Biopolymer-Based Films and Coatings

With the growing concern for the environment and the rising price of crude oil, there is increasing demand for non-petroleum-based polymers from renewable resources. Biopolymer films have been regarded as potential replacements for synthetic films in food packaging due to a strong marketing trend toward environmentally friendly materials. Biopolymer-based films and coatings display good barrier properties, flexibility, transparency, economic profitability, and environmental compatibility. Therefore, they have successfully been used for packaging various food products.

Biopolymer-Based Films and Coatings: Trends and Challenges elaborates on the recent methods and ingredients for making biodegradable films and coatings, as well as the current requirements for food security and environmental issues. This book also explores films and coatings prepared with essential oils, antimicrobial substances, and bioactive components that *make up* this active packaging. Films and coating chapters are based on biopolymers used to prepare films and coatings, that is, carbohydrates, lipids, protein, and so on. This book provides a platform for researchers and industrialists on the basic and advanced concepts of films and coatings.

Key Features

- Provides a comprehensive analysis of recent findings on biopolymers (carbohydrate, protein, and lipid) based films and coatings
- Contains a wealth of new information on the properties, functionality, and applications of films and coatings
- Presents possible active and functional components and ingredients for developing films and coatings
- Guides start-up researchers on where to start the latest research work in packaging

It has been estimated that the global production of bioplastics is set to hike from ~2.11 in 2020 to ~2.87 million tons in 2025. Further, the demand for fresh, ready-to-eat, or semi-finished foods is increasing, and the need to maintain food safety and quality further exacerbates the challenges in the supply chain, especially with the globalization of the food trade and the use of centralized processing facilities for food distribution. It is an urgent requirement to increase shelf life and reduce food product loss. Considering the great market demand for biodegradable material-based packaging systems, this book comes at an opportune time to enable researchers and food scientists to develop suitable solutions considering the sustainability and economic feasibility of the process.

Biopolymer-Based Films and Coatings

Trends and Challenges

Edited by
Sneh Punia Bangar
Anil Kumar Siroha

CRC CRC Press
Taylor & Francis Group
Boca Raton London New York

CRC Press is an imprint of the
Taylor & Francis Group, an **informa** business

First edition published 2023
by CRC Press
6000 Broken Sound Parkway NW, Suite 300, Boca Raton, FL 33487–2742

and by CRC Press
4 Park Square, Milton Park, Abingdon, Oxon, OX14 4RN

CRC Press is an imprint of Taylor & Francis Group, LLC

ISBN: 9781032293387 (hbk)
ISBN: 9781032301549 (pbk)
ISBN: 9781003303671 (ebk)

DOI: 10.1201/9781003303671

Typeset in Times
by Apex CoVantage, LLC

Contents

Preface

Food safety and quality are affected by packaging materials. The majority of commercial packaging contains polymers and materials derived from petroleum. As crude oil reserves decline, environmentalists want natural, biodegradable polymers to replace plastic packaging. Because of the cost of separating polymers and each recycling operation, traditional recycling cannot handle the large volume of collected waste. Biodegradable materials are the most environment-friendly alternative to petroleum-based packaging materials. Polymers derived from renewable resources such as polysaccharides, proteins, lipids, and resins have been used to develop biodegradable packaging materials. This project was undertaken to expand knowledge of biodegradable packaging materials. *Biopolymer-Based Films and Coatings* is based on the principal component for film preparation: polysaccharide-based films, protein-based films, lipid-based films, microorganism-based biopolymers, and nanotechnology-based biopolymers. This book contains 18 chapters, each written by experts in their field.

Polysaccharide-based films and coatings are becoming more appealing due to their wide availability. Siroha and Bangar contributed Chapter 1; this chapter highlights the properties, film preparation methods, and applications of starch-based packaging materials. Millions of tons of cellulose waste are generated yearly by the agricultural industry, which is a pollutant. The use of cellulose in producing environmentally friendly films is addressed in Chapter 2. Chapter 3 was authored by Espinosa-Andrews and colleagues. The properties and applications of chitosan-based films and coatings are discussed in this chapter. Gum sources, properties, and their utilization in films and coatings are addressed in Chapter 4. Chapter 5 focuses on the development of films from alginates and is contributed by Dave and coworkers. In Chapter 6, Nayi and his coworkers focus on green and sustainable technology for packaging material (bio-based) production. The materials generally derived from sustainable or renewable biomass are known as bio-based materials.

Protein-based films and coatings have received a lot of attention because of their ability to form transparent films and their mechanical characteristics. The characteristics of films and coatings made from soy proteins are elaborated in Chapter 7, which was authored by Simmi Deo and her coworkers. It also discusses the mechanisms through which they provide protection to food products and help improve their quality and safety. Gluten is a wheat protein, and its use in the formulation of films and coatings is covered in Chapter 8; which provides an overview of the properties, applications, recent trends, and limitations of gluten-free films.

The incorporation of essential oils into packaging is not only considered eco-friendly, but the antimicrobial agents in these oils also inhibit bacterial and fungal growth, thereby extending the food product's shelf life. Edible films and coatings incorporating lipids are discussed in Chapter 9. This chapter focuses on applications of lipid-based films and coatings. Chapter 10 is contributed by Manjunatha and his coworkers, and discusses essential oil–based films/coatings. Essential oils, the types, and the method of incorporation into packaging materials are described.

Polyhydroxyalkanoates (PHAs), the green plastics produced from microorganisms, are stored in the cytoplasm of microorganisms as energy storage material. PHA production, properties, and applications are discussed in Chapter 11, contributed by Kumari and Singh. Recent advances in the development of PHB (polyhydroxybutyrate)-based packaging materials are elaborated in Chapter 12. Panda and Dash contributed Chapter 13. In this chapter, the preparation, properties, and applications of poly(3-hydroxybutarate-co-3-hydroxyvalerate) (PHBV)-based films and coatings are addressed. Pullulan's basic structure, production methods, and its applications in films and coatings are discussed in Chapter 14, contributed by Mitharwal and Suhag. Shrestha, Bangar, and Ayele contributed Chapter 15, "Bionanocomposites in Food and Medicine." The utilization of nanomaterials in films and coatings is discussed in Chapter 16.

Chapter 17, "Biopolymer Production Methods and Regulatory Aspects," is authored by Kishore and his coworkers. In this chapter, production methods, common biopolymer materials, regulatory aspects, and applications are discussed. Food contamination is a major problem for producing safe and healthy food. Contaminants in food can be incorporated from various sources, such as pesticides sprayed on crops, from food processing equipment, and leaching of chemicals from packaging materials to food products. Soni and his coworkers elaborate the various sources of contaminants from packaging.

Editors

Sneh Punia Bangar, PhD, is Researcher at Clemson University (United States). Earlier, she worked as Assistant Professor (C) in the Department of Food Science and Technology, Chaudhary Devi Lal University, Sirsa. Her research interests include extraction and functional characterization of bioactive compounds, starches, functional foods, and nanocomposite films and coatings. She has presented her research at various national and international conferences and has published more than 100 research papers/book chapters in national and international journals/books. To date, she has authored or coauthored more than 80 (published/ accepted) articles, 1 book series, 2 authored books (CRC/Taylor & Francis), 5 edited books (CRC/Taylor & Francis, Springer, Elsevier), 1 reference book (CRC/ Taylor & Francis), 35 book chapters, 20 conference proceedings/seminars, and 3 guest editor of journals' special issues (*Food Research International, International Journal of Food Science and Technology, Journal of Food Processing and Preservation, Foods and Polymers*).

Anil Kumar Siroha, PhD, is presently working as Assistant Professor (C) in the Food Science and Technology Department at Chaudhary Devi Lal University, Sirsa. He received his doctorate in food science and technology at Chaudhary Devi Lal University, Sirsa, India. His areas of interest include starch, starch modification, and the development of new products. He has published more than 30 research papers, 20 book chapters, 3 edited books, and 1 authored book. He also serves as a reviewer for various national and international journals. Dr. Siroha is an active member of the Association of Food Scientists and Technologists (AFSTI) in Mysore, India. Dr. Siroha has also co-supervised more than 25 research projects of MSc students in the Food Science and Technology Department at Chaudhary Devi Lal University, Sirsa, Haryana.

Contributors

Advaita
University of Delhi
Delhi, India

Aparna Agarwal
Lady Irwin College
Delhi University
New Delhi, India

Sonia Attri
Maharshi Dayanand University
Haryana, India

Abebaw Ayele
Bahir Dar Institute of Technology–Bahir Dar
    University
Bahir Dar, Ethiopia

Sneh Punia Bangar
Clemson University
Clemson, SC, USA

Verbi P. Bhagabati
University of Delhi
Delhi, India

Disha Bhattacharjee
Clemson University
Clemson, SC, USA

Nisha Chaudhary
Agriculture University
Rajasthan, India

Ho-Hsien Chen
National Pingtung University of Science and
    Technology
Neipu, Taiwan

Nidhi Dangi
Maharshi Dayanand University
Haryana, India

Priya Dangi
University of Delhi
New Delhi, India

Pranjyan Dash
National Taipei University of Technology
Taipei City, Taiwan

Pinal K. Dave
Gujarat Vidyapith–Sadra
Gujarat, India

Simmi Deo
Mahidol University
Bangkok, Thailand

Itu Dutta
University of Delhi
New Delhi, India

Hugo Espinosa-Andrews
Centro de Investigación y Asistencia en
    Tecnología y Diseño del Estado de Jalisco
Guadalajara, Mexico

Clara Flores
Clemson University
Clemson, SC, USA

Marisol Verdín García
Universidad Autónoma De Querétaro
Querétaro, México

Eristeo García-Márquez
Centro de Investigación y Asistencia
    en Tecnología y Diseño del Estado
    de Jalisco
Nuevo Leon, Mexico

Prixit Guleria
Maharshi Dayan and University
Rohtak, India

Surangna Jain
University of Tennessee
Knoxville, TN, USA

P. Karthik
Karpagam Academy of Higher Education
Coimbatore, India

Anand Kishore
National Institute of Food Technology and
    Entrepreneurship and Management
Kundli, India

Navneet Kumar
Anand Agricultural University
Godhra, India

Pradeep Kumar
National Institute of Food Technology and
    Entrepreneurship and Management
Sonepat, India

Khushbu Kumari
National Dairy Research Institute
Karnal, India

Purnima Kumari
National Institute of Food Technology
    Entrepreneurship and Management
Sonepat, India

Kinshuk Malik
Indian Institute of Technology Kharagpur
West Bengal, India

Vishal Manjunatha
Clemson University
Clemson, SC, USA

Swati Mitharwal
National Institute of Food Technology
    Entrepreneurship and Management
Sonepat, India

Shalini Mohan
Kalasalingam Academy of Research and
    Education
Krishnankoil, India

Lakshmanan Muthulakshmi
Kalasalingam Academy of Research and
    Education
Krishnankoil, India

Pratik Nayi
National Pingtung University of Science and
    Technology
Neipu, Taiwan

Pradeep Kumar Panda
Yonsei University
Wonju-si, South Korea

Rohan Jitendra Patil
National Institute of Food Technology
    and Entrepreneurship and
    Management
Sonepat, India

Anchita Paul
University of Delhi
Institute of Home Economics
New Delhi, India

Navya Puri
University of Delhi
Institute of Home Economics
New Delhi, India

R. Rajam
Kalasalingam Academy of Research and
    Education
Krishnankoil, India

T. V. Ramana Rao
VIGNAN's Foundation for Science, Technology
    and Research
Vadlamudi, India

Rizwana
Delhi University
New Delhi, India

Rogelio Rodríguez-Rodríguez
Universidad de Guadalajara
Guadalajara, Mexico

Urvi Shah
North Carolina State University
Raleigh, NC, USA

Kritika Sharma
University of Delhi
New Delhi, India

Pratiksha Shrestha
Department of Food Technology and Quality
    Control
Kathmandu, Nepal

Anupama Singh
National Institute of Food Technology
    Entrepreneurship and Management
Kundli, India

Sneha Singhal
University of Delhi
Institute of Home Economics
New Delhi, India

Anil Kumar Siroha
Chaudhary Devi Lal University
Sirsa, India

Ian Blaise Smith
Clemson University
Clemson, SC, USA

Kartik Soni
Delhi University
New Delhi, India

Rajat Suhag
Free University of Bozen-Bolzano
Bolzano, Italy

Anjelina Sundarsingh
Ghani Khan Chaudhary Institute
    of Engineering and
    Technology
West Bengal, India

V. R. Thakkar
Sardar Patel University
Vallabh, Vidyanagar, India

Shivanki Tomar
Maharshi Dayanand University
Rohtak, India

Abhishek Dutt Tripathi
Banaras Hindu University
Varanasi, India

Juan Alberto Resendiz Vazquez
Universidad Autónoma De Querétaro
Querétaro, México

Celso Velásquez-Ordoñez
Universidad de Guadalajara
Guadalajara, Mexico

1

Properties and Applicability of Starch-Based Films

Anil Kumar Siroha and Sneh Punia Bangar

CONTENTS

1.1 Introduction

Packaging materials influence food safety and quality during shelf life. The majority of commercialized packaging in the world is produced from petroleum-based polymers and materials (Henning et al., 2021). As crude oil stocks decrease, environmentalists are increasingly asking for the replacement of plastic packaging with materials made of natural, biodegradable polymers (Łupina et al., 2022). According to the European Bioplastics Organisation (2020), bioplastics account for only 1% of the 368 million tons of plastic manufactured each year globally, with starch-based polymers accounting for 18.7% of the total, mostly used in flexible packaging and consumer goods. By 2025, global bioplastic production is predicted to increase by 36%. Furthermore, conventional recycling procedures cannot handle the vast amount of collected waste due to the inherent costs associated with separating the various polymers and the costs associated with each recycling process. As a result, there is a strong need to develop environmentally friendly, cost-effective packaging materials (Ilyas et al., 2018). Natural and renewable polymers are increasingly being used in place of synthetic packaging sheets in a variety of applications. Polymers derived from renewable resources, including polysaccharides, proteins, lipids, and resins, have been studied to generate biodegradable packaging materials (Martins et al., 2012).

In this context, starch is a well-known green ingredient. This substance is biodegradable, is edible, does not need the use of fossil fuels, and is widely available (Molavi et al., 2021). Because starch has low thermoplastic activity, suitable plasticizers are required to compensate. However, native starch-based films' low moisture barrier and lesser mechanical properties compared to non-biodegradable plastic films have limited their practical adoption in the food industry (Yousefi et al., 2019). Extensive study has focused on developing functional and intelligent starch-based films. Antibacterial, antioxidant, and hydrophobic active substances can be added to starch-based films to provide antibacterial, antioxidant, and barrier capabilities. Furthermore, including a pH indicator into starch-based films can make them pH-responsive (Cui et al., 2021). This chapter includes various methods, ingredients, and properties of starch films.

1.2 Structure and Functions of Starch

Starch is primarily composed of two polysaccharides, amylopectin and amylose, which are both constituted of α-D-glucose linked by α (1–4) glycoside bonds. Amylopectin is highly branched, with the decreasing end side connected by α (1–6), which accounts for 70–80% of the starch. These components composition can alter the starch-based film's qualities in this regard (Henning et al., 2021). According to research, starches from various sources gelatinize at different temperatures due to amylose–amylopectin association–dissociation interactions (Thakur et al., 2019). The starch unit ratio influences the microstructure during heating, and hence the viscosity of the film, generating a suspension that controls network retraction during film drying (Basiak et al., 2017).

1.2.1 Properties of Starches

1.2.1.1 Physicochemical Properties

Amylose and amylopectin, two key components of starch, significantly impact the physicochemical properties of starch-based films (Song et al., 2021). Amylose's linear structure results in a dense network in the film matrix. Compared to amylose, branching amylopectin produces a less compact network (Lee et al., 2020; Song et al., 2021). Amylose content is observed for corn at 20.6% (Siroha et al., 2020a), rice at 19.2% (Dhull et al., 2021b), potato at 21.6% (Siroha et al., 2020a), and millets at 11.01–21.93% (Bangar et al., 2021d; Siroha et al., 2020b; Sandhu & Siroha, 2017). The amylose content has been reported to vary with the botanical source of the starch and is influenced by climatic and soil conditions throughout grain development (Singh et al., 2006). The swelling power (SP) of millets is 14.43–18.83 g/g (Bangar et al., 2021d), potato 24.20 g/g (Yang et al., 2022), corn 17.3 g/g (Siroha et al., 2020a), rice 15.9 g/g (Dhull et al., 2021b), mung bean 15.6 g/g (Punia et al., 2019), etc. The SP of starch granules is related to amylopectin, and a higher SP is related to amylopectin's high water-holding capacity (Yang et al., 2020). The amylose leached during heating affects starch solubility power, which is affected by starch components and granule size (Lin et al., 2015). Solubility is regulated by the crystalline organization, granule size, gelatinization magnitude, and starch granule shape, according to Xia et al. (2019). Intermediate-length amylopectin branches ($13 \leq DP \leq 24$) are thought to limit starch swelling by increasing the stability of molecular accumulation and crystallization areas (Srichuwong et al., 2005). Furthermore, the number of short amylopectin chains was positively associated with starch SP (Li et al., 2020).

1.2.1.2 Pasting Characteristics

Pasting characteristics reveal essential information regarding starch cooking behavior during heating cycles. To determine pasting characteristics, starch is cooked in excess water at a constant shearing rate, and the temperature profile is controlled according to the method selected (Figure 1.1). Water content, shear rate, temperature profile, starch type, and starch concentration utilized for determination all impact pasting qualities. Rapid Visco Analyzer (RVA, Newport Scientific Pvt. Ltd., Australia),

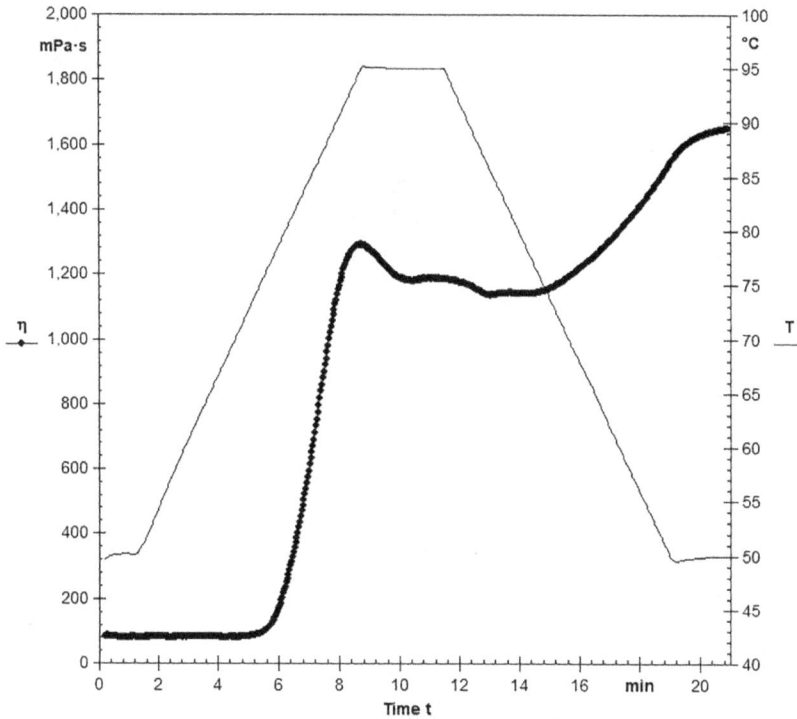

FIGURE 1.1   Pasting graph of starch.

Brabender Visco-Amylo-Graph (BVA) (Brabender, Germany), and rheometer (Anton Paar, Austria) are mainly used for measuring pasting characteristics of starches. Peak viscosity (PV) is the largest viscosity achieved during pasting or heating. The third stage is the holding stage, in which fast heating causes granules to swell, allowing more granules to reach their maximum viscosity before disintegration, known as breakdown viscosity (BV). The holding stage is preceded by a cooling stage, in which viscosity increases as the starch granules cool, resulting in retrogradation, known as setback viscosity (SV) (Bangar et al., 2021a; Punia et al., 2021). Because SV and FV are formed as a result of retrogradation, they play a significant role in the food business when starch is utilized as an ingredient/additive. Many factors influence the pasting behavior and characteristics of starches, including starch concentration, composition (amylose content, amylose-to-amylopectin ratio), cooking and cooling temperatures, and the presence of solutes, such as lipids and sugars, and pH (Mahajan et al., 2021).

1.2.1.3 Thermal Properties

Granular starch exhibits an order–disorder phase shift known as gelatinization when heated with enough water over various temperatures. Gelatinization is characterized by water intake in the amorphous zone, radial expansion of the granules, disintegration of the crystalline region with the breaking of double helices, and starch molecule leaching (Hoover, 2001). Gelatinization of the starch, which plays a vital role in food processing, is used to calculate the cooking capabilities of starches, and starches with low gelatinization temperatures have superior cooking characteristics (Waters et al., 2006). Starch crystal disruption involves the separation of double helices in the crystal register, followed by unwinding (Biliaderis, 2009). Differential scanning calorimetry (DSC) is the preferred method for measuring starch gelatinization temperature and enthalpy, which indicates the energy required to disrupt starch crystallites (Hsieh et al., 2019). DSC thermograms are used to calculate gelatinization temperatures, such as onset (To, onset of gelatinization endotherm; Tp, temperature at peak; Tc, conclusion; and H, enthalpy) (Figure 1.2).

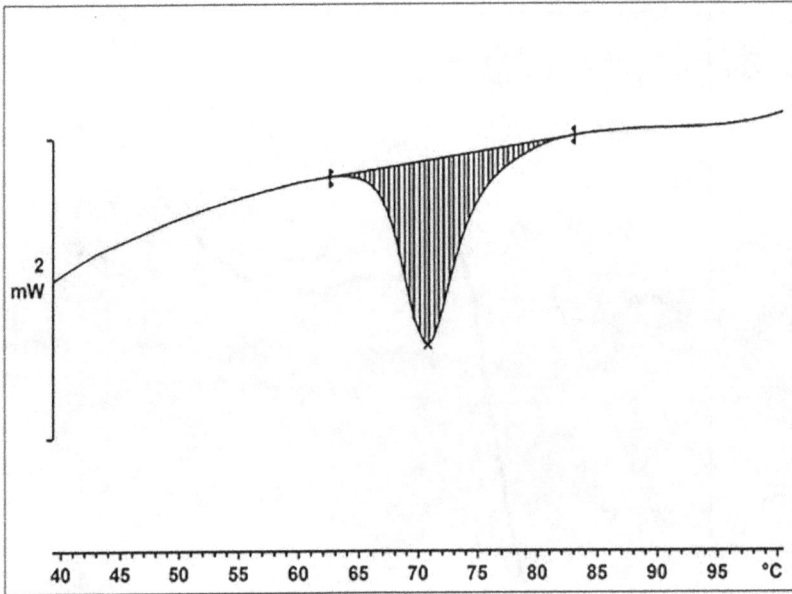

FIGURE 1.2   DSC graph of starch.

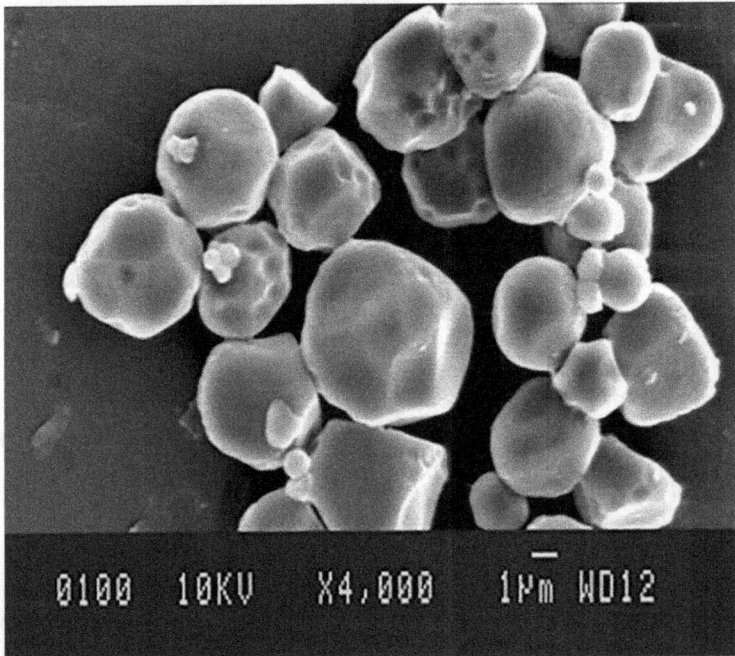

FIGURE 1.3   Scanning electron micrograph of starch.

The melting temperature range (Tc–To) demonstrates amylopectin quality and uniformity (Annor et al., 2014). A small melting range indicates more uniform quality with more amylopectin crystal stability, and vice versa (Ratnayake et al., 2001). The changes in gelatinization temperature can be related to variances in amylose concentration, starch granule size, shape, and distribution, as well as the internal organization of starch fractions within the granule (Singh et al., 2004).

1.2.1.4 Morphological Properties

The size and shape of starch granules differ significantly between plant species (Lindeboom et al., 2004). Scanning electron microscopy (SEM), transmission electron microscopy (TEM), light microscopes, normal light microscopes (NLM), and polarized light microscopes (PLM) are all used to analyze the structural morphology of starch (Bangar et al., 2021a). Morphological properties of starches derived from various plant sources differ depending on genotype and biological origin. The size and form of starch granules differ depending on their biological origin (Svegmark & Hermansson, 1993). Starch granules vary in size (from 1 to 100 µm diameter) and shape (polygonal, spherical, lenticular) and can differ in content, structure, and organization of amylose and amylopectin molecules, amylopectin branching pattern, and crystallinity degree (Lindeboom et al., 2004). Generally, the variation in exterior morphology is adequate to allow unequivocal botanical origin characterization using optical microscopy (Pérez & Bertoft, 2010). A scanning electron micrograph of starch is shown in Figure 1.3.

1.3 Starch as a Sustainable Polymer

Polymers are flexible materials with a wide range of qualities that are easy to manufacture and inexpensive. Polymers have been utilized for decades in various applications, such as in packaging, automotive, civil construction, agriculture, medicine, and health care (Pelissari et al., 2019). When compared to synthetic materials, starch has two primary disadvantages. Water is a key plasticizer for starch because it contains three hydroxyl groups; changes in the physical and mechanical properties of starch-based products can be significant. Second, natural starch does not have thermoplastic properties. As a result, without the addition of plasticizers, starch cannot be melt-processed using standard plastic equipment (Mohammadi et al., 2013). These components' composition can alter the starch-based film's qualities in this regard. Films with a higher amylose content often have better film-forming qualities, such as mechanical strength, elongation, and gas barrier properties (Liu et al., 2005). Because starch's breakdown temperature is lower than its melting point prior to gelatinization, it cannot be thermally treated without a plasticizer or gelatinization agent (Liu et al., 2008). While starch is widely available and flexible, its high water solubility and brittleness limit its ability to make films. This can be reduced by including low-molecular-weight plasticizers (polyhydroxyl compounds) into the starch solution (film-forming recipe) (Menzel et al., 2020). Glycerol (Mallakpour & Khodadadzadeh, 2018) and sorbitol (Abera et al., 2020) are common plasticizers used in the production of thermoplastic starch (TPS) (Martins et al., 2021).

1.4 Preparation of Starch-Based Films

There are three main methods for preparing starch-based films: casting, extrusion/thermocompression, and blow molding.

1.4.1 Casting Method

Because of its simplicity and ability to evaluate the biopolymer's potential to build filmogenic matrices and their properties, direct casting is the most commonly employed approach on a laboratory scale. Following the selection of the biopolymer, the following stages are carried out for the manufacturing of the films, according to Marangoni Júnior et al. (2020); Priyadarshi and Rhim (2020); Umaraw et al. (2020); Wróblewska-Krepsztul et al. (2018): (1) The biopolymer is first dissolved in a suitable solvent, which is commonly water. (2) In the case of protein, the pH is altered to promote solubility. (3) Additional substances, such as plasticizers, are added, and (4) combinations with another biopolymer, if necessary. (5) Then, sieving and heating to produce a homogeneous and viscous solution. (6) After cooling, additional functional components are added and filtered (7) solutions are loaded onto baking trays. (9) Next, drying under temperature and relative humidity (RH) controlled settings. (10) The film is then detached from the plate, and (11) temperature and RH are regulated during storage (Figure 1.4). The main advantage of the casting method of film formation is the ease of production without needing specialist equipment (Chen

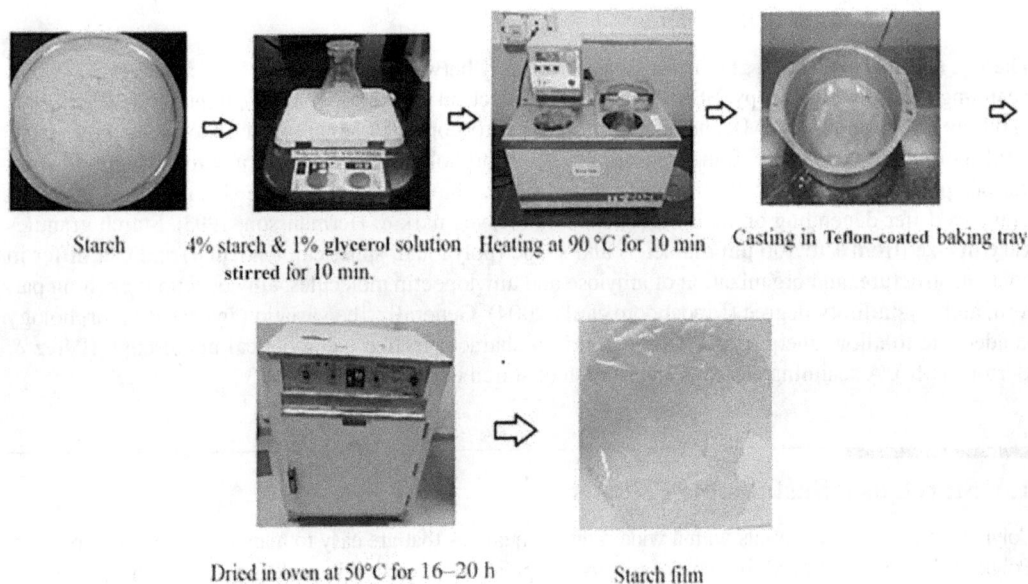

FIGURE 1.4   Method for casting the starch film.

Source: da Rosa Zavareze et al. (2012).

et al., 2008). Because casting is a wet operation, it promotes more significant particle–particle interaction, resulting in more homogeneous particle packaging and fewer flaws (Yang et al., 2011).

1.4.2 Extrusion Procedure

Extrusion is a commercially used polymer processing method that is also widely used and generally favored for producing polymer films. This approach modifies the structural properties and improves the physicochemical properties of extruded materials (Calderón-Castro et al., 2018). Liu et al. (2009) examined standard and novel starch treatment approaches. Thermal processing methods for starch include sheet/film extrusion, foaming extrusion, injection molding, compression molding, and reactive extrusion, according to these authors (a special type of extrusion in which chemical reaction and typical extrusion take place). In general, the extrusion process is separated into three zones: (1) the feeding zone, (2) the kneading zone, and (3) the heating zone at the machine's final part/exit (Hauck & Huber, 1989; Calderón-Castro et al., 2018). While amylose content increases processibility, as seen by higher torque, instability (torque variation), and higher die pressure, these negative effects may be mitigated by increasing the water content, temperature, and screw speed and/or decreasing the feed rate (Li et al., 2011).

1.4.3 Blow Molding

A twin-screw extruder is used to pelletize the starch and additives, followed by a single-screw blown-film extruder to make a starch-based film. Extrusion blow molding is mostly used to manufacture food plastic packaging products. The use of extrusion blow molding to make starch-based films has garnered a lot of interest in recent years (Cui et al., 2021). However, extrusion blow molding for starch-based materials has not yet been investigated since high shear stress causes deterioration (molecular bond breakup) and increases gelatinization even at low moisture levels (Xie et al., 2008; Liu et al., 2009). Because the parison must bear its weight before being trapped by the mold, a polymer with strong melt strength, thermal stability, and restricted swelling is usually suitable for extrusion blow molding (Wagner et al., 2014).

1.5 Role of Plasticizers in Films and Coatings

Plasticizers also have an essential role in preventing starch retrogradation during film storage (Domene-López et al., 2019). The best plasticizers include hydroxyl groups, which limit the interaction of starch hydroxyl groups in the film, lowering molecular density while boosting free volume and molecular mobility (Henning et al., 2021). The bulk of plasticizers is very hydrophilic and hygroscopic, attracting water molecules and forming a huge plasticizer water hydrodynamic complex. Plasticizers that are often utilized include glycerin, propylene glycol, sorbitol, sucrose, polyethylene glycol, and corn syrup (Han, 2013; Kumar & Neeraj, 2019). Dias et al. (2010) found that increasing the plasticizer concentration in starch films with glycerol or sorbitol enhanced water vapor permeability. In the same test settings, starch films plasticized with sorbitol were more efficient moisture barriers than those plasticized with glycerol (temperature and relative humidity gradient). Glycerol ($C_3H_8O_3$) is a common plasticizer used in the production of starch-based biodegradable materials. Because the inclusion of glycerol weakens the internal hydrogen bond force of amylose and increases the fluidity of the starch chain, the EAB of starch-based biodegradable products may be improved (Aghazadeh et al., 2018b; Myllärinen et al., 2002). Waxes and lipids are also employed as additives in the production of starch films and coatings. Because of changes in the inner structure and film surface, lipids increase the barrier, mechanical, and optical characteristics of the film (Jiménez et al., 2013; Jost et al., 2014). The addition of waxes increases the film's moisture barrier qualities. Natural antioxidants such as organic acids, phenolic acids, terpenes, tocopherols, carotenoids, vitamins, and others have been employed in starch films to increase product stability and extend storage term (Ashwar et al., 2014; Inam u Nisa et al., 2015). Changes in starch and plasticizer type, as well as concentration, produce films with varying opacity, thickness, water solubility, and water vapor permeability (Ramos et al., 2022). The inclusion of the plasticizer enhances the hydrophilic content, hence increasing film solubility. Furthermore, it changes the contact between the polymer chains, which may promote polymer disintegration (Haq et al., 2014).

1.6 Incorporation of Functional Compounds

To improve overall protection, edible and biodegradable films and coatings may be formulated with antibacterial and antioxidant compounds (Jiménez et al., 2012a). Plant extracts include natural antibacterial compounds, such as polyphenols, flavonoids, tannins, and alkaloids. These chemicals generally have antibacterial properties and may disrupt cell membranes, interfere with active transport, and decrease enzyme function (Cui et al., 2021). Antimicrobial food packaging comprises coating foods with antimicrobials to keep hazardous germs away. Antimicrobials may be incorporated into packaging materials by directly incorporating antimicrobial agents into polymers or by applying the antimicrobial coating to polymer surfaces (Fu & Dudley, 2021). Khalid et al. (2018) created an antibacterial film for the first time by combining pomegranate peel as an antibacterial ingredient with polycaprolactone and a starch matrix. When the starch/pomegranate peel ratio hits 40%, the area of the antibacterial film's inhibitory zone against *S. aureus* reaches 115.37 mm². In terms of antibacterial properties, chitosan has shown significant promise for application in food preservation packaging against a broad range of pathogens, owing to its antimicrobial activity (Dutta et al., 2009). Particles have been used in food packaging to provide antibacterial activity to the material and enhance mechanical and barrier qualities (Bodirlau et al., 2013; Scarfato et al., 2017; de Moraes Crizel et al., 2018). Fenugreek oil (*T. foenum-graecum*) has potential antimicrobial properties against certain microbe strains that cause foodborne disease and may therefore be used in place of chemical preservatives (Dhull et al., 2021a). Alwhibi et al. (2014) found that fenugreek oil is effective against Gram-positive and Gram-negative bacterial isolates, such as *E. coli*, *Pseudomonas aeruginosa*, *Klebsiella pneumonia*, *Staphylococcus aureus*, and *Salmonella Typhi*. The impact of different curcumin concentrations (0%, 0.5%, 1%, 2%, and 3%) on the characteristics of PMS films was reported. TS dropped and EAB increased as curcumin content increased. Furthermore, ABTS and DPPH radical scavenging activities were observed, indicating that the films had high antioxidant characteristics (Baek & Song, 2019). The lowest recorded essential oil content in a film was 0.05% w/w (Liu et al., 2014), while the maximum was 7.5% w/w (Liu et al., 2014; Almasi et al., 2020). Akhter et al. (2019) found increased

TS, water barrier qualities, and thermal stability after incorporating 0.5% rosemary essential oil, mint essential oil, nisin, and lactic acid into chitosan/pectin/starch films (0.75:1.5:0.75% w/w).

1.7 Film Characteristics

Mechanical properties measurement gives critical information for analyzing the strength and toughness of materials, indicating the applications' potential value. The mechanical parameters of the thermoplastic starch film were determined by measuring the mechanical properties of both the longitudinal and transverse films (Liu et al., 2020). Table 1.1 summarizes the physicochemical characteristics of films.

1.7.1 Mechanical Characteristics

Mechanical characteristics of polymeric films are essential in determining their ability to withstand external pressure before rupture. Mechanical properties like TS, EAB, Young's and storage modulus, and loss factor are all important in packing materials (Bangar et al., 2021c). The mechanical characteristics of the films are determined by the formulation (macromolecule, solvent, plasticizer, pH) and the production procedure. The glass transition temperature is another feature of films that influences their mechanical properties (Zamudio-Flores et al., 2006). The highest stress created in a film during a tensile test is *TS* (Liu et al., 2021), while *EAB* evaluates film flexibility and stretchability; it indicates the film elongation from the original length to the breaking point (Henning et al., 2021). TS and EAB show the potential of food packaging in terms of breaking resistance and preserving integrity under stress during processing, handling, and storage (Santana et al., 2019). The impact of sonication on TS and EAB was studied by Garcia-Hernandez et al. (2017). The film's TS (7.3 MPa for S0 to 5.9 MPa) and EAB (48.8% for S0 and a minimum value of 38.7%) dropped as the sonication period increased.

1.7.2 Barrier Properties

The film's WVP measures the ease with which moisture may permeate the film. The WVP of the film should be low in order to establish a barrier or to reduce moisture transfer between the food and the surrounding environment (Vellaisamy Singaram et al., 2021). Barrier characteristics are crucial determinants since they determine the shelf life of packaged foodstuff. Barrier qualities aid in preserving moisture in packed food and eliminating microbiological contaminations from the surrounding environmental conditions (Hiemenz & Rajagopalan, 1997; Jiménez et al., 2012b). The two most essential barrier properties of biodegradable polymer packaging are water and oxygen. These barrier qualities inhibit moisture and oxygen exchange between the

TABLE 1.1

Physical and mechanical properties of films

Type of starch used	Film thickness (mm)	Solubility (%)	Moisture content (%)	Tensile strength (MPa)	Elongation at break (%)	Reference
Barnyard millet	0.10	41.44	23.17	6.56	—	Sharma et al. (2021a)
Lichi kernel	0.152	43.66	23.64	1.41	19.88	Sharma et al. (2021b)
Maize starch	0.266	—	22.26	1.49	51	Żołek-Tryznowska and Kałuża (2021)
Mango kernel	0.121	—	—	2.99	14.16	Vellaisamy Singaram et al. (2021)
Mango kernel	0.063	34.58	—	26.41	3.82	Nawab et al. (2016)
Millets	0.11–0.15	25.6–31.5	10.1–14.6	3.79–6.95	53.4–73.2	Bangar et al. (2021d)
Oat starch	0.266	—	21.77	0.36	27	Żołek-Tryznowska and Kałuża (2021)
Pearl millet	0.103	38.34	26.08	6.80	62.8	Siroha et al. (2021)
Pearl millet	0.105	34.04	27.11	—	—	Bangar et al. (2021b)
Pearl millet	0.18	14.77	24.16	3.44	56.5	Dhull et al. (2021a)
Rice starch	0.145	—	18.72	1.80	49	Żołek-Tryznowska and Kałuża (2021)
Tapioca starch	0.136	—	17.22	0.78	137	Żołek-Tryznowska and Kałuża (2021)

product and its environment (Bangar & Whiteside, 2021). A substance's permeability is determined by its solubility and diffusion through the film and is determined by the chemical structure, microstructure, degree of crystallinity, and molecular weight of the polymer matrix (Agarwal, 2021). Because starch films have good oxygen barrier qualities, they may be used to preserve food. For example, the oxygen permeabilities of wheat starch film, cornstarch film, and rice starch film are 0.0085, 0.0090, and 0.0086 cc mm/m2 d kPa, respectively (Aghazadeh et al., 2018a). Starch films have lower WVP values than protein and polysaccharide films (García et al., 2004; Parra et al., 2004), but they are greater than synthetic polymers like polyethylene.

1.7.3 Solubility

Solubility influences film's usage in food sectors. Some food materials/products need water-insoluble films to maximize shelf life/integrity and water resistance. Solubility indicates the film structure's integrity and affinity to interact in an aqueous medium (Gutiérrez et al., 2015). According to Ballesteros-Mártinez et al. (2020), water insolubility may be required to retain product integrity and water resistance in packaging materials. On the other hand, high water solubility may be advantageous for the coatings of fresh and less-processed foods. Potential uses of biodegradable films need partial solubility, mainly to retain product integrity and water resistance for packaging materials; nevertheless, solubility before consumption may be advantageous in certain circumstances (Thakur et al., 2018). By incorporating a hydrophobic material into the starch matrix, the water solubility of the starch film is lowered (Cui et al., 2021). Pérez et al. (2021) created active starch films by combining tapioca starch with glycerin, ethanol, zein, natamycin, and nisin. The water solubility of the film with 3% zein reduced from 34.96% to 28.43% when compared to the film without zein.

1.7.4 Thickness of Films

Film thickness is identified as a significant characteristic since it impacts film properties, such as transparency, TS, water vapor, and gas permeability (Vellaisamy Singaram et al., 2021). The larger the thickness of the film, the higher the film's moisture content (Baranzelli et al., 2019). Film thickness may be varied in the casting procedure by adjusting the filmogenic-suspension-weight-to-plate-area ratio. The drying conditions (rate and temperature) define film characteristics (e.g., water content, crystallinity, etc.), which impact the microstructure and properties of the film (Versino et al., 2016). Bader and Goritz (1994) emphasized that by altering drying conditions, the amorphous-crystalline structure, which is highly connected to film barrier and mechanical characteristics, may be adjusted. Because starch is a semi-crystalline molecule, physical and chemical characteristics, such as tensile and gas barrier qualities, are impacted by both amorphous and crystalline zones and their cohesive energy density (Liu, 2005). As previously stated, native starch with a high amylopectin concentration is more crystalline, but starch films with more amylose have greater crystallinity (García et al., 2000). Evaluating thickness is closely connected to evaluating film uniformity (Hosseini et al., 2021).

1.8 Films from Modified Starches

Derivatization, esterification, etherification, cross-linking, decomposition (acid or enzymatic hydrolysis and oxidization of starch), or heat and moisture are used to modify starch. Different ways of modification may aid in improving the mechanical and functional qualities of starch-based biodegradable films (Fonseca et al., 2015). Oxidized starch is prepared by reacting starch with a specific quantity of an oxidizing reagent under temperature and pH control (Zhao et al., 2012). Depolymerization occurs during oxidation, resulting in a decreased dispersion viscosity and the introduction of carbonyl and carboxyl groups, which inhibit recrystallization (Mirmoghtadaie et al., 2009). The films were made using oxidized banana starch and glycerol. Three stages of oxidation were used to create oxidized banana starch (0.5, 1.0, and 1.5% of active chlorine). The TS and WVPc values were highest in films produced using oxidized banana starch, and they increased as the oxidation level in the starch increased (Zamudio-Flores et al., 2006). The impact of dual modification (oxidation and acetylation) was investigated by Zamudio-Flores et al. (2009). It was noticed that oxidation raised the TS of the film, and when the oxidized starch was acetylated, this parameter improved. When the oxidation level increased, the percentage E value declined, and the impact became more noticeable.

WVP increased as oxidation level increased; however, acetylation lowered this parameter. Narváez-Gómez et al. (2021) examined the films prepared from modified yam starch. The WVP of the films ranged from 4.4×10^{-10} to 1.5×10^{-9} g/m*s*Pa, with oxidized yam starch showing a 58.04% decrease over native starch. The TS of oxidized yam starch films decreased by 17.51% as glycerol content increased. Vellaisamy Singaram et al. (2021) investigated the characteristics of chemically modified mango kernel starch (MKS) films (oxidation and benzylation). The TS of modified films was more than double that of the native starch film. Oxidation prevented retrogradation by interrupting starch recrystallization. When compared to the native film, benzylation reduced WVP by 58% and increased UV absorption by 80%. González-Soto et al. (2019) investigated a film made from dual-modified (acetylation and cross-linking) potato starch. It was discovered that modified starch films had a larger elongation percentage (82.81%) than their native counterparts (57.4%), but lower TS (3.51 MPa for native and 2.17 MPa for dual modified) and reduced crystallinity in both fresh and stored films. Oyeyinka et al. (2017) investigated bambara starch that had been treated with varying concentrations of stearic acid (0%, 2.5%, 3.5%, 5%, 7%, and 10%). The change boosted film solubility while decreasing TS, EAB, and WVR. Bambara starch film may be treated with 2.5% stearic acid to increase WVP and thermal stability while having no influence on TS. For example, as compared to the control, maize starch film treated with 15% stearic acid reduced WVP by approximately 27% (Jiménez et al., 2012b). Prachayawarakorn and Kansanthia (2022) investigated the properties and features of oxidized, cross-linked, and dually modified cassava starch films using hydrogen peroxide oxidation and glutaraldehyde cross-linking. Furthermore, glutaraldehyde cross-linked starch (GLC) films had greater stress at maximum load and Young's moduli than hydrogen peroxide oxidized (HPO) and native films. Both dually modified starch films had much reduced swelling and moisture absorption than singly modified starch films. Furthermore, the mechanical characteristics of dually modified starch films were significantly superior to those of singly modified and native starch films, particularly for GLC/HPOS film. STMP, phosphorus oxychloride, epichlorohydrin, and a combination of adipic and acetic mixed anhydrides are among the reagents used to cross-link starch (Wattanachant et al., 2003). Cross-linked (CL) starch resists heat, acid, and shear (Jyothi et al., 2006). Cross-linking joins polymer molecules using covalent, ionic, or hydrogen bonds (Mehboob et al., 2020). Cross-linking limits the starch–water interface, resulting in structural integrity that makes starch more stable under heat, acid, and shearing (El-Tahlawy et al., 2007). Cross-linked (CL) starch films had reduced moisture, solubility, WVP, and EAB values while having greater thickness, opacity, and thermal and TS values, according to Dhull et al. (2021a). The moisture content, water solubility, and TS of the films loaded with fenugreek oil were reduced, whereas thickness, opacity, WVP, and EB were increased. Siroha et al. (2021) created the films by modifying pearl millet starch with epichlorohydrin (EPI) at various concentrations (0.1%, 0.3%, 0.5%, and 0.8%). CL starch films had lower moisture, solubility, and thickness than native starch films and a larger percentage of opacity. It was also discovered that when the concentration of EPI increased, the TS and EAB value similarly increased. Sharma et al. (2021a) investigated the ability of CL barnyard millet starches to create films. CL starches had greater TS but lower moisture content, WVP, and solubility, suggesting superior barrier and mechanical qualities as compared to native starch films. CL (5%) starch film had the lowest WVP (1.28 g/Pa.s.m2) and the greatest TS (13.12 MPa) of all films. Sharma et al. (2021b) investigated the impact of cross-linking on litchi kernel starch as well. It was discovered that films made from CL starch had lower moisture content, thickness, water solubility, WVP, and EAB while increasing opacity and TS. Starch CL increased the mechanical qualities of biofilms, and such films might be employed for low-density material packing (Sandhu et al., 2017). Bangar et al. (2021b) used several physical treatments to modify pearl millet starch (heat moisture treatment, sonication treatment, microwave treatment). After heat treatment, moisture content and thickness decreased, while solubility and opacity increased. Cold plasma is a green method for starch modification, with most studies focusing on cross-linking and depolymerization mechanisms (Sifuentes-Nieves et al., 2019). The chemical, mechanical, and barrier characteristics of films produced from plasma-modified maize starch (MSF) were studied as a function of amylose concentration in this work (30, 50, and 70%). For modified starches, EAB and WVPR decreased. Young's modulus and TS decreased for 30 and 50% amylose content modified starch but increased for 70% amylose content modified starch (Sifuentes-Nieves et al., 2019). Kanatt (2020) created the films using irradiation tapioca starch and gelatin (0, 5, 10, and 20 kGy). The mechanical characteristics of films made with irradiation starch irradiated at 20 kGy were found to be the best. The TS of films created with irradiation starch at 20, 10, and 5 kGy were 58.61%, 26.9%, and 9.6% higher than that of films prepared with non-irradiated starch, respectively. Kim et al. (2008) also found that films made using irradiated cornstarch, PVA, and glycerol have superior mechanical characteristics than those made with native starch. The maximum puncture strength was found in starch gelatin films made with irradiation (20 kGy) starch (2.02 N). Hu et al. (2019) created the films using normal maize starch that has been debranched, hydroxypropylated, and debranched/hydroxypropylated. Debranching reduced EAB while increasing TS of the starch film, while hydroxypropylation enhanced EAB while decreasing TS. Surprisingly, the combination of debranching and hydroxypropylation might overcome their disadvantages, significantly improving the qualities of native maize starch film. The film's EAB and TS were enhanced to 57.2% and 7.35 MPa, respectively, after dual modification. Martins et al. (2022) investigated the effect of starch nanocrystals on the physicochemical, thermal, and structural properties of starch-based films. Adding 0.1% potato starch nanoparticles (PSNCs) to films made from native starch lowered the film's solubility to 16%. The starch nanoparticles (SNCs) raised the film's TS at least twice as much as the control; they also decreased the film's WVP and boosted elongation capacity. Including 0.3% rice nanoparticles (RSNCs) and 0.1% PSNCs improved film cohesiveness while decreasing surface cracks. Roy et al. (2020) made composite films by varying the quantities of nano starch in native mung bean starch (0.5, 1.0, 2.0, 5.0, and 10.0%). Compared to native starch films, composite films were clearer, were more transparent, and had superior textural qualities. Compared to native starch films, composite films had greater thickness and burst strength while having lower moisture content, WVTR, and solubility. Table 1.2 summarizes the properties of films prepared from modified starches.

TABLE 1.2

Physical and mechanical properties of films prepared by modified starches

Starch	Types of modification	Major results	Reference
Bambara groundnut	Stearic acid	Solubility of films increased, whereas TS, EAB, and WVR decreased after modification process.	Oyeyinka et al. (2017)
Barnyard millet	Cross-linking	CL starch films showed higher TS but lower moisture content, WVP, and solubility.	Sharma et al. (2021a)
Corn	Palmitic, stearic, and oleic modification	Decrease in the EM and tensile strength and elongation at break, except in the case of oleic acid.	Jiménez et al. (2012b)
Corn	Plasma modification	EAB and WVPR decreased for modified starches. Young's modulus and TS decreased for 30 and 50% amylose content and increased at 70% amylose content modified starch.	Sifuentes-Nieves et al. (2019)
Litchi	Cross-linking	CL starch showed lesser moisture content, thickness, water solubility, and WVP and EAB decreases, whereas opacity and TS increase.	Sharma et al. (2021b)
Maize	Debranching and hydroxypropylation	Dual modification (debranched/hydroxypropylated), elongation at break and tensile strength of the film were increased to 57.2% and 7.35 MPa, respectively.	Hu et al. (2019)
Pearl millet	Cross-linking	Moisture, solubility power, thickness, and EAB decreased, whereas TS increased.	Siroha et al. (2021)
Pearl millet	Cross-linking	Lower moisture, solubility, WVP, and EB values while having higher thickness, opacity, thermal, and TS values.	Dhull et al. (2021a)
Pearl millet	Heat moisture treatment, microwave treatment, sonication treatment	Films prepared using modified starches showed less moisture and higher solubility and opacity content compared to films prepared using native starch.	Bangar et al. (2021b)
Potato	Acetylation and cross-linking (dual modification)	Films have higher elongation percentage but lower TS and lower crystallinity in fresh and stored films.	González-Soto et al. (2019)
Tapioca	Irradiated starch	Puncture strength and tensile strength increased while elongation strength decreased after modification.	Kanatt (2020)
Yam starch	Oxidized, cross-linked, and dual: oxidized/cross-linked	WVP of films reduced after modification. The TS of oxidized films showed a decrease with an increase in glycerol concentration.	Narváez-Gómez et al. (2021)

1.9 Applications

For food packaging materials, starch-based biodegradable polymers must have a particular mechanical, barrier, antibacterial, and antioxidant properties. To extend the shelf life of foods, several efforts have been made to strengthen the barrier by mixing with other polymers or adding antioxidants/antimicrobial agents in film-forming materials (Bangar et al., 2021c). Sarak et al. (2022) concentrated on manufacturing and characterizing films and coatings made from native cassava starch, cationic cassava starch, and a combination of the two. The films and coatings are designed to be used as coating materials for mango fruits. Furthermore, after ten days of storage, the mangoes covered with the starch mix film lost the least weight and showed only minor changes in physical appearance. Aguilar-Méndez et al. (2008) tested the shelf life of avocado fruits at 6°C using starch–gelatin coatings. Over time, the flesh firmness, color changes, weight loss, and interior quality of the fruit were effectively maintained. Nawab et al. (2017) investigated the use of edible film made from MKS in tomatoes (*Solanum lycopersicum* L.). Over the shelf life, a combination of starch (4% w/v) and sorbitol (50% w/w starch foundation) showed substantial increases in sensory qualities, lesser weight loss, and enhanced stiffness. Another strategy for controlling infections and extending the shelf life of minimally processed fruits and vegetables is to add natural antimicrobials such as essential oils (EOs) to starch coatings. A 3% cassava starch coating containing carvacrol reduced pathogens in minimally processed pumpkins while also preventing weight loss (Santos et al., 2016). Starch-based edible coatings are also used on nuts and baked goods. Due to its greater flexibility than wheat and cornstarch, rice starch was suggested as a walnut coating (Galvão et al., 2018; Syafiq et al., 2020). Rice starch plasticized with 2% glycerol forms a barrier that reduces oxygen, moisture, and heat on walnuts. By removing the walnut husk and shell, the starch-based coating reduced storage space. Moreover, chitosan and red palm oil incorporation provided a smoother layer due to matrix compactness and homogeneity (Galvão et al., 2018). Rahmasari and Yemiş (2022) used ginger starch films for antibacterial meat packaging. Sensory studies showed that ginger starch films with varied concentrations of coconut shell liquid smoke (CSLS) do not impact ground beef's flavor.

1.10 Conclusion

This chapter summarizes the source, properties, and applications of starch films. Biodegradable packaging has been developed using polysaccharides, proteins, lipids, and resins. Starch is a green element in this sense. Amylose and amylopectin, two important starch components, significantly influence the physicochemical characteristics of starch-based films. Starch films are not very strong; thus, different additives are added to increase their mechanical qualities, which are desirable for its uses in many food and packaging industries. Plasticizers are often used to enhance the characteristics of starch films. Starches are modified to improve their ability to produce films, such as solubility, WVTR, TS, etc. Starch films are used for coating and edible films on fruits and vegetables, as well as in the meat, baking, and packaging industries. More research is needed to improve the mechanical quality of starch films.

REFERENCES

Abera, G., Woldeyes, B., Demash, H. D., & Miyake, G. (2020). The effect of plasticizers on thermoplastic starch films developed from the indigenous Ethiopian tuber crop Anchote (*Coccinia abyssinica*) starch. *International Journal of Biological Macromolecules, 155*, 581–587.

Agarwal, S. (2021). Major factors affecting the characteristics of starch based biopolymer films. *European Polymer Journal, 160*, 110788.

Aghazadeh, M., Karim, R., Abdul Rahman, R., Sultan, M. T., Paykary, M., & Johnson, S. (2018a). Effect of glycerol on the physicochemical properties of cereal starch films. *Czech Journal of Food Sciences, 36*, 403–409.

Aghazadeh, M., Karim, R., Sultan, M. T., Paykary, M., Johnson, S. K., & Shekarforoush, E. (2018b). Comparison of starch films and effect of different rice starch-based coating formulations on physical properties of walnut during storage time at accelerated temperature. *Journal of Food Process Engineering, 41*(1), e12607.

Aguilar-Méndez, M. A., Martín-Martínez, E. S., Tomás, S. A., Cruz-Orea, A., & Jaime-Fonseca, M. R. (2008). Gelatine—starch films: Physicochemical properties and their application in extending the postharvest shelf life of avocado (*Persea americana*). *Journal of the Science of Food and Agriculture, 88*(2), 185–193.

Akhter, R., Masoodi, F. A., Wani, T. A., & Rather, S. A. (2019). Functional characterization of biopolymer based composite film: Incorporation of natural essential oils and antimicrobial agents. *International Journal of Biological Macromolecules, 137*, 1245–1255.

Almasi, H., Azizi, S., & Amjadi, S. (2020). Development and characterization of pectin films activated by nanoemulsion and Pickering emulsion stabilized marjoram (*Origanum majorana* L.) essential oil. *Food Hydrocolloids, 99*.

Alwhibi, M. S., & Soliman, D. A. (2014). Evaluating the antibacterial activity of fenugreek (*Trigonella foenum-graecum*) seed extract against a selection of different pathogenic bacteria. *Journal of Pure and Applied Microbiology, 8*(2), 817–821.

Annor, G. A., Marcone, M., Bertoft, E., & Seetharaman, K. (2014). Physical and molecular characterization of millet starches. *Cereal Chemistry, 91*(3), 286–292.

Ashwar, B. A., Shah, A., Gani, A., Shah, U., Gani, A., Wani, I. D., . . . Masoodi, F. A. (2014). Rice starch actixve packaging films loaded with antioxidants-development and characterization. *Starch/Stärke, 66*, 1–9.

Bader, H. G., & Goritz, D. (1994). Investigations on high amylose cornstarch films. Part 3: Stress strain behaviour. *Starch/Stärke, 46*, 453–439.

Baek, S. K., & Song, K. B. (2019). Characterization of active biodegradable films based on proso millet starch and curcumin. *Starch/Stärke, 71*(3–4), 1800174.

Ballesteros-Mártinez, L., Pérez-Cervera, C., & Andrade-Pizarro, R. (2020). Effect of glycerol and sorbitol concentrations on mechanical, optical, and barrier properties of sweet potato starch film. *NFS Journal, 20*, 1–9.

Bangar, S. P., Kumar, M., Whiteside, W. S., Tomar, M., & Kennedy, J. F. (2021a). Litchi (*Litchi chinensis*) seed starch: Structure, properties, and applications-a review. *Carbohydrate Polymer Technologies and Applications, 2*, 100080.

Bangar, S. P., Nehra, M., Siroha, A. K., Petrů, M., Ilyas, R. A., Devi, U., & Devi, P. (2021b). Development and characterization of physical modified pearl millet starch-based films. *Foods, 10*(7), 1609.

Bangar, S. P., Purewal, S. S., Trif, M., Maqsood, S., Kumar, M., Manjunatha, V., & Rusu, A. V. (2021c). Functionality and applicability of starch-based films: An eco-friendly approach. *Foods, 10*(9), 2181.

Bangar, S. P., Siroha, A. K., Nehra, M., Trif, M., Ganwal, V., & Kumar, S. (2021d). Structural and film-forming properties of millet starches: A comparative study. *Coatings, 11*(8), 954.

Bangar, S. P., & Whiteside, W. S. (2021). Nano-cellulose reinforced starch bio composite films-a review on green composites. *International Journal of Biological Macromolecules, 185*, 849–860.

Baranzelli, J., Kringel, D. H., Mallmann, J. F., Bock, E., Mello El Halal, S. L., Prietto, L., . . . Renato Guerra Dias, A. (2019). Impact of wheat (*Triticum aestivum* L.) germination process on starch properties for application in films. *Starch-Stärke, 71*(7–8), 1800262.

Basiak, E., Lenart, A., & Debeaufort, F. (2017). Effect of starch type on the physico-chemical properties of edible films. *International Journal of Biological Macromolecules, 98*, 348–356.

Biliaderis, C. G. (2009). Be Miller, J. and Whistler, R. (Eds). Structural transitions and related physical properties of starch. In *Starch* (pp. 293–372). London: Academic Press.

Bodirlau, R., Teaca, C. A., & Spiridon, I. (2013). Influence of natural fillers on the properties of starch-based biocomposite films. *Composites Part B: Engineering, 44*(1), 575–583.

Calderón-Castro, A., Vega-García, M., Zazueta-Morales, J. D., Fitch-Vargas, P. R., Carrillo-López, A., Gutiérrez-Dorado, R., Limón-Valenzuela, V., & Aguilar-Palazuelos, E. (2018). Effect of extrusion process on the functional properties of high amylose cornstarch edible films and its application in mango (*Mangifera indica* L.) cv. Tommy Atkins. *Journal of Food Science and Technology, 55*(3), 905–914.

Chen, Q., Roether, J., & Boccaccini, A. (2008). Tissue engineering scaffolds from bioactive glass and composite materials. *Topics in Tissue Engineering*, 1–27.

Cui, C., Ji, N., Wang, Y., Xiong, L., & Sun, Q. (2021). Bioactive and intelligent starch-based films: A review. *Trends in Food Science & Technology*, *116*, 854–869.

da Rosa Zavareze, E., Pinto, V. Z., Klein, B., El Halal, S. L. M., Elias, M. C., Prentice-Hernández, C., & Dias, A. R. G. (2012). Development of oxidised and heat—moisture treated potato starch film. *Food Chemistry*, *132*(1), 344–350.

de Moraes Crizel, T., de Oliveira Rios, A., Alves, V. D., Bandarra, N., Moldão-Martins, M., & Flôres, S. H. (2018). Active food packaging prepared with chitosan and olive pomace. *Food Hydrocolloids*, *74*, 139–150.

Dhull, S. B., Bangar, S. P., Deswal, R., Dhandhi, P., Kumar, M., Trif, M., & Rusu, A. (2021a). Development and characterization of active native and cross-linked pearl millet starch-based film loaded with fenugreek oil. *Foods*, *10*(12), 3097.

Dhull, S. B., Punia, S., Kumar, M., Singh, S., & Singh, P. (2021b). Effect of different modifications (physical and chemical) on morphological, pasting, and rheological properties of black rice (*Oryza sativa* L. *Indica*) starch: A comparative study. *Starch/Stärke*, *73*(1–2), 2000098.

Dias, A. B., Müller, C. M. O., Larotonda, F. D. S., & Laurindo, J. B. (2010). Biodegradable films based on rice starch and rice flour. *Journal of Cereal Science*, *51*, 213–219.

Domene-Lopez, D., Delgado-Marín, J. J., Martin-Gullon, I., García-Quesada, J. C., & Montalban, M. G. (2019). Comparative study on properties of starch films obtained from potato, corn and wheat using 1-ethyl-3-methylimidazolium acetate as plasticizer. *International Journal of Biological Macromolecules*, *135*, 845–854.

Dutta, P. K., Tripathi, S., Mehrotra, G. K., & Dutta, J. (2009). Perspectives for chitosan based antimicrobial films in food applications. *Food Chemistry*, *114*(4), 1173–1182.

El-Tahlawy, K., Venditti, R. A., & Pawlak, J. J. (2007). Aspects of the preparation of starch microcellular foam particles cross-linked with glutaraldehyde using a solvent exchange technique. *Carbohydrate Polymers*, *67*(3), 319–331.

European Bioplastics. (2020). *Bioplastics market development update*. Retrieved from https://docs.european-bioplastics.org/conference/Report_Bioplastics_Market_Data_2020_short_version.pdf.

Fonseca, L. M., Gonçalves, J. R., El Halal, S. L. M., Pinto, V. Z., Dias, A. R. G., Jacques, A. C., & da Rosa Zavareze, E. (2015). Oxidation of potato starch with different sodium hypochlorite concentrations and its effect on biodegradable films. *LWT-Food Science and Technology*, *60*(2), 714–720.

Fu, Y., & Dudley, E. G. (2021). Antimicrobial-coated films as food packaging: A review. *Comprehensive Reviews in Food Science and Food Safety*, *20*(4), 3404–3437.

Galvão, A. M. M. T., de Oliveira Araújo, A. W., Carneiro, S. V., Zambelli, R. A., & Bastos, M. D. S. R. (2018). Coating development with modified starch and tomato powder for application in frozen dough. *Food Packaging and Shelf Life*, *16*, 194–203.

García, M. A., Martino, M. N., & Zaritzky, N. E. (2000). Microstructural characterization of plasticized starch-based films. *Starch/Stärke*, *52*, 118–124.

García, M. A., Pinotti, A., Martino, M. N., & Zaritzky, N. E. (2004). Characterization of composite hydrocolloid films. *Carbohydrate polymers*, *56*(3), 339–345.

Garcia-Hernandez, A., Vernon-Carter, E. J., & Alvarez-Ramirez, J. (2017). Impact of ghosts on the mechanical, optical, and barrier properties of cornstarch films. *Starch-Stärke*, *69*(1–2), 1600308.

González-Soto, R. A., Núñez-Santiago, M. C., & Bello-Pérez, L. A. (2019). Preparation and partial characterization of films made with dual-modified (acetylation and cross-linking) potato starch. *Journal of the Science of Food and Agriculture*, *99*(6), 3134–3141.

Gutiérrez, T. J., Morales, N. J., Pérez, E., Tapia, M. S., & Famá, L. (2015). Physico-chemical properties of edible films derived from native and phosphated cush-cush yam and cassava starches. *Food Packaging and Shelf Life*, *3*, 1–8.

Han, J. H. (2013). Edible films and coatings: A review. In *Innovations in Food Packaging: Second Edition*. London: Academic Press. doi: 10.1016/B978-0-12-394601-0.00009-6.

Haq, M. A., Hasnain, A., & Azam, M. (2014). Characterization of edible gum cordia film: Effects of plasticizers. *LWT—Food Science and Technology*, *55*(1), 163–169.

Hauck, B. W., & Huber, G. R. (1989). Single screw vs twin screw extrusion. *Cereal Food World*, *24*, 930–939.

Henning, F. G., Ito, V. C., Demiate, I. M., & Lacerda, L. G. (2021). Non-conventional starches for biodegradable films: A review focussing on characterisation and recent applications in food. *Carbohydrate Polymer Technologies and Applications*, 100157.

Hiemenz, P. C., & Rajagopalan, R. (1997). Surface tension and contact angle. In *Principles of colloid and surface chemistry* (pp. 248–255). New York: Marcel Dekker.

Hoover, R. (2001). Composition, molecular structure, and physicochemical properties of tuber and root starches: A review. *Carbohydrate Polymers, 45*, 253–267.

Hosseini, S. N., Pirsa, S., & Farzi, J. (2021). Biodegradable nano composite film based on modified starch-albumin/MgO; antibacterial, antioxidant and structural properties. *Polymer Testing, 97*, 107182.

Hsieh, C. F., Liu, W., Whaley, J. K., & Shi, Y. C. (2019). Structure, properties, and potential applications of waxy tapioca starches—A review. *Trends in Food Science & Technology, 83*, 225–234.

Hu, X., Jia, X., Zhi, C., Jin, Z., & Miao, M. (2019). Improving properties of normal maize starch films using dual-modification: Combination treatment of debranching and hydroxypropylation. *International Journal of Biological Macromolecules, 130*, 197–202.

Ilyas, R. A., Sapuan, S. M., Ishak, M. R., & Zainudin, E. S. (2018). Development and characterization of sugar palm nanocrystalline cellulose reinforced sugar palm starch bionanocomposites. *Carbohydrate Polymers, 202*, 186–202.

Inam u Nisa, Ashwar, B. A., Shah, A., Gani, A., Gani, A., & Masoodi, F. A. (2015). Development of potato starch based active packaging films loaded with antioxidants and its effect on shelf life of beef. *Journal of Food Science and Technology, 52*, 7245–7253.

Jiménez, A., Fabra, M. J., Talens, P., & Chiralt, A. (2012a). Edible and biodegradable starch films: A review. *Food and Bioprocess Technology, 5*(6), 2058–2076.

Jiménez, A., Fabra, M. J., Talens, P., & Chiralt, A. (2012b). Effect of re-crystallization on tensile, optical and water vapour barrier properties of cornstarch films containing fatty acids. *Food Hydrocolloids, 26*(1), 302–310.

Jiménez, A., Fabra, M. J., Talens, P., & Chiralt, A. (2013). Physical properties and antioxidant capacity of starch—sodium caseinate films containing lipids. *Journal of Food Engineering, 116*, 695–702.

Jost, V., Kobsik, K., Schmid, M., & Noller, K. (2014). Influence of plasticiser on the barrier, mechanical and grease resistance properties of alginate cast films. *Carbohydrate Polymers, 110*, 309–319.

Jyothi, A. N., Moorthy, S. N., & Rajasekharan, K. N. (2006). Effect of cross-linking with epichlorohydrin on the properties of cassava (*Manihot esculenta Crantz*) starch. *Starch-Stärke, 58*(6), 292–299.

Kanatt, S. R. (2020). Irradiation as a tool for modifying tapioca starch and development of an active food packaging film with irradiated starch. *Radiation Physics and Chemistry, 173*, 108873.

Khalid, S., Yu, L., Feng, M., Meng, L., Bai, Y., Ali, A., . . . Chen, L. (2018). Development and characterization of biodegradable antimicrobial packaging films based on polycaprolactone, starch and pomegranate rind hybrids. *Food Packaging and Shelf Life, 18*, 71–79.

Kim, J. K., Jo, C., Park, H. J., Byun, M. W. (2008). Effect of gamma irradiation on the physicochemical properties of a starch-based film. *Food Hydrocolloids, 22*, 248–254.

Kumar, N., & Neeraj (2019). Polysaccharide-based component and their relevance in edible film/coating: A review. *Nutrition & Food Science, 49*(5), 793–823.

Lee, J. S., Lee, E. S., & Han, J. (2020). Enhancement of the water-resistance properties of an edible film prepared from mung bean starch via the incorporation of sunflower seed oil. *Scientific Reports, 10*(1), 13622.

Li, M., Liu, P., Zou, W., Yu, L., Xie, F., Pu, H., . . . Chen, L. (2011). Extrusion processing and characterization of edible starch films with different amylose contents. *Journal of Food Engineering, 106*(1), 95–101.

Li, Q., Li, C., Li, E., Gilbert, R. G., & Xu, B. (2020). A molecular explanation of wheat starch physicochemical properties related to noodle eating quality. *Food Hydrocolloids, 108*, 106035.

Lin, L., Huang, J., Zhao, L., Wang, J., Wang, Z., & Wei, C. (2015). Effect of granule size on the properties of lotus rhizome C-type starch. *Carbohydrate Polymers, 134*, 448–457.

Lindeboom, N., Chang, P. R., & Tyler, R. T. (2004). Analytical, biochemical and physicochemical aspects of starch granule size, with emphasis on small granule starches: A review. *Starch/Stärke, 56* (3–4), 89–99.

Liu, C., Yu, B., Tao, H., Liu, P., Zhao, H., Tan, C., & Cui, B. (2021). Effects of soy protein isolate on mechanical and hydrophobic properties of oxidized cornstarch film. *LWT-Food Science and Technology, 147*, 111529.

Liu, H., Li, J., Zhu, D., Wang, Y., Zhao, Y., & Li, J. (2014). Preparation of soy protein isolate (SPI)-pectin complex film containing cinnamon oil and its effects on microbial growth of dehydrated soybean curd (Dry Tofu). *Journal of Food Processing and Preservation, 38*(3), 1371–1376.

Liu, H., Xie, F., Yu, L., Chen, L., & Li, L. (2009). Thermal processing of starch-based polymers. *Progress in Polymer Science, 34*(12), 1348–1368.

Liu, W., Wang, Z., Liu, J., Dai, B., Hu, S., Hong, R., . . . Zeng, G. (2020). Preparation, reinforcement and properties of thermoplastic starch film by film blowing. *Food Hydrocolloids*, *108*, 106006.

Liu, X. X., Yu, L., Liu, H. S., Chen, L., & Li, L., (2008). In situ thermal decomposition of starch with constant moisture. *Polymer Degradation and Stability*, *93*, 260–262.

Liu, Z. (2005). Edible films and coatings from starch. In J. H. Han (Ed.), *Innovations in food packaging* (pp. 318–332). London: Elsevier Academic Press.

Łupina, K., Kowalczyk, D., Lis, M., Raszkowska-Kaczor, A., & Drozłowska, E. (2022). Controlled release of water-soluble astaxanthin from carboxymethyl cellulose/gelatin and octenyl succinic anhydride starch/gelatin blend films. *Food Hydrocolloids*, *123*, 107179.

Mahajan, P., Bera, M. B., Panesar, P. S., & Chauhan, A. (2021). Millet starch: A review. *International Journal of Biological Macromolecules*, *180*, 61–79.

Mallakpour, S., & Khodadadzadeh, L. (2018). Ultrasonic-assisted fabrication of starch/MWCNT-glucose nanocomposites for drug delivery. *Ultrasonics Sonochemistry*, *40*, 402–409.

Marangoni Júnior, L., Vieira, R. P., & Anjos, C. A. R. (2020). Kefiran-based films: Fundamental concepts, formulation strategies and properties. *Carbohydrate Polymers*, *246*, 116609.

Martins, A. B., Silveira, A. M., Morisso, F. D. P., & Santana, R. M. C. (2021). Gelatinized and nongelatinized starch/pp blends: Effect of starch source and carboxylic and incorporation. *Journal of Polymer Research*, *28*(1), 1–11.

Martins, J. T., Cerqueira, M. A., Bourbon, A. I., Pinheiro, A. C., Souza, B. W., & Vicente, A. A. (2012). Synergistic effects between κ-carrageenan and locust bean gum on physicochemical properties of edible films made thereof. *Food Hydrocolloids*, *29*(2), 280–289.

Martins, P. C., Latorres, J. M., & Martins, V. G. (2022). Impact of starch nanocrystals on the physicochemical, thermal and structural characteristics of starch-based films. *LWT Food Science and Technology*, *156*, 113041.

Mehboob, S., Ali, T. M., Sheikh, M., & Hasnain, A. (2020). Effects of cross linking and/or acetylation on sorghum starch and film characteristics. *International Journal of Biological Macromolecules*, *155*, 786–794.

Menzel, C., González-Martínez, C., Vilaplana, F., Diretto, G., & Chiralt, A. (2020). Incorporation of natural antioxidants from rice straw into renewable starch films. *International Journal of Biological Macromolecules*, *146*, 976–986.

Mirmoghtadaie, L., Kadivar, M., & Shahedi, M. (2009). Effects of cross-linking and acetylation on oat starch properties. *Food Chemistry*, *116*(3), 709–713.

Mohammadi Nafchi, A., Moradpour, M., Saeidi, M., & Alias, A. K. (2013). Thermoplastic starches: Properties, challenges, and prospects. *Starch/Stärke*, *65*(1–2), 61–72.

Molavi, H., Behfar, S., Shariati, M. A., Kaviani, M., & Atarod, S. (2021). A review on biodegradable starch based film. *Journal of Microbiology, Biotechnology and Food Sciences*, *2021*, 456–461.

Myllärinen, P., Partanen, R., Seppälä, J., & Forssell, P. (2002). Effect of glycerol on behaviour of amylose and amylopectin films. *Carbohydrate Polymers*, *50*, 355–361.

Narváez-Gómez, G., Figueroa-Flórez, J., Salcedo-Mendoza, J., Pérez-Cervera, C., & Andrade-Pizarro, R. (2021). Development and characterization of dual-modified yam (*Dioscorea rotundata*) starch-based films. *Heliyon*, *7*(4), e06644.

Nawab, A., Alam, F., Haq, M. A., & Hasnain, A. (2016). Biodegradable film from mango kernel starch: Effect of plasticizers on physical, barrier, and mechanical properties. *Starch/Stärke*, *68*(9–10), 919–928.

Nawab, A., Alam, F., & Hasnain, A. (2017). Mango kernel starch as a novel edible coating for enhancing shelf-life of tomato (*Solanum lycopersicum*) fruit. *International Journal of Biological Macromolecules*, *103*, 581–586.

Oyeyinka, S. A., Singh, S., & Amonsou, E. O. (2017). Physicochemical and mechanical properties of bambara groundnut starch films modified with stearic acid. *Journal of Food Science*, *82*(1), 118–123.

Parra, D. F., Tadini, C. C., Ponce, P., & Lugão, A. B. (2004). Mechanical properties and water vapor transmission in some blends of cassava starch edible films. *Carbohydrate Polymers*, *58*(4), 475–481.

Pelissari, F. M., Ferreira, D. C., Louzada, L. B., dos Santos, F., Corrêa, A. C., Moreira, F. K. V., & Mattoso, L. H. (2019). Starch-based edible films and coatings: An eco-friendly alternative for food packaging. *Starches for Food Application*, 359–420.

Pérez, S., & Bertoft, E. (2010). The molecular structures of starch components and their contribution to the architecture of starch granules: A comprehensive review. *Starch/Stärke*, *62*(8), 389–420.

Pérez, P. F., Ollé Resa, C. P., Gerschenson, L. N., & Jagus, R. J. (2021). Addition of zein for the improvement of physicochemical properties of antimicrobial tapioca starch edible film. *Food and Bioprocess Technology, 14*(2), 262–271.

Prachayawarakorn, J., & Kansanthia, P. (2022). Characterization and properties of singly and dually modified hydrogen peroxide oxidized and glutaraldehyde cross-linked biodegradable starch films. *International Journal of Biological Macromolecules, 194*, 331–337.

Priyadarshi, R., & Rhim, J. W. (2020). Chitosan-based biodegradable functional films for food packaging applications. *Innovative Food Science and Emerging Technologies, 62*.

Punia, S., Kumar, M., Siroha, A. K., Kennedy, J. F., Dhull, S. B., & Whiteside, W. S. (2021). Pearl millet grain as an emerging source of starch: A review on its structure, physicochemical properties, functionalization, and industrial applications. *Carbohydrate Polymers, 260*, 117776.

Punia, S., Siroha, A. K., Sandhu, K. S., & Kaur, M. (2019). Rheological and pasting behavior of OSA modified mungbean starches and its utilization in cake formulation as fat replacer. *International Journal of Biological Macromolecules, 128*, 230–236.

Rahmasari, Y., & Yemiş, G. P. (2022). Characterization of ginger starch-based edible films incorporated with coconut shell liquid smoke by ultrasound treatment and application for ground beef. *Meat Science, 188*, 108799.

Ramos da Silva, L., Velasco, J. I., & Fakhouri, F. M. (2022). Bioactive films based on starch from white, red, and black rice to food application. *Polymers, 14*(4), 835.

Ratnayake, W., Hoover, R., Shahidi, F., Perera, C., & Jane, J. (2001). Composition, molecular structure, and physicochemical properties of starches from four field pea (*Pisum sativum* L.) cultivars. *Food Chemistry, 74*, 189–202.

Roy, K., Thory, R., Sinhmar, A., Pathera, A. K., & Nain, V. (2020). Development and characterization of nano starch-based composite films from mung bean (*Vigna radiata*). *International Journal of Biological Macromolecules, 144*, 242–251.

Sandhu, K. S., Sharma, L., Singh, C., & Siroha, A. K. (2017). Recent advances in biodegradable films, coatings and their applications. In *Plant biotechnology: Recent advancements and developments* (pp. 271–296). Singapore: Springer.

Sandhu, K. S., & Siroha, A. K. (2017). Relationships between physicochemical, thermal, rheological and in vitro digestibility properties of starches from pearl millet cultivars. *LWT-Food Science and Technology, 83*, 213–224.

Santana, J. S., de Carvalho Costa, É. K., Rodrigues, P. R., Correia, P. R. C., Cruz, R. S., & Druzian, J. I. (2019). Morphological, barrier, and mechanical properties of cassava starch films reinforced with cellulose and starch nanoparticles. *Journal of Applied Polymer Science, 136*(4), 47001.

Santos, A. R., Da Silva, A. F., Amaral, V., Ribeiro, A. B., de Abreu Filho, B. A., & Mikcha, J. M. (2016). Application of edible coating with starch and carvacrol in minimally processed pumpkin. *Journal of Food Science and Technology, 53*(4), 1975–1983.

Sarak, S., Boonsuk, P., Kantachote, D., & Kaewtatip, K. (2022). Film coating based on native starch and cationic starch blend improved postharvest quality of mangoes. *International Journal of Biological Macromolecules, 209*, 125–131.

Scarfato, P., Avallone, E., Galdi, M. R., Di Maio, L., & Incarnato, L. (2017). Preparation, characterization, and oxygen scavenging capacity of biodegradable α-tocopherol/PLA microparticles for active food packaging applications. *Polymer Composites, 38*(5), 981–986.

Sharma, V., Kaur, M., Sandhu, K. S., Kaur, S., & Nehra, M. (2021a). Barnyard millet starch cross-linked at varying levels by sodium trimetaphosphate (STMP): Film forming, physico-chemical, pasting and thermal properties. *Carbohydrate Polymer Technologies and Applications, 2*, 100161.

Sharma, V., Kaur, M., Sandhu, K. S., Nain, V., & Janghu, S. (2021b). Physicochemical and rheological properties of cross-linked Litchi Kernel starch and its application in development of bio-films. *Starch/Stärke, 73*(7–8), 2100049.

Sifuentes-Nieves, I., Hernández-Hernández, E., Neira-Velázquez, G., Morales-Sanchez, E., Mendez-Montealvo, G., & Velazquez, G. (2019). Hexamethyldisiloxane cold plasma treatment and amylose content determine the structural, barrier and mechanical properties of starch-based films. *International Journal of Biological Macromolecules, 124*, 651–658.

Singh, N., Kaur, L., Sandhu, K. S., Kaur, J., & Nishinari, K. (2006). Relationships between physical, morphological, thermal, rheological properties of rice starches. *Food Hydrocolloids, 20*, 532–542.

Singh, N., Sandhu, K. S., & Kaur, M. (2004). Characterization of starches separated from Indian chickpea (*Cicer arietinum* L.) cultivars. *Journal of Food Engineering, 63*(4), 441–449.

Siroha, A. K., Bangar, S. P., Sandhu, K. S., Trif, M., Kumar, M., & Guleria, P. (2021). Effect of cross-linking modification on structural and film-forming characteristics of pearl millet (*Pennisetum glaucum* L.) starch. *Coatings, 11*(10), 1163.

Siroha, A. K., Punia, S., Kaur, M., & Sandhu, K. S. (2020a). A novel starch from Pongamia pinnata seeds: Comparison of its thermal, morphological and rheological behaviour with starches from other botanical sources. *International Journal of Biological Macromolecules, 143*, 984–990.

Siroha, A. K., Punía, S., Sandhu, K. S., & Karwasra, B. L. (2020b). Physicochemical, pasting, and rheological properties of pearl millet starches from different cultivars and their relations. *Acta Alimentaria, 49*(1), 49–59.

Song, H. G., Choi, I., Lee, J. S., Chung, M. N., Yoon, C. S., & Han, J. (2021). Comparative study on physicochemical properties of starch films prepared from five sweet potato (*Ipomoea batatas*) cultivars. *International Journal of Biological Macromolecules, 189*, 758–767.

Srichuwong, S., Sunarti, T. C., Mishima, T., Isono, N., & Hisamatsu, M. (2005). Starches from different botanical sources II: Contribution of starch structure to swelling and pasting properties. *Carbohydrate Polymers, 62*(1), 25–34.

Svegmark, K., & Hermansson, A. M. (1993). Microstructure and rheological properties of composites of potato starch granules and amylose: a comparison of observed and predicted structures. *Food Structure, 12*(2), 181–193.

Syafiq, R., Sapuan, S. M., Zuhri, M. Y. M., Ilyas, R. A., Nazrin, A., Sherwani, S. F. K., & Khalina, A. (2020). Antimicrobial activities of starch-based biopolymers and biocomposites incorporated with plant essential oils: A review. *Polymers, 12*(10), 2403.

Thakur, R., Pristijono, P., Golding, J. B., Stathopoulos, C. E., Scarlett, C., Bowyer, M., . . . Vuong, Q. V. (2018). Effect of starch physiology, gelatinization, and retrogradation on the attributes of rice starch-ι-carrageenan film. *Starch/Stärke, 70*(1–2), 1700099.

Thakur, R., Pristijono, P., Scarlett, C. J., Bowyer, M., Singh, S. P., & Vuong, Q. V. (2019). Starch-based films: Major factors affecting their properties. *International Journal of Biological Macromolecules, 132*, 1079–1089.

Umaraw, P., Munekata, P. E. S., Verma, A. K., Barba, F. J., Singh, V. P., Kumar, P., & Lorenzo, J. M. (2020). Edible films/coating with tailored properties for active packaging of meat, fish and derived products. *Trends in Food Science and Technology, 98*, 10–24.

Vellaisamy Singaram, A. J., Guruchandran, S., Bakshi, A., Muninathan, C., & Ganesan, N. D. (2021). Study on enhanced mechanical, barrier and optical properties of chemically modified mango kernel starch films. *Packaging Technology and Science, 34*(8), 485–495.

Versino, F., Lopez, O. V., Garcia, M. A., & Zaritzky, N. E. (2016). Starch-based films and food coatings: An overview. *Starch/Stärke, 68*(11–12), 1026–1037.

Wagner, J. R., Mount, E. M., & Giles, H. F. (2014). Extrusion. In *The definitive processing guide and handbook* (2nd ed.). Amsterdam: William Andrew Publishing.

Waters, D. L., Henry, R. J., Reinke, R. F., & Fitzgerald, M. A. (2006). Gelatinization temperature of rice explained by polymorphisms in starch synthase. *Plant Biotechnology Journal, 4*(1), 115–122.

Wattanachant, S., Muhammad, K. M. A. T., Hashim, D. M., & Rahman, R. A. (2003). Effect of cross-linking reagents and hydroxypropylation levels on dual-modified sago starch properties. *Food Chemistry, 80*(4), 463–471.

Wróblewska-Krepsztul, J., Rydzkowski, T., Borowski, G., Szczypiński, M., Klepka, T., & Thakur, V. K. (2018). Recent progress in biodegradable polymers and nanocomposite-based packaging materials for sustainable environment. *International Journal of Polymer Analysis and Characterization, 23*(4), 383–395.

Xia, W., Chen, J., He, D., Wang, Y., Wang, F., Zhang, Q., . . . Li, J. (2019). Changes in physicochemical and structural properties of tapioca starch after high speed jet degradation. *Food Hydrocolloids, 95*, 98–104.

Xie, F., Yu, L., Chen, L., & Li, L. (2008). A new study of starch gelatinization under shear stress using dynamic mechanical analysis. *Carbohydrate Polymers, 72*(2), 229–234.

Yang, J., Juanli, Y., & Yong, H. (2011). Recent developments in gelcasting of ceramics. *Journal of the European Ceramic Society, 31*(14), 2569–2591.

Yang, S., Dhital, S., Zhang, M. N., Wang, J., & Chen, Z. G. (2022). Structural, gelatinization, and rheological properties of heat-moisture treated potato starch with added salt and its application in potato starch noodles. *Food Hydrocolloids, 31*, 107802.

Yang, Y., Li, T., Li, Y., Qian, H., Qi, X., Zhang, H., & Wang, L. (2020). Understanding the molecular weight distribution, in vitro digestibility and rheological properties of the deep-fried wheat starch. *Food Chemistry, 331*, 127315.

Yousefi, A. R., Savadkoohi, B., Zahedi, Y., Hatami, M., & Ako, K. (2019). Fabrication and characterization of hybrid sodium montmorillonite/TiO$_2$ reinforced cross-linked wheat starch-based nanocomposites. *International Journal of Biological Macromolecules, 131*, 253–263.

Zamudio-Flores, P. B., Bautista-Baños, S., Salgado-Delgado, R., & Bello-Pérez, L. A. (2009). Effect of oxidation level on the dual modification of banana starch: The mechanical and barrier properties of its films. *Journal of Applied Polymer Science, 112*(2), 822–829.

Zamudio-Flores, P. B., Vargas-Torres, A., Pérez-González, J., Bosquez-Molina, E., & Bello-Pérez, L. A. (2006). Films prepared with oxidized banana starch: mechanical and barrier properties. *Starch/Stärke, 58*(6), 274–282.

Zhao, J., Schols, H. A., Chen, Z., Jin, Z., Buwalda, P., & Gruppen, H. (2012). Substituent distribution within cross-linked and hydroxypropylated sweet potato starch and potato starch. *Food Chemistry, 133*(4), 1333–1340.

Żołek-Tryznowska, Z., & Kałuża, A. (2021). The influence of starch origin on the properties of starch films: packaging performance. *Materials, 14*(5), 1146.

2

Cellulose-Based Eco-Friendly Films

Juan Alberto Resendiz Vazquez and Marisol Verdín García

CONTENTS

2.1 Introduction

To solve the high pollution era we are facing and the accumulation of petroleum-based non-biodegradable polymers, like polyethene (PE), polyvinyl chloride (PVC), and polypropylene (PP), and also to achieve a green and renewable future, the development of biodegradable, biocompatible, renewable, low-cost, and non-CO_2-emitting materials has been of great interest for many applications, for example, biofilms, packing, aerogels, bioabsorbents, biofiltration, bioseparation process, energy storage biomaterials, pharmaceuticals, etc. (see Figure 2.1) (Huang et al., 2022a; Ren et al., 2022; Xie et al., 2022b; Xu et al., 2022).

In this way, lignocellulosic materials (lignin, hemicellulose, and cellulose compounds) garnered attention a long time ago, owing to their biodegradable, economic, and renewable characteristics, which are of big interest to the manufacturing sector (Garrido-Romero et al., 2022).

From lignocellulosic, cellulose is of main interest. The main reasons are: (1) it is the most abundant natural polymer on Earth; (2) cellulose can be extracted from almost all waste biomass sources, like wood, agricultural residues, bacteria, etc.; and (3) it as well presents good mechanical properties, nontoxicity, hydrophobicity, etc. (Yang et al., 2022b). Structurally, cellulose $(C_6H_{10}O_5)_n$ is a linear homopolymer (compound of amorphous and crystalline cellulose) of D-glucose units linked with β-D 1, 4 glycosidic bondages (see Figure 2.2a) primarily obtained from the plant cell wall (Mary et al., 2022; Xu et al., 2022). It is essential to say that both structure and properties of cellulose depend on the biomass source and extraction methods (Mary et al., 2022).

Any type of cellulose nanoparticles (<100 nm in width and length) can be named nanocellulose. Its light weight, biodegradability, low density, and high tensile strength make nanocellulose an essential

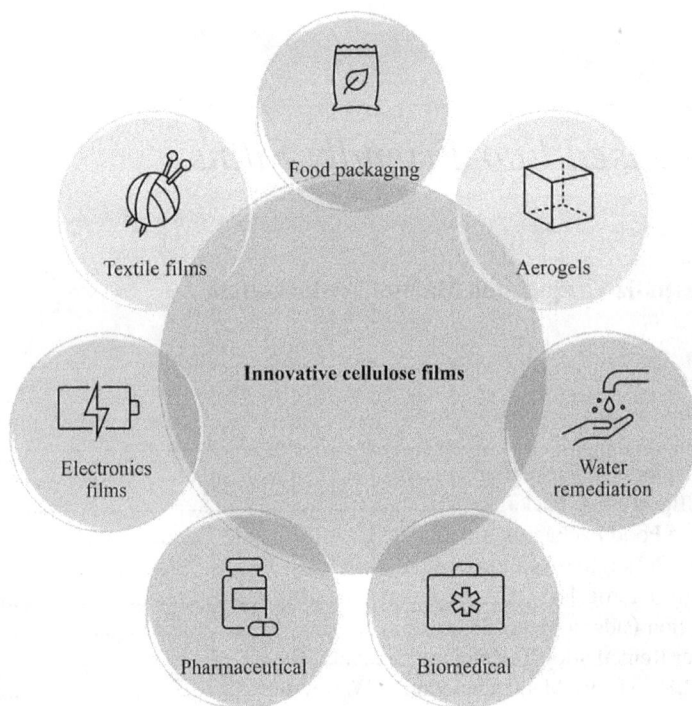

FIGURE 2.1 Main types of applications developed from cellulose and its derivatives.

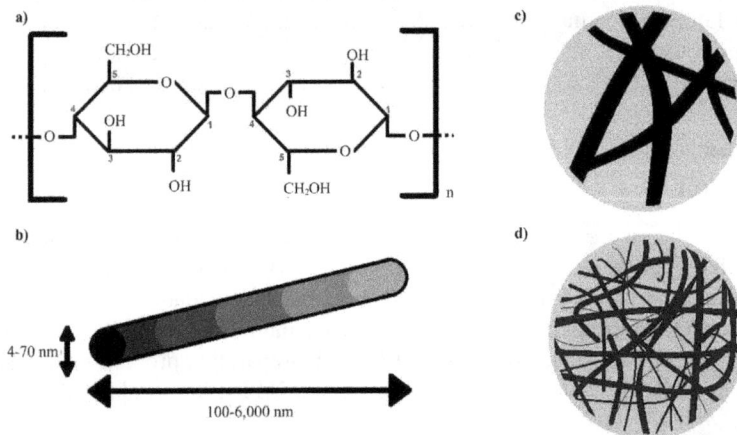

FIGURE 2.2 Main types of micro- and nanocelluloses: (a) cellulose, (b) cellulose nanocrystal, (c) cellulose nanofibrils, and (d) bacterial nanocellulose.

material for many applications. Regularly, nanocellulose is segmented into three types based on the extraction method, crystallinity, particle size, morphology, purity degree, and so on.: (1) cellulose nanocrystals (CNCs), rodlike shape, synthetized mostly from acid hydrolysis; (2) cellulose nanofibrils (CNFs) with cobweb-like interlocking morphologies, obtained by mechanical disintegration (see Figure 2.2c); and (3) bacterial cellulose (BC) with tridimensionality interconnected webs (3D), obtained from microbial fermentation (see Figure 2.2d) (Chen et al., 2022; Mary et al., 2022; Perumal et al., 2022).

From cellulose derivatives, the CNC is of big interest because they are rodlike or whisker-shaped nanoparticles with sizes from 100 to 6,000 nm in length and 4 to 70 nm in width (see Figure 2.2b). Furthermore, it has excellent thermostability and barrier properties (Resendiz-Vazquez et al., 2022). To obtain the CNC, cellulose purification from lignocellulosic biomass is necessary. This process includes sample pretreatment with delignification or bleaching process, lignin removal, and microcellulose collection at this step. Then, more purification steps are crucial, such as acid hydrolysis for the removal of the amorphous part and crystalline fraction fragmentation by mechanical disintegration for the obtention of CNC (Mary et al., 2022; Resendiz-Vazquez et al., 2022).

On the other hand, CNF is characterized by its flexibility in containing both amorphous and crystalline cellulose and is longer than CNC. CNF can be obtained, as previously mentioned, by mechanical disintegration treatments, like ultrasonication, cryo-crushing, high-pressure homogenization, extrusion, electrospinning, etc. The more the mechanical disintegration steps, the smaller the size achieved, with more uniformity (Mary et al., 2022).

Finally, BC is obtained through carbohydrate fermentation. BC is an extracellular metabolite produced by a wide range of Gram-positive and Gram-negative bacteria. This process is divided into two significant events: (1) polymerization and (2) crystallization. Its cellulose has good biocompatibility, higher liquid absorption than other types of plant-derived cellulose, and higher crystallinity for diverse applications, like food packing, electronics, biomedicals, and more (Mary et al., 2022).

In this context, the advancement of nanotechnology has driven the trend of using nanostructured reinforcements in the packaging of materials, incorporating these suitable reinforcement components, resulting in nanocomposite films with improved structural properties by increasing matrix interactions and inducing the formation of a robust three-dimensional network (Xiao et al., 2021). Therefore, the objective of this chapter is to present the most recent developments of biodegradable films based on microcellulose and nanocellulose applied in different complex matrices, such as films or coatings for packing (green, intelligent, and active packaging), aerogels (sorption), water remediation, biomedical applications (biosensors, tissue regeneration), pharmaceutical applications (drug delivery, coatings, drug carriers), electronic (advanced materials and sensors), and textile use (reinforced fibers).

2.2 Packaging

2.2.1 Biodegradable Packaging

Perishable foods that lack protection are highly susceptible to mechanical damage due to handling, transportation, or storage but also are highly prone to decay due to enzymatic activity and microbial metabolites. From the previous aspects, food packaging consists of a basic strategy that positively affects either the protection/preservation of food, such as gases, dust, etc. (Rodrigues et al., 2021). Existing materials applied for the protection of food, like plastic bags and other plastic derivatives, have become essential materials in people's daily lives. Nevertheless, their waste is classified as unsafe materials, disposing annually eight million tons of waste into the open water (Chen et al., 2021). Most of the current synthetic polymers come from petrochemical products. Although some are recyclable, they are not biodegradable; thus, they are recalcitrant in the environment if not treated. Furthermore, they can alter the ecosystems when they are dispersed (Oyeoka et al., 2021).

To amend the pollution caused by synthetic plastic waste, research has been based in recent years on the improvement of eco-friendly materials made by biopolymers and resource derivatives (Chen et al., 2021). The main advantage of biodegradable materials is their ease of degradation by the enzymatic action of living organisms, such as bacteria, yeasts, and molds (Oyeoka et al., 2021). Even though cellulose is the most abundant biopolymer on the planet and is low cost and biodegradable, the mechanical properties of cellulose films are limited in environments with high humidity, which prevents its application as a packaging material for the substitution of traditional synthetic plastics (Chen et al., 2021). Therefore, in the design of food packaging, different forms of the practical application of the biomaterial in food packaging must be taken into account, for example, soaking processes, spreading, wrapping, thermal and pressure treatments, storage, refrigeration, environmental conditions, the interaction of the container with the food, etc. (He et al., 2022).

Currently, ecological food packaging has been developed, capable of matching to a certain extent the physical, chemical, and mechanical properties of synthetic polymers. Among the most recent

TABLE 2.1

Main applications of biodegradable films from cellulose and nanocellulose

Cellulose	Method of elaboration	Modified properties	Polymer matrix	Potential application	Reference
CNC	Extrusion	Renewable material	PBTA/PLA/CNC	Food packaging material	(Andrade et al., 2022)
CNC	Casting	Increase of mechanical properties	PCL/CNC/ZnO	Food packaging	(Gibril et al., 2022)
LCNF	Casting	High resistance, self-cleaning	Chitosan/LCNF	Organic food packaging	(Xu et al., 2022)
CMC	Casting	Biodegradable	CMC/gelatin	Food packaging	(He et al., 2022)
CA	Interfacial molecular film	High density, biodegradable	CA/starch	Biodegradable films	(Yokoyama et al., 2022)
MC	Physicochemical multistage	Ultraviolet shielding	MC/spiropyran	Food packaging	(Qi et al., 2021)
CNC	Casting	Increase in thermal properties (385°C)	PVA/gelatin/CNC	Food packaging	(Oyeoka et al., 2021)
Anisotropic cellulose	Cross-linked	Hydrophobic barrier	Anisotropic cellulose/ myristic acid	Biodegradable waterproof packaging	(Chen et al., 2021)

Abbreviations: *CNC*, cellulose nanocrystals; *MC*, microcellulose; *LCNF*, lignin-contained cellulose nanofibrils; *PVA*, polyvinyl alcohol; *PCL*, polycaprolactone; *CA*, cellulose acetate; *PLA*, poly (lactic acid); *PBTA*, poly (butylene adipate-co-terephthalate); *CMC*, carboxymethyl cellulose; *ZnO*, zinc oxide.

progress in food packaging, there are composite films (biocomposites) based on PBTA/PLA/CNC (poly(butylene adipate-co-terephthalate)/poly (lactic acid)/cellulose nanocrystals) (Perumal et al., 2022); PCL/CNC/ZnO (polycaprolactone/cellulose nanocrystals/zinc oxide) (Gibril et al., 2022); chitosan/LCNF (lignin with cellulose nanofibrils) (Xu et al., 2022) (see Table 2.1), to mention a few examples. The latter (chitosan/LCNF) has competitive properties against petroleum-based polymers, such as ethylene vinyl alcohol (EVOH) and polyvinylidene chloride (PVDC). However, the most important property to highlight is the high degradability of the biopolymers developed, which are under the current trend of generating products that contribute to the reduction of the residuality of the packages and packaging used in the handling and transport of foods. In this sense, the incorporation of nanocellulose (incorporated in the form of CNC and CNF in the biocomposites) is a green and effective strategy to promote the integral properties of composite films due to its inherent biocompatibility (Xu et al., 2022).

2.2.2 Intelligent Food Packaging

Currently, either food security or food waste is an issue of spreading concern for society (Gomes et al., 2022). Each year, 1 in 10 people on Earth falls ill from eating contaminated food. More than 200 diseases are caused by eating contaminated food with undesirable microorganisms, such as bacteria, viruses, and parasites, or chemical substances, such as heavy metals, mycotoxins, and biogenic amines, among others. Antimicrobial-resistant microorganisms can prevail via the food supply chain, through direct contact between animals and people and infected staff, or in the environment (see Figure 2.3). Annually, an estimated 700,000 people die worldwide due to infections with antimicrobial resistance (*A Guide to World Food Safety Day 2022: Safer Food, Better Health*, n.d.).

Moreover, each year globally, betwixt a quarter and a third of the food created for human consumption is lost or spoiled. This is equivalent to 1,300 million tons of food, of which 30% correspond to cereals, 40–50% to roots, oilseeds, vegetables, and fruits, 20% to meat and dairy products, and 35% to seafood. The FAO (Food and Agriculture Organization) has estimated that such food would be enough to feed 2 billion people (*Losses and Food Waste in Latin America and the Caribbean | FAO*, n.d.).

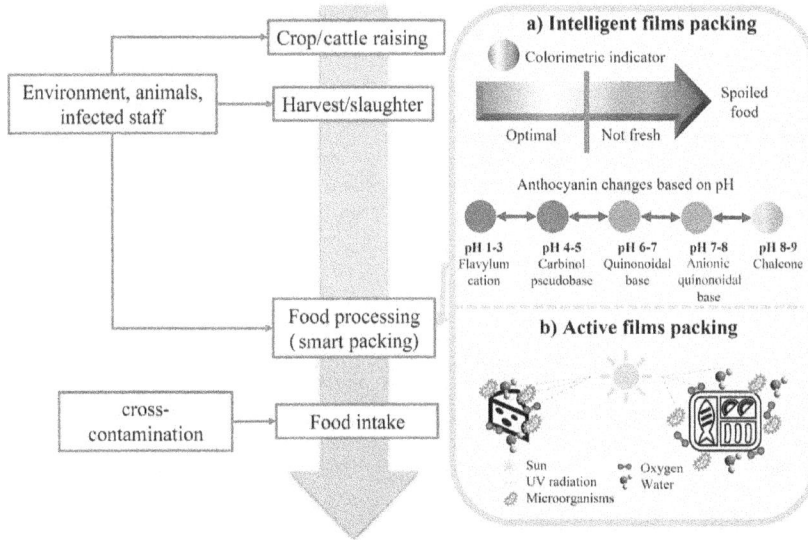

FIGURE 2.3 Schematic illustration of the food supply chain in food processing can be applied smart packing: (a) intelligent packing or (b) active packaging.

Currently, the growing demand and preference of the consumers for fresh and healthy foods, with high quality and long shelf life, has favored the application of state-of-the-art packaging systems to address these problems through the conservation of food and provide information on food quality, maintaining its safety throughout the supply chain to the final consumer (Ghadiri Alamdari et al., 2022; Gomes et al., 2022). Smart packaging is a kind of innovative packaging system whose purpose is to monitor the conditions of the packaged product in real time (Gomes et al., 2022; Shi et al., 2022).

Behind the technologies applied in smart packaging, food spoilage indicators are one of the most studied, since they provide immediate qualitative and semi-quantitative visual information about the packaged food through a color change (see Figure 2.3) (Gomes et al., 2022; Shi et al., 2022).

Within the food industry, these chromatic indicators have been extensively studied with specialized equipment and standardized methods for the analysis, monitoring, and identification of the metabolites involved in the deterioration of food products. This allows for generating an estimate of the shelf life and freshness of the food. However, the main disadvantage is that this evaluation is expensive and can only be carried out *in situ* in the food as quality control.

The measurement of food quality in real time within its packaging can be carried out by monitoring the different quality indicators, such as changes in pH, oxygen levels (O_2), levels of carbon dioxide (CO_2), and temperature indicators, among others (see Table 2.2) (Gomes et al., 2022). In this order of ideas, time–temperature, freshness, and gas leak indicators are the three main devices that are commercially available food freshness indicators (FFI) due to the ability to directly monitor the state of deterioration of food matrices through a color change perceptible to the eye. The FFIs provide benefits that facilitate the incorporation of such smart packaging within the food industry. Some of these benefits are its simplicity, versatility, manufacturing profitability, and sensitivity (Ghadiri Alamdari et al., 2022).

Some pH indicator sensors have been successfully applied to food quality inspection, mainly in pork (Liu et al., 2022b), fish (You et al., 2022), and shrimp (Ghadiri Alamdari et al., 2022). For the evaluation of the modification of the pH in the food, the use of pigments sensitive to pH and solid substrate support is required (You et al., 2022).

The pH change detection in the food matrix is an effective method to identify food degradation since it can indicate the production of various microbial metabolites, such as biogenic amines and organic acids, as part of the degradation of amino acids and glucose, respectively. The preceding leads to an increase in pH due to the accumulation of biogenic amines or the production of acidity caused by the increase in organic acids like carboxylic acid mostly (Gomes et al., 2022).

TABLE 2.2

Trends in the use of smart films from cellulose and nanocellulose applied to food.

Cellulose	Method of elaboration	Properties	Metabolite detected	Composition of the biopolymer	Potential application	Reference
CA	Casting	CRF due to pH change (4 to 8)	Biogenic amines	CA/pyranoflavylium/ glycerol	SIF in food	(Gomes et al., 2022)
BC	Indicator-coated film	CRF due to pH change (3–10)	Biogenic amines	C3G/BC	SIF in tilapia fillets	(Shi et al., 2022)
BC	Indicator-coated film	CRF due to pH change (4–12)	Biogenic amines	BC/ITA	SIF in shrimp	(Ghadiri Alamdari et al., 2022)
CMC	Casting	CRF due to pH change (2–12)	CO_2	CMC/Ovalbumin/ ACN	SIF in mushrooms	(Liu et al., 2022a)
CMC	Physicochemical multistage	CRF due to pH change (3–10)	Biogenic amines	CMC	SIF in pork	(Liu et al., 2022b)
CNF	Casting	CRF due to pH change (2–14)	Not indicated	CNF/PLA/ ACN-thymol	SIF in cherry tomato	(Zabidi et al., 2022)
CMC	Physicochemical multistage	CRF due to pH change (2–12)	Buffer simulation	KGM-CMC/BCA	Real-time monitoring of meat product quality	(You et al., 2022)
CNF	Physicochemical multistage	Thermochromism	NA	CNF/ionic liquids of nickel and chromium/ nanofoams	Real-time monitoring of food packaging and packaging	(Karzarjeddi et al., 2022)

Abbreviations: *CNF*, cellulose nanofibrils; *CA*, cellulose acetate; *C3G*, cyanidin-3-glucoside; *BC*, bacterial cellulose; *ITA*, *Ixiolirion tataricum* anthocyanins; *CMC*, carboxymethyl cellulose; *ACN*, anthocyanins; *PLA*, poly (lactic acid); *KGM*, konjac glucomannan; *BCA*, blackcurrant anthocyanin; *NA*, not applied; *CRF*, chromatic response of the film; *SIF*, smart indicator of freshness.

Currently, research focuses more and more on the rational use of biopolymers and natural pigments, taking into consideration environmental protection and food safety. One of the natural pigments frequently used in the food packaging industry is anthocyanin. Anthocyanins are water-soluble flavonoid compounds commonly found in both fruits (such as berries) and vegetables (like red cabbage) and plants. They offer within their physicochemical properties high sensitivity to changes in pH when the medium is modified. They are characterized by a wide variety of colors, for example, from pH 1-3 it can be seen in red/brown hue; from pH 4-5 a little hue can be noticed; from pH 6-8 purple color can be noticed, and finally pH >11 it is presented in blue tonality (see Figure 2.3a). This characteristic makes its application within the food industry feasible, as a key element in the design and incorporation of smart packaging (Gomes et al., 2022; Rodrigues et al., 2021).

Some researchers developed a biocomposite based on pyranoflavylium as the immobilized colorimetric indicator, at a final concentration of 0.1% (w/w), in cellulose acetate (CA) with different concentrations of glycine (0–40%) as a plasticizer. High sensitivity to pH was found in the films enriched with 20 and 30% glycine after 20 minutes of immersion in an environment rich in the following biogenic amines: cadaverine, putrescine, tyramine, and histamine (0.5–4 g/L) (Gomes et al., 2022). These enriched films showed a remarkable color change (from red to green) in the pH range from 4 to 8, which is characteristic of food deterioration. In the same way, a great response to an environment rich in amines is presented, which suggests a great potential for its application as an indicator of freshness in perishable foods. Similar results were reported in fresh tilapia fillets, but using a polymeric matrix composed of cyanidin-3-glucoside (C3G) bacterial cellulose (BC), presenting a wide detection range of pH change of 3 to 10, just to mention a few examples (see Table 2.2). This represents a rapid, non-invasive, sensitive, and economical detection strategy, based on the change in the pH of the food and determining the degree of freshness in perishable foods, especially in marine products and fresh meat (Shi et al., 2022).

TABLE 2.3

Main applications of active films from cellulose and nanocellulose in food

Cellulose	Method of elaboration	Properties	Composition of the biopolymer	Potential application	Reference
CA	Casting	Antibacterial/ biodegradable	CA/sodium alginate or carrageenan	Biodegradable food/film packaging	(Rajeswari et al., 2020)
Cellulose	Chemical incorporation	High antioxidant and UV blocking properties	Cellulose/tannin	Active food packaging	(Huang et al., 2022a)
CNC	Casting	Antifungal/ antibacterial	PVA/CNC	Active food packaging	(Perumal et al., 2022)
BC	Ultrasound	Antibacterial	BC/G	Nanopaper with antimicrobial activity	(Abral et al., 2022)
CNF	Ultrasound/casting	Antimicrobial/ thermal stability	CNF/AgNPs	Biodegradable food packaging	(Ren et al., 2022)
MC	Cross-link/ dehydration/ etherification	Gas barrier (O_2) and water resistance	MC/PVA	Food packaging	(Xie et al., 2022b)
CNC	Ultrasound-assisted casting	Gas barrier (O_2) and water resistance	Soy protein/CNC	Edible film/ coating	(Xiao et al., 2021)
BC	Thermal cross-linking	Antibacterial activity	BC/gelatin	Candidate for and food packing	(Thongsrikhem et al., 2022)
MFC	Alkaline treatment/ steam explosion/ modification process with ZnO	Gas barrier (CO_2)	PCL/MFC	Food packaging	(Reis et al., 2021)
CNF	Casting	Antifungal/ antimicrobial	CNF/PLA/ ACN-thymol	Active packaging increases shelf life in cherry tomato	(Zabidi et al., 2022)
Cellulose	Chemical cross-linking	Antimicrobial/ biodegradable	Cellulose/ZnO	Food packaging	(Xie et al., 2022a)
CMC	Physicochemical multistage	Antibacterial/ antioxidant	KGM-CMC/BCA	Active packaging of meat products	(You et al., 2022)

Abbreviations: *CNC*, cellulose nanocrystals; *CNF*, cellulose nanofibrils; *MC*, microcellulose; *PVA*, polyvinyl alcohol; *MFC*, microfibrillated cellulose; *PCL*, polycaprolactone; *CA*, cellulose acetate; *G*, *uncaria gambir* leaves extract; *AgNPs*, silver nanoparticles; *KGM*, konjac glucomannan; *BCA*, blackcurrant anthocyanin; *CMC*, carboxymethyl cellulose; *ZnO*, zinc oxide; *PLA*, poly (lactic acid); *ACN*, anthocyanins; *BC*, bacterial cellulose.

However, although there are different packaging systems or smart materials developed for the identification of deterioration in food caused by physical, chemical, and biological changes or their combinations, the use of some smart biocomposites as polymeric systems must be studied to reach their maximum efficiency, practicality, low cost, and evaluation of the smart label-product-consumer interaction.

2.2.3 Active Food Packaging

The progress in ecological, active, environmentally friendly packaging materials has gained further impetus within the packaging industry (Huang et al., 2022a). Food packaging is essential in protecting and preserving all perishable items, such as fresh meat (chicken, fish, pork, and beef), processed fruits, and vegetables. In addition, during the food distribution chain to the markets, this protection is essential to protect the food matrix against dehydration by oxidation, deterioration, and alteration caused by microbial agents to increase the product's shelf life (Reis et al., 2021). In recent years, biofilms have been developed to exert antibacterial and antifungal activity, free radical scavenging activity, barrier properties against UV rays, and barrier properties against gases such as CO_2 and O_2 (see Table 2.3; see Figure 2.3b). Among the most relevant developments in the formulation of biofilms based on cellulose derivatives are those combined with different systems of micro and nanoparticles, such as sodium alginate/carrageenan, silver nanoparticles (AgNPs), zinc oxide (ZnO), essential oils (thymol), and variants of flavonoid compounds (anthocyanins). These composite polymeric systems showed antibacterial activity against some of the leading human pathogens reported in food, such as *Escherichia coli* (*E. coli*), *Salmonella Typhi* (*S. Typhi*), and *Staphylococcus aureus* (*S. aureus*). In addition, in the specific case of cellulose nanofibrils/poly (lactic acid)/anthocyanins/thymol (CNF/PLA/ACN-thymol) and polyvinyl alcohol/cellulose nanocrystals (PVA/CNC) biofilms, an inhibitory antifungal effect was observed, evaluated in cherry tomato and against *Colletotrichum gloeosporioides* and *Lasiodiplodia theobromae*, respectively. These investigations show the potential application of these novel biocomposites in food packaging which can significantly prolong (depending on the food matrix) the freshness and shelf life of high-consumption perishable products in supermarkets, thus reducing food losses caused by food management, in which according to the report on the UN's (United Nations) 2021 food waste index, it is estimated that 931 million tons of food (equivalent to 17% of the total food produced and available to consumers) are lost in the value chain (*El Desperdicio Masivo de Alimentos, Un Problema No Solo de Los Países Ricos | Noticias ONU*, n.d.).

2.2.4 Other Packaging Applications

Cellulose, as a renewable biopolymer, was recognized as a promising alternative to petroleum polymer. However, cellulose's poor thermoplasticity caused a limitation in its full utilization because it cannot be dissolved in common solvents and melted during the heating-extrusion process (Yang et al., 2022c). To date, cellulose-based plastic materials are primarily achieved through solvent-based processing. Contrarily, the processing of thermoplastics is efficient, simple, and green. Therefore, the preparation of melt-processable cellulose (or derivatives) is attractive for eco-friendly manufacturing, which maximizes the economic potential of cellulose (see Table 2.4). In this sense, some researchers developed an easier strategy for the heterogeneous obtention of cellulose-grafted thermoplastic polyurethane with a green approach for the heterogeneous preparation of thermoplastic cellulose-grafted polyurethane (RCP-g-PU) from amorphous regenerated cellulose pulp (RCP) via hydroxyl/isocyanate chemistry, where a series of RCP-g-PU thermoplastics with pour temperatures ranging from 160°C to 226°C were synthesized by adding hexamethylene diisocyanate in RCP without using other organic solvents. Subsequently, the resulting RCP-g-PU can be directly hot-pressed into transparent films with flexibility and foldability (Hou et al., 2020). Furthermore, the solvent-free preparation of thermoplastic biomaterials from MC (micro cellulose) by reactive extrusion has been reported, where the polymeric matrix was developed from MC, kraft lignin (KL), and glycerol. The polymer matrix was formed by screw extrusion and subsequent hot-pressing, obtaining a polymer with hydrophobic characteristics that can remain stable in aqueous solutions for up to seven days, demonstrating its excellent barrier capability, mechanical and thermoplastic properties, with potential application in multiple polymer industries; owing to the main component of the polymer (80 wt%) is the biodegradable lignocellulosic material, unlike most of the polymeric matrices implemented to date (Yang et al., 2022c).

In this same context, the application of soybean hulls has been investigated as a source to extract MFC (microfibrillated cellulose) and used as reinforcement in polymers based on thermoplastic starch (TPS) and PVA (polyvinyl alcohol) for thermoplastic injection, where the addition of MFC and 6% PVA to TPS made the composites more rigid and mechanically reinforced, providing greater dimensional stability after thermal injection (Bortolatto et al., 2022). Recent advances have been reported in the implementation of these TPS materials reinforced with cellulose particles (3–20 wt%) in the e-commerce or delivery industry, where these materials are being used as primary, secondary packaging, or low-grammage product packaging (Ma et al., 2022; Noshirvani et al., 2018). Other examples of cellulose-based thermoplastic materials are presented in Table 2.4.

Another innovative and alternative method to produce biodegradable packaging is the life cycle assessment of cellulose nanofibril films by spray deposition and vacuum filtration pathways for small-scale

TABLE 2.4

Other applications from cellulose and nanocellulose in the packaging industry

Cellulose	Method of elaboration	Properties	Composition of the biopolymer	Potential application	Reference
RCP	Thermoformable	High transparency, flexibility, and foldability through hot-pressing	RCP-g-PU	Thermoformable packaging/ extruded packaging	(Hou et al., 2020)
MFC	Extrusion followed by injection	Increased tensile strength	TPS/PVA/ MFC	Thermoplastic materials	(Bortolatto et al., 2022)
MC	Extrusion followed by hot-pressing	Increased tensile strength and water vapor permeation	TPS/MC/ glycerol	Thermoplastic materials	(Chen et al., 2020)
CNF	Casting	Exceptional tensile strength (~161%), tensile modulus (~167%), thermal stability, and crystallinity	TPS/CNF/ glycerol	Polymer industry	(Dominic et al., 2021)
MC + KL	Extrusion reactive followed by hot-pressing	Excellent thermoplastic and water permeation	MC/KL/ glycerol	Multipurpose polymer industry	(Yang et al., 2022c)
CNF	Spray deposition or vacuum filtration	Reduction of environmental impact	CNF	Polymer industry	(Nadeem et al., 2022)
CNF	Casting	Thermal stability and mechanical resistance	Kc/Alg/CNF	Biodegradable household items and coatings	(Ulrich & Faez, 2022)

Abbreviations: *TPS*, thermoplastic starch; *CNF*, cellulose nanofibrils; *MC*, microcellulose; *PVA*, polyvinyl alcohol; *MFC*, microfibrillated cellulose; *Kc*, κ-carrageenan; *Alg*, Alginate films; *KL*, kraft lignin; *PU*, polyurethane; *RCP*, amorphous regenerated cellulose paste.

production. These alternative film-formation methods aim to minimize water consumption, energy, gas, and pollutant emissions into the environment. However, although films produced through CNF spray deposition have lower environmental impacts than films produced by vacuum filtration, these impacts were approximately 16–21% higher than with a PET film with a similar basis weight. However, to date, no industry benchmark for other nanomaterials or CNF film/coating products can be compared (Nadeem et al., 2022).

2.3 Sorption Films (Aerogels)

The recovery of residual biomass from industrial activities or even natural waste is an attractive approach for the extraction of biopolymers and the production of innovative nanomaterials such as aerogels. The latter are defined as highly porous three-dimensional (3D) materials, extremely light and porous, with a high surface area and low density (Benito-González et al., 2021; Ihsanullah et al., 2022; Zhuang et al., 2022).

Aerogels can be classified as films/membranes, powders, and monoliths based on their structure; based on their appearance, they can be divided into mesoporous, microporous, and mixed; and based on their composition, they can be organic, inorganic, or hybrid, which are useful for an extensive range of applications, owing to their high absorption/adsorption capacity (Ihsanullah et al., 2022). Among the main applications of aerogels is in bioremediation, for example, the absorption or adsorption of some contaminants such as oil, heavy metals, gases, salts, medicines, etc. (Ihsanullah et al., 2022; Peng et al., 2022), which have attracted attention in other areas, such as in food packaging, in the treatment of microplastics in water, and as drug carriers (see Figure 2.4a). However, its application in these areas has been scarcely studied.

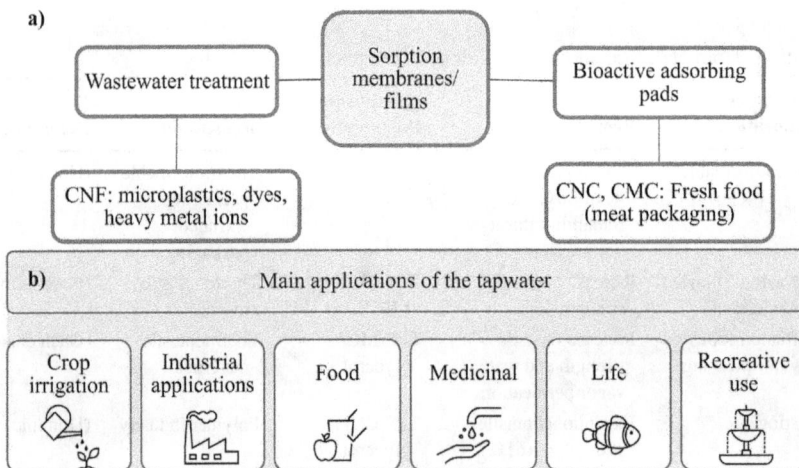

FIGURE 2.4 Main applications of sorption membranes/films and tapwater: (a) Cellulose membrane/films and derivatives; (b) Main applications of fresh water.

2.3.1 Sorption Pads in Meat

In the case of meat packaging, absorption pads are frequently used as humidity control elements, which allow the retention of excess liquids released by the meat during storage. These pads are usually made of a non-stick, non-permeable synthetic polymer, such as polyethene, and a hydrophilic non-woven bottom layer containing active substances that limit microbial growth, such as citric acid and sodium bicarbonate (Fontes-Candia et al., 2019).

Even though these permeable synthetic polymers are predominantly obtained by chemical synthesis, this makes them hard to degrade and effect a negative impact on the environment when disposed of in the trash (Yang et al., 2022a). A sustainable alternative to polyethene-based pads are pads made from cellulose. However, one of the main drawbacks of cellulose-based materials is their highly hydrophilic character, which can be detrimental to their application in food packaging (Benito-González et al., 2021).

Another problem related to the synthesis of cellulose aerogels is that they are typically acquired utilizing a complex synthesis method involving multiple steps: (a) disruption of its crystalline structure for the cellulose dissolution, (b) gelation, (c) cellulose regeneration, (d) solvent exchange, and (e) freeze-drying or supercritical drying. Such a process presents two main drawbacks: high costs and the unsuitability of the produced aerogels for food-grade applications, owing to the usage of organic solvents (Fontes-Candia et al., 2019). To get over this obstacle, often seen is the application of sophisticated strategies based on chemical modification, the reticulation of the coating with hydrophobic compounds not suitable for food usage, compromising the sustainability of the materials obtained (Benito-González et al., 2021).

In the last years, a broadened eco-friendly alternatives and compatible alternatives have been developed for these systems, from the creation of cellulose aerogels with hydrophobic and antioxidant properties to achieve the maximum preservation and shelf life characteristics during meat-chilling storage (Benito-González et al., 2021; Fontes-Candia et al., 2019).

A group of researchers made a *Posidonia oceanica* waste biomass, a hydrophilic and hydrophobic antioxidant extract, with a green technique and incorporated them into cellulose aerogels. These extracts made aerogels less porous and more compact structures. What they found was that all gels were capable of adsorbing considerable quantities of both water in less than three hours (1,500%) or oil in less than seven days (1,900%) (see Table 2.5) (Benito-González et al., 2021).

2.3.2 Water Remediation Technology

Water covers ~70% of our planet's surface and is indispensable for life as we know it; however, nearly 3% of it corresponds to fresh water. For food processing, irrigation of crops, drinking, and most industrial applications fresh water is necessary (see Figure 2.4b). Water pollution can be attributed mainly to the disposition of large amounts of wastewater with non-permissible levels of heavy metal ions, such as Cd(II), Pb(II), Ni(II), Cu(II), Hg(II), Zn(II), Co(II), and Mn(II), and organic and inorganic compounds, petroleum, drugs, salts, and synthetic dyes, without treatment before discharge into open water. Wastewater treatment is essential for the protection of aquatic, soil, and air environments and therefore ensuring both animal and human well-being. The United Nations (UN) in 2010 had established water sanitation as a main goal. In this sense, it required sanitation systems or technologies based on eco-friendly methodologies to safeguard water remediation (James, 2020; Lustenberger & Castro-Muñoz, 2022; Yang et al., 2022b; Zhuang et al., 2022).

A leading wastewater treatment technology is membrane filtration. Commercial membranes for water treatment are made of synthetic polymers like polyvinylidene fluoride (PVDF), polyacrylonitrile (PAN), polysulfone (PS), nylon (NY), polyethylene (PE), and polyethersulfone (PES). The main issue with the usage of these types of membranes is its dirtying, demanding frequent maintenance, and also its difficult degradation in the environment. To face THESE problems, these synthetic polymers have now been replaced, little by little, with natural polymers. Natural cellulose membranes are the most popular ones for water filtration, especially nanocellulose (CNC and CNF), which has gained special focus due to increased surface area and functionality, improving membrane properties (Lim et al., 2020).

For instance, a membrane system has been developed from CNF with ~80% porosity, showing a permeation flux of 127.6 ± 21.8 L·m^{-2}·h^{-1}·bar^{-1}, efficiency separation of 99.9%, excellent flexibility, stability in hot water, pH resistance, wet resistance from 3.5 to 8.0 MPa, and low cost. These shown characteristics make nanocellulose membranes a promising and feasible alternative for ultrafiltration membranes in wastewater treatment (Lim et al., 2020).

This section presents the ultimate developments in cellulose-based adsorption aerogels focused on heavy metal ion chelation, microplastics, and dyes as three of the main contaminants present in wastewater.

2.3.2.1 Heavy Metal Chelation in Wastewater

Technologies to address the metal-contaminated water include adsorption, reverse osmosis, electrodialysis, photodegradation, ultrafiltration, ion exchange, and membrane separation. From these ones, the most economical is adsorption, having the most common adsorbents: activated carbon, cellulose, metal oxides, etc. (Ihsanullah et al., 2022).

Recently established was a novel method to produce high-porosity aerogels from pineapple leaf waste fibers. These aerogels were functionalized with diethylenetriamine, for the Ni (II) ions confiscation from wastewater, showing a maximum adsorption capacity of 0.835 mmol/g (Lim et al., 2020).

To obtain a good aerogel for heavy metal chelation in an aqueous environment, some physicochemical characteristics must be taken into consideration, such as the surface functional groups, porosity, and stability, which are intimately related to the efficiency of heavy metal ion uptake. It is demonstrated that the more hydrophilic groups (containing O, S, P, and N) on the surface, the better the adsorption. It is also seen that a higher surface area can provide more active sites for the adsorption; thus, a better yield can be achieved. A summary of the ultimate investigations for wastewater remediation on the cellulose nanofibrils (CNF)–based adsorbent or absorbent, along with their respective experimental conditions, is depicted in Table 2.5 (Lim et al., 2020).

2.3.2.2 Microplastics

Microplastics, by definition, are plastic particles (<5 mm in size) but also can be present as tiny microplastics (<20 μm). These microplastics are normally found with more ampleness in water and thus present a severe issue of biotoxicity derived from difficulty in degradation. Natural Earth events such as wind, climate, ocean currents, and human activity can help dissipate microplastics, resulting in a negative ecological impact. In humans, it is demonstrated that microplastics can bioaccumulate in the body via intake (drinking water, present at any step in the food chain), settling in the digestive tract, placenta, and blood. Owing to the previous reasons, some countries have forbidden microplastics due to their toxicity (Zheng et al., 2022; Zhuang et al., 2022).

The conventional way for water purification is its sanitization through water plants. The main problem with water plants is that their process is not designed for microplastic removal, so treated water still contains high quantities of microplastic when discharged into the environment. It is estimated that between 15,000 and 4.5 million microplastic particles are disposed every day into open water. Some water plants include rapid sand filtration, disc filters, dissolver air flotation, etc. Filtration has resulted in an effective strategy for large size microplastics; however, this is inefficient for tiny microplastics (due to their size, it can easily block the membrane filter). Membrane bioreactors remove 95% of >20 μm microplastics; nonetheless, its operational cost is relatively high and has constant membrane fouling (Zheng et al., 2022; Zhuang et al., 2022).

To solve this main problem, aerogels can be used for the reasons previously described (high specific surface area, high porosity, low density, high specific surface area, and more); therefore, aerogels have resulted to being very effective and widely used in wastewater bioremediation. Another benefit of the adsorption method is its efficiency, low-cost operation, and easy processing (Zhuang et al., 2022).

To face this problem, some researchers used cellulose nanofibers (CNF), but this matrix has poor adsorption capacity compared to microplastics. To solve it, they incorporated a modifier, 3-epoxypropyl trimethyl ammonium chloride, due to its good cationization efficiency. They used this modified aerogel for the adsorption/removal of microplastics. They informed a maximum value of 146.38 mg/g, which was ~9 times higher when compared with the control (unmodified aerogel) 15.58 mg/g. Before this research, there was no available method for efficient microplastic reduction in water, so this offers a promising application for the anionic, non-desirable molecules present in the wastewater (Zhuang et al., 2022).

To date, the amount of information in this investigation line is almost nonexistent, stating the urgent need for the generation of novel knowledge to mitigate ecological and human health issues.

2.3.2.3 Colorant or Dye Bioremediation in Water

Directly or indirectly, diverse industries like textile, paper, leather, plastics, cosmetics, pharmaceutical, food and drug manufacturing have spread great quantities of dyes in the environment, having, therefore, colorant pollution. These types of molecules present complex aromatic structures with great stability in the environment, thus causing diseases in both humans (e.g., kidney, skin, brain, liver, central nervous, reproductive, and endocrine systems damage and also mutagenic and carcinogenic effects) and animals; nevertheless, their severity depends on the frequency and the pollutant's nature. For this reason, there is an urgent need for dye removal from wastewater before discharge into open water (Dadigala et al., 2022; Sanchez et al., 2022).

There have been developed some strategies to counter these organic dyes, like adsorption and membrane filtration; however, they are of low efficiency (depending on the approach applied). An alternative is the usage of nanozymes. Nanozymes are nanomaterials which can ape the enzyme's natural activity. In this sense, Pd nanoparticles (PdNPs) are widely employed as nanozymes, having the capability of ROS (reactive oxygen species) production in the presence of H_2O_2, essential for dye degradation. Even though there are two main issues of its application in the colloidal form, they aggregate each other, thus reducing catalytic activity and low recovery and recyclability due to its small size; hence, the operational cost increases. To address this limitation, nanocellulose could be used as a support for metal nanoparticles, providing great surface area, biocompatibility, and structural flexibility so it can be prepared in various morphological structures (films, foams, aerogels, and beads), but especially, the nanocellulose presents simple processing (Dadigala et al., 2022; Sanchez et al., 2022).

A research on cellulose nanofibrils–supported PdNPs (PdNPs/PCNF) with outstanding peroxidase and oxidase activities, an optimal activity pH 5.0 for dye degradation, with good recyclability (ten cycles) with ~90% activity at ambient temperature and 0.4 M H_2O_2, has been reported (Dadigala et al., 2022). This offers a novel method for dye pollutant removal from wastewater.

Another magnetic material studied are iron oxide nanoparticles, demonstrating being a good pollutant remover. Other researchers developed in polyvinyl alcohol (PVA) with five different formulations of CNF obtained from wheat straw for PVA/CNF beads creation, containing iron oxides as the magnetic nanoparticle (MNPs). They proved that PVA, CNF, and MNPs had strong interactions with each other through hydrogen bonding; thus, the magnetic beads were able to remove both cationic and anionic dye pollutants from an aqueous solution (Sanchez et al., 2022).

TABLE 2.5

Trends in the use of nanocellulose as sorption films/membranes

Type of cellulose	Method of elaboration	Properties	Composition	Application	Reference
CNF	Cross-linked/ self-cross-linking	Hydrophobic properties	CNF-PAE	Wastewater treatment	(Yang et al., 2022b)
CNF	Liquid nitrogen freezing	Absorption of microplastics	CNF/EPTMAC/ PVA	Bioremediation/ removing microplastics from water	(Zhuang et al., 2022)
CNF	Blending/ freeze-dryer	Hydrophobic and lipophilic properties	CNF/MMT/PEI/ OTS	Wastewater treatment (colorants)	(Fan et al., 2022)
CNF	Ion cross-linked	3D porous membrane/ antimicrobial activity	CNF/curcumin	Drug carrier	(Jose et al., 2022)
CNF	*In situ* physical/ chemical double cross-linking	Electrostatic attraction and chelating effect	CNF/ polyacrylamide	Wastewater treatment (heavy metal, Cu-II)	(Mo et al., 2022)
CNF	Ion cross-linked	Magnetic nanorods	CNF/PVA	Wastewater treatment (pollutants cationic and anionic)	(Sanchez et al., 2022)
CMC	Cross-linked/ ultra-freeze/ freeze-dryer	Absorbent antibacterial surface	CMC/AgNPs/ BC/CA	Bioactive adsorbing pads/fresh packaged foods	(Yang et al., 2022a)
CNF	Cross-linked/ solvent exchange/ supercritical CO$_2$	Porous membrane	Cellulose/pectin	Drug carrier	(Groult et al., 2022)
CNC	Cross-linked/ freeze-dryer/ coated multilayer	High sorption capacities of water/oil (1,500–1,900 %), reduced oxymyoglobin and lipid, oxidation in red meat upon storage	CNC (hydrophilic)/ CNC-PLA (hydrophobic)/ extract phenolic	Bioactive adsorbing pads/fresh packaged foods	(Benito-González et al., 2021)

Abbreviations: CNC, cellulose nanocrystals; CNF, cellulose nanofibrils; PVA, polyvinyl alcohol; AgNPs, silver nanoparticles; CMC, carboxymethyl cellulose; PLA, poly (lactic acid); EPTMAC, 2, 3-epoxypropyl trimethyl ammonium chloride; MMT, montmorillonite; PEI, polyethyleneimine; OTS, octadecyl trichlorosilane.

2.4 Biomedical and Pharmaceutical Applications

In general, cellulose nanoparticles are an alluring class of nanomaterials in a variety of applications, as seen throughout this chapter. Nanocellulose exhibits many attractive properties, including high tensile strength, large surface area, rigidity, colloidal stability, and modification capability. Cellulose nanomaterials are currently being destined as promising eco-friendly biomaterials which can be obtained from diverse plant resources, in substitution of non-biocompatible synthetic materials. These biomaterials, due to their superior intrinsic biological properties, such as non-toxicity, biocompatibility, biodegradability, and low cost of both preparation and processing, are of interest not only for researchers but also for companies in specific advocated pharmacy or biomedical applications (Janmohammadi et al., 2023; Mali & Sherje, 2022; Reshmy et al., 2021). In this section, a brief review of the most recent advances and applications of cellulose nanoparticles is discussed with special emphasis on the bioproperties and their innovational applications in selected biomedical fields (e.g., drug delivery systems, tissue implants, and wound healing), as well as principal current limitations and future potential in these applications.

Cellulose nanoparticles are being used exhaustively because they are one of the main promising eco-friendly material resources in different areas of industrial interest, particularly in the pharmaceutical area, where a variety of applications is being developed for different types of drugs owing to their most envied properties, such as sustainability, biocompatibility, biodegradability, and their physical, structural, and chemical properties. In the last year, researchers have focused on cellulose nanoparticle–based systems as drug delivery vehicles. Controlled drug release represents a variable potential for numerous applications and administration routes (Das et al., 2022; Samanta et al., 2022).

The main advantage of drug delivery systems is that they can be designed for site-specific drug delivery, responding to both environmental and biological stimuli, like temperature, pH, and electric, magnetic, and light fields (Mali & Sherje, 2022).

Thus, cellulose and its derivatives are being studied as an excipient for the modulation of drug release or design of advanced drug carrier systems. Also, CNC have been studied for drug delivery via nanoparticles, nanocomposites, and other systems (Mali & Sherje, 2022).

Despite its long history and knowledge of tableting, research into the usage of cellulose and new types or derivatives of cellulose (e.g., CNF, CNC, and BC) on drug carrier systems is still in sophisticated developments. The main objective is to lower the disintegration rate of the tablets as special excipients, or to extend the release of drugs as novel drug carriers (Das et al., 2022).

At present, the improvement of novel forms of both materials and vehicles is being placed on cellulose nanoparticles, which, at a molecular level, is based on the generation of nanostructured excipients that harbor active ingredients, for example, pigments, drugs, or molecules with beneficial activity on the human body (see Table 2.6). These cellulose-based systems can be established from native or functionalized structures, with the main goal of incrementing or enhancing the effect in blank organs (Das et al., 2022; Schmidt et al., 2022).

To generate a pharmacological systemic effect, the drug can be administered by diverse absorption, such as intake, intravenous, inhalation, dermic, or injection. Nonetheless, among the aforementioned, the preferred route is oral administration, in which the drug or the bioactive compound is taken (Das et al., 2022).

To this extent, it was designed a novel oral controlled-release active carrier system based on chitosan/sodium alginate/ethyl cellulose (CS/SA/EC) using zolmitriptan and etodolac as test bioactives. This was never reported before in studies for its use in film preparation for release in the buccal mucosa (Wang et al., 2022).

Furthermore, established was an emerging 3D printing technology to construct versatile printable materials for drug delivery. However, the 3D printing–nanomaterials relationship, to date, has been scarcely investigated in the pharmaceutical area. The researchers used a nanostructured mesoporous silica matrix (SBA-15) and CMC (SBA-15/CMC) as the carrier matrix for the hydrophobic triamcinolone acetonide drug that was 3D-printed as a hydrophilic polymeric film using the semisolid extrusion (SSE) technique. The 3D-printed films showed complete drug release after 12 hours, and the presence of triamcinolone acetonide–loaded SBA-15 enhanced its *in vitro* mucoadhesion, elucidating their promising application in some oral mucosal treatments. These researchers concluded that this system represents an innovative platform with potential use for the development of water-based mucoadhesive formulations incorporating a hydrophobic drug. This is the first report proposing the development of 3D-SSE-printed nanomedicines containing SBA-15/CMC loaded with the drug (Schmidt et al., 2022).

Another innovative system is the powerful drug delivery system (DDS), which has a great influence on the management of the pharmacokinetic scheme with respect to the rate of drug release, the target site, and the bioavailability of the drug without evading side effects. The purpose of this investigation was to evaluate the effect of CNC-NCG nanofillers with PVA/MeC/PEG-based fiber reinforcement on the release of ketorolac tromethamine (KT) and other properties of drug- and nanofiller-loaded nanofibrous systems. CNF was isolated from jute fibers, while nanocollagen (NCG) was obtained from waste fish scales for the manufacture of the innovative PVA/MeC/PEG cross-linked glutaraldehyde electrospun nanofiber nanocomposites reinforced with drugs. The resulting composite film was characterized, tested *in vitro*, and the drug release results demonstrate that the charged electrospun films showed remarkable sustained release of KT up to ≥16 hours and can work as excellent transdermal DDS (Samanta et al., 2022).

Furthermore, other state-of-the-art pharmaceutical and/or biomedical applications are multifunctional polymeric coatings that can serve as drug delivery vehicles around biofilm formation on prosthetic and implant surfaces. Although CA films are broadly investigated for scaffolds in tissue engineering, their applications as a protective coating and drug delivery vehicle for metallic implants are limited. The goal is that adhesion to stainless steel (SS) substrates is known in detail. In this context, SS-coated CA films were investigated by dipping and electrospun membranes, remarking that SS substrates were successfully bound and subsequently loaded with daptomycin powder. Nonetheless, the drug slightly influences CA fiber morphology but is well-incorporated into composite films. Moreover, release tests confirmed such incorporation. Functionalized implant surfaces with antimicrobial films have the heightened capability to prevent infections caused by the formation of a peripheral bacterial film. Yet a deeper understanding of the complex host–implant interactions and immunomodulatory properties is needed. Presently, advances in functional antimicrobial coatings continue to be forged, focusing their effectiveness on the prevention and mitigation of the severity of bacterial infections (Faria et al., 2022).

Finally, a multifunctional dressing has been developed for cutaneous wounds with antibacterial and sustained release properties using cellulose acetate (CA) as a matrix with silver nanoparticles (AgNPs) as a component and dimethyloxallyl glycine (DMOG) as the drug anchor. *In vitro* release tests performed on the multifunctional bandage confirmed that CA/DMOG/AgNPs nanofibers can gradually release DMOG at about 84 hours, which meets typical diffusion, and the main driving force is the concentration gradient betwixt the nanofibers that carry DMOG and the release medium. Antibacterial testing demonstrates that the material exhibits evident antibacterial performance against *E. coli* and *Bacillus subtilis*, with a little side effect on cell viability, as verified by cell compatibility testing. This multifunctional composite matrix, CA/DMOG/Ag-NPs, has the potential to be a wound dressing material with good prospects for use as a promoter of wound healing in patients with diabetes (Li et al., 2022).

The nanoscale administration system is a favorable drug administration route which is specifically targeted to specific tissues or organs, thus reducing side effects and drug toxicity (Mali & Sherje, 2022).

The usage of cellulose nanoparticles to transport pharmaceuticals remains an attractive idea, but numerous questions remain, in particular about drug–drug interactions, their interaction with other types of cellulose nanoparticles, modulation of drug release and the structure/reduction–destruction relationship, and its influence on pharmacological activity (Das et al., 2022).

Natural polymers have been extensively studied for potential biomedical applications, owing to their intrinsic *in vitro* and *in vivo* biocompatibility with animal cells and tissues (Khan et al., 2022). Natural polymers such as cellulose must have similar physicochemical properties to those of synthetic polymers to be possible viable substitutes with different applications; their biodegradable and biocompatible nature prevents any response triggered when inserted into body tissues, so it is not toxic, but these materials are not recommended for load-bearing bone tissue engineering structural purposes due to their poor mechanical resistance (Mali & Sherje, 2022). In contrast, scaffolds based on synthetic polymers, like PLA, PCL, etc., degrade slowly and tolerate high mechanical forces compared to natural polymers; however, they do not promote cell adhesion and growth (Janmohammadi et al., 2023).

Desirable characteristics of engineered cellulose nanoparticles include a high surface-to-volume ratio, high tensile strength (10 GPa), good flexibility, and high rigidity (110–130 GPa) and can be chemically/structurally modified to provide specific properties for high-value technical and biological applications. For covalent and non-covalent binding of bioactive molecules, the surface chemistry of cellulose nanoparticles is regulated by their hydroxyl groups, which are transformed into other functional groups at high densities and determine their functionality (Mali & Sherje, 2022).

In this order of ideas, the development of engineering biomaterials has evolved into a more advanced technique for the analysis of materials and their physicochemical properties. This knowledge is useful in causing chemical alterations and adaptations to guide the required reaction in a focused biological environment. Due to their ability to reinforce polymeric matrices and promote cell proliferation, cellulose nanoparticles play an important role in tissue engineering (Mali & Sherje, 2022).

Biomaterials created from cellulose nanoparticles have been investigated for use in several applications, for example, 2D or 3D tissue engineering scaffolds, biomedical grafts for blood vessel growth, tissue reconstruction, breast prostheses, strengthening of adhesion of bone implants, urethral catheters, drug release, artificial skin, etc. (see Table 2.6) (Mali & Sherje, 2022; Reshmy et al., 2021).

Recently, a group of scientists manufactured a 3D nanocomposite scaffold from CS/CNC/HN and curcumin as bioactive release. The nanocomposite scaffold represented high biomineralization, great cell proliferation, and a desirable cell attachment. Moreover, the capability of the nanocomposite scaffold for

TABLE 2.6

Development of films carrying active ingredients and trends in use of cellulose-based composite scaffolds for tissue engineering

Cellulose	Method of fabrication	Properties	Composition	Application	Reference
EC	Interfacial reaction solvent-drying	Drug delivery of zolmitriptan and etodolac	CS/SA/EC	Novel buccal mucosal delivery vehicle	(Wang et al., 2022)
CMC	Homogenized/extruded	Drug delivery of triamcinolone acetonide	SBA-15/CMC-TA	Novel 3D buccal mucosal delivery vehicle	(Schmidt et al., 2022)
CNF	Cross-linked/electrospinning nanofiber synthesis	Drug delivery of ketorolac tromethamine	NCG-CNF-KT/PVA-MeC-PEG	Innovative sustained release of ketorolac tromethamine transdermal	(Samanta et al., 2022)
CA	Cross-linked/electrospinning synthesis	Drug delivery of Dimethyloxallyl glycine/antimicrobial properties	CA/DMOG/AgNPs	Multifunctional wound dressing with sustained release/coating	(Li et al., 2022)
CA	Dip-coated electrospinning	Drug delivery of daptomycin/antimicrobial properties	CA/SS/Daptomycin	Multifunctional polymeric coating for implant surfaces	(Faria et al., 2022)
CA	Dip-coated electrospinning	Selective drug delivery of disulfiram/antimicrobial	DS/CA/PEO	Novel selective anticancer agent	(El Fawal et al., 2022)
CNC	Ultrasound/casting/freeze-dryer	Great cell proliferation/good biomineralization/acceptable cell attachment	CS/CNC/HN-curcumin	Bone tissue engineering applications	(Doustdar et al., 2022)
BC	Ultrasound/freeze-dryer	Scaffold/antibacterial/biocompatibility	BC/MWCNT	Novel dressing material for diabetic wounds/coating	(Khalid et al., 2022)

Abbreviations: *EC*, ethyl cellulose; *CA*, cellulose acetate; *CNF*, cellulose nanofibrils; *PVA*, polyvinyl alcohol; *CMC*, carboxymethyl cellulose; *SBA-15*, mesoporous silica nanostructured; *NCG*, nanocollegen; *MeC*, methylcellulose; *SS*, stainless steel; *PEG*, polyethylene glycol; *KT*, ketorolac tromethamine; *TA*, triamcinolone acetonide; *DMOG*, dimethyloxallyl glycine; *AgNPs*, silver nanoparticles; *PEO*, ethylene oxide; *DS*, disulfiram; *HN*, halloysite nanotubes; *CS*, chitosan; *CNC*, cellulose nanocrystals; *BC*, bacterial cellulose; *MWCNT*, multiwalled carbon nanotubes.

curcumin delivery was allowed through cell proliferation, cumulative release, and antibacterial studies. Cell proliferation of the nanocomposite with 10% (wt.) curcumin-loaded halloysite nanotubes reached ~175% at 72 hours. Considering the results, the prepared nanocomposite scaffold holds great potential for being used in bone tissue engineering applications (Doustdar et al., 2022).

In addition, a healing material was developed from BC where the stated matrix was reinforced with multiwalled carbon nanotubes (MWCNT) to develop a dressing for infection control and time reduction in diabetic wound healing process. The BC-MWCNT composite film was evaluated for antibacterial activity and wound healing activity *in vivo*, quantitatively determining the temporal expression of interleukin (IL-1α), tumor necrosis factor (TNF-α), endothelial growth factor vascular growth factor (VEGF), and platelet-derived growth factor (PDGF) by qPCR. The characterization results proved the reinforcement of the BC matrix with MWCNT where the dressing acted as a mechanical and antibacterial barrier for the fragile tissue in the process of healing and contributed to moisture retention, reducing inflammation, which resulted in efficient healing of the wound. The composite film showed antibacterial activity against microorganisms such as *S. aureus*, *E. coli*, *P. aeruginosa*, *S. Typhi*, and *Klebsiella pneumoniae*. In addition, macroscopic wound analysis demonstrated accelerated diabetic wound closure in the BC-MWCNT group (99% healing) compared to the negative control (77%) at 21 days. Then, histological studies supported the results, in which complete re-epithelization of the epidermis and healthy granulation tissue were recognized in the group treated with the reinforced BC-MWCNT matrix. Molecular studies revealed that the BC-MWCNT group had reduced expression of the proinflammatory TNF-α and cytokines IL-1α and increased expression of VEGF, leading to faster healing when compared with the control treatment. These researchers concluded that the adaptive properties of BC were synergistic with MWCNT and are a trigger for creating compounds with potential applications in diabetic wound healing. However, these researchers suggest additional studies focused on establishing the biocompatibility of the composite film before clinical trials (Khalid et al., 2022).

An ideal 3D scaffold should be non-toxic and biocompatible, have a chemical surface that supports cell adhesion, proliferation, and differentiation, and resemble the microscale morphology of the extracellular matrix (ECM) (Khan et al., 2022). For this reason, issues related to the practical obstacles to developing and designing biomedical devices from these biomolecules are crucial to evaluating and implementing commercial products based on cellulose nanoparticles.

2.5 Electronic Films or Coatings

The rapid consumption of non-renewable resources (e.g., petroleum and coal) has brought serious environmental concerns and energy crisis. Therefore, various renewable energy sources (for instance, hydro energy, solar energy, wind energy, tidal energy, and geothermal energy) have attracted rising attention from the scientific and industrial community.

In daily life, the most usual application pattern is to transform these renewable energy sources into electric energy with the help of diverse electric generators (see Table 2.7) (Zhang et al., 2022).

Following this order of ideas, advanced sensor, and energy devices with novel applications (for example, autonomous electric vehicles, intelligent electronic equipment, high-capacity batteries, and other electronic systems) are one of the most tangible foundations of modern intellectual life. Regardless, presently there are scientific and economic limitations that seriously interfere with the development of electronic devices, such as the unsustainability of some materials used in their development; the constant increase in the cost of base materials, such as lithium; the complex manufacturing process; and the performance limitations of the same material.

In the last decade, cellulose nanoparticles have stood out due to their presence and abundance in different natural resources, their renewability, their biodegradability, their low cost, and their remarkable physicochemical properties. These unique properties make cellulose nanoparticles highly competitive as a matrix material for the fabrication and development of advanced functional composites for use in energy-related fields (Chen et al., 2022; Zhang et al., 2022).

In recent years, there has been intense discussion about the progress of nanocellulose for emerging energy storage and harvesting, but also sensor applications. For example, within conductive nanocellulose-based composites, the following stand out: (1) nanocellulose-carbon composites, like carbon nanotubes, single-walled carbon nanotubes, and multiwalled carbon nanotubes; (2) nanocellulose-conductive polymer composites, such as polyaniline, polypyrrole, polyacetylene, polythiophene, poly[3,4-(ethylenedioxy) thiophene], and polyphenylene used in diverse areas, including batteries, electrochemical capacitors, sensors, solar cells, electromagnetic shielding materials, and electrical conductors; (3) nanocellulose-metal nanoparticle composites, for example, nanoparticles of gold, silver, copper, tin, ruthenium, and bismuth, which have promising application potential in flexible electronics; and (4) nanocellulose-metalorganic frameworks (MOFs) composites. For instance, MOFs are highly porous materials linked by metal ions and organic ligands via coordination bonds, which have excellent chemical stability, good pore accessibility, high surface area (up to 7,000 $m^2 g^{-1}$), and versatile functionalities. Then there are (5) nanocellulose-COFs composites, for example, COFs are novel crystalline porous polymers with pre-designable pore structures, highly tunable functionality, large surface area, and exceptional chemical stability, which are constructed by strong covalent bonds from reactions of light elements, including boron, carbon, nitrogen, oxygen, and silicon (Chen et al., 2022).

Among the applications of advanced materials are the supercapacitors areas solar cells, lithium-ion batteries (LIBs), triboelectric nanogenerators, and sensors. Supercapacitors (SCs) have gained great attention due to their high-power density, fast charge–discharge rate, and excellent cycling stability, which bridge the gap between conventional capacitors and rechargeable batteries and can be divided into three types: (1) electrochemical double-layer capacitors (EDLCs), (2) pseudocapacitors, and (3) hybrid capacitors, according to the charge storage mechanism. According to the roles played by nanocellulose in SCs, it can be mainly divided into three types: (1) electrode materials integrated with other conductive materials or converted to carbon materials through pyrolytic processes; (2) separator materials between two SCs electrodes; (3) electrolyte materials that facilitate the electrolyte ions adsorption.

Solar cells where nanocellulose is an ideal substrate have been explored for use in solar cells because its size is smaller than the wavelength of visible light, which makes the prepared nanocellulose-based paper high in transparency, as well as significant scattering along the light transport direction. In lithium-ion batteries, nanocellulose as a building block source can be developed as electrodes, separators, and electrolytes in LIBs.

Triboelectric nanogenerators (TENGs) are an emerging mechanical energy–harvesting technology that can transform environmental mechanical energy (e.g., human movement, water wave, and vibration) into electrical energy through coupled triboelectric effect and electrostatic induction, which have unique merits, including high power density, high efficiency, structural diversity, low cost, light weight, and simplicity. Nanocellulose has lured great interest as a promising alternative, owing to its abundance, biocompatibility, biodegradability, triboelectric effect, and the triboelectric sequence of nanocellulose is higher than in most common polymers, which makes it an ideal positive material for TENGs.

In sensors, nanocellulose-based materials with extraordinary electrical, optical, and mechanical properties as attractive alternatives have been employed in several sensing applications, such as environmental monitoring, food safety, physical sensing, human disease detection, and health care (Chen et al., 2022).

Among the most recent research in the field of advanced materials, presented was a method that demonstrates a non-chemical cross-linking and an ambient drying way to prepare functional aerogels from cellulose nanofibrils (CNFs) and carbon nanotubes (CNTs), building aerogel pore walls with sufficient mechanical properties by repeated freezing and thawing to cross-link CNFs and CNTs. During the analysis, a dual network interpenetration pathway was elaborated that favored the tubular dispersion of CNFs and CNTs, resulting in the development of hybrid dual networks of hydrophilic and hydrophobic nanofibers. As a result, aerogels with tunable densities (0.0519 g cm^3), good conductivity (30.95 S cm^1), and high surface area (157.24 m^2 g cm^1) can be developed. Furthermore, aerogels can be easily recycled due to the absence of chemical cross-linking. Various structural and shape characteristics of aerogels can be modified by 3D printing, indicating the future potential for large-scale production (aerogel diameter up to 8.68 cm). To demonstrate their potential use in various applications, the aerogels were evaluated, and the results indicated high specific electromagnetic shielding properties of 440.9 dB cm^3 g^1 and applications as a stand-alone electrode to load active materials such as manganese dioxide, exhibiting good storage properties of energy (551 F g^1) (Huang et al., 2022b).

Otherwise, evaluated were electrothermal compounds based on MWCNT, reduced graphene oxide (RGO), and cationic cellulose nanofibrils (CCNFs). The electrothermal compound exhibited outstanding cooling, heating response, electrothermal stability, and the efficient heating temperature (102.15°C) was reached when a ratio of MWCNT:RGO of 35:5 (w/w) was carried out at an applied voltage of 18V. Also, mixed use of MWCNT:RGO effectively improved the electrothermal performance; its average electrothermal response time was less than 43.32 seconds. The electrical energy consumed was 32.85 mW/C. This electrothermal compound has promising applications in environmentally friendly and flexible resistance heating electronics (Liang et al., 2022).

Successfully prepared and reported were ZnS nanocomposite films from cellulose nanofibers (CNFs), isolated from bagasse pulp, with different weight ratios of ZnS nanoparticles (ZnS NP). They investigated the electrical properties within the frequency range from 5 to 100 Hz and the temperature range from 25 to 120°C. The results showed that all the ZnS/CNF nanocomposite thin films had semiconductive properties, and the ZnS NP sample had the highest conductivity value (5% wt.). The results indicate that thin films prepared from flexible ZnS/CNF nanocomposites are promising candidates for semiconductor materials used in electronic devices (Abdel-Karim et al., 2022).

TABLE 2.7

Cellulose-based functional materials for applications in advanced energy and/or sensors

Cellulose	Method of elaboration	Properties	Composition	Application as coating	Reference
CCNF	Ultrasonic dispersion and vacuum filtration	Improved the electrothermal performance	CCNF/ MWCNT/ RGO	Flexible resistance heating electronics	(Liang et al., 2022)
CNF	Cycling freeze-thawing to cross-linking	High specific electromagnetic shielding	CNF/CNT	Freestanding electrode	(Huang et al., 2022b)
CNF	Doping of ZnS	Nanocomposite thin films had semiconductor properties and sample had the highest conductivity value	CNF/ZnS	Semiconductor materials used in electronic devices	(Abdel-Karim et al., 2022)
HEC	Multistage	Biocompatible, stretchable, conformable on-skin sensors that can sensitively detect human motions, specifically, skin touch, finger bending, wrist bending, skin wrinkling, breathing, and walking, with excellent stretchability, excellent sensitivity, quick recovery, and conformality	PEDOT/PSS/ HEC	Skin health-monitoring devices	(Han et al., 2022)
HEC	Multistage	High-performing, mechanically stable	PEDOT/ PSS-carbon/ HEC	Printed supercapacitor electrodes	(Belaineh et al., 2022)

Abbreviations: *CNF*, cellulose nanofibrils; *CCNF*, cationic cellulose nanofibril; *MWCNT*, multiwalled carbon nanotube; *RGO*, reduced graphene oxide; *CNT*, carbon nanotubes; *PEDOT*, poly-3,4-ethylenedioxythiophene; *PSS*, poly(styrenesulfonate; *HEC*, hydroxyethyl cellulose.

Likewise, biocompatible skin sensors based on a highly conductive film synthesized from poly-3,4-ethylenedioxythiophene and polystyrenesulfonate (PEDOT/PSS) and hydroxyethyl cellulose (HEC) have been developed. PEDOT/PSS/HEC films were highly biocompatible, stretchable, and conformable films that can sensitively detect human skin touch movements, finger flexion, wrist flexion, skin wrinkling, breathing, walking, with excellent stretchability, excellent sensitivity, fast recovery, and compliance. PEDOT/PSS/HEC have a superior potential for use in skin-worn devices that can detect various human movements due to their excellent performance and high stretchability, conductivity, robustness, and biocompatibility (Han et al., 2022).

On the other hand, it has been reported that the combination of cellulose/PEDOT/PSS-carbon and carbon derivatives for bulk supercapacitor electrodes was adapted for printed electronics. PEDOT/PSS acts as a mixed ion and electron-conducting glue, physically bonding activated carbon particles and, at the same time, facilitating the rapid transport of electrons and ions. A 10% PEDOT/PSS is required for optimal performance. This research showed that cellulose added to PEDOT/PSS-carbon enables high-performance, mechanically stable printed SC electrodes using a combination of printing methods (Belaineh et al., 2022).

2.6 Textile Applications

Due to the increasing population, there is an increasing need for clothing production, but also textile dyeing. Normally, fabrics are stained with water-based synthetic colorants, of which, based on traditional methodologies, only ~80% dye is covalently joined to the cotton fabric. So, the unreacted dye is discharged into the wastewater. Besides, the dyeing industry for clothing requires huge quantities of water that drop hazardous components into the water and eventually into the environment. It is reported that the textile dyeing industry is responsible for 20% of all water pollution (Liyanapathiranage et al., 2020).

As mentioned in the previous sections, natural polymers are an excellent substitution for petroleum-based resources. In this sense, nanofibrillated cellulose hydrogels are of great interest due to their efficiency as carriers for textile dyes. When these nanofibrillated cellulose hydrogels are combined with the dye, it is then produced a colored hydrogel. The benefits of this alternative approach are six times less water used, fewer salts, and fewer alkali requirements. This makes nanofibrillated-cellulose hydrogels a greener approach (Liyanapathiranage et al., 2020).

In this sense, it has been reported that the incorporation of nanofibrillated cellulose hydrogel enhances dye fixation (~30% more) and, even more, reduces up to 60% of its dye discharge, even though its efficiency is mainly dependent on the temperature, type of fibrillated nanocellulose, and dye's chemical structure (Liyanapathiranage et al., 2020). So nanocellulose could be used as a binding agent but also can serve as a dye carrier (Barik et al., 2022).

The military sector is frequently looking for novel approaches, both promising and innovative, and the uniforms are not an exception. Nanocellulose presents great versatility, which fits military applications, for example, antimicrobial properties, self-cleaning and dirty-free fabric, resistant and smart fabric (non-allergic, improved wear, tear resistant, etc.) (Jenol et al., 2022).

2.7 Conclusion

The tendency for nanocellulose usage in the following years will be incremented owing to the economic savings generated when compared to traditional strategies; moreover, environmental benefits are generated. However, despite the technological advances and the actual applications, more focus/punctual research is needed to implement novel and specific applications in which cellulose, or its respective derivatives are applicable or could be implemented.

REFERENCES

Abdel-Karim, A. M., Salama, A. H., & Hassan, M. L. (2022). High dielectric flexible thin films based on cellulose nanofibers and zinc sulfide nanoparticles. *Materials Science and Engineering B: Solid-State Materials for Advanced Technology*, *276*, 115538. https://doi.org/10.1016/j.mseb.2021.115538

Abral, H., Kurniawan, A., Rahmadiawan, D., Handayani, D., Sugiarti, E., & Muslimin, A. N. (2022). Highly antimicrobial and strong cellulose-based biocomposite film prepared with bacterial cellulose powders, uncaria gambir, and ultrasonication treatment. *International Journal of Biological Macromolecules*, *208*, 88–96. https://doi.org/10.1016/j.ijbiomac.2022.02.154

A guide to World Food Safety Day 2022: Safer food, better health. (n.d.). Retrieved June 8, 2022, from www.who.int/publications/i/item/WHO-HEP-NFS-AFS-2022.1

Andrade, M. S., Ishikawa, O. H., Costa, R. S., Seixas, M. V. S., Rodrigues, R. C. L. B., & Moura, E. A. B. (2022). Development of sustainable food packaging material based on biodegradable polymer reinforced with cellulose nanocrystals. *Food Packaging and Shelf Life*, *31*. https://doi.org/10.1016/j.fpsl.2021.100807

Barik, B., Maji, B., Sarkar, D., Mishra, A. K., & Dash, P. (2022). Cellulose-based nanomaterials for textile applications. In *Bio-based nanomaterials: Synthesis protocols, mechanisms and applications*. INC. https://doi.org/10.1016/B978-0-323-85148-0.00009-9

Belaineh, D., Brooke, R., Sani, N., Say, M. G., Håkansson, K. M. O., Engquist, I., Berggren, M., & Edberg, J. (2022). Printable carbon-based supercapacitors reinforced with cellulose and conductive polymers. *Journal of Energy Storage*, *50*, 104224. https://doi.org/10.1016/j.est.2022.104224

Benito-González, I., López-Rubio, A., Galarza-Jiménez, P., & Martínez-Sanz, M. (2021). Multifunctional cellulosic aerogels from Posidonia oceanica waste biomass with antioxidant properties for meat preservation. *International Journal of Biological Macromolecules*, *185*, 654–663. https://doi.org/10.1016/j.ijbiomac.2021.06.192

Bortolatto, R., Bittencourt, P. R. S., & Yamashita, F. (2022). Biodegradable starch/polyvinyl alcohol composites produced by thermoplastic injection containing cellulose extracted from soybean hulls (Glycine max L.). *Industrial Crops and Products*, *176*, 114383. https://doi.org/10.1016/j.indcrop.2021.114383

Chen, J., Wang, X., Long, Z., Wang, S., Zhang, J., & Wang, L. (2020). Preparation and performance of thermoplastic starch and microcrystalline cellulose for packaging composites: Extrusion and hot pressing. *International Journal of Biological Macromolecules, 165*, 2295–2302. https://doi.org/10.1016/j.ijbiomac.2020.10.117

Chen, L., Yassin, S., Abdalkarim, H., Yu, H., Li, Y., Tam, K. C., Chen, X., & Tang, D. (2022). Nanocellulose-based functional materials for advanced energy and sensor applications *Nano Research, 15*, 7432–7452. https://doi.org/10.1007/s12274-022-4374-7

Chen, Q., Chang, C., & Zhang, L. (2021). Surface engineering of cellulose film with myristic acid for high strength, self-cleaning and biodegradable packaging materials. *Carbohydrate Polymers, 269*, 118315. https://doi.org/10.1016/j.carbpol.2021.118315

Dadigala, R., Bandi, R., Alle, M., Park, C.-W., Han, S.-Y., Kwon, G.-J., & Lee, S.-H. (2022). Effective fabrication of cellulose nanofibrils supported Pd nanoparticles as a novel nanozyme with peroxidase and oxidase-like activities for efficient dye degradation. *Journal of Hazardous Materials, 436*, 129165. https://doi.org/10.1016/j.jhazmat.2022.129165

Das, S., Ghosh, B., & Sarkar, K. (2022). Nanocellulose as sustainable biomaterials for drug delivery. *Sensors International, 3*, 100135. https://doi.org/10.1016/j.sintl.2021.100135

Dominic, C. D., dos Santos Rosa, D., Camani, P. H., Kumar, A. S., Neenu, K. V., Begum, P. M. S., Dinakaran, D., John, E., Baby, D., Thomas, M. M., Joy, J. M., Parameswaranpillai, J., & Saeb, M. R. (2021). Thermoplastic starch nanocomposites using cellulose-rich Chrysopogon zizanioides nanofibers. *International Journal of Biological Macromolecules, 191*, 572–583. https://doi.org/10.1016/j.ijbiomac.2021.09.103

Doustdar, F., Olad, A., & Ghorbani, M. (2022). Development of a novel reinforced scaffold based on chitosan/cellulose nanocrystals/halloysite nanotubes for curcumin delivery. *Carbohydrate Polymers, 282*, 119127. https://doi.org/10.1016/j.carbpol.2022.119127

El desperdicio masivo de alimentos, un problema no solo de los países ricos | Noticias ONU. (n.d.). Retrieved June 11, 2022, from https://news.un.org/es/story/2021/03/1489102

El Fawal, G., Abu-Serie, M. M., El-Gendi, H., & El-Fakharany, E. M. (2022). Fabrication, characterization and in vitro evaluation of disulfiram-loaded cellulose acetate/poly(ethylene oxide) nanofiber scaffold for breast and colon cancer cell lines treatment. *International Journal of Biological Macromolecules, 204*, 555–564. https://doi.org/10.1016/j.ijbiomac.2022.01.145

Fan, K., Zhang, T., Xiao, S., He, H., Yang, J., & Qin, Z. (2022). Preparation and adsorption performance of functionalization cellulose-based composite aerogel. *International Journal of Biological Macromolecules, 211*, 1–14. https://doi.org/10.1016/j.ijbiomac.2022.05.042

Faria, J., Dionísio, B., Soares, Í., Baptista, A. C., Marques, A., Gonçalves, L., Bettencourt, A., Baleizão, C., & Ferreira, I. (2022). Cellulose acetate fibres loaded with daptomycin for metal implant coatings. *Carbohydrate Polymers, 276*. https://doi.org/10.1016/j.carbpol.2021.118733

Fontes-Candia, C., Erboz, E., Martínez-Abad, A., López-Rubio, A., & Martínez-Sanz, M. (2019). Superabsorbent food packaging bioactive cellulose-based aerogels from Arundo donax waste biomass. *Food Hydrocolloids, 96*, 151–160. https://doi.org/10.1016/j.foodhyd.2019.05.011

Garrido-Romero, M., Aguado, R., Moral, A., Brindley, C., & Ballesteros, M. (2022). From traditional paper to nanocomposite films: Analysis of global research into cellulose for food packaging. *Food Packaging and Shelf Life, 31*, 100788. https://doi.org/10.1016/j.fpsl.2021.100788

Ghadiri Alamdari, N., Forghani, S., Salmasi, S., Almasi, H., Moradi, M., & Molaei, R. (2022). Ixiolirion tataricum anthocyanins-loaded biocellulose label: Characterization and application for food freshness monitoring. *International Journal of Biological Macromolecules, 200*, 87–98. https://doi.org/10.1016/j.ijbiomac.2021.12.188

Gibril, M. E., Ahmed, K. K., Lekha, P., Sithole, B., Khosla, A., & Furukawa, H. (2022). Effect of nanocrystalline cellulose and zinc oxide hybrid organic—inorganic nanofiller on the physical properties of polycaprolactone nanocomposite films. *Microsystem Technologies, 28*(1), 143–152. https://doi.org/10.1007/s00542-019-04497-x

Gomes, V., Pires, A. S., Mateus, N., de Freitas, V., & Cruz, L. (2022). Pyranoflavylium-cellulose acetate films and the glycerol effect towards the development of pH-freshness smart label for food packaging. *Food Hydrocolloids, 127*. https://doi.org/10.1016/j.foodhyd.2022.107501

Groult, S., Buwalda, S., & Budtova, T. (2022). Tuning bio-aerogel properties for controlling drug delivery. Part 2: Cellulose-pectin composite aerogels. *Biomaterials Advances, 135*, 212732. https://doi.org/10.1016/j.bioadv.2022.212732

Han, J. W., Wibowo, A. F., Park, J., Kim, J. H., Prameswati, A., Entifar, S. A. N., Lee, J., Kim, S., Chan Lim, D., Moon, M. W., Kim, M. S., & Kim, Y. H. (2022). Highly stretchable, robust, and conductive lab-synthesized PEDOT:PSS conductive polymer/hydroxyethyl cellulose films for on-skin health-monitoring devices. *Organic Electronics*, *105*, 106499. https://doi.org/10.1016/j.orgel.2022.106499

He, B., Wang, S., Lan, P., Wang, W., & Zhu, J. (2022). Topography and physical properties of carboxymethyl cellulose films assembled with calcium and gelatin at different temperature and humidity. *Food Chemistry*, *382*, 132391. https://doi.org/10.1016/j.foodchem.2022.132391

Hou, D. F., Liu, Z. Y., Zhou, L., Tan, H., Yang, W., & Yang, M. B. (2020). A facile strategy towards heterogeneous preparation of thermoplastic cellulose grafted polyurethane from amorphous regenerated cellulose paste. *International Journal of Biological Macromolecules*, *161*, 177–186. https://doi.org/10.1016/j.ijbiomac.2020.05.203

Huang, X., Ji, Y., Guo, L., Xu, Q., Jin, L., Fu, Y., & Wang, Y. (2022a). Incorporating tannin onto regenerated cellulose film towards sustainable active packaging. *Industrial Crops and Products*, *180*, 114710. https://doi.org/10.1016/j.indcrop.2022.114710

Huang, Z., Zhang, H., Guo, M., Zhao, M., Liu, Y., Zhang, D., Terrones, M., & Wang, Y. (2022b). Large-scale preparation of electrically conducting cellulose nanofiber/carbon nanotube aerogels: Ambient-dried, recyclable, and 3D-printable. *Carbon*, *194*, 23–33. https://doi.org/10.1016/j.carbon.2022.03.056

Ihsanullah, I., Sajid, M., Khan, S., & Bilal, M. (2022). Aerogel-based adsorbents as emerging materials for the removal of heavy metals from water: Progress, challenges, and prospects. *Separation and Purification Technology*, *291*, 120923. https://doi.org/10.1016/j.seppur.2022.120923

James, S. (2020). Remediation technologies bioventing. *Natural Water Remediation* https://doi.org/10.1016/B978-0-12-803810-9.00008-5

Janmohammadi, M., Nazemi, Z., Salehi, A. O. M., Seyfoori, A., John, J. V., Nourbakhsh, M. S., & Akbari, M. (2023). Cellulose-based composite scaffolds for bone tissue engineering and localized drug delivery. *Bioactive Materials*, *20*, 137–163. https://doi.org/10.1016/j.bioactmat.2022.05.018

Jenol, M. A., Norrrahim, M. N. F., & Nurazzi, N. M. (2022). Nanocellulose nanocomposites in textiles. In *Industrial applications of nanocellulose and its nanocomposites* (Issue 2020). Elsevier. https://doi.org/10.1016/b978-0-323-89909-3.00002-x

Jose, J., Pai, A. R., Gopakumar, D. A., Dalvi, Y., Rubi, V., Bhat, S. G., Pasquini, D., Kalarikkal, N., & Thomas, S. (2022). Novel 3D porous aerogels engineered at nano scale from cellulose nano fibers and curcumin: An effective treatment for chronic wounds. *Carbohydrate Polymers*, *287*, 119338. https://doi.org/10.1016/j.carbpol.2022.119338

Karzarjeddi, M., Ismail, M. Y., Antti Sirviö, J., Wang, S., Mankinen, O., Telkki, V. V., Patanen, M., Laitinen, O., & Liimatainen, H. (2022). Adjustable hydro-thermochromic green nanofoams and films obtained from shapable hybrids of cellulose nanofibrils and ionic liquids for smart packaging. *Chemical Engineering Journal*, *443*. https://doi.org/10.1016/j.cej.2022.136369

Khalid, A., Madni, A., Raza, B., ul Islam, M., Hassan, A., Ahmad, F., Ali, H., Khan, T., & Wahid, F. (2022). Multiwalled carbon nanotubes functionalized bacterial cellulose as an efficient healing material for diabetic wounds. *International Journal of Biological Macromolecules*, *203*, 256–267. https://doi.org/10.1016/j.ijbiomac.2022.01.146

Khan, S., Ul-Islam, M., Ullah, M. W., Zhu, Y., Narayanan, K. B., Han, S. S., & Park, J. K. (2022). Fabrication strategies and biomedical applications of three-dimensional bacterial cellulose-based scaffolds: A review. *International Journal of Biological Macromolecules*, *209*(PA), 9–30. https://doi.org/10.1016/j.ijbiomac.2022.03.191

Li, C., Liu, Z., Liu, S., Tiwari, S. K., Thummavichai, K., Ola, O., Ma, Z., Zhang, S., Wang, N., & Zhu, Y. (2022). Antibacterial properties and drug release study of cellulose acetate nanofibers containing ear-like Ag-NPs and dimethyloxallyl glycine/beta-cyclodextrin. *Applied Surface Science*, *590*, 153132. https://doi.org/10.1016/j.apsusc.2022.153132

Liang, S., Wang, H., & Tao, X. (2022). Effects of multiwalled carbon nanotubes and reduced graphene oxide of different proportions on the electrothermal properties of cationic cellulose nanofibril-based composites. *Journal of Materials Research and Technology*, *17*, 2388–2399. https://doi.org/10.1016/j.jmrt.2022.02.004

Lim, Z. E., Thai, Q. B., Le, D. K., Luu, T. P., Nguyen, P. T. T., Do, N. H. N., Le, P. K., Phan-Thien, N., Goh, X. Y., & Duong, H. M. (2020). Functionalized pineapple aerogels for ethylene gas adsorption and nickel (II) ion removal applications. *Journal of Environmental Chemical Engineering*, *8*(6), 104524. https://doi.org/10.1016/j.jece.2020.104524

Liu, L., Wu, W., Zheng, L., Yu, J., Sun, P., & Shao, P. (2022a). Intelligent packaging films incorporated with anthocyanins-loaded ovalbumin-carboxymethyl cellulose nanocomplexes for food freshness monitoring. *Food Chemistry, 387*, 132908. https://doi.org/10.1016/j.foodchem.2022.132908

Liu, Y., Ma, Y., Liu, Y., Zhang, J., Hossen, M. A., Sameen, D. E., Dai, J., Li, S., & Qin, W. (2022b). Fabrication and characterization of pH-responsive intelligent films based on carboxymethyl cellulose and gelatin/curcumin/chitosan hybrid microcapsules for pork quality monitoring. *Food Hydrocolloids, 124*(PA), 107224. https://doi.org/10.1016/j.foodhyd.2021.107224

Liyanapathiranage, A., Peña, M. J., Sharma, S., & Minko, S. (2020). Nanocellulose-based sustainable dyeing of cotton textiles with minimized water pollution. *ACS Omega, 5*(16), 9196–9203. https://doi.org/10.1021/acsomega.9b04498

Losses and food waste in Latin America and the Caribbean | FAO. (n.d.). Retrieved June 8, 2022, from www.fao.org/americas/noticias/ver/en/c/239392/

Lustenberger, S., & Castro-Muñoz, R. (2022). Advanced biomaterials and alternatives tailored as membranes for water treatment and the latest innovative European water remediation projects: A review. *Case Studies in Chemical and Environmental Engineering, 5*, 100205. https://doi.org/10.1016/j.cscee.2022.100205

Ma, J., He, J., Kong, X., Zheng, J., Han, L., Liu, Y., Zhu, Z., & Zhang, Z. (2022). From agricultural cellulosic waste to food delivery packaging: A mini-review. Chinese Chemical Letters. https://doi.org/10.1016/j.cclet.2022.04.005

Mali, P., & Sherje, A. P. (2022). Cellulose nanocrystals: Fundamentals and biomedical applications. *Carbohydrate Polymers, 275*, 118668. https://doi.org/10.1016/j.carbpol.2021.118668

Mary, R., Tharayil, A., Thresia, R., Antony, T., Kargarzadeh, H., Jose, C., & Thomas, S. (2022). A review on the emerging applications of nano-cellulose as advanced coatings. *Carbohydrate Polymers, 282*, 119123. https://doi.org/10.1016/j.carbpol.2022.119123

Mo, L., Zhang, S., Qi, F., & Huang, A. (2022). Highly stable cellulose nanofiber/polyacrylamide aerogel via in-situ physical/chemical double cross-linking for highly efficient Cu(II) ions removal. *International Journal of Biological Macromolecules, 209*(PB), 1922–1932. https://doi.org/10.1016/j.ijbiomac.2022.04.167

Nadeem, H., Dehghani, M., Garnier, G., & Batchelor, W. (2022). Life cycle assessment of cellulose nanofibril films via spray deposition and vacuum filtration pathways for small scale production. *Journal of Cleaner Production, 342*, 130890. https://doi.org/10.1016/j.jclepro.2022.130890

Noshirvani, N., Hong, W., Ghanbarzadeh, B., Fasihi, H., & Montazami, R. (2018). Study of cellulose nanocrystal doped starch-polyvinyl alcohol bionanocomposite films. *International Journal of Biological Macromolecules, 107*, 2065–2074. https://doi.org/10.1016/j.ijbiomac.2017.10.083

Oyeoka, H. C., Ewulonu, C. M., Nwuzor, I. C., Obele, C. M., & Nwabanne, J. T. (2021). Packaging and degradability properties of polyvinyl alcohol/gelatin nanocomposite films filled water hyacinth cellulose nanocrystals. *Journal of Bioresources and Bioproducts, 6*(2), 168–185. https://doi.org/10.1016/j.jobab.2021.02.009

Peng, H., Xiong, W., Yang, Z., Xu, Z., Cao, J., Jia, M., & Xiang, Y. (2022). Advanced MOFs@aerogel composites: Construction and application towards environmental remediation. *Journal of Hazardous Materials, 432*, 128684. https://doi.org/10.1016/j.jhazmat.2022.128684

Perumal, A. B., Nambiar, R. B., Sellamuthu, P. S., Sadiku, E. R., Li, X., & He, Y. (2022). Extraction of cellulose nanocrystals from areca waste and its application in eco-friendly biocomposite film. *Chemosphere, 287*(P2), 132084. https://doi.org/10.1016/j.chemosphere.2021.132084

Qi, Y., Lin, S., Lan, J., Zhan, Y., Guo, J., & Shang, J. (2021). Fabrication of super-high transparent cellulose films with multifunctional performances via postmodification strategy. *Carbohydrate Polymers, 260*, 117760. https://doi.org/10.1016/j.carbpol.2021.117760

Rajeswari, A., Christy, E. J. S., Swathi, E., & Pius, A. (2020). Fabrication of improved cellulose acetate-based biodegradable films for food packaging applications. *Environmental Chemistry and Ecotoxicology, 2*, 107–114. https://doi.org/10.1016/j.enceco.2020.07.003

Reis, R. S., Souza, D. de H. S., Marques, M. de F. V., da Luz, F. S., & Monteiro, S. N. (2021). Novel bionanocomposite of polycaprolactone reinforced with steam-exploded microfibrillated cellulose modified with ZnO. *Journal of Materials Research and Technology, 13*, 1324–1335. https://doi.org/10.1016/j.jmrt.2021.05.043

Ren, D., Wang, Y., Wang, H., Xu, D., & Wu, X. (2022). Fabrication of nanocellulose fibril-based composite film from bamboo parenchyma cell for antimicrobial food packaging. *International Journal of Biological Macromolecules, 210*, 152–160. https://doi.org/10.1016/j.ijbiomac.2022.04.171

Resendiz-Vazquez, J. A., Roman-Doval, R., Santoyo-Fexas, F., Gómez-Lim, M. A., Verdín-García, M., & Mendoza, S. (2022). Chemical and biological delignification treatments from blue agave and sorghum by-products to obtain cellulose nanocrystals. *Waste and Biomass Valorization, 13*(2), 1157–1168. https://doi.org/10.1007/s12649-021-01547-2

Reshmy, R., Philip, E., Madhavan, A., Arun, K. B., Binod, P., Pugazhendhi, A., Awasthi, M. K., Gnansounou, E., Pandey, A., & Sindhu, R. (2021). Promising eco-friendly biomaterials for future biomedicine: Cleaner production and applications of nanocellulose. *Environmental Technology and Innovation, 24,* 101855. https://doi.org/10.1016/j.eti.2021.101855

Rodrigues, C., Souza, V. G. L., Coelhoso, I., & Fernando, A. L. (2021). Bio-based sensors for smart food packaging—current applications and future trends. *Sensors, 21*(6), 1–24. https://doi.org/10.3390/s21062148

Samanta, A. P., Ali, M. S., Orasugh, J. T., Ghosh, S. K., & Chattopadhyay, D. (2022). cross-linked nano-collagen-cellulose nanofibrils reinforced electrospun polyvinyl alcohol/methylcellulose/polyethylene glycol bionanocomposites: Study of material properties and sustained release of ketorolac tromethamine. *Carbohydrate Polymer Technologies and Applications, 3,* 100195. https://doi.org/10.1016/j.carpta.2022.100195

Sanchez, L. M., Espinosa, E., Mendoza Zélis, P., Morcillo Martín, R., de Haro Niza, J., & Rodríguez, A. (2022). Cellulose nanofibers/PVA blend polymeric beads containing in-situ prepared magnetic nanorods as dye pollutants adsorbents. *International Journal of Biological Macromolecules, 209,* 1211–1221. https://doi.org/10.1016/j.ijbiomac.2022.04.142

Schmidt, L. M., dos Santos, J., de Oliveira, T. V., Funk, N. L., Petzhold, C. L., Benvenutti, E. V., Deon, M., & Beck, R. C. R. (2022). Drug-loaded mesoporous silica on carboxymethyl cellulose hydrogel: Development of innovative 3D printed hydrophilic films. *International Journal of Pharmaceutics, 620,* 121750. https://doi.org/10.1016/j.ijpharm.2022.121750

Shi, C., Ji, Z., Zhang, J., Jia, Z., & Yang, X. (2022). Preparation and characterization of intelligent packaging film for visual inspection of tilapia fillets freshness using cyanidin and bacterial cellulose. *International Journal of Biological Macromolecules, 205,* 357–365. https://doi.org/10.1016/j.ijbiomac.2022.02.072

Thongsrikhem, N., Taokaew, S., Sriariyanun, M., & Kirdponpattara, S. (2022). Antibacterial activity in gelatin-bacterial cellulose composite film by thermally cross-linking with cinnamaldehyde towards food packaging application. *Food Packaging and Shelf Life, 31,* 100766. https://doi.org/10.1016/j.fpsl.2021.100766

Ulrich, G. D., & Faez, R. (2022). Thermal, mechanical and physical properties of composite films developed from seaweed polysaccharides/cellulose nanofibers. *Journal of Polymers and the Environment, 30*(9), 3688–3700. https://doi.org/10.1007/s10924-022-02459-5

Wang, S., Gao, Z., Liu, L., Li, M., Zuo, A., & Guo, J. (2022). Preparation, in vitro and in vivo evaluation of chitosan-sodium alginate-ethyl cellulose polyelectrolyte film as a novel buccal mucosal delivery vehicle. *European Journal of Pharmaceutical Sciences, 168,* 106085. https://doi.org/10.1016/j.ejps.2021.106085

Xiao, Y., Liu, Y., Kang, S., & Xu, H. (2021). Insight into the formation mechanism of soy protein isolate films improved by cellulose nanocrystals. *Food Chemistry, 359,* 129971. https://doi.org/10.1016/j.foodchem.2021.129971

Xie, Y., Pan, Y., & Cai, P. (2022a). Cellulose-brabicatedrobial films incorporated with ZnO nanopillars on surface as biodegradable and antimicrobial packaging. *Food Chemistry, 368,* 130784. https://doi.org/10.1016/j.foodchem.2021.130784

Xie, Y., Pan, Y., & Cai, P. (2022b). Hydroxyl cross-linking reinforced bagasse cellulose/polyvinyl alcohol composite films as biodegradable packaging. *Industrial Crops and Products, 176,* 114381. https://doi.org/10.1016/j.indcrop.2021.114381

Xu, K., Li, Q., Xie, L., Shi, Z., Su, G., Harper, D., Tang, Z., Zhou, J., Du, G., & Wang, S. (2022). Novel flexible, strong, thermal-stable, and high-barrier switchgrass-based lignin-containing cellulose nanofibrils/chitosan biocomposites for food packaging. *Industrial Crops and Products, 179,* 114661. https://doi.org/10.1016/j.indcrop.2022.114661

Yang, J., Zhang, X., Chen, L., Zhou, X., Fan, X., Hu, Y., Niu, X., Xu, X., Zhou, G., Ullah, N., & Feng, X. (2022a). Antibacterial aerogels with nano-silver reduced in situ by carboxymethyl cellulose for fresh meat preservation. *International Journal of Biological Macromolecules, 213*(22), 621–630. https://doi.org/10.1016/j.ijbiomac.2022.05.145

Yang, M., Lotfikatouli, S., Chen, Y., Li, T., Ma, H., Mao, X., & Hsiao, B. S. (2022b). Nanostructured all-cellulose membranes for efficient ultrafiltration of wastewater. *Journal of Membrane Science, 650,* 120422. https://doi.org/10.1016/j.memsci.2022.120422

Yang, P., Yan, M., Tian, C., Huang, X., Lu, H., & Zhou, X. (2022c). International Journal of Biological Macromolecules Solvent-free preparation of thermoplastic bio-materials from microcrystalline cellulose (MCC) through reactive extrusion. *International Journal of Biological Macromolecules*, *217*, 193–202. https://doi.org/10.1016/j.ijbiomac.2022.07.006

Yokoyama, T., Ohashi, T., Kikuchi, N., & Fujimori, A. (2022). Fabrication of cellulose nanofibers by the method of interfacial molecular films and the creation of organized soluble starch molecular films. *Colloids and Surfaces A: Physicochemical and Engineering Aspects*, *643*(January), 128784. https://doi.org/10.1016/j.colsurfa.2022.128784

You, P., Wang, L., Zhou, N., Yang, Y., & Pang, J. (2022). A pH-intelligent response fish packaging film: Konjac glucomannan/carboxymethyl cellulose/blackcurrant anthocyanin antibacterial composite film. *International Journal of Biological Macromolecules*, *204*, 386–396. https://doi.org/10.1016/j.ijbiomac.2022.02.027

Zabidi, N. A., Nazri, F., Tawakkal, I. S. M. A., Basri, M. S. M., Basha, R. K., & Othman, S. H. (2022). Characterization of active and pH-sensitive poly(lactic acid) (PLA)/nanofibrillated cellulose (NFC) films containing essential oils and anthocyanin for food packaging application. *International Journal of Biological Macromolecules*, *212*, 220–231. https://doi.org/10.1016/j.ijbiomac.2022.05.116

Zhang, C., Wang, H., Gao, Y., & Wan, C. (2022). Cellulose-derived carbon aerogels: A novel porous platform for supercapacitor electrodes. *Materials & Design*, *219*, 110778. https://doi.org/10.1016/j.matdes.2022.110778

Zheng, B., Li, B., Wan, H., Lin, X., & Cai, Y. (2022). Coral-inspired environmental durability aerogels for micron-size plastic particles removal in the aquatic environment. *Journal of Hazardous Materials*, *431*. https://doi.org/10.1016/j.jhazmat.2022.128611

Zhuang, J., Rong, N., Wang, X., Chen, C., & Xu, Z. (2022). Adsorption of small size microplastics based on cellulose nanofiber aerogel modified by quaternary ammonium salt in water. *Separation and Purification Technology*, *293*, 121133. https://doi.org/10.1016/j.seppur.2022.121133

3

Chitosan-Based Films and Coatings

Properties and Applications

Hugo Espinosa-Andrews , Eristeo García-Márquez , Celso Velásquez-Ordoñez , and Rogelio Rodríguez-Rodríguez

CONTENTS

3.1 Introduction

The United Nations Environment Program Report reported that about 17% of food products were lost in 2019 due to inappropriate handling, microbial infections, and insufficient packaging (Khan et al., 2021). Minimally processed fresh food products undergo quality losses due to deterioration during storage by oxidation reactions, microbial growth, and environmental stress (Khan et al., 2021).

Moreover, plastic-based packaging of petrochemical origin have usually been used in the food industry. These products cause a serious health problem because they are not biodegradable (Khan et al., 2021). The plastic-based packaging of petrochemical origin can effectively protect food from physical and microbiological damage during storage and transport (Wang et al., 2022c). It is estimated that the industry uses 36% of the synthetic plastic manufactured worldwide, where in 2015, it was projected in 400 million tons, of which only 9% was recycled, and 79% end up as garbage, contaminating the environment. Therefore, it is essential to find a sustainable alternative to substitute synthetic plastics using biodegradable polymers (Caicedo et al., 2022).

A coating is a thin layer that covers the surface of foods. In contrast, a film is a preformed thin layer produced by one or more polymers placed on or between the surface of the food. The coating must be placed in a liquid state on the food surface by immersing the product into the solution of polymer, while

DOI: 10.1201/9781003303671-3

the film is first molded into solid pieces and applied as a food product wrapper (Blancas-Benitez et al., 2022; Maringgal et al., 2020; Tavassoli-Kafrani et al., 2016).

The edible films and coatings have the following properties: (1) barrier against water vapor, gases, and compounds migration; (2) physical and mechanical protection; and (3) appropriate appearance (Blancas-Benitez et al., 2022). In addition, coatings can be carriers of biomolecules, such as nutrients, essential oils, antioxidants, and antimicrobials. The addition of these compounds in films and coating favors their functional properties, increasing the food stability, quality, and safety (Blancas-Benitez et al., 2022; Iñiguez-Moreno et al., 2021).

The film-forming solutions are mainly produced with polysaccharides, proteins, and lipids or a combination of these (Blancas-Benitez et al., 2022). Proteins and polysaccharides are polymers usually used for film and coating applications. They display suitable mechanical and gas barrier (O_2 and CO_2) properties that delay breathing and aging in many fruits. However, they are highly hydrophilic, resulting in a reduced barrier against the loss of humidity. At the same time, the proteins may have allergenic protein fractions, which may cause adverse immunological reactions in sensitive people (Blancas-Benitez et al., 2022).

3.2 Chitosan: Origin, Structure, and Physicochemical and Biological Properties

Nature creates different biopolymers to give the structure of living things. Biological systems can control molecular weight, size, orientation, degree of crystallinity, morphology, microstructure, and phase (Grunenfelder et al., 2014). For example, crustaceans produce nanofibers of different sizes composed of a chitin–protein network with amorphous or crystalline calcium carbonate or calcium phosphate to protect their body, a process known as biomineralization (Agulló et al., 2003; Hild et al., 2008; Mushi et al., 2014). Typically, a crustacean shell contains proteins (20–40%), calcium, and magnesium salts (30–60%), chitin (20–30%), and lipids (0–14%) (Agulló et al., 2003). Chitin is located in invertebrates, including crustacean shells or insect cuticles, and in the mycelium of several fungi, such as *Mucor rouxii*, *Absidia glauca*, *Aspergillus niger*, and some green algae, and yeasts (El Knidri et al., 2018; Huq et al., 2022; Mohan et al., 2020). It has been estimated that 1,560 million tons of chitin are found in marine sources (Alishahi et al., 2011).

Chitin comprises N-acetyl-glucosamine units linked by β-1,4 glycosidic bonds combined in complex polymorphic structures with proteins, minerals, and lipids (Figure 3.1). Their intra- and intermolecular hydrogen bonding network promotes low solubility in organic and inorganic solvents, while it is soluble in inorganic acids, including hydrochloric, sulfuric, and phosphoric acids (El Knidri et al., 2018).

Chitosan is a polysaccharide produced by deacetylation of N-acetyl-glucosamine segments of chitin using concentrated NaOH or KOH solutions (40–50%) at temperatures above 100°C (Rinaudo, 2006; Younes & Rinaudo, 2015). For example, chitosan production from shrimp waste can be produced using the following steps: (1) raw shrimp shells are washed and ground on a 24 mesh; (2) shrimp shells powder

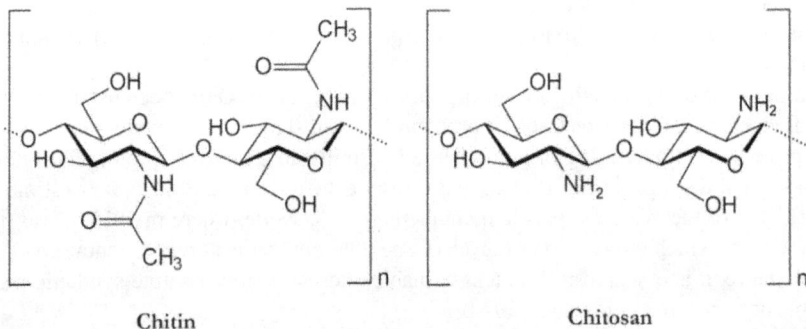

FIGURE 3.1 Chemical structure of chitin and chitosan.

Source: Reprinted from Pita-López et al. (2021), copyright 2021, with permission of Elsevier.

is deproteinized three times using 5% NaOH with a solid-to-solvent ratio of 1:5 (w/v) for 3 h at 95°C; (3) deproteinized shells are demineralized using 0.25 M HCl solution for 30 minutes; (4) the sample is decolorated using 0.315% NaOH solution for 15 minutes; (5) chitin is deacetylated using 50% NaOH solution for 1 h at 120 °C; and (6) finally, the sample is washed to neutral pH and dried. However, alkali or acid procedures are dangerous to the environment, so biological chitosan production has been developed as a more friendly procedure. For example, bacteria lactic acid can be used for demineralization, while proteases can deproteinize (Abdou et al., 2008; De Queiroz Antonino et al., 2017).

Chitosans with different average molecular weights are produced by controlling the hydrolysis rate, addition of free radicals, enzymatic reactions, or the intensity of the energy supplied for their production, including radiation, ultrasound, microwave, and thermal treatments (Chattopadhyay & Inamdar, 2010). It has been reported that chitosan can be obtained by chemical or biological extraction (Chattopadhyay & Inamdar, 2010; Kumari & Rath, 2014; Li et al., 2022; Mahdy Samar et al., 2013; Mohan et al., 2022).

Low-molecular-weight chitosan is produced using enzymatic methods. The enzymes used in these methods include cellulase, lipase pectinase, papain, protease, chitinase, chitosanases, and lysozyme (Mohan et al., 2020, 2022). For example, chitinases can hydrolyze acetylated–acetylated bounds, whereas chitosanase can cleave acetylated–acetylated and deacetylated–deacetylated boundaries.

High-molecular-weight chitosan molecules are disrupted into small ones using high-energy sources (Mohan et al., 2022). For example, microwave extraction reduces the reaction time and molecular weight of chitosan samples. It has been reported that microwave irradiation of chitosan solution, either 100 W for 80 minutes in 0.1 M acetic acid or 400 W for 3 minutes in H_2O_2, did not change the DA, but it can reduce the molecular weight of chitosan (Doan et al., 2021; Wasikiewicz & Yeates, 2013).

Chemically, chitosan is composed of glucosamine and N-acetyl-glucosamine units linked by β-1,4 glycosidic bonds (Figure 3.1) (Islam et al., 2017). Two-mole fractions describe chitosan structures; the degree of acetylation (DA) is determined by the mole fraction of N-acetylated units, and the mole fraction of β-1,4-D-glucosamine units defines the degree of deacetylation (DD). The functionality, polarity, and water solubility of chitosan depend on the DD (Chattopadhyay & Inamdar, 2010; Cheung et al., 2015). Commercial chitosan samples used for pharmaceutical or food applications typically have DD values between 85% and 95%, while lower DD values are employed in agricultural applications (~50%). Depending on the procedure and chitin source, chitosan can be produced with several DD and average molecular weights ranging from 5×10^4 to 2×10^6 Da (Kaczmarek et al., 2019; Mourya & Inamdar, 2008). Reducing the molecular weight of chitosan decreases viscosity and increases water solubility (Mohan et al., 2022).

The typical FTIR spectra of chitin and chitosan from 4,000 to 600 cm^{-1} are as follows (Dahmane et al., 2014): the C-O-C bond stretching vibration bands are observed at 890–1156 cm^{-1}. The C-N bond stretching vibration bands and the C-H binding modes of methylene at 1,330–1,380 cm^{-1}. Chitin displayed typical vibrations bands of the amine groups at 1586–1624 cm^{-1}, while chitosan displayed de-amide bands at 1,643–1,657 cm^{-1}. It has been reported that chitosan absorption bands at 1556 cm^{-1} correspond to amide II (N-H bending), and the vibration band at 1315 cm^{-1} corresponds to amide III (C-N stretching) (Dahmane et al., 2014). Broadband around 3,330–3,440 cm^{-1} corresponds to a dual contribution of O-H stretching and N-H stretching intra- and intermolecular hydrogen bonds. Three different polymorphic forms in nature (α, β, and γ) have been reported, where α-chitin is found in crustaceans, insects, fungi, and yeast cell walls (Dahmane et al., 2014; Espinosa-Andrews et al., 2010).

3.2.1 Physicochemical and Biological Activities of Chitosan

Chitosan is characterized by its excellent biocompatibility, biodegradability, and low toxicity; it possesses antimicrobial, antibacterial, and coagulant activities and acts as a structural, bioadhesive, and wound-healing material for cosmetics, medicine, agriculture, food packing, and wastewater treatment (Kumirska et al., 2011; Pita-López et al., 2021). Chitosan is soluble in numerous organic and inorganic acid solutions, favoring the protonation of the amino groups. The most common chitosan solvents are formic and acetic acids, which display weak polyelectrolyte behavior. It has been reported that the amino groups of chitosan are completely protonated at pH values below 4.5, showing an isoelectric point between pH 6.5 and 7.2 (Espinosa-Andrews et al., 2007; Rodríguez-Rodríguez et al., 2019). Some reports suggest that the antimicrobial properties of chitosan are related to the cationic charge of amino groups because these groups can disrupt the membrane of several micro-organisms (Chang et al., 2015; Chen et al., 2010). The cationic charge of chitosan molecules allows electrostatic interaction with polysaccharides, proteins, or cross-linking agents, including glutaral-dehyde, genipin, or sodium tripolyphosphate. These electrostatic complexes are used in pharmaceutical applications to produce hydrogels or particles that can be used as controlled release systems or build structures in tissue engineering applications (Pita-López et al., 2021; Rodríguez-Rodríguez et al., 2020). The intrinsic viscosity of chitosan is a usual method to calculate its molecular weight. Commonly, low solubility and high viscosity of chitosan are associated with samples of high molecular weight (Mohan et al., 2022). Chitosan displays typical Newtonian behavior in a proper solvent at very low concentration. Acetic acid is commonly used to disperse chitosan at a concentration of 0.1 M or 1% (Rinaudo et al., 1999). It has been reported that the intrinsic viscosity for chitosan in 0.25 M acetic acid and 0.25 M sodium acetate solution (90% DA) and a molecular weight of 654 KDa was 9.4 dL/g, while for a molecular weight of 135.8 KDa, it was 2.55 dL/g (Chattopadhyay & Inamdar, 2010). A Newtonian flow region is reported at low shear rate values; however, as the inter-molecular entanglements decrease in moderated shear regions, a power–law flow region is observed (Hwang, 2000).

Chitosan possesses water- and oil-binding capacities, giving significant technological properties as an emulsifying, thickening, and stabilizing agent. For example, according to its origin, molecular weight, and DD, chitosan has a water-binding capacity between 200% and 750% and an oil-binding capacity of 200% to 450% (Mahdy Samar et al., 2013; Mohan et al., 2022). The hydrogels can be easily produced, exhibit good biocompatibility, and have a low degradation rate. Their mechanical properties can be improved by chemical cross-linking with glutaraldehyde or physical cross-linking with TPP (Berger et al., 2004; Pita-López et al., 2021). Chitosan is recognized for its wound-healing, hypolipid-emic activity, antiaging, antitumoral, and antioxidant activity (Ways et al., 2018; Mohan et al., 2022; Pita-López et al., 2021). The swelling behavior of chitosan hydrogels strongly depends on pH and ionic strength of the environment (Pita-López et al., 2021; Rodríguez-Rodríguez et al., 2020, 2022). Chitosan can interact with anionic molecules like sodium tripolyphosphate or polyanions like alginate, sodium carboxymethyl-cellulose, xanthan gum, pectin, or gum arabic to produce nanoparticles used for con-trolled delivery (Ways et al., 2018). Drugs, proteins, peptides, and nucleic acids are microencapsulated in chitosan nanoparticles for cancer therapy, protein, peptide, or gene delivery. Faster antifungal activ-ity than antibacterial and antityphoid activity has been observed, comparable to the standard antibiotics used in clinical practice (Goy, 2009; Yanat & Schroën, 2021). Higher antimicrobial activity is inter-related with chitosan of low molecular weights (Zheng & Zhu, 2003). Chitosan-based particles have been also used to protect probiotic bacteria (*Lactobacillus* and *Bifidobacteria*) (Barajas-Álvarez et al., 2022). Mucoadhesion of chitosan on the mucin epithelial surface improves the absorption and bioavail-ability of drugs; however, their adhesion strength is impacted by the DD and molecular weight. Some chitosan modifications have been proposed to improve its stability in different environments, includ-ing trimethyl chitosan, carboxymethyl chitosan, thiolated chitosan, acrylated chitosan, half-acetylated chitosan, glycol chitosan, chitosan conjugated, etc. (Ways et al., 2018). Also, chitosan can be used as a chelate for anionic molecules in solution, including phosphorus, heavy metals, and oils from water (Pestov & Bratskaya, 2016).

3.3 Chitosan Film and Coating Properties

Chitosan has been the polysaccharide of choice for the film and coating fabrication. These films and coatings are produced as a barrier to prevent contamination by pathogens, limit the oxidation rate and dehydration of foods, and control the gaseous exchange (oxygen, water vapor, carbon dioxide, and excess gases) (Affes et al., 2022; Cha & Chinnan, 2004; Gasti et al., 2022). The development of pack-aging with a lower polluting effect also involves the incorporation of antimicrobial agents of natural origin, such as essential oils (Soltani Firouz et al., 2021). All these parameters have the purpose of prolonging the shelf life of foods.

3.3.1 Mechanical Properties

Among the important parameters for testing the mechanical properties include thickness, tensile strength, water vapor permeability (WVP), optical properties, and thermal stability analysis of films. These parameters are essential to obtain films and coatings with uniform thickness and the data needed to calculate the parameters of mechanical properties. The thickness of films is frequently determined using micrometric equipment, in which an average thickness is obtained (Affes et al., 2022; Alirezalu et al., 2021; Hosseini et al., 2022). The thickness of plastic films can be measured using the ASTM D6988–21 methodology or the ASTM D8136–17 methodology.

The mechanical tests are determined to obtain data and to know the resistance of the films. Quantitative tensile analysis of a film provides information about their bonding properties. Tensile strength offers information on molecular bonding between polymers. The ASTM D882–91 method is frequently used to determine the tensile strength of films (Alirezalu et al., 2021; Deshmukh et al., 2021; Gasti et al., 2022; Hosseini et al., 2022). The tensile strength (usually, it is measured in MPa) is the result of the breaking load coefficient (N) and the product of the original width (mm) and the original thickness (mm) of the film (Alizadeh-Sani et al., 2021). Tensile strength is usually determined using the methodology described by ASTM. The films are fixed to the ends of the strips, the grip separation is generally 50 mm, and the separation speed is less than 1 mm/s (Vanden Braber et al., 2021). Affes et al. (2022) have suggested that tensile strength can be determined by rheometry to determine the maximum load before failure from the tensile stress versus strain curve. To measure the tensile strength of films, it is suggested to use a rheometer equipped with geometry to measure mechanical properties (Affes et al., 2022).

3.3.2 Barrier Properties

The barrier properties of films depend on the type of polymer, proportion, functional groups, cross-linking agent, heat treatment, time treatment, and other parameters (Affes et al., 2022). During the production of films, the mentioned parameters should be considered. These will provide the best barrier properties against adverse environmental conditions for food products. Usually, water vapor transmission rate (WVTR) and WVP explain the water vapor barrier properties of films, which are in contact with the environment (Cha & Chinnan, 2004; Huang et al., 2022; Zhao et al., 2022). The environmental conditions, internal microstructure, and chemical composition of the polymers used influence the properties of the films. The control of the passage of water vapor seems to depend on the chemical interaction between the hydroxyl groups and functional groups of the polymers. The more significant the number of groups, the higher the interaction of hydrogen bonds and the lower the passage of water through the film (Liu et al., 2017; Zhao et al., 2022). WVP and water vapor transmission rate are critical parameters to quantify when antioxidant or antimicrobial compounds are added. The control of water vapors through the films preserves the quality and prolongs useful life of the product (Gasti et al., 2022; Mittal et al., 2021). These substances modify the mechanical and physicochemical properties of films used for food protection (Flórez et al., 2022; Mittal et al., 2021).

3.3.3 Water Vapor Permeability (WVP)

The WVP of films is evaluated by gravimetry, which has been standardized by ASTM E96–95 (1995). The salt method is usually used. The methodology has been widely described in the literature, with some changes to improve the permeability of films under different conditions (Alirezalu et al., 2021; Alizadeh-Sani et al., 2021; Benbettaïeb et al., 2014; Laboulfie et al., 2013; Mittal et al., 2021). Saturated salts of calcium chloride, sodium chloride, sodium nitrate, and sodium chloride are used to determine WPV in packaging film design.

WVP (g/h*m²) is determined using the water permeation rate through the designed film area. The WVP is obtained by calculation using the following equations.

$$WVPTR = \frac{m}{A} \ [g/(m^2{*}h)] \tag{3.1}$$

where *WVPTR* is the average value of water vapor, *m* is the slope, weight change of water vapor per hour, and *A* is the area of the film.

The data from Equation 3.1 are substituted in the following equation to obtain the rate of water vapor as a function of the film (Equation 3.2).

$$WVT_c = WVTR * \frac{\Delta P_1}{\Delta P_2} \; [\text{g/(m}^2\text{*h)}] \tag{3.2}$$

where WVT_c is water vapor rate corrected value; ΔP_1 is the difference between partial pressure of water vapor in the saturated solution in the container and the partial pressure of water vapor on the outer side of the film (Pa); ΔP_2 is the difference between partial pressure of water vapor in the external saturated solution and the partial pressure of water vapor inside the film (Pa).

Finally, the result is substituted in Equation 3.3, and the WVP is obtained.

$$WVP_m = \frac{WVT_c * L}{\Delta P \; (vapor\; pressure\; of\; water\; trhough\; the\; film)} \; [\text{g/(m*s*Pa)}] \tag{3.3}$$

where *L(mm)* is film thickness and ΔP *(Pa)* is water vapor pressure change.

3.3.4 Optical Properties

The translucency of films influences the appearance and consumer acceptance (Alizadeh-Sani et al., 2021). The translucency affects keeping food products as fresh as possible and avoiding deterioration due to degradation processes (Terzioğlu et al., 2021). Protective films must be selective and prevent the passage of photodegrading light. The translucency of the films is affected by the composition of polymers and plasticizers used (Bi et al., 2021). Both compounds can adsorb specific wavelengths and increase the opacity of the developed films. The light barrier property can be determined by scanning the film sample over a wide range of wavelengths (200 to 800 nm) by UV-vis spectroscopy (Bi et al., 2021; Wang et al., 2021), and it is expressed as a percentage of light transmittance. Light transmittance is negatively correlated to the film UV-vis light barrier ability. This means that a higher light transmittance percentage is equivalent to a lower barrier property to UV-vis light (Yong et al., 2019).

Wang et al. (2019) produced chitosan-based films containing a soybean extract. The films showed higher UV-vis light barrier properties than those of chitosan films. Yong et al. (2019) reported similar results on chitosan-based film with purple and black rice anthocyanins. The addition of curcumin induced a dramatic effect on UV-vis light transmittance. Wang et al. (2022a) developed bio-polyols based on waterborne polyurethane coatings, where chitosan-modified nanoparticles of ZnO were added. The results showed a decrease in optical transmittance with regard to a pure waterborne polyurethane coating, an effect attributed to the aggregation of large particles, and, thus a reduction in film transparency. Alizadeh-Sani et al. (2021) proposed to measure film transparency by spectrophotometry in the visible region. Film transparency is calculated as the ratio of transmittance divided by film thickness (D). The transparency of the films was determined using the ASTM methodology (D1746–09).

Deshmukh et al. (2021) agree on the quantification of transparency of the films, although they reported the reciprocal of transmittance. In contrast, Bi et al. (2021) determined a spectrogram between 200 to 800 nm and reported the light transmission obtained from the sample at the mentioned range using a UV-vis spectrophotometer.

3.3.5 Thermal Stability

Thermal stability is determined via thermogravimetric analysis (TGA) and is considered a parameter of great relevance for the characterization of films and composites stability of packaging materials (Terzioğlu et al., 2021). The technique is based on mass loss due to water evaporation, hydrogen bond breakage, and removal of amino groups in its first two stages, and the third-stage events are related to the decomposition of the backbone (Yuan et al., 2022; Zhang et al., 2021a). Stability under increasing temperatures was improved in films added a $W_1/O/W_2$ emulsion (Yuan et al., 2022); chitosan added with anthocyanins extracted from black soybean seed coat (Wang et al., 2019); waterborne polyurethane reinforced with chitosan-modified ZnO nanoparticles (Wang et al., 2022a); biodegradable films based on chitosan, glycerol, and defatted *Chlorella* biomass, attributed to strong hydrogen bonds between defatted *Chlorella* biomass moieties and chitosan (Deshmukh et al., 2021). The addition of components such as orange peel can increase the thermal stability of films due to the presence of hydroxyl groups participating in strong hydrogen bonding interactions with polymeric compounds present in the composition, which prevents water loss by evaporation and degradation of the polymer chains (Terzioğlu et al., 2021).

However, not all functional molecules would demonstrate an effect on thermal stability, as demonstrated by Pelissari et al. (2009) in the characterization of films made with cassava starch chitosan and oregano essential oils, where there was no significant influence.

3.4 Chitosan-Based Films and Coatings

The general formulation of all bioplastics involves the use of at least one component to produce a structural matrix with suitable cohesiveness. The functional properties of bioplastics depend on the nature of additional constituents incorporated into their composition (Porta et al., 2011).

The goal of polymer blending is to produce materials using an easy and inexpensive method that combines components' features, improving their valuable characteristics and reducing their disadvantages (Haghighi et al., 2020). Films and coatings produced only with chitosan possess disadvantages, such as brittleness, high moisture/oxygen permeability, and low bioactive functionality (Qu & Luo, 2021). Also, the hydrophilic nature of chitosan induces films with weak mechanical properties, like tensile strength and tensile modulus (Qu & Luo, 2021). The properties of films and coating produced by chitosan can be enhanced by combining them with different biomolecules.

3.4.1 Plasticizers

Chitosan is widely used to produce edible films and coatings. However, chitosan films and coating are brittle (Rodríguez-Núñez et al., 2014; Suyatma et al., 2005). For example, Bajdik et al. (2009) produced chitosan films without plasticizer, where their mechanical properties were so poor that they were not appropriate for the preliminary testing. The films formed from the chitosan solution were scarce, and the chitosan films were broken during the drying process. Moreover, Kim et al. (2006) reported that eggs coated with chitosan and glycerol as a plasticizer (0, 0.5, 1.0, 1.5, and 2.0%) showed higher internal quality than those coated only with chitosan. Paul et al. (2018) evaluated the effect of chitosan (0.5–3% w/w) and glycerol (0–3% w/w) concentrations like edible coating for tomatoes using a response surface methodology. The chitosan coating on tomatoes displayed a homogenous structure and a smooth surface, related to an appropriate mixing of chitosan and glycerol at different concentrations. The authors suggested that glycerol can break the hydrogen bonds of the polymer network of chitosan, favoring the formation of a polymer plastic with new hydrogen bonds (Paul et al., 2018).

Furthermore, the brittleness of chitosan films is influenced by the polymer–polymer interactions, which can be controlled by adding plasticizer agents to increase film flexibility (Caicedo et al., 2022; Rodríguez-Núñez et al., 2014). Plasticizers are low-molecular-weight molecules that are used to enhance the plasticity of polymers (Ma et al., 2019; Rodríguez-Núñez et al., 2014; Suyatma et al., 2005). For example, water, oligosaccharides, polyols (glycerol, propylene glycol, sorbitol, xylitol, and maltitol), and lipids are plasticizers commonly used in chitosan-based films (Caicedo et al., 2022; Ma et al., 2019; Rodríguez-Núñez et al., 2014). The ratio and chemical nature of plasticizers have an essential effect on chitosan films. Ma et al. (2019) evaluated the impact of three different polyols as plasticizers on the microstructure and physicochemical properties of chitosan films in wet and dry states. The authors found that the addition of glycerol induced higher hydrophilicity than those obtained with maltitol and xylitol. Also, glycerol plasticized films exhibited the lowest tensile strength and highest elongation break, typical rubber-like behavior. The chitosan films plasticized with maltitol showed higher tensile strength and elongation to break than the xylitol plasticized sample, suggesting that it possesses good mechanical properties. The authors demonstrated by FTIR that the polyols molecules principally interact with the hydroxyl groups rather than the NH_3^+ groups of chitosan. Thakhiew et al. (2010) produced chitosan films plasticized with different glycerol concentrations. Chitosan films with high glycerol concentrations (75 and 125%) required longer drying time than those with low glycerol concentrations (0 and 25%) at 40 and 90°C. Also, thickness (after conditioning at 75% relative humidity for 48 hours) and elongation percent of chitosan films increased with increasing glycerol concentration. These results are attributed to the capacity of glycerol to bind water molecules.

3.4.2 Polysaccharides

Polysaccharides are polymers widely used as film-forming and coating materials. Polysaccharides contain many hydroxyl groups and other polar groups which are essential for film formation (Chen et al., 2021). Generally, polysaccharides do not have antimicrobial properties to prevent bacteria growth in foods. These properties are indispensable for food packaging to increase product safety and extend shelf life. To overcome this deficiency, blending polysaccharides with natural antimicrobial agents is suitable for improving polysaccharide films' structure and characteristics. Polysaccharides can be combined with chitosan to produce chitosan-based films. For example, chitosan has been blended with cellulose (Cazón et al., 2020; dos Santos et al., 2021; Indriyati et al., 2021; Kostag & El Seoud, 2021; Mohammadi sadati et al., 2021; Ozturk et al., 2021; Resende et al., 2018; Santos et al., 2021; Yang et al., 2018; Zhang et al., 2020; Zhou et al., 2021), pectin (Akalin et al., 2022; Elsabee et al., 2008; Esmaeili & Khodanazary, 2021; Gao et al., 2019; Hoagland & Parris, 1996; Machado et al., 2020; Medeiros et al., 2012; Porta et al., 2011; Younis et al., 2020; Younis & Zhao, 2019), carrageenan (Hadiyanto et al., 2019; Ismillayli et al., 2020; Lin et al., 2018; Moller et al., 2022; Olaimat et al., 2014; Park et al., 2001; Pinheiro et al., 2012a, 2012b; Rochima et al., 2018; Volod'ko et al., 2021; Webber et al., 2021; Yu et al., 2018), and sodium alginate (Arroyo et al., 2020; Arzate-Vázquez et al., 2012; Carneiro-da-Cunha et al., 2010; Cook et al., 2013; Dulta et al., 2022; Kim et al., 2018; Kou et al., 2019; Li et al., 2019, 2020; Nair et al., 2018, 2020; Raeisi et al., 2020; Salama et al., 2018; Santos & Machado, 2021; Souza et al., 2015; Zam, 2019; Zhu et al., 2019). An important aspect is that polysaccharide blends are cheaper, more stable, and exhibit better water and thermal stability than lipid- and protein-based films (Nair et al., 2020).

3.4.2.1 Chitosan/Cellulose

Cellulose is the polysaccharide found in the structure of all plants (Kontturi & Spirk, 2019). Cellulose is composed of a linear structure of β-1,4-glucose units organized in fibrils (Szymańska-Chargot et al., 2019). The polymeric structure of cellulose defines its hydrophilicity, degradability, chirality, and chemical variability due to the presence of hydroxyl groups (Cherian et al., 2022). Indriyati et al. (2021) produced bacterial cellulose/chitosan films by casting using glycerol as a plasticizer agent. Also, different chitosan concentrations (1%, 2%, and 3% w/w) were used according on the dried weight of bacterial cellulose. The results showed an effect of chitosan concentration on the surface microstructure of the films. For example, the surface of chitosan-based films was more compact than the bacterial cellulose film. Santos et al. (2021) developed chitosan films reinforced with cellulose nanofibers at 0%, 4%, and 8% w/w. The addition of cellulose nanofibers in chitosan films promoted the formation of porous structures, increasing the barrier properties in the visible light and UV light region. dos Santos et al. (2021) reported that an increase in cellulose concentration favored the solubility of chitosan films and WVP properties. These results were related to the high hydrophilicity of cellulose nanofibers.

Studies have reported that coatings can prevent microbial spoilage and contamination of pathogenic microorganisms, as well as control gasses getting access to fruit and vegetable surfaces by acting as a semipermeable barrier, reducing water loss and maintaining tissue tightness (Ozturk et al., 2021). Zhang et al. (2020) produced a cellulose nanofiber/chitosan coating for cantaloupe rind and fresh-cut pulp. Results showed that adding chitosan favored the formation of chitosan/cellulose films on the surface of cantaloupe rind and fresh-cut pulp and increased the antimicrobial activity of the food. However, incorporating chitosan in the cellulose coating solution made the film appear more transparent, less adhesive, and peeled off more easily. Ozturk et al. (2021) reported that the addition of chitosan improved film formation from cellulose nanofiber as a coating, increasing the gas barrier capacity of the fresh-cut cantaloupe during storage.

3.4.2.2 Chitosan/Pectin

Pectin extracted from apple, citrus, carrot, and hibiscus can be used to prepare edible films (Matta & Bertola, 2020). This heteropolysaccharide is present in many plant and fruits' primary cell walls, and middle lamellae and is commonly associated with cellulose, hemicellulose, and lignin structures (Matta & Bertola, 2020; Valdés et al., 2015). Pectin is widely used for coating applications for its excellent barrier properties to oxygen, aroma preservation, and suitable mechanical properties. However, pectin is highly hydrophilic, which is ineffective against moisture transfer through films (Valdés et al., 2015). Akalin et al. (2022) evaluated the effect of pectin addition on the crystallinity, thermal, mechanical, and antibacterial properties of chitosan-based films. The chitosan films containing pectin displayed a smooth and homogeneous surface and the lowest decomposition temperature (199.7°C). Similar surface structures were reported by Machado et al. (2020). Younis and Zhao (2019) hypothesized that blending pectin and chitosan could result in improved mechanical properties than their individual polymers films. In this sense, Akalin et al. (2022) reported an increase in the tensile strength and elongation of the chitosan films with the addition of pectin.

Chitosan and pectin can produce electrostatic complexes by intermolecular interactions between the cationic amino groups of chitosan (pKa ≈ 6.0–6.5) and anionic carboxylic groups of pectin (pKa ≈ 3.6–4.1) (Machado et al., 2020). Younis and Zhao (2019) reported that pectin film displays higher hydrophilicity than chitosan film, which was related to high WVP, while chitosan/pectin films showed lower WVP values than the pectin film. These results are suitable for food packaging as it reduces moisture loss or uptake of food products over long-term storage, resulting in longer shelf life (Younis & Zhao, 2019).

Several authors have reported pectin/chitosan bilayer coatings as packaging materials (Elsabee et al., 2008; Esmaeili & Khodanazary, 2021; Medeiros et al., 2012). For example, Medeiros et al. (2012) coated mangoes with five successive layers of chitosan and pectin. The multilayer chitosan/pectin coating was effectively adsorbed on the mangoes' skin. The multilayer coating (five layers) displayed a thickness of ~266 nm. Interestingly, the multilayer coating resulted in suitable efficiency for reducing gas flow, which resulted in less mass loss than uncoated mangoes.

3.4.2.3 Chitosan/Carrageenan

Carrageenans are natural hydrophilic polymers derived from red seaweed, composed of a complex mixture of polysaccharides with a linear chain of partially sulfated galactans. These polysaccharides display a high potential for film-forming (Hamzah et al., 2013; Tavassoli-Kafrani et al., 2016). The edible carrageenan coating can be selectively permeable for O_2 and CO_2 gases, modifying the internal atmosphere of the fruit and extending its shelf life (Dwivany et al., 2020).

Pinheiro et al. (2012a) developed chitosan/carrageenan-based nanolayered coating, in which chitosan (cationic polyelectrolyte) and carrageenan (anionic polyelectrolyte) interacted by electrostatic forces. Similar results have been reported by Ismillayli et al. (2020). Volod'ko et al. (2021) produced films using polyelectrolyte complexes based on carrageenan/chitosan. They observed a dense structure of the chitosan film and a layer-by-layer structure of diverse thicknesses for the polyelectrolyte complexes.

Pinheiro et al. (2012a) confirmed the formation of chitosan/carrageenan multilayer coating by the increase in absorbance of the films. The addition of carrageenan layers induced increasing hydrophilicity, whereas the deposition of chitosan led to decreasing hydrophilicity. The chitosan/carrageenan nanolayers showed a lower WVP than those obtained for edible chitosan films. Park et al. (2001) reported an impact of the type of organic acids on the mechanical strength and WVP properties of κ-carrageenan/chitosan composite films. The composite films displayed the highest mechanical properties and lowest WVP in conditions of high acidity regulated with malic acid. Figure 3.2 shows the molecular structures of κ-carrageenan, chitosan, and composite films produced by the solvent cast method.

3.4.2.4 Chitosan/Sodium Talginate

Chitosan and sodium alginate films have numerous limitations, such as inadequate mechanical properties and poor water resistance, which limit their applications in food (Li et al., 2019). However, when

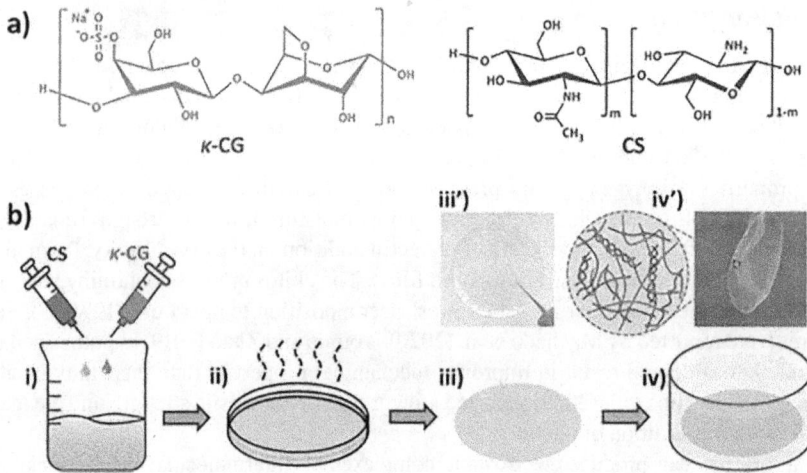

FIGURE 3.2 (a) Molecular structures of κ-carrageenan (κ-CG) and chitosan (CS). (b) Preparation of κ-carrageenan/ chitosan films: (i) solutions of κ-CG and CS were mixed in acid solution; (ii) the mixture was transferred into a petri dish and dried at 70°C; (iii) the casted film was swelled in water to obtain a κ-CG/CS film.

Source: Reprinted with permission from Yu et al. (2018). Copyright 2018. American Chemical Society.

chitosan and sodium alginate solutions are blended, the cationic groups of chitosan can electrostatically interact with the anionic groups of sodium alginate to induce the formation of insoluble complexes. To overcome these limitations, both polymers can be deposited in a layer-by-layer assembly on the surface of the foods (Li et al., 2019; Senturk Parreidt et al., 2018). Li et al. (2019) produced sodium alginate/chitosan films through layer-by-layer deposition, and physicochemical and mechanical properties were compared with films produced by direct blending. Chitosan/sodium alginate films produced by layer-by-layer assembly displayed similar tensile strength, lower thickness, and higher opacity than those produced by direct blending. The authors described that layer-by-layer assembly increased the interaction between polymer structures by increasing the contact area. Santos and Machado (2021) coated alginate particles with chitosan to increase the stability of probiotic yeast. The alginate–chitosan particles displayed lower moisture content, water activity, and hygroscopicity than alginate particles, indicating superior stability during storage. Carneiro-da-Cunha et al. (2010) produced layer-by-layer assembled films of chitosan and sodium alginate. The chitosan/alginate films displayed low water vapor permeability VWP related to their interactions, decreasing permeability. Souza et al. (2015) coated fresh-cut mangoes surfaces with five nanolayers with alternating deposition of alginate and chitosan. The coated fresh-cut mangoes showed lower mass loss compared with uncoated fruit. The results indicated that nanomultilayer coating acted as an effective barrier against water vapor, avoiding water loss.

3.4.3 Proteins

Protein-based edible films have suitable physical and barrier properties due to the formation of tightly packed and ordered structures (Kumar et al., 2022). Protein-based films possess higher mechanical and gas barrier properties than those obtained with lipid and polysaccharide films (Kumar et al., 2022). However, the mechanical properties of protein-based films are poor (Mohamed et al., 2020). Also, proteins like gelatin do not possess antibacterial, antioxidant, and UV-shielding properties, which are extremely important for food preservation (Fu et al., 2022). To avoid these disadvantages, chitosan can be blended with various proteins to produce chitosan-based films, for example, sodium caseinate (Di Giuseppe et al., 2022; Fiore et al., 2021; Hua et al., 2021; Khwaldia et al., 2014; Pereda et al., 2008, 2009; Ríos-de-Benito et al., 2021; Volpe et al., 2017; Zhang et al., 2014), gelatin (Benbettaïeb et al., 2014; Bonilla & Sobral, 2020; Cai et al., 2019; Cardoso et al., 2016; Fu et al., 2022; Gedarawatte et al., 2021;

Koc & Altıncekic, 2021; Mohammed Manshor et al., 2018; Pereda et al., 2011; Pinto Ramos et al., 2019; Poverenov et al., 2014a, 2014b; Wang et al., 2022b, 2021; Zhang et al., 2021b), and collagen (Andonegi et al., 2020; Bhuimbar et al., 2019; Hou et al., 2020; Indrani et al., 2017; Jiang et al., 2020; Jing et al., 2021; Kaczmarek et al., 2018; Qu et al., 2022a, 2022b; Shah et al., 2019; Sionkowska et al., 2006; Socrates et al., 2019), individually in a composite form.

3.4.3.1 Chitosan/Sodium Caseinate

Sodium caseinate is a protein obtained from casein (the main protein in cow's milk) (Kumar et al., 2022; Pereda et al., 2008). Sodium caseinate can form films due to its random coil nature and its ability to generate extensive intermolecular hydrogen, electrostatic and hydrophobic bonds, increasing interchain cohesion (Khwaldia et al., 2004; Kumar et al., 2022). Also, casein-based edible films have high nutritional value and exceptional sensory properties (Schou et al., 2005).

Volpe et al. (2017) studied the effect of chitosan and sodium caseinate concentrations on the physical properties and structure of the films. Increasing the sodium/chitosan ratio increased the film thickness and elastic modulus, while WVP decreased. Di Giuseppe et al. (2022) reported similar results, who produced chitosan/sodium caseinate films using a mixing ratio of 1:1 (w/v) with glycerol as plasticizer. Chitosan/sodium caseinate displayed similar tensile strength and higher elongation break than those films obtained from chitosan and sodium caseinate blends.

Pereda et al. (2008) reported that chitosan/sodium caseinate films could be favored by forming polymeric complexes through electrostatic interactions. Ríos-de-Benito et al. (2021) reported a transparent and flexible film with a smooth and homogeneous microstructure of chitosan/sodium caseinate films. The films containing mesoporous silica nanoparticles filled with oregano essential oil displayed higher surface roughness of sodium caseinate/chitosan (4:1) films than those using the 8:1 ratio, results that may be related to protein aggregation. Pereda et al. (2009) evaluated the effect of glycerol concentration on the physical and mechanical properties of chitosan/sodium caseinate films. The results displayed that increasing glycerol concentrations increased the equilibrium moisture content in a water sorption study at 23°C.

3.4.3.2 Chitosan/Gelatin

Gelatin is produced from the partial hydrolysis of collagen (Lu et al., 2022; Ramos et al., 2016). Gelatin possesses film-forming ability, high water-binding capacity, low cost, and high availability (Gedarawatte et al., 2021; Ramos et al., 2016). Also, gelatin can be blended to improve the physical, functional, and barrier properties of chitosan films. In this sense, the mechanical and barrier properties of protein-based films are better than polysaccharide-based films (Benbettaïeb et al., 2014). Benbettaïeb et al. (2014) evaluated the physical and mechanical properties of chitosan/gelatin films at different proportions and relative humidities. The chitosan/gelatin films displayed homogeneous appearances and flexible structures. Also, pure chitosan and chitosan/gelatin films withhigh chitosan concentration showed higher b* values than pure gelatin films. Their thickness strongly influences the water vapor sorption of biopolymeric films. The authors reported that water vapor sorption increased linearly with increasing thickness of films from 52 to 159 μm.

Cai et al. (2019) compared the properties of chitosan/gelatin films produced by casting, electrospinning, and coaxial electrospinning. The results displayed that the electrospinning processes induced the formation of a surface-visible and uniform granular polymer network structure. The chitosan/gelatin film produced by the solvent cast was soft and transparent, with low water solubility. Koc and Altıncekic (2021) compared the influence of gelatin concentration on the swelling behavior of chitosan/gelatin films. The swelling ratios of films decreased from 458 to 281% with increasing gelatin concentration at pH 8.0. Also, the results indicated that swelling behavior values did not change at pH 1.5 with increasing gelatin concentration.

3.4.3.3 Chitosan/Collagen

Collagen is the central component of the extracellular matrix (Irastorza et al., 2021). Collagen is composed of two identical polypeptide chains (α1) as well as a polypeptide chain (α2), which polymerize with each other to form a single triple-helix structure (Sun et al., 2021; Tang et al., 2022). Collagen-based films possess good rigidity but poor hydrophobicity, and chitosan-based films have moderate antibacterial activity but are brittle, with a low elongation at break value. For these reasons, collagen or chitosan are insufficient to form an eligible edible film using a single polymer. Therefore, collagen and chitosan have been blended to compensate for their disadvantages (Qu et al., 2022b). Chitosan and collagen can be blended to produce films whose properties are influenced by the concentration of polymers (Sionkowska, 2021).

Bhuimbar et al. (2019) fabricated collagen/chitosan films using the skin of medusa (*Centrolophus niger*) to extract soluble acid collagen. The films displayed suitable mechanical strength without cracks or rollover. Also, the films showed an appearance that is shiny, transparent, and smooth. Hou et al. (2020) produced collagen/chitosan films at different ratio of collagen/chitosan (100:0, 60:40, 50:50, 40:60, and 0:100). The collagen/chitosan films were viewed like plastics with a thin, compact, and homogeneous structure. Also, with the addition of chitosan, the collagen films became smoother, more rigid, and less flexible. The mechanical properties (tensile strength), light barrier properties, and WVP of collagen films increased with the addition of chitosan. Qu et al. (2022a) studied the sweep frequency pulsed ultrasound method to improve the fabrication of tuna skin collagen/chitosan films. The tensile strength of collagen/chitosan films increased with increasing ultrasonic frequency. The chitosan/collagen films displayed smooth and uniform structures. Ultrasonic frequencies had no significant effect on water solubility, while WVP values increased with increasing ultrasonic frequency. These results suggest that ultrasonic frequency favored the interaction of collagen and chitosan, making the film more compact. FT-IR analysis confirmed that the ultrasonic treatment enhanced the cross-linking of the films between collagen and chitosan.

3.5 Applications of Chitosan-Based Films and Coatings

Consumers request that vegetables and fruits preserve their beneficial compounds (Tahir et al., 2019). In this sense, scientists and food processing industries have developed polymeric systems that help to improve food quality, safety, freshness, and shelf life by using natural, edible, and biodegradable polymers. These films and coatings can decrease the loss of quality of foods, producing a barrier around vegetables and fruits (Nair et al., 2020).

3.5.1 Chitosan-Based Films

Table 3.1 shows several applications of chitosan-based films in food packing. In general, chitosan-based films promote antimicrobial activity.

Gómez-Estaca et al. (2010) evaluated the antimicrobial activity of gelatin/chitosan-based films with clove essential oil against six microorganisms. The results displayed that gelatin/chitosan-based films containing clove essential oil inhibited the growth of microorganisms, that is, *P. fluorescens*, *S. putrefaciens*, *P. phosphoreum*, *L. innocua*, *E. coli*, and *L. acidophilus*. Qu et al. (2022b) developed collagen/chitosan films by adding polyphenols of pomegranate under ultrasound conditions. The composite films displayed antibacterial activity against *B. subtilis*, *E. coli*, *L. monocytogenes*, and *S. aureus*. Also, tensile strength increased by 47.03%, while antioxidant capacity increased by 24.16 folds compared to the control films (Figure 3.3).

Hua et al. (2021) produced zein-sodium caseinate nanoparticles loaded with clove-essential-oil and incorporated into a chitosan film. The incorporation of nanoparticles increased the tensile strength and break elongation of films. Also, the chitosan/nanoparticles films displayed antibacterial properties against *E. coli* and *S. aureus*.

3.5.2 Chitosan-Based Coatings

Polymer coatings create a semipermeable barrier around food surfaces to control gas exchange, decrease water loss, maintain tissue firmness, and reduce microbial growth (González-Aguilar et al., 2009). Table 3.2 shows several applications of chitosan-based coatings in food packing.

TABLE 3.1

Chitosan-based films for food packing

Chitosan-based films	Plasticizers/ Additives	Bioactive molecules	Properties	References
Collagen/ chitosan	Glycerol	Pomegranate peel extract (PPE)	Antibacterial properties (>5% PPE)	(Bhuimbar et al., 2019)
	Glycerol	Lemon essential oil	Low oxygen permeability, high tensile strength, high elongation at break, lipid oxidation inhibition	(Jiang et al., 2020)
	Glycerol	Pomegranate polyphenol	Antioxidant capacity and antibacterial activity against *B. subtilis*, *E. coli*, *L. monocytogenes*, and *S. aureus*	(Qu et al., 2022b)
Gelatin/ chitosan	Glycerol	Boldo of Chile leaf's extract	Total disintegration on the first day	(Bonilla & Sobral, 2020)
	Sorbitol, glycerol, soya lecithin	Clove essential oil	Antibacterial activity	(Gómez-Estaca et al., 2010)
	Sorbitol, glycerol, soya lecithin	Oregano essential oil	Reduction of mechanical properties and increased light barrier, water vapor barrier, and antimicrobial activity	(Wu et al., 2014)
	Glycerol	Microcapsules of *Pulicaria jaubertii* extract	High barrier property toward UV and visible light, antioxidant activity	(Al-Maqtari et al., 2022)
	Glycerol	N/A	Antibacterial activity against *E. coli*	(Pereda et al., 2011)
	Glycerol	Curcumin	Antibacterial activity against *L. monocytogenes*, *E. coli*, and *S. putrefaciens*	(Wang et al., 2022b)
Pectin/ chitosan	Glycerol, glutaraldehyde, silver nitrate	N/A	Antibacterial activity against *E. coli*	(Akalin et al., 2022)
	Calcium chloride	Tea polyphenols	Antioxidant/antibacterial activity and inhibition of the color deterioration	(Gao et al., 2019)
Cellulose/ chitosan	Glycerol	Saffron anthocyanins	Antibacterial activity	(Alizadeh-Sani et al., 2021)
	Glycerol	N/A	Antibacterial activity	(Indrani et al., 2017)
	N/A	N/A	Antibacterial activity against *E. coli*, *S. epidermidis*, and *M. luteus*	(Szymańska-Chargot et al., 2019)
Sodium caseinate/ chitosan	Zein nanoparticles, glycerol	Clove essential oil	Antibacterial activity with an increase on mechanical properties	(Hua et al., 2021)
	Glycerol	N/A	Suitable water vapor permeability	(Volpe et al., 2017)
Carrageenan/ chitosan	N/A	N/A	Antibacterial activity against *S. aureus* and *E. coli*	(Ismillayli et al., 2020)
	N/A	Nisin Z	Antibacterial activity against *S. aureus*	(Webber et al., 2021)
Sodium alginate	Glycerol	N/A	Low water vapor permeability, suitable color properties, and antibacterial activity	(Salama et al., 2018)

For example, Moller et al. (2022) evaluated the effect of chitosan/carrageenan in combination with allyl isothiocyanate as an edible coating on the microbial load in the chicken breast during storage. *Salmonella Typhimurium* and *Campylobacter coli* were the microorganisms used in this study. The results showed that populations of *Salmonella Typhimurium* in the chicken breast were reduced using chitosan/carrageenan coating with 200 μg/mL of allyl isothiocyanate at 4°C. Similar results were obtained for *Campylobacter coli* using 20 μg/mL of allyl isothiocyanate after 21 days at 4°C.

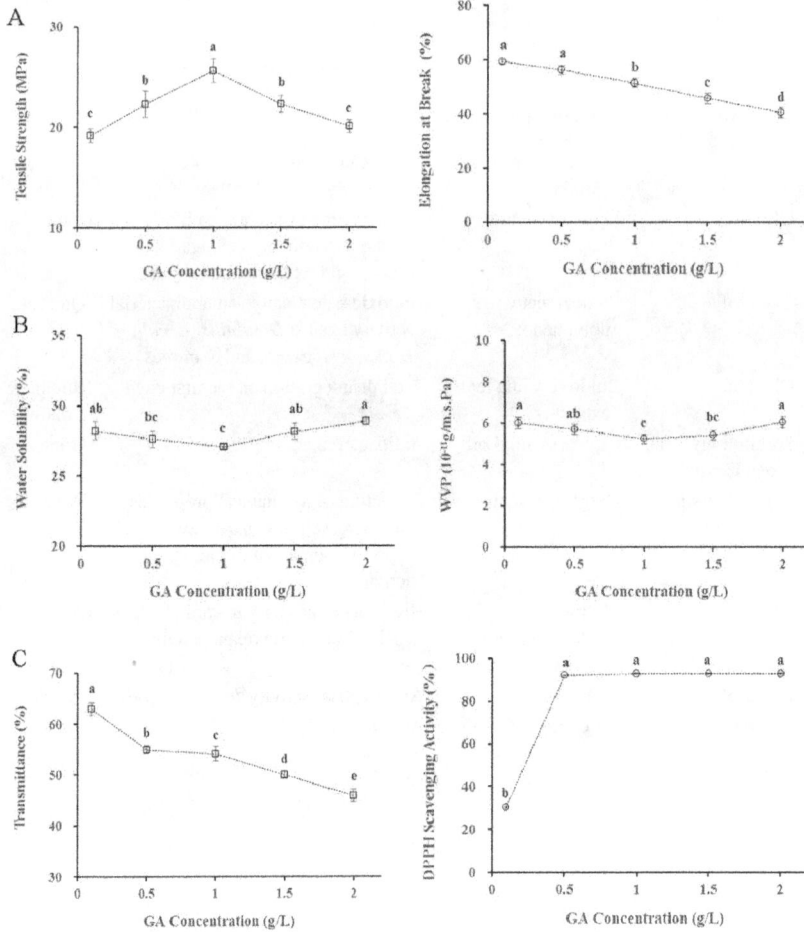

FIGURE 3.3   Mechanical properties (A), water solubility and water vapor permeability (B), and transmittance and DPPH scavenging activity of the films under different gallic acid concentrations (C).

Source: Reprinted with permission from Qu et al. (2022b). Copyright 2020. Elsevier.

Zhang et al. (2021b) reported that carboxymethyl chitosan-gelatin coating with $CaCl_2$ and ascorbic acid positively affected the quality and nutritional characteristics of sweet cherries. The chitosan-based coating reduced weight loss and maintained skin color, peduncle freshness, fruit firmness, total phenolic content, anthocyanins concentration, and antioxidant capacity. Kim et al. (2018) developed an antimicrobial layer-by-layer coating composed of chitosan, sodium alginate, and grapefruit seed extract to improve shrimp stability. The authors reported that the bilayer coating decreased the off-flavor of the shrimp for 15 days at 4°C. Also, the chitosan-based coating displayed antibacterial activity against mesophilic and psychrotrophic microorganisms, reducing bacterial counts.

In recent years, inorganic nanomaterials have been used due to their exceptional characteristics (Yu et al., 2022). For example, the use of nanoparticles in packaging matrix materials improves the performance of food packaging with changes in the physical, mechanical, rheological, barrier, and thermal properties (Taherimehr et al., 2021; Yu et al., 2022). Arroyo et al. (2020) reported the addition of ZnO nanoparticles on the ripening process, water loss, texture, color, rot index, and physicochemical properties of chitosan/alginate coating applied in guavas. The addition of nanoparticles displayed an antibacterial action preventing the apparition of rot in guavas. Also, the chitosan/alginate/nanoparticles coating delayed the ripening process of guavas and decreased the mass loss and the fruit lesions after 20 days of storage. However, inorganic compounds containing metal ions in the film that will inevitably penetrate

TABLE 3.2

Chitosan-based coatings in food packing

Chitosan-based films	Plasticizers/Additives	Foods systems	Properties	References
Chitosan/alginate	Glycerol, ZnO nanoparticles	Guavas	Prevention of rot appearance, antibacterial effect, and delaying the ripening process	(Arroyo et al., 2020)
	Grapefruit seed extract, Tween 80	Shrimp	Antibacterial activity, reduction of off-flavors	(Kim et al., 2018)
	ε-polysine	Fish	Antibacterial activity, reduction of off-flavors	(Li et al., 2020)
	N/A	Mangoes	Decreased weight loss, pH value, browning rate, and control of microbial growth	(Souza et al., 2015)
Chitosan/pectin	Glycerol, tarragon essential oil	Fish muscle	Antibacterial activity, decrease of lipid oxidation	(Esmaeili & Khodanazary, 2021)
	N/A	Mangoes	Reduction of mass loss, antibacterial activity, color stability	(Medeiros et al., 2012)
Chitosan/cellulose	Tween 80 and Span 80	Pears	Reduction in ripening and quality deterioration (weight loss, color, and texture)	(Deng et al., 2017)
	Tween 80	Cantaloupe rind and fresh-cut pulp melon	Antimicrobial activity, decreased weight loss, and lightness maintained	(Ozturk et al., 2021; Zhang et al., 2020)
Chitosan/carrageenan	N/A	Chicken breast	Antibacterial activity	(Moller et al., 2022)
	N/A	Strawberries	Reduction of mass and firmness loss, antibacterial activity	(Resende et al., 2018)
	Oriental mustard extract	Chicken breast	Antibacterial activity	(Olaimat et al., 2014)
Chitosan/gelatin	Glycerol	Beef	Reduction of weight loss and lipid oxidation, color stability, control of microbial growth	(Cardoso et al., 2016, 2019)
	N/A	Cantaloupe melons and fresh-cut pulp	Inhibition of the total microbial growth, reduction of weight loss	(Poverenov et al., 2014a)
	N/A	Red bell peppers	Inhibition of microbial growth, improvement of fruit texture	(Poverenov et al., 2014b)
Carboxymethyl chitosan-gelatin	Glycerol, Tween-20, ascorbic acid	Sweet cherry	Reduction of weight loss, preservation of skin color and fruit firmness	(Zhang et al., 2021b)
Chitosan-caseinate	Rosemary essential oil, Tween 80	Chicken breast	Reduction of oxidation	(Fiore et al., 2021)
	Silica nanoparticles filled with oregano essential oil	Panela cheese	Inhibition of microbial growth	(Ríos-de-Benito et al., 2021)

into the food, posing a potential threat to human health. Therefore, it is essential to replace these inorganic nanoparticles with natural, non-toxic antibacterial agents (Fu et al., 2022).

3.6 Conclusion

This chapter reviews the fabrication, properties, and applications of chitosan-based films and coatings using plasticizers, polysaccharides, and proteins. Chitosan-based films and coatings combined with

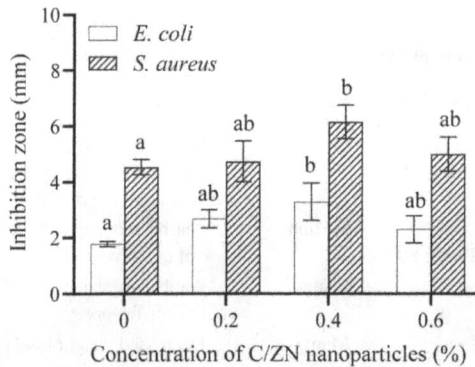

FIGURE 3.4 Inhibition zones of chitosan-based films with chitosan/zein nanoparticles (test bacteria: *E. coli*, the column with white padding; *S. aureus*, the column with a forward slash).

Source: Reprinted from Hua et al. (2021). Copyright 2021, with permission of Elsevier.

polysaccharides and proteins can potentially be used as an exciting alternative to plastic packaging, which has caused severe environmental problems worldwide.

Conflicts of Interest

The authors declare no conflicts of interest.

REFERENCES

Abdou, E. S., Nagy, K. S. A., & Elsabee, M. Z. (2008). Extraction and characterization of chitin and chitosan from local sources. *Bioresource Technology*, *99*(5), 1359–1367. https://doi.org/10.1016/j.biortech.2007.01.051

Affes, S., Maalej, H., Li, S., Abdelhedi, R., Nasri, R., & Nasri, M. (2022). Effect of glucose substitution by low-molecular weight chitosan-derivatives on functional, structural and antioxidant properties of Maillard reaction-cross-linked chitosan-based films. *Food Chemistry*, *366*, 130530. https://doi.org/10.1016/j.foodchem.2021.130530

Agulló, E., Rodríguez, M. S., Ramos, V., & Albertengo, L. (2003). Present and future role of chitin and chitosan in food. *Macromolecular Bioscience*, *3*(10), 521–530. https://doi.org/10.1002/mabi.200300010

Akalin, G. O., Oztuna Taner, O., & Taner, T. (2022). The preparation, characterization and antibacterial properties of chitosan/pectin silver nanoparticle films. *Polymer Bulletin*, *79*(6), 3495–3512. https://doi.org/10.1007/s00289-021-03667-0

Alirezalu, K., Pirouzi, S., Yaghoubi, M., Karimi-Dehkordi, M., Jafarzadeh, S., & Mousavi Khaneghah, A. (2021). Packaging of beef fillet with active chitosan film incorporated with ε-polylysine: An assessment of quality indices and shelf life. *Meat Science*, *176*, 108475. https://doi.org/10.1016/j.meatsci.2021.108475

Alishahi, A., Mirvaghefi, A., Tehrani, M. R., Farahmand, H., Shojaosadati, S. A., Dorkoosh, F. A., & Elsabee, M. Z. (2011). Enhancement and characterization of chitosan extraction from the wastes of shrimp packaging plants. *Journal of Polymers and the Environment*, *19*(3), 776–783. https://doi.org/10.1007/s10924-011-0321-5

Alizadeh-Sani, M., Tavassoli, M., McClements, D. J., & Hamishehkar, H. (2021). Multifunctional halochromic packaging materials: Saffron petal anthocyanin loaded-chitosan nanofiber/methyl cellulose matrices. *Food Hydrocolloids*, *111*, 106237. https://doi.org/10.1016/j.foodhyd.2020.106237

Al-Maqtari, Q. A., Al-Gheethi, A. A. S., Ghaleb, A. D. S., Mahdi, A. A., Al-Ansi, W., Noman, A. E., Al-Adeeb, A., Odjo, A. K. O., Du, Y., Wei, M., & Yao, W. (2022). Fabrication and characterization of chitosan/gelatin films loaded with microcapsules of *Pulicaria jaubertii* extract. *Food Hydrocolloids*, *129*, 107624. https://doi.org/10.1016/j.foodhyd.2022.107624

Andonegi, M., Heras, K. L., Santos-Vizcaíno, E., Igartua, M., Hernandez, R. M., de la Caba, K., & Guerrero, P. (2020). Structure-properties relationship of chitosan/collagen films with potential for biomedical applications. *Carbohydrate Polymers*, *237*, 116159. https://doi.org/10.1016/j.carbpol.2020.116159

Arroyo, B. J., Bezerra, A. C., Oliveira, L. L., Arroyo, S. J., Melo, E. A. D., & Santos, A. M. P. (2020). Antimicrobial active edible coating of alginate and chitosan add ZnO nanoparticles applied in guavas (*Psidium guajava* L.). *Food Chemistry*, *309*, 125566. https://doi.org/10.1016/j.foodchem.2019.125566

Arzate-Vázquez, I., Chanona-Pérez, J. J., Calderón-Domínguez, G., Terres-Rojas, E., Garibay-Febles, V., Martínez-Rivas, A., & Gutiérrez-López, G. F. (2012). Microstructural characterization of chitosan and alginate films by microscopy techniques and texture image analysis. *Carbohydrate Polymers*, *87*(1), 289–299. https://doi.org/10.1016/j.carbpol.2011.07.044

Bajdik, J., Marciello, M., Caramella, C., Domján, A., Süvegh, K., Marek, T., & Pintye-Hódi, K. (2009). Evaluation of surface and microstructure of differently plasticized chitosan films. *Journal of Pharmaceutical and Biomedical Analysis*, *49*(3), 655–659. https://doi.org/10.1016/j.jpba.2008.12.020

Barajas-Álvarez, P., González-Ávila, M., & Espinosa-Andrews, H. (2022). Microencapsulation of Lactobacillus rhamnosus HN001 by spray drying and its evaluation under gastrointestinal and storage conditions. *LWT*, *153*, 112485. https://doi.org/10.1016/j.lwt.2021.112485

Benbettaïeb, N., Kurek, M., Bornaz, S., & Debeaufort, F. (2014). Barrier, structural and mechanical properties of bovine gelatin—chitosan blend films related to biopolymer interactions. *Journal of the Science of Food and Agriculture*, *94*(12), 2409–2419. https://doi.org/10.1002/jsfa.6570

Berger, J., Reist, M., Mayer, J. M., Felt, O., Peppas, N. A., & Gurny, R. (2004). Structure and interactions in covalently and ionically cross-linked chitosan hydrogels for biomedical applications. *European Journal of Pharmaceutics and Biopharmaceutics*, *57*(1), 19–34. https://doi.org/10.1016/S0939-6411(03)00161-9

Bhuimbar, M. V., Bhagwat, P. K., & Dandge, P. B. (2019). Extraction and characterization of acid soluble collagen from fish waste: Development of collagen-chitosan blend as food packaging film. *Journal of Environmental Chemical Engineering*, *7*(2), 102983. https://doi.org/10.1016/j.jece.2019.102983

Bi, F., Qin, Y., Chen, D., Kan, J., & Liu, J. (2021). Development of active packaging films based on chitosan and nano-encapsulated luteolin. *International Journal of Biological Macromolecules*, *182*, 545–553. https://doi.org/10.1016/j.ijbiomac.2021.04.063

Blancas-Benitez, F. J., Montaño-Leyva, B., Aguirre-Güitrón, L., Moreno-Hernández, C. L., Fonseca-Cantabrana, Á., Romero-Islas, L. D. C., & González-Estrada, R. R. (2022). Impact of edible coatings on quality of fruits: A review. *Food Control*, *139*, 109063. https://doi.org/10.1016/j.foodcont.2022.109063

Bonilla, J., & Sobral, P. J. A. (2020). Disintegrability under composting conditions of films based on gelatin, chitosan and/or sodium caseinate containing boldo-of-Chile leafs extract. *International Journal of Biological Macromolecules*, *151*, 178–185. https://doi.org/10.1016/j.ijbiomac.2020.02.051

Cai, L., Shi, H., Cao, A., & Jia, J. (2019). Characterization of gelatin/chitosan ploymer films integrated with docosahexaenoic acids fabricated by different methods. *Scientific Reports*, *9*(1), 8375. https://doi.org/10.1038/s41598-019-44807-x

Caicedo, C., Díaz-Cruz, C. A., Jiménez-Regalado, E. J., & Aguirre-Loredo, R. Y. (2022). Effect of plasticizer content on mechanical and water vapor permeability of maize starch/PVOH/chitosan composite films. *Materials*, *15*(4), 1274. www.mdpi.com/1996-1944/15/4/1274

Cardoso, G. P., Andrade, M. P. D., Rodrigues, L. M., Massingue, A. A., Fontes, P. R., Ramos, A. D. L. S., & Ramos, E. M. (2019). Retail display of beef steaks coated with monolayer and bilayer chitosan-gelatin composites. *Meat Science*, *152*, 20–30. https://doi.org/10.1016/j.meatsci.2019.02.009

Cardoso, G. P., Dutra, M. P., Fontes, P. R., Ramos, A. D. L. S., Gomide, L. A. D. M., & Ramos, E. M. (2016). Selection of a chitosan gelatin-based edible coating for color preservation of beef in retail display. *Meat Science*, *114*, 85–94. https://doi.org/10.1016/j.meatsci.2015.12.012

Carneiro-da-Cunha, M. G., Cerqueira, M. A., Souza, B. W. S., Carvalho, S., Quintas, M. A. C., Teixeira, J. A., & Vicente, A. A. (2010). Physical and thermal properties of a chitosan/alginate nanolayered PET film. *Carbohydrate Polymers*, *82*(1), 153–159. https://doi.org/10.1016/j.carbpol.2010.04.043

Cazón, P., Vázquez, M., & Velazquez, G. (2020). Environmentally friendly films combining bacterial cellulose, chitosan, and polyvinyl alcohol: Effect of water activity on barrier, mechanical, and optical properties. *Biomacromolecules*, *21*(2), 753–760. https://doi.org/10.1021/acs.biomac.9b01457

Cha, D. S., & Chinnan, M. S. (2004). Biopolymer-based antimicrobial packaging: A review. *Critical Reviews in Food Science and Nutrition*, *44*(4), 223–237. https://doi.org/10.1080/10408690490464276

Chang, S.-H., Lin, H.-T. V., Wu, G.-J., & Tsai, G. J. (2015). pH Effects on solubility, zeta potential, and correlation between antibacterial activity and molecular weight of chitosan. *Carbohydrate Polymers*, *134*, 74–81. https://doi.org/10.1016/j.carbpol.2015.07.072

Chattopadhyay, D. P., & Inamdar, M. S. (2010). Aqueous behaviour of chitosan. *International Journal of Polymer Science, 2010,* 939536. https://doi.org/10.1155/2010/939536

Chen, L.-C., Kung, S.-K., Chen, H.-H., & Lin, S.-B. (2010). Evaluation of zeta potential difference as an indicator for antibacterial strength of low molecular weight chitosan. *Carbohydrate Polymers, 82*(3), 913–919. https://doi.org/10.1016/j.carbpol.2010.06.017

Chen, W., Ma, S., Wang, Q., McClements, D. J., Liu, X., Ngai, T., & Liu, F. (2021). Fortification of edible films with bioactive agents: A review of their formation, properties, and application in food preservation. *Critical Reviews in Food Science and Nutrition, 62,* 5029–5055. https://doi.org/10.1080/10408398.2021.1881435

Cherian, R. M., Tharayil, A., Varghese, R. T., Antony, T., Kargarzadeh, H., Chirayil, C. J., & Thomas, S. (2022). A review on the emerging applications of nano-cellulose as advanced coatings. *Carbohydrate Polymers, 282,* 119123. https://doi.org/10.1016/j.carbpol.2022.119123

Cheung, R. C. F., Ng, T. B., Wong, J. H., & Chan, W. Y. (2015). Chitosan: An update on potential biomedical and pharmaceutical applications. *Marine Drugs, 13*(8), 5156–5186. www.mdpi.com/1660-3397/13/8/5156

Cook, M. T., Tzortzis, G., Khutoryanskiy, V. V., & Charalampopoulos, D. (2013). Layer-by-layer coating of alginate matrices with chitosan—alginate for the improved survival and targeted delivery of pro-biotic bacteria after oral administration. *Journal of Materials Chemistry B, 1*(1), 52–60. https://doi.org/10.1039/C2TB00126H

Dahmane, E. M., Taourirte, M., Eladlani, N., & Rhazi, M. (2014). Extraction and characterization of chitin and chitosan from Parapenaeus longirostris from Moroccan local sources. *International Journal of Polymer Analysis and Characterization, 19*(4), 342–351. https://doi.org/10.1080/1023666X.2014.902577

Deng, Z., Jung, J., Simonsen, J., Wang, Y., & Zhao, Y. (2017). Cellulose nanocrystal reinforced chitosan coat-ings for improving the storability of postharvest pears under both ambient and cold storages. *Journal of Food Science, 82*(2), 453–462. https://doi.org/10.1111/1750-3841.13601

De Queiroz Antonino, R. S. C. M., Lia Fook, B. R. P., De Oliveira Lima, V. A., De Farias Rached, R. Í., Lima, E. P. N., Da Silva Lima, R. J., Peniche Covas, C. A., & Lia Fook, M. V. (2017). Preparation and char-acterization of chitosan obtained from shells of shrimp (Litopenaeus vannamei Boone). *Marine Drugs, 15*(5), 141. www.mdpi.com/1660-3397/15/5/141

Deshmukh, A. R., Aloui, H., Khomlaem, C., Negi, A., Yun, J.-H., Kim, H.-S., & Kim, B. S. (2021). Biodegradable films based on chitosan and defatted Chlorella biomass: Functional and physical charac-terization. *Food Chemistry, 337,* 127777. https://doi.org/10.1016/j.foodchem.2020.127777

Di Giuseppe, F. A., Volpe, S., Cavella, S., Masi, P., & Torrieri, E. (2022). Physical properties of active biopoly-mer films based on chitosan, sodium caseinate, and rosemary essential oil. *Food Packaging and Shelf Life, 32,* 100817. https://doi.org/10.1016/j.fpsl.2022.100817

Doan, V. K., Ly, K. L., Tran, N. M.-P., Ho, T. P.-T., Ho, M. H., Dang, N. T.-N., Chang, C.-C., Nguyen, H. T.-T., Ha, P. T., Tran, Q. N., Tran, L. D., Vo, T. V., & Nguyen, T. H. (2021). Characterizations and antibacterial efficacy of chitosan oligomers synthesized by microwave-assisted hydrogen peroxide oxidative depo-lymerization method for infectious wound applications. *Materials (Basel, Switzerland), 14*(16), 4475. https://doi.org/10.3390/ma14164475

dos Santos, T. A., de Oliveira, A. C. S., Lago, A. M. T., Yoshida, M. I., Dias, M. V., & Borges, S. V. (2021). Properties of chitosan—papain biopolymers reinforced with cellulose nanofibers. *Journal of Food Processing and Preservation, 45*(9), e15740. https://doi.org/10.1111/jfpp.15740

Dulta, K., Koşarsoy Ağçeli, G., Thakur, A., Singh, S., Chauhan, P., & Chauhan, P. K. (2022). Development of alginate-chitosan based coating enriched with ZnO nanoparticles for increasing the shelf life of orange fruits (Citrus sinensis L.). *Journal of Polymers and the Environment.* https://doi.org/10.1007/s10924-022-02411-7

Dwivany, F. M., Aprilyandi, A. N., Suendo, V., & Sukriandi, N. (2020). Carrageenan edible coating applica-tion prolongs cavendish banana shelf life. *International Journal of Food Science, 2020,* 8861610. https://doi.org/10.1155/2020/8861610

El Knidri, H., Belaabed, R., Addaou, A., Laajeb, A., & Lahsini, A. (2018). Extraction, chemical modification and characterization of chitin and chitosan. *International Journal of Biological Macromolecules, 120,* 1181–1189. https://doi.org/10.1016/j.ijbiomac.2018.08.139

Elsabee, M. Z., Abdou, E. S., Nagy, K. S. A., & Eweis, M. (2008). Surface modification of polypropylene films by chitosan and chitosan/pectin multilayer. *Carbohydrate Polymers, 71*(2), 187–195. https://doi.org/10.1016/j.carbpol.2007.05.022

Esmaeili, M., & Khodanazary, A. (2021). Effects of pectin/chitosan composite and bi-layer coatings combined with Artemisia dracunculus essential oil on the mackerel's shelf life. *Journal of Food Measurement and Characterization, 15*(4), 3367–3375. https://doi.org/10.1007/s11694-021-00879-w

Espinosa-Andrews, H., Báez-González, J. G., Cruz-Sosa, F., & Vernon-Carter, E. J. (2007). Gum Arabic–chitosan complex coacervation. *Biomacromolecules, 8*(4), 1313–1318. https://doi.org/10.1021/bm0611634

Espinosa-Andrews, H., Sandoval-Castilla, O., Vázquez-Torres, H., Vernon-Carter, E. J., & Lobato-Calleros, C. (2010). Determination of the gum Arabic—chitosan interactions by Fourier Transform Infrared Spectroscopy and characterization of the microstructure and rheological features of their coacervates. *Carbohydrate Polymers, 79*(3), 541–546. https://doi.org/10.1016/j.carbpol.2009.08.040

Fiore, A., Park, S., Volpe, S., Torrieri, E., & Masi, P. (2021). Active packaging based on PLA and chitosan-caseinate enriched rosemary essential oil coating for fresh minced chicken breast application. *Food Packaging and Shelf Life, 29*, 100708. https://doi.org/10.1016/j.fpsl.2021.100708

Flórez, M., Guerra-Rodríguez, E., Cazón, P., & Vázquez, M. (2022). Chitosan for food packaging: Recent advances in active and intelligent films. *Food Hydrocolloids, 124*, 107328. https://doi.org/10.1016/j.foodhyd.2021.107328

Fu, B., Liu, Q., Liu, M., Chen, X., Lin, H., Zheng, Z., Zhu, J., Dai, C., Dong, X., & Yang, D.-P. (2022). Carbon dots enhanced gelatin/chitosan bio-nanocomposite packaging film for perishable foods. *Chinese Chemical Letters, 33*(10), 4577–4582. https://doi.org/10.1016/j.cclet.2022.03.048

Gao, H.-X., He, Z., Sun, Q., He, Q., & Zeng, W.-C. (2019). A functional polysaccharide film forming by pectin, chitosan, and tea polyphenols. *Carbohydrate Polymers, 215*, 1–7. https://doi.org/10.1016/j.carbpol.2019.03.029

Gasti, T., Dixit, S., Hiremani, V. D., Chougale, R. B., Masti, S. P., Vootla, S. K., & Mudigoudra, B. S. (2022). Chitosan/pullulan based films incorporated with clove essential oil loaded chitosan-ZnO hybrid nanoparticles for active food packaging. *Carbohydrate Polymers, 277*, 118866. https://doi.org/10.1016/j.carbpol.2021.118866

Gedarawatte, S. T. G., Ravensdale, J. T., Johns, M. L., Azizi, A., Al-Salami, H., Dykes, G. A., & Coorey, R. (2021). Effectiveness of gelatine and chitosan spray coating for extending shelf life of vacuum-packaged beef. *International Journal of Food Science & Technology, 56*(8), 4026–4037. https://doi.org/10.1111/ijfs.15025

Gómez-Estaca, J., López de Lacey, A., López-Caballero, M. E., Gómez-Guillén, M. C., & Montero, P. (2010). Biodegradable gelatin—chitosan films incorporated with essential oils as antimicrobial agents for fish preservation. *Food Microbiology, 27*(7), 889–896. https://doi.org/10.1016/j.fm.2010.05.012

González-Aguilar, G. A., Valenzuela-Soto, E., Lizardi-Mendoza, J., Goycoolea, F., Martínez-Téllez, M. A., Villegas-Ochoa, M. A., Monroy-García, I. N., & Ayala-Zavala, J. F. (2009). Effect of chitosan coating in preventing deterioration and preserving the quality of fresh-cut papaya 'Maradol'. *Journal of the Science of Food and Agriculture, 89*(1), 15–23. https://doi.org/10.1002/jsfa.3405

Goy, R. C., de Britto, D., & Assis, O. B. G. (2009). A review of the antimicrobial activity of chitosan. *Polímeros, 19*(3).

Grunenfelder, L. K., Herrera, S., & Kisailus, D. (2014). Crustacean-derived biomimetic components and nano-structured composites. *Small, 10*(16), 3207–3232. https://doi.org/10.1002/smll.201400559

Hadiyanto, H., Christwardana, M., Suzery, M., Sutanto, H., Nilamsari, A. M., & Yunanda, A. (2019). Effects of carrageenan and chitosan as coating materials on the thermal degradation of microencapsulated phycocyanin from *Spirulina* sp. *International Journal of Food Engineering, 15*(5–6). https://doi.org/10.1515/ijfe-2018-0290

Haghighi, H., Licciardello, F., Fava, P., Siesler, H. W., & Pulvirenti, A. (2020). Recent advances on chitosan-based films for sustainable food packaging applications. *Food Packaging and Shelf Life, 26*, 100551. https://doi.org/10.1016/j.fpsl.2020.100551

Hamzah, H. M., Osman, A., Tan, C. P., & Mohamad Ghazali, F. (2013). Carrageenan as an alternative coating for papaya (Carica papaya L. cv. Eksotika). *Postharvest Biology and Technology, 75*, 142–146. https://doi.org/10.1016/j.postharvbio.2012.08.012

Hild, S., Marti, O., & Ziegler, A. (2008). Spatial distribution of calcite and amorphous calcium carbonate in the cuticle of the terrestrial crustaceans Porcellio scaber and Armadillidium vulgare. *Journal of Structural Biology, 163*(1), 100–108. https://doi.org/10.1016/j.jsb.2008.04.010

Hoagland, P. D., & Parris, N. (1996). Chitosan/pectin laminated films. *Journal of Agricultural and Food Chemistry, 44*(7), 1915–1919. https://doi.org/10.1021/jf950162s

Hosseini, S. F., Kaveh, F., & Schmid, M. (2022). Facile fabrication of transparent high-barrier poly(lactic acid)-based bilayer films with antioxidant/antimicrobial performances. *Food Chemistry*, *384*, 132540. https://doi.org/10.1016/j.foodchem.2022.132540

Hou, C., Gao, L., Wang, Z., Rao, W., Du, M., & Zhang, D. (2020). Mechanical properties, thermal stability, and solubility of sheep bone collagen—chitosan films. *Journal of Food Process Engineering*, *43*(1), e13086. https://doi.org/10.1111/jfpe.13086

Hua, L., Deng, J., Wang, Z., Wang, Y., Chen, B., Ma, Y., Li, X., & Xu, B. (2021). Improving the functionality of chitosan-based packaging films by cross-linking with nanoencapsulated clove essential oil. *International Journal of Biological Macromolecules*, *192*, 627–634. https://doi.org/10.1016/j.ijbiomac.2021.09.197

Huang, X., Ge, X., Zhou, L., & Wang, Y. (2022). Eugenol embedded zein and poly(lactic acid) film as active food packaging: Formation, characterization, and antimicrobial effects. *Food Chemistry*, *384*, 132482. https://doi.org/10.1016/j.foodchem.2022.132482

Huq, T., Khan, A., Brown, D., Dhayagude, N., He, Z., & Ni, Y. (2022). Sources, production and commercial applications of fungal chitosan: A review. *Journal of Bioresources and Bioproducts*, *7*(2), 85–98. https://doi.org/10.1016/j.jobab.2022.01.002

Hwang, J.-K., & Hae-Hun, S. (2000). Rheological properties of chitosan solutions. *Korea-Australia Rheology Journal*, *12*, 175–179.

Indrani, D. J., Lukitowati, F., & Yulizar, Y. (2017). Preparation of chitosan/collagen blend membranes for wound dressing: A study on FTIR spectroscopy and mechanical properties. *IOP Conference Series: Materials Science and Engineering*, *202*(1), 012020. https://doi.org/10.1088/1757-899x/202/1/012020

Indriyati, D. F., Primadona, I., Srikandace, Y., & Karina, M. (2021). Development of bacterial cellulose/chitosan films: Structural, physicochemical and antimicrobial properties. *Journal of Polymer Research*, *28*(3), 70. https://doi.org/10.1007/s10965-020-02328-6

Iñiguez-Moreno, M., Ragazzo-Sánchez, J. A., & Calderón-Santoyo, M. (2021). An extensive review of natural polymers used as coatings for postharvest shelf-life extension: Trends and challenges. *Polymers*, *13*(19), 3271. www.mdpi.com/2073-4360/13/19/3271

Irastorza, A., Zarandona, I., Andonegi, M., Guerrero, P., & de la Caba, K. (2021). The versatility of collagen and chitosan: From food to biomedical applications. *Food Hydrocolloids*, *116*, 106633. https://doi.org/10.1016/j.foodhyd.2021.106633

Islam, S., Bhuiyan, M. A. R., & Islam, M. N. (2017). Chitin and chitosan: Structure, properties and applications in biomedical engineering. *Journal of Polymers and the Environment*, *25*(3), 854–866. https://doi.org/10.1007/s10924-016-0865-5

Ismillayli, N., Andayani, I. G. A. S., Honiar, R., Mariana, B., Sanjaya, R. K., & Hermanto, D. (2020). Polyelectrolyte Complex (PEC) film based on chitosan as potential edible films and their antibacterial activity test. *IOP Conference Series: Materials Science and Engineering*, *959*(1), 012009. https://doi.org/10.1088/1757-899x/959/1/012009

Jiang, Y., Lan, W., Sameen, D. E., Ahmed, S., Qin, W., Zhang, Q., Chen, H., Dai, J., He, L., & Liu, Y. (2020). Preparation and characterization of grass carp collagen-chitosan-lemon essential oil composite films for application as food packaging. *International Journal of Biological Macromolecules*, *160*, 340–351. https://doi.org/10.1016/j.ijbiomac.2020.05.202

Jing, X., Li, X., Jiang, Y., Zhao, R., Ding, Q., & Han, W. (2021). Excellent coating of collagen fiber/chitosan-based materials that is water- and oil-resistant and fluorine-free. *Carbohydrate Polymers*, *266*, 118173. https://doi.org/10.1016/j.carbpol.2021.118173

Kaczmarek, B., Sionkowska, A., & Skopinska-Wisniewska, J. (2018). Influence of glycosaminoglycans on the properties of thin films based on chitosan/collagen blends. *Journal of the Mechanical Behavior of Biomedical Materials*, *80*, 189–193. https://doi.org/10.1016/j.jmbbm.2018.02.006

Kaczmarek, M. B., Struszczyk-Swita, K., Li, X., Szczęsna-Antczak, M., & Daroch, M. (2019). Enzymatic modifications of chitin, chitosan, and chitooligosaccharides [Review]. *Frontiers in Bioengineering and Biotechnology*, *7*. https://doi.org/10.3389/fbioe.2019.00243

Khan, M. R., Volpe, S., Valentino, M., Miele, N. A., Cavella, S., & Torrieri, E. (2021). Active casein coatings and films for perishable foods: Structural properties and shelf-life extension. *Coatings*, *11*(8), 899. www.mdpi.com/2079-6412/11/8/899

Khwaldia, K., Banon, S., Perez, C., & Desobry, S. (2004). Properties of sodium caseinate film-forming dispersions and films. *Journal of Dairy Science*, *87*(7), 2011–2016. https://doi.org/10.3168/jds.S0022-0302(04)70018-1

Khwaldia, K., Basta, A. H., Aloui, H., & El-Saied, H. (2014). Chitosan—caseinate bilayer coatings for paper packaging materials. *Carbohydrate Polymers*, *99*, 508–516. https://doi.org/10.1016/j.carbpol.2013.08.086

Kim, J.-H., Hong, W.-S., & Oh, S.-W. (2018). Effect of layer-by-layer antimicrobial edible coating of alginate and chitosan with grapefruit seed extract for shelf-life extension of shrimp (Litopenaeus vannamei) stored at 4 °C. *International Journal of Biological Macromolecules*, *120*, 1468–1473. https://doi.org/10.1016/j.ijbiomac.2018.09.160

Kim, S. H., No, H. K., Kim, S. D., & Prinyawiwatkul, W. (2006). Effect of plasticizer concentration and solvent types on shelf-life of eggs coated with chitosan. *Journal of Food Science*, *71*(4), S349–S353. https://doi.org/10.1111/j.1750-3841.2006.00008.x

Koc, F. E., & Altıncekic, T. G. (2021). Investigation of gelatin/chitosan as potential biodegradable polymer films on swelling behavior and methylene blue release kinetics. *Polymer Bulletin*, *78*(6), 3383–3398. https://doi.org/10.1007/s00289-020-03280-7

Kontturi, E., & Spirk, S. (2019). Ultrathin films of cellulose: A materials perspective [review]. *Frontiers in Chemistry*, *7*. https://doi.org/10.3389/fchem.2019.00488

Kostag, M., & El Seoud, O. A. (2021). Sustainable biomaterials based on cellulose, chitin and chitosan composites—A review. *Carbohydrate Polymer Technologies and Applications*, *2*, 100079. https://doi.org/10.1016/j.carpta.2021.100079

Kou, X., He, Y., Li, Y., Chen, X., Feng, Y., & Xue, Z. (2019). Effect of abscisic acid (ABA) and chitosan/nanosilica/sodium alginate composite film on the color development and quality of postharvest Chinese winter jujube (*Zizyphus jujuba* Mill. cv. Dongzao). *Food Chemistry*, *270*, 385–394. https://doi.org/10.1016/j.foodchem.2018.06.151

Kumar, A., Hasan, M., Mangaraj, S., Pravitha, M., Verma, D. K., & Srivastav, P. P. (2022). Trends in edible packaging films and its prospective future in food: A review. *Applied Food Research*, *2*(1), 100118. https://doi.org/10.1016/j.afres.2022.100118

Kumari, S., & Rath, P. K. (2014). Extraction and characterization of chitin and chitosan from (*Labeo rohit*) fish scales. *Procedia Materials Science*, *6*, 482–489. https://doi.org/10.1016/j.mspro.2014.07.062

Kumirska, J., Weinhold, M. X., Thöming, J., & Stepnowski, P. (2011). Biomedical activity of chitin/chitosan based materials—Influence of physicochemical properties apart from molecular weight and degree of N-acetylation. *Polymers*, *3*(4), 1875–1901. www.mdpi.com/2073-4360/3/4/1875

Laboulfie, F., Hémati, M., Lamure, A., & Diguet, S. (2013). Effect of the plasticizer on permeability, mechanical resistance and thermal behaviour of composite coating films. *Powder Technology*, *238*, 14–19. https://doi.org/10.1016/j.powtec.2012.07.035

Li, K., Zhu, J., Guan, G., & Wu, H. (2019). Preparation of chitosan-sodium alginate films through layer-by-layer assembly and ferulic acid cross-linking: Film properties, characterization, and formation mechanism. *International Journal of Biological Macromolecules*, *122*, 485–492. https://doi.org/10.1016/j.ijbiomac.2018.10.188

Li, P., Zhou, Q., Chu, Y., Lan, W., Mei, J., & Xie, J. (2020). Effects of chitosan and sodium alginate active coatings containing ε-polysine on qualities of cultured pufferfish (Takifugu obscurus) during cold storage. *International Journal of Biological Macromolecules*, *160*, 418–428. https://doi.org/10.1016/j.ijbiomac.2020.05.092

Li, Z., Liu, C., Hong, S., Lian, H., Mei, C., Lee, J., Wu, Q., Hubbe, M. A., & Li, M.-C. (2022). Recent advances in extraction and processing of chitin using deep eutectic solvents. *Chemical Engineering Journal*, *446*, 136953. https://doi.org/10.1016/j.cej.2022.136953

Lin, M. G., Lasekan, O., Saari, N., & Khairunniza-Bejo, S. (2018). Effect of chitosan and carrageenan-based edible coatings on post-harvested longan (Dimocarpus longan) fruits. *CyTA—Journal of Food*, *16*(1), 490–497. https://doi.org/10.1080/19476337.2017.1414078

Liu, J., Liu, S., Wu, Q., Gu, Y., Kan, J., & Jin, C. (2017). Effect of protocatechuic acid incorporation on the physical, mechanical, structural and antioxidant properties of chitosan film. *Food Hydrocolloids*, *73*, 90–100. https://doi.org/10.1016/j.foodhyd.2017.06.035

Lu, Y., Luo, Q., Chu, Y., Tao, N., Deng, S., Wang, L., & Li, L. (2022). Application of gelatin in food packaging: A review. *Polymers*, *14*(3), 436. https://doi.org/10.3390/polym14030436

Ma, X., Qiao, C., Wang, X., Yao, J., & Xu, J. (2019). Structural characterization and properties of polyols plasticized chitosan films. *International Journal of Biological Macromolecules*, *135*, 240–245. https://doi.org/10.1016/j.ijbiomac.2019.05.158

Machado, B. R., Facchi, S. P., de Oliveira, A. C., Nunes, C. S., Souza, P. R., Vilsinski, B. H., Popat, K. C., Kipper, M. J., Muniz, E. C., & Martins, A. F. (2020). Bactericidal pectin/chitosan/glycerol films for food pack coatings: A critical viewpoint. *International Journal of Molecular Sciences, 21*(22), 8663. www. mdpi.com/1422-0067/21/22/8663

Mahdy Samar, M., El-Kalyoubi, M. H., Khalaf, M. M., & Abd El-Razik, M. M. (2013). Physicochemical, functional, antioxidant and antibacterial properties of chitosan extracted from shrimp wastes by microwave technique. *Annals of Agricultural Sciences, 58*(1), 33–41. https://doi.org/10.1016/j.aoas.2013.01.006

Maringgal, B., Hashim, N., Mohamed Amin Tawakkal, I. S., & Muda Mohamed, M. T. (2020). Recent advance in edible coating and its effect on fresh/fresh-cut fruits quality. *Trends in Food Science & Technology, 96*, 253–267. https://doi.org/10.1016/j.tifs.2019.12.024

Matta, E., & Bertola, N. (2020). Development and characterization of high methoxyl pectin film by using iso-malt as plasticizer. *Journal of Food Processing and Preservation, 44*(8), e14568. https://doi.org/10.1111/jfpp.14568

Medeiros, B., Pinheiro, A. C., Carneiro-da-Cunha, M. G., & Vicente, A. A. (2012). Development and characterization of a nanomultilayer coating of pectin and chitosan—evaluation of its gas barrier properties and application on 'Tommy Atkins' mangoes. *Journal of Food Engineering, 110*(3), 457–464. https://doi.org/10.1016/j.jfoodeng.2011.12.021

Mittal, A., Singh, A., Benjakul, S., Prodpran, T., Nilsuwan, K., Huda, N., & Caba, K. D. L. (2021). Composite films based on chitosan and epigallocatechin gallate grafted chitosan: Characterization, antioxidant and antimicrobial activities. *Food Hydrocolloids, 111*, 106384. https://doi.org/10.1016/j.foodhyd.2020.106384

Mohamed, S. A. A., El-Sakhawy, M., & El-Sakhawy, M. A.-M. (2020). Polysaccharides, protein and lipid-based natural edible films in food packaging: A review. *Carbohydrate Polymers, 238*, 116178. https://doi.org/10.1016/j.carbpol.2020.116178

Mohammadi Sadati, S. M., Shahgholian-Ghahfarrokhi, N., Shahrousvand, E., Mohammadi-Rovshandeh, J., & Shahrousvand, M. (2021). Edible chitosan/cellulose nanofiber nanocomposite films for potential use as food packaging. *Materials Technology, 37*, 1276–1288. https://doi.org/10.1080/10667857.2021.1934367

Mohammed Manshor, N., Rezali, M. I., Jai, J., & Yahya, A. (2018). Effect of plasticizers on physicochemical and mechanical properties of chitosan-gelatin films. *IOP Conference Series: Materials Science and Engineering, 358*, 012040. https://doi.org/10.1088/1757-899x/358/1/012040

Mohan, K., Ganesan, A. R., Ezhilarasi, P. N., Kondamareddy, K. K., Rajan, D. K., Sathishkumar, P., Rajarajeswaran, J., & Conterno, L. (2022). Green and eco-friendly approaches for the extraction of chitin and chitosan: A review. *Carbohydrate Polymers, 287*, 119349. https://doi.org/10.1016/j.carbpol.2022.119349

Mohan, K., Ganesan, A. R., Muralisankar, T., Jayakumar, R., Sathishkumar, P., Uthayakumar, V., Chandirasekar, R., & Revathi, N. (2020). Recent insights into the extraction, characterization, and bioactivities of chitin and chitosan from insects. *Trends in Food Science & Technology, 105*, 17–42. https://doi.org/10.1016/j.tifs.2020.08.016

Moller, A., Leone, C., Kataria, J., Sidhu, G., Rama, E. N., Kroft, B., Thippareddi, H., & Singh, M. (2022). Effect of a carrageenan/chitosan coating with allyl isothiocyanate on microbial load in chicken breast. *LWT, 161*, 113397. https://doi.org/10.1016/j.lwt.2022.113397

Mourya, V. K., & Inamdar, N. N. (2008). Chitosan-modifications and applications: Opportunities galore. *Reactive and Functional Polymers, 68*(6), 1013–1051. https://doi.org/10.1016/j.reactfunctpolym.2008.03.002

Mushi, N. E., Butchosa, N., Salajkova, M., Zhou, Q., & Berglund, L. A. (2014). Nanostructured membranes based on native chitin nanofibers prepared by mild process. *Carbohydrate Polymers, 112*, 255–263. https://doi.org/10.1016/j.carbpol.2014.05.038

Nair, M. S., Saxena, A., & Kaur, C. (2018). Effect of chitosan and alginate based coatings enriched with pomegranate peel extract to extend the postharvest quality of guava (Psidium guajava L.). *Food Chemistry, 240*, 245–252. https://doi.org/10.1016/j.foodchem.2017.07.122

Nair, M. S., Tomar, M., Punia, S., Kukula-Koch, W., & Kumar, M. (2020). Enhancing the functionality of chitosan- and alginate-based active edible coatings/films for the preservation of fruits and vegetables: A review. *International Journal of Biological Macromolecules, 164*, 304–320. https://doi.org/10.1016/j.ijbiomac.2020.07.083

Olaimat, A. N., Fang, Y., & Holley, R. A. (2014). Inhibition of Campylobacter jejuni on fresh chicken breasts by κ-carrageenan/chitosan-based coatings containing allyl isothiocyanate or deodorized oriental mustard extract. *International Journal of Food Microbiology, 187*, 77–82. https://doi.org/10.1016/j.ijfoodmicro.2014.07.003

Ozturk, S., Zhang, J., Singh, R. K., & Kong, F. (2021). Effect of cellulose nanofiber-based coating with chitosan and trans-cinnamaldehyde on the microbiological safety and quality of cantaloupe rind and fresh-cut pulp. Part 2: Quality attributes. *LWT, 147,* 111519. https://doi.org/10.1016/j.lwt.2021.111519

Park, S. Y., Lee, B. I., Jung, S. T., & Park, H. J. (2001). Biopolymer composite films based on κ-carrageenan and chitosan. *Materials Research Bulletin, 36*(3), 511–519. https://doi.org/10.1016/S0025-5408(01)00545-1

Paul, S. K., Sarkar, S., Sethi, L. N., & Ghosh, S. K. (2018). Development of chitosan based optimized edible coating for tomato (*Solanum lycopersicum*) and its characterization. *Journal of Food Science and Technology, 55*(7), 2446–2456. https://doi.org/10.1007/s13197-018-3162-6

Pelissari, F. M., Grossmann, M. V. E., Yamashita, F., & Pineda, E. A. G. (2009). Antimicrobial, mechanical, and barrier properties of cassava starch–chitosan films incorporated with oregano essential oil. *Journal of Agricultural and Food Chemistry, 57*(16), 7499–7504. https://doi.org/10.1021/jf9002363

Pereda, M., Aranguren, M. I., & Marcovich, N. E. (2008). Characterization of chitosan/caseinate films. *Journal of Applied Polymer Science, 107*(2), 1080–1090. https://doi.org/10.1002/app.27052

Pereda, M., Aranguren, M. I., & Marcovich, N. E. (2009). Water vapor absorption and permeability of films based on chitosan and sodium caseinate. *Journal of Applied Polymer Science, 111*(6), 2777–2784. https://doi.org/10.1002/app.29347

Pereda, M., Ponce, A. G., Marcovich, N. E., Ruseckaite, R. A., & Martucci, J. F. (2011). Chitosan-gelatin composites and bi-layer films with potential antimicrobial activity. *Food Hydrocolloids, 25*(5), 1372–1381. https://doi.org/10.1016/j.foodhyd.2011.01.001

Pestov, A., & Bratskaya, S. (2016). Chitosan and its derivatives as highly efficient polymer ligands. *Molecules, 21*(3), 330. www.mdpi.com/1420-3049/21/3/330

Pinheiro, A. C., Bourbon, A. I., Medeiros, B., da Silva, L. H. M., da Silva, M. C. H., Carneiro-da-Cunha, M. G., Coimbra, M. A., & Vicente, A. A. (2012a). Interactions between κ-carrageenan and chitosan in nanolayered coatings—structural and transport properties. *Carbohydrate Polymers, 87*(2), 1081–1090. https://doi.org/10.1016/j.carbpol.2011.08.040

Pinheiro, A. C., Bourbon, A. I., Quintas, M. A. C., Coimbra, M. A., & Vicente, A. A. (2012b). K-carrageenan/chitosan nanolayered coating for controlled release of a model bioactive compound. *Innovative Food Science & Emerging Technologies, 16,* 227–232. https://doi.org/10.1016/j.ifset.2012.06.004

Pinto Ramos, D., Sarjinsky, S., Alizadehgiashi, M., Möbus, J., & Kumacheva, E. (2019). Polyelectrolyte vs polyampholyte behavior of composite chitosan/gelatin films. *ACS Omega, 4*(5), 8795–8803. https://doi.org/10.1021/acsomega.9b00251

Pita-López, M. L., Fletes-Vargas, G., Espinosa-Andrews, H., & Rodríguez-Rodríguez, R. (2021). Physically cross-linked chitosan-based hydrogels for tissue engineering applications: A state-of-the-art review. *European Polymer Journal, 145,* 110176. https://doi.org/10.1016/j.eurpolymj.2020.110176

Porta, R., Mariniello, L., Di Pierro, P., Sorrentino, A., & Giosafatto, C. V. L. (2011). Transglutaminase cross-linked pectin- and chitosan-based edible films: A review. *Critical Reviews in Food Science and Nutrition, 51*(3), 223–238. https://doi.org/10.1080/10408390903548891

Poverenov, E., Rutenberg, R., Danino, S., Horev, B., & Rodov, V. (2014a). Gelatin-chitosan composite films and edible coatings to enhance the quality of food products: Layer-by-layer vs. blended formulations. *Food and Bioprocess Technology, 7*(11), 3319–3327. https://doi.org/10.1007/s11947-014-1333-7

Poverenov, E., Zaitsev, Y., Arnon, H., Granit, R., Alkalai-Tuvia, S., Perzelan, Y., Weinberg, T., & Fallik, E. (2014b). Effects of a composite chitosan—gelatin edible coating on postharvest quality and storability of red bell peppers. *Postharvest Biology and Technology, 96,* 106–109. https://doi.org/10.1016/j.postharvbio.2014.05.015

Qu, B., & Luo, Y. (2021). A review on the preparation and characterization of chitosan-clay nanocomposite films and coatings for food packaging applications. *Carbohydrate Polymer Technologies and Applications, 2,* 100102. https://doi.org/10.1016/j.carpta.2021.100102

Qu, W., Guo, T., Zhang, X., Jin, Y., Wang, B., Wahia, H., & Ma, H. (2022a). Preparation of tuna skin collagen-chitosan composite film improved by sweep frequency pulsed ultrasound technology. *Ultrasonics Sonochemistry, 82,* 105880. https://doi.org/10.1016/j.ultsonch.2021.105880

Qu, W., Xiong, T., Wang, B., Li, Y., & Zhang, X. (2022b). The modification of pomegranate polyphenol with ultrasound improves mechanical, antioxidant, and antibacterial properties of tuna skin collagen-chitosan film. *Ultrasonics Sonochemistry, 85,* 105992. https://doi.org/10.1016/j.ultsonch.2022.105992

Raeisi, M., Hashemi, M., Aminzare, M., Ghorbani Bidkorpeh, F., Ebrahimi, M., Jannat, B., Tepe, B., & Noori, S. M. A. (2020). Effects of sodium alginate and chitosan coating combined with three different essential oils on microbial and chemical attributes of rainbow trout fillets. *Journal of Aquatic Food Product Technology*, 29(3), 253–263. https://doi.org/10.1080/10498850.2020.1722777

Ramos, M., Valdés, A., Beltrán, A., & Garrigós, M. C. (2016). Gelatin-based films and coatings for food packaging applications. *Coatings*, 6(4), 41. www.mdpi.com/2079-6412/6/4/41

Resende, N. S., Gonçalves, G. A. S., Reis, K. C., Tonoli, G. H. D., & Boas, E. V. B. V. (2018). Chitosan/cellulose nanofibril nanocomposite and its effect on quality of coated strawberries. *Journal of Food Quality*, 2018, 1727426. https://doi.org/10.1155/2018/1727426

Rinaudo, M. (2006). Chitin and chitosan: Properties and applications. *Progress in Polymer Science*, 31(7), 603–632. https://doi.org/10.1016/j.progpolymsci.2006.06.001

Rinaudo, M., Pavlov, G., & Desbrières, J. (1999). Influence of acetic acid concentration on the solubilization of chitosan. *Polymer*, 40(25), 7029–7032. https://doi.org/10.1016/S0032-3861(99)00056-7

Ríos-de-Benito, L. F., Escamilla-García, M., García-Almendárez, B., Amaro-Reyes, A., Di Pierro, P., & Regalado-González, C. (2021). Design of an active edible coating based on sodium caseinate, chitosan and oregano essential oil reinforced with silica particles and its application on panela cheese. *Coatings*, 11(10), 1212. www.mdpi.com/2079-6412/11/10/1212

Rochima, E., Fiyanih, E., Afrianto, E., Subhan, U., Praseptiangga, D., Panatarani, C., & Joni, I. M. (2018). The addition of nanochitosan suspension as filler in carrageenan-tapioca biocomposite film. *AIP Conference Proceedings*, 1927(1), 030042. https://doi.org/10.1063/1.5021235

Rodríguez-Núñez, J. R., Madera-Santana, T. J., Sánchez-Machado, D. I., López-Cervantes, J., & Soto Valdez, H. (2014). Chitosan/hydrophilic plasticizer-based films: Preparation, physicochemical and antimicrobial properties. *Journal of Polymers and the Environment*, 22(1), 41–51. https://doi.org/10.1007/s10924-013-0621-z

Rodríguez-Rodríguez, R., Espinosa-Andrews, H., & García-Carvajal, Z. Y. (2022). Stimuli-responsive hydrogels in drug delivery. In S. Jana & S. Jana (Eds.), *Functional biomaterials: Drug delivery and biomedical applications* (pp. 75–103). Singapore: Springer. https://doi.org/10.1007/978-981-16-7152-4_3

Rodríguez-Rodríguez, R., Espinosa-Andrews, H., Morales-Hernández, N., Lobato-Calleros, C., & Vernon-Carter, E. J. (2019). Mesquite gum/chitosan insoluble complexes: relationship between the water state and viscoelastic properties. *Journal of Dispersion Science and Technology*, 40(9), 1345–1352. https://doi.org/10.1080/01932691.2018.1513848

Rodríguez-Rodríguez, R., Espinosa-Andrews, H., Velasquillo-Martínez, C., & García-Carvajal, Z. Y. (2020). Composite hydrogels based on gelatin, chitosan and polyvinyl alcohol to biomedical applications: A review. *International Journal of Polymeric Materials and Polymeric Biomaterials*, 69(1), 1–20. https://doi.org/10.1080/00914037.2019.1581780

Salama, H. E., Abdel Aziz, M. S., & Sabaa, M. W. (2018). Novel biodegradable and antibacterial edible films based on alginate and chitosan biguanidine hydrochloride. *International Journal of Biological Macromolecules*, 116, 443–450. https://doi.org/10.1016/j.ijbiomac.2018.04.183

Santos, M. A. S., & Machado, M. T. C. (2021). Coated alginate—chitosan particles to improve the stability of probiotic yeast. *International Journal of Food Science & Technology*, 56(5), 2122–2131. https://doi.org/10.1111/ijfs.14829

Santos, T. A., Cabral, B. R., de Oliveira, A. C. S., Dias, M. V., de Oliveira, C. R., & Borges, S. V. (2021). Release of papain incorporated in chitosan films reinforced with cellulose nanofibers. *Journal of Food Processing and Preservation*, 45(11), e15900. https://doi.org/10.1111/jfpp.15900

Schou, M., Longares, A., Montesinos-Herrero, C., Monahan, F. J., O'Riordan, D., & O'Sullivan, M. (2005). Properties of edible sodium caseinate films and their application as food wrapping. *LWT—Food Science and Technology*, 38(6), 605–610. https://doi.org/10.1016/j.lwt.2004.08.009

Senturk Parreidt, T., Müller, K., & Schmid, M. (2018). Alginate-based edible films and coatings for food packaging applications. *Foods*, 7(10), 170. www.mdpi.com/2304-8158/7/10/170

Shah, R., Stodulka, P., Skopalova, K., & Saha, P. (2019). Dual cross-linked collagen/chitosan film for potential biomedical applications. *Polymers*, 11(12), 2094. www.mdpi.com/2073-4360/11/12/2094

Sionkowska, A. (2021). Collagen blended with natural polymers: Recent advances and trends. *Progress in Polymer Science*, 122, 101452. https://doi.org/10.1016/j.progpolymsci.2021.101452

Sionkowska, A., Wisniewski, M., Skopinska, J., Poggi, G. F., Marsano, E., Maxwell, C. A., & Wess, T. J. (2006). Thermal and mechanical properties of UV irradiated collagen/chitosan thin films. *Polymer Degradation and Stability*, 91(12), 3026–3032. https://doi.org/10.1016/j.polymdegradstab.2006.08.009

Socrates, R., Prymak, O., Loza, K., Sakthivel, N., Rajaram, A., Epple, M., & Narayana Kalkura, S. (2019). Biomimetic fabrication of mineralized composite films of nanosilver loaded native fibrillar collagen and chitosan. *Materials Science and Engineering: C, 99,* 357–366. https://doi.org/10.1016/j.msec.2019.01.101

Soltani Firouz, M., Mohi-Alden, K., & Omid, M. (2021). A critical review on intelligent and active packaging in the food industry: Research and development. *Food Research International, 141,* 110113. https://doi.org/10.1016/j.foodres.2021.110113

Souza, M. P., Vaz, A. F. M., Cerqueira, M. A., Teixeira, J. A., Vicente, A. A., & Carneiro-da-Cunha, M. G. (2015). Effect of an edible nanomultilayer coating by electrostatic self-assembly on the shelf life of fresh-cut mangoes. *Food and Bioprocess Technology, 8*(3), 647–654. https://doi.org/10.1007/s11947-014-1436-1

Sun, P.-P., Ren, Y.-Y., Wang, S.-Y., Zhu, H., & Zhou, J.-J. (2021). Characterization and film-forming properties of acid soluble collagens from different by-products of loach (*Misgurnus anguillicaudatus*). *LWT, 149,* 111844. https://doi.org/10.1016/j.lwt.2021.111844

Suyatma, N. E., Tighzert, L., Copinet, A., & Coma, V. (2005). Effects of hydrophilic plasticizers on mechanical, thermal, and surface properties of chitosan films. *Journal of Agricultural and Food Chemistry, 53*(10), 3950–3957. https://doi.org/10.1021/jf048790+

Szymańska-Chargot, M., Chylińska, M., Pertile, G., Pieczywek, P. M., Cieślak, K. J., Zdunek, A., & Frąc, M. (2019). Influence of chitosan addition on the mechanical and antibacterial properties of carrot cellulose nanofibre film. *Cellulose, 26*(18), 9613–9629. https://doi.org/10.1007/s10570-019-02755-9

Taherimehr, M., YousefniaPasha, H., Tabatabaeekoloor, R., & Pesaranhajiabbas, E. (2021). Trends and challenges of biopolymer-based nanocomposites in food packaging. *Comprehensive Reviews in Food Science and Food Safety, 20*(6), 5321–5344. https://doi.org/10.1111/1541-4337.12832

Tahir, H. E., Xiaobo, Z., Mahunu, G. K., Arslan, M., Abdalhai, M., & Zhihua, L. (2019). Recent developments in gum edible coating applications for fruits and vegetables preservation: A review. *Carbohydrate Polymers, 224,* 115141. https://doi.org/10.1016/j.carbpol.2019.115141

Tang, C., Zhou, K., Zhu, Y., Zhang, W., Xie, Y., Wang, Z., Zhou, H., Yang, T., Zhang, Q., & Xu, B. (2022). Collagen and its derivatives: From structure and properties to their applications in food industry. *Food Hydrocolloids, 131,* 107748. https://doi.org/10.1016/j.foodhyd.2022.107748

Tavassoli-Kafrani, E., Shekarchizadeh, H., & Masoudpour-Behabadi, M. (2016). Development of edible films and coatings from alginates and carrageenans. *Carbohydrate Polymers, 137,* 360–374. https://doi.org/10.1016/j.carbpol.2015.10.074

Terzioğlu, P., Güney, F., Parın, F. N., Şen, İ., & Tuna, S. (2021). Biowaste orange peel incorporated chitosan/polyvinyl alcohol composite films for food packaging applications. *Food Packaging and Shelf Life, 30,* 100742. https://doi.org/10.1016/j.fpsl.2021.100742

Thakhiew, W., Devahastin, S., & Soponronnarit, S. (2010). Effects of drying methods and plasticizer concentration on some physical and mechanical properties of edible chitosan films. *Journal of Food Engineering, 99*(2), 216–224. https://doi.org/10.1016/j.jfoodeng.2010.02.025

Valdés, A., Burgos, N., Jiménez, A., & Garrigós, M. C. (2015). Natural pectin polysaccharides as edible coatings. *Coatings, 5*(4), 865–886. www.mdpi.com/2079-6412/5/4/865

Vanden Braber, N. L., Di Giorgio, L., Aminahuel, C. A., Díaz Vergara, L. I., Martín Costa, A. O., Montenegro, M. A., & Mauri, A. N. (2021). Antifungal whey protein films activated with low quantities of water soluble chitosan. *Food Hydrocolloids, 110,* 106156. https://doi.org/10.1016/j.foodhyd.2020.106156

Volod'ko, A. V., Davydova, V. N., Petrova, V. A., Romanov, D. P., Pimenova, E. A., & Yermak, I. M. (2021). Comparative analysis of the functional properties of films based on carrageenans, chitosan, and their polyelectrolyte complexes. *Marine Drugs, 19*(12), 704. www.mdpi.com/1660-3397/19/12/704

Volpe, S., Cavella, S., Masi, P., & Torrieri, E. (2017). Effect of solid concentration on structure and properties of chitosan-caseinate blend films. *Food Packaging and Shelf Life, 13,* 76–84. https://doi.org/10.1016/j.fpsl.2017.07.002

Wang, C., Zhang, J., Chen, J., Shi, J., Zhao, Y., He, M., & Ding, L. (2022a). Bio-polyols based waterborne polyurethane coatings reinforced with chitosan-modified ZnO nanoparticles. *International Journal of Biological Macromolecules, 208,* 97–104. https://doi.org/10.1016/j.ijbiomac.2022.03.066

Wang, F., Wang, R., Pan, Y., Du, M., Zhao, Y., & Liu, H. (2022b). Gelatin/chitosan films incorporated with curcumin based on photodynamic inactivation technology for antibacterial food packaging. *Polymers, 14*(8), 1600. www.mdpi.com/2073-4360/14/8/1600

Wang, H., Ding, F., Ma, L., & Zhang, Y. (2021). Edible films from chitosan-gelatin: Physical properties and food packaging application. *Food Bioscience, 40,* 100871. https://doi.org/10.1016/j.fbio.2020.100871

Wang, Q., Chen, W., Zhu, W., McClements, D. J., Liu, X., & Liu, F. (2022c). A review of multilayer and composite films and coatings for active biodegradable packaging. *NPJ Science of Food*, 6(1), 18. https://doi.org/10.1038/s41538-022-00132-8

Wang, X., Yong, H., Gao, L., Li, L., Jin, M., & Liu, J. (2019). Preparation and characterization of antioxidant and pH-sensitive films based on chitosan and black soybean seed coat extract. *Food Hydrocolloids*, 89, 56–66. https://doi.org/10.1016/j.foodhyd.2018.10.019

Wasikiewicz, J. M., & Yeates, S. G. (2013). "Green" molecular weight degradation of chitosan using microwave irradiation. *Polymer Degradation and Stability*, 98(4), 863–867. https://doi.org/10.1016/j.polymdegradstab.2012.12.028

Ways, T. M. M., Lau, W. M., & Khutoryanskiy, V. V. (2018). Chitosan and its derivatives for application in mucoadhesive drug delivery systems. *Polymers*, 10(3), 267. https://doi.org/10.3390/polym10030267

Webber, J. L., Namivandi-Zangeneh, R., Drozdek, S., Wilk, K. A., Boyer, C., Wong, E. H. H., Bradshaw-Hajek, B. H., Krasowska, M., & Beattie, D. A. (2021). Incorporation and antimicrobial activity of nisin Z within carrageenan/chitosan multilayers. *Scientific Reports*, 11(1), 1690. https://doi.org/10.1038/s41598-020-79702-3

Wu, J., Ge, S., Liu, H., Wang, S., Chen, S., Wang, J., Li, J., & Zhang, Q. (2014). Properties and antimicrobial activity of silver carp (Hypophthalmichthys molitrix) skin gelatin-chitosan films incorporated with oregano essential oil for fish preservation. *Food Packaging and Shelf Life*, 2(1), 7–16. https://doi.org/10.1016/j.fpsl.2014.04.004

Yanat, M., & Schroën, K. (2021). Preparation methods and applications of chitosan nanoparticles; with an outlook toward reinforcement of biodegradable packaging. *Reactive and Functional Polymers*, 161, 104849. https://doi.org/10.1016/j.reactfunctpolym.2021.104849

Yang, J., Kwon, G.-J., Hwang, K., & Kim, D.-Y. (2018). Cellulose-chitosan antibacterial composite films prepared from LiBr solution. *Polymers*, 10(10), 1058. https://doi.org/10.3390/polym10101058

Yong, H., Liu, J., Qin, Y., Bai, R., Zhang, X., & Liu, J. (2019). Antioxidant and pH-sensitive films developed by incorporating purple and black rice extracts into chitosan matrix. *International Journal of Biological Macromolecules*, 137, 307–316. https://doi.org/10.1016/j.ijbiomac.2019.07.009

Younes, I., & Rinaudo, M. (2015). Chitin and chitosan preparation from marine sources. Structure, properties and applications. *Marine Drugs*, 13(3). https://doi.org/10.3390/md13031133

Younis, H. G. R., Abdellatif, H. R. S., Ye, F., & Zhao, G. (2020). Tuning the physicochemical properties of apple pectin films by incorporating chitosan/pectin fiber. *International Journal of Biological Macromolecules*, 159, 213–221. https://doi.org/10.1016/j.ijbiomac.2020.05.060

Younis, H. G. R., & Zhao, G. (2019). Physicochemical properties of the edible films from the blends of high methoxyl apple pectin and chitosan. *International Journal of Biological Macromolecules*, 131, 1057–1066. https://doi.org/10.1016/j.ijbiomac.2019.03.096

Yu, D., Yu, Z., Zhao, W., Regenstein, J. M., & Xia, W. (2022). Advances in the application of chitosan as a sustainable bioactive material in food preservation. *Critical Reviews in Food Science and Nutrition*, 62(14), 3782–3797. https://doi.org/10.1080/10408398.2020.1869920

Yu, H. C., Zhang, H., Ren, K., Ying, Z., Zhu, F., Qian, J., Ji, J., Wu, Z. L., & Zheng, Q. (2018). Ultrathin κ-carrageenan/chitosan hydrogel films with high toughness and antiadhesion property. *ACS Applied Materials & Interfaces*, 10(10), 9002–9009. https://doi.org/10.1021/acsami.7b18343

Yuan, D., Hao, X., Liu, G., Yue, Y., & Duan, J. (2022). A novel composite edible film fabricated by incorporating W/O/W emulsion into a chitosan film to improve the protection of fresh fish meat. *Food Chemistry*, 385, 132647. https://doi.org/10.1016/j.foodchem.2022.132647

Zam, W. (2019). Effect of alginate and chitosan edible coating enriched with olive leaves extract on the shelf life of sweet cherries (*Prunus avium* L.). *Journal of Food Quality*, 2019, 8192964. https://doi.org/10.1155/2019/8192964

Zhang, J., Ozturk, S., Singh, R. K., & Kong, F. (2020). Effect of cellulose nanofiber-based coating with chitosan and trans-cinnamaldehyde on the microbiological safety and quality of cantaloupe rind and fresh-cut pulp. Part 1: Microbial safety. *LWT*, 134, 109972. https://doi.org/10.1016/j.lwt.2020.109972

Zhang, X., Li, Y., Guo, M., Jin, T. Z., Arabi, S. A., He, Q., Ismail, B. B., Hu, Y., & Liu, D. (2021a). Antimicrobial and UV blocking properties of composite chitosan films with curcumin grafted cellulose nanofiber. *Food Hydrocolloids*, 112, 106337. https://doi.org/10.1016/j.foodhyd.2020.106337

Zhang, Y., Niu, Y., Luo, Y., Ge, M., Yang, T., Yu, L., & Wang, Q. (2014). Fabrication, characterization and antimicrobial activities of thymol-loaded zein nanoparticles stabilized by sodium caseinate—chitosan hydrochloride double layers. *Food Chemistry*, 142, 269–275. https://doi.org/10.1016/j.foodchem.2013.07.058

Zhang, Y.-L., Cui, Q.-L., Wang, Y., Shi, F., Fan, H., Zhang, Y.-Q., Lai, S.-T., Li, Z.-H., Li, L., & Sun, Y.-K. (2021b). Effect of edible carboxymethyl chitosan-gelatin based coating on the quality and nutritional properties of different sweet cherry cultivars during postharvest storage. *Coatings, 11*(4), 396. www.mdpi.com/2079-6412/11/4/396

Zhao, S., Jia, R., Yang, J., Dai, L., Ji, N., Xiong, L., & Sun, Q. (2022). Development of chitosan/tannic acid/cornstarch multifunctional bilayer smart films as pH-responsive actuators and for fruit preservation. *International Journal of Biological Macromolecules, 205*, 419–429. https://doi.org/10.1016/j.ijbiomac.2022.02.101

Zheng, L.-Y., & Zhu, J.-F. (2003). Study on antimicrobial activity of chitosan with different molecular weights. *Carbohydrate Polymers, 54*(4), 527–530. https://doi.org/10.1016/j.carbpol.2003.07.009

Zhou, H., Tong, H., Lu, J., Cheng, Y., Qian, F., Tao, Y., & Wang, H. (2021). Preparation of bio-based cellulose acetate/chitosan composite film with oxygen and water resistant properties. *Carbohydrate Polymers, 270*, 118381. https://doi.org/10.1016/j.carbpol.2021.118381

Zhu, J., Wu, H., & Sun, Q. (2019). Preparation of cross-linked active bilayer film based on chitosan and alginate for regulating ascorbate-glutathione cycle of postharvest cherry tomato (*Lycopersicon esculentum*). *International Journal of Biological Macromolecules, 130*, 584–594. https://doi.org/10.1016/j.ijbiomac.2019.03.006

4

Characterization of Novel Films and Coatings Based on Gums

Nidhi Dangi, Sonia Attri, Shivanki Tomar, and Prixit Guleria

CONTENTS

4.1 Introduction

In the past few years, there has been an upsurge in awareness regarding good dietary habits, and more conscious efforts have been put forth to study the harmful effects of chemical-based additives in food products added intentionally for food preservation. Doubtless, the spoilage of food products is a vital problem, causing a significant economic loss, approximately more than 25% of the food before consumption (Huang et al., 2012; Lopez-Carballo et al., 2012). Food packaging systems provide several functions relating to containment, information, and marketing, although the prime function of packaging is to provide a barrier between food and the surrounding environment, resulting in reduced exposure from spoiling factors (water, oxygen, microbes, and off-flavors) and avoiding loss of desirable flavor components, thus enhancing the shelf life of food (Otoni et al., 2017). Therefore, to overcome the ever-growing need and demand for food quality, safety, and preservation, novel food packaging techniques are being explored, being the necessity of the hour (Chawla et al., 2021; Sofi et al., 2018). The usual non-biodegradable packaging materials possess a substantial threat to the environment owing to the higher levels of toxic emissions, changes in the carbon dioxide cycle, and composting problems (Ezeoha and Ezenwanne, 2013). The packaging waste contributes to a significant portion of the solid waste in

DOI: 10.1201/9781003303671-4

urban communities, resulting in increased environmental distress. Plastic is an important source of litter offering substantial challenges to waste management systems and thus indicates the importance of biodegradable raw materials (Zhong et al., 2020). Though complete substitution of prevalent petroleum-based plastics with biodegradable materials is impossible to attain, the application of biodegradable food packaging materials could be a solution, at least in food commodities (Ataei et al., 2020; Pereda et al., 2012). With the rising concerns over the environment and food safety, several researchers have shown their interest in developing biodegradable films and coatings.

Biopolymers have good shear strength and flow characteristics which diversify their applications. Natural polymers have several advantages over synthetic polymers, such as availability, biocompatibility, and biodegradability, thus leading to ecological safety and providing the possibility of preparing different derivatives of biopolymers for specific purposes (Nussinovitch, 2009; McClements, 2009). The prime choice for biodegradable packaging materials are biopolymers, including polysaccharides, proteins, and lipids (Galus and Kadzińska, 2015; Ganiari et al., 2017; Salehi, 2020). Biopolymer films and coatings can also be loaded with antioxidants, anti-microbial ingredients, bioactive components, and flavoring compounds to impart desired characteristics in the food product. Also, these films and coatings are edible, readily available, and biodegradable, making them a suitable option as compared to synthetic packaging materials (Gahruie et al., 2020a, 2020b).

4.2 Classification of Films and Coatings

Films and coatings are broadly classified into four categories based on their principal component.

4.2.1 Polysaccharide-Based Films and Coatings

The most commonly used polysaccharides in the making of films and coatings are starch, chitosan, alginate, cellulose, carrageenan, pullulan, gums, and mucilages (Hashemi and Khaneghah, 2017; Salehi, 2020). Polysaccharide-based films and coatings are carbohydrates in nature, and the films formed by these polymers have hydrophilic structures which react strongly with water, thereby having poor water barrier properties. The number of hydroxyl groups in the main polymer chain is responsible and plays an important function in the formation of films and their characteristics. The moisture content in the film is greatly affected by the moisture barrier properties and water vapor permeability. Further, increased hydration rates lead to an augmentation in elongation properties but decrease mechanical properties (Alizadeh-Sani et al., 2019. The selective permeability of these films toward gases could be used to generate a modified atmosphere (Dehghani et al., 2018; Ramos et al., 2012).

4.2.2 Protein-Based Films and Coatings

Protein-based films and coatings are prepared from proteins derived from plants or animals. The raw material can be obtained from plants or animals, which include animal tissues (collagen and gelatin), milk (casein and whey protein), eggs (albumin), grains (corn zein and wheat gluten), and oilseeds (peanut protein). Proteins are heteropolymers with specific amino acids as their monomer units, and the structure of protein could be modified with heat, pressure, or other reactions to control the physical and mechanical structure of the resultant films and coatings. These films have excellent barriers toward oxygen, carbon dioxide, and lipids at relatively low humidity. The reason behind the excellent barrier of these films toward oxygen is their tightly packed hydrogen-bonded structure (Raghav et al., 2016). The protein-based films have adequate mechanical strength, that is, structural stability, which makes them capable of holding a particular form.

4.2.3 Lipid-Based Films and Coatings

Lipid-based films and coatings provide a shiny appearance to the foods and impart hydrophobicity. Lipids used in the making of films include neutral glycerides which are the esters of glycerol and fatty acids and waxes which are the esters of long-chain monohydric alcohol and fatty acids (Kamal, 2019;

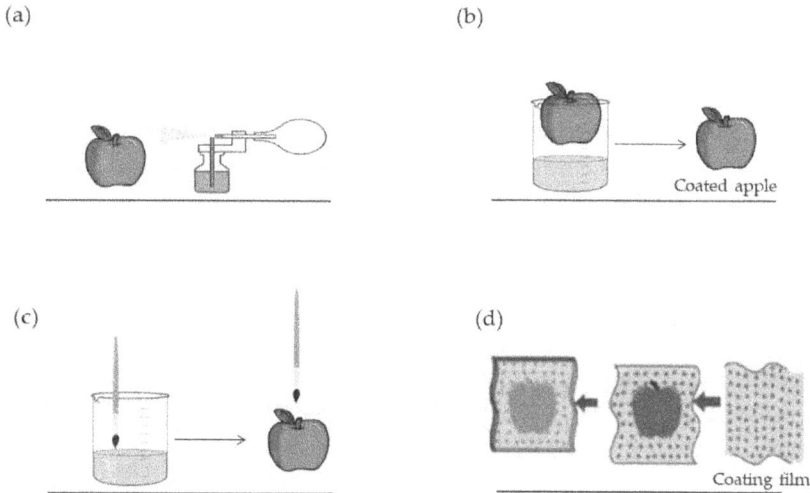

FIGURE 4.1 Representation of the film and coating techniques: (a) spraying, (b) immersion, (c) brushing, and (d) wrapping.

Source: Lazaridou and Biliaderis (2020).

Aydin et al., 2017). Lipid-based films and coatings have excellent water repellent and low penetration properties, and the extent of these properties depends on the type of hydrophobic compound used. The non-polymeric nature of lipids limits their ability to form cohesive films; thus, they are generally used as a coating in most cases. They require hydrocolloids or a supporting matrix in conjugation to form stand-alone films (Rhim and Shellhammer, 2005).

4.3 Preparation Methods of Films and Coatings

A film or coating can be defined as a thin and continuous layer that is placed on the surface of foods for preserving their quality. Films and coatings are often termed interchangeably; however, films are formed before their use and applied on the food surface after formation and include wraps, covers, etc., while coatings are formed directly on the food surface and are considered a part of the final product. In the case of coatings, the method of application on the product surface and the ability of coating material to adhere to the surface are the most crucial factors. In addition, the barrier properties provided by the films and coatings depend on the type of food that is to be protected. For example, in the case of fruits and vegetables, films should have a low water evaporation rate and low oxygen permeability to decrease respiration, but not too low to impart anaerobic conditions, which might cause off-flavor production. This might pose a challenge in developing specific films and coatings for a specific food product (Mohamed et al., 2020; Otoni et al., 2017).

Food products are generally coated by immersing, spraying, and forming a thin layer of film on the food surface, acting as a semipermeable membrane and thus controlling the moisture loss and gas transfer rate (Figure 4.1) (Dhanapal et al., 2012). The functionality and efficacy of films and coatings greatly depend on their barrier, mechanical, and color properties, which in turn depend on film composition and its formation process. Immersion is the most common method of applying coatings on fruits and vegetables. The coating is formed by dipping in a gum solution of desired properties, such as density, viscosity, and surface tension, and the food is immersed directly into the solution for 5 to 30 seconds (Vargas, 2008). However, it has been reported that coatings applied by the brushing method gave good results on beans and highly perishable fruits and vegetables, such as strawberries, berries, etc. (Raghav et al., 2016). The spraying process is generally used for the coating that is formed from less-viscous coating solutions. Highly viscous solutions are not easily sprayed, and thus, the immersion method is suitable for high-viscous coating solutions. The spraying process involves a jet system consisting of a spray gun and nozzle. The spray exits the nozzle as a sheet of liquid covering the product, and the quality of the coating is greatly dependent on the spray gun, nozzle, temperature, air, and flow rate of liquid. The films produced by compression molding, casting, and extrusion are applied by the wrapping method (Diaz-Montes et al., 2021; Tavassoli-Kafrani et al., 2020). The solvent casting method is widely used for the preparation of gum-based films. In this process, gum dispersions are spread on a suitable substrate and dried. The film structure depends on the drying conditions, film thickness, and composition of the casting solution (Dhanapal et al., 2012). During the drying process, the solvent evaporates, leading to a decrease in solubility till polymer chains align themselves to form films. The extrusion method relies on the thermoplastic characteristics of the biopolymers and involves the heating of polymer above their glass transition temperature in limited water conditions. They have certain limitations as they cannot be applied on irregular surfaces of food products (Kamal, 2019). Also, the extrusion process often needs the addition of plasticizers, such as glycerol or sorbitol. Extrusion is mainly employed by industrial preparations since it does not require solvent addition and evaporation.

4.4 Gums for Films and Coatings

Hydrocolloids belong to the class of polysaccharides which includes different mucilages, gums, and glucans consisting of structurally different biological macromolecules with diverse physicochemical properties and are widely used for various applications in food, pharmacy, and medicine (Malviya et al., 2011). Hydrocolloids are high-molar-mass polymers and are composed of hydrophilic string. These can form dispersions when dissolved in water and exhibit colloidal properties due to which they are termed hydrocolloids. Gums can be defined as polymers composed of complex carbohydrates that can form gels due to their solubility in water. They can be formed by galactose, arabinose, rhamnose, xylose, galacturonic acid, and other compounds. Gums are present in high quantities in different plants, animals, seaweeds, fungi, and other microbial sources, where they perform several structural and metabolic functions; plant sources provide the largest amounts. The different available gums are classified in Table 4.1.

TABLE 4.1

Classification of gums

Basis	Class	Example
Origin	Natural seeds	Guar gum, karaya gum, ipomoea, fenugreek, LBG
	Plant exudates	Konjac gum, chicle gum, arabic gum, gum karaya, ghatti gum, tragacanth gum
	Microbial	Xanthan gum, gellan gum, spruce gum
	Seaweeds	Sodium alginate, agar-agar, carrageenans
Charge	Anionic gums	Arabic gum, gum karaya, gellan gum, carrageenans
	Non-ionic gums	Guar gum, LBG, tamarind gum, xanthan gum
Shape	Branch on branch	Gum arabic, tragacanth gum
	Short branch	Xanthan gum, guar gum
Structure	Galactomannans	Guar gum, locust bean gum, fenugreek gum, tara gum, dhaincha gum, cassia gum
	Uronic acid–containing gums	Xanthan gum
	Glucomannans	Konjac gum
	Tri-heteroglycans	Gellan gum
	Tetra-heteroglycans	Arabic gum, psyllium seed gum
	Penta-heteroglycans	Tragacanth gum, gum ghatti
Functionality	Thickening	Xanthan, gum arabic, guar gum, LBG, tara gum, konjac gum, tragacanth gum, β-glucan
	Gelling behavior	Gellan gum, flaxseed gum,
	Emulsification and Stabilization	Konjac gum, arabic gum

Source: Kapoor et al. (2013); Prajapati et al. (2013b); Rana et al. (2011).

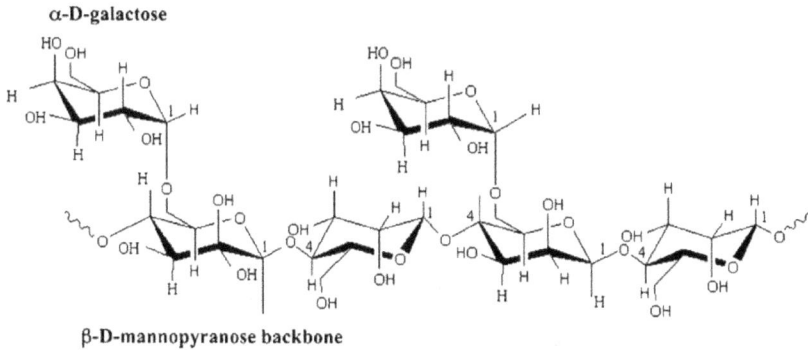

FIGURE 4.2 Intermolecular hydrogen bonding of gums when dissolved in water.

These are classified based on their origin, such as plants, seaweeds, and animal sources, as well as from microbial sources and on nature, that is, natural or modified polymers (Williams and Phillips, 2000).

Gums are naturally occurring polysaccharides possessing complex, branched polymeric structures, because of which they exhibit high cohesive and adhesive properties. They are often termed hydrocolloids and used as thickeners, gelling agents, stabilizers, emulsifiers, films, coatings, and dietary fibers (Barak and Mudgil, 2014; Dick et al., 2015). All the properties of gums are closely related to their chemical structure. They could be formed by different sugar molecules, in the main backbone chain or side chains, or can be branched, which in turn determines their complexity. Gums are high-molecular-weight polymers and can form highly viscous solutions even at low concentrations. The ability of gums to thicken the solution greatly depends on its molecular structure. When dissolved in water, they form intense hydrogen bonds, and in solution, gum molecules arrange themselves in an ordered structure known as a micelle, which is stabilized by intermolecular hydrogen bonding, as shown in Figure 4.2. The micelle entraps and immobilizes water molecules and thus imparts viscosity to the solution or either forms a gel that comprises both solid and liquid-like characteristics and is viscoelastic (Salehi and Kashaninejad, 2015; Williams and Phillips, 2000).

Gums are used for films and coatings because of their texturing abilities. The hydrophilic nature of gums provides good barrier properties against carbon dioxide and oxygen under certain conditions (Cerqueira et al., 2011). In low humid conditions, most gums have good barrier properties, and the quality of films prepared from gums greatly depends on the gum structure, that is, the presence of large hydroxyl groups and other polar groups which form hydrogen bonds. The uniform distribution of the polymer chain enhances the coating's ability to form hydrogen bonds and their participation in ionic interactions. Polysaccharide breaks down the interactions between long-chain polymer segments, form new intermolecular hydrogen and hydrophilic bonds as the solvent evaporates, and thus form a coating matrix (Janjarasskul et al., 2010). High-molecular-weight polymers consist of large-chain polymeric structures required for the formation of polymer matrices possessing the required cohesive strength that provides mechanical strength to the coating (Moncayo et al., 2013). However, gum-based films have low water vapor barrier properties, leading to poor stability and mechanical properties (sensitive to moisture content) (Li and Nie, 2016), since gums can act as potent replacers of synthetic polymers such as plastics due to their functional and edible properties. Therefore, the film-forming characteristics of different gums and their applications in food packaging systems are discussed further.

4.5 Structure and Film-Forming Properties of Gums

4.5.1 Guar Gum

Guar or cluster bean is an annual agricultural crop, *Cyamopsis tetragonolobus*, belonging to the family *Leguminosae*. It is grown in desert zones of India's west and northwest, in Pakistan and Sudan, and in sections of the United States (Thombare et al., 2016). Guar gum is a gel-forming galactomannan

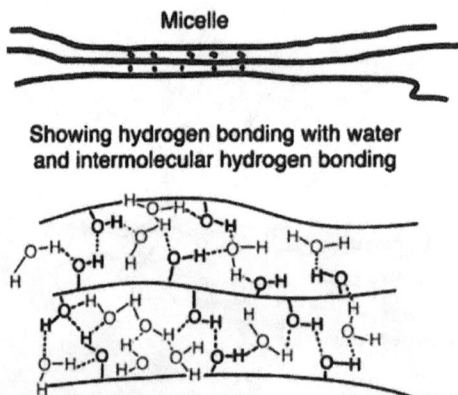

Micelle

Showing hydrogen bonding with water
and intermolecular hydrogen bonding

FIGURE 4.3   Structure of guar gum.

Source: Barak and Mudgil (2014).

obtained by grinding the endosperm of *Cyamopsis tetragonolobus*. This gum is extremely water-soluble, biodegradable, and widely utilized as a thickener, stabilizer, and emulsifier, viscosity builder, and water binder in a variety of culinary products. Guar gum has properties similar to carrageenan, alginate, xanthan gum, and alginate gum; however, it is less expensive. The structure gum is made up of linear backbone chains of $(1 \rightarrow 4)$-β-d-mannopyranosyl units connected by $(1 \rightarrow 6)$ connections to branch points of α -d-galactopyranosyl units. In many studies, the ratio of mannose to galactose units in guar gum was found to be in the range of 1.6:1 to 1.8:1 (Pollard et al., 2006). Guar gum has the greatest molecular weights of any water-soluble polysaccharide found in nature. The average molecular weight was found to be in the range of 10^6 to 2×10^6 using modern techniques, such as size exclusion chromatography and low angle laser light scattering (Thombare et al., 2016). The starch-based edible films were successfully developed by Saberi et al., 2016 from pea starch, glycerol, and guar gum. The findings revealed that pea starch, glycerol, and guar gum significantly affected viscosity, solubility, moisture content, transparency, and color of the films. The optimum formulation of pea starch–based films was found to be guar gum 0.3 g, glycerol of 25% w/w, and pea starch of 2.5 g, respectively. This research discovered that pea starch and guar gum may be combined to create a composite film with acceptable packaging capabilities, which can then be used to coat fruits and vegetables to extend their shelf life. Rao et al. (2010) has developed the chitosan- and guar gum-based composite films and studied the physical, mechanical, and antimicrobial properties. The optical, mechanical, and antibacterial properties of the films were all impacted by the concentration of the two polysaccharides in the film. When compared to chitosan film without guar gum, the film created with a combination of 85% CH and 15% GG (v/v) was found to be the best since it had lower oxygen permeability, superior mechanical properties, and equivalent antibacterial capabilities of chitosan. The potential of guar gum/Ag-Cu nanocomposite films as an active food packing material was studied by Arfat et al. (2017). The effects of nano particles loadings (0.5–2%) on the thermomechanical, optical, spectral, oxygen barrier, and antimicrobial characteristics of the GG/Ag-Cu NC films were examined. The results revealed that the tensile strength and thermal properties of the film were improved. The nano composite films showed excellent UV, light, and oxygen barrier capability. Nanocomposite films demonstrated strong antibacterial activity against both Gram-positive and Gram-negative bacteria; as a result, the film could be regarded as an active food packaging.

4.5.2 Fenugreek Gum

Fenugreek (Trigonella foenum-graecum) is a leguminous plant in the Fabaceae family that is widely grown as a semiarid crop in northern Africa, the Mediterranean, India, and Canada (Kasran et al., 2013). For many years, fenugreek seed, which grows in pods at the plant's extremities, has been used medicinally and as a food ingredient. As shown in Figure 4.4, fenugreek gum is made up of the heterogeneous

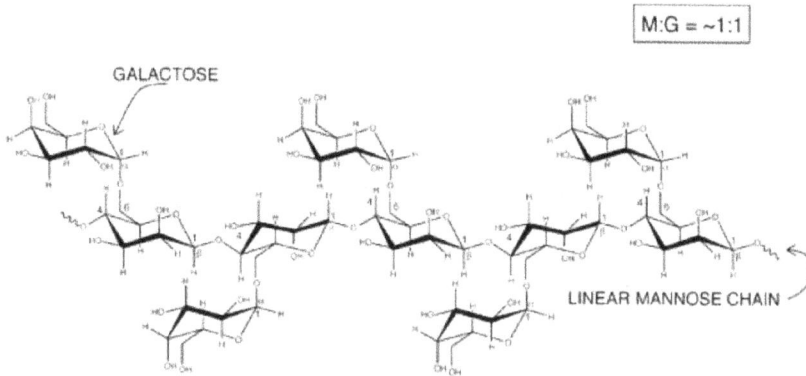

FIGURE 4.4 Structural features of fenugreek gum.

Source: Salarbashi et al. (2019).

polysaccharide galactomannan, which has a linear (1,4)-D-mannan backbone with various degrees of D-galactosyl substituents connected via 1,6-glycosidic linkage and is generated from the seed endosperm. The molar ratio of mannose to galactose in fenugreek gum is about 1:1 (Iurian et al., 2017; Youssef et al., 2009). Fenugreek gum has been used as a thickener, stabilizer, and emulsifier in many food products (Youssef et al., 2009). Galactomannan had a molecular weight M_w of 3.23×10^5 g mol^{-1} and intrinsic viscosity of 235 mL g^{-14} (Jiang et al., 2007). Fenugreek seed gum is the most soluble among seed gums. Because of its high thickening qualities, fenugreek gum can be used in food industries. This gum also has a high emulsification capability, making it suitable for use in the food, cosmetics, and pharmaceutical industries. The effect of reinforcement of fenugreek gum–based films with different types and amounts of nanoclays on the physicochemical, mechanical, barrier, antimicrobial, and microstructural properties was investigated by Memiş et al. (2017). The results revealed that increasing the amount of nanoclays decreased the moisture content and oxygen permeability of films. The nanocomposite film results showed favorable mechanical and antibacterial activity when tested against foodborne pathogens. Because of the clays' poor dispersion and high surface energy, nanoclays concentrations >5% did not improve the TS. It was discovered that the surfaces of the films containing up to 5% nanoclays were largely uniform and smooth. Jayaprada and Umapathy (2020) have reported the properties of microfibrillated cellulose (MFC)–reinforced pectin/fenugreek gum biocomposite. In comparison to the neat-and-blend film, the solubility and water vapor permeability of the films containing plant extracts significantly decreased while the elongation and tensile strength increased. The thermal property of the MFC films containing plant extracts showed little weight loss.

4.5.3 Locust Bean Gum

Locust bean gum or carob gum is a white to creamy white powder obtained after milling of seed endosperm of fruit pod of the carob tree, a member of the legume family, botanically known as *Ceratonia siliqua L.*, which is found in Mediterranean regions (Barak et al., 2014). Galactomannan makes up the majority of locust bean endosperm, accounting for about 80% of the total weight, with proteins and contaminants accounting for the remaining 20% (Prajapati et al., 2013b). Galactomannan is a high-molecular-weight polymer found in locust bean gum. It is made up of two units: galactose and mannose. The linear chain of (1→4)-linked β-d-mannopyranosyl units with (1→6)-linked α-d-galactopyranosyl residues as side chains, as shown in Figure 4.5. The galactose-to-mannose ratio of locust bean gum is approximately 1:3.1–1:3.9 (McCleary, 1985). The molecular weight of locust galactomannan is reported as approximately 535–826 kDa (Haddarah et al., 2014)

In cold water, locust bean gum is somewhat soluble. The complete solubilization of locust galactomannan in water necessitates heating. The viscosity of locust bean gum solution reduces with increasing

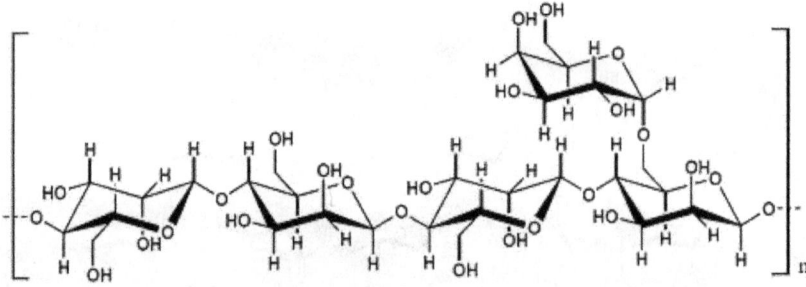

FIGURE 4.5 Structural features of locust bean gum.

Source: Barak and Mudgil (2014).

pH above 9 and decreasing below 4. This gum is utilized in a variety of foods as a thickener, stabilizer, and gelling agent, including baked goods, drinks, dairy products, and processed fruit products. Locust bean gum is utilized for its stabilizing, thickening, and fat-replacing characteristics and is designated as GRAS (generally recognized as safe) by the FDA (Food and Drug Administration) (Soma et al., 2009). Due to its dietary fiber action, it is also helpful in the management of numerous health issues, like diabetes, bowel movements, heart disease, and colon cancer (Barak and Mudgil, 2014). Mostafavi et al. (2016) investigated the films prepared from blends of solution of gum tragacanth and locust bean gum. The surface tension of the two polysaccharides was reduced when they were blended, which could improve their spreadability and coating integrity when applied to food products. Physical observations revealed that the gum blending ratio had no effect on film thickness, density, or retraction ratio; adding locust bean gum to the film increased transparency, water barrier, and mechanical qualities. The effect of xanthan and locust bean gum synergistic interaction on characteristics of the biodegradable edible film was studied by Kurt et al. (2017). Concentrations of locust bean gum, xanthan gum, and glycerol in the optimized film sample were found to be 89.6%, 10.4%, and 20%, respectively. The film's water vapor permeability, tensile strength, elongation at break %, and elastic modulus values were found to be 0.22 g mm h^{-1} m^2 kPa, 86.97 MPa, 33.34%, and 177.25 MPa, respectively, at the optimum point. The optimized film was characterized by its physical, thermal, and structural behavior. The analyses performed using scanning electron microscopy (SEM), X-ray diffraction (XRD), differential scanning calorimetry (DSC), and Fourier transform infrared spectroscopy (FTIR) revealed polymer miscibility and the presence of interaction. In conclusion, the interaction of xanthan gum and locust bean gum was successfully used to produce biodegradable films and coatings with enhanced properties. Application of locust bean gum edible coating to extend shelf life of sausages and garlic-flavored sausages was investigated by Dilek et al. (2011). Sausage that was coated and flavored with garlic had a longer shelf life than the control sample under storage conditions. Increased amounts of plasticizers, which are liquid at room temperature, had a diluting effect on the film solution and slightly increased the moisture loss values of coated samples. These finding suggests that LBG film may be suitable as edible packaging for sausages and garlic-flavored sausage, as well as for convenient consumer use to reduce food packaging waste.

4.5.4 Tara Gum

Tara gum, commonly known as Peruvian carob, is a white powder obtained from the seed endosperm of the *Caesalpinia Spinosa* tree, which is native to Peru but widely grown in China's Yunnan and Sichuan regions. As shown in Figure 4.6, galactomannan polysaccharides are the principal component of tara gum, consisting of a linear main chain of (1–4)-β-d-mannopyranose units linked by (1–6) linkage with α-d-galactopyranose units (Wu et al., 2015). In terms of structure and functional characteristics, tara gum is comparable to guar gums and locust bean gum. The ratio of mannose/galactose molecules in tara gum is 3:1, whereas it is 2:1 and 4:1 in guar gum and locust bean gum, respectively (Wu et al., 2012). It is a neutral and reserved polysaccharide with an approximate molecular weight of (Mw) of 1.0×10^6 g·mol^{-1}

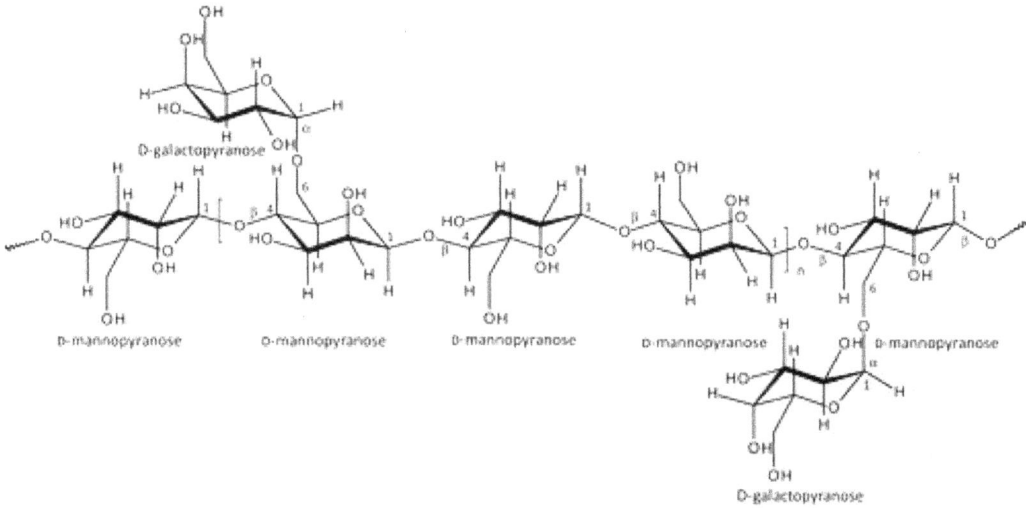

FIGURE 4.6 Structural features of tara gum.

Source: Rigano et al. (2019).

(Pollard et al., 2006). To avoid syneresis, tara gum has been employed in foods as a thickener, binder, stabilizer, and moisture retainer. It has the benefits of being colorless, insipid, and stable; it works well with other gums and acts as a gelling agent (Viebke et al., 1996). Antoniou et al. (2014) evaluated tara gum as edible film material and the influence of polyols as plasticizers on the properties of the films. The study showed that tara gum has good mechanical properties and can be utilized to make freestanding edible films using the casting method. The addition of polyol plasticizers influenced the film's thermomechanical and barrier properties but did not affect the polysaccharides. Tara gum edible film incorporated with oleic acid was studied by Ma et al. (2016). The effect of different oleic acid concentrations (0%, 5%, 10%, 15%, and 20%) on moisture, color, opacity, mechanical properties, water vapor permeability, and thermal stability of tara gum was examined. The study revealed that the polymer matrix became discontinuous as a result of the addition of oleic acid, lowering the tara gum film's tensile strength and light transmittance. Oleic acid has also reduced the water vapor permeability and moisture of film and improved the surface contact angle and thermal stability of the film. Ma et al. (2017) has investigated the rheology of film-forming solutions and physical properties of tara gum–reinforced polyvinyl alcohol. Due to the high concentration of hydroxyl groups, the results showed that TG and polyvinyl alcohol (PVA) interacted strongly via hydrogen bonds. The mechanical, hydrophobic, and barrier properties were improved by the addition of PVA. However, there was an optimum level for the interaction between TG and PVA, where TG was the dominant phase in the film system and the film had better properties, as indicated by the tensile strength, oxygen permeability, and water vapor permeability values for various ratios of films. As a result, the two polymers combined to create a high-quality blend film with potential applications in the packaging and film industries.

4.5.5 ARABIC GUM

Gum arabic is an edible, dried, gummy exudate from the stems and branches of Acacia senegal and A. seyal that is rich in non-viscous soluble fiber. Arabic gum is a branched-chain, complex polysaccharide that is found as mixed calcium, magnesium, and potassium salt of a polysaccharide acid and is either neutral or slightly acidic. The backbone is composed of 1,3-linked β-d-galactopyranosyl units. The side chains are made up of two to five 1,3-linked β-d-galactopyranosyl connected by 1,6-linkages to the main chain. Units of α -l-arabinofuranosyl, α-l-rhamnopyranosyl, β-d-glucuronopyranosyl, and 4-O-methyl-β-d-glucuronopyranosyl can be found in the main and side chains, the latter two typically as end units, as shown in Figure 4.7 (Islam et al., 1997; Anderson et al., 1966). Arabic gum is a highly heterogeneous

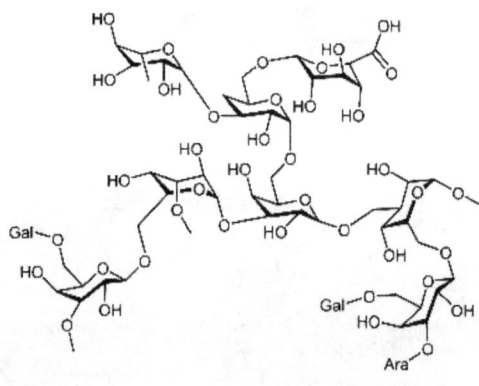

FIGURE 4.7   Structural features of arabic gum.

Source: Jahandideh et al. (2021).

material but was separated into three major fractions by hydrophobic affinity chromatography. The majority of the gum (88.4%) was an arabinogalactan with a molecular mass of 3.8×10^5 Da and very low protein content (0.35%). The second fraction (10.4% of total), an arabinogalactan-protein complex, contained 11.8% protein and had a molecular mass of 1.45×10^6 Da. The third fraction (1.2% of total gum), referred to as a low-molecular-weight glycoprotein (GP), had a protein content of 47.3% and a molecular mass of 2.5×10^5 Da (Ray et al., 1995). In Middle Eastern countries, gum arabic is used as a treatment for chronic renal disease. It has a wide range of applications in the food sector due to its edibility, high water solubility, generally recognized as safe (GRAS) status, absence of aftertaste, and other favorable qualities. It is used in food compositions, such as in ice creams, jellies, sweets, soft drinks, beverages, syrups, and chewing gums, because of its emulsifying, stabilizing, thickening, and binding properties. It's perfect for confectionery coatings and glazes because of its film-forming capabilities (Patel et al., 2015). Kang et al. (2021) developed a bionanocomposite film based on gum arabic and reinforced with cellulose nanocrystals. The effect of cellulose nanocrystals on the physicochemical, functional, and rheological of the gum arabic–based films was evaluated. The results revealed that the film incorporated with 4% cellulose nanocrystals showed better tensile strength and elongation at break when compared to gum arabic film. The film containing 4% cellulose nanocrystals also reduced the oxygen and water vapor permeability and improved the ultraviolet light barrier and thermal stability of the film. Chu et al. (2019) have developed active films based on chitosan and gum arabic incorporated with cinnamon essential oil. The results revealed that electrostatic interactions existed between chitosan and gum arabic and led to the formation of an entangled structure. In terms of mechanical properties, the addition of gum arabic reduced the film resistance to breaking and stretching, but it sharply reduced the water vapor permeability and moisture content of composite films. The antioxidant effectiveness was greatly enhanced when the ratio of gum chitosan/gum arabic changed from 1:0 to 1:20. The effects of gum arabic–based edible coating on guava fruit characteristics during storage were studied by El-Gioushy et al. (2022). The findings showed that coatings made with gum arabic (10%) alone or in combination with natural plant extracts significantly reduced weight loss, decay, and rot ratio. The application of gum arabic (10%) + moringa extract (10%) was the best treatment out of all those put to the test for the majority of the parameters examined, and it showed the highest values for maintaining firmness, total soluble solids, total sugars, and total antioxidant activity.

5.5.6 Gum Karaya

Also known as sterculia gum, it is a plant exudate that is a dry exudate of the *Sterculia urens* tree, which is often found in India. Gum karaya is a hydrophilic polysaccharide with an anionic charge. It

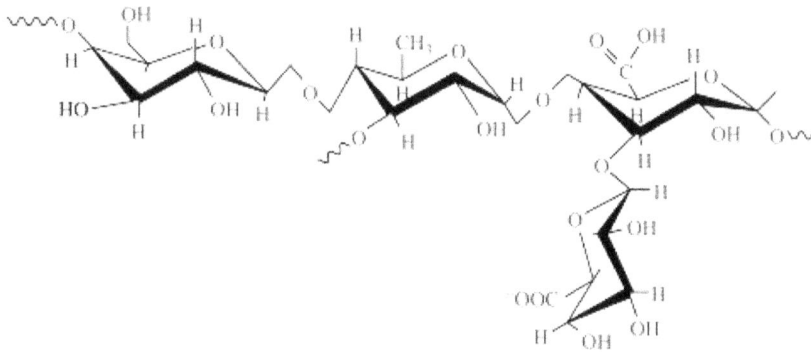

FIGURE 4.8 Structural features of karaya gum.

Source: Kander et al. (2021).

has a molecular weight of 1.6 107 Da and is made up of 55–60% galactose and rhamnoses, 37–40% glucuronic and galacturonic acids, and 8% acetyl functional groups (Mirhosseini and Amid, 2012). As shown in Figure 4.8, the backbone of the gum is made up of D-galacturonic acid and L-rhamnose residues, with side chains linked to the main chain via a 1,2-linkage of D-galactose or a 1,3-linkage of D-glucuronic acid. In addition, half of the main chain's rhamnose residues are 1, 4-linked to D-galactose units (Anderson et al., 1983; Chauhan et al., 2019).

Gum karaya and its derivatives have several appealing properties, including increased acidic stability, high swelling, good viscosity, and water retention (Preetha and Vishalakshi, 2020). Importantly, gum karaya has no toxicity, allergenic, mutagenic, or teratogenic effects (Silva et al., 2020). Furthermore, gum karaya and its modified polysaccharides have antibacterial and antioxidant properties and can aid in the anticancer (Pooja et al., 2015) and antihypertensive activity of other drug molecules in various formulations (Laha et al., 2019). Gum karaya is utilized as a stabilizer, emulsifier, thickener, and adulterate for tragacanth gum in the food business because the two have similar physical qualities (Anderson, 1989). Cao and Song (2019) studied the development of active gum karaya/Cloisite Na+ nanocomposite films containing cinnamaldehyde and revealed that the gum karaya film displayed an amorphous condition, low tensile strength, high degree of elasticity, and high water solubility. The amorphous condition allowed the addition of Cloisite Na+ to increase the physical and water barrier qualities of the gum karaya films by up to 11%. The addition of cinnamaldehyde to the EB of the gum karaya films lessened the unfavorable effects of the Cloisite Na+ addition. Additionally, the 0.75% cinnamaldehyde-containing gum karaya nanocomposite film demonstrated antibacterial activity against the tested pathogens. Yousuf et al. (2021) prepared a novel biodegradable film by combining karaya gum with *S. chinensis* oil or oleogel. The water vapor permeability, which is thought to have a significant impact on the practical applicability of biodegradable films, was decreased by the incorporation of oil/oleogel. The addition of oil/oleogel into films somewhat decreased the tensile strength but enhanced the elongation at break. The addition of oil or oleogel also made films thicker. The phenolic content and DPPH radical scavenging activity of the films were also improved by the addition of oil/oleogel.

4.5.7 Gum Ghatti

Gum ghatti, also known as Indian gum, is a non-starch polysaccharide that is native to India. The major species is *Anogeissus latifolia* (Combretaceae, Myrtales), a huge deciduous tree that grows in dry places. It is a long-used plant whose name comes from the term Ghat, which means "mountain pass," and was given to the gum due to its historic mountain transportation routes (Deshmukh et al., 2012). The dried product is sifted and processed as usual, and the gum seeps naturally, typically colored by non-carbohydrate impurities. Its quality and quantity could not be guaranteed in the past; hence, it was not widely used as a tree gum in food goods (Amar et al., 2006). Since 1976, the United

$$
\begin{array}{ccc}
\text{R} & & \text{R} \\
\uparrow & & \uparrow \\
6 & & 6 \\
\rightarrow 4)\text{-}\beta\text{-D-GlcA - } (1\rightarrow 2)\text{- D-Man}p\text{-}(1\rightarrow 4)\text{-}\beta\text{-GlcA- }(1\rightarrow 2)\text{-D-Man}p\text{-}(1\text{-} \\
3 & & 3 \\
\uparrow & & \uparrow \\
1 & & 1 \\
\text{L-Ara}p & & \text{L-Ara}p \\
3 & & 3 \\
\uparrow & & \uparrow \\
\text{R'} & & \text{R'}
\end{array}
$$

$$\text{R = L-Ara}f\text{-or-L-Ara}f\text{-}(1\rightarrow 2,\ 3\ \text{or}\ 5)\text{-L-Ara}f\text{-}(1\text{-}$$

$$
\text{R'} = \begin{array}{c} \beta\text{-Glc}p\text{A-} \\ \text{or} \\ \beta\text{-Gal}p\text{-} \end{array} \left[(1\rightarrow 6)\text{-D-Gal}p \atop 3 \right]_n \!\!\!\!\!-\!(1
$$

$$\uparrow \atop \text{R}$$

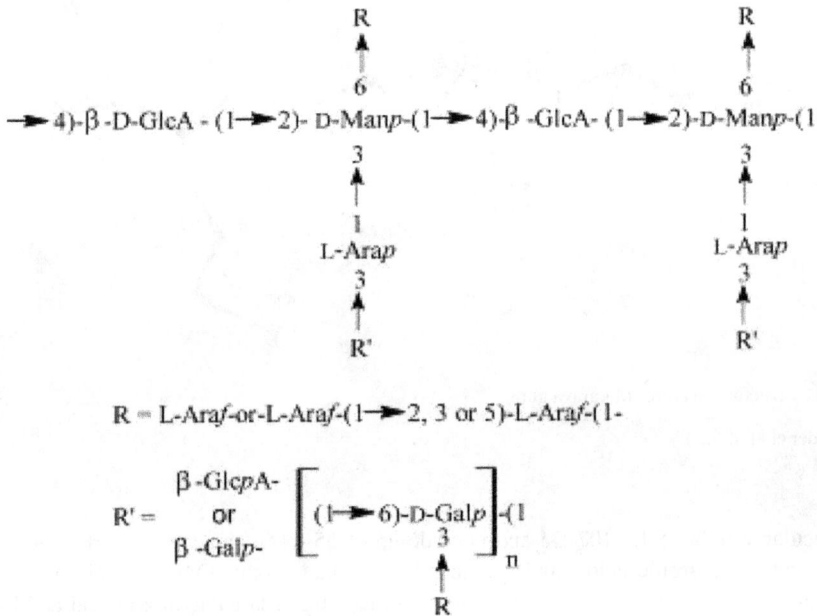

FIGURE 4.9   Structural features of gum ghatti.

Source: Deshmukh et al. (2012).

States regulatory status as "generally recognized as safe" (GRAS) has been based on tests for toxicity, mutagenicity, and teratogenicity. However, the European Union has demanded a more detailed evaluation of the safety of these gums as food additives, and the United States has failed to provide the necessary information. As a result, gum ghatti has been removed from European lists of authorized additives (Glicksman, 1983). Gum ghatti is a polysaccharide that forms naturally as a mixture of calcium and magnesium salts of uronic acid. It contains acid and/or ghattic acids. It's made up of L-arabinose, D-galactose, D-mannose, D-xylose, and D-glucuronic acid in a 48:29:10:5:10 molar ratio, with 1% rhamnose as non-reducing end groups, as shown in Figure 4.9 (Aspinall et al., 1965; Tischer et al., 2002). The gum contains alternating 4-O-substituted and 2-O-substituted-D-mannopyranose units, as well as chains of 1–6 connected D-galactopyranose units with L-arabinose units on the side chains (Aspinall et al., 1965; Jefferies et al., 1977).

A nonchemical physical procedure involving dissolving, filtering, sterilization, and spray-drying was used to create "Gatifolia," a novel gum ghatti product. It has better emulsification ability, acid resistance, and salt tolerance (Pszczola and Banasiak, 2006). Gatifolia has been shown to have a broader spectrum of proteinaceous molecular components for binding oil, making it superior to other natural emulsifiers. In comparison to many other polysaccharides, the gum has a low viscosity but is more water-soluble. Gum ghatti was also investigated for its film-forming property. The physical, barrier, mechanical, optical, and thermal properties were assessed as a function of gum ghatti content (0.75 and 1.0%, w/v), plasticizer type (glycerol and sorbitol), and concentration (15%, 30%, and 45%, w/w). These properties are crucial for their applications in food preservation or packaging. As a function of gum ghatti content (0.75% and 1.0%, w/v), plasticizer type (glycerol and sorbitol), and concentration (15%, 30%, and 45%, w/w), the physical, barrier, mechanical, optical, and thermal properties—which are crucial to their applications on food preservation or packaging—were also determined. Overall, gum ghatti demonstrated good potential to produce edible films that met the requirements. The inclusion of essential oils should be studied further to strengthen additional features, like antioxidant activity, antibacterial activity, and reduction of water vapor permeability. It's important to discuss how well gum ghatti edible film preserves perishable foods, particularly those that are susceptible to microbial contamination and oxidation processes (Zhang et al., 2016). In order to enhance the functionality of biodegradable sodium alginate film, Cheng et al. (2021) studied the interaction between sodium alginate and gum ghatti. Strong electrostatic interactions as well as hydrogen bonds have been created between sodium alginate and gum ghatti. High mechanical strength, light barrier, and thermal stability of sodium alginate/gum ghatti blend films are promising biomaterials and packaging materials with numerous application possibilities.

4.5.8 Xanthan Gum

Xanthan gum is an extracellular anionic heteropolysaccharide made from pure cultures of the genus *Xanthomonas* fermenting carbohydrates. Because of its outstanding rheological qualities, it is employed in a variety of industries. Scientists at the Northern Regional Research Laboratory, Department of Agriculture, United States of America, discovered xanthan in 1950 while looking for microorganisms capable of generating water-soluble gums with economic potential. Among the polysaccharides generated in xanthan, the most fascinating features to compete with other natural and synthetic gums were discovered during the research. As a result, xanthan gum became the first industrially manufactured microbial polymer (Rosalam and England, 2006). The first commercially available xanthan gum was produced in 1960 using *Xanthomonas campestris* NRRL B-1459, and the product was marketed in 1964 (García-Ochoa et al., 2000). Xanthan gum is a heteropolysaccharide made up of a long chain of glucose, mannose, and glucuronic acid pentasaccharide units in a 2:2:1 molar ratio (Figure 4.10). The main chain is made up of β-D-glucose units linked at the 1 and 4 positions, similar to cellulose, while the trisaccharide side chain is made up of a D-glucuronic acid unit sandwiched between two D-mannose units attached to the main chain's alternative glucose unit (Jansson et al., 1975). The terminal D-mannose unit is linked to a pyruvic acid residue, while the D-mannose unit belonging to the main chain is linked to an acetyl group (Rosalam and England, 2006). The inclusion of acetic acid and pyruvic acid moieties in xanthan gives it an anionic character. With different strains of *X. campestris*, the structural feature of xanthan gum varies (Cottrell and Kang, 1978).

Xanthan gum is a free-flowing white to cream-colored powder. It is soluble in both cold and hot water, but most organic solvents are insoluble. The attributes will vary depending on the product's intended purpose; for example, microbial loads and heavy metal levels for food applications must be very low, whereas this is not the case for other industrial applications (Sandford, 1979). Most commercially available thickeners, such as cellulose derivatives, starch, pectin, gelatin, dextrin, alginate, and carrageenan, are compatible with xanthan gum. It exhibits a synergistic increase in viscosity with galactomannan, meaning, that the measured viscosity is more than the sum of the separate gum viscosities. Very specific and defined qualities can be obtained by combining different gums with xanthan gum in varying quantities—for example, viscosity, pseudo-plasticity, and "mouth feel." Zheng et al.

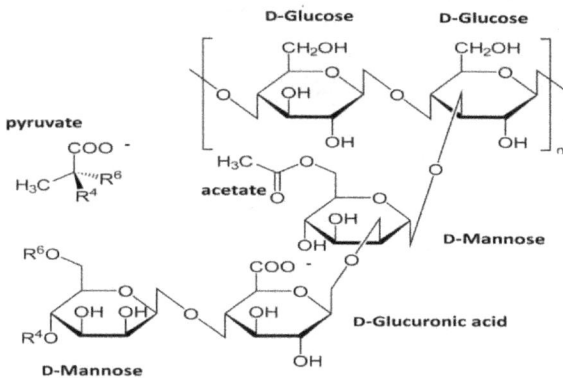

FIGURE 4.10 Chemical structure of xanthan gum.

Source: Petri (2015).

(2022) prepared the hydroxypropyl methylcellulose (HPMC) film incorporating xanthan gum. He concluded that the water vapor transmission rate (WVTR), mechanical characteristics, and light transmittance all showed that xanthan gum and hydroxypropyl methylcellulose worked well together due to the hydrogenbond interaction, and that xanthan gum had a large impact on the microstructure, crystalline texture, and chemical makeup of the xanthan-hydroxypropyl methylcellulose composite film. Bananas were coated with the best xanthan-hydroxypropyl methylcellulose sample, which had an ideal xanthan gum concentration of 2 g/L, and the weight loss rate was able to be reduced from 25 to 3% (without xanthan-hydroxypropyl methylcellulose coating) to 16 to 4% (with xanthan-hydroxypropyl methylcellulose coating). As a result, there was a reduction in the release of flavoring agents. With xanthan-hydroxypropyl methylcellulose composite film for food preservation, banana shelf life has qualitatively enhanced, confirming its applicability in food packaging applications. The effect of xanthan coating supplemented with cinnamic acid on freshly cut "Nashpati" and "Babughosha" pears was studied by Sharma and Rao (2015), and they found that the coating prevented the activity of browning-related enzymes, delayed the onset of browning, and increased the shelf life of the fruit by up to four days and eight days, respectively. The effects of xanthan and locust bean gum synergistic interaction on the properties of biodegradable edible film were studied by Kurt et al. (2017). When these polysaccharides and glycerol are combined in various ratios, the mechanical and barrier properties of the films network are altered. The interaction of xanthan and locust bean gum was effectively exploited to produce biodegradable films and coatings with better properties.

4.5.9 Konjac Gum

Konjac gum, also known as konjac mannan or konjac glucomannan, is a hydrocolloid produced from the root of the plant *Amorphophallus spp.* Konjac is a root vegetable that has been used in Chinese cuisine for over 2,000 years. In Asian markets, they're also popular health foods. Konjac is a heteropolysaccharide made up of 5:8 monosaccharide units of glucose and mannose (Shimahara et al., 1975), with β—(1–4) glycosidic linkages connecting them (Figure 4.11). In addition, one acetyl ester group per 19 sugar residues is found in konjac mannan. The molecular weight of konjac mannan varies according to the amorphophallus species or even the variety from which it is derived, as well as the extraction method used. Sugiyama et al. (1972) estimated average molecular weights of 0.67–1.9 million Daltons, depending on the amorphophallus type, while Li et al. (2006a) found 1.04 million Daltons.

One of the intriguing features of konjac gum is its ability to work in combination with other hydrocolloids. Takigami (2000) found a synergistic relationship between konjac mannan and xanthan, resulting in an elastic gel, as well as a considerable increase in the gel strength of konjac mannan–carrageenan and konjac mannan–agar mixes. Furthermore, konjac glucomannan has high water absorbency, absorbing up to 100 g of water per gram of sample, and its water absorbency is reduced as the degree of acetylation on its chains increases (Koroskenyi and McCarthy, 2001). The limited presence of the acetyl group in the chain contributes to the gelation behavior of konjac glucomannan. At low solid concentration, an alkali

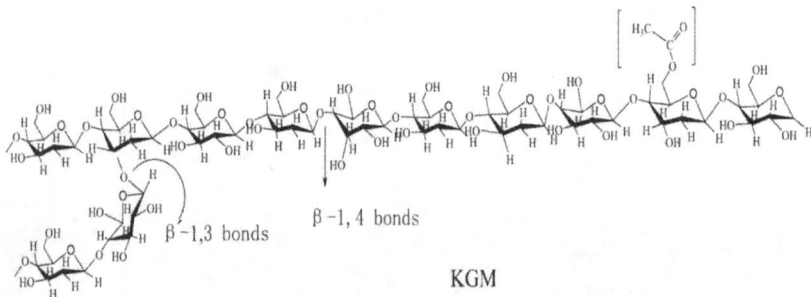

FIGURE 4.11 Structure of konjac glucomannan.

Source: Zhou et al. (2022).

solution is necessary for the gelation of konjac glucomannan (Gao and Nishinari, 2004a); however, at high solid content, konjac glucomannan can gel without alkali. Xiao et al. (2022) studied the interactions between the four polysaccharides—konjac glucomannan, agar, gum arabic, and virgin coconut oil—and concluded that the mixture improved the water barrier properties of the emulsified films compared to pure konjac glucomannan film, with decreased water vapor permeability, water vapor adsorption, water swelling, water solubility, and a larger water contact angle. Additionally, the addition of virgin coconut oil enhanced the elongation ratio at break (EAB) of konjac glucomannan film while having just a little effect on the tensile strength. Due to its superior water barrier qualities, the emulsified film has the potential to be used in the preservation of cucumber freshness. Liu et al. (2021) developed the edible films made from konjac glucomannan that include thyme essential oil and studied their properties. The addition of thyme essential oil greatly improved the physical characteristics and antioxidant–antibacterial activities of konjac glucomannan films, with the best results being shown in konjac glucomannan–based films loaded with 1.2% of thyme essential oil. Therefore, the concentrations of thyme essential oil had significant influence on the characteristics of konjac glucomannan–based films loaded with thyme essential oil, and that the thyme essential oil–loaded konjac glucomannan films have a great deal of promise for use in food packaging.

4.5.10 Gellan Gum

Sphingomonas elodea produces gellan gum, which is a type of fermentation polysaccharide (previously identified as *Pseudomonas elodea*, but later reclassified). CP Kelco discovered this microbe while conducting a global screening effort to find new and unusual gums that could be produced through fermentation. Gellan gum was supposed to be a "universal" gelling agent at the time of its discovery. Oddly, the texture of gellan gum gel can range from soft and elastic to rigid and brittle. Gellan gum may form gels with both monovalent and divalent cations, which adds to its attraction (Valli and Clark, 2009). As shown in Figure 4.12, gellan gum has a linear chemical structure made up of repeating glucose, rhamnose, and glucuronic acid units. Two acyl substituents—acetate and glycerate—are present in their native or high-acyl form. Both substituents are found on the same glucose residue, with one glycerate and one acetate per every two repeating units on average (Kuo et al., 1986). It has a repeating unit of β-1,3-D-glucose, β-1,4-D-glucuronic acid, α-1,4-L-rhamnose, and two acyl groups, acetate and glycerate, attached to the glucose residue next to the glucuronic acid (Figure 4.12) (Prajapati et al., 2013c). Based on the acyl substituents, native gellan gum is divided into two types: glyceryl gellan gum and acetyl gellan gum. Low-acyl gellan gum is produced by partially hydrolyzing both of them.

Gellan gum hydrates in hot water, and the low acyl form hydrates with sequestrants in cold water. When native high-acyl gellan gum is cooled, it produces soft, stretchy gels. Low-acyl gellan gum gels with monovalent and divalent cations at very low concentrations create stiff, brittle textures with exceptional thermal stability. Gellan gum blends can be used to manage syneresis and create a variety of textures, from soft and supple to stiff and brittle (Valli and Clark, 2009). The structural alteration of the blend films (konjac glucomannan and gellan gum) was examined by Xu et al. (2007) using FT-IR, XRD, DSC, and transparency. The results revealed that intermolecular hydrogen bonding occurred between

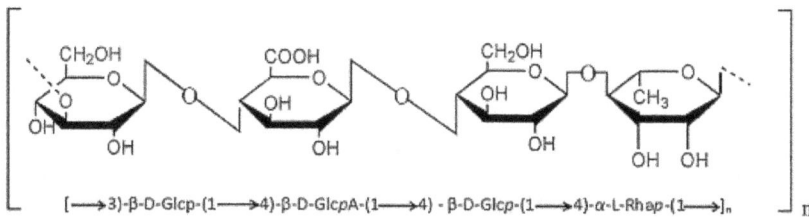

FIGURE 4.12 Chemical structure of gellan gum.

Source: Prajapati et al. (2013c).

konjac glucomannan and gellan gum, and the interaction of the blend film was significantly stronger than that of the others when the konjac glucomannan content in the blend films was approximately 70% wt percent. A higher maximum tensile strength and a reduced moisture uptake value are other features of the mixed film. According to research on the antimicrobial properties of konjac glucomannan–gellan gum films integrating nisin at different ratios against food pathogenic bacteria, specifically *S. aureus*, the inhibitory action has a favorable relationship to the gellan gum concentration.

4.6 Application of Gum-Based Films and Coatings

Nowadays, there is an increase in the production rate of non-degradable petroleum-based plastic materials for the packaging of foods which causes serious health issues as well as environmental concerns for their degradation (Gahruie et al., 2019). Therefore, many efforts have been made by various researchers to find non-toxic, environmentally friendly materials. Proteins, lipids, polysaccharides (starch, tapioca, corn, cellulose, and cellulose derivatives), and other naturally occurring plant-based components have received much attention due to their biodegradability (Gahruie et al., 2017; Galus and Kadzińska, 2015; Ganiari et al., 2017; Salehi and Kashaninejad, 2014; Salehi et al., 2014). The main reason for less use of synthetic packaging is their non-degradability, and hence, this will make more interest in natural recourses to making biodegradable edible coatings and their use as a packaging material. Edible coatings and gums are widely used in meat, fish, and poultry, bakery, and confectionery products as a shelflife enhancer and decreasing in respiration rate and transpiration in fruit and vegetables (Salehi, 2020). Edible coatings act as physical barriers on fruit and vegetable surfaces and decrease permeability to oxygen, carbon dioxide, and water vapor, which decrease respiration rate and transpiration. Edible films and coating also act as carriers of active ingredients, such as antioxidants, flavors, antimicrobial agents, fortified nutrients, and spices (Gennadios and Weller, 1990). Except for the barrier properties, these are also used to control adhesion, cohesion, and durability and improve the physical appearance of coated foods.

Fresh fruits and vegetables are highly perishable, and the spoilage of fruit and vegetables mainly occurred during harvest, handling, transportation, and storage. Edible coatings play a very important role to handle this situation. Edible coatings are applied on whole and fresh-cut fruits and vegetables (Krochta et al., 1994; Dhall, 2013; Youssef et al., 2015). Fruits and vegetables which have been coated include orange, apple, grapefruit, cherry, papaya, lemon, strawberry, mango, peach, tomato, cucumber, capsicum, cantaloupe, and minimally processed carrot, fresh-cut potato, fresh-cut cabbage, fresh-cut tomato slices, fresh-cut onion, lettuce, etc. Edible coatings are very useful for the protection of fruit and vegetable quality, with the additional benefit of reducing the weight of non-biodegradable packaging material for their packaging. Therefore, the maintenance of fruit and vegetable quality has been achieved by using some edible coatings and films based on hydrocolloids (gums), such as chitosan, gum arabic, pectin, xanthan gum, and alginate (Table 4.1.) (Najafi et al., 2021; Ghosh et al., 2021; Priyadarshi et al., 2022).

In meat and poultry products, edible films and protective coatings have been used for a long time to prevent their quality. They can improve the quality of fresh, frozen, and processed meat and poultry products by delaying moisture loss and reducing lipid oxidation and discoloration (Gennadios, 2002) because meat and meat products are highly perishable food commodities, mainly due to their enriched nutrient composition, high pH, and high water activity. These types of characteristics increase the chances of the growth of several pathogenic and spoilage-causing microorganisms (Samelis, 2006). Nowadays, active edible films and coatings made with antimicrobial properties have been developed. The most important antimicrobial agents added to edible coating formulations are bacteriocins and organic acids (Cagri et al., 2004). Polysaccharide-based (chitosan-based) coatings are also used to preserve meat products because of the inherent antimicrobial properties of this biopolymer (Dutta et al., 2009).

Nuts, cereal grains, cereal-based products, bakery products, and sweets are low-moisture-content products and prone to chances of spoilage. These types of spoilages are reduced with the use of edible coating and films and enhanced the product's quality. Nuts are used in different types of foods like ice cream, biscuits, ready-to-eat kheer, and high-fat foods. The most common types of spoilage in nuts and nut-based foods are sogginess due to moisture uptake, rancidity due to lipid oxidation, and loss of flavor

(Trezza and Krochta, 2002). The protection of nuts from oxidative rancidity can be achieved by edible coating. Firstly, the oxygen level within the nut by a high oxygen barrier coating and then use of coating as a carrier of antioxidants on the surface of nuts. Almonds and hazelnuts were also protected from oxidation with low methoxyl pectin or cellulose derivative coatings (Debeaufort, 1998). Padua and Wang (2002) prepared zein coatings and found them effectively used as oxygen, lipid, and moisture barriers for nuts, candies, and confectionery products.

Bread is a highly perishable product because of its high water activity, and its marketing period is too short and a maximum of seven days best before. The main problem with bread is its crumb hardness due to the leakage in the package, giving a sensation of drying product by ingestion, and

TABLE 4.2

Effects of polysaccharide-based coatings on fruits and vegetables quality

Polysaccharide used	Additives used	Coated fruits/ Vegetables	Effects on fruits and vegetables	References
Gum arabic	Glycerol	Ponkan orange (*Citrus poonensis*)	Reduction in postharvest decay and membrane lipid peroxidation Retention of nutritional quality Decreased fruit quality deterioration	(Huang et al., 2021)
	Glycerol	Strawberry (*Fragaria ananassa*)	Complete inhibition offungal growth	(Tahir et al., 2018)
	Glycerol	Tomato (*Solanum lycopersicum*)	Reduced water activity during storage of tomatoes	(Oladipupo et al., 2019)
	Aloe vera gel, ethanolic	Mango (*Mangifera indica*)	Prevention of weight loss, acidity loss Delay in the ripening process	(Ebrahimi et al., 2020)
Alginate	Citric acid, ascorbic acid	Apple (*Malus pumila*)	Reduced microbial growth and weight loss	(Najafi et al., 2021)
	Orange essential oil	Tomato (*Solanum lycopersicum*)	Retard the growth of bacteria Stop ripening and spoilage	(Das et al., 2020)
	Ascorbic acid	Pineapple (*Ananas comosus*)	Retained color Inhibited the polyphenol oxidase enzyme	(Lopez-Cordoba and Aldana-Usme, 2019)
Chitosan	Curcumin	Kiwi fruit (*Actinidia deliciosa*)	Reduced weight loss, firmness loss, respiration rate, and microbial count for 10 days of storage at a temperature of 10°C	(Ghosh et al., 2021)
	Acetic acid	Cucumber (*Cucumis sativus*)	Increased fresh-cut cucumber freshness and shelf life up to 12 days Fungal count reduced	(Olawuyi et al., 2019)
	Calcium chloride	Papaya (*Carica papaya L.*)	Storage time increased Reduced the growth of spoilage-causing fungi	(Romanazzi et al., 2017)
Pectin, pullulan	Grape seed extract (*Vitis vinifera*)	Peanut (*Arachis hypogaea*)	Reduction in bacterial growth and rancidity Lipid oxidation prevented and prolonged shelf life	(Priyadarshi et al., 2022)
Xanthan gum	Citric acid, glycerol	Lotus root (*Nelumbo nucifera*)	Decreased enzymatic browning Stopped the growth of *Bacillus subtilis*	(Lara et al., 2020)

another one is the growth of mold by moisture uptake (Stauffer, 1994). These types of shortcomings are noticed and solved by various researchers with the use of active packaging methods, modified atmospheric packaging, and the use of films prepared by different types of materials like polysaccharides, proteins, and lipids (Falguera, 2011). Swathi et al. (2019) prepared an edible coating with pectin, alginate, and whey protein and applied it to bread as a barrier to evaluate its effectiveness for retaining the quality of bread by reducing moisture loss and textural changes during storage. They found that the coatings displayed a retaining effect, with the percentage increase in crumb hardness lower for coated samples (1 L: 126.5% and 2 L: 231.2%) than for control samples (Control: 271.8%). In bakery products, microencapsulation has been extensively used for protecting iron salts and increasing their bioavailability (Cocato et al., 2007). Altamirano-Fortoul et al. (2012) prepared functional bread with starch-based coatings and microencapsulation of *Lactobacillus acidophilus*. Different types of probiotic coatings were applied on the surface of partially baked bread, and it was found that in all treatments, microencapsulated *L. acidophilus* survived after baking and storage time, although the reduction was higher in the sandwich treatment (starch solution/sprayed microcapsules/starch solution).

4.7 Conclusion

In this chapter, edible films and coatings sources, classification, and applications were reviewed. A growing amount of attention has been paid to biopolymer films and coatings in recent years due to their biodegradable characteristics. Due to the removal of plastic waste and the environmental harm it causes, the development of alternative, edible biodegradable films made of different gums to replace synthetic polymers has increased. Biopolymer-based packaging materials are typically made from various gum sources, such as arabic gum, tara gum, guar gum, fenugreek gum, etc., and can act as lipid, gas, aroma, and moisture barriers to improve food quality and properties, by reducing food deterioration and extending shelf life. Food industries like fruit and vegetable, meat, bakery, and confectionery use edible coatings and films to make healthy products with the incorporation of bioactive components in them. In addition, the use of an edible coating also helps in the reduction of packaging waste and makes an eco-friendly environment for society.

REFERENCES

Alizadeh-Sani, M., Ehsani, A., Kia, E. M. and Khezerlou, A. 2019. Microbial gums: Introducing a novel functional component of edible coatings and packaging. *Applied Microbiology and Biotechnology* 17: 6853–6866.

Altamirano-Fortoul, R., Moreno-Terrazas, R., Quezada-Gallo, A. and Rosell C. M. 2012. Viability of some probiotic coatings in bread and its effect on the crust mechanical properties. *Food Hydrocolloids* 29: 166–174.

Amar, V., Al-Assaf, S. and Phillips, G. O. 2006. An introduction to gum ghatti: Another proteinaceous gum. *Foods and Food Ingredients Journal of Japan* 211: 275.

Anderson, D. M. W. 1989. Evidence for the safety of gum karaya (Sterculia spp.) as a food additive. *Food Additives & Contaminants* 6: 189–199.

Anderson, D. M. W., McNab, C. G. A., Anderson, C. G., Brown, P. M. and Pringuer, M. A. 1983. Studies of uronic acid materials, Part 58: Gum exudates from the genus Sterculia (gum karaya). *International Tree Crops Journal* 2: 147–154.

Anderson, D. M. W. and Stoddart, J. F. 1966. Studies on uronic acid materials: Part XV. The use of molecular-sieve chromatography in studies on acacia senegal gum (Gum Arabic). *Carbohydrate Research* 2: 104–114.

Antoniou, J., Liu, F., Majeed, H., Qazi, H. J. and Zhong, F. 2014. Physicochemical and thermomechanical characterization of tara gum edible films: Effect of polyols as plasticizers. *Carbohydrate Polymers* 111: 359–365.

Arfat, Y. A., Ejaz, M., Jacob, H. and Ahmed, J. 2017. Deciphering the potential of guar gum/Ag-Cu nanocomposite films as an active food packaging material. *Carbohydrate Polymers* 157: 65–71.

Aspinall, G. O., Bhavanandan, V. P. and Christensen, T. B. 1965. Gum ghatti (Indian gum). Part V. Degradation of the periodate-oxidized gum. *Journal of the Chemical Society (Resumed)*, 2677–2684.

Ataei, S., Azari, P., Hassan, A., Pingguan-Murphy, B., Yahya, R. and Muhamad, F. 2020. Essential oils-loaded electrospun biopolymers: A future perspective for active food packaging. *Advanced Polymer Technology*. Article ID 9040535.

Aydin, F., Kahve, H. I. and Ardic, M. 2017. Lipid-based edible films. *Journal of Scientific and Engineering Research* 9: 86–92.

Barak, S. and Mudgil, D. 2014. Locust bean gum: Processing, properties and food applications. A review. *International Journal of Biological Macromolecules* 66: 74–80.

Cagri, A., Ustunol, Z. and Ryser, E. T. 2004. Antimicrobial edible films and coatings. *Journal of Food Protection* 67: 833–848.

Cao, T. L. and Song, K. B. 2019. Effects of gum karaya addition on the characteristics of loquat seed starch films containing oregano essential oil. *Food Hydrocolloids* 97: 105198.

Cerqueira, M. A., Bourbon, A. I., Pinheiro, A. C., Martins, J. T., Souza, B. W. S., Teixeira, J. A. and Vicente, A. A. 2011. Galactomannans use in the development of edible films/coatings for food applications. *Trends in Food Science & Technology* 22: 662–671.

Chauhan, G., Verma, A., Doley, A. and Ojha, K. 2019. Rheological and breaking characteristics of Zr-cross-linked gum karaya gels for high-temperature hydraulic fracturing application. *Journal of Petroleum Science and Engineering* 172: 327–339.

Chawla, R., Sivakumar, S. and Kaur, H. 2021. Antimicrobial edible films in food packaging: Current scenario and recent nanotechnological advancements-a review. *Carbohydrate Polymer Technologies and Applications* 2: 100024.

Cheng, T., Xu, J., Li, Y., Zhao, Y., Bai, Y., Fu, X. and Mao, X. 2021. Effect of gum ghatti on physicochemical and microstructural properties of biodegradable sodium alginate edible films. *Journal of Food Measurement and Characterization* 15: 107–118.

Chu, Y., Xu, T., Gao, C., Liu, X., Zhang, N. I., Feng, X., Xingxun, L., Xingxun, S. and Tang, X. (2019). Evaluations of physicochemical and biological properties of pullulan-based films incorporated with cinnamon essential oil and Tween 80. *International Journal of Biological Macromolecules* 122: 388–394.

Cocato, M., Maria, R., Messias, N., Helena, C. and Celia, C. 2007. In vitro and in vivo evaluation of iron bio-availability from microencapsulated ferrous sulfate. *Revista de Nutrição* 20: 239–247.

Cottrell, I. W. and Kang, K. S. 1978. Xanthan gum, is a unique bacterial polysaccharide for food applications. *Developments in Industrial Microbiology* 19: 117–131.

Das, S., Vishakha, K., Banerjee, S., Mondal, S. and Ganguli, A. 2020. Sodium alginate-based edible coating containing nanoemulsion of Citrus sinensis essential oil eradicates planktonic and sessile cells of food-borne pathogens and increased quality attributes of tomatoes. *International Journal of Biological Macromolecules* 162: 1770–1779.

Debeaufort, F., Quezada-Gallo, J. A. and Voilley, A. 1998. Edible films and coatings: tomorrow's packagings: A review. *Critical Reviews in Food Science and Nutrition* 38: 299–313.

Dehghani, S., Hosseini, S. V. and Regenstein, J. M. 2018. Edible films and coatings in seafood preservation. A review. *Food Chemistry* 240: 505–513.

Deshmukh, A. S., Setty, C. M., Badiger, A. M. and Muralikrishna, K. S. 2012. Gum ghatti: A promising polysaccharide for pharmaceutical applications. *Carbohydrate Polymers* 87: 980–986.

Dhall, R. K. 2013. Advance in edible coating for fresh fruits and vegetables: A review. *Critical Reviews in Food Science and Nutrition* 53: 435–450.

Dhanapal, A., Sasikala, P., Rajamani, L., Kavitha, V., Yazhini, G. and Banu, M. S. 2012. Edible films from polysaccharides. *Food Science and Quality Management* 3: 9–17.

Diaz-Montes, E. and Castro-Munoz, R. 2021. Edible films and coatings as food-quality preserves: An overview. *Foods* 10: 249.

Dick, M., Costa, T. M. H., Gomaa, A., Subirade, M., Rios, A. D. O. and Flores, S. H. 2015. Edible film production from chia seed mucilage: Effect of glycerol concentration on its physicochemical and mechanical properties. *Carbohydrate Polymers* 130: 198–205.

Dilek, M., Polat, H., Kezer, F. and Korcan, E. 2011. Application of locust bean gum edible coating to extend shelf life of sausages and garlic-flavoured sausage. *Journal of Food Processing and Preservation* 35: 410–416.

Dutta, P. K., Tripathi, S., Mehrotra, G. K. and Dutta, J. 2009. Perspectives for chitosan based antimicrobial films in food applications. *Food Chemistry* 114: 1173–1182.

Ebrahimi, F. and Rastegar, S. 2020. Preservation of mango fruit with guar-based edible coatings enriched with Spirulina platensis and Aloe vera extract during storage at ambient temperature. *Scientia Horticulturae* 265: 109258.

El-Gioushy, S. F., Abdelkader, M. F. M., Mahmoud, M. H., Abou El Ghit, H. M., Fikry, M., Bahloul, A. M. E., Morsy, A. R., Abdelaziz, A. M. R. A., Alhaithloul, H. A. S., Hikal, D. M., Abdein, M. A., Hassan, K. H. A. and Gawish, M. S. 2022. The effects of a gum Arabic-based edible coating on guava fruit characteristics during storage. *Coatings* 12: 90.

Ezeoha, S. L. and Ezenwanne, J. N. 2013. Production of biodegradable plastic packaging film from cassava starch. *IOSR Journal of Engineering* 10: 14–20.

Falguera, V. Quintero, J. P. Jiménez, A., Munoz, J. A. and Ibarz, A. 2011. Edible films and coatings: Structures, active functions and trends in their use. *Trends in Food Science and Technology* 22: 292–303.

Gahruie, H. H., Eskandari, M. H., Khalesi, M., Van der Meeren, P. and Hosseini, S. M. H. 2020a. Rheological and interfacial properties of basil seed gum modified with octenyl succinic anhydride. *Food Hydrocolloids* 101: 105489.

Gahruie, H. H., Eskandari, M. H., Vander Meeren, P. and Hosseini, S. M. H. 2019. Study on hydrophobic modification of basil seed gum-based (BSG) films by octenyl succinate anhydride (OSA). *Carbohydrate Polymers* 219: 155–161.

Gahruie, H. H., Mostaghimi, M., Ghiasi, F., Tavakoli, S., Naseri, M. and Hosseini, S. M. H. 2020b. Effects of fatty acids chain length on the techno-functional properties of basil seed gum-based edible films. *International Journal of Biological Macromolecules* 160: 245–251.

Gahruie, H. H., Ziaee, E., Eskandari, M. H. and Hosseini, S. M. H. 2017. Characterization of basil seed gum-based edible films incorporated with *Zataria multiflora*essential oil nanoemulsion. *Carbohydrate Polymers* 166: 93–103.

Galus, S. and Kadzińska, K. 2015. Food applications of emulsion-based edible films and coatings. *Trends in Food Science and Technology* 45: 273–283.

Ganiari, S., Choulitoudi, E. and Oreopoulou, V. 2017. Edible and active films and coatings as carriers of natural antioxidants for lipid food. *Trends in Food Science & Technology* 68: 70–82.

Gao, S. and Nishinari, K. 2004a. Effect of deacetylation rate on gelation kinetics of konjac glucomannan. *Colloids and Surfaces B: Biointerfaces* 38: 241–249.

Gao, S. and Nishinari, K. 2004b. Effect of degree of acetylation on gelation of konjac glucomannan. *Biomacromolecules* 5: 175–185.

García-Ochoa, F., Santos, V. E., Casas, J. A. and Gómez, E. 2000. Xanthan gum: Production, recovery and properties. *Biotechnology Advances* 18: 549–579.

Gennadios, A. 2002. *Protein-based films and coatings. 1st edition.* Boca Raton, FL: CRC Press. 367–392.

Gennadios, A. and Weller, C. L. 1990. Edible films and coatings from wheat and corn proteins. *Food Technology* 44: 63–69.

Ghosh, T., Nakano, K. and Katiyar, V. 2021. Curcumin doped functionalized cellulose nanofibers based edible chitosan coating on kiwifruits. *International Journal of Biological Macromolecules* 184: 936–945.

Glicksman, M. (Ed.). 1983. *Food hydrocolloids. 1st edition. Gum ghatti (Indian gum).* Boca Raton, FL: CRC Press.

Haddarah, A., Bassal, A., Ismail, A., Gaiani, C., Ioannou, I., Charbonnel, C., Hamieh, T. and Ghoul, M. 2014. The structural characteristics and rheological properties of Lebanese locust bean gum. *Journal of Food Engineering* 120: 204–214.

Hashemi, S. M. B., Mousavi Khaneghah, A., Ghaderi Ghahfarrokhi, M., and Ismail, E. Ş. 2017. Basil seed gum containing *Origanum vulgare* sub sp. *viride* essential oil as edible coating for fresh-cut apricots. *Postharvest Biology and Technology* 125: 26–34.

Huang, Q., Wan, C., Zhang, Y., Chen, C. and Chen, J. 2021. Gum Arabic edible coating reduces postharvest decay and alleviates nutritional quality deterioration of Ponkan fruit during cold storage. *Frontiers in Nutrition* 8: 717596.

Huang, W., Xu, H., Xue, Y., Huang, R., Deng, H. and Pan, S. 2012. Layer-by-layer immobilization of lysozyme—chitosan—organic rectorite composites on electrospun nanofibrous mats for pork preservation. *Food Research International* 48: 784–791.

Islam, A. M., Phillips, G. O., Sljivo, A., Snowden, M. J. and Williams, P. A. 1997. A review of recent developments on the regulatory, structural and functional aspects of gum Arabic. *Food Hydrocolloids* 11: 493–505.

Iurian, S., Dinte, E., Iuga, C., Bogdan, C., Spiridon, I., Barbu-Tudoran, L., Bodoki, A., Tomuta, L. and Leucuţa, S. E. 2017. The pharmaceutical applications of a biopolymer isolated from Trigonella foenum-graecum seeds: Focus on the freeze-dried matrix-forming capacity. *Saudi Pharmaceutical Journal* 25: 1217–1225.

Jahandideh, A., Ashkani, M. and Moini, N. 2021. Biopolymers in textile industries. In *Biopolymers and their industrial applications. 1st edition.* Amsterdam, Netherlands, Elsevier. 193–218.

Janjarasskul, T. and Krochta, J. 2010. Edible packaging materials. *Annual Review of Food Science and Technology* 1: 415–448.

Jansson, P. E., Kenne, L. and Lindberg, B. 1975. Structure of the extracellular polysaccharide from Xanthomonas campestris. *Carbohydrate Research* 45: 275–282.

Jayaprada, M. and Umapathy, M. J. 2020. Preparation and properties of a microfibrillated cellulose reinforced pectin/fenugreek gum biocomposite. *New Journal of Chemistry* 44: 18792–18802.

Jefferies, M., Pass And, G. and Phillips, G. O. 1977. The viscosity of aqueous solutions of gum ghatti. *Journal of the Science of Food and Agriculture* 28: 173–179.

Jiang, J. X., Zhu, L. W., Zhang, W. M. and Sun, R. C. 2007. Characterization of galactomannan gum from fenugreek (*Trigonella foenum-graecum*) seeds and its rheological properties. *International Journal of Polymeric Materials* 56: 1145–1154.

Kamal, I. 2019. Edible films and coatings: Classification, preparation, functionality and applications. A review. *Archive of Organic and Inorganic Chemical Science* 4: 501–510.

Kandar, C. C., Hasnain, M. S. and Nayak, A. K. 2021. Natural polymers as useful pharmaceutical excipients. In *Advances and Challenges in Pharmaceutical Technology.* London: Academic Press. 1–44.

Kang, S., Xiao, Y., Guo, X., Huang, A. and Xu, H. 2021. Development of gum Arabic-based nanocomposite films reinforced with cellulose nanocrystals for strawberry preservation. *Food Chemistry* 350: 129199.

Kapoor, M., Khandal, D., Seshadri, G., Aggarwal, S. and Khandal, R. K. 2013. Novel hydrocolloids: Preparations & applications—A review. *International Journal of Recent Research and Applied Studies* 16: 432–482.

Kasran, M., Cui, S. W. and Goff, H. D. 2013. Covalent attachment of fenugreek gum to soy whey protein isolate through natural Maillard reaction for improved emulsion stability. *Food Hydrocolloids* 30: 552–558.

Koroskenyi, B. and McCarthy, S.P. 2001 Synthesis of acetylated konjac glucomannan and effect of degree of acetylation on water absorbency. *Biomacromolecules* 2: 824–826.

Krochta, J. M., Baldwin, E. A. and Nisperos-Carriedo, M. O. 1994. *Edible coatings and films to improve food quality.* Lancaster, PA: Technomic Publishing. 1–379.

Kuo, M. S., Mort, A. J. and Dell, A. 1986. Identification and location of L-glycerate, an unusual acyl substituent in gellan gum. *Carbohydrate Research* 156: 173–187.

Kurt, A., Toker, O. S. and Tornuk, F. 2017. Effect of xanthan and locust bean gum synergistic interaction on characteristics of biodegradable edible film. *International Journal of Biological Macromolecules* 102: 1035–1044.

Laha, B., Goswami, R., Maiti, S. and Sen, K. K. 2019. Smart karaya-locust bean gum hydrogel particles for the treatment of hypertension: Optimization by factorial design and pre-clinical evaluation. *Carbohydrate Polymers* 210: 274–288.

Lara, G., Yakoubi, S., Villacorta, C.M., Uemura, K., Kobayashi, I., Takahashi, C., Nakajima, M. and Neves, M. A. 2020. Spray technology applications of xanthan gum-based edible coatings for fresh-cut lotus root (Nelumbo nucifera). *Food Research International* 137: 109723.

Lazaridou, A. and Biliaderis, C. G. 2020. Edible films and coatings with pectin. In V. Kontogiorgos (Ed.), *Pectin: Technological and physiological properties.* Cham: Springer. 99–123.

Li, B., Xie, B. and Kennedy, J. F. 2006a. Studies on the molecular chain morphology of konjac glucomannan. *Carbohydrate Polymers* 64: 510–515.

Li, B., Xie, B., and Kennedy, J. F. 2006b. RETRACTED: Studies on the molecular chain morphology of konjac glucomannan. *Carbohydrate Polymers* 64: 510–515.

Li, J. M. and Nie, S. P. 2016. The functional and nutritional aspects of hydrocolloids in foods. *Food Hydrocolloids* 53: 46–61.

Liu, Z., Lin, D., Shen, R., Zhang, R., Liu, L. and Yang, X. 2021. Konjac glucomannan-based edible films loaded with thyme essential oil: Physical properties and antioxidant-antibacterial activities. *Food Packaging and Shelf Life* 29: 100700.

Lopez-Carballo, G., Gomez-Estaca, J., Catala, R., Hernandez-Munoz, P. and Gavara, R. 2012. Active anti-microbial food and beverage packaging. In K. L. Yam and D. S. Lee (Eds.), *Emerging food packaging technologies*. Cambridge: Woodhead Publishing. 27–54.

Lopez-Cordoba, A. and Aldana-Usme, A. 2019. Edible coatings based on sodium alginate and ascorbic acid for application on fresh-cut pineapple (Ananas comosus (L.) Merr). *Agronomia Colombiana* 37: 317–322.

Ma, Q., Du, L., Yang, Y. and Wang, L. 2017. Rheology of film-forming solutions and physical properties of tara gum film reinforced with polyvinyl alcohol (PVA). *Food Hydrocolloids* 63: 677–684.

Ma, Q., Hu, D., Wang, H. and Wang, L. 2016. Tara gum edible film incorporated with oleic acid. *Food Hydrocolloids* 56: 127–133.

Malviya, R., Srivastava, P. and Kulkarni, G. T. 2011. Applications of mucilages in drug delivery. A review. *Advances in Biological Research* 5: 1–7.

McCleary, B. 1985. The fine structures of carob and guar galactomannans. *Carbohydrate Research* 139: 237–260.

McClements, D. J. 2009. Biopolymers in food emulsions. *Modern Biopolymer Science* 4: 129–166.

Memiş, S., Tornuk, F., Bozkurt, F. and Durak, M. Z. 2017. Production and characterization of a new biodegradable fenugreek seed gum-based active nanocomposite film reinforced with nano clays. *International Journal of Biological Macromolecules* 103: 669–675.

Mirhosseini, H. and Amid, B. T. 2012. A review study on chemical composition and molecular structure of newly planted gum exudates and seed gums. *Food Research International* 46: 387–398.

Mohamed, S. A. A., El-Sakhawy, M. and El-Sakhawy, M. A. M. 2020. Polysaccharides, protein and lipid-based natural edible films in food packaging: A review. *Carbohydrate Polymers* 238: 116178.

Moncayo, D., Buitrago, G. and Algecira, N. 2013. The surface properties of biopolymer-coated fruit: A review. *Ingeniería e Investigacion* 33: 11–16.

Mostafavi, F. S., Kadkhodaee, R., Emadzadeh, B. and Koocheki, A. 2016. Preparation and characterization of tragacanth—locust bean gum edible blend films. *Carbohydrate Polymers* 139: 20–27.

Mudgil, D., Barak, S. and Khatkar, B. S. 2014. Guar gum: Processing, properties and food applications. A review. *Journal of Food Science and Technology* 51: 409–418.

Najafi Marghmaleki, S. Mortazavi, S. M., Saei, H. and Mostaan, A. 2021. The effect of alginate-based edible coating enriched with citric acid and ascorbic acid on texture, appearance and eating quality of apple fresh-cut. *International Journal of Fruit Science* 21: 40–51.

Nussinovitch, A. 2009. Biopolymer films and composite coatings. *Modern Biopolymer Science* 10: 295–326.

Oladipupo, Q. A., Yousif, A. I. E. and Babiker. E. E. 2019. Effects of gum arabic edible coatings and sun-drying on the storage life and quality of raw and blanched tomato slices. *Journal of Culinary Science & Technology* 17(1): 45–58.

Olawuyi, I. F., Park, J. J., Lee, J. J. and Lee, W. Y. 2019. Combined effect of chitosan coating and modified atmosphere packaging on fresh-cut cucumber. *Food Science and Nutrition* 7: 1043–1052.

Otoni, C. G., Avena-Bustillos, R. J., Azeredo, H. M. C., Lorevice, M. V., De Moura, M. R., Mattoso, L. H. C. and McHugh, T. H. 2017. Recent advances on edible films based on fruits and vegetables—A review. *Comprehensive Reviews in Food Science and Food Safety* 16: 1151–1169.

Padua, G. W. and Wang, Q. 2002. Formation and properties of corn zein films and coatings. In A. Gennadios (Ed.), *Protein-based films and coatings*. Boca Raton, FL: CRC Press. 43–68

Patel, S. and Goyal, A. 2015. Applications of natural polymer gum Arabic: A review. *International Journal of Food Properties* 18: 986–998.

Pereda, M., Amica, G. and Marcovich, N. E. 2012. Development and characterization of edible chitosan/olive oil emulsion films. *Carbohydrate Polymers* 87: 1318–1325.

Petri, D. F. 2015. Xanthan gum: A versatile biopolymer for biomedical and technological applications. *Journal of Applied Polymer Science* 132: 42035.

Pollard, M. A. and Fischer, P. 2006. Partial aqueous solubility of low-galactose-content galactomannans. *Current Opinion in Colloid & Interface Science* 11: 184–190.

Pooja, D., Panyaram, S., Kulhari, H., Reddy, B., Rachamalla, S. S. and Sistla, R. 2015. Natural polysaccharide functionalized gold nanoparticles as biocompatible drug delivery carriers. *International Journal of Biological Macromolecules* 80: 48–56.

Prajapati, V. D., Jani, G. K., Moradiya, N. G. and Randeria. N. P. 2013a. Pharmaceutical applications of various natural gums, mucilages and their modified forms. *Carbohydrate Polymers* 92: 1685–1699.

Prajapati, V. D., Jani, G. K., Moradiya, N. G., Randeria, N. P. and Nagar, B. J. 2013b. Locust bean gum: A versatile biopolymer. *Carbohydrate Polymers* 94: 814–821.

Prajapati, V. D., Jani, G. K., Zala, B. S. and Khutliwala, T. A. 2013c. An insight into the emerging exopolysaccharide gellan gum as a novel polymer. *Carbohydrate Polymers* 93: 670–678.

Preetha, B. K. and Vishalakshi, B. 2020. Microwave-assisted synthesis of karaya gum-based montmorillonite nanocomposite: Characterization, swelling and dye adsorption studies. *International Journal of Biological Macromolecules* 154: 739–750.

Priyadarshi, R., Riahi, Z. and Rhim, J. W. 2022. Antioxidant pectin/pullulan edible coating incorporated with Vitis vinifera grape seed extract for extending the shelf life of peanuts. *Postharvest Biology and. Technology* 183: 111740.

Pszczola, D. E. and Banasiak, K. 2006. Enter IFT's magic ingredient kingdom. *Food Technology-Chicago* 60: 45–94.

Raghav, P. K., Agarwal, N. and Saini, M. 2016. Edible coating of fruits and vegetables: A review. *International Journal of Scientific Research and Modern Education* 1: 188–204.

Ramos, O. L., Fernandes, J. C., Silva, S. I., Pintado, M. E. and Malcata, F. X. 2012. Edible films and coatings from whey proteins: A review on formulation, and on mechanical and bioactive properties. *Critical Reviews in Food Science and Nutrition* 52: 533–552.

Rana, V., Rai, P., Tiwary, A. K., Singh, R. S., Kennedy, J. F. and Knill. C. J. 2011. Modified gums: Approaches and applications in drug delivery. *Carbohydrate Polymers* 3: 1031–1047.

Rao, M. S., Kanatt, S. R., Chawla, S. P. and Sharma, A. 2010. Chitosan and guar gum composite films: Preparation, physical, mechanical and antimicrobial properties. *Carbohydrate Polymers* 82: 1243–1247.

Ray, A. K., Bird, P. B., Iacobucci, G. A. and Clark Jr, B. C. 1995. Functionality of gum Arabic. Fractionation, characterization and evaluation of gum fractions in citrus oil emulsions and model beverages. *Food Hydrocolloids* 9: 123–131.

Rhim, J. W. and Shellhammer, T. H. 2005. Lipid-based edible films and coatings. In J. H. Han (Ed.), *Innovations in food packaging*. San Diego, CA: Elsevier Academic Press. 362–383.

Rigano, L., Deola, M., Zaccariotto, F., Colleoni, T. and Lionetti, N. 2019. A new gelling agent and rheology modifier in cosmetics: Caesalpinia spinosa gum. *Cosmetics* 6: 34.

Romanazzi, G., Feliziani, E., Baños, S. B. and Sivakumar, D. 2017. Shelf life extension of fresh fruit and vegetables by chitosan treatment. *Critical Reviews in Food Science and Nutrition* 57: 579–601.

Rosalam, S. and England, R. 2006. Review of xanthan gum production from unmodified starches by *Xanthomonas campestris* sp. *Enzyme and Microbial Technology* 39: 197–207.

Saberi, B., Thakur, R., Vuong, Q. V., Chockchaisawasdee, S., Golding, J. B., Scarlett, C. J. and Stathopoulos, C. E. 2016. Optimization of physical and optical properties of biodegradable edible films based on pea starch and guar gum. *Industrial Crops and Products* 86: 342–352.

Salarbashi, D., Bazeli, J. and Fahmideh-Rad, E. 2019. Fenugreek seed gum: Biological properties, chemical modifications, and structural analysis—A review. *International Journal of Biological Macromolecules* 138: 386–393.

Salehi, F. 2020. Edible coating of fruits and vegetables using natural gums: A review. *International Journal of Fruit Science* 20: S570–S589.

Salehi, F. and Kashaninejad, M. 2014. Kinetics and thermodynamics of gum extraction from wild Sage seed. *International Journal of Food Engineering* 10: 625–632.

Salehi, F. and Kashaninejad, M. 2015. Static rheological study of ocimum basilicum seedgum. *International Journal of Food Engineering* 11: 97–103.

Salehi, F., Kashaninejad, M. and Behshad, V. 2014. Effect of sugars and salts on rheological properties of Balangu seed (Lallemantia royleana) gum. *International Journal of Biological Macromolecules* 67:16–21.

Sandford, P. A. 1979. Exocellular, microbial polysaccharides. *Advances in Carbohydrate Chemistry and Biochemistry* 36: 265–313.

Sharma, S. and Rao, T. R. 2015. Xanthan gum based edible coating enriched with cinnamic acid prevents browning and extends the shelf-life of fresh-cut pears. *LWT-Food Science and Technology* 62: 791–800.

Shimahara, H., Suzuki, H., Sugiyama, N. and Nisizawa, K. 1975. Isolation and characterization of oligosaccharides from an enzymic hydrolysate of konjac glucomannan. *Agricultural and Biological Chemistry* 39: 293–299.

Silva, S. C. C. C., de Araujo Braz, E. M., de Amorim Carvalho, F. A., de Sousa Brito, C. A. R., Brito, L. M., Barreto, H. M. and da Silva, D. A. 2020. Antibacterial and cytotoxic properties from esterified Sterculia gum. *International Journal of Biological Macromolecules* 164: 606–615.

Sofi, S. A., Singh, J., Rafiq, S., Ashraf, U., Dar, B. N. and Nayik, G. A. 2018. A comprehensive review on antimicrobial packaging and its use in food packaging. *Current Nutrition & Food Science* 14: 305–312.

Soma, P. K., Williams, P. D. and Lo, Y. M. 2009. Advancements in non-starch polysaccharides research for frozen foods and microencapsulation of probiotics. *Frontiers of Chemical Engineering in China* 3: 413–426.

Stauffer, C. E. 1994. Frozen bakery products. In C. P. Mallett (Ed.), *Frozen food technology*. Cambridge: Chapman & Hall.

Sugiyama, N., Shimahara, H., Andoh, T., Takemoto, M. and Kamata, T. 1972. Molecular weights of konjac mannans of various sources. *Agricultural and Biological Chemistry* 36: 1381–1387.

Swathi, S.N.C., Chiara, C., Federica B., Angelo, F. and Marco D. R. 2019. Evaluation of drying of edible coating on bread using NIR spectroscopy. *Journal of Food Engineering* 240: 29–37.

Tahir, H. E., Xiaobo, Z., Jiyong, S., Mahunu, G. K., Zhai, X. and Mariod, A. A. 2018. Quality and postharvest-shelf life of cold-stored strawberry fruit as affected by gum Arabic (Acacia senegal) edible coating. *Journal of Food Biochemistry* 42: 12527.

Takigami, S. 2000. Konjac mannan. In *Handbook of hydrocolloids*. Boca Raton, FL: CRC/Woodhead Publishing. 413–424.

Tavassoli-Kafrani, E., Gamage, M. V., Dumee, L. F., Kong, L. and Zhao, S. 2020. Edible films and coatings for shelf life extension of mango: A review. *Critical Reviews in Food Science and Nutrition* 62: 1–26.

Thombare, N., Jha, U., Mishra, S. and Siddiqui, M. Z. (2016). Guar gum as a promising starting material for diverse applications: A review. *International Journal of Biological Macromolecules* 88: 361–372.

Tischer, C. A., Iacomini, M., Wagner, R. and Gorin, P. A. 2002. New structural features of the polysaccharide from gum ghatti (Anogeissus latifola). *Carbohydrate Research* 337: 2205–2210.

Trezza, T. A. and Krochta, J. M. 2002. Application of edible protein coatings to nuts and nut-containing food products. In A. Gennadios (Ed.), *Protein-based films and coatings*. Boca Raton, FL: CRC Press LLC. 527–550.

Valli, R. and Clark, R. 2009. Gellan gum. In A. Imeson (Ed.), *Food stabilizers, thickeners and gelling agents*. Chichester: Blackwell Publishing Ltd. 145–166.

Vargas, M., Pastor, C., Chiralt, A., McClements, J. and Gonzalez-Martinez, C. 2008. Recent advances in edible coatings for fresh and minimally processed fruits. *Critical Reviews in Food Science and Nutrition* 48: 496–511.

Viebke, C. and Piculell, L. 1996. Adsorption of galactomannans onto agarose. *Carbohydrate Polymers* 29: 1–5.

Williams, P. A. and Phillips, G. O. 2000. Introduction to food hydrocolloids. In G. O. Phillips and P. A. Williams (Eds.), *Handbook of hydrocolloids*. New York: CRC Press. 1–19.

Wu, Y., Ding, W., Jia, L. and He, Q. 2015. The rheological properties of tara gum (*Caesalpinia Spinosa*). *Food Chemistry* 168: 366–371.

Wu, Y., Li, W., Cui, W., Eskin, N. A. M. and Goff, H. D. 2012. A molecular modeling approach to understand conformation—functionality relationships of galactomannans with different mannose/galactose ratios. *Food Hydrocolloids* 26: 359–364.

Xiao, M., Luo, L., Tang, B., Qin, J., Wu, K. and Jiang, F. 2022. Physical, structural, and water barrier properties of emulsified blend film based on konjac glucomannan/agar/gum Arabic incorporating virgin coconut oil. *LWTFood Science and Technology* 154: 112683.

Xu, X., Li, B., Kennedy, J. F., Xie, B. J. and Huang, M. 2007. Characterization of konjac glucomannan—gellan gum blend films and their suitability for release of nisin incorporated therein. *Carbohydrate Polymers* 70: 192–197.

Youssef, A. R. M., Ali E. A. A. and Emam, H. E. 2015. Influence of postharvest application of some edible coating on storage life and quality attributes of novel Orange fruits during cold storage. *International Journal of ChemTechResearch* 8: 2189–2200.

Youssef, M. K., Wang, Q., Cui, S. W. and Barbut, S. 2009. Purification and partial physicochemical characteristics of protein-free fenugreek gums. *Food Hydrocolloids* 23: 2049–2053.

Yousuf, B., Wu, S. and Gao, Y. 2021. Characteristics of karaya gum based films: Amelioration by inclusion of *Schisandra chinensis* oil and its oleogel in the film formulation. *Food Chemistry* 345: 128859.

Zhang, P., Zhao, Y. and Shi, Q. 2016. Characterization of a novel edible film based on gum ghatti: Effect of plasticizer type and concentration. *Carbohydrate Polymers* 153: 345–355.

Zheng, M., Chen, J., Tan, K. B., Chen, M. and Zhu, Y. 2022. Development of hydroxypropyl methylcellulose film with xanthan gum and its application as an excellent food packaging bio-material in enhancing the shelf life of bananas. *Food Chemistry* 374: 131794.

Zhong, Y., Godwin, P., Jin, Y. and Xiao, H. 2020. Biodegradable polymers and green-based antimicrobial packaging materials: A mini-review. *Advanced Industrial and Engineering Polymer Research* 3: 27–35.

Zhou, N., Zheng, S., Xie, W., Cao, G., Wang, L. and Pang, J. 2022. Konjac glucomannan: A review of structure, physicochemical properties, and wound dressing applications. *Journal of Applied Polymer Science* 139: 51780.

5

Development of Films and Coatings from Alginates

Pinal K. Dave, T. V. Ramana Rao, and V. R. Thakkar

CONTENTS

DOI: 10.1201/9781003303671-5

5.1 Introduction

Alginate is a non-digestible polymer found in a variety of brown seaweeds (Phaeophyceae), including *Macrocystis pyrifera*, *Laminaria hyperborean*, *Laminaria digitate*, *Ascophyllum nodosum*, *Laminaria japonica*, *Lesonianegrescens*, *Eclonia maxima*, and *Sargassum* sp. (Skurtys et al., 2014). Bacteria like *Azotobacter vinelandii* and *Pseudomonas aeruginosa mucoid* strains may also create alginate-like polymers, and they produce alginate as bacterial exopolysaccharides (Emmerichs et al., 2004). As shown in Table 5.1, there is a vast difference between the structure of bacterial and algal alginates. Alginate is a water-soluble white or yellow-colored fibrous powder substance with excellent adhesion and filming capabilities (Raghav et al., 2016). Alginates are non-toxic, biodegradable, and naturally occurring since they are made from sea algae. A hydrocolloid is a substance made up of water and water-absorbing colloidal substances, which are big, water-soluble molecules that reduce viscosity and are used as a texturizer (Lee and Rogers, 2012). Hydrocolloids aid in the absorption of dietary fiber.

Nowadays, edible films and coatings made from alginate are gaining popularity because they improve fruit properties, reduce weight loss, improve appearance and color, maintain quality, extend shelf life, inhibit microbial growth (including yeast, mold, *E. coli*, and *Listeria*), and also act as an antioxidant and anti-browning agent (Hershko and Nussinovitch, 1998).

5.2 Extraction of Alginate from Seaweed

Brown seaweeds possess alginate in their cell walls (Chee et al., 2011). The collected biomass (seaweed) will be washed with tap water, dried in the shade (except for *M. pyrifera*, which is treated while wet), and then they will be chopped into minute pieces before being subjected to further process (Gombotz and Wee, 1998). After removal or degradation of homopolysaccharides such as laminarin and fucoidan with dilute mineral acid, alginate is recovered from dried and milled algal material. Alkaline Earth cations and H+ are swapped out at the same time. Adding sodium carbonate to the insoluble protonated alginate at a pH below 10 converts it to the soluble sodium salt. Before being turned to salt or acid, alginate might be further treated (Fertah, 2017).

5.3 Derivatives of Alginates

The most common kind of alginate is sodium alginate. Alginic acid, calcium, ammonium, and potassium salts, as well as an ester, propylene glycol alginate, are formed in smaller amounts (McHugh, 1987). During the calcium alginate process, alginic acid and calcium alginate are produced; both may be extracted at an appropriate stage, dried, and milled to the desired particle size after thorough washing. The other salts are made by neutralizing wet alginic acid with an alkali, most often ammonium hydroxide or potassium carbonate; water or alcohol may be added to keep the product workable, and they are processed similarly to sodium alginate.

TABLE 5.1

Difference between bacterial alginate and algal alginate

Bacterial alginate	Algal alginate
They have higher molecular weight.	Low molecular weight compared to bacterial alginate.
O-acetyl group is present.	O-acetyl group is absent.
The oligomeric guluronic sequences are absent.	Alternating sequences of M and G and homopolymeric monomers are present.

Source: Sachdeva et al. (2020).

Propylene glycol alginate, an ester of alginic acid, differs from sodium alginate in terms of characteristics and usage. It was initially patented in 1947, and as manufacturing processes improved, further patents were issued. It's created by combining liquid propylene oxide with wet alginic acid (20% or more solids) that's been partly reacted with sodium carbonate in a pressure vessel for 2 hours at about 80°C. The finished product is ready once it has been dried and milled (McHugh, 1987).

5.4 Preparation of Films and Coatings from Alginates

5.4.1 Cross-Linking

From film creation through coating manufacture, Parreidt et al. (2018a) presented the whole cross-linking procedure utilizing alginate. Alginate is a kind of polyuronide, and it is a well-known natural ion exchanger (Yoo and Krochta, 2011). The advantage of natural ion-exchange feature of alginate is that its charged state assists in image production. Only if bivalent ions are lacking that alginate may be used to increase viscosity (Lee and Rogers, 2012). When a bivalent cation is introduced to an alginate solution, ion exchange occurs, resulting in the formation of a gel (Lu et al., 2006). The affinity of alginate for alkaline earth metals rises in this order: Ca^{2+}, Sr^{2+}, and Ba^{2+} (Kohn, 1975). Monovalent cations cannot gel with Mg^{2+} ions (Sutherland, 1991). Pb^{2+}, Cu^{2+}, Cd^{2+}, Co^{2+}, Ni^{2+}, Zn^{2+}, and Mn^{2+} can all gel, but their toxicity limits their application. Making an alginate gel is a challenging process. The number and length of guluronic acid blocks (G-blocks) in the polymeric chain, the capacity to bind the number of divalent ions, the kind of gelling ions, and the gelling conditions all influence the hydrogel properties of alginate (Soazo et al., 2015). Alginate exhibits confirmation changes, such as G-block alignment and the creation of the egg-box structure when divalent cations (usually Ca^{2+} ions) are added to the solution (Grant, 1973). When calcium ions are bound between two chains, divalent salt bridges form (Tapia et al., 2008).

In a linear binary copolymer, alginates are glycosidic linkages (Figure 5.1) that link the residues of -D-mannuronic acid (M) and -L-guluronic acid (G) (Nisperos-Carriedo, 1994). G-blocks make stiff, thick gels, while M-blocks make porous, flexible gels (Kierstan, 1982). As a consequence, in gels with a high polyguluronic alginate concentration, high-molecular-weight molecules have a high diffusional resistance. According to Olivas and Barbosa-Canovas (2008), films with higher G-block proportions (and hence a lower M/G ratio) have better moisture barrier qualities. External and internal gelling modes (Papajova et al., 2012) are two methods for incorporating gelling ions into an alginate solution to produce a hydrogel. (1) External gelation (the traditional method) entails exposing an alginate solution to a solution of gelling ions, where Ca^{2+} binds quickly with the carboxylic groups of guluronic acid residues, resulting in an irreversible hydrogel (Liu et al., 2002). (2) The internal gelling approach, entails mixing an insoluble source of gelling ions with alginate solutions and then releasing the gelling ions by

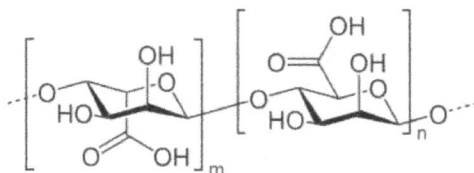

FIGURE 5.1 The supply of calcium ions influences gel formation (calciumchloride, calciumlactate, calcium gluconate, calcium nitrate, and calcium propionate). Brownlee et al. (2005) discovered that cross-linking calcium chloride with calcium gluconate, calcium nitrate, or calcium propionate made alginate gels stronger. Because the rate of gelation and the concentration of Ca^{2+} are inversely proportional, calcium supply has an impact on gelation kinetics (Lee and Rogers, 2012). Calcium chloride (75 g/100 mL), calcium lactate (8 g/100 mL), and calcium gluconate (3 g/100 mL) are the three most soluble calcium compounds at 20°C (Soazo et al., 2015). The first calcium salts to acquire steady-state gel strength were calcium chloride, calcium lactate, and calcium gluconate (Soazo et al., 2015). In a study conducted by Chrastil (1991), it was discovered that the gelation kinetic constants are unaffected by calcium availability. The calcium delivery mechanism has no impact on the gel's strength or resistance to diffusion of calcium.

pH-lowering processes, such as organic acid addition or gradual lactone hydrolysis. Internal gelation creates more homogenous, less-dense gel matrices with larger pore widths than external gelation because Ca2+ is replaced by H+ when acid is injected (Chan et al., 2006). According to Mancini and McHugh (2000), the third (3) method of starting controlled alginate gelation is to chill the heated solution, which comprises all the components. Due to the thermal energy of the alginate solution, calcium-induced hydrogel synthesis can only occur after cooling.

Despite its high solubility, calcium chloride is a poor calcium source because it imparts a harsh flavor to food (Lee and Rogers, 2012). In coatings where flavor is a consideration, calcium gluconate and calcium lactate may be used. Alginate gelation in situ and the creation of extremely homogeneous structures are also of interest to the biotechnology industry (e.g., tissue engineering, immobilization of cell and enzyme systems, carriers for drug delivery system). It has been reported how to make homogeneous alginate gels with delayed calcium ion release (Draget, 1989). Kuo and Ma (2001) used calcium carbonate D-glucono-lactone ($CaCO_3$ -GDL) or calcium sulfate dihydrate ($CaSO4.2H2O$ -$CaCO_3$ -GDL) as a gelation agent because of their low solubilities, which allowed for more controlled and homogeneous gel formation. The gel's dimensional stability (swelling tendency due to an increased tendency of the carboxyl and hydroxyl groups to interact with water molecules and osmotic pressure) was controlled by adjusting the calcium ion concentration in the external aqueous environment, cross-linking density, polymer concentration, and chemical composition of alginate (Kuo and Ma, 2008). When an alginate film or coating is immersed in a cross-linking calcium solution, two types of reactions occur: diffusion of the multivalent ion, which results in formation of calcium linkage with the carboxyl groups, resulting in insolubilization of the alginate film, and dissolution of alginate by the solution, as Pavlath et al. pointed out (1999). Because of the larger concentration of the multivalent ion, the dissolving process loses its supremacy. Rhim (2004) observed a similar effect, noting that the method of $CaCl_2$ treatment affected the thickness of the film that is, immediately applying the cross-linking agent to the alginate films resulted in thicker films, rather than soaking the alginate films in a cross-linking solution resulting in thicker films. Pavlath et al. (1999) reported that mixing at 20°C yielded an immediate gel that could not be cast, mixing at 50°C yielded a thick solution that could be poured into the frames.

The capacity of the alginate's 3D structure to expand in the solvent is affected by the degree of cross-linking, resulting in lower permeability to various solutes and usage in pharmaceutical controlled-release systems (Zactiti and Kieckbusch, 2009). They also discovered that when the concentration of Ca+2 ions in the cross-linking solution grew, the degree of swelling decreased (i.e., cross-linking increased), reducing alginate film solubility and elongation while improving tensile strength. Increasing $CaCl_2$ concentration enhanced tensile strength while decreasing percent elongation at break, according to Rhim (2004). According to research on the effects of calcium chloride dipping alone (without coating), calcium has the capacity to bind cell wall polymers, retain structure, and reduce the water solubility of pectic compounds by forming calcium pectate (Howard, 1994). The amount of $CaCl_2$ in the solution increases the hardness, whereas the dipping duration has no effect (Luna-Guzman, 1999).

5.4.2 Synthesis of Biofilms from Alginates

According to Castro-Yobal et al.'s (2021) publications, there are three main procedures for producing films using alginates.

Three steps were used to create the biofilms. An alginate matrix, glycerol (a plasticizer), oleic acid (a controller of hydrophilicity/hydrophobicity), and calcium chloride ($CaCl_2$) were among the ingredients (a cross-linking agent). Each component was introduced at a varied concentration in each example, and the Teflon molds used had a diameter of 15 cm. At room temperature, biofilms were collected and sealed in bags.

5.4.2.1 Biofilms (Alginate-Glycerol) Added with Oleic Acid without Cross-Linking

The polymer matrix solution was produced by dissolving 1% sodium alginate in distilled water, heating at 70°C for 20 minutes, and continually stirring. Then 0.25%, 0.5%, and 1% (v/v) glycerol and 0.25%

(v/v) oleic acid were added and stirred for 20 minutes at 600 rpm. The slurry was put into the molds to make the films when the waiting period had passed. After that, the films were dried for 12 hours on a drying stove at temperatures ranging from 40 to 45°C.

5.4.2.2 Biofilms (Alginate-Glycerol) without Addition of Oleic Acid and Cross-Linking

$CaCl_2$ was used to cross-link the films in this instance. The films were retained in the molds under the same circumstances as before, but without the use of oleic acid, after they had been formed and dried. They were then cross-linked for 2 minutes with 0.75% and 1% (v/v) $CaCl_2$ solutions. They were taken from the mold when the time had passed, and dried in a drying oven for 5 minutes at 40°C.

5.4.2.3 Biofilms (Alginate-Glycerol) Added with Oleic Acid and Cross-Linking

To make these films, oleic acid was mixed into a sodium alginate solution, and glycerol was added during the stirring stage at 50°C, where it was constantly swirled for 4 hours. After 4 hours, the gel-filled container was put in a desiccator, and a vacuum of 51 kPa was applied for 15 minutes to remove any bubbles that had formed during component mixing. After that, it was put into circular molds and allowed to solidify for 12 hours at 38°C. The $CaCl_2$ solution was added when the films were dry, and the cross-linking agent and the films were allowed in contact for an additional minute. The films were ultimately taken from the mold after drying for 5 minutes at 40°C.

5.5 Use of Additives to Make Films and Coatings from Alginates

The inclusion of different natural or artificial additives may impact the mechanical, functional, organoleptic, and nutritional characteristics of alginate films and coatings (Parreidt et al., 2018b), which are listed in what follows.

5.5.1 Nutraceuticals and Nutritional Improvements, Flavors, Pigments

Nutraceuticals are health supplements that improve the nutritional content of the coated food product by including them in the coating/film composition (Baldwin et al., 2011). Zinc, vitamin E, calcium, and other nutraceuticals are used in the coating/film composition. Vitamin E is used to fight cancer, arthritis, diabetes, cataracts, cardiovascular disease, and Alzheimer's disease. The vitamin E content of sunflower oil is high. It also contains a significant amount of unsaturated and polyunsaturated fats (EFSA, 2005). Calcium is a common intracellular messenger as well as a cofactor for extracellular enzymes and proteins; thus, it's included. It's also important for bone and tooth formation (Skurtys et al., 2014). Carotene (-carotene) is a food and dietary supplement. It's also a dye (Garcia et al., 2016). Anthocyanin is a naturally occurring pigment with antibacterial properties and the potential to lower cancer and cardiovascular disease risks. Essential oils add flavor and odor to the food. Menthol is also used as a flavor that may be employed (Embuscado et al., 2009). The sensory attributes of coated food is enhanced by added color and flavor (Skurtys et al., 2014). Vegetable oils are available in a variety of sizes and forms, each with its own set of health advantages.

5.5.2 Antioxidants and Antimicrobials

Antioxidants are essential in the manufacture of alginates coatings, according to Sachdeva et al. (2021). Degradation, oxidative rancidity, discoloration, dehydration, and oil diffusion are all prevented by using antioxidants in the coating material. Antioxidants increase food shelf life, stability, quality, and nutritional value (Garcia et al., 2016). Because oxidation is a surface–air phenomena, antioxidants work better on the product's surface (Embuscado and Huber, 2009). Antioxidants are alkaline or phenolic chemicals. Antioxidants such as butylated hydroxytoluene (BHT), tert-butylhydroquinone (TBHQ), butylated hydroxyanisole (BHA), propyl gallate (PG), and tocopherols protect lipids from being oxidized (Shahidi, 2000). Some of the antioxidants included in this tea include ascorbic acid, pomegranate peel extract, citric acid, rosemary oleoresin, horsemint essential oil, clove essential oil, tea polyphenols, and lycopene.

In contrast to their direct application to food items, adding antibacterial and antioxidant chemicals to edible coatings and films allows them to slowly release and retain a critical concentration for a longer time (Appendini et al., 2002). Antimicrobials added directly to food, rather than those that migrate from the coating, will reduce bacteria quickly, but the repair of wounded cells and subsequent reproduction of undestroyed cells may result in quality losses and/or foodborne diseases (Quintavalla and Vicini, 2002).

Antimicrobial agents have been used in a variety of combinations in edible films and coatings made from alginate. Alginate might be used as a substrate for antimicrobials to reduce the microbial load on coated foods, according to the findings. Essential oils (Eos) have been studied extensively for their antibacterial characteristics, mechanism of action, and prospective applications (Burt and Reinders, 2003). To add antibacterial qualities to edible films and coatings, several natural preservatives have been employed. Their unpleasant taste (abietane diterpenes, carnosol, and ursolic acid) discourages people from using those (Cagri et al., 2004).

5.5.3 Anti-Browning Agents

Cutting fruits and vegetables may modify their look and color in unfavorable ways (Baldwin et al., 2011), and as a result brown, crimson, or black coloration develops in the fruit or vegetable (Rocha and Morais, 2002). Color is highly significant in product marketing, as appearance plays a key role for the customer. The oxidation of phenolic chemicals, both enzymatic and non-enzymatic, is responsible for the change in color (Martinez and Whitaker, 1995). Polyphenol oxidase is an enzyme that converts phenolic chemicals into very unstable quinones, which are then polymerized to produce the red, brown, and black colors (Dong et al., 2000). The activity of polyphenol oxidase rises when a food item is peeled and sliced. The sensory characteristics of the food are harmed by browning (Martinez and Whitaker, 1995). The amount of active polyphenol oxidase present, the amount of phenolic compounds present, the temperature, pH, and oxygen availability all impact the rate of browning. Because polyphenol oxidase requires oxygen to initiate the enzymatic activity that causes browning, an oxygen barrier must be present to prevent browning.

When phenols and heavy metals come into contact, browning might develop (Rocha and Morais, 2002). Anti-browning chemicals are employed in edible coatings and film material because it efficiently transports the agent. In food storage, transportation, and distribution, edible films and coatings inhibit oxygen diffusion due to mechanical damage (Garcia et al., 2016). In most cases, anti-browning agents are included into the cross-linking solution. Ascorbic acid, onion extracts, erythorbic acid, L-cysteine, 4-hexylresorcinol, and sucrose ester are some of the most often used anti-browning substances (Baldwin et al., 2011). Sulfite is also used to prevent food browning, both enzymatic and non-enzymatic.

5.5.4 Plasticizers

Plasticizer is important for preparing coatings, mostly for polysaccharide- and protein-based coatings, as the structure of such coatings is hard and rigid due to extensive connections between polymer molecules (Krochta, 2002). In addition to refining mechanical properties of the coating solutions, the plasticizer also affects the resistance of coatings to the permeation of gases and vapors (Sothornvit and Krochta, 2001), where the hydrophilic plasticizers usually improve the water vapor permeability of the coatings.

Plasticizers are used to increase the free volume or molecular mobility of polymers, decrease intermolecular tensions, provide flexibility, reduce brittleness, improve tear impact resistance, and control the flow of the coating material (Rojas-Grau et al., 2007). Water is the most frequent and effective plasticizer, yet owing to the sensitivity of water's plasticizing effects on hydrophilic biopolymers to environmental factors such as relative humidity and temperature, it is difficult to acquire (Guilbert et al., 1995). In food coating research, plasticizers such as glycerol, sorbitol, acetylated monoglyceride, polyethylene glycol, sucrose, and others have been utilized in addition to water (Jost et al., 2014). Hydrophilic plasticizers enhance the coating's water vapor permeability (WVP) and change its mechanical characteristics. As a result, the kind and amount of plasticizer utilized in the formulation of an edible coating are critical. According to Parris et al. (1995), lower doses induce brittleness, whereas greater doses produce stickiness. In contrast to sorbitol, which was stiffer, glycerin and sodium lactate produced stronger and more elastomeric alginate-based films. Because it is poor at inhibiting intermolecular hydrogen bonding between polymer molecules, sorbitol-enhanced films have better water vapor barrier characteristics at the same concentration. The impact of glycerol and sorbitol on the mechanical characteristics of alginate films were explored by Jost et al. (2014), who discovered that both plasticizers reduced equilibrium moisture content and porosity. Glycerol improved WVP and oxygen permeability, whereas sorbitol had little effect. Olivas and Barbosa-Canovas (2008) investigated the effects of a variety of plasticizers (glycerol, sorbitol, polyethylene glycol (PEG), and fructose) on the WVP and mechanical properties of calcium alginate films at two different RH values. Fructose-, sorbitol-, glycerol-, and PEG-embedded films improved the order of WVP. Plasticizers caused a rapid rise in moisture content and a change in mechanical characteristics by raising tensile strength in the moisture sorption isotherm graphs.

Glycerol has been utilized as a plasticizer in a number of studies, the most notable of which is in alginate films and coatings. The quantity utilized, on the other hand, might vary. The water vapor resistance (WVR) of alginate coatings rises with increasing glycerol content in the formulation up to 1.75% (v/v), according to Rojas-Grau et al. (2007); however, the WVR decreases at higher glycerol concentrations. When optimizing an alginate coating formulation, Azarakhsh et al. (2012) saw a similar impact and calculated the quantity to be 1.16% (w/v). According to Tapia et al. (2008), glycerol concentrations over 1.5% (w/v) reduced WVR. When designing the formulations, edible coatings with outstanding wettability and consistent spreading ability on the target food product are required (Ribeiro et al., 2007). As a result, the components' effects on the surface tension of the coating solution are vital. Because they are not tensio-active chemicals, glycerol and sorbitol have no influence on the surface tension of solutions (Senturk et al., 2018). To combat the brittleness of alginate films and coatings, various attempts mixed multiple plasticizers. Glycerol, palmitic acid, -cyclodextrin, and glycerolmonostearate were the film components chosen by Fan et al. (2009). Su Cha et al. (2002) employed a 1:1 mixture of glycerol and polyethylene glycol.

5.5.5 Surfactants

The coating solution and the solid surface of fruits and vegetables must be compatible in order for an edible coating solution to function. To minimize interfacial tension and improve adhesion, the coating solution's surface tension must be lowered to match the lower surface tension of the surface. Surface-active ingredients in the film-forming solution, such as emulsifiers and other amphiphilic compounds, reduce the surface tension of the coating solution, reducing the difference between the solid surface energy and the surface tension of the coating solution, thereby enhancing adhesion work (Han and Gennodio, 2005).

Because of low surface free energy, adhesion to hydrophobic, uneven surfaces, and homogenous food coatings might be challenging (Pavlath and Orts, 2009). Surfactants (surface-active agents) are required to improve a product's wettability and coating material adherence (Senturk et al., 2018). Surfactants and emulsifiers exhibited a reduced rate of moisture loss in coating formulations because their surface water activity was lowered (Baldwin et al., 1995). Surfactants are found in greater numbers on the surface of interfaces (liquid–air, liquid–liquid, and liquid–solid) than in the liquid itself (Porter, 1991). The charge type of the surface-active component and the chemical structure of the hydrophilic groups are two factors that may be used to classify surfactants. There are four different kinds of ions: anionic (negatively charged), non-ionic (uncharged), cationic (positively charged), and amphoteric (uncharged) (may be positively or negatively charged, or both, depending on the conditions). Surfactants are divided into four categories. An important efficacy indication is the ability of an edible coating to distribute uniformly across the target product. As a result, while developing coating formulations, researchers took into account food surface free energy, edible coating surface tension, and spreading coefficient. Senturk et al. (2018) investigated alginate-based coating formulations with various surfactant concentrations and types (0–5% Tween 40, Tween 80, Span 80, Span 60, and soy lecithin).

5.6 Application Methods

The physical characteristics of the final product are heavily influenced by the film formation methods and coating process conditions. To maximize the film's functions, it must be homogenous and defect-free (no air bubbles or mechanical damage) (Skurtys et al., 2010).

5.6.1 Mechanisms of Film Formation

Following is a list of the factors that contribute to the formation of edible films (Parreidt et al., 2018a).

Simple coacervation: (1) solvent evaporation (drying); (2) addition of a hydrosoluble non-electrolyte (in which the hydrocolloid is insoluble); and (3) pH adjustment with the addition of electrolyte, which activate salting out or cross-linking.

Polymer complex coacervation: A polymer complex is formed by mixing two hydrocolloid solutions with opposite electron charges.

A *macromolecule* (e.g., ovalbumin proteins) is dissolved or a hydrocolloid dispersion is chilled to produce precipitation or gelation (e.g., agar, gelatin). Despite environmental changes, shelf-standing film is made using the same processes as thermoplastic film: solvent casting and extrusion (Skurtys et al., 2010).

5.6.2 Coating

The application technique and the coating's capacity to attach to the food surface are two of the most important features of edible coatings (Dhanapal et al., 2012).

5.6.2.1 Dipping Technique

The method's main advantage is that it may be used on any surface, even the most difficult or abrasive ones (Andrade, 2012). Because the sliced surfaces of food items are hydrophilic, immersing them in the coating solution will not provide adequate adhesion. To minimize moisture loss and gas transfer, food is often coated by dipping or spraying a thin coating on the surface that acts as a semipermeable barrier (Lin and Zhao, 2007). The sample is removed and the surplus film-forming solution covering the product is drained after soaking for a few minutes in alginate dispersions. After that, the alginate-coated sample is placed in the cross-linking bath for a second time to complete the gel-formation process and drain any surplus solution (Nussinovitch, 2009a). The evaporation of solvents from the coating and cross-linking solutions is often disregarded since the dipping technique is generally rather short (Cisneros-Zevallos and Krochta, 2003). The dipping and draining times vary every research, but they generally last 30 seconds to 5 minutes. A multilayer coating or layer-by-layer (LbL) technique was used to address the challenges of the coating adhering to the hydrophilic surface of the cut surfaces (Sipahi et al., 2013). Food items are dipped in coating solutions containing oppositely charged polyelectrolytes to create physical and chemical interaction (Sipahi et al., 2013). Fruits and vegetables are now generally treated using the LbL method (Arnon-Rips and Poverenov, 2018).

5.6.2.2 Spraying

Spraying a semipermeable barrier over the surfaces of food products is another popular technique (Zapata et al., 2008). The coating solution is applied using a spraying device that employs nozzles to create droplets that are spread throughout the necessary food surface area (Andrade et al., 2012). The spraying method requires less coating material to provide efficient coverage because of the high spraying pressure (about 60–80 psi) (Tharanathan, 2003). Uniform coating, thickness control, multilayer applications, coating solution contamination avoidance, solution temperature control, and the capacity to handle large surface areas are some of the other advantages of spraying. It is important that the spraying solution isn't too thick. Earle and McKee (1976) used a water-soluble algin dispersion, followed by a gelling agent, to coat newly slaughtered hanging animal carcasses. The viscosity of the solution was discovered to be the most important feature in the vertical spraying application for consistent adherence to the alginate layer. Papajova et al. (2012) created an externally gelling method by airbrushing gelling solution aerosols onto planar alginate hydrogels. The spraying was done in a different way by Amanatidou et al. (2000). After dipping carrot slices in $CaCl_2$ solutions, they were sprayed with an alginate-based solution after drying.

5.6.2.3 Vacuum Impregnation

In order to add vitamins and minerals to foods, researchers have used the vacuum impregnation approach. This coating has recently shown that it can produce a thicker, more effective film when solutes are incorporated into the air, including porous food matrices like fruits and vegetables (Vargas et al., 2009). This method employs the same dipping and draining procedures as the dipping method. The samples are put in two airtight vacuum chambers that are driven by vacuum pumps rather than dipping tanks. The items are then immersed in the coating solution and subjected to atmospheric restoration once the vacuum application is completed.

5.7 Coatings with Sodium Alginate

Edible films and coatings based on sodium alginate have a wide range of applications, which are listed next.

5.7.1 Fruits and Vegetables

Hydrocolloid-based coatings have been used to reduce fruit and vegetable respiration because of their unique permeability to the gases O_2 and CO_2 (Nisperos-Carriedo et al., 1994).

A composite covering of sodium alginate and olive oil protected the quality of ber fruit, according to Rao et al. (2016). Antioxidant levels rose, and cell wall hydrolase activity reduced during storage when ber fruits were covered with this composite coating enriched with ascorbic and citric acids. Edible coatings based on sodium alginate and its mixes with ascorbic acid have demonstrated to be a viable solution for preserving the aesthetic appeal of minimally processed pineapple during refrigerated storage, according to Cordoba and Usme (2019). These coatings formed a thick layer on top of fresh-cut pineapple samples, giving them a bright, translucent, and youthful appearance. Coated samples appeared better after storage than uncoated ones. Overall, our data suggest that coating fresh-cut pineapple with sodium alginate may be a viable postharvest approach. Moreover, Gol et al. (2015) noticed the use of alginate as a coating on jamun fruit reduced the weight loss and decay percentage while retaining the total soluble solids and acidity.

According to Parreidt et al. (2019), fruits were covered with an alginate-based coating layer to protect them from the environment. On two kinds of coated meals as well as solo films, the alginate-based coating's water vapor barrier effectiveness was shown. The findings help us better understand the water barrier characteristics of alginate-based coatings. Adding a calcium lactate solution to the coating process at the beginning encourages the production of gels on the fruit surface while also enhancing coating uniformity. Researchers may employ the new coating approach to enhance the adhesion of their planned alginate-based coatings, and it might be used to improve the effects of edible coatings on quality metrics. Freshly cut melon slices with a covering lost less water than those without. However, more water was lost in the strawberry samples. The use of an alginate-based coating approach combined with an additional calcium dipping step enhances hydrophilic cut surfaces, particularly in porous food samples, while the influence on waxy surfaces requires further investigation. The vapor sorption analyzer may be set up in a variety of ways to assist researchers in better understanding what is going on.

As a consequence of rising awareness of excellent eating habits and a lack of time to prepare meals, consumers are becoming more interested in fresh-cut fruits and vegetables (Baldwin et al., 1995).

Consumer preferences are shifting toward convenience while maintaining quality (Wong et al., 1994). In minimally processed fresh-cut fruits and vegetables, washing, peeling, cutting, slicing, coring, and other preparation activities remove these barriers, form tissue lesions, degrade the fruit's integrity, and cause wounding stress (Rojas-Grau et al., 2007). Contamination, enzymatic browning, unwelcome volatile generation, and texture alterations are all potential outcomes (Olivas et al., 2007). Degradative effects, as well as the growth of spoilage and dangerous microorganisms, may be reduced by enhancing natural barriers or replacing them with artificial barriers surrounding the product, such as edible films and coating applications (Forney et al., 2009). Fruits and vegetables must have a delay in respiration and physiological activity to extend their shelf life. Coatings and films that change gas flow might be used to cover fresh vegetables (Ncama et al., 2018). The swelling ratio and water solubility of alginate films are important issues in the case of fresh-cut fruits with high moisture surfaces. According to Tapia et al. (2008) alginate films resist dissolving in water and may therefore cover fresh-cut fruits with a high moisture content.

5.7.2 Effect of Sodium Alginate as a Multilayered Edible Coating on Fruits

As a coating, cohesion and adhesion are two types of interactions that must occur throughout the film manufacturing process. Controlling adhesion and cohesion coefficients is important for optimizing a coating solution since the former supports liquid spreading while the latter encourages liquid contraction (Ribeiro et al., 2007).

A coating solution is put to the fruit's surface to begin the procedure. An adequate coating solution is spontaneously applied to the fruit's surface (Mittal, 1997). Forming a better coating might be challenging. "As a solution's interfacial tension falls, so does interfacial activity" (Gaonkar, 1991). The surface tension of water is quite strong (72.8 dyn cm1), while the surface tension of most solid surfaces is much lower (Nussinovitch, 2009b).

In this regards, attempt has been made by us using a rheometer. Three different coating materials were examined in different quantities: chitosan, sodium alginate, and hydroxipropyl methylcellulose (0.5%, 1%, and 1.5%). The surface tension of several edible coating solutions was determined using a tensiometer. Because all the coating solutions in this study had a higher surface tension, Tween-80 (0.1%) was added to all of them to reduce surface tension. Based on the results, intermediate viscous concentrations of the various coatings were chosen to evaluate the effect of these coatings as a multilayered postharvest coating on the postharvest shelf life of phalsa, plum, custard apple, and cape gooseberry fruit, as well as efforts to preserve their nutritional quality attributes. Multilayer coatings of (sodium alginate–$CaCl_2$-Chitosan) boosted antioxidant activity, texture, and lowered respiration rate in plums but reduced browning, respiration, and improved antioxidant potential in custard apples.

5.7.3 Frozen and Fried Foods

Fontes et al. (2011) evaluated the effects of an alginate-coated fried sweet potato chip. Based on its chemical and physical properties, alginate film would be an excellent choice for coating sweet potato chips, since it has the highest tensile strength and elongation, ensuring that the coating would cover the material uniformly and with minimal risk of rupturing during processing. The color of the product did not change throughout the frying process.

Khanedan et al. (2011) looked at how different alginate concentrations acted as a film on kilka fish during storage. During the experiment, he noted that as the concentration of sodium alginate grew, the moisture value decreased. Adding sodium alginate to the edible coating reduced total volatile nitrogen, stopped lipid oxidation, increased the shelf life of kilka fish, and reduced moisture.

A variety of cheeses are wrapped with edible coatings and films to prevent quality loss. Zhong et al. (2014) used four different application procedures on mozzarella cheese to examine the effectiveness of three different coating materials (sodium alginate, chitosan, and soy protein isolate). Alginate-coated samples were shown to have improved overall characteristics due to their enhanced wettability on the product surface.

5.7.4 Poultry Products

Meat processing has long been considered one of the most polluting industries in the food sector. In terms of pollution, packaging materials are a major contributor. Manufacturers prefer plastics over other materials because of its qualities (Gheorghita et al., 2020). When a single applied layer isn't enough to offer all the desired properties, several laminate films are often used. Apart from the toxicity of the components when used alone, the use of several compositions and adhesives for the overlap makes these intricate films very hazardous; all these activities make them difficult to separate and recycle after they have been discarded. Application techniques for edible films and coatings on meat and poultry include foaming, dipping, spraying, casting, brushing, individual wrapping, rolling, or vacuum impregnation. When the films are thin or the coating is only sprayed on one side of the product, compressed air is used to apply the coating; brushing is used to apply the coating entirely to the product surface, and spraying is used when the films are thin or the coating is only sprayed on one side. Spraying is used for film-forming solutions that are in the form of an emulsion and require the presence of a foaming agent, brushing is used for complete coating application on the product surface, and spraying is used for film-forming solutions that are in the form of an emulsion and require the presence of a foaming agent (Huber and Embuscado, 2009). The vacuum impregnation method is used to improve a product by adding vitamins or minerals (Parreidt et al., 2018b). To apply composite or multilayer films, casting and extrusion are used (Skurtys et al., 2010). Packaging in the meat business is designed to keep the product safe from gas and vapor leakage (Troy and Kerry, 2010). This should help the meat retain its weight and color (Haile et al., 2013). Studies have demonstrated that edible films created from biopolymers, particularly polysaccharides, may help meat and meat products last longer by reducing dehydration, rancidity, and browning of muscle tissue (Sanchez-Ortega et al., 2014). On the top of smoked or steamed meats, edible coatings break down, increasing structure and texture while inhibiting moisture transfer (Cutter, 2006). Edible packaging may include organic acids, essential oils or plant extracts, bacteriocins, antioxidants, colorings, flavorings, vitamins, and other natural ingredients. The FDA regulates the amount of antibacterial compounds that may be added to food and the materials used in food packaging (Malhotra et al., 2015).

Antimicrobial packaging is a critical weapon in the fight against harmful microorganism contamination in meat. The microbial load caused by inappropriate raw material processing and handling, as well as incorrect packaging, causes contamination of meat products to appear on the outside surface. Antimicrobial films may assist in reducing cross-contamination and extending the shelf life of meat and meat products by inhibiting germs on the surface of newly processed goods. Antimicrobial chemicals released slowly through the structure of antimicrobial foils are an essential and successful approach for protecting meat from hazardous germs. The microbial load caused by inappropriate raw material processing and handling, as well as incorrect packaging, causes contamination of meat products to appear on the outside surface. Antimicrobial films may assist in reducing cross-contamination and extending the shelf life of meat and meat products by inhibiting germs on the surface of newly processed goods (Gheorghita et al., 2020).

Different types of additives have been incorporated into alginate-based edible films and coatings. The summarized results (Table 5.2) show that alginate forms an effective base for antimicrobials and antioxidants to decrease the microbial load and to increase the antioxidant capacity, respectively, of coated food products.

5.7.5 Sodium Alginate as a Functional Food Ingredient

Alginate's exceptional ability to gel at low temperatures and strong heat stability make it ideal for use as thickeners, stabilizers, and restructuring agents. Alginate is also gaining popularity as a vehicle for protective coatings on packed, diced, or prepared fruits and vegetables, as well as encapsulating active enzymes and biological microbes. With appropriate chemical and biological alterations to modify their structures and features, novel applications of certain alginates in the food industry with high bioactivities at low concentrations may be conceivable. Food manufacturers, alginate producers, and food and nutrition specialists will need to work together more closely to develop innovative alginate applications in the food and beverage industry (Qin et al., 2018).

TABLE 5.2

The application of alginate-based (along with additives) edible films and coatings on different foods

Foods	Coatings/Additives/ Cross-linking	Effects	Sources
Fresh-cut pineapple	Alginate, sunflower oil/CaCl$_2$ (EC) lemongrass EO	Shelf life of fresh-cut pineapple was prolonged, and microbes were significantly reduced.	Azarakhsh et al. (2014)
Ground beef	Alginate/CaCl$_2$ nisin, acetic acid, lactic acid, potassium sorbate	Immobilization in alginate/CaCl$_2$ enhanced the activity of only some of the antimicrobial agent/combination. Only acetic and lactic acid inhibited *E. coli*.	Fang and Tsai (2003)
Guava	Alginate/CaCl$_2$ pomegranate peel extract	Antioxidant activity was increased.	Nair et al. (2018)
Silver carp fillet	Alginate/CaCl$_2$/clove oil	Use of clove oil along with the alginate/CaCl$_2$ significantly decreased lipid oxidation.	Jalali et al. (2018)
Apple pieces	Alginate/lipid	Pieces of apple were coated with double layers of (alginate/acetylated monoglyceride) decreased respiratory activity.	Wong et al. (1994)
Strawberry	Alginate	Yeast antagonist along with the alginate increased the quality.	Fan et al. (2009)
Blueberry	Alginate/CaCl$_2$	Quality loss was controlled by the combination of the coatings.	Duan et al. (2004)
Cherry	Alginate	Coating of alginate increased the total antioxidant activity, phenolic activity, and also enhanced storage life.	Diaz-Mula et al. (2012)
Pork chops	Starch-alginate with rosemary oleoresin	Use of composite coating inhibited lipid oxidation and formation of hexanal, pentane, and total volatiles.	Handley et al. (1997)
Cheese	*Pimpinella saxifrage*	The addition of *Pimpinella saxifrage* in sodium alginate coating was effective in improving the oxidative and bacterial stability of the coated cheese and in reducing the weight loss, preserving pH and color.	Ksoudaa et al. (2019)
Fresh-cut papaya	Alginate, oregano, and thyme	The coated papaya had highest sensory evaluation scores forstored for 12 days at 4°C and also improved microbiological food safety.	Tabassum and Khan (2020)
Chicken breast meat	Alginate-based film with black cumin	Use of black cumin indicated antimicrobial activity against *E. coli* and also had less variation in pH, and color changes for chicken breast.	Takma and Korel (2019)
Sausages	Sodium alginate	Use of sodium alginate decreased the growth of bacterial population for up to 7 days and enhanced the shelf life of minced beef meat.	Shahbazi and Shavisi (2019)
Fish fillet	Alginate-based nanocomposite films with several oils	Marjoram and clove alginate–based nanocomposite films coating inhibited the growth of *Listeria monocytogenes*.	Alboofetileh et al. (2016)

5.7.6 Sodium Alginate in the Food Industry

Food, drinks, medicines, health and personal care items, agriculture/animal feed/poultry, and agriculture/animal feed/poultry products may all benefit from sodium alginate E401. Salad dressings, puddings, jams, tomato juices, tinned goods, ice creams, yoghurts, creams, and cheeses all include sodium alginate E401 as a thickening and emulsifying agent (Newseed, 2015).

5.7.6.1 Thickener and Emulsion

Salad dressings, pudding, jams, tomato ketchups, and canned goods all benefit from sodium alginate E401, which helps increase product stability while reducing liquid outflow.

5.7.6.2 Stabilizer

Sodium alginate E401, an ice cream stabilizer that replaces starch and carrageenan, may aid in the prevention of ice crystals while also increasing the flavor. Ice lollies, iced fruit juice, and iced milk are examples of mixed beverages. The finished product does not stick to the box when something is added to dairy goods, like refined cheese, canned cream, and dry cheese. Additionally, when used as a cover for mild meals, sodium alginate may maintain the product fine and prevent it from breaking apart.

5.7.6.3 Hydration

Noodles, vermicelli, and ice powder all benefit from sodium alginate's high cohesiveness, which makes them less prone to pull, bend, or break. It is highly beneficial when wheat flour has a low gluten concentration. Sodium alginate E401 may also aid with the product's internal form equalization and water retention, allowing it to be stored for extended periods of time. When sodium alginate E401 is added to iced sweets, it acts as a protective barrier against overheating while also speeding up flavor release and raising the melting point.

5.7.6.4 Coacervation

Sodium alginate E401 may be used to create fine, non-leaky, contractible gel goods. As a result, it may be used on both frozen and non-frozen goods. It may be used to keep fruit, meat, and seafood fresher for longer by exploiting its unique role as a cover. It may be used as a sugar coating on bread, a covering for fillings and cakes, and an auto-coagulant in canned foods, since it is unaffected by temperature or acidity. It may be used to create a crystal soft sugar that is elastic, non-sticky, and transparent by replacing carrageenan for it.

5.7.7 Alginates in Food Packaging

The primary objective of packaging is to protect and secure the contents. The packaging protects the included food product from physical/mechanical damage, physicochemical changes produced by light, oxygen, moisture, and odors, and biological changes caused by germs and pests; all the aforementioned characteristics result in a decline in product quality and safety (Kontominas, 2020). Due to the detrimental environmental effect of synthetic packaging materials, researchers and the food industry are looking into using biodegradable and renewable natural-source materials. Proteins (whey, wheat, maize, and soy proteins, gelatin), lipid derivatives (waxes, acetylated triglycerides), and carbohydrates (whey, wheat, maize, and soy proteins, gelatin) are among the biopolymers used in food packaging (starch, cellulose and its derivatives, carrageenan, pectin, chitosan, alginates). Alginates are biopolymers made up of natural hydrophilic polysaccharide biopolymers that are derived mostly from sea brown algae. They have excellent film-forming qualities, low permeability to O_2 and vapors, flexibility, water solubility, and gloss, and they are tasteless and odorless when used in the form of films or coatings. They help conserve moisture, delay oxidation, prevent color and texture deterioration, reduce microbial load, boost sensory acceptance, and minimize cooking losses when used in combination with additives such as organic acids, essential oils, plant extracts, bacteriocins, and nanomaterials. Alginates were first employed to reduce the rate of respiration in perishable fresh fruits and vegetables, but they're now being utilized to extend the shelf life of a variety of foods, including beef, chicken, fish, and cheese. Alginates may be utilized to create a multifunctional food packaging system that satisfies the ultimate purpose of food packaging technology when used in combination with the concepts of active, intelligent, and green packaging technologies (Kontominas, 2020).

5.7.8 Application of Alginates during Transportation

Moisture content, water vapor permeability (WVP), water vapor resistance (WVR), and water vapor transmission rate (WVTR) of alginate-based edible films and coatings have all been studied extensively in the literature. The epidermal cell layer and cuticles minimize weight loss in raw, untreated fruits and vegetables, but edible coatings and films provide an extra barrier layer on the stomata, reducing transpiration (and hence weight loss) (Appendini et al., 2002). Vapor-phase diffusion, which is aided by the differential in water vapor pressure between the product and the surrounding air, is the principal mechanism for moisture loss from food products. The mass transfer rate is influenced by the thickness of the formed film, moisture permeability, temperature, and relative humidity of the surrounding medium (Yaman and Bayondrl, 2002). The second most researched mode of transportation is gas barrier properties (particularly O_2). The two techniques employed in gas transport are capillary diffusion (which occurs more often in porous, imperfect materials) and activated diffusion (which comprises gas solubilization in the film, diffusion through the film, and release on the opposite side of the film) (Donhowe and Fennema, 1994).

The relative humidity (RH) of a product has a big impact on the rates of O_2 and CO_2 transit (Bonilla et al., 2012). When the RH rises, more water and film/coating molecules interact, forming a plasticized structure that allows mass to be transferred. Because of the high RH of a freshly cut surface, predicting a coating's permeability attributes is particularly challenging. Because these gases are water-soluble, increasing the gas diffusivity and solubility in edible films formed from hydrophilic gelling matrix results in increased gas permeability (Guilbert et al., 1995).

5.7.9 Alginates in the Pharmaceutical Industry

Biopolymers are biocompatible, biodegradable, non-antigenic, and non-toxic, with molecular masses ranging from 32,000 to 40,000 g/mol in commercial grades. Commercial alginates are utilized in biomedical engineering, biotechnology, environmental remediation, and medicine, among other sectors. Wound dressings, bone regeneration, neovascularization, protein distribution, cell transport, theranostic agents, oral medication administration, controlled release systems, raft formulations, biological agent immobilization, and environmental contamination remediation are just a few of the applications for alginates. Hydrogels, tablets, microcapsules, films, matrices, microspheres, liposomes, nanoparticles, beads, cochleate, floating, and supersaturated drug delivery systems are some of the applications for alginates (Kothale et al., 2020).

5.7.9.1 Alginate Microparticles

Because of its intrinsic properties, such as better biocompatibility and biodegradability for improved transport, stability, and long-term release of encapsulated medications, sodium alginate–based particles have become one of the most sought-after drug delivery systems (Yoo et al., 2011). In pharmacological research, tissue engineering, and regenerative medicine, they're also often used to encapsulate living cells. Individual cells may be watched and controlled individually using microgels like these, which can be used to investigate the impact of confinement on cell destiny or distribute cells for tissue repair (Utech et al., 2015).

5.8 Other Uses of Alginates

Alginate is also utilized in textile and paper manufacturing, as well as in immobilized biocatalysts, welding rods, and fish feed binders.

5.9 Future Trends and Challenges

Developing formulations for each type of food, ensuring biochemical, physicochemical, and antimicrobial stability without affecting food sensory quality, being completely safe for human consumption, having sufficient mechanical strength and a competitive cost to replace current synthetic materials, and eliminating the use of chemical or synthetic preservatives are just a few of the major research challenges to consider for alginate use as films and coatings. These characteristics are the source of the term "future packaging technology." Despite the numerous advantages of edible coatings for prolonging shelf life, improving quality, and assuring microbiological safety in fresh fruits and vegetables, their commercial application on a wide variety of different foods is still restricted.

Recent research has expanded on the use of edible biopolymers to make food films and coverings. To retain the product's quality throughout its shelf life, edible/biodegradable films and coatings may be utilized. On fresh-cut fruits and vegetables as well as meat items coated with alginate solutions, including additives, promising results have been found. New antibacterial, antioxidant, and anti-browning chemicals, on the other hand, may provide even higher food safety and quality advantages. It will be possible to better understand any potential synergistic effects between alginate coating and active substances. One way to enhance the structure's characteristics is to combine film-forming biopolymers. Another research need has been identified: developing new synergistic gelling technologies. Variable quantities of alginate and cross-linking agents may be used to investigate the diffusion capabilities of active compounds from the alginate gel matrix, as well as its structure. It's possible to do a comparative research. Techniques indicated the impact of various calcium salts (e.g., calcium chloride, calcium lactate, calcium gluconate) on quality indicators may be evaluated in detail. The great majority of alginate coating research has been done in the lab, and commercial applications are still rare. Further research with practical applications should be focused on industrial implementation in order to commercialize alginate-coated food items with prolonged shelf lives. In order to avoid the disadvantages of the application methods, coating application methods can be redesigned to include a recycle process that does not waste too much coating solution, reduce the microbial load of the solution during recycling, design spraying methods for irregular surfaces, design industrial-size vacuum tanks, and so on. As a result, edible films and coatings based on sodium alginate might be utilized much more often than they are now.

5.10 Conclusion

Plastic containers might be replaced with edible films and alginate coatings. Alginate-based edible coatings and films extend shelf life, prevent water and moisture loss, restrict microbiological development, and postpone ripening. Coatings and films containing additives improve the nutritional demands of food products, increase flexibility, improve adhesion, and add flavor, color, and other characteristics that today's customers want. Alginate-based edible films and coatings do not have much widespread acceptance in the marketplace. The focus of future study should be on commercial applications. To overcome this issue, coating application methods may be altered; spraying techniques can be enhanced to make them suitable for coating application, especially on rough surfaces of food products.

REFERENCES

Alboofetileh, M., Rezaei, M., Hosseini, H. & Abdollahi, M. (2016). Efficacy of activated alginate-based nano-composite films to control *Listeria monocytogenes* and spoilage flora in rainbow trout slice. *Journal of Food Science and Technology. 53*, 521–530.

Amanatidou, A., Slump, R. A., Gorris, L. G. M. & Smid, E. J. (2000). High oxygen and high carbondioxide modified atmospheres for shelf-life extension of minimally processed carrots. *Journal of Food Science. 65*, 61–66.

Andrade, R. D., Skurtys, O. & Osorio, F. A. (2012). Atomizing spray systems for application of edible coatings. *Comprehensive Reviews in Food Science and Food Safety. 11*, 323–337.

Appendini, P. & Hotchkiss, J. H. (2002). Review of antimicrobial food packaging. *Innovation Food Science Emerging Technology. 3*, 113–126.

Arnon-Rips, H. & Poverenov, E. (2018). Improving food products' quality and storability by using layer by layer edible coatings. *Trends in Food Science and Technology. 75*, 81–92.

Azarakhsh, N., Osman, A., Ghazali, H. M., Tan, C. P. & Mohd Adzahan, N. (2014). Lemongrass essential oil incorporated into alginate-based edible coating for shelf-life extension and quality retention of fresh-cut pineapple. *Postharvest Biology Technology. 88*, 1–7.

Azarakhsh, N., Osman, A., Tan, C. P., Mohd Ghazali, H. & Mohd Adzahan, N. (2012). Optimization of alginate and gellan-based edible coating formulations for fresh-cut pineapples. *International Food Research Journal. 19*, 279–285.

Baldwin, E. A., Hagenmaier, R. & Bai, J. (2011). *Edible coatings and films to improve food quality.* Boca Raton, FL: CRC Press.

Baldwin, E., Nisperos-Carriedo, M. & Baker, R. (1995). Edible coatings for lightly processed fruits and vegetables. *Horticulture Science. 30*, 35–38.

Bonilla, J., Atares, L., Vargas, M. & Chiralt, A. (2012) Edible films and coatings to prevent the detrimental effect of oxygen on food quality: Possibilities and limitations. *Journal of Food Engineering. 110*, 208–213.

Brownlee, I. A., Allen, A., Pearson J. P., Dettmar, P. W., Havler, M. E., Atherton, M. R. & Onsoyen, E. (2005). Alginate as a source of dietary fiber. *Critical Reviews in Food Science and Nutrition. 45*(6), 497–510.

Burt, S. A. & Reinders, R. D. (2003). Antibacterial activity of selected plant essential oils against *Escherichia coli* O157:H7. Letter. *Applied Microbiology. 36*, 162–167.

Cagri, A., Ustunol, Z. & Ryser, E.T. (2004). Antimicrobial edible films and coatings. *Journal Food Protection. 67*, 833–848.

Castro-Yobal, M. A., Contreras-Oliva, A., Saucedo-Rivalcoba, V., Rivera-Armenta, J. L., Hernandez-Ramirez, G., Salinas-Ruiz, J. & Herrera-Corredor, A. (2021). Evaluation of physicochemical properties of film based alginate for food packing applications. *e-Polymers. 21*, 82–95.

Chan, L. W., Lee, H. Y. & Heng, P. W. S. (2006). Mechanisms of external and internal gelation and their impact on the functions of alginate as a coat and delivery system. *Carbohydrate Polymer. 63*, 176–187.

Chee, S. Y., Wong, P. K. & Wong, C. L. (2011). Extraction and characterisation of alginate from brown seaweeds (Fucales, Phaeophyceae) collected from Port Dickson, Peninsular Malaysia. *Journal of Applied Phycology. 23*(2), 191–196.

Chrastil, J. (1991). Gelation of calcium alginate. Influence of rice starch or rice flour on the gelation kinetics and on the final gel structure. *Journal of Agricultural Food Chemistry. 39*, 874–876.

Cisneros-Zevallos, L. & Krochta, J. M. (2003). Dependence of coating thickness on viscosity of coating solution applied to fruits and vegetables by dipping method. *Journal of Food Science. 68*, 503–510.

Cordoba, L. A. & Usme, A. A. (2019). Edible coatings based on sodium alginate and ascorbic acid for application on fresh-cut pineapple (*Ananas comosus* (L.) Merr) Recubrimientos comestibles a base de alginato de sodio y ácido ascórbico para aplicación sobre piña (*Ananas comosus* (L.) Merr) fresca cortada. *Agronomia Colombiana. 37*(3), 317–322.

Cutter, C. N. (2006). Opportunities for bio-based packaging technologies to improve the quality and safety of fresh and further processed muscle foods. *Meat Science. 74*, 131–142.

Dhanapal, A., Rajamani, L. & Shakila Banu, M. (2012). Edible films from polysaccharides. *Food Science Quality Management. 3*, 9–17.

Diaz-Mula, H. M., Serrano, M. & Valero, D. (2012). Alginate coatings preserve fruit quality and bioactive compounds during storage of sweet cherry fruit. *Food Bioprocess Technology. 5*, 2990–2997.

Dong, X., Wrolstad, R. E. & Sugar, D. (2000). Extending shelf life of fresh-cut pears. *Journal of Food Science. 65*(1), 181–186.

Donhowe, I. G. & Fennema, O. (1994). Edible films and coatings: Characteristics, formation, definitions, and testing methods. In *Edible coatings and films to improve food quality.* Krochta, J. M., Baldwin, E. A. & Nisperos-Carriedo, M. O., Eds. Lancaster, NH: Technomic Publication, pp. 1–24.

Draget, K. I., Ostgaard, K. & Smidsrod, O. (1989). Alginate-based solid media for plant tissue culture. *Applied Microbiology and Biotechnology. 31*, 79–83.

Duan, J., Wu, R., Strik, B. C. & Zhao, Y. (2004). Effect of edible coatings on the quality of fresh blueberries (duke and elliott) under commercial storage conditions. *Postharvest Biology and Technology. 59*, 71–79.

Earle, R. D. & McKee, D. H. (1976). Process for treating fresh meats. U.S. Patent 3991218, 9 November.

Embuscado, M. E. & Huber, KC. (2009). *Edible films and coatings for food applications.* New York: Springer.

Emmerichs, N., Wingender, J., Flemming, H. C. & Mayer, C. (2004). Interaction between alginates and manganese cations: Identification of preferred cation binding sites. *International Journal of Biological Macromolecules. 34*(1–2), 73–79.

European Food Safety Authority (EFSA). (2005) Opinion of the Scientific Panel on Dietetic Products, Nutrition and Allergies [NDA] related to nutrition claims concerning omega-3 fatty acids, monounsaturated fat, polyunsaturated fat and unsaturated fat. *EFSA Journal. 3*, 253.

Fan, Y., Xu, Y., Wang, D., Zhang, L., Sun, J., Sun, L. & Zhang, B. (2009). Effect of alginate coating combined with yeast antagonist on strawberry *(Fragaria × ananassa)* preservation quality. *Postharvest Biology and Technology. 53*, 84–90.

Fang, T. J. & Tsai, H. C. (2003). Growth patterns of *Escherichia coli* O157:H7 in ground beef treated with nisin, chelators, organic acids and their combinations immobilized in calcium alginate gels. *Food Microbiology. 20*, 243–253.

Fertah, M. (2017). Venkatesan, J., Anil S., and Kim, S. (Eds). Isolation and characterization of alginate from seaweed. In *Seaweed polysaccharides*. Amsterdam, The Netherlands: Elsevier, pp. 11–26.

Fontes, L. C. B., Bamos, K. K., Sivi, T. C. & Queiroz F. P. C. (2011). Biodegradable edible films from renewable sources potential for their application in fried foods. *American Journal of Food Technology. 6*(7), 555–567.

Forney, C. F., Mattheis, J. P. & Baldwin, E.A. (2009). Effects on flavor. In *Modified and controlled atmosphere for the storage, transpiration and packaging of horticultural commodities.* Yahia, E. M., Ed. Boca Raton, FL: CRC Press, pp. 119–158.

Gaonkar, A. G. (1991). Surface and interfacial activities and emulsion of some food hydrocolloids. *Food Hydrocolloids. 5*, 329–337.

Garcia, M. P., Gomez-Guillen, M. C., Lopez-Caballero, M. E. & Barbosa-Canovas, G. V. (2016). *Edible films and coatings: Fundamentals and applications.* Boca Raton, FL: CRC Press.

Gheorghita, R., Gutt, G. & Amariei, S. (2020). The use of edible films based on sodium alginate in meat product packaging: an eco-friendly alternative to conventional plastic materials. *Coatings. 10*, 166.

Gol, N. B., Vyas, P. B. & Ramana Rao T. V. (2015). Evaluation of polysaccharide-based edible coatings for their ability to preserve the postharvest quality of indian blackberry *(Syzygium cumini* l.), *International Journal of Fruit Science, 1–25.* DOI: 10.1080/15538362.2015.1017425.

Gombotz, W. R. & Wee, S. (1998). Protein release from alginate matrices. *Advanced Drug Delivery Reviews. 31*(3), 267–285.

Grant, G. T., Morris, E. R., Rees, D. A., Smith, P. J. C. & Thom, D. (1973). Biological interactions between polysaccharides and divalent cations: The egg-box model. *FEBS Letters. 32*, 195–198.

Guilbert, S., Gontard, N. & Cuq, B. (1995). Technology and applications of edible protective films. *Packaging Technology Science. 8*, 339–346.

Haile, D., De Smet, S., Claeys, E. & Vossen, E. (2013). Effect of light, packaging condition and dark storage durations on colour and lipid oxidative stability of cooked ham. *Journal of Food Science and Technology. 50*, 239–247.

Han, J. H. & Gennodio, A., (2005). Edible films and coating: A review. In *Innovation in food packaging.* Han, J. H., Ed. London: Elsevier, pp. 239–262 and Technology Books, pp. 239–259.

Handley, D., Ma-Edmonds, M., Hamouz, F., Cuppett, S., Mandigo, R. & Schnepf, M. (1997). Controlling oxidation and warmed-over flavor in precooked pork chops with rosemary oleoresin and edible film. In *Natural antioxidants chemistry, health effects, and applications.* Shahidi, F., Ed. Champaign, IL: AOCS Press, pp. 311–318.

Hershko, V. & Nussinovitch, A. (1998). Physical properties of alginate coated onion *(Allium cepa)* skin. *Food Hydrocolloid.* doi: 10.1016/S0268-005X(98)00029-0.

Howard, L. R., Burma, P. & Wagner, A. B. (1994). Firmness and cell wall characteristics of pasteurized jalapeño pepper rings affected by calcium chloride and acetic acid. *Journal of Food Science and Technology. 59*, 1184–1186.

Huber, K. & Embuscado, E. (2009). *Edible films and coatings for food applications.* New York: Springer-Verlag, pp. 245–268.

Jalali, N., Ariiai, P. & Fattahi, E. (2018). Effect of alginate/carboxyl methyl cellulose composite coating incorporated with clove essential oil on the quality of silver carp fillet and *Escherichia coli* O157:H7 inhibition during refrigerated storage. *Journal of Food Science and Technology. 53*, 757–765.

Jost, V., Kobsik, K., Schmid, M. & Noller, K. (2014). Influence of plasticizer on the barrier, mechanical and grease resistance properties of alginate cast films. *Carbohydrate Polymer 110*, 309–319.

Khanedan, N., Motalebi, A. A., Khanipour, A. A., Koochekian sabour A., Seifzadeh, M. & Hasanzati, R. (2011). Effects of different concentrations of Sodium alginate as an edible film on chemical changes of dressed Kilka during frozen storage. *Iranian Journal of Fisheries Sciences. 10*(4), 654–662.

Kierstan, M., Darcy, G. & Reilly, J. (1982). Studies on the characteristics of alginate gels in relation to their use in separation and immobilized applications. *Biotechnology and Bioenergy. 24*, 1507–1517.

Kohn, R. (1975). Ion binding on polyuronates-alginate and pectin. *Pure Applied Chemistry. 42*, 371–397.

Kontominas, M. G. (2020). Use of alginates as food packaging materials. *Foods. 12*(10), 1440.

Kothale, D., Verma, U., Dewagan, N., Jana, P., Jain, A. & Jain, D. (2020). Alginate as a promising polymer for pharmaceutical, food and biomedical applications. *Current Drug Delivery. 17*(9), 755–775.

Krochta, J. M. (2002). Protein as raw materials for films and coatings: Definitions, current status, and opportunities. In *Protein-based films and coatings*. Gennadios, A., Ed. Boca Raton, FL: CRC Press, pp. 1–41.

Ksoudaa, G., Sellimia, S., Merlierb, F., Falcimaigne, A., Thomassetb, B., Nasria, M. & Hajjia, M. (2019). Composition, antibacterial and antioxidant activities of Pimpinella saxifrage essential oil and application to cheese preservation as coating additive. *Food Chemistry. 288*, 47–56.

Kuo, C. K. & Ma, P. X. (2001). Ionically cross-linked alginate hydrogels as scaffolds for tissue engineering: Part 1. Structure, gelation rate and mechanical properties. *Biomaterials. 22*, 511–521.

Kuo, C. K. & Ma, P. X. (2008). Maintaining dimensions and mechanical properties of ionically cross-linked alginate hydrogel scaffolds in vitro. *Journal of Bio-mediated Material Research. Part A. 84A*, 899–907.

Lee, P. & Rogers, M. A. (2012). Effect of calcium source and exposure-time on basic caviar spherification using sodium alginate. *International Journal of Gastronomy and Food Science. 1*(2), 96–100.

Lin, D. & Zhao, Y. (2007). Innovations in the development and application of edible coatings for fresh and minimally processed fruits and vegetables. *Comprehensive Reviews in Food Science and Food Safety. 6*, 60–75.

Liu, X. D., Yu, W.Y., Zhang, Y., Xue, W. M., Yu, W. T., Xiong, Y., Ma, X. J., Chen, Y. & Yuan, Q. (2002). Characterization of structure and diffusion behavior of ca-alginate beads prepared with external or internal calcium sources. *Journal of Microencapsulation. 19*, 775–782.

Lu, J., Zhu, W., Guo, Y. L., Hu, Z. X. & Yu, P., (2006). Electro-spinning of sodium alginate with poly (ethylene oxide). *Polymer. 47*, 8026–8031.

Luna-Guzman, I., Cantwell, M. & Barrett, D. M. (1999). Fresh-cut cantaloupe: Effects of $CaCl_2$ dips and heat treatments on firmness and metabolic activity. *Postharvest Biology and Technology. 17*, 201–213.

Malhotra, B., Keshwani, A. & Kharkwal, H. (2015). Antimicrobial food packaging: Potential and pitfalls. *Frontier Microbiology. 6*, 1–9.

Mancini, F. & McHugh, T. H. (2000). Fruit-alginate interactions in novel restructured products. *Food/ Nahrung. 44*, 152–157.

Martinez, M. V. & Whitaker, J. R. (1995). The biochemistry and control of enzymatic browning. *Trends in Food Science & Technology. 6*(6), 195–200.

McHugh, D. J. (1987). Production, properties and uses of alginates. *Food and Agriculture Organization. 288*.

Mittal, K. L. (1997). The role of the interface in adhesion phenomena. *Polymer Engineering and Science. 17*, 467–473.

Nair, M. S., Saxena, A. & Kaur, C. (2018). Effect of chitosan and alginate based coatings enriched with pomegranate peel extract to extend the postharvest quality of guava (*Psidium guajava* L.). *Food Chemistry. 240*, 245–252.

Ncama, K., Magwaza, L., Mditshwa, A. & Zeray Tesfay, S. (2018). Plant-based edible coatings for managing postharvest quality of fresh horticultural produce: A review. *Food Packaging and Shelf Life. 16*, 157–167.

Newseed (2015). Application and uses of Sodium Alginate. (Sodium Alginate Applications, Sodium Alginate Uses. foodsweeteners.com.

Nisperos-Carriedo, M. O. (1994). Edible coatings and films based on polysaccharides. In *Edible coatings and films to improve food quality*. Volume 1. Krochta, J. M., Baldwin, E. A., Nisperos-Carriedo, M. O., Eds. Lancaster, NH: Technomic Publishing, pp. 322–323.

Nussinovitch, A. (2009a). Biopolymer films and composite coatings. In *Modern biopolymer science*. Kasapis, S., Norton, I. T. & Ubbink, J. B. London: Academic Press, pp. 295–326.

Nussinovitch, A. (2009b). Hydrocolloids for coatings and adhesives. In *Handbook of Hydrocolloids*, 2nd ed. Shaston: Woodhead Publishing, pp. 760–806.

Olivas, G. I. & Barbosa-Canovas, G. V. (2008). Alginate—calcium films: Water vapor permeability and mechanical properties as affected by plasticizer and relative humidity. *LWT Food Science and Technology. 41*, 359–366.

Olivas, G. I., Mattinson, D. S. & Barbosa-Canovas, G. (2007). Alginate coatings for preservation of minimally processed 'Gala' apples. *Postharvest Biology and Technology. 45,* 89–96.

Papajova, E., Bujdos, M., Chorvat, D., Stach, M. & Lacik, I. (2012). Method for preparation of planar alginate hydrogels by external gelling using an aerosol of gelling solution. *Carbohydrate Polymer.* 90, 472–482.

Parreidt, T. S., Lindner, M., Rothkopf, I., Schmid, M. & Muller, K. (2019). The development of a uniform alginate-based coating for cantaloupe and strawberries and the characterization of water barrier properties. *Foods.* 8, 203. DOI: 10.3390/foods80602033.

Parreidt, T. S., Muller, K. & Schmid, M. (2018a), Alginate-based edible films and coatings for food packaging applications. *Foods.* 7, 170.

Parreidt, T. S., Schmid, M. & Muller, K. (2018b). Effect of dipping and vacuum impregnation coating techniques with alginate based coating on physical quality parameters of cantaloupe melon. *Journal of Food Science.* 929–936.

Parris, N., Coffin, D. R., Joubran, R. F. & Pessen, H. (1995). Composition factors affecting the water vapor permeability and tensile properties of hydrophilic films. *Journal of Agricultural Food Chemistry.* 43, 1432–1435.

Pavlath, A. E., Gossett, C., Camirand, W. & Robertson, G. H. (1999). Ionomeric films of alginic acid. *Journal of Food Science and Technology.* 64, 61–63.

Pavlath, A. E. & Orts, W. (2009). Edible films and coatings: Why, what, and how? In *Edible films and coatings for food applications.* Huber, K. C. & Embuscado, M. E., Eds. New York: Springer, pp. 1–23.

Porter, M. R. (1991). *Handbook of surfactants.* Glasgow: Blackie.

Qin, Y., Jiang J., Zhao, L., Zhang, J. & Wang F. (2018). Grumezescu, A.M., and Holaban, A.M. (Eds.). Application of alginates as a functional food ingredients. In *Biopolymers for food design. Hand book of food Bioengineering.* Academic Press, Cambridge, Massachusetts pp. 409–429.

Quintavalla, S. & Vicini, L. (2002). Antimicrobial food packaging in meat industry. *Meat Science.* 62, 373–380.

Raghav, P. K., Agarwal, N. & Saini, M. (2016). Edible coating of fruits and vegetables: A review. *International Journal of Scientific Research and Modern Education.* 1(1), 188–204.

Rao, T. V. R., Baraiya, N. S., Vyas, P. B. & Patel, D. M. (2016). Composite coating of alginate-olive oil enriched with antioxidants enhances postharvest quality and shelf life of Ber fruit (*Ziziphus mauritiana* Lamk. Var. Gola). *Journal of Food Science and Technology.* 53(1), 748–756.

Rhim, J. W. (2004). Physical and mechanical properties of water resistant sodium alginate films. *LWT Food Science and Technology.* 37, 323–330.

Ribeiro, C., Vicente, A. A., Teixeira J. A. & Miranda, C. (2007). Optimization of edible coating composition to retard strawberry fruit senescence. *Postharvest Biology and Technology.* 44, 63–70.

Rocha, A. M. & Morais, A. M. (2002). Polyphenoloxidase activity and total phenolic content as related to browning of minimally processed 'Jonagored' apple. *Journal of the Science of Food and Agriculture.* 82(1), 120–126.

Rojas-Grau, M. A., Raybaudi-Massilia, R. M., Soliva-Fortuny, R. C., Avena-Bustillos, R. J., McHugh, T. H. & Martin-Belloso, O. (2007). Apple puree-alginate edible coating as carrier of antimicrobial agents to prolong shelf-life of fresh-cut apples. *Postharvest Biology and Technology.* 45, 254–264.

Sachdeva, A., Gupta, V., Rahi, R., Neelam, D. & Devki. (2021). Seaweed polysaccharides based edible coatings and films: An alternative approach. *International Journal of Recent Scientific Research.* 12(3), 41198–41206.

Sanchez-Ortega, I., Garcia-Almendarez, B. E., Santos-Lopez, E. M., Amaro-Reyes, A., Barboza-Corona, J. E. & Regalado, C. (2014). Antimicrobial edible films and coatings for meat and meat products preservation. *Science World Journal.* 1–18.

Senturk Parreidt, T., Schott, M., Schmid, M. & Muller, K. (2018). Effect of presence and concentration of plasticizers, vegetable oils, and surfactants on the properties of sodium-alginate-based edible coatings. *International of Molecular Science.* 19, 742.

Shahbazi, Y. & Shavisi, N. (2019). Effects of sodium alginate coating containing Mentha spicata essential oil and cellulose nanoparticles on extending the shelf life of raw silver carp (*Hypophthalmichthys molitrix*) fillets. *Food Science and Biotechnology.* 28, 433–440. DOI: 10.1007/s10068-018-0486-y.

Shahidi, F. (2000). Antioxidants in food and food antioxidants. *Food/Nahrung.* 44(3), 158–163.

Sipahi, R. E., Castell-Perez, M. E., Moreira, R. G., Gomes, C. & Castillo, A. (2013). Improved multilayered antimicrobial alginate-based edible coating extends the shelf life of fresh-cut watermelon (*Citrullus lanatus*). *LWT Food Science and Technology.* 51, 9–15.

Skurtys, O., Acevedo, C., Pedreschi, F., Enrione, J., Osorio, F. & Aguilera, J. M. (2010). Food hydrocolloid edible films and coatings. In *Food hydrocolloids characteristics, properties and structures.* Hollingworth, C. S., Ed. New York: Nova Science Publishers, pp. 41–80.

Skurtys, O., Acevedo, C., Pedreschi, F., Enronoe, J., Osorio, F. & Aguilera, J. M. (2014). *Food hydrocolloid edible films and coatings*. Hauppauge, NY: Nova Science Publishers.

Soazo, M., Baez, G., Barboza, A., Busti, P. A., Rubiolo, A., Verdini, R. & Delorenzi, N. J. (2015). Heat treatment of calcium alginate films obtained by ultrasonic atomizing: Physicochemical characterization. *Food Hydrocolloids. 51*, 193–199.

Sothornvit, R. & Krochta, J. M. (2001). Plasticizer effect on mechanical properties of beta-lactoglobulin films. *Journal of Food Engineering, 50*, 149–55.

Su Cha, D., Choi, J. H., Chinnan, M. S. & Park, H. J. (2002). Antimicrobial films based on Na-alginate and κ-carrageenan. *LWT Food Science and Technology. 35*, 715–719.

Sutherland, I. W. (1991). Alginates. In *Biomaterials: Novel materials from biological sources*, 1st ed. Byrom, D., Ed. Basingstoke: Palgrave Macmillan, pp. 307–331.

Tabassum, N. & Khan, M. A. (2020). Modified atmosphere packaging of fresh-cut papaya using alginate based edible coating: Quality evaluation and shelf life study. *Scientia Horticulturae. 259*, 108853.

Takma, D. K. & Korel, F. (2019). Active packaging films as a carrier of black cumin essential oil: Development and effect on quality and shelf-life of chicken breast meat. *Food Packaging and Shelf Life. 19*, 210–217.

Tapia, M. S., Rojas-Grau, M. A., Carmona, A., Rodriguez, F. J., Soliva-Fortuny, R. & Martin-Belloso, O. (2008). Use of alginate- and gellan-based coatings for improving barrier, texture and nutritional properties of fresh-cut papaya. *Food Hydrocolloids. 22*, 1493–1503.

Tharanathan, R. N. (2003). Biodegradable films and composite coatings: Past, present and future. *Trends Food Science and Technology. 14*, 71–78.

Troy, D. & Kerry, J. (2010) Consumer perception and the role of science in the meat industry. *Meat Science. 86*, 214–226.

Utech, S., Prodanovic, R., Mao, A. S., Ostafe, R., Mooney, D. J. & Weitz, D. A. (2015). Microfluidic generation of monodisperse, structurally homogeneous alginate microgels for cell encapsulation and 3D cell culture. *Advance Healthcare Material. 4*(11), 1628–1633.

Vargas, M., Chiralt, A., Albors, A. & Gonzalez-Martinez, C. (2009). Effect of chitosan-based edible coatings applied by vacuum impregnation on quality preservation of fresh-cut carrot. *Postharvest Biology and Technology. 51*, 263–271.

Wong, D. W., Tillin, S. J., Hudson, J. S. & Pavlath, A. E. (1994). Gas exchange in cut apples with bilayer coatings. *Journal of Agricultural Food Chemistry. 42*, 2278–2285.

Yaman, O. & Bayondrl, L. (2002). Effects of an edible coating and cold storage on shelf-life and quality of cherries. *LWT Food Science and Technology. 35,* 146–150.

Yoo, J. W., Irvine D. J., Discher, D. E. & and Mitragotri, S. (2011). Bio-inspired, bioengineered and biomimetic drug delivery carriers. *Nature Reviews Drug Discovery. 10*, 7. DOI: 10.1038/nrd3499.

Yoo, S. & Krochta, J. M. (2011). Whey protein—polysaccharide blended edible film formation and barrier, tensile, thermal and transparency properties. *Journal of Science and Food Agriculture. 91*, 2628–2636.

Zactiti, E. M. & Kieckbusch, T. G. (2009). Release of potassium sorbate from active films of sodium alginate cross-linked with calcium chloride. *Packaging Technology Science. 22*, 349–358.

Zapata, P. J., Guillen, F., Martinez-Romero, D., Castillo, S., Valero, D. & Serrano, M. (2008). Use of alginate or zein as edible coatings to delay postharvest ripening process and to maintain tomato (*Solanum lycopersicon* Mill) quality. *Journal of the Science of Food and Agriculture. 88*, 1287–1293.

Zhong, Y., Cavender, G. & Zhao, Y. (2014). Investigation of different coating application methods on the performance of edible coatings on mozzarella cheese. *LWT Food Science Technology. 56*, 1–8.

6

Bio-Based Materials for Food Packaging

A Green and Sustainable Technology

Pratik Nayi, Navneet Kumar, and Ho-Hsien Chen

CONTENTS

DOI: 10.1201/9781003303671-6

6.1 Introduction

This chapter aims to provide current knowledge about different sustainable or green materials for food packaging and various applications. It also presents the other properties of sustainable packaging materials and their processing methods to develop green packaging materials for food packaging. Environmental concerns over food packaging motivated the development of new environmentally friendly alternatives. The use of bio-based materials has great potential in the development of such packaging materials. This chapter demonstrates that different bio-based polymers can protect food and keep it well in sensory quality. The bio-based material–based packages showed excellent gas barrier properties for ensuring the extended storage life of the modified active packaged foods even using the material with poorer barrier qualities (Peelman et al., 2014).

Using bio-based packaging materials also helps reduce the consumption of crude oil and CO_2 emissions, reducing the industrial sector's dependency on fluctuating oil prices. The food by-product utilization in bio-based material makes the technology more lucrative and beneficial in attempting the environmental concerns generated by food wastes. However, bio-based and/or biodegradable packaging, like conventional packaging, must perform a variety of critical roles, including food confinement and protection, sensory quality and safety, and consumer communication. Generally, several bio-based packaging options are already available for short-term applications of dried items, which are hardly affected by oxygen and water vapor transmission. Polylactide (PLA) packaging, presently used for yogurt and pasta, might be used for green peppers, fresh-cut romaine lettuce, blueberries, etc. (Almenar et al., 2010). Starch-based packaging materials can be utilized for different food products, for example, organic tomatoes, milk chocolates, fresh-cut beef steaks, whole fresh celery. Cellulose packaging might be utilized for modified atmosphere packaging of fresh food and is readily accepted for packaging potato chips and desserts. Although, the wider application of bio-based packaging for different food products under modified atmospheric conditions is still to be explored. The modified atmosphere packaging is a widely accepted technology for suppressing biochemical and microbiological activities and enhancing the shelf life of foods (Peelman et al., 2013).

Excellent quality and safe food have become the topmost priority of consumers, among other parameters, which are being met by packaging manufacturing units to provide a solution to enhance the shelf life of foods, retention of nutrients, and reduction in microbial load during storage and transport of the food. Food packaging is primarily used to preserve the quality of food and to protect food safety. The production of food packaging waste out of the used packaging material has become a serious concern nowadays. Green or bio-based packaging material has paved the way for a sustainable solution for effective packaging and minimum environmental impact, using renewable and sustainable biomass rather than limited petrochemicals (Wang et al., 2021a).

Several factors responsible for food deterioration include microbial spoilage, oxidation of ingredients, and metabolic activities of food. These factors depend on environmental conditions viz light, temperature, humidity, microorganisms, physical damage, dust, and aromas (Han et al., 2018). The storage requirement for different types of food remains different, for example, fruit and vegetable preservation necessitates lowering the rate of respiration and transpiration, which is normally accomplished by controlling gas (O_2, CO_2, ethylene) environment, temperature, humidity, light, etc. Oxidation and microbiological development, and external factors such as light, oxygen, and moisture, should be carefully monitored for dairy products, mainly milk, cream, and cheese. Meat products are prone to discoloration, which can be avoided by using vacuum packing or modified environment packaging. Biopolymer packaging, which is ecologically benign, has been widely employed to ensure the safety and quality of meat products (Chen et al., 2019).

Diverse methods of food packaging have been adopted in terms of aims and operating mechanisms, which include the following:

  (i) Active packaging
 (ii) Passive packaging
(iii) Sustainable or green packaging
 (iv) Intelligent or smart packaging

Passive packaging prioritized thermal stability, barrier performance, and mechanical strength.

Green packaging is becoming more and more relevant and important with growing concern about the environment due to the generation of huge amounts of food packaging waste. The development of food packaging materials with minimal environmental effects remains the goal for sustainable packaging. The impact is typically determined by how the production and processing of packaging material are carried out. Further, the disposal, incineration, composting, and recycling of the product are also a matter of concern.

There are three main aspects of sustainable or green packaging, mainly raw material, production, and management of waste (Han et al., 2018). Green packaging mainly prefers the materials which renewable sources can produce to eradicate CO_2 emission, by decreasing petrochemical utilization, lighter and thinner packaging with cost and energy-effective processes. Material that reduces the effect on the environment, such as recyclable, biodegradable, and reusable, is preferred, though different approaches have been used for the packaging of food. The safety concerns are widespread, particularly with active and nano packaging technologies. Bio-based materials are chemically organized in singular polymer molecules. Polymer sizes, at the molecular level, depend on the origin and category of polymers. The bio-based nanomaterials discussed here mainly consist of nanocrystals and nanofibers, including numerous polymer chains. Nanocrystals typically have diameters ranging from several nanometers to several micrometers. Natural fibers are usually thick, which contain one or more chemical components (Wang et al., 2021b).

Food packaging is mainly characterized by food preservation, and it's about increasing food products' shelf life. It results in limiting food waste and sustainable food value chains. Plastic materials or polymers have been widely used as packaging materials for several decades, which are derived from traditional fossil-based resources, and their widespread usage in the food packaging industry is mainly due to their excellent features, technical advances, and low cost. Non-renewable resources are used primarily to produce conventional plastic materials, are not biodegradable, and are not always totally recyclable. The widespread use of such materials adds to environmental issues, such as depletion of natural resources, trash, and global warming (Nilsen-Nygaard et al., 2021).

6.2 An Overview of Bio-Based Packaging Materials: Global Scenario

The global packaging market is expanding all the time. The worldwide packaging market is valued at about $914.7 billion in the year 2019, with consistent growth of about 8.4% over the last five years. The world market growth was made slower due to the impact of the COVID-19 pandemic, but still, it is expected to cross the mark of $1.05 trillion in the year 2025, which is also expected to further increase by 1.8% per year after 2025 to reach the product of $1.15 trillion in 2030. The worldwide packaging materials industry utilized paper/paperboard, flexibles, rigid plastics, metals (steel/aluminum), glass, and others. Paper/paperboard and plastics are the most common packaging materials and contribute about

78% of packaging materials. About 70% of paper-based packaging and 14% of plastic packaging were reported to be recycled globally (Stark & Matuana, 2021).

Most of these materials are easily recyclable when utilized as a single material in package types, such as uncoated paperboard or corrugated cardboard. However, packaging materials are frequently integrated into multiple structures, mainly coated materials, multilayers, laminates, and so on, making recycling impractical and, in most cases, economically inconvenient. These packaging materials have caused severe environmental problems since they are either landfilled or burnt at the end of their lifetimes. Sustainable packaging minimizes the environmental effect and footprint over time. It is a rapidly growing packaging industry and is thus recognized as an important concern of consumers and industries. In recent years, some significant brand firms have included a sustainability plan in their industrial strategy. For example, PepsiCo announced its 2025 sustainability plan in 2015, aiming to use recoverable or recyclable packaging completely (Stark & Matuana, 2021). Petrochemical industries produce more than 95% of plastic material from non-renewable sources (Mangaraj et al., 2019). Plastics are used in the packaging of foods, construction material, medical gadgets, etc., while about 43% of the synthetic polymers are being used in the packaging industry in India.

Packaging is crucial in contemporary life since it helps minimize the loss of food material and waste generated. Still, it helps reduce energy usage for the transportation of food materials. In the packaging industry, the quantity of waste materials is increasing nowadays, so any sound solution must be needed from the industry. Packaging is an essential component of the manufacturing of food, starting from storage, preservation, distribution, and other unit operations (Ivonkovic et al., 2017). In recent years, traditional plastics are now being replaced with bioplastics. A *bioplastic* is a plastic that is biodegradable or has a bio-based origin (Shaikh et al., 2021).

6.3 Processing of Bio-Based Packaging Materials

Three-step methods mainly develop green or sustainable packaging materials, break down intermolecular linkage, synthesize new molecular arrangements, and develop a 3D polymeric network (Galić et al., 2011). Polymer shape and processing conditions affect the formation of new molecular bonds. Hydrophobic, electrostatic, covalent, and H-bonding interactions maintain the newly formed films (Zubair & Ullah, 2020). The bio-based plastic materials are prepared using two methods: the wet processing method and the dry processing method, as shown in Figure 6.1. Wet processing is concerned with dissolution, solvent type, and pH of the solvent, which can affect polymer conformation. Dry processing is concerned with the polymer's thermoplastic properties, which affect the inhibition of disulfide conversion reaction (Asgher et al., 2020).

6.3.1 Wet Processing

The wet processing method is used to prepare bio-based plastic from renewable resources, like carbohydrates, proteins, lipids, etc. The wet processing method depends on the dissolution of biopolymer in a suitable solvent to prepare the film-forming solution. As additions to bioplastic films, various compounds such as antibacterial agents, plasticizers, micro/nanostructures, cross-linking agents, fillers, and antioxidants, are employed (Felix et al., 2017). The casting of film-forming solution and solvent evaporation characterize this process. Using a plasticizer is beneficial because it decreases intermolecular links and hardness, resulting in a smoother and more flexible substance (Sanyang et al., 2016). Moreover, wet processing is preferred in the packaging industry due to the improved mechanical properties of prepared bioplastic sheets (Asgher et al., 2020).

6.3.2 Dry Processing

Dry processing is based on the thermoplastic properties of polymers, which play an essential part in developing packaging materials. This process is linked to the glass transition concept, which states that at a specific temperature, a glassy substance transforms into a semi-solid state. This semi-solid transition state modifies the polymers' mechanical and physicochemical properties. Intermolecular connections between protein

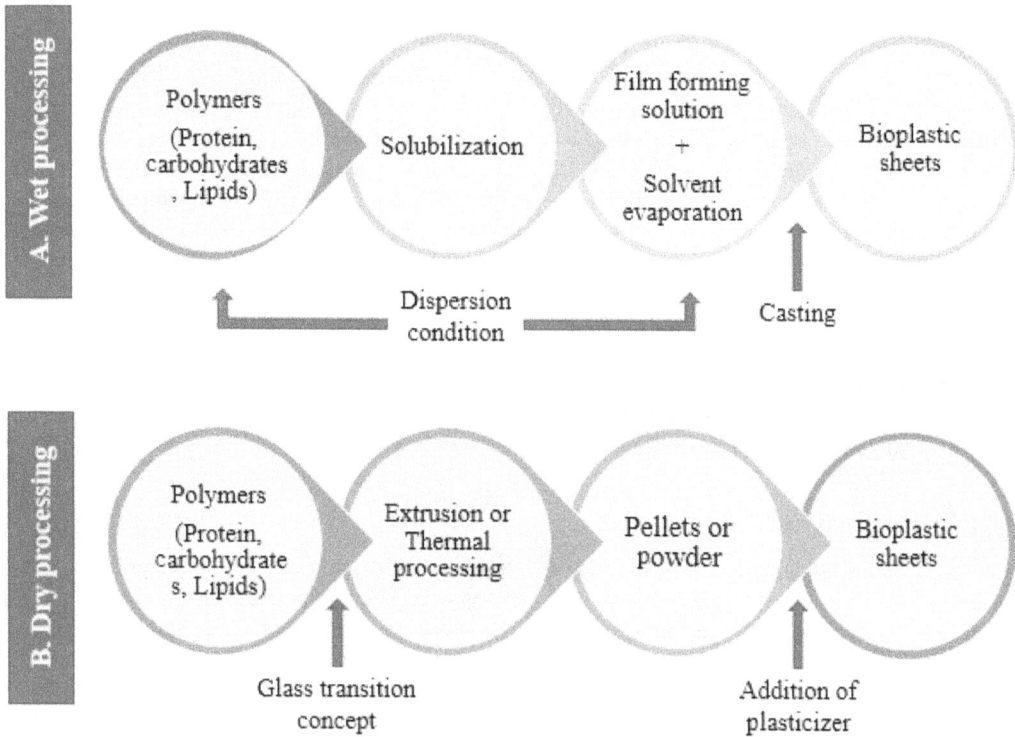

FIGURE 6.1 Wet processing and dry processing for sustainable packaging materials.

molecules are broken, producing denaturation, while new linkages and bonds form, causing changes in material characteristics. Polymer-based packaging is mainly made by several approaches, like thermal processing and extrusion. Both procedures can be employed separately or concurrently, with extrusion utilized for minor adjustments and mixing and heat processing for final product development (Asgher et al., 2020).

6.4 Bio-Based Food Packaging Materials

Bio-based polymers, nanomaterials, fibers, and their combinations have been proven sustainable or green packaging materials. Bio-based polymers are generally divided into four categories:

(i) Polymers mainly extracted from biomass
(ii) Synthetic polymers obtained from biomass monomers
(iii) Bio-based polymers produced by microorganisms
(iv) Biodegradable polymers developed from petrochemical monomers

Because food packaging materials must often meet several criteria, including permeability, mechanical characteristics, and antibacterial capabilities, generally, food packaging materials are usually made of bio-based composites (Wang et al., 2021a).

6.4.1 Bio-Based Polymers

6.4.1.1 Cellulose

Cellulose is a linear polymer which is semi-rigid in nature and available abundantly in nature. The interaction of hydrogen at intermolecular or intramolecular makes it difficult to dissolve directly. Because the cellulose chain contains multiple hydroxyl groups, several processes can be used to produce cellulose derivatives. It has also been stated that cellulose derivatives, such as carboxymethyl cellulose (CMC), are mixed with polyvinyl alcohol for creating a mechanically stable packaging film. The freshness of strawberries is protected by applying a two-layer edible coating of CMC and chitosan due to a decrease in metabolite level (Yan et al., 2019). The antimicrobial activity potential in packaging material is also observed in acetate cellulose (Gouvêa et al., 2015). Other cellulose derivatives, such as cellulose cinnamate, have been described as helpful packing polymers for fruits such as strawberries and mini-date vine tomatoes due to superior mechanical qualities, thermal stability, and limited water vapor, oxygen, and oil permeability (Wang et al., 2021b).

Cellulose is a linear polymer made up of cellobiose repeating units, which remain insoluble in organic solvents. It has a crystalline structure. It can be converted into various forms due to its insolubility and poor fluency. This change is accomplished by varying levels of replacement. Mechanical characteristics and degradation rate decrease as substitution degree rises. Cellulose acetate (CA) is a cellulose derivative with a tensile strength comparable to polypropylene. CA's applicability in thermal processing is limited due to its high glass transition temperature (Tg) (Shaikh et al., 2021).

6.4.1.2 Hemicellulose

Pentosans and hexosans merge to produce hemicelluloses, which are branched polymers. Generally, hemicelluloses are extremely water-soluble due to several side chains, which provide good film-forming properties. In the cell wall of hardwoods, xylans are the most abundant type of hemicellulose. It has been observed that xylan films have strong barrier characteristics against oxygen and grease. Without employing chemical processes, galactomannan and xyloglucan were isolated from *Caesalpinia pulcherrima* and *Tamarindus indica*. Their blended films were revealed to have high barrier qualities and thermal stability, allowing them to be potential environment-friendly, biodegradable, and edible packaging films (Mendes et al., 2017).

Nonetheless, due to the low mechanical strength and sensitivity to moisture, hemicellulose films are generally less preferable, which can be improved by combining with other polymers and chemical modification. Hemicellulose may be chemically modified to produce materials with a wide range of unique attributes.

Chitosan added with intercalated quaternized hemicellulose (QH) in montmorillonite resulted in better water vapor and oxygen barrier qualities in food packaging material with compact and mechanically robust films. The compact nacre-like structure was primarily responsible for the increase in mechanical capabilities, whereas the high viscosity of chitosan was mainly responsible for improving barrier qualities (Chen et al., 2016). An active food packaging material can also be prepared using a hemicellulose-based composite film (Mugwagwa & Chimphango, 2020).

Plasticizer addition enhances flexibility, toughness, and low oxygen permeability because hemicellulosic-based films are brittle. Because of low oxygen permeability, these films always play an essential role in packaging applications. Softwood glucuronoxylan and hemicelluloses are being used to synthesize the packaging materials with desired flexibility and oxygen permeability using bioplasticizers (Silva & Lucia, 2015). However, the films made may remain susceptible to water uptake. Decreasing film flexibility and the addition of more polymer content (alginate or CMC) may enhance the resistance to moisture absorption and mechanical strength.

6.4.1.3 Chitosan/Chitin

Cellulose chitin also remains one of the abundantly available biopolymers on Earth. It is derived from the exoskeletons of marine insects and invertebrates and fungi cell walls. Deacetylation of chitin makes chitosan a cationic biopolymer. Antibacterial properties against both Gram-positive and Gram-negative bacteria are also observed due to the presence of amino and hydroxyl groups in chitosan. It also shows antioxidant properties along with antibacterial properties in food packaging (Kumar et al., 2020). Synthetic polymers, mainly polyvinyl alcohol and polylactic acid, nanomaterials such as metal and metal oxide nanomaterials, functional extracts such as beeswax and other biopolymers, mainly protein and polysaccharides, are blended with chitosan to modify the thermal, barrier, and mechanical properties for food packaging (Wang et al., 2018).

Chitin is a linear copolymer. It is composed of N-glucosamine and N-acetylglucosamine, which are linked with a β-1,4 linkage. The organization of monomers in the polymers depends on the processing method. Chitin is typically found in bug, crab, and shrimp shells, and it's also widely accessible and is classified as amino cellulose. Chitin may also be obtained by fungal culture, which has a protein content ranging from 10% to 15%. Because chitin has poor solubility, it is frequently mixed for use in packaging (Shaikh et al., 2021).

6.4.1.4 Lignin

Lignin is another major complex biopolymer in the world. It is found primarily on the plant cell wall and has cross-linked structures. Ester bonds are commonly used to connect them to hemicellulose. Lignin comprises hydroxyl, carboxyl, carbonyl, methoxy, and benzene active groups. With its antibacterial, UV-shielding, and anti-oxidation qualities, lignin is extensively employed in the packaging of foods due to its particular structures and groups. The antibacterial characteristics of lignin were examined using hydroxypropyl methylcellulose (HPMC)-lignin-chitosan and HPMC-lignin composite films, and it was discovered that lignin concentration is a prime factor for its antimicrobial activity (Alzagameem et al., 2019).

The antimicrobial properties of lignin are achieved by forming composites by combining them with PVA or gelatin. The cell lysis occurs due to the interaction of the hydroxyl group of lignin phenolic compounds. To improve mechanical and wettability properties, lignin may be introduced into various matrixes of polymers, like chitosan, latex, PLA, alginate, and polyvinyl alcohol (PVA). Lignin has also been used as an antioxidant agent in active food packaging in various ways. Because of the added methoxyl groups, lignin-containing syringyl groups were shown to have UV-shielding capabilities and high antioxidant activity (Guo et al., 2019). Lignin acts as a nucleating agent and antioxidant through the addition of polyhydroxy alkynoate (PHA) and 3-hydroxybutyrate (PHB). Greater stiffness and lower permeability to carbon dioxide and oxygen of the food packaging film are also reported (Wang et al., 2021a).

6.4.1.5 Starch

Starch comprises amylose (linear/helical) and branching amylopectin. The flexibility of starch film can be enhanced by enzymatic, chemical, physical, and plasticization treatment. Citric acid and gelatin are added to improve the characteristics and structure of starch films. Superior water vapor and gas barrier qualities in the paper coating are obtained by the addition of acetylated starch. The sulfur hexafluoride plasma treatment was used to counteract starch's hydrophilicity. The additives, like nanoclay, ZnO, and MgO, are added with starch to produce a better barrier, UV shield, and antimicrobial properties (Nechita & Roman, 2020). Good mechanical, thermal, and high barrier properties are obtained by adding pectin and lemongrass oils to starch to make the composite film (Mendes et al., 2020).

Various forms of starch are generally used in food packaging applications, such as commercial starch, cornstarch, and thermoplastic starch (TPS).

Guarás et al. (2015) used PCL to improve the matrix stability of thermoplastic starch to provide improved mechanical strength. The level of 15% PCL on a weight basis exhibited a noticeable impact on the stabilization of starch blends. Basil and green tea extract were also added to cassava starch film for the preparation of packaging material to make the packaging intelligent and biodegradable (Medina-Jaramillo et al., 2017). The oxidative process during food preservation is also slowed down by adding solid polyphenolic components. Carotenoids and chlorophyll components make the films pH-sensitive (Din et al., 2020).

6.4.1.6 Pectin

Pectin is a complex heteropolysaccharide. It is obtained from plant cell walls. It is commonly used in food packaging. Pectin films are prepared by thermocompression molding or casting for use in food packaging. Pectin is also compatible with other biopolymers, such as polysaccharides, lipids, proteins, etc. Plasticizers such as glycerol were commonly utilized in pectin films.

Direct coating of pectin is also used to encapsulate cut fruits and vegetables to provide active packaging and the release of active components at a controlled rate—for example, oxidation of soybean oil can be delayed using pectin films till 30 days. Pectin films containing essential oil obtained from clove were added with barrier, antioxidant, antibacterial, and mechanical attributes (Nisar et al., 2018). Corn flour composites also effectively maintain the freshness of tomatoes when used to coat the surface of tomatoes (Mellinas et al., 2020). The shelf life of food can be enhanced by adding marjoram essential oil by encapsulating it with a pectin matrix (Almasi et al., 2020). The antibacterial characteristics, barrier qualities, tensile strength, and thermal stabilities of pectin/starch/chitosan films increased with the addition of essential oils and nisin (Akhter et al., 2019). Another study was reported by Lei et al. (2019), and it was found that tea extracts were effectively helpful for enhancing the antimicrobial, antioxidant, and mechanical properties of pectin.

The film is designed to fulfill the functional requirement of various preferable characteristics of packaging material. A copolymer synthetization is popular to provide the functionality and strength of pectin films. Active films are formed when citrus pectin is combined with marjoram essential oil. Mechanical strength was enhanced due to the creation of hydrogen bonds between pectin and essential oil (Almasi et al., 2020). Similarly, pectin films are added with thyme essential oil to lower elongation at break value (Lin et al., 2020). The antimicrobial activity and pharmacological and therapeutic potential of edible pectin films containing essential oil have been studied (Ghasemi et al., 2017). Some studies have sought to improve pectin films' hydrophobicity. The hydrophilicity of the modified films was lowered by including curcumin/gamma-aminobutyric acid in pectin. Compared to pure pectin films, the inclusion of curcumin boosted thermal stability by 112.42% (Meerasri & Sothornvit, 2020).

6.4.1.7 Alginate

Alginate is also known as a biopolymer (anionic) that is primarily derived from brown algae. It is regarded as safe for use in direct contact with the material. Alginate and calcium chloride were generally used to apply to paperboard surface to improve oil resistance and barrier attributes. A composite made from sodium alginate/carboxymethyl cellulose and sodium propylene glycol alginate/alginate demonstrates excellent paper coating with materials having increased oil barrier capabilities, water resistance, and mechanical properties modification (Wang et al., 2021b).

Alginate is a polymer found naturally in the cell walls of certain brown algae (Senturk Parreidt et al., 2018). Alginate has excellent thickening stability, colloidal characteristics, moisture retention, and gel-forming ability (Dou et al., 2018). Generally, alginate is utilized for various purposes, like as stabilizer, gelling agent, and thickener in the food sector. For example, a well-known method of grape preservation is by applying CMC and brewer yeast for coating grapes. The application of sodium alginate with lemongrass and glycerol is also used to keep cut fruits fresh for about 16 days (Azarakhsh et al., 2014). Similarly, alginate with thymol also resulted in the preservation of cut fruits and vegetables (Chen et al., 2021). Alginate films added with salicylic acid inhibit microbial growth (Kurczewska et al., 2021). Alginate-based films are effective and can be used in the packaging of foods on a small scale or large scale (Atta et al., 2022).

6.4.1.8 Proteins

Some proteins can make free- or self-standing films, which are also good in film-forming characteristics to package food materials. Protein sources commonly used in protein-based films are soybean, whey, and wheat gluten. Other proteins studied and reported for food packaging applications include milk protein, maize zein, fish gelatin, and myofibrillar proteins. Proteins have inherent disadvantages, such as poor barrier, mechanical, thermal, and physicochemical attributes for use as food packaging materials. The quality attributes of protein films can be enhanced by adding nanofillers to the formulations.

Nanofillers such as titanium dioxide, zinc oxide, silica, carbon nanotubes/nanoparticles, layered silicates, as well as organic nanofillers like nanocellulose, and nanochitin are used to enhance the barrier and mechanical properties of the films (Zubair & Ullah, 2020). Plasticizers and other active substances were also used for protein-based packaging to enhance microbial delay development and water barrier qualities (Chen et al., 2019). Furthermore, proteins can be changed chemically, physically, and enzymatically to produce films with improved water resistance and mechanical strength. For their actual uses, proteins were utilized as coatings over the surfaces of cellulose-based packaging because protein-based films' mechanical strength was often lower (Coltelli et al., 2015).

6.4.1.9 Collagen and Gelatin

Proteins of numerous polypeptides, such as proline, lysine, glycine, and hydroxyproline, form collagen in the form of connective tissue protein. The glycine content offers flexibility for the collagen. They are used in PVA films and cellulose.

The chemical breakdown of collagen forms gelatin, and it is a significant polypeptide of more molecular weight which is formed due to the chemical breakdown of the collagen. It contains 19 amino acids and has a robust film-making ability. The packaging material's water vapor barrier and mechanical strength are greatly influenced by its amino acid composition, molecular weight distribution, and some plasticizers. The film's applicability in packaging is limited due to the thermal instability of the collagen. To obtain great film attributes for food packaging, several additives are used to increase or change the barrier qualities and mechanical stability of the film. A film of 4% gelatin with 2.5 and 5.0% maize and olive oil, respectively, is utilized for the packaging of sausages (Ramos et al., 2016). The enzyme protease is responsible for gelatin degradation (Shaikh et al., 2021).

Collagen as well as gelatin are derived from animals. Collagen is an abundantly available protein in nature. It accounts for around 20 to 25% of body mass in animals. Its structure comprises three cross-linked α-chains, whereas gelatin is a denatured collagen derivative of numerous polypeptides. Collagen is high in hydroxyproline/proline, methionine, and glycine (Asgher et al., 2020). The extrusion process produces collagen-based bioplastics. They have a wide range of uses, whereas gelatin-based films need a wet process. The collagen-based film offers excellent mechanical characteristics; for example, hydrolyzed collagen films are popular due to their higher tensile strength (Fadini et al., 2013), while gelatin films have poor mechanical and barrier characteristics, indicating their hydrophilic nature (Ciannamea et al., 2018).

6.4.1.10 Wheat Gluten and Soy Protein

Wheat gluten is the by-product of the starch production process and is usually available at a lower cost. They degrade faster than conventional polymers and produce no hazardous by-products. It is proved to be a superior film-forming agent, although it needs a plasticizer due to its brittleness. Water-soluble carbohydrates are not present in soy protein concentrate. Textured soy protein (TSP) is produced from soy protein concentrate and has a 70% protein concentration. The TSP films show lower mechanical strength and poor barrier qualities due to the hydrophilic nature of the protein. The films made from isolated soy protein remain sensitive to moisture. Stearic acid is added to enhance the tensile and thermal properties and reduce moisture sensitivity (Lodha & Netravali, 2005). For biodegradable soybean-based packaging containers, κ-carrageenan, gellan gum, and glycerol are used to provide soy protein film.

Compression, extrusion, and molding are used with plasticized wheat gluten to prepare a biodegradable film (Zubeldía et al., 2015). Hydrophobic, hydrogen, and disulfide interactions are synthesized during film production, and sulfhydryl groups help stabilize disulfide bonds (Sharma et al., 2017). Heating causes polymer denaturation and the breakdown of natural disulfide and hydrophobic groups. The oxidation of gluten results in the formation of new disulfide linkages during drying. The clarity of the film is determined by the gluten mass purity and the casting media used, which might be acidic or alkaline (Chiou et al., 2020). The films developed high uniformity, good mechanical strength, and gas barrier properties (Mojumdar et al., 2011). Soy proteins can also be used in the production of bioplastic sheets for use in packaging. Soy protein–based films are more translucent, smoother, flexible, and less expensive than other protein-based bioplastics (Otoni et al., 2016).

Furthermore, they have strong oxygen barrier characteristics in low-moisture situations. Compared to low-density polyethylene, the main disadvantages refer to heat instability, poor mechanical strength, and allergenicity. Stainless steel plates at high temperatures are used from films using soy protein isolates in an aqueous solution. Several researchers have reported that film creation utilizes different protein-based raw materials, such as pistachio globulin protein, canola protein cake, pumpkin oil, pea protein, etc. (Acquah et al., 2020).

6.4.2 Bio-Based Nanomaterials

6.4.2.1 Cellulose Nanofibers

Nanofibers of cellulose are used for manufacturing film to enhance the water vapor barrier capabilities studied by Bedane et al. (2015). Cellulose nanofibers (CNFs) were used for soybean-based polymers to enhance the water vapor barrier characteristics (Lu et al., 2014). The techniques like sol-gel technique, composite extrusion, electrospinning, and layer-by-layer assembly can be employed to enhance water resistance and water vapor barrier properties of cellulose nanofiber–based films (Nechita & Roman, 2020). Polyethylene glycol (PEG) and polylactic acid (PLA) are added with CNFs to develop composites for food packaging. The cellulose nanofibers were chemically treated to achieve optimal compatibility and improvement in hydrophobic properties with PLA. The biodegradable composites are coated on the surface of packing paper to reduce water vapor permeability (Song et al., 2014). The active chemicals through the nanoporous networks of the CNFs, tensile strength, air resistance, and water retention properties were enhanced by coating CNFs on paper substrates (Jin et al., 2021). The carboxymethyl cellulose nanofibers, along with silver/chitosan nanoparticles, were also used to prepare cushioning and antibacterial bifunctional aerogels in 3D-printing inks (Zhou et al., 2021).

6.4.2.2 Cellulose Nanocrystals

Cellulose nanocrystals (CNC) are extracted from several sources. CNC is frequently used as a reinforcing agent or to provide chemical compatibility for numerous packaging materials. It is joined with polypropylene, polyethylene, PVA, PLA, PET, and CMC to prepare food packaging materials (Huang et al., 2020). The resulting PVA/CNC/CMC composite films demonstrated better thermal, barrier, and mechanical characteristics and promote transparency for packaging material (El Achaby et al., 2017). PLA/nanocellulose composite materials may have antibacterial capabilities when combined with antimicrobial ingredients. Organic materials and inorganic materials are the two types of antibacterial materials. Organic acids, polymers, and enzymes are categorized under organic antimicrobial materials, whereas metal and metal oxide nanoparticles are inorganic antimicrobial agents (Gan & Chow, 2018). Cellulose nanocrystals obtained from the baggage of sugarcane were added with zinc ions to form hydrogels. Ripening was delayed, and the coated fruits, such as papaya, avocado, banana, and strawberry, were effectively kept for a longer duration (Jung et al., 2020).

6.4.2.3 Bacterial Cellulose

Some bacteria rather than plants are helpful in producing bacterial cellulose. Bacterial cellulose possesses a high water-holding capacity, high strength, excellent purity, and high degree of polymerization. Bacterial cellulose obtained using bovine lactoferrin was employed as antimicrobial edible packaging (Padrão et al., 2016). To develop food packaging, antimicrobial peptides were immobilized in bacterial cellulose (Malheiros et al., 2018). Polymer composite films using bacterial cellulose have also been prepared for food packaging (Bandyopadhyay et al., 2018). Bacterial cellulose nanowhiskers are used as reinforcement for PVA/gelatin-based food packaging (Haghighi et al., 2021). Edible packaging material with antimicrobial and excellent mechanical characteristics can be made using bacterial cellulose nanofibers, starch, and chitosan (Abral et al., 2021). The presence of silver nanoparticles and bacterial cellulose nanocrystals in the chitosan matrix makes the packaging antimicrobial. Furthermore, the mechanical characteristics and water vapor permeation properties were enhanced (Salari et al., 2018). To develop food packaging materials, nanofibrillated bacterial cellulose and fruit purees were mixed with pectin.

The food packaging prepared with the incorporation of fruit purees does not possess a desirable mechanical strength, barrier attributes, etc., while nanofibrillated bacterial cellulose incorporation enhanced mechanical and barrier properties (Viana et al., 2018). The edible composite film created by combining chitosan, tapioca starch, and bacterial cellulose has superior mechanical qualities, high heat stability, and antibacterial capabilities (Abral et al., 2021). Aside from CNF, CNC, and bacterial cellulose, additional celluloses, such as microfibrillated cellulose, cellulose acetate, and microcrystalline cellulose, have been employed as PLA additives for packaging food with a better barrier, mechanical, and transparent properties (Khosravi et al., 2020).

6.4.3 Bio-Based Fibers

6.4.3.1 Paper

Cellulose fibers are used to make paper. As compared to other synthetic polymers, it has benefits, like biodegradability, recyclability, and compostability. Paper-based packaging includes juice boxes, tea bags, egg trays, etc. These are obtained from wood or agroresidual fibers using mechanical or chemical pulping. The composition and shape of fibers are prime factors for determining packaging needs. A fiber coarseness of 1.2 mg/m is desirable for the preparation of tea bags (Sood & Sharma, 2021). The complete replacement of wood fiber with agroresidue fiber is not possible due to the short length of fiber and the presence of non-fibrous components in agroresidue-based fibers. However, replacement of soft wood fiber with up to 85% is possible for the agroresidue-based fiber (Bhardwaj et al., 2019). Pine needles are used as a lignocellulosic source with an ethylene scavenger (Kumar et al., 2021). Paper's hydrophilic nature and barrier qualities are inadequate. A high-relative-humidity environment promotes hydro expansion, which provides a deleterious impact on product qualities like dimensional stability and compressive strength (Niini et al., 2021). Mineral oil-free or bio-based inks are recommended for the production of recycled fibers to overcome the difficulty in handling impurities (Buist et al., 2020).

6.4.3.2 Other Natural Fibers and Their Composites

Natural fibers can be formed mainly in three ways, namely, (1) geological processes, such as asbestos; (2) plants, cotton lint, or sisal, hemp, flax fibers; or (3) animals, such as silk and wool. Natural fibers provide reinforcement as fillers in biocomposites, where they are added to the matrix of proteins and polysaccharides in full bioplastics. They can be manufactured in the same way as traditional plastics, such as resin transfer molding, compression/injection molding, and extrusion. Aside from the obvious advantages of degradability and renewability, natural fiber possesses good strength and lower density and is inexpensive. Moreover, the hydrophilic characteristic of natural fiber composites, like with paper-based packaging, poses a severe difficulty (Berthet et al., 2016). Research has been conducted to investigate fibers of various origins and prepare standard operating procedures for manufacturing food packaging material. Polyhydroxyalkanoate was used as bioplastics and added with 10% coconut fibers for antibacterial activity along with essential oregano oil. It was observed that the ideal fiber size was greater than 1,500 m, with an aspect ratio of 7. There was no significant disadvantage in terms of processing or material qualities while employing coconut fibers, which would save a significant amount of expensive bioplastics. Oregano essential oil treatment for coating coconut fibers results in a bacteriostatic effect (Torres-Giner et al., 2018).

6.4.4 Synthetic Bio-Based, Biodegradable Polymers for Food Packaging

6.4.4.1 Polylactic Acid (PLA)

Because of its availability, compostability, biocompatibility, and qualities similar to traditional fossil-based polymers, it is one of the promising bio-based polymers. Polylactic acid is biodegradable, but it requires about 55–60°C for composting due to its higher melting point and high temperature of glass transition (Meereboer et al., 2020; Nilsen-Nygaard et al., 2021). The advantages of utilizing PLA for food packaging include biodegradability in industrial circumstances, making it recyclable, biocompatible, and a promising substitute to replace traditional plastic materials (Mangaraj et al., 2019). The US Food and Agriculture Agency has also approved it as safe. Moreover, because of poor mechanical and barrier qualities, PLA's use in food packaging is currently limited. PLA's qualities may be designed and balanced by modifying its chemical composition and molecular features. Furthermore, mixing it with other materials makes it ready for the specific needs of various food items.

PLA is aliphatic polyester derived, which is derived through lactide monomer polymerization. Fermentation of renewable resources is used to produce lactic acid monomers. Due to its higher molecular weight, it degrades quickly and becomes biodegradable or recyclable (Singla & Mehta, 2012). PLA's characteristics may be modified from crystalline to amorphous by varying the monomeric ratio. Commercially available PLA glass transition temperatures range from 63.0 to 63.8°C (Briassoulis, 2004). The polymer degradation rate is influenced by initial monomer and crystallinity concentration. Higher monomer content exhibits a lower degradation rate due to its higher crystalline nature (Shaikh et al., 2021). The characteristics can be modified by the addition of PBA to PLA and transforming it from brittle to ductile. The system's elongation capacity was eventually enhanced. Other mechanical qualities were improved as a result of the PBA acting as a reinforcing component for the PLA matrix. When PBA is added to PLA, toughness may be increased about three times from 3.9 to 12.6 kJ/m². The mechanical and morphological qualities of PLA can be improved by incorporating silver nanoparticle aggregates and microcrystalline cellulose (Fortunati et al., 2010). All three components display distinct qualities. The addition of microcrystalline cellulose improves permeability and thermal properties and enhances crystallinity. Ali et al. (2014) reported that the chain synthesis of PLA-based polyurethane (PLAPU) films with PCL diol was used as a softening/flexibility agent. The mechanical and gas barrier qualities of an acquired mix with a PLAPU/PCL ratio of 1:3 were the best.

6.4.4.2 Polyhydroxyalkanoates

Polyhydroxyalkanoates (PHAs) are bio-based polyesters which are generally produced by bacterial fermentation (Samui & Kanai, 2019). Pure microbe cultures cultivated under sterile circumstances on various renewable sources such as glucose can be used to produce these biogenic polyesters. Wastewater containing organic acids and sugars from industrial applications can also be used to produce PHAs (Colombo et al., 2019). The popularity of PHAs follows the increasing trend not only due to a potential alternative for fossil-based plastics but also because they possess similar physicochemical qualities and biodegradability in many conditions (Chan et al., 2019).

Biopol™, a commercially available PHA product aimed toward food packaging applications, is manufactured by Metabolix Inc. (USA) (Bajpai, 2019). Biopol™ has an outstanding coating and film-forming capabilities, and it is used to prepare disposables to keep or serve foods. A PHA bioplastic is also developed by a collaboration of Metabolix Inc. and Archer Daniels Midland Co., which is suitable for high melting and strength-grade bioplastics. Bioplastic works well for trays, tubs, cold cups, hot cups, lids, etc., especially for single-use serving (Nilsen-Nygaard et al., 2021).

The biodegradability of these PHAs also shows thermoplastic biopolymer. These can be combined with other copolymers to provide excellent solutions and versatile use. The addition of zein, pullulan, and WPI enhanced high barrier property to observed oxygen barrier capabilities (Fabra et al., 2014). Improvements in WVPC and OPC values of 38–48% and 28–35%, respectively, have been reported due to the addition to the matrix. The fibrous morphology of PHAs can be obtained by adding pullulan and zein, while beaded morphology can be obtained by the addition of WPI (Din et al., 2020). *Posidonia oceanica* (PO) was added to PHA to improve tensile strength by 80%, while mechanical strength was enhanced from 1.63 to 3.8 kJ/m² (Seggiani et al., 2017).

6.4.4.3 Polycaprolactone (PCL)

PCL is a fossil-based polymer that is biodegradable, easy to produce, non-toxic, flexible, and hydrophobic. It has many characteristics, mainly as a biodegradable, semi-crystalline, easy-to-process, and low-cost fossil-based polymer. It expands applicability as a polyurethane compatibilizer because of its solubility in inorganic and organic solvents with 60°C as glass transition temperature (Vroman &

Tighzert, 2009). The hydrophobicity of hydrophilic chitosan polymer is also boosted by the incorporation of PCL. When compared to pure films, the combination has a lower rate of water vapor transmission. Food preserved in such films has a longer shelf life as a result of this feature. A number of these films are also commercially available (Shaikh et al., 2021). Guarás et al. (2015) synthesized TPS and PCL blends for use in food packaging. PCL in pure form has a good crystalline ability which can be improved through synthetization (Din et al., 2020).

6.4.4.4 Polybutylene Succinate (PBS)

Polybutylene succinates are polyalkene dicarboxylates formed by polycondensing glycols. It is a crystalline polymer with white color and easier processibility, with a glass transition and melting point temperature of −45 to −10°C and 90 to 120°C, and possesses elongation of about 330% at the break. It has similar mechanical properties to polypropylene and polyethylene (Wang et al., 2007). These were initially developed in Japan in 1990. Many new copolymers have been developed since then, such as polybutylene succinate-co-adipate (PBSA), which was created by adding adipic acid at a specified quantity. A tiny amount of coupling agents can be used to raise the polymer's molecular weight. Various businesses sell various PBS by modifying the monomeric components too. The characteristics and breakdown rate of these polymers are influenced by the type of diols and diacids utilized during condensation (Shaikh et al., 2021).

6.4.4.5 Polyglycolide (PGA)

Polyglycolic acid or polyglycolide are synthesized using polycondensation of glycolic acid. It has a glass transition and melting point temperature in the range of 35 to 40°C and 220 to 250°C and remains as basic aliphatic polyesters. Because of its high crystallinity of 40–55%, it is insoluble in water. However, it is soluble in fluorinated solvents and can be used to prepare polymer films of higher molecular weight. Within five to six months, the polymer is reabsorbed by the body (Tiberiu, 2011). The Chemours Company also produces a low-molecular-weight film (Shaikh et al., 2021).

6.4.4.6 Polyvinyl Alcohol (PVA)

It is primarily available in amorphous phases with minor crystallinity, which makes it a semicrystalline polymer. The molecular weight and degree of hydrolysis usually vary between 20,000 to 400,000 and 80 to 99%, depending on the vinyl acetate length used in PVA preparation (Abdullah et al., 2017).

6.5 Sustainable Packaging Materials

6.5.1 Based on Wood Fiber

Because cellulose is the most prevalent biopolymer, it is a useful and easily accessible bioresource for preparing sustainable packaging solutions. Cellulose may be obtained from algae, forestry waste, agricultural waste, wood, plants, and even microorganisms. Other key wood components include cellulose, lignin, and hemicellulose, which vary in quantity depending on the parent material. Cellulose may be extracted from the wood cell containing about 40 to 50% cellulose for the preparation of fiber-based packaging. Isolation methods for cellulose affect the chemical content and shape of cellulose fibers, resulting in varied packaging uses. Cellulose fibers may help with sustainable packaging in a variety of ways, from 100% cellulose-based goods like paperboard or paper to transforming into less than 1% nanocellulose in biopolymers to enhance the barrier properties of packaging (Stark & Matuana, 2021)

6.5.2 Paper and Paperboard

Packaging applications are a significant market for produced paper and paperboard. In 2000, packaging accounted for around 47% of total paper and paperboard production. These are considered sustainable while prepared from easily recyclable components. Inks and coatings, on the other hand, might have a detrimental influence on recyclability. Plants have been used to make paper for millennia. The majority of paper today is made from the wood pulp of various trees, for example, coniferous (James et al., 2002). Because of its weak barrier qualities, low heat sealability, and strength, protection as packaging is limited. The protection level and quality of paper packaging can be improved by adding strengthening agents, brighteners, and colorants. Another way to improve the strength is by laminating with plastic or aluminum to obtain the desired quality. Paper is coated with wax, or its density increased, to provide better barrier qualities and grease resistance. The functional qualities of the paper can be enhanced by a change in processing parameters too (Stark & Matuana, 2021).

6.5.3 Cellulose Nanomaterials (CNs)

In the last decade, there has been a remarkable increase in interest in CNs. The word *CN* refers to a kind of cellulose particle with at least one dimension on the nanoscale. They are inexpensive, light in weight, and ecologically beneficial when compared to other nanoparticles. Wood is the most prevalent source of CNs. The addition of CNs provides higher surface area, lower density, good transparency, better strength and barrier qualities, and lower thermal expansion than the other packaging material, which make these suitable for food packaging applications desiring batter barrier properties (Stark, 2016; Vilarinho et al., 2018). The manufacturing of CNs is anticipated to have an environmental impact due to the usage of huge volumes of acid/water for hydrolysis. CNs, on the other hand, may be utilized to make films for preparing polymeric-based composites or laminates. Extensive research on CNs paves the way for commercial utilization in the packaging of food materials (Yousseff & El-Sayed, 2018). Furthermore, the safe use of this material also provides strong justification for the commercialization of these materials.

6.5.4 Cellulose Nanomaterial Films

CNFs have substantially larger aspect ratios than CNCs, which allow the production of CNs from aqueous suspension directly. A robust, stiff, transparent film can be prepared by an entanglement of fibers between hydrogen bonding, which are not redisbursable during the removal of water. Because of the similarities in production procedures with cellulosic-based paper, CNF films have been dubbed "nanopaper." Mechanical characteristics can be influenced by elements such as raw material, manufacturing process, film-making techniques, and testing environment (Qing et al., 2015). The oxygen transfer rates (OTRs) of CNs are superior to those of traditional packaging materials. Because of the lower porosity in cellulose nanomaterial films, diffusion of oxygen occurs in place of transfer for appropriate thick sheets due to the non-linking of pores (Minelli et al., 2010). The OTR of CNF films might vary depending on the treatment, manufacturing technique, and thickness. Increasing film thickness (Aulin et al., 2010) or lowering void volume with glycerol plasticizer (Minelli et al., 2010) can increase OTR. However, because CNF films are hydrophilic, their OTRs rise as humidity increases.

6.5.5 Bioplastic Materials

Polymers derived from petroleum have traditionally been used in the packaging sector. However, plastics derived from biological material provide sustainable solution due to the rise in oil prices, landfill problems, environmental concerns, regulation, and consumer preferences for finding out more biological material for use in packaging. The packaging industry is shifting its focus toward bio-based or degradable materials that will assist the sector to achieve its future sustainability demands for a wide range of applications. Bioplastics have a lower carbon footprint and are more compostable than petroleum-based plastics (Boonniteewanich et al., 2014). Furthermore, bio-based plastics exhibit several physicomechanical qualities equivalent to petroleum-based polymers, making them an excellent substitute (Türünç et al., 2011).

6.6 Properties

Several properties of biodegradable or non-biodegradable polymers are studied to understand the characteristics of polymers in a specific set of conditions, uses, etc. Some of the important properties are illustrated in Figure 6.2.

6.6.1 Tensile Strength

The tensile strength of a material is the highest amount of stress that it can bear before failing. This property is the quite-popular mechanical property for determining the strength of any polymer. Mechanical properties remain critical for determining the protection using packaging material. Tensile strength is affected by polymer type, manufacturing conditions, chemical modification, additives, etc. Tensile strength also varies with the processing method, time, and storage (Briassoulis & Giannoulis, 2018). The tensile strength of PLA can be enhanced by preparing bio-nanocomposites with the incorporation of nanoparticles (Shaikh et al., 2021).

6.6.2 Water Vapor Transmission Rate

It can be defined as the amount of water vapor passed through the material per unit of time and area (kg/mm²s) (Auras et al., 2006). Food material is very sensitive to moisture and degrades faster due to the absorption of moisture and an increase in the moisture level of food during storage. Water vapor transmission rate (WVTR) is a prime factor for storing meat, dairy foods, seafood, etc., where critical control of moisture level is desirable. The WVTR of any packaging material is measured at a temperature of 38°C and 90% relative humidity condition. Biodegradable plastics show better control on water vapor permeability in comparison to thermoplastic polymers, allowing them to be utilized for dry product storage, and because PBS has low water vapor retention, its use is limited. When mixed with PLA at a 20:80 ratio, PBS produces a composite with excellent water vapor retention (Shaikh et al., 2021).

6.6.3 Oxygen Transmission Rate

Oxygen transmission rate is indicated through oxygen permeability coefficients (OPC), which can be defined as the amount of oxygen that passes through a substance per unit area and time under pressure

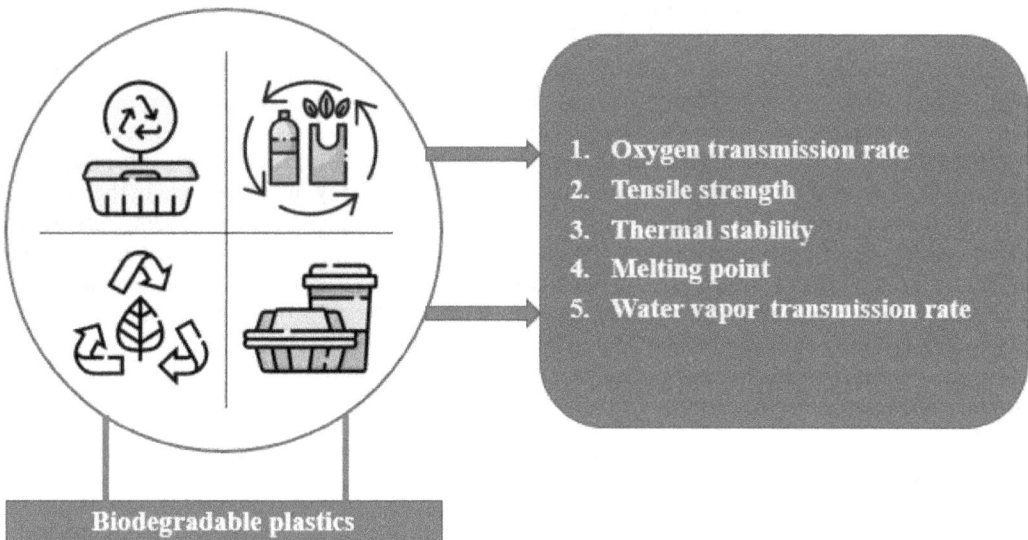

1. Oxygen transmission rate
2. Tensile strength
3. Thermal stability
4. Melting point
5. Water vapor transmission rate

Biodegradable plastics

FIGURE 6.2 Illustration of key properties of biodegradable polymers.

(kg/mm²sPa). The oxidation process can be restricted using low-OPC polymers, which provide the benefit of extending the shelf life of food material (Oliveira et al., 2004). The oxygen transmission rate in biodegradable polymers is very low, allowing only a limited quantity of oxygen to infiltrate. Blends are created in some circumstances by combining biodegradable polymers to improve barrier characteristics. The hydrophobicity of packaging material was also enhanced due to the addition of chitosan in 20:80 chitosan–starch films as compared to native chitosan films (Akter et al., 2014).

6.6.4 Melting Point

Melting temperature is defined as the temperature at which a material begins to change from solid to liquid phase (Tm). It is also the highest temperature that a polymer may withstand before deforming (thermodynamic characteristic). The melting temperatures of PC and PET have the greatest values of 240 to 280°C and 245 to 270°C, respectively. Chitin has a Tm of 290–300°C, which is comparable to PET or PC in the case of biodegradable polymers. However, chitin is frequently combined with other materials due to its inability to shape alone (Shaikh et al., 2021).

6.6.5 Thermal Stability

Thermal characteristics are important when considering the possible usage of polymeric materials for various packaging applications. These properties are important for the development of materials with better characteristics (Begum et al., 2020). *Thermal stability* refers to the retention of various properties, like toughness, strength, elasticity, etc., of polymeric material at a particular temperature. The thermogravimetric study of polymers is commonly used to measure their heat stability (TGA). Polymer thermal stability is determined by its molecular weight, degree of crystallinity, and chemical structure. The cross-linking activities and aromatic features increase the heat stability of polymers. Polymers are less resistant to high temperatures in case of having double bonds or the presence of oxygen in the main chain of the structure (Shaikh et al., 2021).

6.7 Applications

6.7.1 Modified Atmosphere Packaging (MAP)

It refers to the packing of a perishable product in an environment that has been changed such that its composition is different from that of air. It is a frequently used technology in packaging primarily fruits and vegetables by altering the composition of gasses in package headspace. The gas composition for a specific food is decided by several elements, including commodity, cultivar, respiration rate, storage temperature, and product weight (Briano et al., 2015). By changing the O_2 and CO_2 concentrations, MAP has shown a reduction in respiration rate and delay in fruit ripening. Shrinkage due to water loss can also be minimized by maintaining relative humidity in the package in the range of 90 to 95% (Giacalone & Chiabrando, 2013). Biodegradable packaging is currently being researched for the packaging of fresh food items (Peelman et al., 2013). The application of an oxygen absorber enhances the shelf life of strawberries in bio-based containers. At a specific temperature range, cabbage, sweet corn, broccoli (entire or chopped), lettuce (shredded or head), tomatoes, and blueberries are successfully stored using biodegradable laminates. Berries have a longer shelf life due to the changed atmosphere, and also, an enclosed container protects food from diseases and other environmental toxins (Briano et al., 2015).

6.7.2 Edible Packaging

Edible packaging is a good solution for food applications as it also has good barrier properties, mechanical strength, control release of chemicals and other sensory attributes. They are an essential component of the cuisine and are consumed with it. Films, sheets, coatings, and pouches are the most common types of edible packaging (Janjarasskul & Krochta, 2010). Edible films are made as soil sheets, while coating remains available in liquid form (Galus & Kadzińska, 2015). Proteins, polysaccharides, and lipids are the key components of edible packaging. The two most common components are chitosan and gelatin/collagen. The coatings minimize O_2 concentration, prevent oil migration, and maintain moisture level. Collagen sausage casing is proved to be an effective edible film available commercially.

6.7.3 Active Packaging

Active packaging is defined as "deliberately including components that would release or absorb chemicals into or from the packed food or the surrounding environment." It is a novel method for ensuring the safety of food products and extending their shelf life. The system comprises O_2 and CO_2, ethylene, and odor absorbers for scavenging activity, and antimicrobial, antioxidant, and flavor emitters for releasing components in packaging systems (Yildirim et al., 2018). The interaction among materials blocks or diminishes the activity of some active compounds when they are directly absorbed into the food. As a result, the regulated release of active ingredients is a prime focus for an effective bulk food system. A thin biopaper is made by adding zinc oxide and oregano essential oil to provide current active packaging solutions, which act as antibacterial to *Staphylococcus aureus* and *E. coli*. If emitting sachets containing eugenol, carvacrol, or trans-anethole are placed within mixed cellulose/PP pillow bundles, they can be utilized to enhance the shelf life of iceberg lettuce. The sachets progressively release the natural antibacterial ingredient and aid with food preservation (Wieczyńska & Cavoski, 2018).

6.8 Future Perspectives

Future research on bio-based materials still needs to focus on improving its effectiveness in meeting the demand for food preservation during storage and transportation. Several things must be addressed when it comes to the actual use of these bio-based products. Suitably available local bio-based materials are to be worked out, and tailor-made solutions should be prepared to address the physical or chemical routes based on the storage conditions of specific foods for their packaging. The commercialization of developed bio-based materials still faces several barriers that need to be overcome to get a real advantage. The energy used for the preparation of bio-based materials, such as biopolymer extraction, nanomaterial isolation, and preparation, should also be minimized to reduce the cost of bio-based packaging material. More flexible commercial procedures and equipment capable of matching large-scale production of bio-based products are in great demand. Another future trend will be to have a better knowledge of the biodegradability and sustainability of these materials to reduce their environmental effect after usage. More work is needed to create greener procedures that use fewer or no harmful organic solvents. The issue of food safety should also be considered critical in the development of various bio-based food packaging. The last issue remains the use of harmful chemicals or procedures in food packaging materials.

6.9 Conclusion

Consumer demands and new legislation provide the transition to sustainable packaging. Biodegradable polymers contribute to lowering the environmental effect of plastic manufacturing and processing. Because biodegradable polymers are generated from renewable feedstocks such as agricultural waste, there is a tremendous possibility for research to be done to capitalize on this economic opportunity. The usage of bio-based polymers for food packaging and other applications is rapidly expanding. However, comprehensive research into the interactions between biopolymers and food components in processing and storage is required before using any food packaging. Future research should concentrate on using nanotechnology and sensors to aid in communicating information to customers. Biodegradable polymers can contribute to long-term environmental sustainability.

REFERENCES

Abdullah, Z. W., Dong, Y., Davies, I. J., & Barbhuiya, S. (2017). PVA, PVA blends, and their nanocomposites for biodegradable packaging application. *Polymer-Plastics Technology and Engineering*, *56*(12), 1307–1344.

Abral, H., Pratama, A. B., Handayani, D., Mahardika, M., Aminah, I., Sandrawati, N., . . . Ilyas, R. A. (2021). Antimicrobial edible film prepared from bacterial cellulose nanofibers/starch/chitosan for a food packaging alternative. *International Journal of Polymer Science*, *2021*.

Acquah, C., Zhang, Y., Dubé, M. A., & Udenigwe, C. C. (2020). Formation and characterization of protein-based films from yellow pea (*Pisum sativum*) protein isolate and concentrate for edible applications. *Current Research in Food Science*, *2*, 61–69.

Akhter, R., Masoodi, F. A., Wani, T. A., & Rather, S. A. (2019). Functional characterization of biopolymer based composite film: Incorporation of natural essential oils and antimicrobial agents. *International Journal of Biological Macromolecules*, *137*, 1245–1255.

Akter, N., Khan, R. A., Tuhin, M. O., Haque, M. E., Nurnabi, M., Parvin, F., & Islam, R. (2014). Thermomechanical, barrier, and morphological properties of chitosan-reinforced starch-based biodegradable composite films. *Journal of Thermoplastic Composite Materials*, *27*(7), 933–948.

Ali, F. B., Kang, D. J., Kim, M. P., Cho, C. H., & Kim, B. J. (2014). Synthesis of biodegradable and flexible, polylactic acid based, thermoplastic polyurethane with high gas barrier properties. *Polymer International*, *63*(9), 1620–1626.

Almasi, H., Azizi, S., & Amjadi, S. (2020). Development and characterization of pectin films activated by nanoemulsion and Pickering emulsion stabilized marjoram (*Origanum majorana* L.) essential oil. *Food Hydrocolloids*, *99*, 105338.

Almenar, E., Samsudin, H., Auras, R., & Harte, J. (2010). Consumer acceptance of fresh blueberries in bio-based packages. *Journal of the Science of Food and Agriculture*, *90*(7), 1121–1128.

Alzagameem, A., Klein, S. E., Bergs, M., Do, X. T., Korte, I., Dohlen, S., . . . Schulze, M. (2019). Antimicrobial activity of lignin and lignin-derived cellulose and chitosan composites against selected pathogenic and spoilage microorganisms. *Polymers*, *11*(4), 670.

Asgher, M., Qamar, S. A., Bilal, M., & Iqbal, H. M. (2020). Bio-based active food packaging materials: Sustainable alternative to conventional petrochemical-based packaging materials. *Food Research International*, *137*, 109625.

Atta, O. M., Manan, S., Shahzad, A., Ul-Islam, M., Ullah, M. W., & Yang, G. (2022). Biobased materials for active food packaging: A review. *Food Hydrocolloids*, *125*, 107419.

Aulin, C., Gällstedt, M., & Lindström, T. (2010). Oxygen and oil barrier properties of microfibrillated cellulose films and coatings. *Cellulose*, *17*(3), 559–574.

Auras, R., Singh, S. P., & Singh, J. (2006). Performance evaluation of PLA against existing PET and PS containers. *Journal of Testing and Evaluation*, *34*(6), 530–536.

Azarakhsh, N., Osman, A., Ghazali, H. M., Tan, C. P., & Adzahan, N. M. (2014). Lemongrass essential oil incorporated into alginate-based edible coating for shelf-life extension and quality retention of fresh-cut pineapple. *Postharvest Biology and Technology*, *88*, 1–7.

Bajpai, P. (2019). *Biobased polymers: Properties and applications in packaging* (pp. 25–111). Netherlands: Elsevier.

Bandyopadhyay, S., Saha, N., Brodnjak, U. V., & Saha, P. (2018). Bacterial cellulose based greener packaging material: A bioadhesive polymeric film. *Materials Research Express*, *5*(11), 115405.

Bedane, A. H., Eić, M., Farmahini-Farahani, M., & Xiao, H. (2015). Water vapor transport properties of regenerated cellulose and nanofibrillated cellulose films. *Journal of Membrane Science*, *493*, 46–57.

Begum, S. A., Rane, A. V., & Kanny, K. (2020). Applications of compatibilized polymer blends in automobile industry. In *Compatibilization of polymer blends* (pp. 563–593). Netherlands: Elsevier.

Berthet, M. A., Angellier-Coussy, H., Guillard, V., & Gontard, N. (2016). Vegetal fiber-based biocomposites: Which stakes for food packaging applications?. *Journal of Applied Polymer Science*, *133*(2).

Bhardwaj, S., Bhardwaj, N. K., & Negi, Y. S. (2019). Cleaner approach for improving the papermaking from agro and hardwood blended pulps using biopolymers. *Journal of Cleaner Production*, *213*, 134–142.

Boonniteewanich, J., Pitivut, S., Tongjoy, S., Lapnonkawow, S., & Suttiruengwong, S. (2014). Evaluation of carbon footprint of bioplastic straw compared to petroleum based straw products. *Energy Procedia*, *56*, 518–524.

Briano, R., Giuggioli, N. R., Girgenti, V., & Peano, C. (2015). Biodegradable and compostable film and modified atmosphere packaging in postharvest supply chain of raspberry fruits (cv. G randeur). *Journal of Food Processing and Preservation*, *39*(6), 2061–2073.

Briassoulis, D. (2004). An overview on the mechanical behaviour of biodegradable agricultural films. *Journal of Polymers and the Environment*, *12*(2), 65–81.

Briassoulis, D., & Giannoulis, A. (2018). Evaluation of the functionality of bio-based food packaging films. *Polymer Testing*, *69*, 39–51.

Buist, H., van Harmelen, T., van den Berg, C., Leeman, W., Meima, M., & Krul, L. (2020). Evaluation of measures to mitigate mineral oil migration from recycled paper in food packaging. *Packaging Technology and Science*, *33*(12), 531–546.

Chan, C. M., Pratt, S., Halley, P., Richardson, D., Werker, A., Laycock, B., & Vandi, L. J. (2019). Mechanical and physical stability of polyhydroxyalkanoate (PHA)-based wood plastic composites (WPCs) under natural weathering. *Polymer Testing*, *73*, 214–221.

Chen, G. G., Qi, X. M., Guan, Y., Peng, F., Yao, C. L., & Sun, R. C. (2016). High strength hemicellulose-based nanocomposite film for food packaging applications. *ACS Sustainable Chemistry & Engineering*, *4*(4), 1985–1993.

Chen, H., Wang, J., Cheng, Y., Wang, C., Liu, H., Bian, H., Pan, Y., Sun, J., & Han, W. (2019). Application of protein-based films and coatings for food packaging: A review. *Polymers*, *11*(12), 2039.

Chen, J., Wu, A., Yang, M., Ge, Y., Pristijono, P., Li, J., . . . Mi, H. (2021). Characterization of sodium alginate-based films incorporated with thymol for fresh-cut apple packaging. *Food Control*, *126*, 108063.

Chiou, B. S., Cao, T., Bilbao-Sainz, C., Vega-Galvez, A., Glenn, G., & Orts, W. (2020). Properties of gluten foams containing different additives. *Industrial Crops and Products*, *152*, 112511.

Ciannamea, E. M., Castillo, L. A., Barbosa, S. E., & De Angelis, M. G. (2018). Barrier properties and mechanical strength of bio-renewable, heat-sealable films based on gelatin, glycerol and soybean oil for sustainable food packaging. *Reactive and Functional Polymers*, *125*, 29–36.

Colombo, B., Calvo, M. V., Sciarria, T. P., Scaglia, B., Kizito, S. S., D'Imporzano, G., & Adani, F. (2019). Biohydrogen and polyhydroxyalkanoates (PHA) as products of a two-steps bioprocess from deproteinized dairy wastes. *Waste Management*, *95*, 22–31.

Coltelli, M. B., Wild, F., Bugnicourt, E., Cinelli, P., Lindner, M., Schmid, M., . . . Lazzeri, A. (2015). State of the art in the development and properties of protein-based films and coatings and their applicability to cellulose based products: An extensive review. *Coatings*, *6*(1), 1.

Din, M. I., Ghaffar, T., Najeeb, J., Hussain, Z., Khalid, R., & Zahid, H. (2020). Potential perspectives of biodegradable plastics for food packaging application-review of properties and recent developments. *Food Additives & Contaminants: Part A*, *37*(4), 665–680.

Dou, L., Li, B., Zhang, K., Chu, X., & Hou, H. (2018). Physical properties and antioxidant activity of gelatin-sodium alginate edible films with tea polyphenols. *International Journal of Biological Macromolecules*, *118*, 1377–1383.

El Achaby, M., El Miri, N., Aboulkas, A., Zahouily, M., Bilal, E., Barakat, A., & Solhy, A. (2017). Processing and properties of eco-friendly bio-nanocomposite films filled with cellulose nanocrystals from sugarcane bagasse. *International Journal of Biological Macromolecules*, *96*, 340–352.

Fabra, M. J., López-Rubio, A., & Lagaron, J. M. (2014). On the use of different hydrocolloids as electrospun adhesive interlayers to enhance the barrier properties of polyhydroxyalkanoates of interest in fully renewable food packaging concepts. *Food Hydrocolloids*, *39*, 77–84.

Fadini, A. L., Rocha, F. S., Alvim, I. D., Sadahira, M. S., Queiroz, M. B., Alves, R. M. V., & Silva, L. B. (2013). Mechanical properties and water vapour permeability of hydrolysed collagen—cocoa butter edible films plasticised with sucrose. *Food Hydrocolloids*, *30*(2), 625–631.

Felix, M., Perez-Puyana, V., Romero, A., & Guerrero, A. (2017). Development of protein-based bioplastics modified with different additives. *Journal of Applied Polymer Science*, *134*(42), 45430.

Fortunati, E., Armentano, I., Iannoni, A., & Kenny, J. M. (2010). Development and thermal behaviour of ternary PLA matrix composites. *Polymer Degradation and Stability*, *95*(11), 2200–2206.

Galić, K., Ščetar, M., & Kurek, M. (2011). The benefits of processing and packaging. *Trends in Food Science & Technology*, *22*(2–3), 127–137.

Galus, S., & Kadzińska, J. (2015). Food applications of emulsion-based edible films and coatings. *Trends in Food Science & Technology*, *45*(2), 273–283.

Gan, I., & Chow, W. S. (2018). Antimicrobial poly (lactic acid)/cellulose bionanocomposite for food packaging application: A review. *Food Packaging and Shelf Life*, *17*, 150–161.

Ghasemi, S., Jafari, S. M., Assadpour, E., & Khomeiri, M. (2017). Production of pectin-whey protein nano-complexes as carriers of orange peel oil. *Carbohydrate Polymers, 177*, 369–377.

Giacalone, G., & Chiabrando, V. (2013). Modified atmosphere packaging of sweet cherries with biodegradable films. *International Food Research Journal, 20*(3), 1263.

Gouvêa, D. M., Mendonça, R. C. S., Soto, M. L., & Cruz, R. S. (2015). Acetate cellulose film with bacteriophages for potential antimicrobial use in food packaging. *LWT-Food Science and Technology, 63*(1), 85–91.

Guarás, M. P., Alvarez, V. A., & Ludueña, L. N. (2015). Processing and characterization of thermoplastic starch/polycaprolactone/compatibilizer ternary blends for packaging applications. *Journal of Polymer Research, 22*(9), 1–12.

Guo, Y., Tian, D., Shen, F., Yang, G., Long, L., He, J., . . . Deng, S. (2019). Transparent cellulose/technical lignin composite films for advanced packaging. *Polymers, 11*(9), 1455.

Haghighi, H., Gullo, M., La China, S., Pfeifer, F., Siesler, H. W., Licciardello, F., & Pulvirenti, A. (2021). Characterization of bio-nanocomposite films based on gelatin/polyvinyl alcohol blend reinforced with bacterial cellulose nanowhiskers for food packaging applications. *Food Hydrocolloids, 113*, 106454.

Han, J. W., Ruiz-Garcia, L., Qian, J. P., & Yang, X. T. (2018). Food packaging: A comprehensive review and future trends. *Comprehensive Reviews in Food Science and Food Safety, 17*(4), 860–877.

Huang, S., Liu, X., Chang, C., & Wang, Y. (2020). Recent developments and prospective food-related applications of cellulose nanocrystals: A review. *Cellulose, 27*(6), 2991–3011.

Ivonkovic, A., Zeljko, K., Talic, S., & Lasic, M. (2017). Biodegradable packaging in the food industry. *Journal of Food Safety and Food Quality, 68*, 26–38.

James, R., Jewitt, M., Matussek, H., Moohan, M., & Potter, J. (2002). *Pulp and paper international facts and price book*. Brussels: Paperloop Publications.

Janjarasskul, T., & Krochta, J. M. (2010). Edible packaging materials. *Annual Review of Food Science and Technology, 1*(1), 415–448.

Jin, K., Tang, Y., Liu, J., Wang, J., & Ye, C. (2021). Nanofibrillated cellulose as coating agent for food packaging paper. *International Journal of Biological Macromolecules, 168*, 331–338.

Jung, S., Cui, Y., Barnes, M., Satam, C., Zhang, S., Chowdhury, R. A., . . . Ajayan, P. M. (2020). Multifunctional bio-nanocomposite coatings for perishable fruits. *Advanced Materials, 32*(26), 1908291.

Khosravi, A., Fereidoon, A., Khorasani, M. M., Naderi, G., Ganjali, M. R., Zarrintaj, P., . . . Gutiérrez, T. J. (2020). Soft and hard sections from cellulose-reinforced poly (lactic acid)-based food packaging films: A critical review. *Food Packaging and Shelf Life, 23*, 100429.

Kumar, A., Gupta, V., Singh, S., Saini, S., & Gaikwad, K. K. (2021). Pine needles lignocellulosic ethylene scavenging paper impregnated with nanozeolite for active packaging applications. *Industrial Crops and Products, 170*, 113752.

Kumar, M., Tomar, M., Saurabh, V., Mahajan, T., Punia, S., del Mar Contreras, M., . . . Kennedy, J. F. (2020). Emerging trends in pectin extraction and its anti-microbial functionalization using natural bioactives for application in food packaging. *Trends in Food Science & Technology, 105*, 223–237.

Kurczewska, J., Ratajczak, M., & Gajecka, M. (2021). Alginate and pectin films covering halloysite with encapsulated salicylic acid as food packaging components. *Applied Clay Science, 214*, 106270.

Lei, Y., Wu, H., Jiao, C., Jiang, Y., Liu, R., Xiao, D., . . . Li, S. (2019). Investigation of the structural and physical properties, antioxidant and antimicrobial activity of pectin-konjac glucomannan composite edible films incorporated with tea polyphenol. *Food Hydrocolloids, 94*, 128–135.

Lin, D., Zheng, Y., Wang, X., Huang, Y., Ni, L., Chen, X., . . . Wu, D. (2020). Study on physicochemical properties, antioxidant and antimicrobial activity of okara soluble dietary fiber/sodium carboxymethyl cellulose/thyme essential oil active edible composite films incorporated with pectin. *International Journal of Biological Macromolecules, 165*, 1241–1249.

Lodha, P., & Netravali, A. N. (2005). Thermal and mechanical properties of environment-friendly 'green' plastics from stearic acid modified-soy protein isolate. *Industrial Crops and Products, 21*(1), 49–64.

Lu, P., Xiao, H., Zhang, W., & Gong, G. (2014). Reactive coating of soybean oil-based polymer on nanofibrillated cellulose film for water vapor barrier packaging. *Carbohydrate Polymers, 111*, 524–529.

Malheiros, P. S., Jozala, A. F., Pessoa-Jr, A., Vila, M. M., Balcao, V. M., & Franco, B. D. (2018). Immobilization of antimicrobial peptides from *Lactobacillus sakei* subsp. *sakei* 2a in bacterial cellulose: Structural and functional stabilization. *Food Packaging and Shelf Life, 17*, 25–29.

Mangaraj, S., Yadav, A., Bal, L. M., Dash, S. K., & Mahanti, N. K. (2019). Application of biodegradable polymers in food packaging industry: A comprehensive review. *Journal of Packaging Technology and Research*, *3*(1), 77–96.

Medina-Jaramillo, C., Ochoa-Yepes, O., Bernal, C., & Famá, L. (2017). Active and smart biodegradable packaging based on starch and natural extracts. *Carbohydrate Polymers*, *176*, 187–194.

Meerasri, J., & Sothornvit, R. (2020). Characterization of bioactive film from pectin incorporated with gamma-aminobutyric acid. *International Journal of Biological Macromolecules*, *147*, 1285–1293.

Meereboer, K. W., Misra, M., & Mohanty, A. K. (2020). Review of recent advances in the biodegradability of polyhydroxyalkanoate (PHA) bioplastics and their composites. *Green Chemistry*, *22*(17), 5519–5558.

Mellinas, C., Ramos, M., Jiménez, A., & Garrigós, M. C. (2020). Recent trends in the use of pectin from agrowaste residues as a natural-based biopolymer for food packaging applications. *Materials*, *13*(3), 673.

Mendes, F. R., Bastos, M. S., Mendes, L. G., Silva, A. R., Sousa, F. D., Monteiro-Moreira, A. C., . . . Moreira, R. A. (2017). Preparation and evaluation of hemicellulose films and their blends. *Food Hydrocolloids*, *70*, 181–190.

Mendes, J. F., Norcino, L. B., Martins, H. H. A., Manrich, A., Otoni, C. G., Carvalho, E. E. N., . . . Mattoso, L. H. C. (2020). Correlating emulsion characteristics with the properties of active starch films loaded with lemongrass essential oil. *Food Hydrocolloids*, *100*, 105428.

Minelli, M., Baschetti, M. G., Doghieri, F., Ankerfors, M., Lindström, T., Siró, I., & Plackett, D. (2010). Investigation of mass transport properties of microfibrillated cellulose (MFC) films. *Journal of Membrane Science*, *358*(1–2), 67–75.

Mojumdar, S. C., Moresoli, C., Simon, L. C., & Legge, R. L. (2011). Edible wheat gluten (WG) protein films: Preparation, thermal, mechanical and spectral properties. *Journal of Thermal Analysis and Calorimetry*, *104*(3), 929–936.

Mugwagwa, L. R., & Chimphango, A. F. (2020). Enhancing the functional properties of acetylated hemicellulose films for active food packaging using acetylated nanocellulose reinforcement and polycaprolactone coating. *Food Packaging and Shelf Life*, *24*, 100481.

Nechita, P., & Roman, M. (2020). Review on polysaccharides used in coatings for food packaging papers. *Coatings*, *10*(6), 566.

Niini, A., Leminen, V., Tanninen, P., & Varis, J. (2021). Humidity effect in heating and cooling of press-formed paperboard food packages: Comparison of storing and heating conditions. *Packaging Technology and Science*, *34*(8), 517–522.

Nilsen-Nygaard, J., Fernández, E. N., Radusin, T., Rotabakk, B. T., Sarfraz, J., Sharmin, N., . . . Pettersen, M. K. (2021). Current status of biobased and biodegradable food packaging materials: Impact on food quality and effect of innovative processing technologies. *Comprehensive Reviews in Food Science and Food Safety*, *20*(2), 1333–1380.

Nisar, T., Wang, Z. C., Yang, X., Tian, Y., Iqbal, M., & Guo, Y. (2018). Characterization of citrus pectin films integrated with clove bud essential oil: Physical, thermal, barrier, antioxidant and antibacterial properties. *International Journal of Biological Macromolecules*, *106*, 670–680.

Oliveira, N. S., Oliveira, J., Gomes, T., Ferreira, A., Dorgan, J., & Marrucho, I. M. (2004). Gas sorption in poly (lactic acid) and packaging materials. *Fluid Phase Equilibria*, *222*, 317–324.

Otoni, C. G., Avena-Bustillos, R. J., Olsen, C. W., Bilbao-Sáinz, C., & McHugh, T. H. (2016). Mechanical and water barrier properties of isolated soy protein composite edible films as affected by carvacrol and cinnamaldehyde micro and nanoemulsions. *Food Hydrocolloids*, *57*, 72–79.

Padrão, J., Gonçalves, S., Silva, J. P., Sencadas, V., Lanceros-Méndez, S., Pinheiro, A. C., . . . Dourado, F. (2016). Bacterial cellulose-lactoferrin as an antimicrobial edible packaging. *Food Hydrocolloids*, *58*, 126–140.

Peelman, N., Ragaert, P., De Meulenaer, B., Adons, D., Peeters, R., Cardon, L., . . . Devlieghere, F. (2013). Application of bioplastics for food packaging. *Trends in Food Science & Technology*, *32*(2), 128–141.

Peelman, N., Ragaert, P., Vandemoortele, A., Verguldt, E., De Meulenaer, B., & Devlieghere, F. (2014). Use of biobased materials for modified atmosphere packaging of short and medium shelf-life food products. *Innovative Food Science & Emerging Technologies*, *26*, 319–329.

Qing, Y., Sabo, R., Wu, Y., Zhu, J. Y., & Cai, Z. (2015). Self-assembled optically transparent cellulose nanofibril films: Effect of nanofibril morphology and drying procedure. *Cellulose*, *22*(2), 1091–1102.

Ramos, M., Valdés, A., Beltran, A., & Garrigós, M. C. (2016). Gelatin-based films and coatings for food packaging applications. *Coatings*, *6*(4), 41.

Salari, M., Khiabani, M. S., Mokarram, R. R., Ghanbarzadeh, B., & Kafil, H. S. (2018). Development and evaluation of chitosan based active nanocomposite films containing bacterial cellulose nanocrystals and silver nanoparticles. *Food Hydrocolloids, 84*, 414–423.

Samui, A. B., & Kanai, T. (2019). Polyhydroxyalkanoates based copolymers. *International Journal of Biological Macromolecules, 140*, 522–537.

Sanyang, M. L., Sapuan, S. M., Jawaid, M., Ishak, M. R., & Sahari, J. (2016). Effect of plasticizer type and concentration on physical properties of biodegradable films based on sugar palm (Arenga pinnata) starch for food packaging. *Journal of Food Science and Technology, 53*(1), 326–336.

Seggiani, M., Cinelli, P., Mallegni, N., Balestri, E., Puccini, M., Vitolo, S., . . . Lazzeri, A. (2017). New bio-composites based on polyhydroxyalkanoates and posidonia oceanica fibres for applications in a marine environment. *Materials, 10*(4), 326.

Senturk Parreidt, T., Müller, K., & Schmid, M. (2018). Alginate-based edible films and coatings for food packaging applications. *Foods, 7*(10), 170.

Shaikh, S., Yaqoob, M., & Aggarwal, P. (2021). An overview of biodegradable packaging in food industry. *Current Research in Food Science, 4*, 503–520.

Sharma, N., Khatkar, B. S., Kaushik, R., Sharma, P., & Sharma, R. (2017). Isolation and development of wheat based gluten edible film and its physicochemical properties. *International Food Research Journal, 24*(1), 94–101.

Silva, T. C. F., Silva, D., & Lucia, L. A. (2015). The multifunctional chemical tunability of wood-based polymers for advanced biomaterials applications. In *Green biorenewable biocomposites: From knowledge to industrial applications* (pp. 427–459). New York: Apple Academic Press.

Singla, R., & Mehta, R. (2012). Preparation and characterization of polylactic acid-based biodegradable blends processed under microwave radiation. *Polymer-Plastics Technology and Engineering, 51*(10), 1014–1017.

Song, Z., Xiao, H., & Zhao, Y. (2014). Hydrophobic-modified nano-cellulose fiber/PLA biodegradable composites for lowering water vapor transmission rate (WVTR) of paper. *Carbohydrate Polymers, 111*, 442–448.

Sood, S., & Sharma, C. (2021). Study on fiber furnishes and fiber morphological properties of commonly used Indian food packaging papers and paperboards. *Cellulose Chemistry and Technology, 55*, 125–131.

Stark, N. M. (2016). Opportunities for cellulose nanomaterials in packaging films: A review and future trends. *Journal of Renewable Materials, 4*(5), 313–326.

Stark, N. M., & Matuana, L. M. (2021). Trends in sustainable biobased packaging materials: A mini review. *Materials Today Sustainability, 15*, 100084.

Tiberiu, N. (2011). Concepts in biological analysis of resorbable materials in oro-maxillo facial surgery. *Oro-Maxillo-Facial Implantology (Romania), 2*(1), 33–38.

Torres-Giner, S., Hilliou, L., Melendez-Rodriguez, B., Figueroa-Lopez, K. J., Madalena, D., Cabedo, L., . . . Lagaron, J. M. (2018). Melt processability, characterization, and antibacterial activity of compression-molded green composite sheets made of poly (3-hydroxybutyrate-co-3-hydroxyvalerate) reinforced with coconut fibers impregnated with oregano essential oil. *Food Packaging and Shelf Life, 17*, 39–49.

Türünç, O., Montero de Espinosa, L., & Meier, M. A. (2011). Renewable polyethylene mimics derived from castor oil. *Macromolecular Rapid Communications, 32*(17), 1357–1361.

Viana, R. M., Sá, N. M., Barros, M. O., de Fátima Borges, M., & Azeredo, H. M. (2018). Nanofibrillated bacterial cellulose and pectin edible films added with fruit purees. *Carbohydrate Polymers, 196*, 27–32.

Vilarinho, F., Sanches Silva, A., Vaz, M. F., & Farinha, J. P. (2018). Nanocellulose in green food packaging. *Critical Reviews in Food Science and Nutrition, 58*(9), 1526–1537.

Vroman, I., & Tighzert, L. (2009). Biodegradable polymers. *Materials, 2*(2), 307–344.

Wang, H., Qian, J., & Ding, F. (2018). Emerging chitosan-based films for food packaging applications. *Journal of Agricultural and Food Chemistry, 66*(2), 395–413.

Wang, J., Cao, Y., Jaquet, B., Gerhard, C., Li, W., Xia, X., . . . Zhang, K. (2021a). Self-compounded nanocomposites: Toward multifunctional membranes with superior mechanical, gas/oil barrier, UV-shielding, and photothermal conversion properties. *ACS Applied Materials & Interfaces, 13*(24), 28668–28678.

Wang, J., Euring, M., Ostendorf, K., & Zhang, K. (2021b). Biobased materials for food packaging. *Journal of Bioresources and Bioproducts, 7*(1), 1–13.

Wang, X., Zhou, J., & Li, L. (2007). Multiple melting behavior of poly (butylene succinate). *European Polymer Journal, 43*(8), 3163–3170.

Wieczyńska, J., & Cavoski, I. (2018). Antimicrobial, antioxidant and sensory features of eugenol, carvacrol and trans-anethole in active packaging for organic ready-to-eat iceberg lettuce. *Food Chemistry*, *259*, 251–260.

Yan, J., Luo, Z., Ban, Z., Lu, H., Li, D., Yang, D., . . . Li, L. (2019). The effect of the layer-by-layer (LBL) edible coating on strawberry quality and metabolites during storage. *Postharvest Biology and Technology*, *147*, 29–38.

Yildirim, S., Röcker, B., Pettersen, M. K., Nilsen-Nygaard, J., Ayhan, Z., Rutkaite, R., . . . Coma, V. (2018). Active packaging applications for food. *Comprehensive Reviews in Food Science and Food Safety*, *17*(1), 165–199.

Youssef, A. M., & El-Sayed, S. M. (2018). Bionanocomposites materials for food packaging applications: Concepts and future outlook. *Carbohydrate Polymers*, *193*, 19–27.

Zhou, W., Fang, J., Tang, S., Wu, Z., & Wang, X. (2021). 3D-printed nanocellulose-based cushioning—antibacterial dual-function food packaging aerogel. *Molecules*, *26*(12), 3543.

Zubair, M., & Ullah, A. (2020). Recent advances in protein derived bionanocomposites for food packaging applications. *Critical Reviews in Food Science and Nutrition*, *60*(3), 406–434.

Zubeldía, F., Ansorena, M. R., & Marcovich, N. E. (2015). Wheat gluten films obtained by compression molding. *Polymer Testing*, *43*, 68–77.

7

Soy Protein–Based Films and Coatings

Functionality and Characterization

Simmi Deo, Anjelina Sundarsingh, and Surangna Jain

CONTENTS

7.1 Introduction

Foods come in a variety of forms and go through a number of procedures before reaching the consumer. Postharvesting processes often have an impact on the appearance and functionality of fruits and vegetables. It is possible to apply the same logic to other foods that may be harmed during distribution and marketing. As a result, food packaging plays a critical role in safeguarding and storing foods. A variety of packing materials is used to make it more convenient for consumers. Consumer preferences for moderately processed food products with extended shelf life and convenience have resulted in novel and innovative food packaging strategies. While novel packaging technologies such as active packaging, bioactive packaging, and intelligent packaging are broadly acknowledged and widely deployed, edible films and coatings are still considered a future trend (Falguera et al., 2011).

A film or coating, interchangeably often termed an edible film for foods, is a coating of a fine layer of material with a thickness of less than 0.3 mm that acts as a barrier for the stuff and can be ingested along with it (Debeaufort & Voilley, 2009; Diaz-Montes & Castro-Munoz, 2021). These layers are frequently applied to the food surface in liquid form by keeping the product in a film-forming solution created by the structural matrix. In nature, edible films are self-supporting structures, whereas edible coatings cling to the food surface (Bourtroom, 2008; Guimarães et al., 2018). Apart from giving protection to the stuff from mechanical, physical, chemical, and microbiological damage, coatings are believed to act as encapsulating agents and preservatives, making them more desirable (Miller & Krochta, 1997).

A lot of potential is there for edible films and coatings in the future to meet customer demand for environmentally friendly and natural foods. They are not totally preferred over traditional food packaging materials, but they do add to the functionality of the food. Because these packing materials are made from agricultural wastes and/or industrial food processing commodities, they add value to biomass. Edible films and coatings can improve the process of food preservation while also reducing the cost and bulk of standard packaging. Biopolymers are based on hydrocolloids, such as starch, cellulose, chitosan, alginates, pectin, and gums, and proteins, such as corn zein, gelatin, soybean proteins, sunflower proteins, wheat gluten, whey, casein, and keratin, which are used to create edible coatings and films. The latest revolutionary invention includes the use of composites and blends to allow for the controlled release of food additives and nutrients, along with the basic functional features of creating a gas or moisture barrier (Campos et al., 2011; Majid et al., 2018).

Two of the most studied biopolymers that can be used to extrude edible coatings are polysaccharides and proteins, among many others (Figure 7.1). Although polysaccharides are readily available, low-cost, non-toxic, and thermostable resources, protein's hydrophobicity, which is a disadvantage of polysaccharides, makes it a preferred source (Hassan et al., 2018; Mikkonen et al., 2007). They are a preferable option for films and coatings because they have good film-forming qualities, a higher nutritional index, biodegradability, gas-barrier properties, and improved mechanical properties. According to Bourtroom (2008), the oxygen permeability of soy protein coatings was 670, 540, 500, and 260 times lower than that of pectin, starch, polyethylene, and methylcellulose, respectively. Proteins with strong barrier qualities, such as maize zein, whey proteins, wheat gluten, soy protein, casein, keratin, egg white, gelatin,

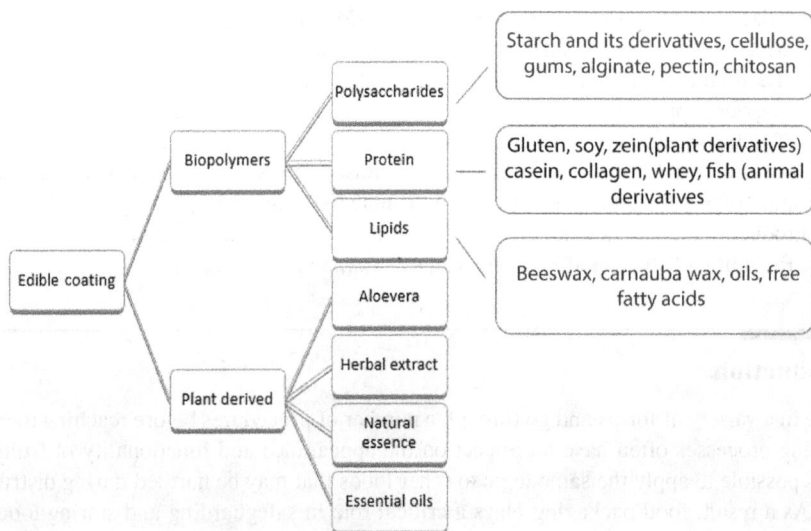

FIGURE 7.1   Edible films from various sources.

collagen, and myofibrillar proteins, have been widely used in the development of edible films and coatings via the solvent-casting technique. The fabrication of protein-based coatings and films utilizing thermo-plasticization and extrusion processes has received very little attention. Controlling the molecular architecture and spatial organization of the natural macromolecule is a big challenge when using thermo-plasticization and extrusion processes with high repeatability (Majid et al., 2018; Mensitieri et al., 2011).

Soy protein is a promising contender for film and coating materials since it has great functionality among proteins. The majority of soy protein films are prepared from soy protein isolate (SPI), which is soy protein that has been highly refined and contains at least 90% protein and is manufactured from soy flour, eliminating the majority of non-protein components. Aside from that, soy flour and soy milk can be used to make it (Cho & Rhee, 2004; Tian et al., 2018). Emulsification, cohesiveness, water and fat absorption, dough and fiber formation, and texture have been improved in films and coatings made from these. In addition, numerous experiments on soy protein have been conducted to improve its film-forming capacity, such as the incorporation of sodium dodecyl sulfate (SDS) and carboxymethyl cellulose (CMC), which increase the film's extendibility and permeability, respectively (Soares et al., 2005; Swain et al., 2004).

7.2 Chemistry of Films

The use of edible coatings dates back to centuries, when meat products were coated by cellulose and fruits and vegetables by waxes. While literature says the Chinese people were one of the groups that used edible coatings during the 12th century, further worldwide use was popularized during the 20th century to prevent water loss from fruits and vegetables. They were especially popularized since having a coating or film helped in retaining the freshness and appearance of fruits and vegetables. Other than that, they were also helpful in improving organoleptic characteristics, such as color, flavor, sweetness, and antimicrobial properties (Han & Gennadios, 2005; Hassan et al., 2018; Vasconez et al., 2009).

Edible coatings are mostly prepared from any biopolymer or chemical which has film-forming properties. The preparation can either be wet or dry processing in food processing. Stages like spraying, dipping, and brushing are some of the common steps to add a coat or film to the foodstuffs (Arshad et al., 2016; Cutter, 2006). The commonly used components for preparation of coating and films are polysaccharides, proteins, lipids, and composites. These biopolymers are mostly eco-friendly, since they are derived from plants (Cutter, 2006). Proteins are one of the biopolymers which have more advantage of being a coating material due to certain functionalities. Among protein biopolymers, plant proteins, like gluten, zein, and soy protein, and animal protein, such as whey, casein, collagen, and fish protein, are significant materials (Chen et al., 2019).

Soy protein (SP), a by-product of the soy oil industry, is currently used in animal feed and as a food supplement. Soy flour (SF), soy protein concentrate (SPC), and soy protein isolate are the three types of commercial soy protein products made from soybean (SPI) (Tian et al., 2018). SPIs are used to make the majority of SP-based films. SPI is a highly refined or purified form of SP that has a minimum moisture-free protein concentration of 90%. Soy proteins are made up of a mix of albumins and globulins, with 90% of them being globular storage proteins. SP comprises 18 amino acids, including those with polar functional groups, such as carboxyl, amine, and hydroxyl, which can chemically react and allow for easy modification of soy protein (Kinsella, 1979). Even though soy proteins are a good choice for films and coatings, their moisture absorption property is a drawback. As a result, SP is modified using enzymatic, physical, and chemical ways to make it more stable and firmer in order to get high-performance SP (Thakur et al., 2016).

Solution-based or melt processing–based technologies are commonly used to produce SP-based products. Extrusion, casting, heating, spinning, and thermal compacting are all used to make SP-based films and coatings. In solution-based processing, a dispersion medium is used to create a solution of SP, which is then used to cast the SP-based film, whereas in thermal processing, SP is melted at a high temperature, allowing it to denature and form a coating. These techniques have some disadvantages, yet they are still the most extensively utilized method of producing SP-based products, particularly in melt processing–based methods, when heat is employed to cause protein denaturation. Because SP is brittle by nature, it is frequently combined with a plasticizer in the thermal processing industry to improve melt flow ability

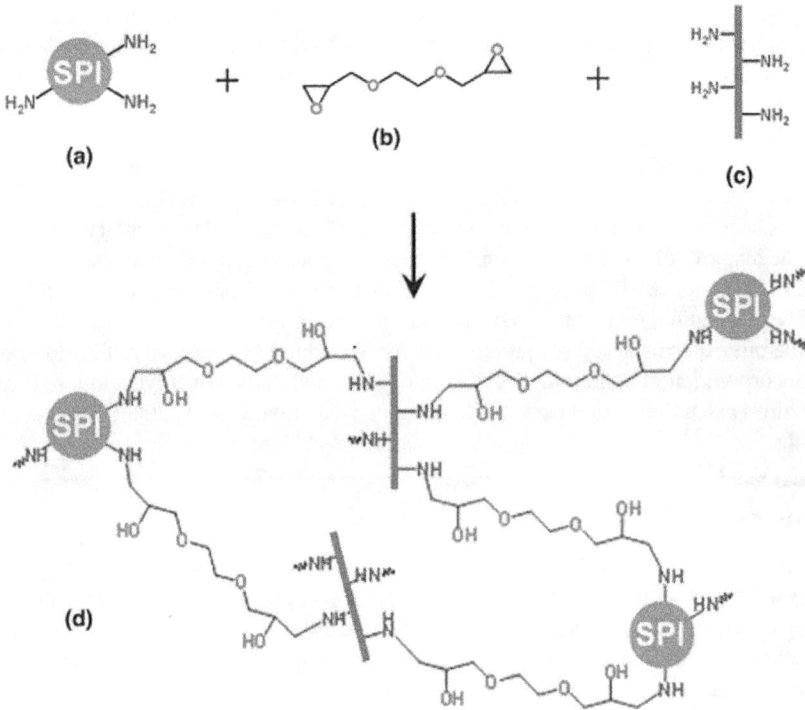

FIGURE 7.2   Reaction leading to cross-linking of SPI based films. (a) SPI, (b) EDGE, (c) MCNC, (d) SPI based film.

Source: Adapted from (Zhang et al., 2016).

and enable the creation of films and coatings. In other situations, different types of biopolymers are added to SP to improve mechanical and thermal durability, resulting in SP composites with improved properties compared to films and coatings made only from SP and SPI (Tian et al., 2018).

The SPI-EDGE-MCNC film successfully displayed higher mechanical property and lower water absorption rate in a study by Zhang et al. (2016) to improve the performance of SPI-coated film by adding modified cellulose nanocrystals (MCNC). The cross-linking of SPI with EDGE and MCNC, as shown in Figure 7.2, led to a considerable improvement in tensile strength from 3.13 MPa to 4.79 MPa, according to the study. According to Xu et al. (2015), the formation of hydrogen bonds and physical cross-linking structures between the CNC and matrix was responsible for the increased tensile strength and decreased elongation at break of the CNC-modified film. Similarly, the moisture content (water absorption test) of the CNC and MCNC samples decreased by 9.7% and 24%, respectively. Several further research found that adding different plasticizers improved functional properties.

7.3 Development of Soy Protein–Based Edible Films and Their Functionalization

SP- and SPI-based films are becoming increasingly popular due to their abundance, biodegradability, biocompatibility, low cost, and nutritional value, as well as their ability to form polymers because of the 7 and 11s protein fractions, which aid in the formation of a three-dimensional cross-linking network (Galus et al., 2020). Despite having exceptional properties that make them a good candidate for films and coatings, their high affinity for water puts them at a disadvantage. Plasticizers are added to SP to improve their functionality for this reason. Wet and dry techniques are primarily used to develop SP-based films and coatings. In wet procedures, casting is primarily employed to create the films and coatings, whereas in dry methods, molding and extraction are typically used (Calva-Estrada et al., 2019; Garrido et al.,

2013). In both the processing, to allow larger extension of their required structure, it is important that denaturation of the native proteins takes place using heat, acid or base, and/or a solvent in both of these procedures. Greater extension allows for more interaction and association between protein chains, which forms the protein film's cohesive matrix through electrostatic interactions, Van der Waals forces, hydrogen bonds, covalent and disulfide bonds, and other interactions (Dhall, 2013).

One of the two procedures widely employed for producing films and coatings at the laboratory scale is wet processing, also known as solution casting. The dispersion or solubilization of proteins in a solvent media is the basis for this approach. Plasticizers are required in this process since the films would be brittle if they were not. Plasticizers are low-molecular-weight compounds with varying degrees of hydrophilicity. The hydroxyl groups of polyol-based plasticizers generate hydrogen bonds with polymers, disrupting polymer–polymer interactions (Garrido et al., 2013).

In the solution-casting technique, SPI is combined with water and glycerol at 30% (w/w) of SPI added to generate a film-forming solution (Eswaranandam et al., 2004; Rani & Kumar, 2019). Depending on the nature of the additions, liquid additives in suitable amounts might be introduced in this phase. Based on the type of plasticizers employed, the solution is next heated in a water bath at a temperature of 60–70°C. This is followed by the casting of a Teflon or glass surface solution. While the basic principle remains the same, the time and temperature combinations can be modified based on the plasticizers used to produce a film with the best functionality (Rani & Kumar, 2019).

One more method which is industrially used to fabricate SP films and coatings is compression molding, also known as dry processing. Majority of SP films are produced using this method as it is simple and less time-consuming compared to solution casting. Since compression molding involves the use of thermal treatment, fabrication of protein-based films might be a hurdle as glass transition temperature of proteins is almost near to their thermal degradation temperature. But the addition of plasticizers during the processing has helped overcome this problem, and compression molding has become much simpler as compared to other processing methods (Garrido et al., 2013).

Initially, SP or SPI is completely mixed with desired plasticizers before being hot-pressed under pressure at a specific temperature. A quantity of 30% (w/w) glycerol was employed as a plasticizer in a study to create SPI films, which were hot-pressed for 15 minutes at 140°C under 15 MPa. Glycerol has been identified as one of the most effective plasticizers, and it can be coupled with soy protein to improve film functionality. But mostly, compression molding involving thermal treatment is known for its additive free processing to get water-stable films and coatings. Technologies like nanoimprint lithography (NIL) is currently being used to develop protein-based films without any additives, but in literatures, SP-based films are mostly developed with the help of a plasticizer. In comparison to solution casting, which has low hydrophobicity and water resistance, compression-molded films have higher transmittance, smoother surfaces, higher elongation, and higher tensile strength (Garrido et al., 2013; Rani & Kumar, 2019).

In order to improve cross-linking and functionality, it is frequently necessary to change the structure of proteins. Chemical or enzymatic alteration, for example. Protein covalent cross-linking processes can be catalyzed by enzymes, resulting in high-molecular-weight biopolymers. Enzymatic cross-linking procedures are more popular and advantageous than chemical cross-linking approaches in terms of safety (Song et al., 2011). Protein films with a range of capabilities can be made using modification processes that are both efficient and reliable. Surface changes allow for the creation of multifunctional films, while additions help overcome native protein film constraints, such as stiffness and mechanical instability. The cytotoxicity of the reagents used for alteration or the ensuing by-products are a key drawback of both techniques. Furthermore, certain modification procedures require the use of organic solvents, which might cause protein denaturation and loss of function (Gopalakrishnan et al., 2021).

7.4 Development of Edible Coatings Derived from Soy Proteins

Edible coatings are thin layers of edible materials that are used for coating different food products and their surfaces and providing protection to them from light, moisture, oxygen, and different microorganisms (Tripathi, 2021). Proteins, mainly globular proteins, such as soy proteins, as coating materials is of huge interest because they are soluble in water as well as in different acidic, basic, and saline solutions (Hassan et al., 2018; Lopez-Polo et al., 2021). Not only that, but these proteins can also generate stable foams and gels and enhance the transport of different bioactive compounds (Sahraee et al., 2019).

The process and mechanism of the development of edible coatings are very close to that of films. The difference between them is that films are formed as thin layers that can be wrapped around or be incorporated in between different foods. On the other hand, coatings are prepared directly on different food products in the form of a coat and are suitable for human consumption (Zink et al., 2016). The process of formation of edible coatings involves a dilute solution of proteins, such as soy proteins, which is put on the food product surface, and eventually, the solvent gets evaporated and leads to the formation of a coat on different food products (Dangaran et al., 2009). The different methods that have been used in various studies to form edible coatings using soy proteins include dipping, spraying, electrostatic spraying, and fluidized-bed coating (Table 8.1).

7.4.1 Dipping

The dipping method is easy to use and the most used method to produce edible coatings using soy proteins, and it includes submerging the food product into a vessel comprising of the coating solution (Andrade et al., 2012). Coating a food product using this method with edible coatings involves a total of three steps that include immersion, deposition, and solvent evaporation (Suhag et al., 2020). During immersion, the food product is kept in a coating solution under a particular speed till there is a complete interaction between the food product and coating matrix (Valdes et al., 2017). During deposition, there is the formation of thin layers of the coating on the food product surface and the draining out of the excess liquid (Suhag et al., 2020). The final and third step involves solvent evaporation from the food product surface using a drying method which could either include drying in air at room temperature or using a dryer (Andrade et al., 2012).

This method is quite good for food coatings that are irregularly shaped or those that require a full coverage, as it leads to a uniform coating to develop on the food product. Alves et al. (2017) employed the dipping method to coat cut apples with soy protein isolate–based edible coatings incorporated with ferulic acid that helps control their weight loss. Similarly, fresh-cut eggplants were dipped in soy protein-cysteine–based edible coatings which provided a uniform coating and prevented enzymatic browning (Ghidelli et al., 2014). Kang et al. (2013) used soy protein edible coatings to coat walnut kernels using the dipping method that helped improved their fat stability and protected them against lipid oxidation. A soy protein isolate–chitosan edible coating has also been applied on apricot fruits during storage using the dipping method that helped reduce their firmness and weight loss (Zhang et al., 2018). Along with these, many different studies have been done using this method on developing edible coatings using soy protein isolate for coating different fruits and vegetables and preventing their degradation (Ghidelli et al., 2015; Li et al., 2019; Shon, 2011; Xu et al., 2001; Zhong et al., 2014).

7.4.2 Spray Coating

The spray coating is one more method that can be used for formation of a thin or thick uniform layer of protein-based edible coating on a food product. The benefit of using this method over others is that it can help coat the food products that have large surface areas (Andrade et al., 2012). This is because it involves the formation of droplets that are then sprayed over the food product using nozzles and allows the application of the coating on a food product several times without contaminating the coating solution (Zhong et al., 2014).

Spray coating to produce an edible coating using soy protein has been used in a few studies. A soy protein and glycerol coatings were applied to successfully coat low-moisture Mozzarella cheese, and it was observed that their physicochemical properties during storage were enhanced (Zhong et al., 2014). Previously, edible coatings prepared from soy proteins using the spray coating method was prepared, and they were found to be chemically durable and exhibited antifouling and self-cleaning property (Liu et al., 2019).

7.4.3 Electrostatic Spraying

Electrostatic spraying is a coating method to develop edible coatings and is more superior to the traditional spray coating method (Zhong et al., 2014). It is commonly used in the paint industry but now is also being used in the food industry too. This method employs electrostatic sprayers which are electrically charged, and they allow the protein solutions to enclose around and evenly coat all types of food products. Its advantage is that is allows the homogenous distribution of the coating, helps increase droplet coverage, and control the droplet size and reduce all types of waste (Maski & Durairaj, 2010). Zhong et al. (2014) indicated that when soy protein solution was electrostatically sprayed on Mozzarella cheese, it enhanced their properties, such as hardness, color, and weight loss during storage. Unfortunately, only very few works have been done using this method to produce soy protein–based edible coatings.

7.4.4 Fluidized-Bed Coating

This method is utilized mainly by the pharmaceutical industries for coating tablets, but now this technique is gaining attention for incorporating a film coating on different foods. It involves the deposition of a thin film coating on a food surface by spraying the coating solution containing protein into a fluidized bed of the food product on which the coating is to be done. This technology is being used by few foods and confectionary industries because this method allows the solvents to be dried much faster than the other conventional coating techniques, such as pan coating (Lin & Krochta, 2006). Not only that, the fluidized-bed coating technique leads to the formation of a uniform coating which is not much agglomerated, as it provides efficient mass and heat transfer between the atomized solvent stream and solid phase (food product). He et al. (2013) have utilized the fluidized-bed coating method to coat pellets using soy protein–stabilized indomethacin nanosuspensions. Other than that, not many studies have been done to develop edible coatings from soy protein using fluidized-bed coating.

TABLE 7.1

Development of soy protein isolate edible coatings and films and improvement in the characteristics of the final product

Formulation	Method	Characteristics	References
Soy protein isolate and ferulic acid	Dipping	Controlled weight loss of apples	(Alves et al., 2017)
Soy protein and cysteine	Dipping	Prevented enzymatic browning of eggplants	(Ghidelli et al., 2014)
Soy protein isolate	Dipping	Enhanced lipid oxidation in walnut kernels	(Kang et al., 2013)
Soy protein and chitosan	Dipping	Reduced weight loss in apricots	(Zhang et al., 2018)
Soy protein isolate, cinnamaldehyde, and zinc oxide	Dipping	Improved postharvest banana quality	(Li et al., 2019)
Soy protein isolate	Dipping	Improved the physicochemical properties of Mozzarella cheese	(Zhong et al., 2014)
Soy protein isolate	Dipping	Reduced browning in artichokes	(Ghidelli et al., 2015)
Soy protein and carboxymethyl cellulose	Dipping	Reduced browning and improved antioxidant activity	(Shon, 2011)
Soy protein isolate, stearic acid, and pullulan	Dipping	Retarded the senescence process of kiwi fruits	(Xu et al., 2001)
Soy protein isolate and glycerol	Spray coating	Improved the physicochemical properties of Mozzarella cheese	(Zhong et al., 2014)
Soy protein isolate	Spray coating	Chemically durable and exhibited antifouling and self-cleaning property	(Liu et al., 2019)
Soy protein isolate and glycerol	Electrospray coating	Properties such as hardness, color, and weight loss during storage were enhanced of cheese	(Zhong et al., 2014)
Soy protein and indomethacin	Fluidized-bed coating	Enhanced properties	(He et al., 2013)

7.5 Functionalization and Modification of Films and Coatings Derived from Soy Proteins

Proteins are found in several structures in many plant and animal tissues and can carry out various functions. One of these functions for which they are most probably known is in the building of tissues and being involved in different biochemical reactions. Due to the wide range of functional properties, they are great options to prepare renewable materials of high performance. Films prepared using proteins can be subjected to various modifications that include various methods that can help improve their mechanical strength and poor resistance to water vapor (Mihalca et al., 2021). To prepare different functionalized and edible films and coatings, the proteins that are frequently utilized and preferred include soy protein, along with gelatin, casein, and zein (Coltelli et al., 2016). Due to the hydrophilic nature of soy proteins, films and coatings produced using them have high water vapor permeability values and poor barrier properties and are weaker, with lesser elongation properties (Wihodo & Moraru, 2013). Therefore, various studies are being done to enhance the functionality of soy protein–based films and coatings using various modifications.

7.5.1 Physical Modifications

7.5.1.1 Plasticization (Thermoplastic Processing)

This method employs the use of a plasticizer that includes glycerol, polyethylene glycerol (PEG), and sorbitol to hinder protein chain interactions and enhance the flexibility of the formed films (Wihodo & Moraru, 2013). This is because the films made only by proteins are brittle and fragile because of the interactions and bonding between the protein chains. But the addition of plasticizers to the film-forming solutions leads to them getting incorporated in between the three-dimensional network of the proteins which help enhance free volume and allow better mobilization of the different polymer chains (Wihodo & Moraru, 2013).

Various studies have been done where different types of plasticizers in different amounts have been added to the different protein films that have enhanced the physical properties of these films in different ways. Wan et al. (2005) have incorporated PEG, sucrose, and sorbitol, respectively, to soy protein isolate–based films and have observed reduced water vapor permeability that led to better barrier properties. They have also observed better tensile strength of the soy protein isolate–based films.

7.5.1.2 Ultrasonic Processing

This method utilizes high-intensity ultrasound that is acoustic waves with frequencies higher than 20 kHz. High-intensity ultrasound generally ranges between 16 and 100 kHz, with intensities greater than 1 W cm^{-2} (Chemat et al., 2011). These high-intensity waves generate a lot of shear forces which can break the covalent bonds of different proteins.

Wang et al. (2013) assessed the impact of ultrasound on soy protein–based films and saw that the barrier properties of these films to oxygen and water vapor were enhanced. Similar results were seen, where they observed that when films made with soy protein isolate were treated with ultrasound, they had better hydrophobicity and film density (Wang et al., 2014). Sun et al. (2012) also found that the soy protein films became more homogenous, with improved mechanical properties, following ultrasound.

7.5.1.3 Compression Molding

During compression molding, proteins are mixed with different plasticizers and placed in an open mold, where a plunger makes these materials to achieve a particular form through the application of pressure and heat (Mihalca et al., 2021). This successive application of both heat and pressure leads to protein denaturation, which is followed by subsequent cooling, where they assume a particular form that is dependent on hydrogen, covalent, ionic, and hydrophobic and hydrophilic interactions (Balny et al., 2002). In this method, the blends are molded to viscoelastic melts at high temperature and pressure.

Soy protein films were formed by compression molding with glycerol at high temperatures (Guerrero & de la Caba, 2010). These films were found to have higher elongation and tensile strength when compared to conventional casting. Similar results were seen when soy protein films were compressed at high temperatures in another study (Ciannamea et al., 2014). They also reported that hydrogen bonds and hydrophobic interactions are mainly responsible for the compression-molded soy protein films. Also, films were prepared using soy protein and exopolysaccharides prepared from lactic acid bacterial strains by compression molding at high temperature (Uranga et al., 2020). The resulting films were found to be homogenous and transparent, indicating a good compatibility between soy protein and exopolysaccharides, and possessed antifungal activity.

7.5.1.4 Extrusion

Extrusion is a method where different raw materials are continuously incorporated into a hopper which is catering to a horizontal barrier (Zink et al., 2016). Firstly, the raw materials are added to the feeding section, where they are mixed, degassed, and compressed. This is followed by the transition phase, where, with the help of different screws, the mixture is moved to the transition section. Here, the mixture is converted to extrudates as the temperature and pressure are increased (Zink et al., 2016). In the final metering section stage, the extrudates are then subjected to very high pressure, temperature, and shear rate, leading them to achieve their final configuration as it is pressed through a die (Hernandez-Izquierdo & Krochta, 2008). Some studies have reported the modification of soy protein–based films using this method.

A mixture of soy protein isolate and montmorillonite was extruded with a twin-screw co-rotating extruder at a temperature of 50°C (Kumar et al., 2010). A huge improvement was observed in the tensile strength, storage modulus, thermal stability, glass transition temperature, and water vapor permeability of the films. Zhang et al. (2001) studied the consequences of different plasticizers on the mechanical and thermal properties of soy protein films during extrusion. The films were found to be strong and elastic. In another study done, a blend of soy protein isolate and cassava starch was subjected to extrusion, and the films were found to have higher water vapor permeability (Ferreira et al., 2021).

7.5.2 Chemical Modifications

7.5.2.1 Acetylation

This method involves the incorporation of an acetyl group to proteins where the acetyl groups form covalent bonds with the amino group of the proteins (Zink et al., 2016). The amino acid side chains of the proteins are known to attract one another, but when the amino groups are bound with the acetyl groups, there is lesser number of amino groups remaining. Hence, this results in the partial unfolding of the protein structure that can bring changes in the functionality of the proteins.

The acetylation of soy proteins was studied, and it was found that the acetylated protein isolates demonstrated changes in their functional properties that included their solubility, foaming and emulsifying properties, and water and oil holding capacity (Kim & Rhee, 1989). In another study, soy protein isolate was acetylated with pectin to prepare conjugates which demonstrated enhanced functional properties, such as solubility and emulsifying properties (Ma et al., 2020). This is important, as proteins and their functional properties help determine the behavior of soy-based films and coatings during food processing, preparation, and storage.

7.5.2.2 Succinylation

Succinylation is a type of acylation reaction and is a preferred chemical modification method, as they provide steric hindrance and electronegative charge to the proteins (Basak & Singhal, 2022). In this method, the amino acid residues in a protein are replaced with dicarboxylic acid anhydrides, such as succinic acid anhydride (SA), octenyl succinic anhydride (OSA), and dodecenylsuccinic anhydride (DDSA) (Basak & Singhal, 2022; Zink et al., 2016). The degree of succinylation is then determined to understand the extent to which the proteins are altered. This type of reaction is mainly known to take place at the lysine ε-amino group, where the cationic amino group is converted into an anionic residue which enhances the solubility and functionality of the proteins (Xu et al., 2021; Zink et al., 2016).

The attributes of succinylated soy protein–based films and their applications were investigated for preserving apples (Wu et al., 2019). They found that succinylated soy protein–based films showed better film-forming properties as the succinylation reaction destroyed the globular molecules of the protein and expanded its peptide chain that led to better tensile strength and elongation properties of the films formed. These films were also able to efficiently improve the quality of the apples. Similarly, succinic anhydride modification of soy protein isolate changed their flexibility and enhanced their interfacial functional properties (He et al., 2021; Lian et al., 2022).

7.5.2.3 Grafting

The grafting method is also a type of acylation reaction where fatty acids are incorporated into the protein structures in alkaline conditions (Zink et al., 2016). This results in long alkyl chains to be incorporated in the protein structure, which behave like internal plasticizers, which leads to lesser intermolecular interactions among the protein chains and changes in protein folding. This results in modifications in the thermal and functional properties of the proteins.

Plant proteins, including soy proteins, were modified by acylation reactions using palmitic acid chloride, and thermoplastic materials formed using these proteins showed enhanced water resistance (Brauer et al., 2007). Soy proteins have also been modified by grafting with methyl methacrylate, and the films prepared using these proteins were more hydrophobic, and their mechanical properties were enhanced (Gonzalez & Alvarez Igarzabal, 2017). Fatty acid chains were also grafted to soy protein isolate by the grafting reaction, and it was seen that the amphiphilic nature of the soy proteins improved and led to their better functional properties (Nesterenko et al., 2014).

7.5.3 Biochemical Modification by Enzymatic Hydrolysis

Enzymatic hydrolysis involves the cleavage of the peptide bonds, which are known to link the amino acids together in a protein. This method involves the use of enzymes that are specific for the hydrolysis of the peptide bonds, such as peptidases, and is known to enhance the solubility of the different proteins (Zink et al., 2016). There are two types of peptidases that are involved in enzymatic hydrolysis. One is exopeptidase, which cleaves the terminal peptide bonds, and the other is endopeptidase, which is known to cleave the nonterminal amino acids (Belitz, 2013). This results in the protein chain shortening and reduced molecular weight that leads to lesser intermolecular interactions between the protein chains and better flexibility. Transglutaminase-treated soy protein films were prepared, and their properties were found to enhance, which included their mechanical and surface hydrophobic properties (Jiang et al., 2007; Zadeh et al., 2018).

7.6 Characterization of Soy Protein–Based Films and Coatings

7.6.1 Mechanical Properties

Packaging of different food products in films and edible coatings is done to protect them during processing, storage, and transportation and to protect them from different mechanical stresses (Hadidi et al., 2022). Therefore, the mechanical properties of protein-based films and coatings are of high importance, as that is known to affect the food product texture during transportation and consumption (Hermawan et al., 2019). These include the tensile strength, elongation, and elastic modulus of the films. The mechanical strength of protein-based films and coatings is known to be affected by amino acids in proteins that influence the protein structure and film properties. Also, intra- and intermolecular interactions between the protein chains are known to affect their mechanical properties (Mihalca et al., 2021).

It was reported that when appropriate concentrations of soy protein isolate and plasticizers such as glycerol are used, it results in enhanced tensile strength, Young's modulus, and elongation at break of the films (Nandane & Jain, 2015). It was also suggested that the addition of graphene to films prepared using soy protein isolate demonstrated enhanced tensile strength and Young's modulus in comparison to films using only soy protein isolate (Han et al., 2017). It has been reported that soy protein isolate–based films containing zinc oxide and cinnamaldehyde showed better mechanical strength with its tensile strength and breaking elongation, increasing by about 1.2 times compared to soy protein isolate films alone (Wu et al., 2019). Similarly, it was found that incorporating licorice residue enhanced the tensile strength of soy protein–based films (Han et al., 2018). Amado et al. (2020) found that the addition of pectin to soy protein isolate–based films led to better tensile strength of the films. The main strategies that can help enhance the properties of soy protein isolate films include the addition of plasticizers or bioactive compounds or cross-linking them with different chemical compounds.

7.6.2 Thermal Properties

The thermal properties of plant protein–based films and coatings are very important, as they are subjected to different temperatures and heat changes during processing and storage (Hadidi et al., 2022). Many different techniques are used to visualize the thermal properties of protein-based films, which include differential scanning calorimetry (DSC), thermogravimetric analysis (TGA), hermomechanical analysis (TMA), and dynamic mechanical, thermal analysis (DMTA) (Garavand et al., 2022).

Soy proteins are globular proteins, which comprise of hydrogen, hydrophobic, and disulfide bonds, which make them good candidates for protein–carbohydrate and protein–protein interactions that can help enhance the thermal properties of prepared films (Hadidi et al., 2022). Also, their thermal stability could be enhanced following different physical and chemical modifications as well as combining them with different biopolymers (Gupta & Nayak, 2015). Filler addition, such as graphene, was found to successfully modify and enhance the thermal properties of soy protein–based films through π–π interactions and hydrogen bonding between soy protein and graphene (Han et al., 2017). Soy protein films have also been plasticized with triethanolamine, and it has been observed that these films had higher thermal stability because of strong interactions between soy protein and triethanolamine (Tian et al., 2009). Silicon dioxide nanoparticles, nanoboron nitride, and polycaprolactone-triol have also been added to soy protein films, which were found to enhance their thermal stability (Dash & Swain, 2013; Han & Wang, 2016; Schmidt & Soldi, 2006).

7.6.3 Barrier Properties

Barrier properties are very important characteristics for edible films and coatings, and this refers to oxygen and carbon dioxide permeability, water vapor permeability, and diffusion of aroma through the film and coating material (Hadidi et al., 2022). Proteins are known to have good gas barrier properties and high permeability that are desired, which is due to their polar nature. Therefore, they are considered potential materials for food packaging applications. But because of their hydrophilicity, they lead to poor water vapor and moisture barrier properties but good oxygen barrier properties (Hadidi et al., 2022). This becomes one major obstacle in the application of proteins to develop film packaging. However, certain modifications in proteins, such as soy proteins, can help enhance their barrier properties for better applications.

Soy protein isolate films were added with zinc oxide nanoparticles, and it was observed that their oxygen barrier properties were enhanced (Tang et al., 2019). Carbon nanoparticles of different sizes were used to modify soy protein isolate films, and it was found that when carbon nanoparticles of the smallest size were used, it led to the best water barrier property with a reduction in water vapor permeability in comparison to soy protein isolate films alone (Li et al., 2016). Han et al. (2018) incorporated licorice residue in soy protein isolate films, and their water vapor and oxygen permeability were found to reduce.

7.6.4 Thickness

The thickness of soy protein–based films is important, as this property is correlated with other properties of protein-based films, such as mechanical strength and barrier properties (Mihalca et al., 2021).

The methods by which the soy protein films are prepared, as well as the drying conditions of the films, are known to affect the thickness of the protein-based films. Hence, it is also an important characteristic that needs to be analyzed when developing soy protein–based films.

The addition of primrose flours to soy protein–based films led to an increased thickness in comparison to the control soy protein–based films, and this difference in thickness was not found to affect its mechanical and barrier properties (Mikus et al., 2021).

7.7 Mechanisms of Food Protection by Soy Protein–Based Films and Edible Coatings

Consumers have gained a greater awareness over the last several decades about plastic's effect on the environment. Therefore, the food packaging industry needs alternative packaging materials with a longer shelf life, higher quality, and fewer environmental impact. Edible packaging has been used to make food look better and keep it fresh for a long time. In the last few decades, it has gotten a lot of attention because it could be used to replace non-biodegradable synthetic packaging materials partially (Hassan et al., 2018). Edible coatings and films give food products better safety and quality, nutritional value, and cost-effective production in the food industry (Bharadwaj et al., 2019). Because of their promising barrier properties against gas and moisture, lipid oxidation, enzymatic activities, as well as contamination, it helps to safeguard the quality, appearance, and shelf life of food products, such as bakery, fruits and vegetables, fish, dairy products, etc. (Suhag et al., 2020).

Proteins have several vital properties for the formation of the film due to their structural and intrinsic properties. The diversity of amino acids, functional groups, and charged regions found throughout protein chains provides multiple locations for chemical interactions, which helps improve film stability (Carpine et al., 2015). Proteins are utilized to make biodegradable films because of their film-forming capabilities, nutritional value, quantity, high gas barrier qualities, and superior mechanical properties compared to polysaccharides and lipids (Chen et al., 2019). Soy protein isolates (SPI) are isolated mainly from soy flour, which is used in edible packaging films and coatings. SPIs are excellent for use as a precursor in producing edible film wraps due to their wide availability, affordability, biodegradability, and absence of toxicological chemicals. These protein derivatives have water, fat absorption, emulsifying, and excellent film-forming characteristics (Rangaraj et al., 2021).

SPIs with a high molecular weight boost the resulting films' mechanical strength without negatively impacting their ability to carry water vapor. Soy protein–based films have oxygen permeabilities that are 260, 500, 540, and 670 times lower than low-density methylcellulose, polyethylene, starch, and pectin, respectively (Chen et al., 2019). Because of their hydrophilic nature, SPI-based edible films have a high water vapor permeability, which prevents them from being used to preserve crispy food products. Fatty acid modifiers are often used as a common technique to lower the water vapor permeability value of SPI-based edible (Rangaraj et al., 2021). The International Union of Pure and Applied Chemistry (IUPAC) council described a plasticizer as "a substance which when added in a material can increase its flexibility or workability" (Chen et al., 2019). Because of the strong connections between polymeric chains, plasticizers are usually added to edible films and coatings, specifically polysaccharides and proteins. Plasticizers are used in polymeric film-forming materials to increase polymer thermoplasticity. They may bind to polymers and prevent polymer–polymer interactions, making them more flexible and processable. Many of the plasticizers used are hydrophilic and hygroscopic, attracting water molecules to create a hydrodynamic complex. Plasticizers include polyethylene glycol, propylene glycol, sorbitol, glycerin, sucrose, and corn syrup (Suhag et al., 2020).

7.8 Application of Soy Protein–Based Films and Edible Coatings in Improving Shelf Life of Food Products

Traditional packing materials are mostly non-recyclable and ecologically unfriendly. Research has recently focused on developing completely biodegradable packaging materials composed entirely of

TABLE 7.2

Application of soy protein–based film and coatings in food

Source of films and coatings	Food products	Functions	References
Soy protein hydrolysates/ whey protein hydrolysates	Pork patties	Inhibition of lipid oxidation	(Pena-Ramos & Xiong, 2003)
Soy protein isolate/ montmorillonite/clove essential oil	Bluefin tuna	Inhibits the microbial growth and lipid oxidation	(Echeverria et al., 2018)
Soy protein isolates/glycerol/ ferulic acids	Apple	Color and texture of the apple maintained, oxidative degradation reduced	(Alves et al., 2017)
Soy protein isolates/cysteine	Eggplant	Shelf life enhanced, control enzymatic browning	(Ghidelli et al., 2014)
Soy protein isolates/chitosan	Apricot	Shelf life enhances, textural property maintained, inhibited pectin degradation	(Zhang et al., 2018)
Soy protein isolates/ carboxymethyl cellulose	Potato pellet chips	Fat absorption reduced, sensorial properties enhanced	(Angor, 2016)
Soy protein isolates gellan gum	Doughnut Potato fries	Fat absorption reduced	(Rayner et al., 2000)
Soy protein isolates/ cinnamaldehyde/tea polyphenols/polylactic acid polybutylene adipate/ wheat gluten	Meat analogs	Effective against *Escherichia coli* (*E. coli*) and *Staphylococcus aureus* (*S. aureus*), prevents moisture loss, maintains textural properties	(Wang et al., 2022)
Soy protein isolates/cellulose nanocrystals	Pork Strawberry	Shelf life enhanced, water mobility decreased	(Xiao et al., 2021)

natural raw materials like proteins and polysaccharides. They are harmless to the environment and may also be eaten with the product they are packaging (Tkaczewska, 2020). Bioactive peptides and protein hydrolysates have a lot of potential as antioxidant food additives, since they may reduce hydroperoxides, scavenge free radicals, chelate prooxidative transition metals, and change the physical features of the foods (Sohaib et al., 2017). Table 7.2 depicts the application of soy protein–based film and coatings on food products. In cooked pork patties, the inhibition of lipid oxidation was reported by the whey protein hydrolysates and soy protein hydrolysates (Pena-Ramos & Xiong, 2003). It was reported that soy protein isolates (SPI) were combined with cortex *Phellodendron* extract with different concentration (0%, 10%, 12.5%, 15%, 17.5%, 20%, or 22.5%, w/w, based on SPI) to create a natural antibacterial and antioxidant coating (Liang & Wang, 2018). This coating has the characteristics of water vapor, oxygen, and light. The SPI film's antioxidant activity was also enhanced. The films were successful in eradicating *Staphylococcus aureus* (Gram-positive bacteria). These findings indicate that the SPI and cortex *Phellodendron* extract coating may increase food's shelf life. Coatings made of protein have been investigated for their use as prospective coating materials for reduced fat uptake (Ananey-Obiri et al., 2018). It was seen that when soy protein isolate was coated on potato chips, there was a reduction in fat than when different concentrations of carboxymethyl cellulose (2%, 6%, 10%, and 14% w/v) were used (Angor, 2016). The use of soy protein in doughnut mix at a concentration of 10% as an edible covering led to a decrease in fat content of 55.12% (Rayner et al., 2000).

7.9 Future Recommendations and Research Gaps

Worldwide, the preservation of foods is a major challenge. Using natural food components with antibacterial and antioxidant properties that don't harm human health to minimize chemical usage in the food sector is gaining popularity. A significant solution to this problem is to use edible coatings and films, which are helpful for both consumers and the environment. These edible coatings are now employed as nutraceutical properties (Hassan et al., 2018). It is found that protein hydrolysates and peptides enhance the shelf life of the food. These compounds have been found to significantly inhibit the oxidation of lipids in food items as well as the development of microbes, even at low concentration (Tkaczewska, 2020).

Moreover, SPI films provide only minimal resistance to water vapor movement because of the intrinsic hydrophilicity of proteins (Nandane & Jain, 2015). On the other hand, edible coatings have several drawbacks, including the ability to prevent oxygen exchange and the production of off-flavors. Some edible coatings have a high hygroscopicity, which aids microbial development. Extensive research is required to reduce the difficulties associated with the use of edible coatings and to make these coatings more helpful and practical in all aspects of food protection (Hassan et al., 2018).

REFERENCES

Alves, M. M., Goncalves, M. P., & Rocha, C. M. R. (2017). Effect of ferulic acid on the performance of soy protein isolate-based edible coatings applied to fresh-cut apples. *LWT–Food Science and Technology, 80*, 409–415. https://doi.org/10.1016/j.lwt.2017.03.013

Amado, L. R., Silva, K. D., & Mauro, M. A. (2020). Effects of interactions between soy protein isolate and pectin on properties of soy protein-based films. *Journal of Applied Polymer Science, 137*(21). https://doi.org/10.1002/app.48732

Ananey-Obiri, D., Matthews, L., Azahrani, M. H., Ibrahim, S. A., Galanakis, C. M., & Tahergorabi, R. (2018). Application of protein-based edible coatings for fat uptake reduction in deep-fat fried foods with an emphasis on muscle food proteins. *Trends in Food Science & Technology, 80*, 167–174. https://doi.org/10.1016/j.tifs.2018.08.012

Andrade, R. D., Skurtys, O., & Osorio, F. A. (2012). Atomizing spray systems for application of edible coatings. *Comprehensive Reviews in Food Science and Food Safety, 11*(3), 323–337. https://doi.org/10.1111/j.1541-4337.2012.00186.x

Angor, M. M. (2016). Reducing fat content of fried potato pellet chips using carboxymethyl cellulose and soy protein isolate solutions as coating films. *Journal of Agricultural Science, 8*(3), 162–168.

Arshad, M., Huang, L. L., & Ullah, A. (2016). Lipid-derived monomer and corresponding bio-based nanocomposites. *Polymer International, 65*(6), 653–660. https://doi.org/10.1002/pi.5107

Balny, C., Masson, P., & Heremans, K. (2002). High pressure effects on biological macromolecules: From structural changes to alteration of cellular processes. *Biochimica Et Biophysica Acta-Protein Structure and Molecular Enzymology, 1595*(1–2), 3–10. https://doi.org/Pii S0167-4838(01)00331-4

Basak, S., & Singhal, R. S. (2022). Succinylation of food proteins–A concise review. *LWT-Food Science and Technology, 154*. https://doi.org/10.1016/j.lwt.2021.112866

Belitz, H. D., & Grosch, W. (2013). *Lehrbuch der lebensmittelchemie.* Springer-Verlag.

Bharadwaj, A., Alam, T., & Talwar, N. (2019). Recent advances in active packaging of agri-food products: A review. *Journal of Postharvest Technology, 7*(1), 33–62.

Bourtroom, T. (2008). Edible films and coatings: Characteristics and properties. *International Food Research Journal, 15*(3), 237–248.

Brauer, S., Meister, F., Gottlober, R. P., & Nechwatal, A. (2007). Preparation and thermoplastic processing of modified plant proteins. *Macromolecular Materials and Engineering, 292*(2), 176–183. https://doi.org/10.1002/mame.200600364

Calva-Estrada, S. J., Jimenez-Fernandez, M., & Lugo-Cervantes, E. (2019). Protein-based films: Advances in the development of biomaterials applicable to food packaging. *Food Engineering Reviews, 11*(2), 78–92. https://doi.org/10.1007/s12393-019-09189-w

Campos, C. A., Gerschenson, L. N., & Flores, S. K. (2011). Development of edible films and coatings with antimicrobial activity. *Food and Bioprocess Technology, 4*(6), 849–875. https://doi.org/10.1007/s11947-010-0434-1

Carpine, D., Dagostin, J. L. A., Bertan, L. C., & Mafra, M. R. (2015). Development and characterization of soy protein isolate emulsion-based edible films with added coconut oil for olive oil packaging: Barrier, mechanical, and thermal properties. *Food and Bioprocess Technology, 8*(8), 1811–1823. https://doi.org/10.1007/s11947-015-1538-4

Chemat, F., Zill-e-Huma, & Khan, M. K. (2011). Applications of ultrasound in food technology: Processing, preservation and extraction. *Ultrasonics Sonochemistry, 18*(4), 813–835. https://doi.org/10.1016/j.ultsonch.2010.11.023

Chen, H. B., Wang, J. J., Cheng, Y. H., Wang, C. S., Liu, H. C., Bian, H. G., Pan, Y. R., Sun, J. Y., & Han, W. W. (2019). Application of protein-based films and coatings for food packaging: A review. *Polymers*, *11*(12). https://doi.org/10.3390/polym11122039

Cho, S. Y., & Rhee, C. (2004). Mechanical properties and water vapor permeability of edible films made from fractionated soy proteins with ultrafiltration. *Lebensmittel-Wissenschaft Und-Technologie-Food Science and Technology*, *37*(8), 833–839. https://doi.org/10.1016/j.lwt.2004.03.009

Ciannamea, E. M., Stefani, P. M., & Ruseckaite, R. A. (2014). Physical and mechanical properties of compression molded and solution casting soybean protein concentrate based films. *Food Hydrocolloids*, *38*, 193–204. https://doi.org/10.1016/j.foodhyd.2013.12.013

Coltelli, M. B., Wild, F., Bugnicourt, E., Cinelli, P., Lindner, M., Schmid, M., Weckel, V., Muller, K., Rodriguez, P., Staebler, A., Rodriguez-Turienzo, L., & Lazzeri, A. (2016). State of the art in the development and properties of protein-based films and coatings and their applicability to cellulose based products: An extensive review. *Coatings*, *6*(1). https://doi.org/10.3390/coatings6010001

Cutter, C. N. (2006). Opportunities for bio-based packaging technologies to improve the quality and safety of fresh and further processed muscle foods. *Meat Science*, *74*(1), 131–142. https://doi.org/10.1016/j.meatsci.2006.04.023

Dangaran, K., Tomasula, P. M., & Qi, P. (2009). Structure and function of protein-based edible films and coatings. *Edible Films and Coatings for Food Applications*, 25–56. https://doi.org/10.1007/978-0-387-92824-1_2

Dash, S., & Swain, S. K. (2013). Effect of nanoboron nitride on the physical and chemical properties of soy protein. *Composites Science and Technology*, *84*, 39–43. https://doi.org/10.1016/j.compscitech.2013.05.004

Debeaufort, F., & Voilley, A. (2009). Edible films and coatings for food applications, *9*, New York: Springer.

Dhall, R. K. (2013). Advances in edible coatings for fresh fruits and vegetables: A review. *Critical Reviews in Food Science and Nutrition*, *53*(5), 435–450. https://doi.org/10.1080/10408398.2010.541568

Diaz-Montes, E., & Castro-Munoz, R. (2021). Edible films and coatings as food-quality preservers: An overview. *Foods*, *10*(2). https://doi.org/10.3390/foods10020249

Echeverria, I., Lopez-Caballero, M. E., Gomez-Guillen, M. C., Mauri, A. N., & Montero, M. P. (2018). Active nanocomposite films based on soy proteins-montmorillonite-clove essential oil for the preservation of refrigerated bluefin tuna (*Thunnus thynnus*) fillets. *International Journal of Food Microbiology*, *266*, 142–149. https://doi.org/10.1016/j.ijfoodmicro.2017.10.003

Eswaranandam, S., Hettiarachchy, N. S., & Johnson, M. G. (2004). Antimicrobial activity of citric, lactic, malic, or tartaric acids and nisin-incorporated soy protein film against *Listeria monocytogenes*, *Escherichia coli* O157: H7, and *Salmonella gaminara*. *Journal of Food Science*, *69*(3), M79–M84.

Falguera, V., Quintero, J. P., Jimenez, A., Munoz, J. A., & Ibarz, A. (2011). Edible films and coatings: Structures, active functions and trends in their use. *Trends in Food Science & Technology*, *22*(6), 292–303. https://doi.org/10.1016/j.tifs.2011.02.004

Ferreira, L. F., de Oliveira, A. C. S., Begali, D. D., Neto, A. R. D., Martins, M. A., de Oliveira, J. E., Borges, S. V., Yoshida, M. I., Tonoli, G. H. D., & Dias, M. V. (2021). Characterization of cassava starch/soy protein isolate blends obtained by extrusion and thermocompression. *Industrial Crops and Products*, *160*. https://doi.org/10.1016/j.indcrop.2020.113092

Galus, S., Kibar, E. A. A., Gniewosz, M., & Krasniewska, K. (2020). Novel materials in the preparation of edible films and coatings-a review. *Coatings*, *10*(7). https://doi.org/10.3390/coatings10070674

Garavand, F., Cacciotti, I., Vahedikia, N., Salara, A. R., Tarhan, O., Akbari-Alavijeh, S., Shaddel, R., Rashidinejad, A., Nejatian, M., Jafarzadeh, S., Azizi-Lalabadi, M., Khoshnoudi-Nia, S., & Jafari, S. M. (2022). A comprehensive review on the nanocomposites loaded with chitosan nanoparticles for food packaging. *Critical Reviews in Food Science and Nutrition*, *62*(5), 1383–1416. https://doi.org/10.1080/10408398.2020.1843133

Garrido, T., Etxabide, A., Penalba, M., de la Caba, K., & Guerrero, P. (2013). Preparation and characterization of soy protein thin films: Processing-properties correlation. *Materials Letters*, *105*, 110–112. https://doi.org/10.1016/j.matlet.2013.04.083

Ghidelli, C., Mateos, M., Rojas-Argudo, C., & Perez-Gago, M. B. (2014). Extending the shelf life of fresh-cut eggplant with a soy protein-cysteine based edible coating and modified atmosphere packaging. *Postharvest Biology and Technology*, *95*, 81–87. https://doi.org/10.1016/j.postharvbio.2014.04.007

Ghidelli, C., Mateos, M., Rojas-Argudo, C., & Perez-Gago, M. B. (2015). Novel approaches to control browning of fresh-cut artichoke: Effect of a soy protein-based coating and modified atmosphere packaging. *Postharvest Biology and Technology*, *99*, 105–113. https://doi.org/10.1016/j.postharvbio.2014.08.008

Gonzalez, A., & Alvarez Igarzabal, C. I. (2017). Study of graft copolymerization of soy protein-methyl methacrylate: Preparation and characterization of grafted films. *Journal of Polymers and the Environment*, *25*(2), 214–220. https://doi.org/10.1007/s10924-016-0797-0

Gopalakrishnan, S., Xu, J. L., Zhong, F., & Rotello, V. M. (2021). Strategies for fabricating protein films for biomaterial applications. *Advanced Sustainable Systems*, *5*(1). https://doi.org/10.1002/adsu.202000167

Guerrero, P., & de la Caba, K. (2010). Thermal and mechanical properties of soy protein films processed at different pH by compression. *Journal of Food Engineering*, *100*(2), 261–269. https://doi.org/10.1016/j.jfoodeng.2010.04.008

Guimarães, A., Abrunhosa, L., Pastrana, L. M., & Cerqueira, M. A. (2018). Edible films and coatings as carriers of living microorganisms: A new strategy towards biopreservation and healthier foods. *Comprehensive Reviews of Food Science and Food Safety*, *17*(3), 594–614.

Gupta, P., & Nayak, K. K. (2015). Characteristics of protein-based biopolymer and its application. *Polymer Engineering and Science*, *55*(3), 485–498. https://doi.org/10.1002/pen.23928

Hadidi, M., Jafarzadeh, S., Forough, M., Garavand, F., Alizadeh, S., Salehabadi, A., Khaneghah, A. M., & Jafari, S. M. (2022). Plant protein-based food packaging films: Recent advances in fabrication, characterization, and applications. *Trends in Food Science & Technology*, *120*, 154–173. https://doi.org/10.1016/j.tifs.2022.01.013

Han, J. H., & Gennadios, A. (2005). Edible films and coatings: A review. *Innovations in Food Packaging*, 239–262. https://doi.org/Doi 10.1016/B978–012311632–1/50047–4

Han, Y. F., Li, K., Chen, H., & Li, J. Z. (2017). Properties of soy protein isolate biopolymer film modified by graphene. *Polymers*, *9*(8). https://doi.org/10.3390/polym9080312

Han, Y. Y., & Wang, L. (2016). Improved water barrier and mechanical properties of soy protein isolate films by incorporation of SiO2 nanoparticles. *RSC Advances*, *6*(113), 112317–112324.

Han, Y. Y., Yu, M., & Wang, L. J. (2018). Preparation and characterization of antioxidant soy protein isolate films incorporating licorice residue extract. *Food Hydrocolloids*, *75*, 13–21. https://doi.org/10.1016/j.foodhyd.2017.09.020

Hassan, B., Chatha, S. A. S., Hussain, A. I., Zia, K. M., & Akhtar, N. (2018). Recent advances on polysaccharides, lipids and protein based edible films and coatings: A review. *International Journal of Biological Macromolecules*, *109*, 1095–1107. https://doi.org/10.1016/j.ijbiomac.2017.11.097

He, M. Y., Li, L. J., Wu, C. L., Zheng, L., Jiang, L. Z., Huang, Y. Y., Teng, F., & Li, Y. (2021). Effects of glycation and acylation on the structural characteristics and physicochemical properties of soy protein isolate. *Journal of Food Science*, *86*(5), 1737–1750. https://doi.org/10.1111/1750-3841.15688

He, W., Lu, Y., Qi, J. P., Chen, L. Y., Yin, L. F., & Wu, W. (2013). Formulating food protein-stabilized indomethacin nanosuspensions into pellets by fluid-bed coating technology: physical characterization, redispersibility, and dissolution. *International Journal of Nanomedicine*, *8*, 3119–3128. https://doi.org/10.2147/Ijn.S46207

Hermawan, D., Lai, T. K., Jafarzadeh, S., Gopakumar, D. A., Hasan, M., Owolabi, F. A. T., Aprilia, N. A. S., Rizal, S., & Khalil, H. P. S. A. (2019). Development of seaweed-based bamboo microcrystalline cellulose films intended for sustainable food packaging applications. *Bioresources*, *14*(2), 3389–3410. https://doi.org/10.15376/biores.14.2.3389-3410

Hernandez-Izquierdo, V. M., & Krochta, J. M. (2008). Thermoplastic processing of proteins for film formation—SA review. *Journal of Food Science*, *73*(2), R30–R39. https://doi.org/10.1111/j.1750-3841.2007.00636.x

Jiang, Y., Tang, C. H., Wen, Q. B., Li, L., & Yang, X. Q. (2007). Effect of processing parameters on the properties of transglutaminase-treated soy protein isolate films. *Innovative Food Science & Emerging Technologies*, *8*(2), 218–225. https://doi.org/10.1016/j.ifset.2006.11.002

Kang, H. J., Kim, S. J., You, Y. S., Lacroix, M., & Han, J. (2013). Inhibitory effect of soy protein coating formulations on walnut (Juglans regia L.) kernels against lipid oxidation. *Lwt-Food Science and Technology*, *51*(1), 393–396. https://doi.org/10.1016/j.lwt.2012.10.019

Kim, K. S., & Rhee, J. S. (1989). Effects of acetylation on physicochemical properties of 11s soy protein. *Journal of Food Biochemistry*, *13*(3), 187–199. https://doi.org/10.1111/j.1745–4514.1989.tb00393.x

Kinsella, J. E. (1979). Functional properties of soy proteins. *Journal of the American Oil Chemists Society*, *56*(3), 242–258.

Kumar, P., Sandeep, K. P., Alavi, S., Truong, V. D., & Gorga, R. E. (2010). Preparation and characterization of bio-nanocomposite films based on soy protein isolate and montmorillonite using melt extrusion. *Journal of Food Engineering*, *100*(3), 480–489. https://doi.org/10.1016/j.jfoodeng.2010.04.035

Li, J. M., Sun, Q., Sun, Y., Chen, B. R., Wu, X. L., & Le, T. (2019). Improvement of banana postharvest quality using a novel soybean protein isolate/cinnamaldehyde/zinc oxide bionanocomposite coating strategy. *Scientia Horticulturae, 258.* https://doi.org/10.1016/j.scienta.2019.108786

Li, Y., Chen, H., Dong, Y. M., Li, K., Li, L., & Li, J. Z. (2016). Carbon nanoparticles/soy protein isolate bio-films with excellent mechanical and water barrier properties. *Industrial Crops and Products, 82,* 133–140. https://doi.org/10.1016/j.indcrop.2015.11.072

Lian, Z. T., Yang, S., Dai, S. C., Tong, X. H., Liao, P. L., Cheng, L., Qi, W. J., Wang, Y. J., Wang, H., & Jiang, L. Z. (2022). Relationship between flexibility and interfacial functional properties of soy protein isolate: Succinylation modification. *Journal of the Science of Food and Agriculture.* https://doi.org/10.1002/jsfa.12012

Liang, S. M., & Wang, L. J. (2018). A natural antibacterial-antioxidant film from soy protein isolate incorporated with cortex phellodendron extract. *Polymers, 10*(1). https://doi.org/10.3390/polym10010071

Lin, S. Y., & Krochta, J. M. (2006). Fluidized-bed system for whey protein film coating of peanuts. *Journal of Food Process Engineering, 29*(5), 532–546. https://doi.org/10.1111/j.1745–4530.2006.00081.x

Liu, X. R., Wang, K. L., Zhang, W., Zhang, J. Z., & Li, J. Z. (2019). Robust, self-cleaning, anti-fouling, super-amphiphobic soy protein isolate composite films using spray-coating technique with fluorinated HNTs/SiO2. *Composites Part B-Engineering, 174.* https://doi.org/10.1016/j.compositesb.2019.107002

Lopez-Polo, J., Monasterio, A., Cantero-Lopez, P., & Osorio, F. A. (2021). Combining edible coatings technology and nanoencapsulation for food application: A brief review with an emphasis on nanoliposomes. *Food Research International, 145.* https://doi.org/10.1016/j.foodres.2021.110402

Ma, X. B., Chen, W. J., Yan, T. Y., Wang, D. L., Hou, F. R., Miao, S., & Liu, D. H. (2020). Comparison of citrus pectin and apple pectin in conjugation with soy protein isolate (SPI) under controlled dry-heating conditions. *Food Chemistry, 309.* https://doi.org/10.1016/j.foodchem.2019.125501

Majid, I., Ahmad Nayik, G., Mohammad Dar, S., & Nanda, V. (2018). Novel food packaging technologies: Innovations and future prospective. *Journal of Saudi Society of Agricultural Science, 17*(4), 454–462.

Maski, D., & Durairaj, D. (2010). Effects of charging voltage, application speed, target height, and orientation upon charged spray deposition on leaf abaxial and adaxial surfaces. *Crop Protection, 29*(2), 134–141. https://doi.org/10.1016/j.cropro.2009.10.006

Mensitieri, G., Di Maio, E., Buonocore, G. G., Nedi, I., Oliviero, M., Sansone, L., & Iannace, S. (2011). Processing and shelf life issues of selected food packaging materials and structures from renewable resources. *Trends in Food Science & Technology, 22*(2–3), 72–80. https://doi.org/10.1016/j.tifs.2010.10.001

Mihalca, V., Kerezsi, A. D., Weber, A., Gruber-Traub, C., Schmucker, J., Vodnar, D. C., Dulf, F. V., Socaci, S. A., Farcas, A., Muresan, C. I., Suharoschi, R., & Pop, O. L. (2021). Protein-based films and coatings for food industry applications. *Polymers, 13*(5). https://doi.org/10.3390/polym13050769

Mikkonen, K. S., Rita, H., Helen, H., Talja, R. A., Hyvonen, L., & Tenkanen, M. (2007). Effect of polysaccharide structure on mechanical and thermal properties of galactomannan-based films. *Biomacromolecules, 8*(10), 3198–3205. https://doi.org/10.1021/bm700538c

Mikus, M., Galus, S., Ciurzynska, A., & Janowicz, M. (2021). Development and characterization of novel composite films based on soy protein isolate and oilseed flours. *Molecules, 26*(12). https://doi.org/10.3390/molecules26123738

Miller, K. S., & Krochta, J. M. (1997). Oxygen and aroma barrier properties of edible films: A review. *Trends in Food Science & Technology, 8*(7), 228–237. https://doi.org/10.1016/S0924-2244(97)01051-0

Nandane, A. S., & Jain, R. (2015). Study of mechanical properties of soy protein based edible film as affected by its composition and process parameters by using RSM. *Journal of Food Science and Technology-Mysore, 52*(6), 3645–3650. https://doi.org/10.1007/s13197-014-1417-4

Nesterenko, A., Alric, I., Silvestre, F., & Durrieu, V. (2014). Comparative study of encapsulation of vitamins with native and modified soy protein. *Food Hydrocolloids, 38,* 172–179. https://doi.org/10.1016/j.foodhyd.2013.12.011

Pena-Ramos, E. A., & Xiong, Y. L. L. (2003). Whey and soy protein, hydrolysates inhibit lipid oxidation in, cooked pork patties. *Meat Science, 64*(3), 259–263. https://doi.org/10.1016/S0309-1740(02)00187-0

Rangaraj, V. M., Rambabu, K., Banat, F., & Mittal, V. (2021). Natural antioxidants-based edible active food packaging: An overview of current advancements. *Food Bioscience, 43.* https://doi.org/10.1016/j.fbio.2021.101251

Rani, S., & Kumar, R. (2019). A review on material and antimicrobial properties of soy protein isolate film. *Journal of Polymers and the Environment, 27*(8), 1613–1628. https://doi.org/10.1007/s10924-019-01456-5

Rayner, M., Ciolfi, V., Maves, B., Stedman, P., & Mittal, G. S. (2000). Development and application of soy-protein films to reduce fat intake in deep-fried foods. *Journal of the Science of Food and Agriculture*, *80*(6), 777–782. https://doi.org/10.1002/(Sici)1097-0010(20000501)80:6<777::Aid-Jsfa625>3.0.Co;2-H

Sahraee, S., Milani, J. M., Regenstein, J. M., & Kafil, H. S. (2019). Protection of foods against oxidative deterioration using edible films and coatings: A review. *Food Bioscience*, *32*. https://doi.org/10.1016/j.fbio.2019.100451

Schmidt, V., & Soldi, V. (2006). Influence of polycaprolactone-triol addition on thermal stability of soy protein isolate based films. *Polymer Degradation and Stability*, *91*(12), 3124–3130. https://doi.org/10.1016/j.polymdegradstab.2006.07.016

Shon, J. H., & Choi, Y.H.. (2011). Effect of edible coatings containing soy protein isolate (SPI) on the browning and moisture content of cut fruit and vegetables. *Journal of Applied Biological Chemistry*, *54*(3), 190–196.

Soares, R. M. D., Scremin, F. F., & Soldi, V. (2005). Thermal stability of biodegradable films based on soy protein and cornstarch. *Macromolecular Symposia*, *229*, 258–265. https://doi.org/10.1002/masy.200551132

Sohaib, M., Anjum, F. M., Sahar, A., Arshad, M. S., Rahman, U. U., Imran, A., & Hussain, S. (2017). Antioxidant proteins and peptides to enhance the oxidative stability of meat and meat products: A comprehensive review. *International Journal of Food Properties*, *20*(11), 2581–2593. https://doi.org/10.1080/10942912.2016.1246456

Song, F., Tang, D. L., Wang, X. L., & Wang, Y. Z. (2011). Biodegradable soy protein isolate-based materials: A review. *Biomacromolecules*, *12*(10), 3369–3380. https://doi.org/10.1021/bm200904x

Suhag, R., Kumar, N., Petkoska, A. T., & Upadhyay, A. (2020). Film formation and deposition methods of edible coating on food products: A review. *Food Research International*, *136*. https://doi.org/10.1016/j.foodres.2020.109582

Sun, Y., Sun, C. Y., & Chen, G. (2012). Effect of ultrasound treatment on properties of soy proteins film. *Materials and Computational Mechanics*, *117–119*, 513–516. https://doi.org/10.4028/www.scientific.net/AMM.117-119.513

Swain, S. N., Biswal, S. M., Nanda, P. K., & Nayak, P. L. (2004). Biodegradable soy-based plastics: Opportunities and challenges. *Journal of Polymers and the Environment*, *12*(1), 35–42. https://doi.org/10.1023/B:Jooe.0000003126.14448.04

Tang, S. Y., Wang, Z., Li, W., Li, M., Deng, Q. H., Wang, Y., Li, C. Y., & Chu, P. K. (2019). Ecofriendly and biodegradable soybean protein isolate films incorporated with ZnO nanoparticles for food packaging. *ACS Applied Bio Materials*, *2*(5), 2202–2207. https://doi.org/10.1021/acsabm.9b00170

Thakur, M. K., Thakur, V. K., Gupta, R. K., & Pappu, A. (2016). Synthesis and applications of biodegradable soy based graft copolymers: A review. *ACS Sustainable Chemistry & Engineering*, *4*(1), 1–17. https://doi.org/10.1021/acssuschemeng.5b01327

Tian, H. F., Guo, G. P., Fu, X. W., Yao, Y. Y., Yuan, L., & Xiang, A. M. (2018). Fabrication, properties and applications of soy-protein-based materials: A review. *International Journal of Biological Macromolecules*, *120*, 475–490. https://doi.org/10.1016/j.ijbiomac.2018.08.110

Tian, H. F., Liu, D. G., & Zhang, L. (2009). Structure and properties of soy protein films plasticized with hydroxyamine. *Journal of Applied Polymer Science*, *111*(3), 1549–1556. https://doi.org/10.1002/app.29160

Tkaczewska, J. (2020). Peptides and protein hydrolysates as food preservatives and bioactive components of edible films and coatings—a review. *Trends in Food Science & Technology*, *106*, 298–311. https://doi.org/10.1016/j.tifs.2020.10.022

Tripathi, A. D., Sharma, R., Agarwal, A., & Haleem, D. R. (2021). Nanoemulsions based edible coatings with potential food applications. *International Journal of Biobased Plastics*, *3*(1), 112–125.

Uranga, J., Llamas, M. G., Agirrezabala, Z., Duenas, M. T., Etxebeste, O., Guerrero, P., & de la Caba, K. (2020). Compression molded soy protein films with exopolysaccharides produced by cider lactic acid bacteria. *Polymers*, *12*(9). https://doi.org/10.3390/polym12092106

Valdes, A., Ramos, M., Beltran, A., Jimenez, A., & Garrigos, M. C. (2017). State of the art of antimicrobial edible coatings for food packaging applications. *Coatings*, *7*(4). https://doi.org/10.3390/coatings7040056

Vasconez, M. B., Flores, S. K., Campos, C. A., Alvarado, J., & Gerschenson, L. N. (2009). Antimicrobial activity and physical properties of chitosan-tapioca starch based edible films and coatings. *Food Research International*, *42*(7), 762–769. https://doi.org/10.1016/j.foodres.2009.02.026

Wan, V. C. H., Kim, M. S., & Lee, S. Y. (2005). Water vapor permeability and mechanical properties of soy protein isolate edible films composed of different plasticizer combinations. *Journal of Food Science, 70*(6), E387–E391.

Wang, L., Xu, J. G., Zhang, M. M., Zheng, H., & Li, L. (2022). Preservation of soy protein-based meat analogues by using PLA/PBAT antimicrobial packaging film. *Food Chemistry, 380.* https://doi.org/10.1016/j.foodchem.2021.132022

Wang, Z., Sun, X. X., Lian, Z. X., Wang, X. X., Zhou, J., & Ma, Z. S. (2013). The effects of ultrasonic/microwave assisted treatment on the properties of soy protein isolate/microcrystalline wheat-bran cellulose film. *Journal of Food Engineering, 114*(2), 183–191. https://doi.org/10.1016/j.jfoodeng.2012.08.004

Wang, Z., Zhou, J., Wang, X. X., Zhang, N., Sun, X. X., & Ma, Z. S. (2014). The effects of ultrasonic/microwave assisted treatment on the water vapor barrier properties of soybean protein isolate-based oleic acid/stearic acid blend edible films. *Food Hydrocolloids, 35*, 51–58. https://doi.org/10.1016/j.foodhyd.2013.07.006

Wihodo, M., & Moraru, C. I. (2013). Physical and chemical methods used to enhance the structure and mechanical properties of protein films: A review. *Journal of Food Engineering, 114*(3), 292–302. https://doi.org/10.1016/j.jfoodeng.2012.08.021

Wu, J., Sun, Q., Huang, H., Duan, Y., Xiao, G., & Le, T. (2019). Enhanced physico-mechanical, barrier and antifungal properties of soy protein isolate film by incorporating both plant-sourced cinnamaldehyde and facile synthesized zinc oxide nanosheets. *Colloids and Surfaces B-Biointerfaces, 180*, 31–38. https://doi.org/10.1016/j.colsurfb.2019.04.041

Xiao, Y. Q., Liu, Y. N., Kang, S. F., & Xu, H. D. (2021). Insight into the formation mechanism of soy protein isolate films improved by cellulose nanocrystals. *Food Chemistry, 359.* https://doi.org/10.1016/j.foodchem.2021.129971

Xu, F. J., Dong, Y. M., Zhang, W., Zhang, S. F., Li, L., & Li, J. Z. (2015). Preparation of cross-linked soy protein isolate-based environmentally-friendly films enhanced by PTGE and PAM. *Industrial Crops and Products, 67*, 373–380. https://doi.org/10.1016/j.indcrop.2015.01.059

Xu, F. Y., Wen, Q. H., Wang, R., Li, J., Chen, B. R., & Zeng, X. A. (2021). Enhanced synthesis of succinylated whey protein isolate by pulsed electric field pretreatment. *Food Chemistry, 363.* https://doi.org/10.1016/j.foodchem.2021.129892

Xu, S. Y., Chen, X. F., & Sun, D. W. (2001). Preservation of kiwifruit coated with an edible film at ambient temperature. *Journal of Food Engineering, 50*(4), 211–216. https://doi.org/10.1016/S0260-8774(01)00022-X

Zadeh, E. M., O'Keefe, S. F., Kim, Y. T., & Cho, J. H. (2018). Evaluation of enzymatically modified soy protein isolate film forming solution and film at different manufacturing conditions. *Journal of Food Science, 83*(4), 946–955. https://doi.org/10.1111/1750-3841.14018

Zhang, J., Mungara, P., & Jane, J. (2001). Mechanical and thermal properties of extruded soy protein sheets. *Polymer, 42*(6), 2569–2578. https://doi.org/10.1016/S0032-3861(00)00624-8

Zhang, L. F., Chen, F. S., Lai, S. J., Wang, H. J., & Yang, H. S. (2018). Impact of soybean protein isolate-chitosan edible coating on the softening of apricot fruit during storage. *LWT–Food Science and Technology, 96*, 604–611. https://doi.org/10.1016/j.lwt.2018.06.011

Zhang, S. F., Xia, C. L., Dong, Y. M., Yan, Y. T., Li, J. Z., Shi, S. Q., & Cai, L. P. (2016). Soy protein isolate-based films reinforced by surface modified cellulose nanocrystal. *Industrial Crops and Products, 80*, 207–213. https://doi.org/10.1016/j.indcrop.2015.11.070

Zhong, Y., Cavender, G., & Zhao, Y. Y. (2014). Investigation of different coating application methods on the performance of edible coatings on Mozzarella cheese. *LWT–Food Science and Technology, 56*(1), 1–8. https://doi.org/10.1016/j.lwt.2013.11.006

Zink, J., Wyrobnik, T., Prinz, T., & Schmid, M. (2016). Physical, chemical and biochemical modifications of protein-based films and coatings: An extensive review. *International Journal of Molecular Sciences, 17*(9). https://doi.org/10.3390/ijms17091376

8

Development and Structural Characterization
of Gluten-Based Films and Coatings

Priya Dangi, Nisha Chaudhary, Itu Dutta, Sneha Singhal, Navya Puri, and Anchita Paul

CONTENTS

8.1 Introduction

Films and coatings have been widely used since the 12th century to prevent moisture loss from vegetables and fruits and maintain their strength and quality during the storage period (Hammam, 2019). Plastic films used to be a popular food packaging material. Unfortunately, these are pretty much difficult

DOI: 10.1201/9781003303671-8

to identify, separate, and decompose, resulting in the significant amounts of wastes and, hence, outlasting its usefulness. As a result, the development of biodegradable and renewable packaging film was required. Later on, it has been found out that there can be a development of two types of films; one is biodegradable, and another one is non-biodegradable films. In the category of biodegradable films, they can be further classified as non-edible, like paper films and edible ones, like films made from proteins, such as whey protein, casein, and gluten (Cousineau, 2012).

Edible films are developing at an exponential rate due to their increasing demand for eco-friendly products and rising awareness toward the environment (Guillaume et al., 2010). They are the alternative to conventional synthetic packaging, cause less harm, and offer numerous advantages over polymeric packaging, like the use of few renewable raw materials, providing nutrient supplementation, avoiding impurities, being cost-effective, and also showing enhanced organoleptic properties in the packaged foods with the controlled release of food additives. Food packaging films and coatings made from edible materials can mechanically protect and also slow down the rate of deterioration and spoilage of food products (Gennadios et al., 1993a). Edible films are considered a cohesive layer that is formed from matrix of bioactive polymers having 0.050–0.250 mm thickness. These layers of edible film and coatings are used as roll, immersion, coat, or spray to provide mechanical protection to food products from gases, water vapor, and dissolved solids or fats (Hammam, 2019). For the production of these type of films, they often use agricultural-based raw materials (Guillaume et al., 2010). These films are composed of highly biodegradable proteins and polysaccharides which are commonly used to preserve sausages, resins, wax, and gelatin capsules (Hammam, 2019; Zuo et al., 2009). Preparations of these films often employ the use of film-forming agents, such as proteins, lipids, and polysaccharides. Among these, proteins are the thermoplastic heteropolymers having polar and non-polar amino acids, which are responsible for the formation of intermolecular linkages and manifest large number of functional properties (Tanada-Palmu et al., 2000). The amino acid content is often responsible for the enhancement of nutritional value of these biofilms. Most commonly, they include caseins, whey proteins, and gluten (Hammam, 2019).

Among the edible films and coatings, gluten films are used quite frequently in various food industries. Gluten, a three-dimensional viscoelastic complex derived from wheat, is one of such renewable and ecological options (Cousineau, 2012). It is one of the extraordinary polymeric materials which show effective oxygen barrier as compared to the non-protein edible coatings or packaging materials. Apart from some wheat gluten-specific characteristics such as allergenicity, they can be recognized as biodegradable and biocompatible (Mojumdar et al., 2011). When formed into film, wheat gluten protein can form a fibrous network exhibiting strength, elasticity, and plasticity (Cousineau, 2012). Film based on wheat gluten is homogenous, transparent, mechanically strong, and relatively water-insoluble (Mojumdar et al., 2011). This is because wheat gluten contains more than 75% of proteins, and the combination of protein with polysaccharide improves the physical properties and, thus, the overall performance. These films and coatings are generally made under laboratory conditions. However, due to their hydrophilic nature, they fail to provide good barrier properties against humidity (Zuo et al., 2009). It was also observed that gluten-based films are very brittle and become difficult to handle (Irissin-Mangata et al., 2001). Further developments are employed with the use of plasticizers like glycerol to reduce intermolecular forces and brittleness and increase the mobility of polymeric chains. This led to the improvement in the flexibility and extensibility properties of the film which prevent the chipping and cracking of the film during the handling process (Tanada-Palmu et al., 2000). Moreover, gluten films, when combined with unmodified and modified polypeptides, showed high tensile strength but were water-soluble. Similar to polyamides, wheat gluten films are good oxygen barriers at low relative humidity but poor water barrier properties in highly humid environment (Maningat et al., 1994). Also, packaging material must be time-stable in order to protect and provide a longer shelf life. However, biopolymers, including wheat gluten films, deteriorate with age. To slow aging, it is critical to identify and comprehend the mechanism and causes of time-dependent physical and chemical changes, but the studies are limited. Currently, there is a need to reduce the limiting factors and find new uses for gluten through the development of non-conventional applications, such as the manufacture of the films that could be used in food packaging, ensuring that they are strong and relatively transparent. These gluten films and coatings are typically not intended to be the replacements for existing non-edible films and coatings. They may, at the very least, reduce the use of currently available synthetic and non-biodegradable products. Gluten films, as edible films, provide an alternative packaging material to those synthetic films and coatings. Because of their biodegradability and renewable natural resources, they have the potential to play a significant role in the packaging and coating industries (Mojumdar et al., 2011).

8.2 Structure of Gluten Films

Gluten films and coatings are the heteropolymers comprising of more than 20 different amino acids, each of them having specific sequences and structures. This molecular diversity helps proteins have a considerable potential for the formation of such linkages that differ with respect to their position, nature, and/or energy (Capezza et al., 2019). Evaluating the structure and morphology of extruded wheat gluten films showed wheat gluten protein assemblies elucidated ranging from nano (4.4 Å and 9 to 10 Å, up to 70 Å) to micrometers (10 μm). The presence of sodium hydroxide in these gluten films induced a tetragonal structure with unit cell parameters as a = 51.85 Å and c = 40.65 Å, whereas ammonium hydroxide resulted in a bi-dimensional hexagonal close-packed (HCP) structure with a lattice parameter of 70 Å. In wheat gluten films with ammonium hydroxide, a highly polymerized protein pattern with intimately mixed glutenins and gliadins bound through SH/SS interchange reactions was found. A large content of β-sheet structures is also found in these films, and the film structure was oriented in the extrusion direction (Kuktaite et al., 2011).

Protein molecular weights have a substantial effect on protein network structure. They also help detect the presence of molecular overlapping, which leads to the formation of physical nodes. As is the case of synthetic macromolecules, overlapping could occur beyond a critical molecular weight (Mc). But a high mean molecular weight restricts polymer flow during material formation that can further lead to defects in the end product. The alpha-helix and beta pleated-sheet secondary structures of proteins are highly stabilized by cooperative hydrogen bonds and can resemble crystalline zones. These zones are important in these film proteins, where many repetitive sequences are seen to be present. They could be responsible for the formation of regular "crystal" lattice-type arrangements, which are known to have a marked effect on the final material properties of conventional synthetic polymers (Krochta, 2002).

WG protein-based materials could, thus, be defined as mainly amorphous three-dimensional arrangements stabilized by low-energy interactions that are partially reinforced by regular "crystal lattice"–type arrangements and strengthened by some covalent bonds (Krochta, 2002).

8.3 Composition of Gluten Films

Gluten is the cohesive and elastic mass that is left over after starch is washed away from wheat flour dough. Wheat gluten is composed of a mixture of complex protein and molecules which are segregated into glutenins and gliadins on the basis of their extractability in aqueous ethanol. Neither of these groups consists of pure proteins, and there are chances of a considerable overlap depending on the extraction conditions (Krochta, 2002). Gliadin refers to the viscous component of gluten and constitutes a heterogeneous group of proteins characterized by single polypeptide chains associated by hydrogen bonding and hydrophobic interactions, having intramolecular disulfide bonds and being soluble in a 70% ethanol or water solution. These proteins have been classified into four groups: α-type and γ-gliadins of relative molecular weight masses (Mr) between 30,000 and 50,000 and ω-gliadins with Mr of about 44,000–74,000 by sodium dodecyl sulfate-polyacrylamide gel electrophoresis (SDS-PAGE) (Khatkar & Schofield, 1997). Glutenin's form an extensive network of intermolecular disulfide bonds as with synthetic polymers (Krochta, 2002). The larger the subunits involved in the cross-linking, the greater the contribution to the elastic properties of the gluten matrix. Glutenin's comprise of a diverse number of protein molecules grouped into high-molecular-weight glutenin subunits (HMW-GS) with Mr in the range of 95,000–145,000 and low-molecular-weight glutenin subunits (LMW-GS) with Mr around 44,000 by SDS-PAGE (Chaudhary et al., 2016). Hydrogen bonding between the repeated regions of

HMW-GS has been found to be responsible for the elasticity of gluten according to the "loop and train" theory of Belton (Krochta, 2002). LMW-GS are partially soluble in 70% ethanol and are analogous, with some high-molecular-weight gliadins (Dangi et al., 2019). Gliadins and glutenins are present in almost equal quantities in wheat gluten and have similar amino acid compositions, being high in terms of quantities of both glutamine and proline. They also have a considerable number of non-polar amino acids containing aliphatic or aromatic groups. These groups, together with a few readily ionizable amino acids, are responsible for the insolubility of gluten in water (Kuktaite et al., 2011). The amount, size distribution, and molecular architecture of glutenins and gliadins greatly influence the rheological, processing, mechanical, and physicochemical properties of gluten (Chaudhary et al., 2021; Dangi et al., 2019).

Commercially, gluten is an industrial by-product with an annual production of 400,000 metric tons worldwide. Wheat gluten is suitable for furnishing numerous food and non-food uses. Its main application is in the bakery industry, where it is used to strengthen weak flours, rendering them suitable for bread-making. It is widely accepted that the unique rheological properties of wheat gluten films are derived from its protein composition, with lipids and carbohydrates being contaminants entangled in the protein matrix (Buffo & Han, 2005).

8.4 Development of Gluten Films and Coatings

The methods to produce edible and biodegradable films from wheat gluten have been reported by several authors (Kuktaite et al., 2011). However, the information and knowledge regarding the separation of gluten into its main components, gliadins and glutenins, as purer sources for films, are limited. Essentially, the development of films and coatings that are sourced from agricultural materials can be described as (1) disruption of intermolecular bonds of the native form of polymer chains and making them mobile by usage of certain chemical or physical agents, (2) achieving required spatial configuration, and (3) formation of intermolecular bonds and removal of cleaving agents which stabilize the required structure obtained in the previous step. This is usually achieved through solvent dispersion or thermoplastic process (Krochta, 2002). The principle and process of both the methods are further elaborated in Figure 8.1.

The conditions in the formation of film ideally should be modulated by keeping in mind the predetermined applicate and usage of the films. For instance, properties defining mechanical attributes and appearance are of prime importance while making a film that will be used as a thin coating over products. If the film is to be used as a moisture barrier between components of non-homogenous packaged foods, a film solution of pH 5 to 6 as well as a low ethanol concentration (20%) can be utilized to achieve the lowest water vapor permeability. To obtain a resistant, homogeneous, and transparent film without insoluble particles or excessively increased water vapor permeability, a high concentration of gluten (12.5 g/100 mL), pH around 4, and ethanol concentration around 32.5% would be recommended in solvent dispersion method (Krochta, 2002). In an investigative study, the following method (Figure 8.2) was used to prepare gluten films derived from Brazilian soft and semi hard wheat varieties. Films prepared from soft wheat flour fall short in comparison between its properties of mechanical strength, oxygen permeability, water vapor permeability, etc. when it comes to films prepared from "semi-hard" wheat mixed with low concentrations of glycerol (Tanada-Palmu & Grosso, 2003).

Protein-based films and coatings have remarkable gas barrier properties and sufficient mechanical attributes than those made from lipids and polysaccharides. The oxygen permeability of the former is 670 times lower than that of polyethylene, polypropylene, and low-density methyl cellulose. Aside from being useful for the preparation of edible films, protein-based biopolymers are also known to have various properties that make them an ideal material for food packaging. These include their ability to prevent the loss of flavor and moisture, as well as their ability to transport active nutrients and chemicals. It is also important that nitrogen sources are available during the degradation of protein-based films to prevent them from degrading. However, these protein-based films, owing to their hydrophilicity, are more prone to moisture and thus allowing a passage of water. Plasticizers, when utilized along with certain complementary post-treatments, such as mechanical treatment, heat, pressure, lipid interfaces, and metal ions, have been effective in overcoming these drawbacks. Incorporating oil into the dispersion being used for film formation is effective in increasing the water barrier properties by 25% when

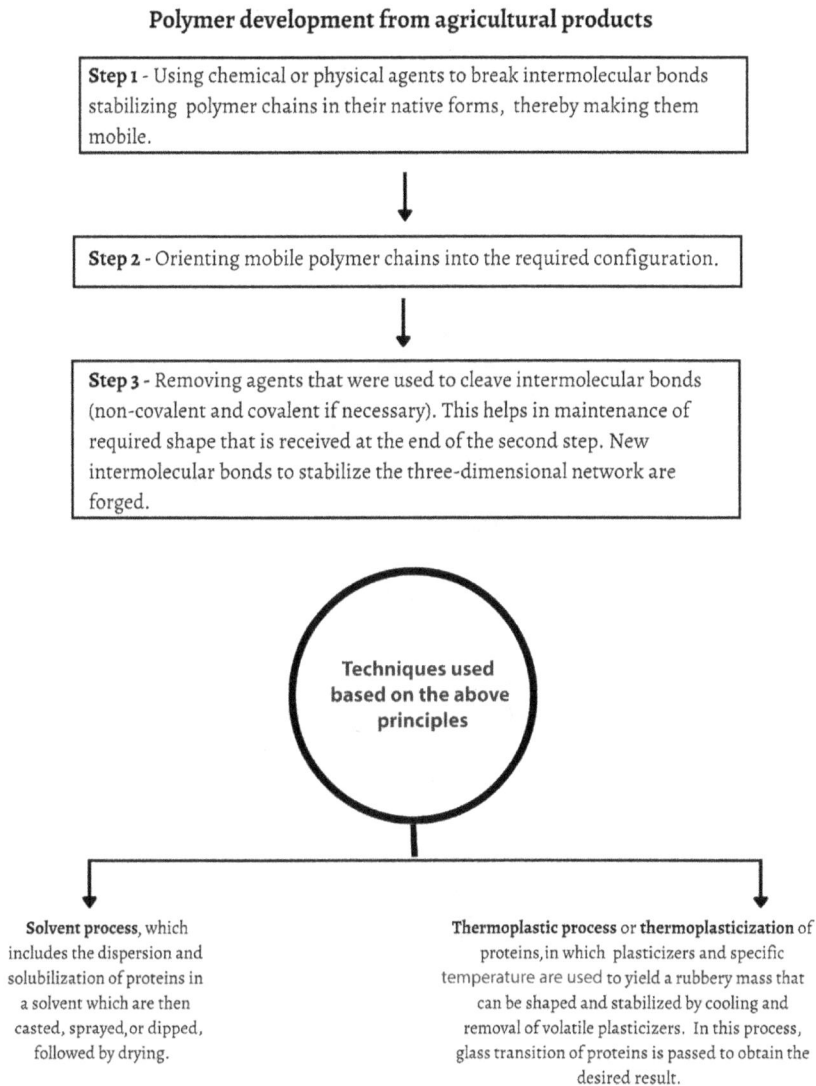

Polymer development from agricultural products

Step 1 - Using chemical or physical agents to break intermolecular bonds stabilizing polymer chains in their native forms, thereby making them mobile.

↓

Step 2 - Orienting mobile polymer chains into the required configuration.

↓

Step 3 - Removing agents that were used to cleave intermolecular bonds (non-covalent and covalent if necessary). This helps in maintenance of required shape that is received at the end of the second step. New intermolecular bonds to stabilize the three-dimensional network are forged.

Techniques used based on the above principles

Solvent process, which includes the dispersion and solubilization of proteins in a solvent which are then casted, sprayed, or dipped, followed by drying.

Thermoplastic process or **thermoplasticization** of proteins, in which plasticizers and specific temperature are used to yield a rubbery mass that can be shaped and stabilized by cooling and removal of volatile plasticizers. In this process, glass transition of proteins is passed to obtain the desired result.

FIGURE 8.1

compared to the film dispersions without oil. Further, mechanical properties in thermal methods for preparing films by casting can be improved by covalent cross-linking of gliadin polypeptide chains (Chen et al., 2019).

The configuration of protein chains as well as the type, distribution, and density of intermolecular and intramolecular linkages and the availability of residues play important roles in the operational properties of the final three-dimensional systems. They are strongly reliant on film polymerization conditions, such as the sort, amount, and order of solvent addition; drying conditions; as well as the type and number of plasticizers (Krochta, 2002). For example, as noted by Gontard et al. (1992), the film attributes were reliant on film-forming conditions. Strong interactions were observed between pH and ethanol concentration, thereby affecting opacity, solubility, and water vapor permeability of the film. The concentration of wheat gluten protein and the pH of the solution used in film formation were crucial parameters defining mechanical attributes of the film. The sturdiest film with a puncture strength of 4.5 N was obtained at higher gluten concentrations, such as 12.5 g/100 mL, and pH of the solution was above 5 (Krochta, 2002).

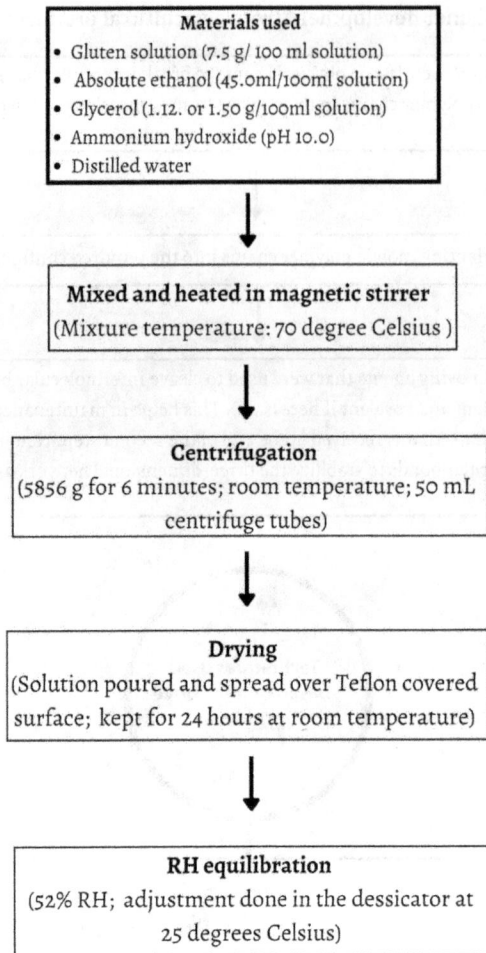

```
┌─────────────────────────────────────────────┐
│              Materials used                  │
│  • Gluten solution (7.5 g/ 100 ml solution)  │
│  • Absolute ethanol (45.0ml/100ml solution)  │
│  • Glycerol (1.12. or 1.50 g/100ml solution) │
│  • Ammonium hydroxide (pH 10.0)              │
│  • Distilled water                           │
└─────────────────────────────────────────────┘
                     │
                     ▼
┌─────────────────────────────────────────────┐
│      Mixed and heated in magnetic stirrer    │
│    (Mixture temperature: 70 degree Celsius )  │
└─────────────────────────────────────────────┘
                     │
                     ▼
┌─────────────────────────────────────────────┐
│              Centrifugation                  │
│  (5856 g for 6 minutes; room temperature; 50 mL │
│              centrifuge tubes)               │
└─────────────────────────────────────────────┘
                     │
                     ▼
┌─────────────────────────────────────────────┐
│                 Drying                       │
│ (Solution poured and spread over Teflon covered │
│  surface; kept for 24 hours at room temperature) │
└─────────────────────────────────────────────┘
                     │
                     ▼
┌─────────────────────────────────────────────┐
│              RH equilibration                │
│ (52% RH; adjustment done in the dessicator at │
│            25 degrees Celsius)               │
└─────────────────────────────────────────────┘
```

FIGURE 8.2

During the course of film formation, sometimes there is an extensive intermolecular interaction which may lead to the brittle nature of the films. To counteract this brittleness, plasticizers are employed, which are small molecules with low volatility. These plasticizers are successful in changing the three-dimensional structures, increasing free volume and mobility of polypeptide chains and decreasing the attractive forces between molecules in the films. In essence, plasticizers are crucial in altering the functional film properties and are efficient in increasing extensibility and the flexible nature of film. A decrease in cohesion and mechanical resistance is noted upon the addition of plasticizers. Preferred plasticizers used in wheat gluten protein films are glycerol and ethanolamine. Some others used are diglycerol, sorbitol, mannitol, propylene glycol, triethylene glycol, maltitol, etc. Upon their addition, these small molecules insert themselves between peptide chains and undergo hydrogen bonding by the presence of hydroxyl groups. Moreover, they have been effective in reducing viscosity and processing time of gluten. Usually, plasticizers are incorporated at levels of 15–40% by the weight of protein, while 20–30% is preferred. The amount of plasticizer used is a matter of the fact that glass transition temperature should be lowered to ambient temperature. Polar plasticizers, however, increase the hygroscopicity and diffuse easily upon contact with a moisture-rich substrate; hence, non-polar plasticizers are preferred. Amphipolar plasticizers, such as fatty acids and derivatives, are therefore promoted for utilization (Krochta, 2002). Water is an optimized and natural plasticizer in hydrocolloid-based films and coatings, with widespread application, as it is effective in reducing the glass transition temperature and increasing free volume of the biopolymers. Saturated fatty acids and glycerin have been used as natural plasticizers in wheat gluten films and coatings (Chen et al., 2019). Composite films made up of wheat protein gluten and fatty substances have exhibited good texture and low transparency (Hammam, 2019). In a study, the influence of plasticizer type and its concentration was found to have a greater effect on mechanical properties and water vapor permeability; however, other physical attributes did not get impacted much. The pressing temperature impacted the final film properties since the mixing time had a degree of influence over cross-linking. Hence, the films with a good balance of attributes and with lighter color were obtained by mixing humid or as received proteins which have no more than 20 weight percent glycerol for 5 minutes at 80°C and then pressing the paste at 100°C (Zubeldía et al., 2015).

The two development methods of preparation of gluten films and coatings are elaborated next.

8.4.1 Solvent Dispersion (Casting)

A light coating of protein matrix should be created and dried during this procedure. Wheat gluten protein films have also been created by accumulating the surface "skin" created during the thermal treatment of film-forming wheat gluten solutions to near-boiling temperatures (Krochta, 2002). In the presence of alcohol and disulfide bond–reducing agents, wheat gluten proteins are immiscible with water and necessitate a complicated solvent system of basic or acidic conditions (Bishnoi et al., 2022). Modifying the pH levels, in broad sense, severely impairs hydrogen and ionic interactions, whilst ethanol interrupts hydrophobic interactions, which are primarily taking place among gliadin molecules. When subjecting gluten to alkaline conditions, intermolecular (in glutenins) and intramolecular (in gliadins and glutenins) covalent disulfide linkages are severely affected and reduced to sulfhydryl groups (Krochta, 2002). Bases, including sodium hydroxide, potassium hydroxide, and ammonium hydroxide, are being used, mainly due to their volatility. Volatile disruptive agents usually get eliminated during drying (Mojumdar et al., 2011). In acidic environments, reducing agents such as sodium sulfite, cysteine, or mercaptoethanol should be used. Acids such as citric acid, phosphoric acid, lactic acid, and propionic acid may be used (Janjarasskul & Krochta, 2010). New interactions due to the freeing of active linkage sites and their increased proximity are now possible. A three-dimensional structure is thus developed as a result of the synergistic interactions of new hydrogen bonds, hydrophobic interactions, and disulfide bonds. Fractionation of solvent removal can help achieve film-forming solutions with very low levels of alcohol (Krochta, 2002).

Colloidal wheat gluten protein dispersions are formed by dissolving wheat gluten in an aqueous alcohol medium also containing acid. Alcohol amount will then be reduced by adding water or diafiltration to the stage in which the proteins turn insoluble (Lacroix & Vu, 2014). Precipitation of wheat gluten proteins as microparticle occur, forming homogeneous colloidal dispersion that is physically and microbiologically stable. When these particles are cast onto a substrate and dried at room temperature, they fuse to form consistent lustrous coatings. The quality attributes of wheat gluten protein–constituted films can be enhanced by selectively choosing the casting requirements or by treating the film-forming solutions prior to casting. The other approach would be to treat preformed films. Some of the pretreatments can be enzymatic (for example, transglutaminase), chemical (by utilizing cations and aldehydes), thermal, or radiative (ultraviolet or radiation) (Krochta, 2002). Further, it has been noted that upon reducing the pH from 6 to 4, an increment in film puncture strength was observed as an effect of acid on the protein structure. Enough protein unfolding, caused by pH levels just below the isoelectric point, has been required to increase the likelihood of cross-bridge formation between protein chains throughout drying, thereby improving the film hardness (Gontard et al., 1992). Further, gluten films, through casting, were prepared, and effects of wheat components spelt and bran on the film parameters were studied. Here, the minimum water permeability was noted at lowest glycerol concentration (2.0%), and at the instance where concentration of spelt bran was highest (6.8%). Bran presence decreased water vapor permeability, whereas glycerol presence increased water vapor permeability. This is a result of the inherent water resistance property of the bran (Mastromatteo et al., 2008).

The process of casting has widespread applications, and is a popular method in research. This process involves creating a thin layer of film using alcoholic aqueous solution. After casting, the resulting films are then collected by boiling the protein solutions. The characteristics of thermos pressed and cast films are different. For instance, the former has a stronger rupture resistance and higher elongation properties. The stress-strain relationship between films produced through different methods is different. This suggests that the production process affects the structure of proteins. The most common solvent used in film-forming is water ethanol. The film's consistency can be controlled by altering the conditions of the solution (Chen et al., 2019). Coatings can be made using processes alike to film production using solvent casting. Diluted protein solution is coated on food surface, and a coating is formed as solvent evaporates. Some common methods to achieve this are dipping, panning, and fluidized-bed processing (Dangaran et al., 2009).

8.4.2 Thermoplastic Process

The regulation of thermoplastic mechanisms, such as thermoforming and extrusion, involves an analysis of the rheological and flow properties of pliable plasticized wheat protein gluten. The chemical structure of wheat gluten protein is one in which the amino acids of the polypeptide chains can provide a broad variety of physical cross-links, primarily hydrogen bonds and/or hydrophobic interactions, which support the final rheological property of the films. Temperature has an impact on the film formation (Krochta, 2002). A common practice of forming film through extrusion is seen under this method.

Plasticized wheat gluten protein can be extruded in a co-rotating twin-screw extruder (Krochta, 2002). Following a dead stop procedure, the extracted screws display three distinct zones, as seen in twin-screw extruder: (1) powder solid conveying, (2) moderately filled "melt" flow, and (3) fully filled "melt" flow. Increased cross-linking inside the material seems to be the cause behind extrudate breakup (Krochta, 2002). Configuration of twin-screw extruder differs from model to model, but the procedure followed is similar. The extruded material is either passed through a die or thermo-molded to form a film. In one case, the barrel or cylinder of the twin-screw extruder (two co-rotating screws with length-to-diameter ratio of 48:1) is subdivided into 11 heating zones, starting from the hopper to the die. A flat sheet die can be used for extrusion of the films; all extrudates are picked up by the conveyor belt and cooled through air ventilation, as seen in practice in Rasel et al. (2016). Depending on the model, the twin-screw extruder may also constitute six different heating zones and two feeding zones, with temperatures ranging from 90 to 120°C in all the six zones. Extrusion in this case was carried out at a screw rotation of 30 Hz and a feed rate of 15 g/min. Extruded materials are cooled to room temperature or 25°C, then pelletized using an automatic pelletizer, and these pellets were then thermo-molded using a hydraulic press (Gutiérrez et al., 2021).

For basic operations, a homogeneous mixture of vital gluten (flour with minimal starch; in this case, crude protein content may be as high as 77.5% on dry basis), glycerol, and distilled water is used as the starting feed material. The twin-screw extruder comprises of different zones and can also be distinguished as feed screw, mixing elements, and extrusion screw. At lower barrel temperatures, sheet produced is sticky, is matte in color, and has uneven surface. At temperatures above 80°C, the sheet tended to rupture and lost its stickiness. A continuous sheet could be formed at set temperatures of barrel and die around 130°C. The die gap when decreased below 0.35 mm hindered film formation as the die got blocked. Hence, it was determined that at a screw speed of 60 rpm, total feed rate of 8 g/min, with gluten-glycerol-and-water ratio of 1.65:1:1.19, die gap of 0.35 mm, suitable gluten sheet could be prepared (Hochstetter et al., 2006).

Extrusion method used to make gluten films reinforced with carbon fibers (Wei et al., 2022) can be described as: (1) Mixing of wheat gluten, glycerol (mass ratio of glycerol to gluten = 30/70), carbon fiber (10% by volume of final composite), and water, which is added as a processing aid, for five minutes in a dough maker. (2) With 30 revolutions per minute at 90°C, the samples were extruded in twin-screw mini extruder. (3) Extrusion was repeated once more for improvement in dispersion of fiber. (4) Water was removed by drying overnight in oven at 50°C and further in silica gel–containing desiccator for 3 days. (5) Hot press was used to produce films from the dried samples. The conclusions showcased that carbon fibers exhibited adherence to gluten matrix when coupled with glycerol. These fibers were effective in reducing moisture uptake of gluten. It has been observed that the degree of aggregation, cross-linking, and conformational changes at molecular level can be determined by modulating certain parameters, such as the amount of plasticizer, shear force applied, input of mechanical energy, and the operating time, temperature, and pressure. Keeping this in mind, adequate temperature range which facilitates moderate viscosity, ensuring optimal aggregation of protein molecules without crossing the threshold beyond which degradation may occur, is vital when it comes to film formation through extrusion (Jiménez-Rosado et al., 2019).

8.5 Development of Composite Films and Coatings

Multilayer agro-materials (wheat gluten/paper WG/other biopolymers such as poly, or lactic acid), as well as composite agro-materials combining wheat gluten with fibers of cotton, sisal, coconut, or straw, have given promising test results and could possibly be used for different applications. One of the main developments in multicomponent wheat gluten–based materials involves the incorporation of lipid compounds into the film structure (Gennadios & Weller, 1990).

Films made from wheat gluten and lipids can combine the water vapor and the resistance of the lipids with the relatively good mechanical properties of wheat gluten. Lipids can either be incorporated in wheat gluten film-forming solutions to obtain composite films by emulsion technique or deposited as layers onto the surfaces of preformed gluten films to obtain bilayer films by coating technique. The effects of lipids incorporated into wheat gluten film-forming solutions can vary from destruction of the protein network to improvement of the mechanical and barrier properties of the cast films. These effects are complex and dependent on the nature and structure of the lipids, along with the chemical interactions between the lipids and the protein. Gontram incorporated 22 different lipids (at 20% w/w dry matter) into wheat gluten film-forming solutions. Among these lipids, 11 were selected because of their ability to form composite films with satisfactory appearance and mechanical properties. Water vapor permeability, opacity, solubility in water, and mechanical properties were determined for each type of gluten–lipid film as a function of lipid concentration and compared with a control gluten film without any lipid component. For all tested lipids, film dispersion in water increased sharply above a lipid content of about 20% (w/w dry matter). Above this threshold is the presence of lipids which could have probably and partially destabilized the protein structural matrix (reduction of intermolecular interactions among protein chains) or decreased water resistance when the lipid component carried polar, hydrophilic groups. The film opacity is highly intensified as lipid concentration increases. This opacity increase was quite low with diacetyl tartaric ester of monoglyceride. Beeswax, a solid and highly hydrophobic lipid, was the most effective lipid in improving the moisture barrier ability of films. As beeswax concentration increases from 0 to 36.8% (w/w dry matter), water vapor permeability decreases sixfold, from 5.08 to 0.83×10^{-12} mol/m s Pa. However, films with beeswax were opaque and brittle and disintegrated easily into water. These drawbacks were reduced when sucroglycerides or diacetyl tartaric esters of monoglycerides were used as the lipid component in the composite films (Gontard et al., 1992). Combining wheat gluten with a diacetyl tartaric ester of monoglycerides (at 20% w/w dry matter) produced films that, when compared to the control, glycerol-plasticized wheat gluten films, were as transparent and had higher mechanical resistance and about 50% lower water vapor permeability. However, improving moisture barrier properties of such composite films by increasing lipid concentration proved to be of limited value, because the hydrophilic protein matrix altered the resistance of the lipid components to that of water vapor transmission (Heralp et al., 1995).

Bilayer films composed of gluten film and modified polyethylene film were prepared. The objective was to combine the gas selectivity of the gluten films with that of excellent mechanical properties and moisture resistance of the polyethylene films. Wheat gluten films were prepared first using a casting procedure and then hot-pressed with that of polyethylene films at different temperatures. Because of chemical incompatibility, no adhesion between the wheat gluten layer and the synthetic layer could be obtained using simple polyethylene films. Three types of modified polyethylene films with different reactive groups were tested for their ability to offer enhanced adhesion with wheat gluten films, such as ethylene/acrylic ester/maleic anhydride terpolymer (EAMAT); ethylene/glycidyl methacrylate copolymer (EGMC); and polyethylene grafted with maleic anhydride (PGMA). Various time and temperature combinations for hot-pressing were tested, and the adhesion of the wheat gluten films on the synthetic films was evaluated. At optimum conditions for hot-pressing (110°C and six minutes), good adhesion was obtained only with PGMA. The adhesion strength (0.04 N/mm width) was equivalent to that of WG/WG or PGMA/PGMA bilayer films hot-pressed at the same conditions. Adhesion of wheat gluten films to EAMAT or EGMC films was not sufficient, probably due to the lower chemical compatibility of these synthetic materials with WG than with PGMA. Tensile strength and percent elongation at break values of PGMA/WG bilayer films were not affected by RH, unlike those of the mentioned gluten films. Also, the gas and water vapor permeability values of the PGMA/WG films were considerably reduced as compared to those of control wheat gluten films. However, the high selectivity of these gluten-based films to oxygen and carbon dioxide was highly reduced when these films were combined with PGMA films (Rayas et al., 1997).

8.6 Properties of Gluten Films

The high-molecular-weight-possessing non-polar character, complexity, and diversity of subunits are some of the unique features of wheat gluten proteins that can be utilized to make films having novel functional properties, such as selective gas barrier properties and rubber-like mechanical properties. The materials are homogeneous, transparent, mechanically strong, and relatively water-resistant. They are biodegradable and biocompatible apart from some wheat gluten–specific characteristics, such as allergenicity. These films could be useful for packaging cheese, fruits, and vegetables or for films for agricultural uses or cosmetic applications. The barrier properties of such films can be substantially improved to approximate those of polyethylene films by incorporating fatty compounds (e.g., beeswax and paraffin) in the film structure. Many mechanical properties of these films have been determined and modelled. For the most resistant gluten film–based materials, critical fracture deformation (Dc 0.7 mm) values are slightly lower than those of synthetic reference materials, such as low-density polyethylene (Dc 2.3 mm) and cellulose films (Dc 3.3 mm) (Gennadios et al., 1993a).

8.6.1 Water Vapor Permeability (WVP)

An edible film or coating should have water vapor permeability as low as possible so as to reduce the transfer of moisture between the surrounding environment and the food (Gennadios et al., 1996). WVP of gluten films is determined by the protein network structure (including tightness) and the water affinity of the film constituents. WVP can be increased by not only composition but also bubbles, pinholes, or points of weakness caused by material folding. Although reducing WVP in gluten films is difficult, some progress has been made in this regard. Soaking gluten films in an aqueous solution with a pH of 7.5 (the isoelectric point of gluten) increased tensile strength and decreased WVP by tightening the film structure. When sodium sulfide was added to the formulation, WVP increased in gluten films. This is because sodium sulfide cleaves disulfide bonds, which increases chain mobility and bonding during film drying while decreasing the polypeptide chain length. It was also reported that adding mineral oil and hydrolyzed keratin to the formulation reduced WVP of gluten films by 23 to 25%. Because of its non-polar and hydrophobic properties, mineral oil improved water barrier properties while also lowering film tensile strength. The introduction of keratin is believed to improve WVP by forming links with gluten units. It is also noted that in gluten films cast in aqueous ethanol and with ammonium hydroxide addition, increasing the plasticizer content increases WVP. Furthermore, the hydrophilic nature of glycerol is thought to improve water adsorption and permeation. It is also stated that as film thickness increases, so does WVP. The impact of film thickness on WVP was associated with structural changes or film swelling (Cousineau, 2012). Hence, WVP values of the various films should be compared with caution, because there were substantial thickness differences among the film (Gontard et al., 1993).

8.6.2 Oxygen Permeability

Oxygen barrier properties are highly needed in gluten films and coating in order to prevent lipid oxidation. Oxygen permeability of all wheat gluten films was low due to their polar nature and linear structure, leading to high cohesive energy density and low free volume. However, with increasing plasticizer concentration, oxygen permeability increases. Water, glycerol, and many other plasticizers increase the free volume of the polymer, thereby increasing polymer mobility and permeability. In one of the studies, wheat gluten films were found to have oxygen permeability of the same magnitude as those of polyvinylidene chloride–based films (Hernández-Muñoz et al., 2003). Oxygen permeability could not be measured for films made with acetic acid because they were too fragile and cracked in the test cell.

8.6.3 Tensile Strength and Percentage Elongation at Break

Tensile strength was tested to determine the resistance of gluten film. The tensile strength of unplasticized gluten films and those made with acetic acid could not be measured because of their high fragility. The effect of glycerol on the mechanical properties of films is observed often. Tensile strength decreases and percentage elongation at break increases with an increase in glycerol content. Similarly, decreased tensile strength and an increased percentage of elongation at break with increasing amounts of plasticizer have been reported for other protein films, such as glycerol-plasticized whey protein isolate and egg albumen films (Hernández-Muñoz et al., 2003). The plasticization effect of glycerol can be explained by the great number of hydrogen bonds between protein chains resulting from the high glutamine content, which is about 45% of wheat gluten proteins. Films prepared from wheat gluten with ammonium hydroxide had low oxygen permeability and high tensile strength at low water activity.

8.6.4 Mechanical Properties

The mechanical properties of materials are largely associated with distribution and density of intermolecular and intramolecular interactions allowed by the primary and spatial structures. Cohesion of wet gluten–based materials depends mainly on the type and density of intramolecular and intermolecular interactions but also results from interactions with other constituents. Cooperative phenomena generally allow optimal thermodynamical stability of systems. The effects of interactions depend on interaction probabilities and interaction energies. From a simplistic point of view, when covalent bonds stabilize the network or when density of bond energy is high, films are very resistant and relatively elastic. Thermoplastic-processed WG films plasticized with glycerol behaved like rubber, and their tensile properties could be modelled well using the statistical theory of rubber-like elasticity. Increasing processing temperature induced cross-linking reactions that were reflected in an increase of elastic modulus and a decrease of solubility in a 2% SDS buffer. A very close relationship could be evidenced between the elastic modulus and the protein solubility in 2% SDS (Weegels & Hamer, 1997).

8.7 Modifications of Gluten Films

Wheat gluten films show altered properties when they are subjected to various chemical and physical treatments. Chemical treatments include the treatment with formaldehyde vapors, while the physical treatment employs the use of heat, UV rays, and γ-rays. Effects of these treatments are observed on mechanical properties, water vapor permeability and solubility, along with color characteristics. Gluten-based films are observed as promising protective coating materials for perishable foods, such as cherry, tomatoes, and Sharon fruits (Hernández-Muñoz et al., 2003). Future research should explore possibilities of improving wheat gluten film properties by promoting cross-linking through enzymatic and chemical protein treatments. Development of edible films from wheat gluten will enhance its value and may provide new markets for wheat gluten (Tsiami et al., 1997). The various treatments subjected to gluten films and their effects (Table 8.1) are discussed in detail in what follows.

TABLE 8.1

Various treatments subjected to gluten films and their effects

Treatment	Experimental conditions	Properties observed	References
Aging	• Freshly cast films were conditioned for 48, 144, and 360 hours at 20°C and 60% RH for studies on the effect of aging.	• Increase in tensile strength and Young's modulus (stiffness) and decrease in elongation when the aging time increases from 48 to 360 hours. • Changes in mechanical properties due to the increase in thiol oxidation, which results in the formation of protein polymers with large molecular sizes.	(Micard et al., 2000)
Chemical treatment			
Formaldehyde	• The gluten film was placed in a hermetic box containing one liter of an ethanolic solution of formaldehyde (10% v/v) for 24 hours.	• Increase in tensile strength. • Increase in Young's modulus. • Increase in mechanical properties by improving the mechanical resistance. • Increase in inherent toxicity. • Lower elongation due to the reduction of mobility of the polymer chains in the film. • Increase in surface hydrophobicity. • Very little effect on water vapor permeability and solubility. • No color change.	(Cui et al., 2017; Micard et al., 2000)
Physical treatment			
Temperature	• Films were heated for 15 minutes on a Teflon sheet at 80, 95, 110, and 125°C, and 1.5 and 15 minutes at 140°C. • High-pressure liquid chromatography was used to quantify glycerol in the film before and after heating to assess the potential loss of glycerol during thermal treatments. • Refractometry was then used for detection.	• Slightly darker and more vivid, with a yellow-brown hue (at temperature higher than 55°C) and becomes more intense at higher temperature. • Obtaining cross-linked and more compact network at 70°C or more. • Decrease slightly with curing temperature, attributed to a partial loss of glycerol. • No change in the moisture-holding capacity (between temperature 40 and 115°C. • Increase in tensile strength and decrease in film extensibility and elongation. • Denser and tighter polymer network was achieved (in temperature 70–85°C or more). • Changes in viscoelastic properties. • Decrease in water vapor permeability when the temperature increased from 65 to 95°C for 2 hours.	(Hernández-Muñoz et al., 2004; Micard et al., 2000)

8.7.1 Aging

Prior to property evaluations, freshly cast films were conditioned for 48, 144, and 360 hours at 20°C and 60% RH for studies on the effect of aging. To test films under the same aging conditions, the time between the end of casting and property measurement was set to 144 hours at 60% RH and 20°C for all other treatments. Post-treatments were always performed on films that had been aged for 48 hours and then stored again at the same temperature and RH for a total aging of 144 hours. When the aging time of the film increases from 48 to 360 hours, there is a 75% increase in tensile strength, a 314% increase in stiffness (Young's modulus), and a 36% decrease in elongation. It is also showed that the storage period and conditions (temperature and relative humidity) of the films have a significant impact on thiol oxidation. The changes in mechanical properties of films may be caused by an increase in thiol oxidation during aging, which results in the formation of protein polymers with large molecular sizes. Thus, aging time influences the film properties to a greater extent, and therefore, an important factor needs to be considered (Micard et al., 2000).

8.7.2 Chemical Treatment

Formaldehyde treatment is one of the most effective treatments in terms of mechanical properties, followed by heat treatment. As a model chemical cross-linker, formaldehyde was chosen. Because of the difficulty in removing excess formaldehyde from films soaked in an ethanolic solution of formaldehyde, vapor post-treatment was chosen. Furthermore, soaking may remove plasticizer from the films, causing them to become brittle. As a result, the film was placed in a hermetic box containing one liter of an ethanolic solution of formaldehyde (10% v/v) for 24 hours. The elongation of wheat gluten films is decreased by 62%, the stiffness (Young's modulus) is increased by 438%, and the tensile strength is increased by 376% after the treatment with formaldehyde. Compared with the soaking method, vapor exposition procedures tend to produce effective results (Micard et al., 2000).

8.7.3 Physical Treatment

8.7.3.1 Treatment Temperature

Heat treatment was performed with a molder containing two heating plates modified as per the exact film thickness but without any applied pressure. Films were heated for 15 minutes on a Teflon sheet at 80, 95, 110, and 125°C, and 1.5 and 15 minutes at 140°C (Micard et al., 2000). Thermal treatments at temperatures higher than 55°C caused significant changes in the color coordinates of gliadin- and glutenin-rich films. Heat-cured films' colors became slightly darker and more vivid (higher chroma), with a yellow-brown hue. When the curing temperature was raised, these effects became more pronounced (Hernández-Muñoz et al., 2004). At atmospheric pressure, the boiling point of pure glycerol is 290°C, but it drops significantly in the presence of water. To assess the potential loss of glycerol during thermal treatments, high-pressure liquid chromatography (HPLC) was used to quantify glycerol in the film before and after heating. A water extract was obtained by immersing the film in water for 24 hours with continuous stirring, followed by 24 hours in a cold room without stirring, and was dosed on an RSil-R5C18–25F HPLC column and eluted by water (1 mL/min). Refractometry was then used for detection (Micard et al., 2000). As an effect of heat treatment, the protein's conformation changed due to the rupturing of non-covalent bonding, and the exposure of hydrophilic groups allowed more interactions between protein chains and solvent. Thus, glycerol distribution was favored, resulting in a decrease in intermolecular forces (Marcuzzo et al., 2011). When the temperature was raised from 80 to 140°C, there was an increase in tensile strength (41 to 329%) but a decrease in elongation (22 to 66%). Tensile strength increased twofold when the temperature was raised from 110 to 125°C, whereas elongation decreased gradually when the temperature was raised from 80 to 140°C. Heat treatment caused protein aggregation via cross-linking (hydrophobic and disulfide bonding). SDS solubility measurements were used to assess the aggregation of proteins in heat-cured films. Heating gluten film at 140°C for 15 minutes resulted in nearly total protein insolubility. When gluten was heated at 80°C for 30 minutes, protein extractability in SDS decreased by 11–32% by virtue of glutelin aggregation. When the treatment time was reduced from 15 to 1.5 minutes at high temperature (140°C), the tensile strength and elongation of wheat gluten film increased significantly when compared to the values of the control film. The 140°C 1.5 min treatment produced results comparable to the 110 °C 15 min treatment (Micard et al., 2000).

8.7.3.2 UV Radiation

A UV oven at 254 nm was used to apply UV treatments to films at 0.25 and 1 J/cm^2 doses. UV-post-treated films showed no or few changes in mechanical properties across the dose range used. UV radiation efficiency is generally determined by the protein source, specifically the amino acid compositions and molecular structures (Micard et al., 2000).

8.7.3.3 Gamma Rays

The Commissariat a' L'Energie Atomique (Cadarache, France) used a cobalt-60 source to treat films that had previously been produced. There were three irradiation doses used for testing, namely, 10, 20, and

40 KGy. Gamma radiation increased the tensile strength while decreasing break elongation. Except for elongation at break, no clear effect of dose irradiation was observed. A 10 KGy dose resulted in a 32% reduction in elongation. The effect was reduced when the radiation dose was increased above 10 KGy. This could be explained by a reduction in insoluble glutenin polymers in wheat flour. It was discovered that increasing the radiation dose from 2.5 to 20 KGy caused a progressive decrease in the number of largest polymers of glutenin, resulting in lower average molecular size via depolymerization and/or covalent linkage breakdown. The formation of dityrosine during irradiation could explain the positive effect of radiation on mechanical properties observed on gluten films irradiated with 10 KGy. As a result, these cross-linkages within polypeptide macromolecules would result in mechanical property changes (Micard et al., 2000).

Conclusively, the mechanical properties of wheat gluten films can be significantly altered by formaldehyde treatment, particularly when used as vapors. Heat curing for a short period of time (15 minutes) at temperatures above 110°C improves the mechanical properties of films. The high aggregation of network proteins causes such changes in mechanical properties. Radiation had little or no effect on the mechanical properties of the films (Hamer & Vliet, 2000).

8.8 Applications of Gluten Films

Packaging usually acts only as a method to contain the food item; some types may facilitate functions such as cooking or consumption. When paper is used as a primary packaging material, it comes into direct contact with the foodstuff and is therefore configured to extend shelf life in some manner. Usually, the design is coupled with other materials in layers that are impervious to gases such as oxygen and vapors (for example, water or aroma). Synthetic materials have long posed many concerns, namely, environmental and health; hence, alternatives in the form of edible coatings have become a crucial checkpoint in the promotion of sustainable and efficient packaging (Rovera et al., 2020).

Edible films have been observed to efficiently control mass transfer between product components as well as product and the external surrounding environment. Wheat protein gluten has exhibited good film-forming capacity due to its cohesive and elastic properties. In packaging, a film composed of gluten would not only be flexible, heat-sealable, and strong but also transparent/translucent, further not impacting the visual perception of the coated products (Tanada-Palmu & Grosso, 2003). Certain differences in applications may also arise as a property of an edible film being a single material intended to be wrapped around food, whereas an edible coating is an external covering that remains adhered to the surface of the coated food (Dhaka & Upadhyay, 2018). In fruits and vegetables, spoilage reduction as a result of decline of respiration rate is observed when they have a coating or film of gluten. Edible films have been used to (1) control mass transfer, (2) reduce oxidative reactions, (3) better surface gloss, and (4) enhance textural properties in fresh and processed fruits and vegetables (Dhaka & Upadhyay, 2018). The widespread use of environment-friendly gluten coatings that are derived from renewable sources in the sector of fruit preservation is due to its remarkable property in regulating mass transfer (Chen et al., 2022).

Keeping in mind gluten sensitivity, many applications are targeted toward packaging. This is not only limited to flexible films for agriculture (for instance, mulching and banana culture films), packaging films, and coatings for paperboards but also can be expanded to composite isolating materials in automobile industries made from wheat gluten–based biodegradable materials reinforced with fibers. Through careful selection and conceptualization of raw materials, careful use of separation and purification techniques and rheological modifying additives, and modification of film-forming processes, there is substantial scope for modifying the properties of wheat gluten–based films (e.g., characteristics and concentration of solvents, dispersing and plasticizing agents, temperature, protein concentration, and drying conditions). In particular, some useful applications in the field of active coatings, active packaging, modified atmosphere packaging, and drug delivery systems can be seen as a result of the moisture, gas, and solute barrier properties exhibited by wheat gluten–based films (Krochta, 2002).

Antimicrobial and antioxidant agents can be carried by polypeptide-based edible polymers. Another potential application for polypeptide-based edible polymers is their use in multiple-layer food packaging

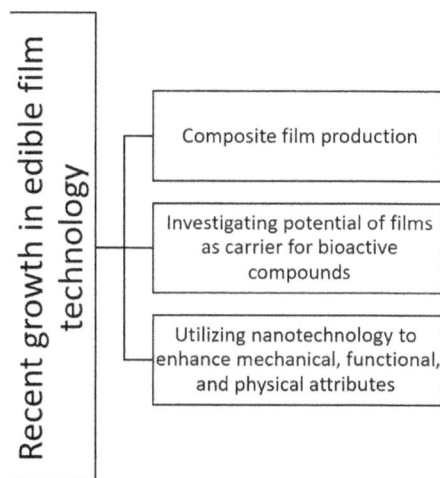

FIGURE 8.3

materials in conjunction with non-edible polymer. The internal layers which are in direct contact with food substances would be the protein-based edible polymer in this case. Edible polypeptide films have a potential application in individual packaging of small portions of food, especially those whose individual packaging has been restricted due to practical reasons, such as nuts and beans. After 36 days of aerobic fermentation and 50 days in farmland soil, wheat gluten films are readily biodegraded without releasing any toxic by-products in the due process. Milling, bakery products, meats, pasta, and breads are just few other examples of where these films can be used (Shit & Shah, 2014). Wheat gluten has been used in eggshell coating based on physical and barrier properties exhibited by gluten proteins. It has also been derived and used as a coating in starch-based packing material (Heralp et al., 1995). Composite gluten, when coated on peanuts, has been seen to retard fat degradation (Chen et al., 2019). In a study comparing wheat gluten–coated strawberries with those without coating (Figure 8.3), noteworthy extension of shelf life and retardation of senescence were observed in coated samples. The coated samples also attained consumer favorability as through sensory evaluation, as approval on taste, flavor, and visual quality was given. Incorporation of lipids into the film matrix proved to be advantageous for retaining firmness and reducing weight loss as a consequence of moisture losses (Tanada-Palmu & Grosso, 2005).

8.9 Trends and Recent Developments in Gluten Films

Food packaging films and coatings made from edible materials have been long proposed as alternatives to conventional polymeric packaging, inspired by the plethora of natural protective coatings found on horticulture and biological produce. Foods could be mechanically protected, slowing the rate of deterioration and spoilage. The current tilt toward edible films and coatings over traditional packaging materials have risen due to some well-established advantages which include the sustainable nature of the raw materials used in edible film production, nutritional supplementation and organoleptic enhancement of packaged foods, application in the interior of foods to control intercomponent moisture and solute migration, individual packaging of small food portions and small-size food products, used in as an internal layer in multilayered packages and controlled food preservative release mechanisms. Based on the nature of biopolymers used in film formation, a general classification of these films into polysaccharide, lipid, protein, and composite (combination of substances from other categories) films has been done (Gennadios et al., 1993b). The idea behind using numerous different ingredients is to take advantage of the synergistic interactions that occur when they are combined. For instance, the physical and barrier properties of tragacanth–locust gum bean composite film were reported to be superior to those of made using only the individual counterparts (Dhaka & Upadhyay, 2018).

At present, the usage of recyclable and sustainably derived agricultural products for packaging, films, and coatings production is on the rise for optimizing food storage. In the sector of vaccine drug delivery, edible polymer vaccines offer a lucrative avenue for providing long-term therapy in smaller doses, with the application of oral vaccine delivery. Edible polymer technology is therefore not only an advantageous mechanism that can further the agenda of decreasing petroleum-based packaging in the agricultural sector but also can actively address forthcoming needs that may arise in the sector of therapy (Shit & Shah, 2014).

In a study by Gennadios et al. (1993b), the following has come forward. Water vapor permeability was high in wheat gluten–based films, and this is still a significant constraint when considering commercial applications. The permeability of oxygen in wheat gluten–based films has been observed to be extremely low. The water vapor permeability of films made with acetylated monoglycerides wasn't significantly lower than that of standard films; however, the tensile strength of all modified films had been found to be significantly higher (by 42–69%) than the standard film. The water vapor permeability of a film made from wheat gluten and corn zein in a 4:1 ratio (weight basis) was significantly lower (by about 23%) than the standard. The oxygen permeability of a film made from wheat gluten and soy protein isolate in a 2.3:1 weight ratio (weight basis) was significantly lower (by about 40%) than the standard film. The use of edible polymers on a variety of food products is becoming more common. Many new industries are eyeing and investigating the possible impact of edible films and coatings as carriers for antimicrobials, flavors, antioxidants, coloring agents, vitamins, probiotics, nutraceuticals, etc. (Shit & Shah, 2014).

Nanotechnology is an effective tool that can be used to improve the nutritional properties of food through the use of nanoscale ingredients. Study on nanocomposites presents a compelling framework for designing novel innovative materials, including safe-to-eat polymers. Micro- and nanoencapsulation of bioactive components with edible coating materials may assist in monitoring their release under particular circumstances, safeguarding them from external factors, such as moisture, heat, as well as other extreme conditions, while significantly giving a boost to their reliability and viability. Through the means of dipping into a series of solutions or by spraying the food surface or by layer-by-layer electro-deposition, nanolaminate coatings derived from food-grade components could be utilized to incorporate and impart the effect of functional ingredients, namely, antimicrobial agents, flavorings, anti-browning agents, etc. (Shit & Shah, 2014).

Further, it has also been noted that enzymatic and chemical treatments on wheat gluten proteins may potentially improve film properties as a result of better cross-linking (Tanada-Palmu & Grosso, 2003). Future research might well encourage the development of solutions to enhance the delivery properties of safe-to-eat polymers. At the time, many tests on food applications are carried out on a laboratory level. However, more investigations on a commercial level are needed to provide more reliable information that will be instrumental in commercializing fresh-cut products encased in edible polymers. Food manufacturers are looking for edible polymers which could be used for a wide variety of foods that are important not only from the aspect of value addition but also from the aspect of increasing shelf life (Shit & Shah, 2014). Current developments in edible polymer films technology can be summarized, as described in Figure 8.4.

A few recent studies on composite gluten films and their properties are discussed in the following.

- Bionanocomposite composed of gluten and birch powder (lignocellulose) was developed, and these films were effective in maintaining appearance, freshness, and aroma of the fruits, thereby augmenting the shelf life of the fruits. These low-cost coatings exhibited favorable oxygen and water resistance, with remarkable UV-blocking and antimicrobial properties. These films also showcased reusable attributes (Chen et al., 2022)

- Wheat gluten coating on paperboard packaging is sensitive to water absorption; hence, in this study, a silica network was integrated by sol-gel chemistry into the gluten matrix. In these hybrid coatings, through spectrometric analysis, it was observed that both the organic and inorganic phases interact through hydrogen bonding. The intrinsic toughness of the inorganic

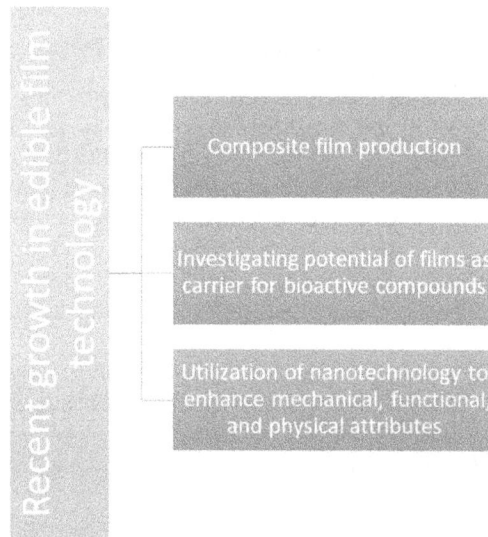

FIGURE 8.4 Developments in gluten films and coatings.

Source: Dhaka & Upadhyay (2018).

phase, specifically, resulted in greater extent of brittleness, which negatively affected the final water vapor barrier performance of the composite film. However, this can be mended by the use of appropriate plasticizers (Rovera et al., 2020)

- In a study focused on improving the aspects of wheat gluten films in the areas of poor mechanical strength, water resistance, and puncture resistance, a combination with other materials, such as apple pectin, carboxymethyl cellulose, egg white proteins, and tartaric acid, was carried out. The first composite film with apple pectin showcased better mechanical properties and film consistency. Upon combination with carboxymethyl cellulose, thermal stability was increased and egg white proteins increased opacity of the films, further improving the ability of the films to block light. Tartaric acid not only boosted thickness but also bestowed UV resistance to the film. All these films had higher mechanical strength and better water resistance (Dong et al., 2022)

- A high-pressure and high-temperature thermoplastic process was used to produce active films made from wheat gluten protein with different concentrations of thyme oil. A comprehensive analysis of the various properties of these films was performed. The results of the antimicrobial activity tests revealed that the addition of the essential oil of thyme to the formulations prepared from wheat gluten protein resulted in a significant increase in the overall antimicrobial activity. It was also shown that this oil can be used to produce both edible and biodegradable films. The addition of oil to various films results in the reduction of their mechanical performance, but they also enhance their flexibility. In vitro results indicated that a minimum of 10 wt.% TO was required to establish an effective inhibition zone against Gram-positive bacteria. Films with 15 wt.% TO exhibited better antimicrobial activity against two types of microflora and two selected vegetables (Ansorena et al., 2016)

Consumer preferences have managed to drive research and development for petroleum-free packaging materials, such as those with biodegradable or edible properties, as well as those derived from sustainable agricultural products. Edible films, gels, or coatings are biopolymers with many such desirable properties. They can be fashioned from an array of materials, which would include polysaccharides, lipids, and proteins, alone or in combined application with other components (Shit & Shah, 2014). Hence, the general consensus is driving for research.

8.10 Challenges toward the Development of Gluten Films

A major cause for concern is the harmful effects of wheat-based foods on celiac disease patients. In this disease, lesions of the small intestinal mucosa (villi atrophy) form, thus undermining nutrient absorption. This disease is linked to specific histocompatibility cell antigens (HCA system) and can begin as early as the young age of 2, as wheat gets incorporated into the diet. Enterocyte injury is caused by peptide sequences with increased levels of proline and glutamine (Pro-Ser-Gln-Gln and Gln-Gln-Gln-Pro), which activate the cellular immune system (T cells). Naturally, a diet without gluten relieves symptoms and restores standard histology and operation of the intestinal mucosa (Krochta, 2002). Non–celiac disease gluten insensitivity also further hampers widespread acceptance of gluten films and coatings.

As described in Dhaka and Upadhyay (2018), films derived from proteins have exhibited unsatisfactory physical properties, such as tensile strength, elongation at break, puncture resistance, strength, etc. Protein-based edible polymer films may be able to replace synthetic polymer films due to mechanical and barrier properties. At low relative humidity, polypeptides were thought to be effective oxygen barriers. However, proteins' moisture–barrier properties are limited by the fact that they are not completely hydrophobic and contain predominantly hydrophilic amino acid residues. WVP of wheat gluten plasticized with glycerol has been found to be $7.00 \times 10-10$ (g m^{-1} s^{-1} Pa^{-1}), as reported in Shit & Shah (2014). Gluten's cohesiveness and elasticity give wheat dough stability and speed up film formation. The hydrophilic nature of wheat and the significant portion of hydrophilic plasticizer added to impart acceptable film flexibility resulted in poor resistance to water vapor in wheat gluten films. Using a cross-linking agent like glutaraldehyde or heat curing can improve the properties of wheat gluten films. Commercial plastic wraps used often in food packaging have been compared with gluten film, and the findings point out to the lower water vapor permeability exhibited by gluten films, which has proven to be a major limitation for large-scale usage of protein-based biodegradable films (Heralp et al., 1995).

8.11 Limitations of Wheat Gluten Films

Although there are numerous benefits of wheat gluten films, significant obstacles remain, such as retailer and consumer skepticism, material costs, and the additional costs of switching technologies. Furthermore, the properties of those gluten films still have some limitations and drawbacks that limit their use in large-scale applications. Aside from mechanical properties that are not yet competitive, these films are still difficult to process in comparison to synthetic polymers. The following are some of the major disadvantages and limitations of gluten films (Shit & Shah, 2014).

8.11.1 Water Sensitivity

The hydrophilic nature of many gluten-based films poses a significant challenge for material manufacturers, as many food applications require materials that are resistant to moist conditions. Most bio-based materials, including WG, are insoluble or difficult to dissolve in water, but they exhibit high water uptake (swelling) and permeability. Besides that, gluten films change mechanical and barrier properties in high-moisture conditions, which is a significant disadvantage. Under the influence of bacteria, this water sensitivity can establish a more unexpected, rapid, and unmanageable degradation, which is a major disadvantage of biodegradability (Mojumdar et al., 2011).

8.11.2 Aging

Controlling the lifetime of wheat gluten films is one of the challenges for their successful use. Products must be stable and functional during storage and intended use, but they must also biodegrade efficiently after that. Package integrity and microbial stability can only be ensured by carefully controlling water activity, pH, nutrients, temperature, oxygen levels, and time. Thus, biodegradable polymer films like wheat gluten films can be safely stored in dry environments and used with dry food products for an extended period, whereas acceptable storage time in moist environments or time of use with moist foods is limited (Krochta & Mulder-Johnston, 1997). Aging causes gluten films' properties to change, rendering them unsuitable for commercial applications. A gluten film's aging can be caused by physical or chemical reactions in the polymer matrix. The most common aging processes that films go through are explained next (Mojumdar et al., 2011).

8.11.3 Physical Aging: Migration of Additives from the Matrix

This is a physical process that occurs when plasticizers migrate to the surface of the film. Migration of these low molecular compounds generally results in stiffer and less-extendable polymers, which may reduce the protective function of the packaging and, as a result, the shelf life of the packaged food. The migration is influenced by molecular mass, concentration, and the hydrophobic or hydrophilic character of the plasticizer. Water, which is one of the most powerful plasticizers for these types of films, migrates from the matrix to the surface over time and evaporates, leaving more brittle films (Gontard & Ring, 1996).

8.11.4 Chemical Aging: Oxidation

Several study results have shown that disulfide bond formation by thiol oxidation occurred during WG film storage, even under conditions (temperature and RH) that reduce molecule mobility. As a result, the mechanical properties of the films may change with age, depending on the rate of thiol oxidation during drying and storage (Micard et al., 2001; Morel et al., 2000). During oxidation, the sulfhydryl in cysteine amino acid is responsible for the formation of disulfide cross-links (Gällstedt, 2004). This is a process that is accelerated when the film solution is heated to form films, and it is critical for achieving good mechanical properties in the finished product (Roy et al., 1999). During storage, unreacted thiol groups can be oxidized, and intramolecular disulfide bonds can be reorganized to intermolecular disulfide bonds via thiol–disulfide exchange reactions. This causes an increase in protein aggregation as well as brittleness of the film structure (Lindsay & Skerrit, 1999).

8.12 Strategies to Enhance Film Properties and Lifetime

Various researchers are attempting to improve the properties of biopolymer films in a variety of ways. The majority of them use their modifications as "pretreatments," which means that the changes take place in the film-forming solution. Other attempts are made with "post-treatments" that are applied to the film (Micard et al., 2000). The following are summaries of the most commonly used methods.

8.12.1 Plasticizers

To address the issues caused by the use of a low-molecular-weight plasticizer such as glycerol (increased diffusion of gas and water vapor through the film and plasticizer migration), glycerol can be replaced with a higher-molecular-weight compound with hydrophobic substituents. This may aid in reducing WVP and plasticizer migration, thereby improving the fracture strain of gluten films. Amphiphilic substances such as fatty acids (lauric, stearic, and oleic acids), octanoic and palmitic acids, dibutyl phthalate, and tartrate have been investigated as possible replacements for glycerol. Many other studies have been conducted in an attempt to reduce plasticizer migration, comparing the effects of various hydrophilic plasticizers (varying in chain length) on the mechanical properties of protein films (Olabarrieta, 2005).

8.12.2 Cross-Linking

Another way to improve the properties of protein films is to modify the polymer network by cross-linking the polymer chains. Because proteins contain reactive side groups, it is possible to cross-link the polypeptide chains using chemical, enzymatic, or physical treatments to improve the functionality of derived protein films. Proteins have been cross-linked with aldehydes in several studies. For this type of reaction, various aldehydes have been used, most notably formaldehyde, glutaraldehyde, and glyoxal. However, despite being highly reactive, these components have a significant disadvantage: toxicity. This must be considered when creating biopolymer materials.

Enzymatic cross-linking with transgluminase was performed on whey protein and wheat gluten, with mechanical properties, water vapor transferability, solubility, and hydrolyzability of the resulting films analyzed.

Gamma irradiation was also used as a pretreatment on the film-forming solution. The use of gamma irradiation to induce cross-linking was discovered to be an effective method for improving the barrier and mechanical properties of some edible films and coatings. UV radiation is another weaker form of electromagnetic radiation. This method of irradiation has been used to cross-link protein films.

Another method can also be described when the solution is thermally treated. Controlled thermal treatments have significantly altered the properties of several proteins. These studies have shown that heating at high temperatures increases film strength and water resistance by forming the intermolecular disulfide bonds required to form an intact film (Olabarrieta, 2005).

8.12.3 Blending with Other Constituents

The incorporation of hydrophobic compounds such as lipids into the film-forming solution has been a widely used method to improve the water vapor barrier properties of films. The incorporation of lipid compounds (waxes and fatty acids) aids in the limitation of moisture migration because the lipids reduce water vapor transmission while the proteins or polysaccharides give the films strength and thus help improve structural integrity. Several studies have been conducted to examine and confirm the effectiveness of composite edible films containing fatty acids (stearic or palmitic acids), paraffin or beeswax, fatty acid esters, and carnauba wax. Blends containing lipids remained brittle, necessitating the addition of a plasticizer (Olabarrieta, 2005).

The unfavorable properties of biopolymer films continue to prevent their direct market application. However, using these films in multilayer packaging appears to be a viable option. The lack of structural integrity and characteristic functionality of biopolymer films can be addressed by laminating them between other synthetic films (Makino & Hirata, 1997). Another method of attempting to improve film mechanical properties is the preparation of composite films by combining compatible polysaccharides and proteins (Olabarrieta, 2005). Fiber-reinforced composites are an appealing research area because they are eco-friendly, sustainable, low in density, and have acceptable mechanical properties and biodegradability. Furthermore, these fibers have excellent thermal and acoustic insulation properties. Natural fibers from grass, hemp, and ramie have been reported as polymer film reinforcements (Espert, 2005). Nanocomposites are a relatively new class of composites in which the reinforcing material has dimensions less than or equal to 1 nanometer. Nanoclays, cellulose nanowhiskers, ultrafine titanium dioxide, and carbon nanotubes are some examples of nano reinforcements that are currently being developed (Olabarrieta, 2005).

8.13 Conclusion

The gradual shift toward widespread application of edible films and coatings is a conscious effort to reduce the burden caused by synthetic packaging films and pushing forward sustainable and biodegradable alternatives. These films have been derived from polysaccharides or lipids or plants/animal proteins or a combination—composite film—of these categories where the individual components in combination displayed synergistic action. One such plant protein used is wheat protein gluten due to its viscoelastic properties and ability to form elastic and strong films with good barrier properties to gaseous exchange. Water vapor permeability is a factor that becomes a disadvantage when it comes to gluten films and coatings which can be overcome by composite film production. Mass transfer has been observed to be effectively controlled between the components of the products and also between its external surrounding environment upon application of gluten coatings and films in packaging. Application in areas of individual packaging of small items such as nuts, internal coating in multiple layered packaging, and controlled release of certain preservative or drug release has been seen. Formed either through thermoplastic methods such as extrusion or solvent dispersion followed by casting, these films are, to a great extent, impacted by the conditions used during their manufacture, and hence, these parameters are adjusted keeping in mind the end use of the film. Plasticizers are utilized to counteract the brittleness that is inherent to the stand-alone film. Wheat gluten, however, also falls short of acceptability when it comes to people suffering from celiac disease or non-celiac gluten insensitivity. Current trends in production of these films are spearheaded by the use of nanotechnology and composite film production.

REFERENCES

Ansorena, M. R., Zubeldía, F., & Marcovich, N. E. (2016). Active wheat gluten films obtained by thermoplastic processing. *LWT—Food Science and Technology*, *69*, 47–54. https://doi.org/10.1016/j.lwt.2016.01.020

Bishnoi, S., Trifol, J., Moriana, R., & Mendes, A. C. (2022). Adjustable polysaccharides-proteins films made of aqueous wheat proteins and alginate solutions. *Food Chemistry*, *391*, 133196. https://doi.org/10.1016/j.foodchem.2022.133196

Buffo, R. A., & Han, J. H. (2005). Edible films and coatings from plant origin proteins. In J. H. Han (Ed.), *Innovations in food packaging* (pp. 277–300). London: Academic Press.

Capezza, A. J., Glad, D., Özeren, H. D., Newson, W. R., Olsson, R. T., Johansson, E., & Hedenqvist, M. S. (2019). Novel sustainable superabsorbents: A one-pot method for functionalization of side-stream potato proteins. *ACS Sustainable Chemistry & Engineering*, *7*(21), 17845–17854. https://doi.org/10.1021/acssuschemeng.9b04352

Chaudhary, N., Dangi, P., & Khatkar, B. S. (2016). Assessment of molecular weight distribution of wheat gluten proteins for chapatti quality. *Food Chemistry*, *199*, 28–35.

Chaudhary, N., Virdi, A. S., Dangi, P., Khatkar, B. S., Mohanty, A. K., & Singh, N. (2021). Protein, thermal and functional properties of α-, γ- and ω-gliadins of wheat and their effect on bread making characteristics. *Food Hydrocolloids*, 107212. https://doi.org/10.1016/j.foodhyd.2021.107212

Chen, H., Wang, J., Cheng, Y., Wang, C., Liu, H., Bian, H., . . . Han, W. (2019). Application of protein-based films and coatings for food packaging: A review. *Polymers (Basel)*, *11*(12). https://doi.org/10.3390/polym11122039

Chen, Y., Li, Y., Qin, S., Han, S., & Qi, H. (2022). Antimicrobial, UV blocking, water-resistant and degradable coatings and packaging films based on wheat gluten and lignocellulose for food preservation. *Composites Part B: Engineering*, *238*, 109868. https://doi.org/10.1016/j.compositesb.2022.109868

Cousineau, J. (2012). Production and characterization of wheat gluten films. *UWSpace*. Retrieved from http://hdl.handle.net/10012/7158

Cui, L., Yuan, J., Wang, P., Sun, H., Fan, X., & Wang, Q. (2017). Facilitation of α-polylysine in TGase-mediated cross-linking modification for gluten and its effect on properties of gluten films. *Journal of Cereal Science*, *73*, 108–115. https://doi.org/10.1016/j.jcs.2016.12.006

Dangaran, K., Tomasula, P. M., & Qi, P. (2009). Structure and function of protein-based edible films and coatings. In K. C. Huber & M. E. Embuscado (Eds.), *Edible films and coatings for food applications* (pp. 25–56). New York: Springer.

Dangi, P., Chaudhary, N., & Khatkar, B. S. (2019). Rheological and microstructural characteristics of low molecular weight glutenin subunits of commercial wheats. *Food Chemistry*, *297*, 124989. https://doi.org/10.1016/j.foodchem.2019.124989

Dhaka, R. K., & Upadhyay, A. (2018). Edible films and coatings: A brief overview. *Pharma Innovation*, *7*(7), 331–333.

Dong, M., Tian, L., Li, J., Jia, J., Dong, Y., Tu, Y., . . . Duan, X. (2022). Improving physicochemical properties of edible wheat gluten protein films with proteins, polysaccharides and organic acid. *LWT*, *154*, 112868. https://doi.org/10.1016/j.lwt.2021.112868

Espert, A. (2005). *Strategies for improving mechanical properties of polypropylene/cellulose composites.* (2005:12 Doctoral thesis, comprehensive summary), KTH, Stockholm (DiVA database). Retrieved from http://urn.kb.se/resolve?urn=urn:nbn:se:kth:diva-179

Gällstedt, M. (2004). *Films and composites based on chitosan, wheat gluten or whey proteins -Their packaging related mechanical and barrier properties.* (2004:12 Doctoral thesis, comprehensive summary), Fiber- och polymerteknologi, Stockholm (DiVA database).

Gennadios, A., Park, H. J., & Weller, C. L. (1993a). Relative humidity and temperature effects on tensile strength of edible protein and cellulose ether films. *Transactions of the ASAE, 36*(6), 1867–1872. https://doi.org/10.13031/2013.28535

Gennadios, A., & Weller, C. L. (1990). Edible films and coatings from wheat and corn proteins. *Food Technology, 44*, 63–69.

Gennadios, A., Weller, C. L., Hanna, M.A., & Froning, G.W. (1996). Mechanical and barrier properties of egg albumen films. *Journal of Food Science, 61*(3), 585–589. https://doi.org/10.1111/j.1365-2621.1996.tb13164.x

Gennadios, A., Weller, C. L., & Testin, R. F. (1993b). Property modification of edible wheat, gluten-based films. *Transactions of the ASAE, 36*(2), 465–470. https://doi.org/10.13031/2013.28360

Gontard, N., Guilbert, S., & Cuq, J.-L. (1992). Edible wheat gluten films: Influence of the main process variables on film properties using response surface methodology. *Journal of Food Science, 57*(1), 190–195. https://doi.org/10.1111/j.1365-2621.1992.tb05453.x

Gontard, N., Guilbert, S., & Cuq, J.-L. (1993). Water and glycerol as plasticizers affect mechanical and water vapor barrier properties of an edible wheat gluten film. *Journal of Food Science, 58*(1), 206–211. https://doi.org/10.1111/j.1365-2621.1993.tb03246.x

Gontard, N., & Ring, S. (1996). Edible wheat gluten film: Influence of water content on glass transition temperature. *Journal of Agricultural and Food Chemistry, 44*(11), 3474–3478. https://doi.org/10.1021/jf960230q

Guillaume, C., Pinte, J., Gontard, N., & Gastaldi, E. (2010). Wheat gluten-coated papers for bio-based food packaging: Structure, surface and transfer properties. *Food Research International, 43*(5), 1395–1401. https://doi.org/10.1016/j.foodres.2010.04.014

Gutiérrez, T. J., Mendieta, J. R., & Ortega-Toro, R. (2021). In-depth study from gluten/PCL-based food packaging films obtained under reactive extrusion conditions using chrome octanoate as a potential food grade catalyst. *Food Hydrocolloids, 111*, 106255. https://doi.org/10.1016/j.foodhyd.2020.106255

Hamer, R. J., & Vliet, T. V. (2000). Understanding the structure and properties of gluten: An overview. In P. R. Shewry & A. S. Tatham (Eds.), *Wheat gluten* (pp. 125–131). The Royal Society of Chemistry, Cambridge, UK

Hammam, A. R. A. (2019). Technological, applications, and characteristics of edible films and coatings: A review. *SN Applied Sciences, 1*(6), 632. https://doi.org/10.1007/s42452-019-0660-8

Heralp, T. J., Gnanasambandam, R., Mcguire, B. H., & Hachmeister, K. A. (1995). Degradable wheat gluten films: Preparation, properties and applications. *Journal of Food Science, 60*(5), 1147–1150. https://doi.org/10.1111/j.1365-2621.1995.tb06311.x

Hernández-Muñoz, P., Kanavouras, A., Ng, P. K. W., & Gavara, R. (2003). Development and characterization of biodegradable films made from wheat gluten protein fractions. *Journal of Agricultural and Food Chemistry, 51*(26), 7647–7654. https://doi.org/10.1021/jf034646x

Hernández-Muñoz, P., Villalobos, R., & Chiralt, A. (2004). Effect of thermal treatments on functional properties of edible films made from wheat gluten fractions. *Food Hydrocolloids, 18*(4), 647–654. https://doi.org/10.1016/j.foodhyd.2003.11.002

Hochstetter, A., Talja, R. A., Helén, H. J., Hyvönen, L., & Jouppila, K. (2006). Properties of gluten-based sheet produced by twin-screw extruder. *LWT—Food Science and Technology, 39*(8), 893–901. https://doi.org/10.1016/j.lwt.2005.06.013

Irissin-Mangata, J., Bauduin, G., Boutevin, B., & Gontard, N. (2001). New plasticizers for wheat gluten films. *European Polymer Journal, 37*(8), 1533–1541. https://doi.org/10.1016/S0014-3057(01)00039-8

Janjarasskul, T., & Krochta, J. M. (2010). Edible packaging materials. *Annual Review of Food Science and Technology, 1*(1), 415–448. https://doi.org/10.1146/annurev.food.080708.100836

Jiménez-Rosado, M., Zarate-Ramírez, L. S., Romero, A., Bengoechea, C., Partal, P., & Guerrero, A. (2019). Bioplastics based on wheat gluten processed by extrusion. *Journal of Cleaner Production, 239*, 117994. https://doi.org/10.1016/j.jclepro.2019.117994

Khatkar, B. S., & Schofield, J. D. (1997). Molecular and physicochemical basis of bread making properties of wheat gluten proteins. *Journal of Food Science Technology, 34*(2), 85–103.

Krochta, J. M. (2002). Proteins as raw materials for films and coatings: Definitions, current status and opportunities. In A. Gennadios (Ed.), *Protein based films and coatings*. Boca Raton, FL: CRC Press.

Krochta, J. M., & Mulder-Johnston, C. (1997). Edible & biodegradable polymer films: Challenges and opportunities. *Food Technology, 51*, 61–74.

Kuktaite, R., Plivelic, T. S., Cerenius, Y., Hedenqvist, M. S., Gällstedt, M., Marttila, S., . . . Johansson, E. (2011). Structure and morphology of wheat gluten films: From polymeric protein aggregates toward superstructure arrangements. *Biomacromolecules, 12*(5), 1438–1448. https://doi.org/10.1021/bm200009h

Lacroix, M., & Vu, K. D. (2014). Edible coating and film materials: Proteins. In J. H. Han (Ed.), *Innovations in food packaging* (2nd ed., pp. 277–304). San Diego, CA: Academic Press.

Lindsay, M. P., & Skerrit, J. H. (1999). The glutenin macropolymer of wheat flour doughs: Structure-function perspectives. *Trends in Food Science and Technology, 10,* 247–253.

Makino, Y., & Hirata, T. (1997). Modified atmosphere packaging of fresh produce with a biodegradable laminate of chitosan-cellulose and polycaprolactone. *Postharvest Biology and Technology, 10,* 247–254.

Maningat, C. C., Bassi, S., & Hesser, J. M. (1994). Wheat gluten in food and non-food systems. *Research Department Technical Bulletin, 16,* 1–8.

Marcuzzo, E., Peressini, D., Debeaufort, F., & Sensidoni, A. (2011). Effect of process temperature on gluten film properties. *Italian Journal of Food Science, 23,* 202–207.

Mastromatteo, M., Chillo, S., Buonocore, G. G., Massaro, A., Conte, A., & Del Nobile, M. A. (2008). Effects of spelt and wheat bran on the performances of wheat gluten films. *Journal of Food Engineering, 88*(2), 202–212. https://doi.org/10.1016/j.jfoodeng.2008.02.006

Micard, V., Belamri, R., Morel, M., & Guilbert, S. (2000). Properties of chemically and physically treated wheat gluten films. *Journal of Agricultural and Food Chemistry, 48*(7), 2948–2953. https://doi.org/10.1021/jf0001785

Micard, V., Morel, M. H., Bonicel, J., & Guilbert, S. (2001). Thermal properties of raw and processed wheat gluten in relation with protein aggregation. *Polymer, 42*(2), 477–485. https://doi.org/10.1016/S0032-3861(00)00358-X

Mojumdar, S. C., Moresoli, C., Simon, L. C., & Legge, R. L. (2011). Edible wheat gluten (WG) protein films: Preparation, thermal, mechanical and spectral properties. *Journal of Thermal Analysis and Calorimetry, 104*(3), 929–936. https://doi.org/10.1007/s10973-011-1491-z

Morel, M. H., Bonicel, J., Micard, V., & Guilbert, S. (2000). Protein insolubilization and thiol oxidation in sulfite-treated wheat gluten films during aging at various temperatures and relative humidities. *Journal of Agricultural and Food Chemistry, 48*(2), 186–192. https://doi.org/10.1021/jf990490i

Olabarrieta, I. (2005). *Strategies to improve the aging, barrier and mechanical properties of chitosan, whey and wheat gluten protein films.* (2005:14 Doctoral thesis, comprehensive summary), KTH, Stockholm (DiVA database).

Rasel, H., Johansson, T., Gällstedt, M., Newson, W., Johansson, E., & Hedenqvist, M. (2016). Development of bioplastics based on agricultural side-stream products: Film extrusion of crambe abyssinica/wheat gluten blends for packaging purposes. *Journal of Applied Polymer Science, 133*(2). https://doi.org/10.1002/app.42442

Rayas, L. M., Hernandez, R. J., & Ng, P. K.W. (1997). Development and characterization of biodegradable/edible wheat protein films. *Journal of Food Science, 62*(1), 160–162. https://doi.org/10.1111/j.1365-2621.1997.tb04390.x

Rovera, C., Türe, H., Hedenqvist, M. S., & Farris, S. (2020). Water vapor barrier properties of wheat gluten/silica hybrid coatings on paperboard for food packaging applications. *Food Packaging and Shelf Life, 26,* 100561. https://doi.org/10.1016/j.fpsl.2020.100561

Roy, S., Weller, C. L., Gennadios, A., Zeece, M. G., & Testin, R. F. (1999). Physical and molecular properties of wheat gluten films cast from heated film-forming solutions. *Journal of Food Science, 64*(1), 57–60. https://doi.org/10.1111/j.1365-2621.1999.tb09860.x

Shit, S. C., & Shah, P. M. (2014). Edible polymers: Challenges and opportunities. *Journal of Polymers, 2014,* 427259. https://doi.org/10.1155/2014/427259

Tanada-Palmu, P. S., & Grosso, C. R. F. (2003). Development and characterization of edible films based on gluten from semi-hard and soft Brazilian wheat flours (development of films based on gluten from wheat flours). *Ciência e Tecnologia de Alimentos, 23.* https://doi.org/10.1590/S0101-20612003000200027

Tanada-Palmu, P. S., & Grosso, C. R. F. (2005). Effect of edible wheat gluten-based films and coatings on refrigerated strawberry (Fragaria ananassa) quality. *Postharvest Biology and Technology, 36*(2), 199–208. https://doi.org/10.1016/j.postharvbio.2004.12.003

Tanada-Palmu, P., Heihn, H., & Hyvonen, L. (2000). Preparation, properties and applications of wheat gluten edible films. *Agricultural and Food Science, 9*(1), 23–35. https://doi.org/10.23986/afsci.5650

Tsiami, A. A., Bot, A., & Agterof, W. G. M. (1997). Rheology of mixtures of glutenin subfractions. *Journal of Cereal Science, 26*(3), 279–287. https://doi.org/10.1006/jcrs.1997.0131

Weegels, P., & Hamer, R. (1997). Temperature induced changes of wheat products. In R. J. Hamer & R. C. Hoseney (Eds.), *Interactions: The key to cereal chemistry* (pp. 95–130). Manhatten, NY: AACC-International.

Wei, X.-F., Ye, X., & Hedenqvist, M. S. (2022). Water-assisted extrusion of carbon fiber-reinforced wheat gluten for balanced mechanical properties. *Industrial Crops and Products, 180*, 114739. https://doi.org/10.1016/j.indcrop.2022.114739

Zubeldía, F., Ansorena, M. R., & Marcovich, N. E. (2015). Wheat gluten films obtained by compression molding. *Polymer Testing, 43*, 68–77. https://doi.org/10.1016/j.polymertesting.2015.02.001

Zuo, M., Song, Y., & Zheng, Q. (2009). Preparation and properties of wheat gluten/methylcellulose binary blend film casting from aqueous ammonia: A comparison with compression molded composites. *Journal of Food Engineering, 91*(3), 415–422. https://doi.org/10.1016/j.jfoodeng.2008.09.019

9

Edible Films and Coatings with Incorporation of Lipids

Verbi P. Bhagabati, Kinshuk Malik, and Advaita

CONTENTS

9.1 Introduction

9.1.1 Overview

Films are thin layers of material that stand alone. They're frequently made out of polymers that can give a thin structural mechanical strength. Sheets are a type of thick film. As a result, other than thickness, there is no discernible variation in composition of material between sheets and films. Further manufacturing operations can turn films into bags, pouches, capsules, wraps, or casings. Coatings are a kind of film that are placed directly to the material's surface (Hun & Cennadios, 2005).

DOI: 10.1201/9781003303671-9

Food packaging materials are critical for confining foods and maintaining food quality and safety along the supply chain until consumption (López-Córdoba et al., 2017; Raybaudi-Massilia et al., 2016). The foodstuffs purchased are usually wrapped in metal foils or plastic or synthetic polymers which are removed and thrown away (Dubey & Dubey, 2020). Aforementioned polymers are not biodegradable, are hard to recycle, hence reuse, and are not made from renewable resources (Fernández et al., 2007). The food processing industry is constantly working to reduce the amount of plastic packaging used in their goods whenever possible. Plastic manufacturing in the world increased from 1.5 million metric tons in 1950 to 348 million metric tons in 2017, with just near about 7% of that being recycled (Shahidi & Hossain, 2022). According to the present plastic pollution outline, around 400,000 plastic bottles will be used. Every minute, over 700,000 plastic bags are eaten around the world (Porta, 2017). Because of the negative impacts of synthetic packaging on human and environmental health, researchers were prompted to create greener alternatives (Gao et al., 2017). Natural origin foods, partially or entirely replacing synthetic foods, are a good option for customers (Fernández et al., 2007). As a result, novel films made from agricultural food waste and natural resources that are low-cost have been considered as a viable and cost-effective alternative to conventional day-to-day plastics (Alzate et al., 2016; Balti et al., 2017; Nisar et al., 2018; Valdés et al., 2014) because few of such substances are eatable and can be combined with other foods, however, being environmentally favorable and confirming that the package serves the primary purpose of preserving the meal, or the product, or being in direct contact with the food (Raybaudi-Massilia et al., 2016).

9.1.2 Historical Background

Healthy-to-eat films and coatings have been utilized to guard food goods for millennia in the food industry; therefore, this is not a new method of protection. For example, coatings of cellulose in meat products and waxing on fruits and vegetables (Raghav et al., 2016). Edible coatings have been utilized since the 12th century in China (Hassan et al., 2018). The first edible film was prepared from soymilk in 15th-century Japan (Yuba) to preserve food (Sánchez-Ortega et al., 2014).

9.1.3 Edible Films/Packing and Its Importance

Edible packaging refers to edible coatings and films in general. Edible coatings and films are thin protective layers developed around food surfaces using edible polymers like polysaccharides, lipids, proteins, or mixtures of these polymers, and they must be made with ingredients that are acceptable under current food rules and must not obstruct with the coated product's sensory profile or other qualitative qualities (Baldwin et al., 2011), while edible coatings are crafted as slim layers and incorporated into food products (Shahidi & Hossain, 2022).

In addition to being environmentally benign, it can be employed to improve food items' organoleptic, nutritional, and microbiological aspects (Lopez-Rubio et al., 2017). Also, they are a viable vehicle for bioactive materials, as they are consumed with meals, guaranteeing that bioactive chemicals, such as antioxidants, vitamins, and probiotics, are delivered as intended. These bioactive substances not only give nutritional value to coated foods but also aid in improving the functional properties of films and coatings; for example, antioxidants aid in extending the shelf life of food items (Piermaria et al., 2015).

The gas and moisture restriction qualities, sensory quality, mechanical properties, and nutritional aspects of coated/wrapped eatables are all improved by this protective layer (Galus & Kadzińska, 2015). Furthermore, edible coatings and films can deliver physical and mechanical shielding for food items that are vulnerable to damage during transportation (Dubey & Dubey, 2020).

Rising awareness of conservation of the environment and its protection, increased demand for greater class of foods, desire for innovative processing of food and preservation methodologies, and scientific discovery of the utility of novel materials are all driving forces behind this ardent endeavor (Chen, 1995). Today, edible films are consumed with a vast diversity of foods, and its total annual revenue is around 100 million US dollars (Dehghani et al., 2018).

9.1.4 Chemistry of Edible Films

Fats, oils, vegetal and natural animal waxes, fatty alcohols, and fatty acids are the most common hydrophobic compounds employed as barrier to the moisture in edible coatings and films. The film's response to these hydrophobic components is determined by their chemical properties and emulsion-forming conditions (Fernández et al., 2007).

Various studies have highlighted the benefits and applications of edible packaging (Guilbert et al., 1996; Krochta, 2002). Proteins, lipids, and polysaccharides are the prime components of edible coatings. Because of their hydrophilic nature, polysaccharides and protein films have great mechanical qualities but are permeable to H_2O (Anker et al., 2002; Kristo et al., 2007). Lipids possess the polar opposite properties (Pérez-Gago & Krochta, 2001). As a result, lipid incorporation to hydrophilic films is an intriguing option for obtaining films with enhanced properties (Quezada Gallo et al., 2000). Since lipids are not biopolymers, they cannot make cohesive films. Due to their low polarity, they are incorporated into biopolymers and form composite films or are used as a coating material that provides a barrier to water vapor (De Azeredo, 2012)

9.1.5 Incorporation of Lipids

Lipids carry a long history of being used as an edible coating for food protection (Dhaka & Upadhyay, 2018). Lipids are attractive applicants for edible coatings and films due to the need to prevent loss of moisture from packaged and nonpackaged food items. Because of their polar nature, lipids create good moisture barriers (Morillon et al., 2010); but lipids are not biopolymers, so they are unable to produce cohesive and self-supporting films and coatings (Shahidi & Hossain, 2022). Instead, coatings are made from a mixture of lipids, proteins, and polysaccharides so as to have better mechanical and barrier qualities. Various lipid suspensions have been the topic of a lot of research in near years (Dubey & Dubey, 2020).

Lipid-based edible coatings and films behave differently in terms of moisture transfer, depending on their unique characteristics. Lipid polarity is influenced by the length of aliphatic chains, the existence and extent of unsaturation, and the distribution of chemical groups. Unsaturated fatty acids have a higher polarity, allowing for more moisture transfer (De Azeredo, 2012).

Coatings made of mono-, di-, and triglycerides are also available. Their chemical structure, particularly water vapor permeability, could substantially impact their functional qualities. Short-chain triglyceride molecules are H_2O-soluble to some extent, while long-chain molecules are not. When equated to saturated fatty acids, unsaturated fatty acids possess lower melting points and high moisture transfer rates. As a result of the greater mobility of hydrocarbon chains and the less-efficient lateral packing of acyl chains, branching of acyl chains increases water vapor permeability (Stuchell & Krochta, 1995).

At low relative humidity, whey proteins generate water-based edible coatings and films that are clear with good fragrance, O_2, and lipid restriction attributes (Krochta, 2002). These films, however, are easy to break and, to increase flexibility, require the addition of plasticizers. A plasticizer improves water vapor permeability (WVP) as a result (Mahmoud & Savello, 1992; Sothornvit & Krochta, 2000). Glycerol, sorbitol, monoglycerides, glucose, and polyethylene are all common plasticizers. Water acts as a plasticizer and has a significant consequence on films' characteristics. The WVP of WPI-based edible films is considerably improved when lipids are added (Banerjee et al., 1996; Krochta, 1999). Whey protein emulsifies the lipid in these films, generating a gel.

Different lipids have effectively been included in edible films based on proteins. Ma et al. (2012) used a microfluidic emulsification approach to include olive oil into gelatin fibers, which had good water and UV light barrier qualities. Lactic acid and oil (Guerrero et al., 2011) or beeswax and oleic acid combinations were used to alter the restrictive, optical, and mechanical properties of edible films based on soya protein (Monedero et al., 2009). Pereda et al. (2010) discovered that incorporating tung oil to composite sodium caseinate films improved mechanical characteristics. Lipid and protein components might network well, resulting in edible films with better functional and structural qualities.

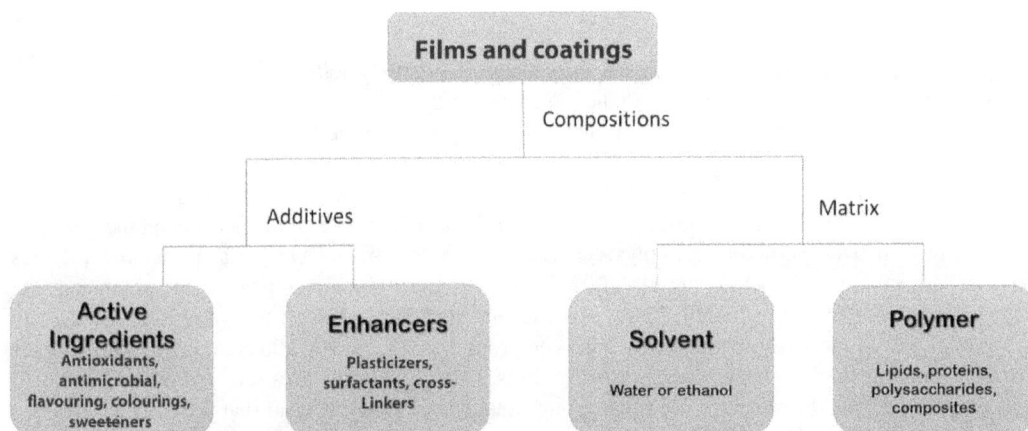

FIGURE 9.1 Composition of edible films and coatings

When liquid oils are added in excess of 40%, lipid incorporation is hampered (Pérez-Gago & Krochta, 2001), and the development of stable emulsion films generally necessitates the incorporation of emulsifying agents (Kokoszka et al., 2010; Ma et al., 2012; Shaw et al., 2002). Lipids that are used for incorporating in edible coatings and films are sunflower oil, paraffin, coconut oil, palm oil, tung oil, cocoa butter, beeswax, herbal wax, acetylated monoglycerides, cinnamon and ginger essential oils, etc. (Dhaka & Upadhyay, 2018; Dubey & Dubey, 2020; Atarés et al., 2010a).

Besides, before beginning edible coating or film production, manufacturers must adhere to national regulatory requirements and legislation. All film-developing constituents and additives should be composed of food-grade substances, with all production requirements satisfying stringent hygiene standards (Pavli et al., 2018).

Even though edible coatings and films are not supposed to substitute old packaging methods completely, the quality of food shielding can be improved by compiling primary edible packaging and secondary nonedible packaging, which is used for hygienic and handling goals (De Azeredo, 2012). This chapter provides a complete update on recent developments in formerly published and ongoing investigations on edible active films for their future uses.

9.2 Synthesis

9.2.1 Edible Films

Just like synthetic plastics are processed, edible films are also manufactured similarly, mainly in three steps:

a) Stabilization of native state of the polymer
b) Polymer chain arrangement
c) New interactions and bonds forming three-dimensional networks

Further, these procedures can be divided into wet processes that work on solubilization/dispersion, followed by drying of ingredients and the dry processes that are solvent-free and, hence, are environmentally favorable (Calva-Estrada et al., 2019; Murrieta-Martínez et al., 2018).

Film-developing substance should produce a spatially rearranged gel system with all added film-developing materials, like plasticizers, biopolymers, and solvents in the case of wet casting, regardless of the film-developing technique, whether casting is wet or dry. To make film-developing solutions, biopolymer film-forming ingredients are usually gelatinized. Excess solvents are removed from the gel structure as the hydrogels are dried further.

Most biopolymers' complete film-forming mechanisms following gelation have yet to be established. Polymer chemistry lab techniques, such as FTIR spectrometry, electrophoresis, NMR spectrometry, X-ray diffraction, polarizing microscopy, and others, are required to detect them. Many thermoplastic attributes, including polymer melting, gelatinization, polymer rearrangement, flow profile, and others, should be researched for extrusion casting (dry process) to forecast film-forming mechanisms.

Lamination, composite creation, particle or emulsion addition, perforation, overcoating, and heat curing are examples of physical alterations of edible films and coatings (Micard et al., 2000; Miller et al., 1997; Gennadios, 1996), ultrasound treatment (Banerjee et al., 1996), radiation, and orientation (Gennadios et al., 1998; Micard et al., 2000).

9.2.2 Film Preparation Using Different Lipid Incorporation, Additives, Emulsifiers, and Plasticizers

TABLE 9.1

Film preparation using different lipid incorporation, additives, emulsifiers, plasticizers, etc.

SL no.	Edible film	Lipids	Additives	Major results	References
1.	Wheat gluten films	Blend of beeswax and acetic esters of mono- and diglycerides	Gluten as plasticizer Glycerol monostearate as emulsifier	Gluten showed high tolerance to UT. At mesoscopic level, there was breakdown in protein without changes in molecular level, indicating existence of a colloidal system represented by glutenins and gliadins chains.	(Marcuzzo et al., 2010) (Greenspan, 1977)
2.	Hydroxypropyl di starch phosphate (HP), oxidized hydroxypropyl starch (OS), and gelatin	Rapeseed oil, shortening, and beeswax	Glycerol	With incorporation of lipids, it was observed that G′ and |η*| values decreased, crystallinity increased, molecular interactions weakened. RO promoted starch plasticization compared to BW and ST. Good water resistance, water vapor barrier property, and surface hydrophobicity were observed, especially in the case of BW.	(Cheng et al., 2021)
3.	Fish water-soluble protein	Beeswax and stearic acid	Glycerol	High tensile strength, elongation at break, and water vapor permeability were observed when oleic acid was incorporated.	(Tanaka et al., 2001)
4.	Sodium caseinate	Ginger and cinnamon essential oil	Glycerol	Limited impact on mechanical properties and slight increase in water vapor permeability. Films with regular surfaces were obtained that affected their optical properties. There was surface roughness and loss of gloss because of lipid aggregation at the time of drying.	(Atarés et al., 2010a)
5.	Sodium alginate	Oleic acid, polyglycerol esters of edible fatty acids (PGE)	Green tea extract (GTE) or grape seed extract (GSE), Tween 80	Films showed antiviral and antioxidant properties. WVP was improved; modifications in transmittance and color parameters were seen.	(Fabra et al., 2018, 2009a,b)
6.	Whey protein isolate	Rapeseed oil	Glycerol	Changes in film thickness and protein structure and increased hydrophobicity were observed.	(Kokoszka et al., 2010; Galus & Kadzińska, 2016a)

(Continued)

TABLE 9.1 Film preparation using different lipid incorporation, additives, emulsifiers, plasticizers, etc. (*Continued*)

SL no.	Edible film	Lipids	Additives	Major results	References
7.	Whey protein Isolati	Almond or walnut oil	Glycerol	More hydrophobicity, decreased WVP and increase in contact angle values, weak mechanical resistance; almond oil had greater plasticizing effect and tendency of changing the functional properties of the protein than walnut oil.	(Galus & Kadzińska, 2016b)
8.	Chitosan	*T. moroderi* (TMEO) or *T. piperella* (TPEO)	Chitosan glycerol, Tween 80	The incorporation of lipids in these cases has proved to enhance the antibacterial and antioxidant properties of chitosan. Highly perishable foods like fish and poultry will find these films very useful for resistance to external environment.	(Ojagh et al., 2010; Ruiz-Navajas et al., 2013)
9.	Whey protein isolate	Butterfat and candelilla wax	Sorbitol and glycerol	At high butterfat and glycerol amounts, films were very soft and unpeelable from the casting surfaces, whereas at low sorbitol amounts, they were very brittle. By changing the amounts of lipid and plasticizers, EMC and solubility of WPI could be changed.	(Kim & Ustunol, 2001)
10.	Gelatin	Olive oil	Glycerol	Lipid droplet size or distribution of films decides the WVP and tensile property. Mechanical resistance and water barrier ability may weaken by lipid agglomerates.	(Ma et al., 2012)

9.3 Edible Film Properties

Edible coatings and films are a kind of packaging used primarily in the food sector. Coatings and films must meet specific requirements due to their direct interaction with food, including acceptable sensory qualities, suitable physicochemical, biochemical, and microbiological consistency, safety, and lack of toxicity (Murrieta-Martínez et al., 2018).

They ought to be elastic and strong, with a little permeability, in order to enable the barrier construction that regulates the pace of passage of food molecular elements, such as flavor, volatile water, O_2, and CO_2 (Calva-Estrada et al., 2019; Parreidt et al., 2018).

Mechanical qualities, chemical and biological barrier features, and thermic behavior must be carefully considered when manufacturing coatings and films, because they greatly affect product quality. The inherent properties of the substances utilized for coating and film preparation, as well as extrinsic property variables, are intimately related to these final qualities (Murrieta-Martínez et al., 2018).

9.3.1 Mechanical Properties

The elongation, flexibility, and hardness of the coating and film polymer are related to the mechanical assets of the food product, which are connected to the package's behavior during storage and handling (Murrieta-Martínez et al., 2018). Compared to those tested on standard packaging, these physical qualities are generally lacking, which has restricted the use of edible coatings and films. As a result, polymers' physical structure should be explored to create edible coatings and films with superior mechanical properties, including greater resilience and strength (Mkandawire & Aryee, 2018). Mechanical metrics (tensile strength, Young's modulus, and elongation at break) were calculated. It can be seen that adding 0.5% almond oil enhanced the tensile strength considerably. However, the reverse effect was noticed when 1.0% almond oil was incorporated into the films. The lower the obtained tensile strength values, the higher the oil content. This is due to the plasticizing impact of almond oil at greater concentrations in films.

For both oils, Young's modulus was greater after oils were added at 0.5% and lesser when oils were added at 1.0%. For gelatin films, Ma et al. observed that at lower olive oil concentrations, there was a surge in elastic modulus and tensile strength and a loss at the greatest olive oil addition (Ma et al., 2012). Boosted cinnamon oil concentration increased soy protein isolate film strength, according to Atarés et al. (2010b). Fang et al. (2002) observed that when the amount of soybean oil in whey protein isolate films increased, the tensile strength decreased. Han et al. (2006) likewise discovered a decrease in pea starch film tensile strength, although the variations were only apparent when beeswax concentration was greater than 20%.

Almond oil had a greater plasticizing result because of its chemical arrangement (high oleic and linoleic acids), which provided a smoother and softer structure, as seen by the water-solubility values. Such behavior was caused by the lipid content and concentration, as well as a cross-linking impact between lipid particles and the whey protein. Ma et al. (2012) found that adding 5 and 10% olive oil to gelatin films reduced elongation at break, whereas adding 15 and 20% olive oil increased it. A similar trend was observed with soy protein films including flaxseed oil at concentrations ranging from 1 to 10% (Hopkins et al., 2015) and incorporation of olive oil in chitosan films at different concentrations (Pereda et al., 2012). However, Fang et al. found that for WPI films, there has been 15% increase in elongation break of soybean oil, whereas higher oil content (30–50%) decreased this parameter. Javanmard and Golestan (2008) found a similar trend in whey films made with olive oil at different oil–protein ratios. The plasticizing impact of lipids in hydrocolloid films is greatly influenced by lipid concentration, as demonstrated by these examples. However, the results showed that a minor amount of oils (0.5 and 1.0%) had an impact on the mechanical characteristics of whey films when compared to a larger degree of oil incorporation. It could possibly be linked to the homogenization process, in which the stresses forced the protein matrix to reorganize.

Elongation at break (EAB) is the greatest length alteration of a test sample just before breaking, while elastic modulus (EM) is an indicator of the film's stiffness. Tensile strength (TS) represents the highest tensile stress that the film can withstand (Pereda et al., 2012). For a film to keep its structure intact and endure external stress, it must be mechanically stronger and extensible (Yang & Paulson, 2000). In an oil-to-protein-ratio-dependent fashion, the tensile strength of the films is more in emulsified film than in the control sample. Olive oil addition increased the EM of the films as well. As a result, the addition of oil in the gelatin film matrix greatly enhanced protein–protein interactions. As a result, the TS values of the resulting films were larger than the films with no incorporation of oil. Alternatively, the presence of antioxidants in olive oil may cause cross-linking of the protein during the film-casting process, which could explain the improvement in stiffness and strength of gelatin films, owing to olive oil addition. Mechanical factors have been shown to be altered differently based on the material utilized, concentration of additional lipid materials, and particle size distribution in several investigations. Because of a lipid phase, the whey protein matrix order was disrupted, resulting in variations in tensile characteristics (Galus & Kadzińska, 2016b).

9.3.2 Barrier Properties

Because the chemical and physical deterioration of food is intricately tied to the transfer of gases between the environment and the product, the barrier qualities of edible films and coatings are critical in estimating product shelf life (Ribeiro et al., 2021).

The primary function of an edible coating or film is to prevent the transfer of moisture. Lipids show good water-repellent properties and, hence, are integrated into the films. Best water barrier characteristics are shown by lipids produced from saturated, straight-chain fatty acids, esters and fatty alcohols having 16 or more carbon atoms. The barrier efficacy order is observed to be highest for stearic alcohol, then tristearin, followed by beeswax, then acetylated monoglycerides; next to it were stearic acid and alkanes. These differences can be explained on the basis of different pore size on the surface layers of lipids and the homogeneity of the network's composition, which is connected to the polymorphic form and alignment of the chains in the network.

Water vapor barrier characteristics are considerably improved when lipids are added to protein-based films by producing a steady emulsion or lipid layer covering the film. Barrier efficacy of composite films is greatly impacted by polarity of particles and the uniform dispersal of hydrophobic compounds in the film. The impact of affinity (polarity difference) between the dispersed and continuous phases on transfer of water was demonstrated; the weaker the affinity, the lesser the film homogeneity. According to some research, bilayer films offer 10 to 1,000 times greater barrier efficacy for transfer of H_2O than emulsified films.

The H_2O vapor efficiency of emulsion-based packaging is determined by the lipid type, quantity, and structure. Because of their reduced water solubility, long fatty acid chains in lipids have a lower WVP. Increase in the weight of a nonpolar molecule leads to stronger solvent–solute interactions than solute–solute interactions, resulting in solubility. The discrepancy in particle sizes could be due to the chemical composition of lipids (Pérez-Gago & Krochta, 2001). The barrier effectiveness of the most hydrophobic lipids is the highest.

Liquid lipids are less effective at blocking moisture transport than crystalline lipids. As a result, edible lipid obstacles in a composite film include fatty acids that are high melting, and hydrogenated fats, monoglycerides, and waxes. Several authors have demonstrated that orthorhombic lipid crystals, such as those discovered in paraffin wax and beeswax, have a higher restrictive performance than hexagonal crystal forms, like those observed in acetylated monoglycerides or tristearin. This is due to the orthorhombic lipid crystals' more agglomerated structure, which contains less-accessible volume for H_2O molecule movement. Lipids can act as effective moisture barriers, but they have a number of drawbacks in terms of applicability, stability of mechanics and chemicals, and organoleptic quality (Khwaldia et al., 2004).

9.3.3 Water Barrier Properties and Glass Transition Temperature

The moisture barrier qualities of a coating are determined by the width of the produced film, temperature, moisture permeability, and RH of the medium around it. The moisture content of a product is determined by the diffusion of water vapor, which is accelerated by the pressure variation between the air around it and the food (Parreidt et al., 2018). Emulsion-based films are not as effective water barriers as bilayer films because lipid circulation is not uniform.

They do, still, have considerable mechanical strength and need simple manufacturing and application technique, while multilayer films necessitate a complicated series of activities based on the different coatings. In emulsion-based films, the tinier the globules of lipid and the greater the homogeneity, the lesser the H_2O vapor permeability (Debeaufort & Voilley, 1995; Pérez-Gago & Krochta, 2001). When the solvent evaporates, the rate of flocculation drops and creaming is not allowed, promoting emulsion stability. The degree of homogenization influences the reduction of size of lipid particle in emulsions, which then is linked to a reduction in water vapor permeability of dried films. The primary advantage of barrier coatings and films might be increased by boosting coating qualities by fabricating nanocomposite films with greater barrier, mechanical, and functional capabilities, hence enhancing the shelf life of fresh good (Kalia & Parshad, 2013).

The glass transition temperature that determines the durability of films is greatly prompted by the makeup and moisture content of the film (Jiménez et al., 2013a). In general, when the amount of water in an amorphous material grows, the glass transition temperature falls (Hambleton et al., 2012). For pistachio globulin protein film, the glass transition temperature was not highly altered by fatty acids addition, according to Zahedi et al. (2010). However, when fatty acids were incorporated to starch films, Jiménez et al. found that the glass transition decreased. Composition of film and also measuring conditions can explain the discrepancies in these phenomena (Jiménez et al., 2013a).

Composite membranes or coatings have a non-homogeneous structure, meaning, coatings are made up of a continuous matrix with certain additions, like lipid globules in emulsions or solid particles in insoluble materials (Taylor, 2010).

The functional features of emulsion-based films generated by casting processes may differ from those of food-surface coatings. However, as an intervening layer, edible films or coating should stick well to the surface and not break or acquire unpleasant sensory qualities in the course of storage (Galus & Kadzińska, 2015).

9.3.4 Functional Properties

Edible coatings and films perform many of the same tasks as traditional packaging, such as serving as barriers to H_2O vapor, flavor substances, and gases, and also increasing the structural reliability and handling capabilities of meals. Even though edible coatings and films are not intended to completely substitute old packaging methods, combining secondary non-edible packaging with primary edible packaging can improve food protection efficiency (Taylor, 2010). Different film-forming behaviors can arise with film formation due to the great diversity of chemical structures and compositions of emulsion-based components (proteins, lipids, and polysaccharides). Lipids' functional qualities are determined by their structure and polarity, which is determined by group distribution, aliphatic chain length, and the existence and degree of unsaturation (Rhim & Shellhammer, 2005).

The qualities of emulsified films should, without a doubt, be managed. Proteins and polysaccharides, even though because of their hydrophilic character, they are bad moisture guards, but they often produce films with strong mechanical qualities. Lipids, on the other hand, produce effective moisture barriers (Hambleton et al., 2012).

In comparison to pure lipid layers, emulsified films have greater mechanical properties. Vegetable oils and other liquid lipid compounds, in general, form bilayer or multilayer layers on the hydrocolloid surface and are unable to form distinct films. The emulsification process consents for the creation of films with certain mechanical properties. Because coating forming emulsified films does not require the mechanical characteristics of films that are stand-alone and utilized as pouches, castings, or wraps, the majority of research describes the advancement in moisture barrier efficiency of films by including lipid in film-forming substances. Generally, emulsified films have mechanical properties similar to a pure hydrocolloid matrix, with lipid type and concentration modifying the mechanical properties (Hopkins et al., 2015). The mechanical characteristics of emulsified films are heavily influenced by the structure and stability of an emulsion.

9.3.5 Light Absorption and Opacity

The transparency or clarity of edible coatings has a straight effect on the coated food's appearance and, hence, is an important criterion for customer approval. Emulsions that form films are impacted in the optical characteristics of dried films by the droplet size distribution and the lipid fraction. In general, emulsified films have greater lightness values that decrease somewhat with increasing lipid concentration (Hopkins et al., 2015; Monedero et al., 2009; Pereda et al., 2010).

The elevated clearance of emulsion-based films is linked to improved structural uniformity (Ortega-Toro et al., 2014) and is affected by droplets of lipid scattered in the matrix of the film, which impede transmission of light through the film (Pereda et al., 2012). The inclusion of an immiscible lipid phase enhances opacity (Guerrero et al., 2011; Ma et al., 2012; Pereda et al., 2010, 2012; Shaw et al., 2002; Yang & Paulson, 2000). The refractive indexes of the phases differ with particle size distribution and concentration, according to the considerable rise in opacity values reported, owing to the incorporation of lipids to hydrocolloid films.

9.3.6 Permeability of Gas

The primary permeating agents engaged in the transportation of film phenomenon are H_2O vapor, O_2, and CO_2, the first of which is related to moisture of the product and the others to respiration of the fresh product (Murrieta-Martínez et al., 2018; Parreidt et al., 2018).

When relative humidity is low, oxygen permeability improves marginally, but when relative humidity is high (>0.75%), it decreases dramatically. Because of their higher chemical affinity and solubility, hydrophobic substances are more permeable to gases, according to Miller and Krochta (Miller et al., 1997). Because of the incorporation of lipid to hydrocolloid films, oxygen permeability was seen to have increased (Hambleton et al., 2012; Jiménez et al., 2013b; Navarro-Tarazaga et al., 2011).

The permeability of carbon dioxide of emulsion-based films, on the other hand, is highly influenced by the chemical constitution of the lipids. Ayranci et al. discovered that adding stearic and palmitic acid to cellulose films reduced carbon dioxide transmission, whereas adding lauric acid kind of had the reverse outcome (Ayranci & Tunc, 2001). Due to shorter length of lauric acid hydrocarbon chain length, the attractive interactions among acid molecules were found to be weak.

The H_2O vapor permeability of films is the most widely researched feature of films because it plays such a crucial part in food deterioration reactions (Rao et al., 2010). Most articles claim a reduction in permeability of water vapor because of the water-repellent nature of lipids, with waxes being the finest moisture obstacles among them (Bourtoom, 2009). On addition of acetylated monoglyceride to WP films, Anker et al. found that water vapor permeability was reduced by 0.5 (emulsion) and 70 (bilayer) (Anker et al., 2002).

Edible films and coatings are often hydrophilic, that is, permeable to H_2O vapor, limiting their capability to fulfill specific purpose in specific applications. As a result, properties of moisture must always be assessed and regulated, considering the package's ultimate food application: although dehydration should be prevented in fresh meals, water permeability is crucial in items like bread and might lead to food deterioration (Murrieta-Martínez et al., 2018). Coatings and films, during respiration, also act as obstacle to exchange of gas, decreasing O_2 intake and, thus, CO_2 formation. This changing atmosphere ought to not induce anaerobic conditions inside the product to prevent anaerobic development (Parreidt et al., 2018).

Even though the aforementioned characteristic properties are incomparable to the traditional packaging, a significant growth in perception of the structure/process/property relationships of substances is allowing the development of quite a few good approaches which allow for the upgradation of characteristics of the coating material without bargaining the main food characters, particularly appearance, taste, and grade (Calva-Estrada et al., 2019; Mkandawire & Aryee, 2018). The major is known as active packaging, and it involves adding natural antibacterial and antioxidant agents to film and coating solutions. They contain compounds that are active and could be very promising in packing as they allow concentration of target ingredient to be maintained at high level that can be slowly released over time (Calva-Estrada et al., 2019; Parreidt et al., 2018).

9.4 Disadvantages of Using Lipids

a) The food has a poor oxygen barrier, making it susceptible to oxidative deterioration.

b) They have a negative impact on food's sensory characteristics (Dhaka & Upadhyay, 2018).

c) Lipid films do not have the structural stability of protein or polysaccharide films in general. The inclusion of lipids causes a diverse structure of film with discontinuities in the network of polymer. As a consequence, adding lipids to polysaccharide or protein films to reduce their barrier qualities can negatively impact the film's strength as measured by puncture or tensile strength (Galus & Kadzińska, 2015).

To address the brittleness of edible films induced by large intermolecular pressures, a plasticizing agent must be added. Plasticizers minimize these stresses and improve mobility, resulting in more flexible films and increased mobility of polymer chains. They also improve their permeability. Plasticizers that impede protein chain hydrogen bonding, like PEG, Gly, Sor, and H2O, are commonly used to produce protein films (Khwaldia et al., 2004).

9.5 Applications

9.5.1 Vegetables and Fruits

Consumers nowadays demand products that are unusual from the market with local products. Regrettably, the condition of such products is harmed by prolonged storage and transportation circumstances. Tropical and exotic fruits and vegetables are dried, candied, or dehydrated osmotically to extend their shelf life. Plums, apricots, raisins, and dactyls are examples of preserved fruit with a high carbohydrate content and a low water content (due to evaporation). Consumers regard stickiness and aggregation as a disadvantage with such product qualities. Furthermore, changes in water content have a negative impact on product qualities.

Water absorption promotes the growth of microorganisms, increases enzyme activity, and increases the flexibility of oxidation reactants, whereas desorption of water tends to make the result less adaptable and rigid. According to certain research, a coating made up of cellulose, waxes, pectin, starches, derivatives, or proteins can provide enough defense (Hagenmaier & Baker, 1993).

Vegetables and fruits are coated and combined with lipids and chemical compounds like fungicides and growth regulators to preserve for a long period. However, these contaminants should be eliminated by washing before being distributed or processed for the market (Hall, 2012). Lipid-based edible coatings minimize moisture loss, resulting in significant financial losses for growers. Most applications, however, rely on spraying or dipping the coating solution directly onto the product's surface to generate the film layer (Dea et al., 1995).

Fresh, unprocessed fruits and vegetables are seen to be the healthiest option. Unfortunately, because of the seasonality and perishability of fruits and vegetables, and also consumer needs and eating habits, producers are increasingly interested in processing and preserving goods and not fresh products. As a result, edible coatings and films are being used in numerous processing processes. Controlling the interior gas composition is critical to the advancement of edible coatings for fresh items (Dhall, 2013). Life activities such as transpiration, respiration, ripening, and over-ripeness continue to occur in raw resources throughout storage. Under typical settings, all these processes reduce the shelf life of vegetables and fruits to a little number or possibly numerous days.

The addition of lipids coating fresh fruits and vegetables has the effect of reducing loss of moisture. In research literature, loss of moisture of apple pieces was found to be greatly reduced when coated with lipid-incorporated protein or polysaccharide compared to control slices. Water loss causes weight loss and shriveling in fresh fruits and vegetables, resulting in loss of economy due to loss of sale. As a result, utilization of edible lipid-based coatings is an intriguing technique for limiting undesired changes and extending the shelf life of foods (Galus & Kadzińska, 2015).

It was found that using a casein and carnauba wax coating forms a hindrance around the product and thus reduces the water content loss in papaya, which further limits water vapor leakage. Others have experienced similar outcomes. During respiration, carbohydrates are oxidized to form CO_2, H_2O, and heat. This unfavorable course results in a reduction in content of carbohydrate as well as loss of weight. There is, again, the possibility of a detrimental change in color, an unpleasant odor and flavor, and a decline in nutritional content, all of which could lead to product deterioration (Galus & Kadzińska, 2015).

The water vapor resistance and the rate of respiration of items that are coated can be used to measure the barrier qualities of edible coatings and films. When they are applied to minimally processed vegetables and fruits, a modified atmosphere is generated for the food, which reduces the respiration rate and, hence, metabolic reactions. Protein and multicomponent coatings are used in this scenario because they are hydrophilic and provide a great barrier to nonpolar molecules like O_2 and CO_2 (Galus & Kadzińska, 2015). Velickova et al. (2013) had shown that the respiration rate of strawberries was reduced when incorporated with beeswax into chitosan-based coatings. Likewise, coatings made of hydroxypropylmethylcellulose coated with chitosan-based, oleic acid-based, caseinate-based, etc. films on plums, strawberries, and carrots show decrease in respiration rate.

Hydrocolloids might lose their activity in water, as they are soluble; hence, lipid-based coatings are a superior choice due to their hydro-repellence. But even lipid has a limitation—it provides a waxy appearance on fruits. In a study, the impact of coatings on apples made of whey protein isolate and beeswax was examined (Perez-Gago et al., 2003). To whey protein coated with beeswax emulsion, **ascorbic acid** or **L-cysteine** (0.5%) was added, which significantly decreased the enzymatic browning of "Golden Delicious apples," according to later research (Perez-Gago et al., 2006). This was in contrast to when antioxidants were used alone. A sensory analysis of coated apples was also conducted. According to the findings, emulsion coating viscosity has to be reduced by lowering emulsion amount and increasing beeswax amount. Strawberry coatings with chitosan-based coatings had a seven-day shelf life at 293K and 53% relative humidity (Velickova et al., 2013). Beeswax used as coating emulsion has a lot of advantages and effect on reduced loss of weight, rate of respiration, sugar retention, color change prevention, protection against fungus, etc. Fruits like apples and citrus fruits have a glossy appearance because of edible coatings made of shellac, which is a natural resin, that also serve as an extra attractive feature (Hall, 2012). Apple pieces layered with a layer product containing carnauba wax, cassava starch, and stearic acid for

30 days at 279K were reported to be limited in color, firmness, and microbial changes (Chiumarelli & Hubinger, 2012), or coated with caseinate, alginate, carrageenan, or pectin and integrated with monoglycerides that are acetylated. When candelilla wax, mesquite gum, and mineral oil coverings were applied, a tendency to lower the normal decay speed of Persian limes was discovered. Similar designs, enhanced by calcium and glycerol addition, led to loss of weight of guava for 15 days. Similar to this, banana, apple, and avocado pieces covered with a mineral oil and candelilla wax mixture maintained suitable quality for 6 days under refrigeration (Saucedo-Pompa et al., 2007). After being stored for 11 days, grapes covered with some mixture of sorbitol, pea protein, and candelilla wax had a considerably higher ascorbic acid and lower sugar content and saw less weight loss, extending their freshness. An additional advantage of coating was the fruit surface's appealing gloss. Kiwi, held at 50% and 288K relative humidity, received emulsion coatings made from pullulan, pea protein, and stearic acid (Xu et al., 2001). Fruits that had been treated saw 1.78% less moisture loss after storage of 54 days than uncoated fruit. If kept at 298K and 84% relative humidity, apricots and green peppers lost moisture at rates of 19.8 to 7.82% and 5.02 to 2.87%, respectively, when coatings on the basis of methylcellulose, polyethylene glycol, and stearic acid were applied. When the same composition was used on cauliflower and mushrooms moisture loss was likewise reduced. Brussels sprouts were coated with substances made of pea protein, sorbitol, and candelilla wax. This lessened mass loss and preserved the vegetables' firmness, vitamin C content, and polyphenols throughout storage. The same recipe applied to heads of broccoli revealed increased vitamin C levels and less texture weakening over time. In studies on zucchini, the usage of acetylated monoglycerides with caseinate coatings considerably decreased loss of moisture during storage (Galus & Kadzińska, 2015). When hydroxypropylmethylcellulose beeswax, which is antifungal, was employed, Fagundes et al. (2014) noticed a decline in moisture loss and an upkeep of firmness of cherry tomatoes that are cold-stored. Physicochemical characteristics and sensory features showed no harmful effects either. Throughout 42 days of preservation, temperature was at 273K. Brussels sprouts coated with a cornstarch–sunflower oil mixture demonstrated reduced moisture loss, preservation of polyphenols and vitamin C, conservation of color, and rigidness. According to Tzoumaki et al. (2009), coatings of whey protein–stearic acid have a positive effect on asparagus spear quality characteristics by delaying moisture loss, lowering basal hardening, and delaying the onset of purple color. The color, hardness, flavor, and general acceptability of kiwi fruits were successfully maintained throughout storage using a composite coat made of rice bran oil and concentrate of whey protein, and for plasticizer, glycerol is used. *Lasiodiplodia theobromae* disease severity and respiration rate were dramatically slowed down in coated avocados with composite coating of pectin, beeswax, and sorbitol. According to Maftoonazad et al. (2007), coated fruits had much lower related quality alterations in texture and color compared to the control.

9.5.2 Nuts

Using an efficient wall, like edible coatings that keep nuts away from O_2 in the air, can stop lipid oxidation on nuts. The efficiency of emulsion coatings made of WPI, carnauba wax, and pea starch in avoiding food spoilage, that is, rancidity of pine nuts and walnuts, was mentioned by Mehyar et al. (2012). Additionally, all covered nuts offered an affirmative sensory preference compared to the control. When whey protein and emulsion of coatings of olive oil were applied to pistachio kernels, Javanmard (2008) reported similar outcomes to increase the quality or maintain its integrity. Chemicals like benzoic acid, calcium lactate, lysozymes, sorbic acid, and plant-derived secondary metabolites with various antibacterial and antimicrobial properties are incorporated directly into the matrix or encapsulated, which stops the degradation of food.

Polysaccharides and essential oils containing nanoemulsions were used to create edible films with functional qualities (Acevedo-Fani et al., 2015).

To prevent the oxidative rancidity, deterioration, and discoloration of some foods, antioxidants are introduced to emulsified coatings. To prevent enzymatic browning of veggies and fruits, ascorbic acid is frequently employed. When methylcellulose and stearic acid emulsions were used, together with potassium sorbate and ascorbic acid, Olivas et al. (2003) reported that the browning of wedges of pear was delayed. A comparable mixture of methylcellulose, ascorbic acid, polyethylene glycol, stearic acid, and green peppers prevented loss of weight and vitamin C (Galus & Kadzińska, 2015).

9.5.3 Meat and Their Products

Meat is frequently subjected to the growth of bacteria like monocytogenes, which are immune to sub-threshold doses because of their propensity to form biofilms and manufacture stress proteins. This growth can be prevented by coatings made from derivatives of cellulose, alginates, and gums, with also the addition of acetic acid and lactic acid having antimicrobial activities. Edible coatings can be added to meat products by various methods, such as dipping, casting, foaming, brushing, spraying, etc.

Before processing, aqueous emulsion coatings with saturated fatty acids or alcohols chain were observed to reduce water loss and freezer burn in chilled or frozen meats during storage. To increase the shelf life of cold-stored hamburgers, acetic or lactic acid with edible films of chitosan–sunflower oil were used, and they showed a decrease in antibacterial efficiency.

Water loss affects the texture of meat products; hence, the quality and freshness are compromised. This was controlled by addition of whey proteins and monoacylglycerols (Stuchell & Krochta, 1995). Also, permeability was found to be reduced by the addition of chitosan matrix of oil of sunflower.

Alternate to vacuum packaging, coatings using pectin and blends of sodium alginates and gelatin were used with corn and olive oil, which helped in weight loss control of dry sausages. Study suggested improvement of quality and longevity of films and casings. Beeswax was added to the mixture to replace half of the lard, which decreased adhesion while also enhancing its barrier qualities (Kester & Fennema, 1986).

Lard, which makes the meat product susceptible to oxidation, can be reduced in amount by the addition of sorbitol or beeswax. But there was some cracking observed; hence, emulsions using glycerol and carrageenan were adopted.

The appearance of the product is also an important consideration for the consumer. When sunflower oil was added to films based on chitosan, pork burger appeared glossy.

This therefore directed a decline in the rise in coarseness of surface and specular reflection in air–film boundary that was achieved by filling the tiny holes with oil produced during the film development.

Chitosan-based coated burgers have higher concentration of metmyoglobin, and low oxygen is required for its formation. Thus, evidence of chitosan being a good oxygen barrier can be seen.

According to yet another study, sunflower oil–coated hamburgers had lower levels of metmyoglobin. Oxygen permeability improved by the incorporation of a component of lipid (Galus & Kadzińska, 2015).

9.5.4 Bakery Products

Snacks like chips, nuts, biscuits, baked products, and morning cereals have a crispy texture due to their reduced moisture content; it is the characteristic feature of these products. When these products are stored under greater relative humidity, water content increases, as a result of which there is a reduction in crispness and increase in softness. Emulsified coatings prepared from soybean oil, cornstarch, and methylcellulose, when incorporated with edible coatings, saw a decrease in hydration of the food products; thus, reduction in crispness was somewhat limited.

A study was conducted on the efficacy of caseinates or chitosan-based films on dry fruit cereal with pineapple. Different emulsion coatings were used on pineapples viz beeswax/oleic acid/sodium caseinate, chitosan/oleic acid, and beeswax/oleic acid/calcium and sodium caseinate to prevent hydration of cereals during storage. The emulsion coatings based on caseinate showed improvement in their shelf life (Galus & Kadzińska, 2015).

9.5.5 Cheese

Cheese comes in a wide range of textures, scents, taste, and composition. Because of the growth of fungus and bacteria, their shelf life is less and thus deteriorates the quality of the product. This can be limited by the addition of antimicrobial coatings on its surface during storage. Consumable hydrocolloid coatings like WPI, guar gum, glycerol, and sunflower oil, along with lactic acid, chitooligiosaccharides, or natamycin as antimicrobial substances, can replace commercial non-edible coatings like polyvinyl acetate or paraffin. In addition to their antimicrobial property restricting the growth of pathogens, they also help minimize water loss, reduce hardness, and prevent change in color during its storage (Galus & Kadzińska, 2015).

9.6 Trends

9.6.1 Recent Trends

New edible coatings and films have developed in recent years with the addition of different herbs that may be eaten and antibacterial chemicals to preserve fresh produce. One of the most promising and appealing research areas in the food sector today is nanotechnology. Given that these systems have a larger surface area, nanoparticles and nanoemulsions may also contribute to layers' functioning and barrier qualities for fruit protection. Undoubtedly, submicronic structures allow for greater homogeneity and distribution on the fruit skin, which has several benefits for various uses.

SLNs (solid lipid nanoparticles) are colloidal lipid submicronic models developed to enclose and transport lipophilic functional components. A surfactant and lipid solution are homogenized at a temperature over the melting point of lipid to create an oil–water nanoemulsion, which is how SLNs are normally made. After being cooled to normal temperature, this hot nanoemulsion forms solid particles. Systems such as SLNs have significant technological potential across a variety of industries, including the food industry.

However, from plants and animal resources, it is expensive and requires specialized methods to extract carbohydrate polymers. Due to their widespread effectiveness in the food industry, polysaccharides derived from microbial resources are currently receiving greater attention. Regarding commercialization, the success of a novel coating system depends on its ease of accessibility, low manufacturing cost, eco-friendliness, potency at low concentrations, and potential for protection against biological and chemical harm. The uses of renewable resources, along with their biodegradable nature and biocompatibility, have ramped up microbial gums' usage, like curdlan, pullulan, dextran, scleroglucan, as well as xanthan. In addition, the production of microbial gums from chosen species can be ascended through changing the conditions needed for the most suitable application.

A novel technology in the food sector is herbal edible coating. The most popular herbs used in edible coats include lemongrass, aloe vera gel, tulsi, neem, turmeric, and rosemary. It is manufactured using herbs or a mixture of many different consumable films. Herbs provide vitamins, antioxidants, and vital minerals in addition to having antibacterial qualities. Aloe vera gel is now frequently used as a coating for vegetables and fruits due to its antibacterial properties, which also help prevent moisture and water loss. Also, essential oils and extracts are employed in the eatable film of vegetables and fruits, such as mint oil, ginger EO, turmeric neem extract, clove bud oil, and others. Herbs serve as a natural nutraceutical and are a good source of minerals, antioxidants, and vitamins that are great for health.

Recent years have seen several attempts to change the fibers of chicken feathers, either by grafting of surface polymers that are synthetic or by blending with plasticizer, in order to convert them into films utilizing compression or casting molding methods. Keratin, a type of protein, makes up around 90% of feathers. Feather keratin is a biopolymer that is renewable, is biodegradable, and may be useful. Feather keratin is a tiny (10 kDa) semi-crystalline protein with a balanced hydrophobic and hydrophilic amino acid makeup, with the primary amino acids being proline, serine, valine, glycine, and cysteine. Using graft copolymerization upon methyl methacrylate and a malic acid/potassium permanganate redox system managed to transform chicken feather fiber. The thermal behavior of ungrafted and grafted fibers showed just very little difference, according to these authors. By employing $K_2S_2O_8/NaHSO_3$ as a redox system and grafting methyl acrylate onto native chicken feather fiber, Jin et al. changed the fiber. They then created films by compression-molding the modified feathers using glycerol as a plasticizer.

Some examples of lipid-based recent advances include the following:

1. Using **lauric acid** on additives phenolic-gum-lipid complex like oligomeric procyanidins, flaxseed gum with properties like water vapor permeability, mechanical properties, and peroxide value was changed.
2. Glycerol and vinegar were used as plasticizers on cornstarch, to which **olive oil** was added, which had properties like tensile strength and water barrier properties that were enhanced.
3. Using **yuba (beeswax as lipid)** on sodium carboxymethyl cellulose and glycerol and sorbitol as plasticizer oxidized ferulic acid as cross-linking agent and sodium pyrophosphate as emulsifier.

It exhibits incomparable properties of edible film, which includes elongation, water resistance, and water vapor permeability.

4. Using **carnauba wax or beeswax** on additive gelatin to change for the better the antioxidant properties, barrier properties, and thermal stability.

9.6.2 Future Trends

The possible use of materials that are edible on food products is currently a hot topic of discussion. Finding the ideal pairing in between food and composition of coating is typically the major focus. Emulsified layers have many benefits, particularly in preventing covered fruits and vegetables from losing moisture. In order to generate multicomponent layers with integrated properties, new formulations should be examined for the integration of functional chemicals. However, further study is required to create goods that are stable for longer periods of storage. A few days, according to some studies, is the maximum amount of time that coated products can experience detrimental changes. However, by employing the right packing method, this period can be successfully extended. Information on emulsion stability, lipid oxidation, and their potential impact on the sensory qualities of coated items is lacking. It is important to evaluate new hydrocolloids derived from unconventional sources as well as new, highly stable lipids as potential components of emulsified coatings. The crucial element is also how coating technology is adapted to industrial tools and possibilities. It is necessary to prepare an emulsion that forms films and contains lipids with high melting points under specified circumstances, which can result in additional expenses for producers that must be considered.

Additionally, a sensory study should be looked into throughout time storage to determine if the coating doesn't quite alter taste and flavor at the end product that is coated so as to evaluate the final attribute of coated items and asses the true efficacy of emulsified coatings and films. But according to current research, nanoemulsions' small lipid size of the particles improves the physicochemical film qualities and may increase the bioactivity of lipophilic drugs by increasing their surface area/mass, which results in lower dosages of active chemicals. As a future group of active packaging, nanoemulsions could be employed for edible coating and film formulations.

9.7 Conclusion

Food, from the garden to the table, takes a lot of time, a lot of travel and storage. In the course of it, food products get deteriorated; hence, packaging is an urgent aspect of the food industry. As an alternative to synthetic packaging substances, edible coatings and films are quickly becoming more popular among consumers and have numerous applications and properties that can be altered for the benefit of the good. Many new types of edible coatings and films have been created as a consequence of research-and-development activities, and they are on par in functionality with their synthetic equivalents. They are a prominent alternative for packaging food commodities due to their edibility and biodegradability. Lipid incorporation to edible films has proved to be a boon to the food packaging sector as there has been enhancement in many properties, like barrier properties, water permeability, mechanical properties, etc. Edible lipid films have acquired many usages in the food industry in processed and unprocessed foods. New advancements target providing the touch of nature like that of its purest form while also increasing shelf life. To enhance the commercial features and sensory aspects of edible coatings and films, however, there are still many murky areas that require attention.

REFERENCES

Acevedo-Fani, A., Salvia-Trujillo, L., Rojas-Graü, M. A., & Martín-Belloso, O. (2015). Edible films from essential-oil-loaded nanoemulsions: Physicochemical characterization and antimicrobial properties. *Food Hydrocolloids, 47,* 168–177. https://doi.org/10.1016/j.foodhyd.2015.01.032

Alzate, P., Miramont, S., Flores, S., & Gerschenson, L. (2016). Effect of the potassium sorbate and carvacrol addition on the properties and antimicrobial activity of tapioca starch – hydroxypropyl methylcellulose edible films. *Starch*, *69*(5–6), 1600261. https://doi.org/10.1002/star.201600261

Anker, M., Berntsen, J., Hermansson, A., & Stading, M. (2002). Improved water vapor barrier of whey protein films by addition of acetylated monoglyceride. *Innovative Food Science & Emerging Technologies*, *3*(1), 81–92.

Atarés, L., Bonilla, J., & Chiralt, A. (2010a). Characterization of sodium caseinate-based edible films incorporated with cinnamon or ginger essential oils. *Journal of Food Engineering*, *100*(4), 678–687. https://doi.org/10.1016/j.jfoodeng.2010.05.018

Atarés, L., De Jesús, C., Talens, P., & Chiralt, A. (2010b). Characterization of SPI-based edible films incorporated with cinnamon or ginger essential oils. *Journal of Food Engineering*, *99*(3), 384–391. https://doi.org/10.1016/j.jfoodeng.2010.03.004

Ayranci, E., & Tunc, S. (2001). The effect of fatty acid content on water vapour and carbon dioxide transmissions of cellulose-based films. *Food Chemistry*, *72*(2), 231–236.

Baldwin, E. A., Hagenmaier, R. D., & Bai, J. (2011). *Edible coatings and films to improve food quality* (2nd ed.). CRC Press/Taylor & Francis.

Balti, R., Mansour, M. B., Sayari, N., Yacoubi, L., Rabaoui, L., Brodu, N., & Massé, A. (2017). Development and characterization of bioactive edible films from spider crab (Maja crispata) chitosan incorporated with Spirulina extract. *International Journal of Biological Macromolecules*, *105*, 1464–1472. https://doi.org/10.1016/j.ijbiomac.2017.07.046

Banerjee, R., Chen, H., & Wu, J. (1996). Milk protein-based edible film mechanical strength changes due to ultrasound process. *Journal of Food Science*, *61*(4), 824–828.

Calva-Estrada, S. J., Jiménez-Fernández, M., & Lugo-Cervantes, E. (2019). Protein-based films: Advances in the development of biomaterials applicable to food packaging. *Food Engineering Reviews*, *11*, 78–92. https://doi.org/10.1007/s12393-019-09189-w

Chen, H. (1995). Functional properties and applications of edible films made of milk proteins. *Journal of Dairy Science*, *78*(11), 2563–2583. https://doi.org/10.3168/jds.S0022-0302(95)76885-0

Cheng, Y., Sun, C., Zhai, X., Zhang, R., Zhang, S., Sun, C., Wang, W., & Hou, H. (2021). Effect of lipids with different physical state on the physicochemical properties of starch/gelatin edible films prepared by extrusion blowing. *International Journal of Biological Macromolecules*, *185*, 1005–1014. https://doi.org/10.1016/j.ijbiomac.2021.06.203

Chiumarelli, M., & Hubinger, M. D. (2012). Stability, solubility, mechanical and barrier properties of cassava starch—carnauba wax edible coatings to preserve fresh-cut apples. *Food Hydrocolloids*, *28*(1), 59–67. https://doi.org/10.1016/j.foodhyd.2011.12.006

De Azeredo, H. M. C. (2012). Edible coatings. In *Advances in fruit processing technologies*. CRC Press, Taylor & Francis Group (pp. 345–362). https://doi.org/10.1201/b12088

Dea, S., Ghidelli, C., & Plotto, A. (1995). Baldwin, E.A., Hagenmaier, R., and Bai, J. (Eds.), Coatings for minimally processed fruits and vegetables. In *Edible coatings and films to improve food quality*, CRC Press, Boca Raton, (pp. 243–289).

Debeaufort, F., & Voilley, A. (1995). Effect of surfactants and drying rate on barrier properties of emulsified edible films. *International Journal of Food Science + Technology*, *30*, 183–190.

Dehghani, S., Hosseini, S. V., & Regenstein, J. M. (2018). Edible films and coatings in seafood preservation: A review. *Food Chemistry*, *240*, 505–513. https://doi.org/10.1016/j.foodchem.2017.07.034

Dhaka, R. K., & Upadhyay, A. (2018). Edible films and coatings: A brief overview. *The Pharma Innovation Journal*, *7*(7), 331–333.

Dhall, R. K. (2013). Advances in edible coatings for fresh fruits and vegetables: A review. *Critical Reviews in Food Science and Nutrition*, *53*, 435–450. https://doi.org/10.1080/10408398.2010.541568

Dubey, N. K., & Dubey, R. (2020). Edible films and coatings: An update on recent advances. In *Biopolymer-based formulations: Biomedical and food applications*. Elsevier Inc. https://doi.org/10.1016/B978-0-12-816897-4.00027-8

Fabra, M. J., Falcó, I., Randazzo, W., Sánchez, G., & López-Rubio, A. (2018). Antiviral and antioxidant properties of active alginate edible films containing phenolic extracts. *Food Hydrocolloids*, *81*, 96–103. https://doi.org/10.1016/j.foodhyd.2018.02.026

Fabra, M. J., Jiménez, A., Atarés, L., Talens, P., & Chiralt, A. (2009a). Effect of fatty acids and beeswax addition on properties of sodium caseinate dispersions and films. *Biomacromolecules*, *10*(6), 1500–1507. https://doi.org/10.1021/bm900098p

Fabra, M. J., Talens, P., & Chiralt, A. (2009b). Microstructure and optical properties of sodium caseinate films containing oleic acid-beeswax mixtures. *Food Hydrocolloids*, *23*(3), 676–683. https://doi.org/10.1016/j.foodhyd.2008.04.015

Fagundes, C., Palou, L., Monteiro, A. R., & Pérez-Gago, M. B. (2014). Effect of antifungal hydroxypropyl methylcellulose-beeswax edible coatings on gray mold development and quality attributes of cold-stored cherry tomato fruit. *Postharvest Biology and Technology*, *92*, 1–8. https://doi.org/10.1016/j.postharvbio.2014.01.006

Fang, Y., Tung, M. A., Britt, I. J., Yada, S., & Dalgleish, D. G. (2002). Tensile and barrier properties of edible films made from whey proteins. *Journal of Food Science*, *67*(1), 188–193. https://doi.org/10.1111/j.1365-2621.2002.tb11381.x

Fernández, L., De Apodaca, E. D., Cebrián, M., Villarán, M. C., & Maté, J. I. (2007). Effect of the unsaturation degree and concentration of fatty acids on the properties of WPI-based edible films. *European Food Research and Technology*, *224*(4), 415–420. https://doi.org/10.1007/s00217-006-0305-1

Galus, S., & Kadzińska, J. (2015). Food applications of emulsion-based edible films and coatings. *Trends in Food Science and Technology*, *45*(2), 273–283. https://doi.org/10.1016/j.tifs.2015.07.011

Galus, S., & Kadzińska, J. (2016a). Moisture sensitivity, optical, mechanical and structural properties of whey protein-based edible films incorporated with rapeseed oil. *Food Technology and Biotechnology*, *54*(1), 78–89. https://doi.org/10.17113/ftb.54.01.16.3889

Galus, S., & Kadzińska, J. (2016b). Whey protein edible films modified with almond and walnut oils. *Food Hydrocolloids*, *52*, 78–86. https://doi.org/10.1016/j.foodhyd.2015.06.013

Gao, P., Wang, F., Gu, F., Ning, J., Liang, J., Li, N., & Ludescher, R. D. (2017). Preparation and characterization of zein thermo-modified starch films. *Carbohydrate Polymers*, *157*, 1254–1260. https://doi.org/10.1016/j.carbpol.2016.11.004

Gennadios, A., Ghorpade, V. M., Weller, C. L., & Hanna, M. A. (1996). Heat curing of soy protein films. *Biological Systems Engineering Papers and Publications, 39*(2), 575–579.

Gennadios, A., Rhim, J. W., Handa, A., Weller, C. L., & Hanna, M. A. (1998). Ultraviolet radiation affects physical and molecular properties of soy protein films. *Journal of Food Science*, *63*(2), 3–6.

Greenspan, L. (1977). Humidity fixed points of binary saturated aqueous solutions. *Journal of Research of the National Bureau of Standards, Section A: Physics and Chemistry, 81*(1), 89–96.

Guerrero, P., Nur Hanani, Z. A., Kerry, J. P., & De La Caba, K. (2011). Characterization of soy protein-based films prepared with acids and oils by compression. *Journal of Food Engineering*, *107*(1), 41–49. https://doi.org/10.1016/j.jfoodeng.2011.06.003

Guilbert, S., Gontard, N., & Gorris, L. G. M. (1996). Prolongation of the shelf life of perishable food products using biodegradable films and coatings. *Lebensmittel-Wissenschaft & Technologie*, *29*, 10–17.

Hagenmaier, R. D., & Baker, R. A. (1993). Reduction in gas exchange of citrus fruit by wax coatings. *Journal of Agricultural and Food Chemistry*, *41*, 283–287.

Hall, D. J. (2012). *Edible coatings from lipids, waxes, and resins.* Taylor and Francis Group, pp. 79–101.

Hambleton, A., Perpiñan-Saiz, N., Fabra, M. J., Voilley, A., & Debeaufort, F. (2012). The Schroeder paradox or how the state of water affects the moisture transfer through edible films. *Food Chemistry*, *132*(4), 1671–1678. https://doi.org/10.1016/j.foodchem.2011.03.009

Han, J. H., Seo, G. H., Park, I. M., Kim, G. N., & Lee, D. S. (2006). Physical and mechanical properties of pea starch edible films containing beeswax emulsions. *Journal of Food Science*, *71*(6), 290–296. https://doi.org/10.1111/j.1750-3841.2006.00088.x

Hassan, B., Chatha, S. A. S., Hussain, A. I., Zia, K. M., & Akhtar, N. (2018). Recent advances on polysaccharides, lipids and protein based edible films and coatings: A review. *International Journal of Biological Macromolecules*, *109*, 1095–1107. https://doi.org/10.1016/j.ijbiomac.2017.11.097

Hopkins, E. J., Chang, C., Lam, R. S. H., & Nickerson, M. T. (2015). Effects of flaxseed oil concentration on the performance of a soy protein isolate-based emulsion-type film. *FRIN*, *67*, 418–425. https://doi.org/10.1016/j.foodres.2014.11.040

Hun, J. H., & Cennadios, A. (2005). Edible films and coatings: A review. In *Innovations in food packaging* (pp. 239–262). Academic Press.

Javanmard, M. (2008). Shelf life of whey protein-coated pistachio kernel (Pistacia Vera L.). *Journal of Food Process Engineering*, *31*(2), 247–259. https://doi.org/10.1111/j.1745-4530.2007.00150.x

Javanmard, M., & Golestan, L. (2008). Effect of olive oil and glycerol on physical properties of whey protein concentrate films. *Journal of Food Process Engineering*, *31*(5), 628–639. https://doi.org/10.1111/j.1745-4530.2007.00179.x

Jiménez, A., Fabra, M. J., Talens, P., & Chiralt, A. (2013a). Phase transitions in starch based fi lms containing fatty acids. Effect on water sorption and mechanical behaviour. *Food Hydrocolloids, 30,* 408–418. https://doi.org/10.1016/j.foodhyd.2012.07.007

Jiménez, A., Fabra, M. J., Talens, P., & Chiralt, A. (2013b). Physical properties and antioxidant capacity of starch—sodium caseinate films containing lipids. *Journal of Food Engineering, 116,* 695–702. https://doi.org/10.1016/j.jfoodeng.2013.01.010

Kalia, A., & Parshad, V. R. (2013). Novel trends to revolutionize preservation and packaging of fruits/fruit products: Microbiological and nanotechnological perspectives. *Critical Reviews in Food Science and Nutrition, 55,* 159–182. https://doi.org/10.1080/10408398.2011.649315

Kester, J. J., & Fennema, O. R. (1986). Edible films and coatings: A review. *Food Technology, 40*(12), 47.

Khwaldia, K., Ferez, C., Banon, S., Desobry, S., & Hardy, J. (2004). Milk proteins for edible films and coatings. *Critical Reviews in Food Science and Nutrition, 44*(4), 239–251. https://doi.org/10.1080/10408690490464906

Kim, S. J., & Ustunol, Z. (2001). Solubility and moisture sorption isotherms of whey-protein-based edible films as influenced by lipid and plasticizer incorporation. *Journal of Agricultural and Food Chemistry, 49*(9), 4388–4391. https://doi.org/10.1021/jf010122q

Kokoszka, S., Debeaufort, F., Lenart, A., & Voilley, A. (2010). Liquid and vapour water transfer through whey protein/lipid emulsion films. *Journal of the Science of Food and Agriculture, 90*(10), 1673–1680. https://doi.org/10.1002/jsfa.4001

Kristo, E., Biliaderis, C. G., & Zampraka, A. (2007). Water vapour barrier and tensile properties of composite caseinate-pullulan films: Biopolymer composition effects and impact of beeswax lamination. *Food Chemistry, 101*(2), 753–764. https://doi.org/10.1016/j.foodchem.2006.02.030

Krochta, J. M. (1999). Water vapor permeability of whey protein. *Journal of Food Science, 64,* 695–698.

Krochta, J. M. (2002). Proteins as raw materials for films and coatings: Definitions, current status, and opportunities. *Protein-Based Films and Coatings,* 1–41. https://doi.org/10.1201/9781420031980-4

López-Córdoba, A., Medina-Jaramillo, C., Piñeros-Hernandez, D., & Goyanes, S. (2017). Cassava starch films containing rosemary nanoparticles produced by solvent displacement method. *Food Hydrocolloids, 71,* 26–34. https://doi.org/10.1016/j.foodhyd.2017.04.028

Lopez-Rubio, A., Fabra, M. J., Martinez-Sanz, M., Mendoza, S., & Vuong, Q. V. (2017). Biopolymer-based coatings and packaging structures for improved food quality. *Journal of Food Quality, 2017,* 2–4. https://doi.org/10.1155/2017/2351832

Ma, W., Tang, C. H., Yin, S. W., Yang, X. Q., Wang, Q., Liu, F., & Wei, Z. H. (2012). Characterization of gelatin-based edible films incorporated with olive oil. *Food Research International, 49*(1), 572–579. https://doi.org/10.1016/j.foodres.2012.07.037

Maftoonazad, N., Ramaswamy, H. S., Moalemiyan, M., & Kushalappa, A. C. (2007). Effect of pectin-based edible emulsion coating on changes in quality of avocado exposed to Lasiodiplodia theobromae infection. *Carbohydrate Polymers, 68*(2), 341–349. https://doi.org/10.1016/j.carbpol.2006.11.020

Mahmoud, R., & Savello, P. A. (1992). Mechanical properties of and water vapor transferability through whey protein films. *Journal of Dairy Science, 75*(4), 942–946. https://doi.org/10.3168/jds.S0022-0302(92)77834-5

Marcuzzo, E., Peressini, D., Debeaufort, F., & Sensidoni, A. (2010). Effect of ultrasound treatment on properties of gluten-based film. *Innovative Food Science and Emerging Technologies, 11*(3), 451–457. https://doi.org/10.1016/j.ifset.2010.03.002

Mehyar, G. F., Al-Ismail, K., Han, J. H., & Chee, G. W. (2012). Characterization of edible coatings consisting of pea starch, whey protein isolate, and carnauba wax and their effects on oil rancidity and sensory properties of walnuts and pine nuts. *Journal of Food Science, 77*(2). https://doi.org/10.1111/j.1750-3841.2011.02559.x

Micard, V., Belamri, R., Morel, M., & Guilbert, S. (2000). Properties of chemically and physically treated wheat gluten films. *Journal of Agricultural and Food Chemistry, 48,* 2948–2953.

Miller, K. S., Chiang, M. T., & Krochta, J. M. (1997). Heat curing of whey protein films. *Journal of Food Science, 62*(6), 1189–1193.

Mkandawire, M., & Aryee, A. N. A. (2018). Resurfacing and modernization of edible packaging material technology. *Current Opinion in Food Science, 19,* 104–112. https://doi.org/10.1016/j.cofs.2018.03.010

Monedero, F. M., Fabra, M. J., Talens, P., & Chiralt, A. (2009). Effect of oleic acid-beeswax mixtures on mechanical, optical and water barrier properties of soy protein isolate based films. *Journal of Food Engineering, 91*(4), 509–515. https://doi.org/10.1016/j.jfoodeng.2008.09.034

Morillon, V., Debeaufort, F., Capelle M., Voilley A., & Blond, G. (2010). Factors affecting the moisture permeability of lipid-based edible films: A review. *Critical Reviews in Food Science and Nutrition, 42*(2), 67–89.

Murrieta-Martínez, C. L., Soto-Valdez, H., Pacheco-Aguilar, R., Torres-Arreola, W., Rodríguez-Felix, F., & Márquez Ríos, E. (2018). Edible protein films: Sources and behavior. *Packaging Technology and Science, 31*(3), 113–122. https://doi.org/10.1002/pts.2360

Navarro-Tarazaga, M. L., Massa, A., & Pérez-Gago, M. B. (2011). Effect of beeswax content on hydroxypropyl methylcellulose-based edible film properties and postharvest quality of coated plums (Cv. Angeleno). *LWT—Food Science and Technology, 44*(10), 2328–2334. https://doi.org/10.1016/j.lwt.2011.03.011

Nisar, T., Wang, Z. C., Yang, X., Tian, Y., Iqbal, M., & Guo, Y. (2018). Characterization of citrus pectin films integrated with clove bud essential oil: Physical, thermal, barrier, antioxidant and antibacterial properties. *International Journal of Biological Macromolecules, 106*, 670–680. https://doi.org/10.1016/j.ijbiomac.2017.08.068

Ojagh, S. M., Rezaei, M., Razavi, S. H., & Hosseini, S. M. H. (2010). Development and evaluation of a novel biodegradable film made from chitosan and cinnamon essential oil with low affinity toward water. *Food Chemistry, 122*(1), 161–166. https://doi.org/10.1016/j.foodchem.2010.02.033

Olivas, G. I., Rodriguez, J. J., & Barbosa-Cánovas, G. V. (2003). Edible coatings composed of methylcellulose, stearic acid, and additives to preserve quality of pear wedges. *Journal of Food Processing and Preservation, 27*(4), 299–320. https://doi.org/10.1111/j.1745-4549.2003.tb00519.x

Ortega-Toro, R., Jiménez, A., Talens, P., & Chiralt, A. (2014). Effect of the incorporation of surfactants on the physical properties of cornstarch films. *Food Hydrocolloids, 38*, 66–75. https://doi.org/10.1016/j.foodhyd.2013.11.011

Parreidt, T. S., Müller, K., & Schmid, M. (2018). Alginate-based edible films and coatings for food packaging applications. *Foods, 7*(10), 1–38. https://doi.org/10.3390/foods7100170

Pavli, F., Tassou, C., Nychas, G. J. E., & Chorianopoulos, N. (2018). Probiotic incorporation in edible films and coatings: Bioactive solution for functional foods. *International Journal of Molecular Sciences, 19*(1). https://doi.org/10.3390/ijms19010150

Pereda, M., Amica, G., & Marcovich, N. E. (2012). Development and characterization of edible chitosan/olive oil emulsion films. *Carbohydrate Polymers, 87*(2), 1318–1325. https://doi.org/10.1016/j.carbpol.2011.09.019

Pereda, M., Aranguren, M. I., & Marcovich, N. E. (2010). Caseinate films modified with tung oil. *Food Hydrocolloids, 24*, 800–808. https://doi.org/10.1016/j.foodhyd.2010.04.007

Pérez-Gago, M. B., & Krochta, J. M. (2001). Lipid particle size effect on water vapor permeability and mechanical properties of whey protein/beeswax emulsion films. *Journal of Agricultural and Food Chemistry, 49*(2), 996–1002. https://doi.org/10.1021/jf000615f

Perez-Gago, M. B., Serra, M., Alonso, M., Mateos, M., & Del Río, M. A. (2003). Effect of solid content and lipid content of whey protein isolate-beeswax edible coatings on color change of fresh-cut apples. *Journal of Food Science, 68*(7).

Perez-Gago, M. B., Serra, M., & Del Río, M. A. (2006). Color change of fresh-cut apples coated with whey protein concentrate-based edible coatings. *Postharvest Biology and Technology, 39*(1), 84–92. https://doi.org/10.1016/j.postharvbio.2005.08.002

Piermaria, J., Diosma, G., Aquino, C., Garrote, G., & Abraham, A. (2015). Edible kefiran films as vehicle for probiotic microorganisms. *Innovative Food Science and Emerging Technologies, 32*, 193–199. https://doi.org/10.1016/j.ifset.2015.09.009

Porta, R. (2017). Plastic pollution and the challenge of bioplastics. *Journal of Applied Biotechnology & Bioengineering, 2*(3), 15406. https://doi.org/10.15406/jabb.2017.02.00033

Quezada Gallo, J. A., Debeaufort, F., Callegarin, F., & Voilley, A. (2000). Lipid hydrophobicity, physical state and distribution effects on the properties of emulsion-based edible films. *Journal of Membrane Science, 180*(1), 37–46. https://doi.org/10.1016/S0376-7388(00)00531-7

Raghav, K., Agarwal, N., & Saini, M. (2016). Edible coating of fruits and vegetables: A review. *International Journal of Scientific Research and Modern Education, I*(I), 188–204. www.researchgate.net/publication/331298687_EDIBLE_COATING_OF_FRUITS_AND_VEGETABLES_A_REVIEW

Rao, M. S., Kanatt, S. R., Chawla, S. P., & Sharma, A. (2010). Chitosan and guar gum composite films: Preparation, physical, mechanical and antimicrobial properties. *Carbohydrate Polymers, 82*(4), 1243–1247. https://doi.org/10.1016/j.carbpol.2010.06.058

Raybaudi-Massilia, R., Mosqueda-Melgar, J., Soliva-Fortuny, R., & Martín-Belloso, O. (2016). Combinational edible antimicrobial films and coatings. *Antimicrobial Food Packaging*, 633–646. https://doi.org/10.1016/B978-0-12-800723-5.00052-8

Rhim, J. W., & Shellhammer, T. H. (2005). Lipid-based edible films and coatings. In *Innovations in food packaging: Overview*. Elsevier Ltd. https://doi.org/10.1016/B978-0-12-311632-1.50053-X

Ribeiro, A. M., Estevinho, B. N., & Rocha, F. (2021). Preparation and incorporation of functional ingredients in edible films and coatings. *Food and Bioprocess Technology*, *14*(2), 209–231. https://doi.org/10.1007/s11947-020-02528-4

Ruiz-Navajas, Y., Viuda-Martos, M., Sendra, E., Perez-Alvarez, J. A., & Fernández-López, J. (2013). In vitro antibacterial and antioxidant properties of chitosan edible films incorporated with Thymus moroderi or Thymus piperella essential oils. *Food Control*, *30*(2), 386–392. https://doi.org/10.1016/j.foodcont.2012.07.052

Sánchez-Ortega, I., García-Almendárez, B. E., Santos-López, E. M., Amaro-Reyes, A., Barboza-Corona, J. E., & Regalado, C. (2014). Antimicrobial edible films and coatings for meat and meat products preservation. *Scientific World Journal*, *2014*. https://doi.org/10.1155/2014/248935

Saucedo-Pompa, S., Jasso-Cantu, D., Ventura-Sobrevilla, J., Sáenz-Galindo, A., Rodríguez-Herrera, R., & Aguilar, C. N. (2007). Effect of candelilla wax with natural antioxidants on the shelf life quality of fresh-cut fruits. *Journal of Food Quality*, *30*(5), 823–836. https://doi.org/10.1111/j.1745-4557.2007.00165.x

Shahidi, F., & Hossain, A. (2022). Preservation of aquatic food using edible films and coatings containing essential oils: A review. *Critical Reviews in Food Science and Nutrition*, *62*(1), 66–105. https://doi.org/10.1080/10408398.2020.1812048

Shaw, N. B., Monahan, F. J., Riordan, E. D. O., & Sullivan, M. O. (2002). Effect of soya oil and glycerol on physical properties of composite WPI films. *Journal of Food Engineering*, *51*, 299–304.

Sothornvit, R., & Krochta, J. M. (2000). Oxygen permeability and mechanical properties of films from hydrolyzed whey protein. *Journal of Agricultural and Food Chemistry*, *48*, 3913–3916.

Stuchell, Y. M., & Krochta, J. M. (1995). Edible coatings on frozen king salmon: Effect of whey protein isolate and acetylated monoglycerides on moisture loss and lipid oxidation. *Journal of Food Science*, *60*(1), 28–31. https://doi.org/10.1111/j.1365-2621.1995.tb05599.x

Tanaka, M., Ishizaki, S., Suzuki, T., & Takai, R. (2001). Water vapor permeability of edible films prepared from fish water. *Journal of Tokyo University of Fisheries*, *87*, 31–37. http://www2.kaiyodai.ac.jp/~toru/websuzuki/images/lab/115.lab.pdf

Taylor, P. (2010). Edible films and coatings: Tomorrow's packagings: A review. *Critical Reviews in Food Science and Nutrition*, 37–41.

Tzoumaki, M. V., Biliaderis, C. G., & Vasilakakis, M. (2009). Impact of edible coatings and packaging on quality of white asparagus (Asparagus officinalis, L.) during cold storage. *Food Chemistry*, *117*(1), 55–63. https://doi.org/10.1016/j.foodchem.2009.03.076

Valdés, A., Mellinas, A. C., Ramos, M., Garrigós, M. C., & Jiménez, A. (2014). Natural additives and agricultural wastes in biopolymer formulations for food packaging. *Frontiers in Chemistry*, *2*, 1–10. https://doi.org/10.3389/fchem.2014.00006

Velickova, E., Winkelhausen, E., Kuzmanova, S., & Alves, V. D. (2013). Impact of chitosan-beeswax edible coatings on the quality of fresh strawberries (Fragaria ananassa cv Camarosa) under commercial storage conditions. *LWT—Food Science and Technology*, *52*(2), 80–92. https://doi.org/10.1016/j.lwt.2013.02.004

Xu, S., Chen, X., & Sun, D. W. (2001). Preservation of kiwifruit coated with an edible film at ambient temperature. *Journal of Food Engineering*, *50*(4), 211–216. https://doi.org/10.1016/S0260-8774(01)00022-X

Yang, L., & Paulson, A. T. (2000). Mechanical and water vapour barrier properties of edible gellan films. *Food Research International*, *33*(7), 563–570. https://doi.org/10.1016/S0963-9969(00)00092-2

Zahedi, Y., Ghanbarzadeh, B., & Sedaghat, N. (2010). Physical properties of edible emulsified films based on pistachio globulin protein and fatty acids. *Journal of Food Engineering*, *100*(1), 102–108. https://doi.org/10.1016/j.jfoodeng.2010.03.033

10

Essential Oil–Based Eco-Friendly Coating Materials

Vishal Manjunatha, Ian Blaise Smith, Urvi Shah, Clara Flores, and Disha Bhattacharjee

CONTENTS

10.1 Introduction

Essential oils (EOs) have an array of beneficial properties and have been used by humans since ancient times for their healing properties. EOs have become a billion-dollar global industry, with an estimated market size valued at 18.6 billion USD in 2020. Revenue from this market is expected to grow from 2021 to 2028 at a compound annual growth rate (CAGR) of 7.4% (Essential oils market size 2022). But what exactly are EOs, and what are they used for? EOs are secondary metabolites produced by a variety of plants, including those with known medicinal properties. These secondary metabolites are produced from primary plant metabolites, including carbohydrates, amino acids, and lipids. They are produced by plants for a variety of reasons, including defense, protection against UV radiation, and as attractants for pollinators. Constituents of essential oil (EO) secondary metabolites include monoterpenes (10 carbons), oxygenated monoterpenes and monoterpene hydrocarbons, sesquiterpenes (15 carbons), aldehydes, ketones, alcohols, esters, epoxides, and other oxygenated derivatives (Bhuyan et al., 2015). Different compositions of these compounds produce characteristic odors, and the oils are most times colorless, with a high refractive index.

EOs, also known as volatile oils, possess a range of biological activities as well as antimicrobial, antifungal, and antioxidant properties (Mancianti & Ebani, 2020). Other properties of EOs that have been documented include anti-inflammatory, antimutagenic, antiseptic, and antibiofilm properties. Since ancient times, EOs have been used for their antiseptic properties and as preservatives in food. *In vivo* and *in vitro* studies have shown that EOs have demonstrated extraordinary healing properties in wounds in part due to their anti-inflammatory, antioxidant, and antimicrobial effects (Pérez-Recalde et al., 2018). To date, over 3,000 EOs are extracted from angiosperms. Oils are most often harvested from flowering plants in the families *Lamiaceae, Rutaceae, Mytaceae,* and *Zingiberaceae* (Yeshi & Wangchuk, 2022). Extraction processes vary and include steam distillation, expression, and mechanical pressing of citrus fruits specifically (Sadgrove et al., 2021). Depending on the extraction method, there may be variability in the amount of total extract. Inappropriate extraction procedures can change or even damage the chemicals present in the EO being processed.

In some cases, a single aromatic plant species can be partitioned into chemotypes which are chemically distinct groups. Chemotypes tend to be highly consistent in their chemistry; however, their chemistry can change due to external factors, such as temperature, moisture content, and soil composition (Sadgrove et al., 2021; Bakkali et al., 2008). For this reason, to maintain EOs with a constant composition, they must be extracted under the exact same conditions. This includes the same climate, same soil, and plants harvested in the same season. Presently, EOs are used in the pharmaceutical, food, and cosmetic industries. Examples of EOs or their components used in the food and cosmetic industry are d-limonene, geranyl acetate, and d-carvone. Recently, more attention has been directed at EOs for their use in aromatherapy, a therapy practice using EOs as the main therapeutic agent to heal an array of complications for mind and body (Ali et al., 2015). In the food industry, EOs can be incorporated into packaging materials to help improve the longevity of foods. We have set forth a detailed summary of the types of essential oils, their components, and their use as biopolymer-based films and coatings.

10.2 Essential Oils and Types of Essential Oils

Thousands of EOs have been identified, and over 300 EOs are employed in the food and fragrance market. Of the known plant species with both aromatic and medicinal properties, around 2,000 species belong to almost 60 botanical families of plants bearing EOs. Plant families that have plants bearing EOs include *Apiaceae, Asteraceae, Burseraceae, Cupressaceae, Fabaceae, Lamiaceae, Laurenceae, Myrataceae,* and *Zingiberaceae.* Examples of EOs extracted from these families are listed in Table 10.1. Essentially, every part of a plant can yield EOs, including flowers, leaves, fruits, peel, bark, seeds, and roots.

Depending on the species of plant, different parts of the plant material are used to extract the EO. Basil, bay leaf, lemongrass, citronella, mint, oregano, peppermint, pine, rosemary, tea tree, thyme,

TABLE 10.1

Examples of plant families containing plants bearing essential oils

Family	Essential oils (EOs)
Apiaceae	Coriander, dill, fennel, anise, cumin, cilantro
Asteraceae	Roman chamomile, German chamomile, blue tansy
Burseraceae	Myrrh, frankincense, palo santo
Cupressaceae	Cypress, juniper berry, arborvitae
Fabaceae	Copaiba, Peru balsam
Lamiaceae	Lavender, patchouli, rosemary, peppermint, basil, thyme, oregano
Laurenceae	Cinnamon, rosewood, camphor, cassia, laurel leaf
Myrataceae	Tea tree, eucalyptus, clove bud, allspice
Zingiberaceae	Cardamon, ginger, turmeric

TABLE 10.2

Therapeutic properties and extraction methods of common essential oils (EOs)

Essential oils	Therapeutic properties	Extraction method(s)
Lavender oil	Antimicrobial, antifungal	Steam distillation
Eucalyptus oil	Antimicrobial, anti-inflammatory	Steam distillation
Peppermint oil	Antibacterial, antifungal, antibiofilm	Steam distillation
Lemon oil	Antifungal	Expression
Clove oil	Antioxidant, antifungal, antibacterial	Hydrodistillation, Steam distillation
Cinnamon oil	Antimicrobial, antimutagenic	Steam distillation, Organic solvent
Frankincense	Antimutagenic, antiseptic	Steam distillation
Grapefruit oil	Antimicrobial, antioxidant, antimutagenic	Expression
Rosemary oil	Antimicrobial, anti-inflammatory, antioxidant	Steam distillation
Lemongrass oil	Antibacterial, antifungal, anti-inflammatory	Steam distillation
Oregano oil	Antimicrobial, antioxidant, anti-inflammatory	Hydrodistillation
Ylang-ylang oil	Antimicrobial, antioxidant, antibiofilm, anti-inflammatory	Hydrodistillation, Steam distillation

kaffir, lantana, and cypress EOs are all collected from the leaves of plants. Seeds are used to extract EOs of anise, cardamom, coriander, cumin, nutmeg, and parsley. Bark contains EOs of cinnamon and sassafras (Tongnuanchan & Benjakul, 2014). While some EOs are known to have a pleasant aroma, such as lavender and peppermint EOs, others, such as neem oil, have a characteristic unpleasant sulfur aroma. Table 10.2 lists some common essential oils, their therapeutic properties, and their most common extraction methods.

10.2.1 Components and Functional Activities of Essential Oils

EOs are not a simple mixture of compounds. Rather, EOs can contain anywhere from 20–60 components in different concentrations, depending on the oil. Typically, an essential oil (EO) will contain two to three components at very high concentrations, along with other components making up a minority composition. EOs are characterized by the major components making up the greatest percentage of the oil. For instance, lavender (*Lavandula angustifolia*) oil is largely composed of linalyl acetate (51%) and linalool (35%) (Prashar et al., 2004). Ylang-ylang (*Cananga odorata*) is mainly composed of linalool (28%) and germacrene D and β-caryophyllene representing 63% of the total hydrocarbon fraction of the oil (Tan et al., 2015). Carvacol (30%) and thymol (27%) are the main components of oregano (*Organum compactum*) EO.

Often, biological activity is determined by the major components of Eos, which include two groups. Of the two groups, terpenes and terpenoids compose the main group, and aromatic and aliphatic components,

TABLE 10.3

Function and structure of monoterpenes

Functions	Acyclic	Monocyclic	Bicyclic
Alcohols	Geraniol, linalol, citronellol, lavandulol, nerol	Menthol, α-terpineol, carveol	Borneol, fenchol, chrysanthenol, thuyan-3-ol
Aldehydes	Geranial, neral, citronellal	—	—
Carbures	Myrcene, ocimene	Terpinenes, p-cimene, phellandrenes	Pinenes, -3-carene, camphene, sabinene, etc.
Ethers	—	1,8-cineole	—
Esters	Linalyl acetate or propionate, citronellyl acetate	Menthyl or α-terpinyl acetate	Isobornyl acetate
Ketone	Tegetone	Menthones, carvone, pulegone, piperitone	Camphor, fenchone, thuyone, ombellulone, pinocamphone, pinocarvone
Peroxides	—	—	Ascaridole
Phenols	—	Thymol, carvacrol	—

Source: Parvin et al. (2014).

of low molecular weight, compose the second group. Terpenes or isoprenoids are classified by the organization and number of isoprene, 5-carbon-base, units they contain. Terpenoids are derived from terpenes and contain multiple cyclic groups and oxygen. The inclusion of oxygen is what differentiates a terpenoid from a terpene. While terpenes are typically volatile, terpenoids can be semi- or non-volatile due to the contribution of other polar moieties. The functions of terpenes are varied in plants and include signaling functions, thermoprotection, and flavoring. Terpenes are traditionally classified as hemiterpenes, monoterpenes, sesquiterpenes, diterpenes, triterpenes, tetraterpenes, and polyterpenes. The coupling of two isoprene units (C10) creates a monoterpene, the most representative molecules in EOs. Monoterpenes are usually linear (acyclic) or contain rings (mono- and bicyclic). Because of the linear or ring assembly they can assume, monoterpenes permit a variety of structures and consist of several functions, some of which are listed in Table 10.3.

Plant phenols are produced from L-phenylalanine and L-tyrosine through the shikimate pathway. They contain at least one hydroxyl group attached to an aromatic ring (Kumar & Goel, 2019). Plant phenolics exhibit several functions, from plant growth to plant development and defense. Their biological activities include anti-inflammatory, anticancer, and antimicrobial properties, to name a few. Oxygenated derivatives of hydrocarbons, compounds of terpenes, include oxides, alcohols, aldehydes, ketones, and acids. Aldehydes are a class of electrophilic carbonyl compounds containing at least one hydrogen substituent on the carbonyl carbon atom. Alcohols are a class of compounds characterized by addition of one or more hydroxyl (-OH) groups attached to a carbon atom of hydrocarbon chain (alkyl group).

The assembly of three isoprene molecules forms a sesquiterpenes. Sesquiterpenes have a similar structure to monoterpenes, including the functions carbures, alcohols, ketones, and epoxide. Sesquiterpene carbure molecules include cadinenes, elemenes, zingiberene, logifolene, etc. Alcohols in this category contain bisabol, cedrol, carotol, viridiflorol, etc. Common EOs containing these compounds are bergamot, lavender, lemongrass, mint, orange, peppermint, rosemary, thyme, sage, and eucalyptus oils.

Minor components in EOs are the aromatic compounds which are derived from phenylpropane. Aromatic compounds include aldehydes, alcohols, phenols, methoxy, derivatives, and methylenedioxy compounds. Examples of plant sources for these minor EO components are cinnamon, clove, parsley, sassafras, and tarragon. Aromatic notes offered by alcohols, aldehydes, and ketones include fruity notes ((E)-nerolidol), floral notes (linalool), citrus notes (limonene), and herbal notes (γ-selinene).

10.3 Essential Oil–Based Active Film/Coating Systems

The demand for safer, more sustainable packaging has risen particularly in the food packaging sector with consumers becoming more socially conscious. Active packaging technologies are those that incorporate additives into the package with the intention of maintaining and/or extending a product's shelf life and quality. While traditional food packaging aims to protect and contain foodstuffs, active packaging takes these concepts a step further. With active packaging, interactions between food and the packaging environment are taken into consideration to provide active protection to the packaged foods. If a package provides additional roles other than an inert barrier to the external environment, it may be considered active. Currently, there are several commercially available active packaging systems on the market based on different principles. Active packaging systems rely on different methodologies, including oxygen scavengers, ethylene scavengers, antimicrobial agents, interleavers, or moisture absorbers (Biji et al., 2015).

Active packaging can be categorized into chemoactive and bioactive, depending on the additives incorporated into the packaging materials. Chemoactive packaging utilizes chemicals as an active agent to influence the chemical composition of the food and/or the gaseous atmosphere within the package. Oxygen and ethylene scavengers tend to remove gases, dehydrating food products and creating unfavorable conditions for microbial growth. The presence of oxygen facilitates growth of aerobic bacteria, while ethylene gas acts as a ripening agent. The removal of these gases helps prolong the life of the food (Sharma et al., 2021). Bioactive active packaging incorporates antimicrobial agents to inhibit bacterial and fungal growth. Studies have shown the addition of certain EOs in packaging products had a great antibacterial effect (Azadbakht et al., 2018).

While there are advantages to active packaging, one drawback with chemoactive packaging is the use of synthetic additives that may cause adverse health effects or lead to excess waste due to unsuitable packaging for recycling. To combat the negative effects of synthetic additives, bioactive compounds from natural sources can be used. In particular, EOs can be added to packaging to prevent food spoilage and extend shelf life.

The antimicrobial and antioxidant nature of many EOs can be used in food packaging to extend shelf life. Foods ranging from fruits and vegetables to meats and dairy can be packaged with materials containing EOs to keep products fresh. While free EOs could be used in packaging, their volatile nature does not allow them to remain for long. To combat this issue, technologies have been developed to encapsulate EOs within certain particles to improve stability (Fernández-López & Viuda-Martos, 2018). Without encapsulation, EOs applied directly to food degrade quickly.

EOs can be incorporated into active packaging via films or coatings. Films are thin sheets applied to food as covers or wrappers. Coatings are films applied directly onto the surface of the food being packaged (Sharma et al., 2021). Edible active film/coating systems are typically produced using edible biopolymers. These biopolymers are derived from numerous natural plant or animal sources, including lipids, proteins, and polysaccharides. Within edible film matrices, EOs can be incorporated via methods such as emulsification or homogenization.

10.4 Methods of Application of Essential Oil Films/Coatings

Several methods may be employed in the application of films/coating impregnated with EOs. The quality of the coating is dependent on the method used. The type of food being packaged should also be taken into consideration when choosing one of the following methods used in active packaging. Methods of application include dipping, spraying, spreading, vacuum impregnation, layer-by-layer method, foaming, and thin-film hydration. Each of these methods will be described in more detail in the following sections.

10.4.1 Dipping

One common application method is dipping, which utilizes a semipermeable covering by dipping a sample into a coating solution for a set amount of time. This coating resists moisture loss and gas exchange. To employ the dipping method, first, the sample is submerged in a coating and cross-linking solution for a predetermined amount of time (Pandey et al., 2022). The time food must remain submerged in the solution ranges anywhere from 30 seconds to 5 minutes. Once sufficient time has elapsed, the sample is removed from the solution and excess solution is allowed to drain. With this method, the sample gets evenly coated with the protective solution even if the surface of the food is uneven or rough.

10.4.2 Spraying

Like dipping, spraying coats food samples with a semipermeable membrane. This method uses nozzles to spray coating solutions onto foodstuffs in the form of droplets. Pressures from spraying systems can reach from 60 to 80 psi. These high pressures allow for small amounts of coating solution to be used. Equal distribution of coating is one advantage to this method, although unlike dipping, some portions of the sample being treated may be missed.

Spraying can also be used to coat packages where food will be placed. Researchers have sprayed PET trays with α-tocopherol, a citrus fruit extract. Once it was sprayed, the antioxidant activity of the package was assessed. Findings from this study indicated that lipid oxidation in the packages coated with α-tocopherol was significantly lower than trays that had not been coated (Anwar et al., 2018).

10.4.3 Spreading

Another method to incorporate an active compound into packaging materials is spreading. This method spreads a thin layer of solution onto a film that is then allowed to dry. The layer containing the active compound is then encapsulated into the film. This film can then be applied to a sample. Previous studies have used this method to produce antimicrobial films using oleoresin extracted from chili peppers for packaging (Anwar et al., 2018).

10.4.4 Vacuum Impregnation Technique

Vacuum impregnation is a packaging technique that relies on diffusion. With this method, mass transfer occurs because of mechanically induced pressure differences. Injection of an external solution into the tissue of fruits and vegetables is accomplished with the aid of a vacuum. Typically, the compounds introduced to the sample target the intracellular spaces and capillaries to change or improve the food's properties. Vacuum impregnation can modify physicochemical properties and increase nutritive value via addition of probiotics or micronutrients. Additional advantages include longer shelf life or modification of sensory attributes.

Vacuum impregnation is a two-step process. The first step in the process consists of a phase where pressure is reduced. After a reduced pressure, the sample is brought back to atmospheric pressure (Pandey et al., 2022). Material becomes impregnated during this process due to hydrodynamic mechanism (HDM) and deformation–relaxation phenomena, leading to the filling of intracellular capillaries (Radziejewska-Kubzdela et al., 2014).

10.4.5 Layer-by-Layer Method

In the layer-by-layer methods, layers are formed onto the sample one by one. Each layer is either physically or chemically bonded together, forming a final multilayer film. Thin films can be developed on foods with this method. Additionally, multilayer edible coatings and biopolymer-based films can be created on food with this technique. A main advantage to this technique is controlled release of certain compounds to the food packaged, extending its shelf life. In one study, by enclosing sorbic acid in a multilayer film, researchers were able to control the release rate of this compound, thereby extending its antimicrobial effects (Jipa et al., 2012).

10.4.6 Foaming

Foams have a great versatility in the food packaging industry and can be used as an edible material or as packaging material. Traditionally, foams were non-biodegradable, synthetic, and polymer based. Foams in this category are petroleum-derived and include polyethylene, polystyrene, polyurethane, and polypropylene. However, alternatives to non-biodegradable foams have come into the market. Biodegradable foams are more environmentally friendly and degrade into carbon dioxide, water, and inorganic compounds. Biodegradable alternative foams can be produced from plant polymers and be used in conjunction with EOs to increase food longevity. Foams begin in a wet stage, in which they are applied to samples. Once applied in the wet stage, the foam is them transitioned to a solid state by cooling, heating, or curing. Some examples of plants that can be used to create foams are potatoes, corn, cassava, and oca (Jarpa-Parra & Chen, 2021). Plant polymers from these plants can be used to create solid foams.

10.4.7 Thin-Film Hydration

Encapsulating Eos is necessary to prolong their bioactive and antimicrobial effects due to their volatile nature. Liposomes are one way that the food industry can encapsulate Eos. Liposomes are sphere-shaped vesicles that consist of at least one phospholipid bilayer enclosing an aqueous or lipid-soluble material. The thin-film hydration method can produce liposomes. With this method, a solvent is removed from a lipid solution, leaving behind a thin film. Addition of a dispersion medium to this film spontaneously forms liposomes (Mukurumbira et al., 2022). While this is a fairly simple method for encapsulating EOs, it is unsuitable for large-scale production (Sherry et al., 2013). Additionally, solvent residues are left behind during this process.

10.5 Types of Essential Oil Films/Coatings

Essential oils have numerous potential applications in films and coatings that can be further utilized by medium of coating used. EOs can turn a film or coating into an active packaging material with antimicrobial properties (Hassan et al., 2018) and decrease the oxidation rate of the coated food, increasing the shelf life of said product (Ju et al., 2019). Typical coating mediums are polysaccharide, lipid, protein based, with a current trend both in research and industry toward the use of composite coating mediums (Hassan et al., 2018).

10.5.1 Polysaccharide-Based Essential Oil Films/Coatings

Polysaccharide-based EO films are especially useful in the creation of films to extend shelf life (Cazón et al., 2017). The majority of polysaccharide films are easily processed and formed into the desired shape and thickness. They also promote high levels of bioavailability for EOs and other compounds added to the film. Because polysaccharides are generally hydrophilic, they do have a number of undesirable qualities for a food film. These include high water solubility, poor barrier properties (especially water vapor permeability), and low extendibility. The inherent properties of most polysaccharide films make them unsuitable for applications involving a liquid product (Kong et al., 2022).

There are several commonly used polysaccharides in edible films and coatings, which can be grouped into starch and non-starch polysaccharides (Anis et al., 2021). Starch-based coatings are composed primarily of amylase, a linear polymer, and amylopectin, a branched polymer. Cornstarch is typically around 25% amylose and 75% amylopectin, while modified cornstarches are up to 85% amylose (Hassan et al., 2018). In order to create a film, the starch must be gelatinized into a homologous mixture. Starch films and coatings have very high oxygen barrier properties, which is highly desirable in many applications and influenced by the ratio of amylopectin to amylose (Cazón et al., 2017). Following proper processing and application, starches will also be tasteless, colorless, and odorless (Hassan et al., 2018). Due to their hydrophilic nature, starch films are susceptible to moisture and relative humidity of the environment. This can be controlled by the crystalline structure of the film (Cazón et al., 2017). Examining the applications of a starch-based film with EO, rosemary EO extract can be embedded into cassava starch films to increase antioxidant activity and the UV-resistant properties of the film (Piñeros-Hernandez et al., 2017).

Another common polysaccharide-based coating comes from cellulose and its derivatives, which include methylcellulose (MC) and hydroxypropyl methylcellulose (HPMC) (Hassan et al., 2018). Cellulose is commonly used due to the ease with which it can be sourced, either from plants or bacteria. Cellulose is inherently insoluble in water but can be chemically treated to increase solubility, or MC and HPMC can be used. MC and HPMC have very poor water barrier properties but do have high oxygen, carbon dioxide, and lipid barrier properties, which make these compounds ideal for coating fried products.

HPMC coatings infused with oregano EO have been shown to increase shelf life, storage quality, and sensory attributes following storage of plums, by decreasing the total microbial growth, decreasing weight loss, and increasing firmness (Choi et al., 2016).

Chitosan is another commonly used polysaccharide in edible coatings and films. Chitosan is the deacetylated form of chitin, containing N-acetyl-d-glucosamine units (Hassan et al., 2018). Chitin is readily available from the exoskeletons of insects and crustaceans, making it an affordable and renewable option. Chitosan itself is desirable for its antimicrobial abilities. The solubility of chitosan varies and depends on the degree of acetylation and the molecular weight, being insoluble in water and organic solvents but soluble in acidic solutions. The applications of chitosan are limited by its properties under heat. Chitosan degrades before its melting point, making it unsuitable for high-heat applications and heat-sealing a film. Like other polysaccharide films and coatings, a chitosan-based coating has high oxygen and carbon dioxide barrier properties but poor water vapor barrier properties (Cazón et al., 2017). Chitosan is a common coating for fruits and vegetables as a microbiological barrier on strawberries, cucumbers, and more or as a gasoline barrier for apples, peaches, and pears (Richardson & Meheriuk, 1982).

Chitosan films with marjoram and basil EO were successful in inhibiting the growth of *E. coli* and *S. aureus*. The dosage of the EO was found to balance the chitosan, as all had measurable antimicrobial benefits but negative organoleptic qualities (De Souza et al., 2020). Studies have also shown antimicrobial and antioxidant properties of chitosan films with clove EO and nisin on ground pork patties. The shelf life was prolonged through decreased microbial activity and quality loss (Venkatachalam & Lekjing, 2020). Another application of a chitosan-based film is the inclusion of *Mentha piperita* or *M. villosa*, commonly known as peppermint and mojito mint, onto table grapes as an antifungal application without negative organoleptic effects (Guerra et al., 2016). The use of oregano EO has shown antimicrobial benefits in the reduction of *Listeria monocytogenes* and *Yersinia enterocolitica* during the storage of cooked chicken products (Ju et al., 2019). The application of a film with lemon EO on strawberries caused decreased respiration rates and increased antifungal abilities (Perdones et al., 2012). As evidenced by the variety of applications, on fruits, vegetables, and meat products, chitosan films and coatings are highly applicable.

Alginate films are made from the brown seaweed family, *Phaeophyceae*. Alginates are salts formed from mannuronic acid and guluronic acid (Cazón et al., 2017). The ratio of mannuronic and guluronic acid is highly important to the properties of the film. Alginate forms when cations (Ca^{+2}) interact with guluronic acid to form a three-dimensional network. The chain pattern of mannuronic and guluronic acid, the molecular weight, and the size of the calcium–guluronic blocks affect the properties of alginate (Hassan et al., 2018). Alginate has differing abilities than many polysaccharide films, especially as a colloid. Like other polysaccharide films, alginate has high oxygen and carbon dioxide barrier properties (Kong et al., 2022).

Alginate is used by itself and with other film-forming polysaccharides. Alginate-only-based films have been infused with cinnamon, palmarosa, clove, and lemongrass EO to increase shelf life and extend organoleptic properties, as well as improve the antimicrobial properties of coated Fuji apples and fresh-cut melons (Raybaudi-Massilia et al., 2008a, 2008b).

Carrageenan is produced from the galactan groups of red seaweed. Carrageenan produces useful films and coatings due to its gel-forming properties. There are multiple subtypes of carrageenan based on the 3,6-anhydro-d-galactose, with slightly different abilities (Kong et al., 2022). With a poor vapor barrier property, carrageenan is similar to many polysaccharide films but is able to be applied in a thicker coating which can provide fleeting protection against water loss (Cazón et al., 2017). A study has shown the potential for the inclusion of cinnamon EO in carrageenan films to increase the barrier and mechanical properties of the product (Praseptiangga et al., 2016).

While the application of a single polysaccharide can form a usable and effective film or coating, multiple polysaccharides are often used in conjunction with EOs to increase the film's properties and enhance the positive benefits of the EOs.

10.5.2 Lipid-Based Essential Oil Films/Coatings

Lipids provide a useful source of edible films and coatings. Typically made from waxes, acyl glycerol, or fatty acids, lipid coatings are hydrophobic or amphiphilic, which causes different functional properties than polysaccharide coatings. This allows the coatings to decrease the negative effects of oxygen, light, and more when infused with EOs, in addition to lowering the water transmission rate (Ju et al., 2019).

One potential base for a lipid-based coating are liposomes. Included in liposomes are phospholipids, which are amphiphilic, with high surface activity (Fathi et al., 2012). This makes them a highly biocompatible material to use in a film or coating, allowing for the protection and controlled release and release of active compounds in the film. Some important functional properties include wettability, the ability for a fluid to spread and remain on a surface, and their amphiphilic ability to encapsulate both hydrophobic and hydrophilic active ingredients (Shishir et al., 2018). Included in phospholipids are the common food ingredients of lecithin, including soy and dairy (Pandey et al., 2022).

Currently used or researched uses for phospholipid coatings with EOs include spray-drying products with a mixture of soy lecithin and clove oil. This has shown an increased ability to encapsulate clove oil and increased antimicrobial effect against *E. coli* and *L. innocua* (Talón et al., 2019). Other researched uses include the encasing of livestock drugs in lecithin with lavandin EO, obtained from lavender, to increase drug efficacy. The lavandin provides additional antimicrobial benefits, while the lecithin film protects the lavandin and increases the ability of the lavandin to remain bioavailable (Varona et al., 2011). The use of thymol and carvacrol, a major constituent of thyme, isolated from *Origanum dictamus*, showed increased antimicrobial abilities against Gram-positive, Gram-negative, and pathogenic fungi when incorporated with phosphatidylcholine-based liposomic films. Antioxidant ability was also greatly increased in these samples (Liolios et al., 2009). Eucalyptus EO has shown potential antifungal properties when entrapped in a liposome-based film. The potential from this research allows additional exploration for the combination of EOs and liposomes for food films and coatings (Moghimipour et al., 2012).

Waxes have very useful applications, especially when incorporated with EOs. Commonly used waxes include paraffin, carnauba, candelilla, and beeswax. They have the ability to greatly increase water barrier ability while also maintaining high barrier properties for oxygen and carbon dioxide. Wax coating can also improve the exterior appearance of the coated product but may decrease the glossiness or appearance of others. With the addition of EOs, waxes can also have antimicrobial benefits. Because of waxes' properties, they can be applied in much thicker coatings than other materials. These uses can be seen in products such as cheese. When applied in these thick coatings, they are often not consumed. However, when applied thinly, wax coatings are still considered edible and have many uses in the food industry (Hassan et al., 2018).

Applications of waxes with EOs have mostly been investigated for use on fruits and vegetables. A study showed the coating of brown rice with paraffin wax and peppermint essential oil. This coating extended the shelf life of the rice during long storage by not decreasing the physical or organoleptic properties of the rice, while inhibiting the growth of molds, specifically *Aspergillus flavus*, during storage (Chaemsanit et al., 2019). Another product with potential increased shelf life and organoleptic properties when coated with carnauba wax and ginger EO is papaya. While the results were not as conclusive for antimicrobial properties, there were still potential applications of the coating (Miranda et al., 2022). When lemongrass EO was combined with carnauba wax and applied to plums, there were definite benefits seen. Ethylene production and weight loss were reduced; the coating had an antimicrobial effect versus *Salmonella* and *E. coli*, without altering the appearance or flavor of the plum (Kim et al., 2013).

10.5.3 Protein-Based Essential Oil Films/Coatings

Numerous proteins have applications as films and coatings for food. The sources are primarily plant proteins, such as soy, gluten, and zein, but also include animal proteins: gelatin and milk proteins, casein, and whey (Ju et al., 2019). Most of the proteins used in coating and film applications are globular and denatured prior to use. The bonding nature of proteins makes these coating highly effective at decreasing the oxygen transmission rate, regardless of relative humidity (Hassan et al., 2018). Plant proteins are often more hydrophobic than animal proteins, expanding their applications (Shishir et al., 2018). The limit of use for protein films and coatings is caused by being hydrophilic, which creates a low water vapor barrier property and high sensitivity to moisture. These drawbacks can be countered by combining them with other film-forming polymers (Seifari & Ahari, 2020).

The milk proteins of whey and casein are highly applicable coatings. Both are excellent emulsifiers, particularly for oil in water emulsions (Pandey et al., 2022) and carriers of bioactive additives, such as

EOs. Casein is an especially good coating due to its hydrophobic nature, carrying EOs and other active ingredients with hydrophobic tendencies with ease (Shishir et al., 2018).

Applications of whey and casein are broad due to their functional properties. Whey protein with clove EO was shown to have high encapsulation of the clove EO, allowing for strong antioxidant and antimicrobial activity (Talón et al., 2019). Another study examined the ability of whey protein films containing eucalyptus, cinnamon, and rosemary EOs and found that the EOs were able to migrate into the salami, potentially increasing the antioxidant and antimicrobial properties of the salami. Ideally, these properties could also be proven in other food products (Ribeiro-Santos et al., 2017). Again, using cured sausage, *Organum virens* EO, from a close relative to oregano, was included in a whey film on traditional Portuguese sausages. Higher acidity, color protection, and decreased lipid oxidation were observed, in addition to an extended shelf life due to lower microbial activity (Catarino et al., 2017). Many of the applications of casein involved the inclusion of either lipid or polysaccharide film-forming agents.

Gelatin is a film-forming protein obtained from animal sources. Gelatin films are sought after due to their low cost of production, useful functional properties, and ability to not alter the appearance of foods (Hassan et al., 2018). Gelatin is highly bioactive and biocompatible. It is also very good at retaining water within a product but also becomes very soluble in aqueous solution. This negates its water-retention abilities, depending on the application (Shishir et al., 2018). Ideal EOs to be encapsulated in gelatin films are those sensitive to pH changes or extremes, hydrophobic, and thermosensitive EOs (Dajic Stevanovic et al., 2020).

Gelatin films are versatile in their applications. One study examined the use of bergamot EO and lemongrass EO. They were both effective antimicrobial films and improved heat stability and water vapor transmission (Ahmad et al., 2012). To improve the shelf life of shrimp, orange leaf EO was included in a gelatin film. This combination increased antimicrobial activity, without negative organoleptic effects, to extend the shrimp's shelf life (Alparslan et al., 2016). The addition of clove and ginger EO to tuna skin gelatin had positive effects on the physical and functional properties of the skin, creating a potential active film packaging system to be further explored (Ningrum et al., 2021).

The primary protein in corn is zein, a hydrophobic prolamin. Most films from zein are formed using an aqueous solution with ethanol as the base. One benefit of a zein coating is their moisture barrier abilities (Hassan et al., 2018). The solubility of zein in different substances makes it extremely useful for the delivery of functional ingredients, such as EO, as well as provides EOs protection. This allows the functional properties of the EO to remain bioactive longer (Shishir et al., 2018). Having antioxidant and antimicrobial benefits itself, the addition of EOs to zein films magnifies these properties (Moradi et al., 2016).

Like other film matrixes, zein is often used in composite to amplify or rectify functional properties and shortcomings. Rose hip EO infused into zein is one example of a purely zein coating. Studied for use on bananas, the EO film showed potential for improving shelf life, therefore reducing food waste of fruits and vegetables (Yao et al., 2016). *Zataria multiflora* Boiss. EO added to a zein film was studied for use on minced meat. *Zataria multiflora* Boiss. is a plant native to Iran and Pakistan with similarities to thyme. The film showed decreased growth of *L. monocytogenes* and *E. coli* plus increased antioxidant abilities to increase shelf life and food safety (Moradi et al., 2016).

Soy protein has similar properties to the other discussed plant proteins. Research has shown that soy protein is more effective when combined with polysaccharides as the majority of the film. When EOs are added, the functional properties of the film have the potential to be greatly increased (Shishir et al., 2018).

Soy protein has been researched with multiple EOs and for multiple food applications. One such study examined the benefits of cinnamon and clove EO incorporation into a soy protein film. There were organoleptic benefits in addition to the typical shelf-life benefits with the addition of EOs (Atarés et al., 2010).

10.5.4 Composite-Based Essential Oil Films/Coatings

Composite films and coatings allow the positive functional properties of the matrixes to interact, potentially even enhancing them. The combination of a composite film or coating plus the addition of EOs offer many functional property advantages, both physical and organoleptic. This is likely an area of research, especially for direct application into industry, where further discussion will occur. Some of the composite films, such as polysaccharide–polysaccharide, polysaccharide–lipid, and polysaccharide–protein, with essential oil coatings are listed in Table 10.4.

TABLE 10.4

Composite-based essential oil films/coatings

Film matrix	Essential oil	Application description/Benefits	Source
Polysaccharide–polysaccharide			
Guar gum, gum arabic	Peppermint	Investigated remaining bioactivity of the peppermint oil	(Sarkar et al., 2013)
Pectin and alginate	Lemon	Reduced raspberry weight and color loss, Reduced microorganism growth	(Guerreiro et al., 2015)
HPMC and chitosan	Bergamot, lemon, tea tree	Decreased water vapor solubility and surface charge of the film	(Sánchez-González et al., 2011)
Polysaccharide–Lipid			
Carnauba HPMC	Ginger	Slowed ripening process, decreased respiration, maintained color to improve papaya shelf life	(Miranda et al., 2022)
Polysaccharide–Protein			
Maltodextrin and whey	Rosemary	Structure of maltodextrin and whey maintained the bioactivity of the rosemary EO with potential to be used in food coatings	(Turasan et al., 2015)
Gum arabic and gelatin	Peppermint	Elongated bioactive window for the peppermint EO in the film	(Dong et al., 2007)
Chitosan and gelatin	Oregano	Inhibits *S. aureus* and *L. monocytogenes* in sweet potato products	(Hosseini et al., 2015)
	Rosemary	Extended shelf life of carbonado chicken through antimicrobial and antioxidant properties, with potential for other RTE meat products	(Huang et al., 2020)

10.6 Effect of Incorporating EOs on Properties of Films/Coatings

Microbial growth on food surfaces results in food spoilage and lowered nutritional value of food due to potential pathogenesis. Excessive amount of synthetic packaging and preservatives causes a massive impact on the environment, such as aquatic and terrestrial pollution and persistence, greenhouse gas emissions, etc. (Atiwesh et al., 2021), and human health, such as disruption of thyroid function, carcinogenicity, etc. (De Groef et al., 2006), implementing a serious demand of spearheaded research into the development of biological polymers made from a variety of agricultural and plant-based commodities as an alternative to conventional plastic films and coatings (Pelissari et al., 2019). Naturally occurring compounds, such as essential and vegetal oils, have gained increased attention for their antimicrobial activity and modest toxicity to nontarget species, including humans, and for possessing antioxidant potential. Essential oils (EOs) have shown antimicrobial activity against various food and spoilage pathogens. Due to their hydrophobic, volatile nature and complex composition, EOs interfere in the functioning of the lipid cell membrane, cell wall, membrane proteins to disrupt proton motive force, active transport, coagulating the cell contents of various pathogens (Burt, 2004; Figueroa-Lopez et al., 2020; Ju et al., 2018; Xing et al., 2019).

Most antimicrobial active compounds in EOs are phenolics, oxygenated terpenoids, and aldehydes, along with quinines, tannins, flavonoids, saponins, coumarins, and alkaloids (Savoia, 2012). Recent studies have examined the implications of EOs from sources including oregano, thyme, basil, cinnamon, sage, coriander, ginger, rosemary, garlic, cumin, nutmeg, turmeric, clove, mace, citrus, and fennel, used either solitary or in amalgamation with other EOs in order to improve the taste and smell and extend the life span of food products (Goulas & Kontominas, 2007; Kaul et al., 2003; Mustafa et al., 2020; Negi & Jayaprakasha, 2001; Oussalah et al., 2006; Takma & Korel, 2019). Incorporation of EOs in active food packaging materials such as chitosan, cellulose, and zein (Arcan & Yemenicioğlu, 2011; Cran et al., 2010; Kanatt et al., 2012) can aid in releasing the bioactive constituents in a more controlled process into the food, increasing the shelf life of the final product by increasing inhibition of bacterial growth.

However, incorporation of EOs can fundamentally influence the functioning of active food packaging compounds. Here, we discuss the differences in properties after EOs are assimilated.

10.6.1 Mechanical Properties

Mechanical and physical properties of food packaging material depend on the polymer matrix and EO interaction. These properties are estimated through elongation at break (EB), tensile strength (TS), and Young's modulus (YM). These toughness parameters essentially evaluate strength, rigidity, the rate at which the extension of the packaging material will break, and the strength versus time or distance parameters (Atarés & Chiralt, 2016). Multiple previous studies published on tensile strength of EO-incorporated films and coatings show an overall decrease in strength of the polysaccharide and protein films in the presence of EOs, tabulated in Table 10.5. Incorporation of EOs into polymer matrix effects a discontinuous, heterogenous film structure, replacing the stronger polymer–polymer synergy with fragile polymer–oil interactions in the matrix, thereby reducing the overall tensile strength (Shojaee-Aliabadi et al., 2013; Zinoviadou et al., 2009). This is evidenced by studying basil EO– or thyme EO–incorporated chitosan, where close-contact polymer–oil interaction produced a more fragile film with increased stretchability due to reduction in chain interaction forces (Bonilla et al., 2012). A similar effect is seen in citrus oil–incorporated fish skin gelatin edible films, where the presence of the EO reduced interaction between the gelatin molecules (Tongnuanchan et al., 2012). Contrastingly, an Atarés study demonstrates that the consolidation of cinnamon EO in soy protein isolate films increased tensile strength (Atarés et al., 2010b). This increase in strength could be attributed to the activated displacement of polymer matrix caused by the EO. EOs contain phenolic compounds in high concentrations. Phenols have the ability to interact with multiple protein site, resulting in more protein crosslinking (Limpisophon et al., 2010).

Along with tensile strength, changes can be observed in elongation break, which is a metric of film's ability to stretch at break, quantified in percent. This parameter is influenced by the type and proportion of EO used in the film (Pranoto et al., 2005). For example, an increase in stretchability is observed in *Satureja hortensis* EO–incorporated K-carrageenan film even in small quantities of EO, whereas *Zataria multiflora* EO–incorporated chitosan films had an increase in pore size, creating possible rupture points, resulting in the reduction of film stretchability (Moradi et al., 2012; Shojaee-Aliabadi et al., 2013). Young's modulus measures the ease of stretchability; an increase in the modulus indicates less ability to bend under applied force (Castro et al., 2021). Infusion with phenolic compounds demonstrably leads to significant increase in Young's modulus, providing resistance against plastic deformation (Xue et al., 2021).

10.6.2 Barrier Properties

Prevention of moisture loss to preserve food quality of the food product is an important criterion that determines packaging for moist products. Alternatively, prevention of moisture entering dry food products is also essential for determining the type of packaging to be used (Pranoto et al., 2005). Water vapor transmission rate (WVTR) is an indicator of permeability of moisture across packaging material. The moisture transmission in films and coatings depends on the hydrophilic-to-hydrophobic ratio of film composition, and the incorporation of mostly hydrophobic EO into film improves water vapor barrier properties, as seen in the incorporation of tea tree, thyme, cinnamon, and ginger EOs in HPMC, hake protein, and sodium caseinate films (Table 10.6) (Atarés et al., 2010a; Pires et al., 2013; Sánchez-González et al., 2009). While EOs are expected to reduce WVTR for better preservation of the product, there are studies on EOs-incorporated films that demonstrate no difference or impair the moisture barrier. Addition of oregano and garlic oil EOs did not significantly affect whey and fish protein films, respectively (Teixeira et al., 2014; Zinoviadou et al., 2009). Incorporation of cinnamon, ginger, bergamot, lemongrass, and garlic oil EOs in soy protein, HPMC, gelatin, and alginate edible films further enhance the function of the films (Ahmad et al., 2012; Atarés et al., 2010b, 2011; Pranoto et al., 2005). Addition of EOs to films can contribute to extended intermolecular interactions and molecular mobility of structural matrix, producing a heterogenous film matrix, contributing to reduction in WVTR.

TABLE 10.5

Mechanical properties of various essential oil–infused films or coatings

Film/coating	EO	TS (MPa)	EB (%)	YM (MPa)	References
Chitosan	Thyme	32.94 ± 3.32	1.8 ± 0.22	19.32 ± 2.44	(Altiok et al., 2010)
HPMC	Ginger	41 ± 13	6 ± 3	—	(Atarés et al., 2011)
HPMC	Bergamot	39 ± 3	2.9 ± 0.7	444 ± 75	(Sánchez-González et al., 2011)
HPMC	Lemon	40 ± 4	3.9 ± 0.4	397 ± 139	(Sánchez-González et al., 2011)
HPMC	Tea tree	34 ± 5	4.2 ± 0.2	365 ± 124	(Sánchez-González et al., 2011)
Chitosan	Bergamot	50 ± 8	6 ± 2	747 ± 225	(Sánchez-González et al., 2011)
Chitosan	Lemon	37 ± 3	6.4 ± 0.2	954 ± 113	(Sánchez-González et al., 2011)
Chitosan	Tea tree	54 ± 5	8 ± 2	652 ± 157	(Sánchez-González et al., 2011)
Alginate	Garlic	38.67	2.73	—	(Pranoto et al., 2005)
κ-carrageenan	*Satureja hortensis*	9.52 ± 0.94d	44.77 ± 3.49	—	(Shojaee-Aliabadi et al., 2013)
Hake protein	Thyme	4.33 ± 0.94	111.2 ± 44.8	—	(Pires et al., 2011)
Fish protein	Clove	7.3 ± 2.3	53.3 ± 21.1	—	(Teixeira et al., 2014)
Fish protein	Garlic	6.6 ± 2.7	55.7 ± 31.7	—	(Teixeira et al., 2014)

The most optimum barrier characteristic for food packaging is dependent on a balance between WVTR and oxygen permeability (OP). OP is impacted by relative humidity and temperature, as diffusion of molecules through the film/coating depends on the fluctuation of temperature and moisture content, which can greatly influence the formulation of edible films/coatings to reduce oxidation of lipids (Miller & Krochta, 1997). EO addition increases OP due to the hydrophobic nature of EOs, such as ginger EO–incorporated HPMC films having a significant OP increase (Atarés et al., 2011).

10.6.3 Optical Properties

Optical properties have an economic impact on product appearance due to consumer desirability (Sivarooban et al., 2008). The type of EO incorporated determines the color changes to the film. This can be demonstrated by addition of tarragon and coriander EO to hake protein films, which made the films appear more yellow than the control (Pires et al., 2013), while no significant color changes were observed when garlic EO was incorporated in alginate films or thyme EO was incorporated into hake protein films (Pires et al., 2011; Pranoto et al., 2005). Transparency of food packaging material is measured using the light transmittance either by the "transparency value," which is quantified by dividing the absorbance measured at 600 nm by thickness of the film or coating (Shiku et al., 2004) or by using the Kubelka-Munk multiple scattering theory, calculated by collecting internal transmittance (T_i) (Hutchings, 2011). Higher "transparency value" indicates lower transparency, whereas an increase in internal transmittance indicates increase in transparency. Sánchez-González et al. (2009) observed transparency reduction caused by addition of tea tree EO in HPMC films, attributed to light scattering due to the difference in refractive index of EO droplets (Sánchez-González et al., 2009).

Film thickness, a parameter influencing film transparency, shows a general increase in the EO-incorporated films over the controls. An increase in thickness was observed on incorporation of mint EO in gelatin edible coatings, which was ascribed to interfered ability of the peptide chains in gelatin to produce a dense film matrix due to infusion of oil droplets or an increase in solid content from the EO itself, as suggested by Ahmad et al. (2012) and Tongnuanchan et al. (2015), where similar results were observed for multi-EO blend–incorporated fish gelatin films formulated at different concentrations. Similarly, ginger EO–incorporated cornstarch-based films or oregano EO–incorporated sweet potato starch films also show a significant increase in thickness (Li et al., 2018; Wang et al., 2022). *Gloss* is defined as the luminous flux reflected from the film to luminous flux reflected of a surface similar to the film. Morphology of the film or coating surface during the film drying stage determines the final gloss, and addition of EO generally results in a decrease in gloss, due to composite heterogeneity and surface roughness (Ward & Nussinovitch, 1996). This was confirmed in bergamot, tea, or lemon EOs

TABLE 10.6

Optical and barrier properties of various EO-infused films and coatings

Films/ Coatings	EO	WVTR ($\times 10^{10}$ g/min. cm.kPa)	OP (cc/ m²/day)	Thickness (µm)	Color	Gloss (60)	T_i (%)	References
Chitosan	Thyme	8.10	4.61	27	White	—	—	(Altiok et al., 2010)
HPMC	Ginger	17	122 ± 15	62 ± 14	Clear	8 ± 2	83.7 ± 1.9	(Atarés et al., 2011)
HPMC	Bergamot	31 ± 4	—	31.1 ± 1.7	—	11 ± 2	0.789 ± 0.035	(Sánchez-González et al., 2011)
HPMC	Lemon	41 ± 6	—	25.6 ± 2.8	—	8.2 ± 1.2	0.842 ± 0.005	(Sánchez-González et al., 2011)
HPMC	Tea tree	57.3 ± 1.5	—	22.3 ± 1.2	—	16 ± 3	0.83 ± 0.02	(Sánchez-González et al., 2011)
Chitosan	Bergamot	92 ± 9	—	36 ± 3	—	8.8 ± 1.5	0.744 ± 0.012	(Sánchez-González et al., 2011)
Chitosan	Lemon	77 ± 3	—	36 ± 3	—	9.9 ± 1.8	0.782 ± 0.009	(Sánchez-González et al., 2011)
Chitosan	Tea tree	74.7 ± 1.8	—	24.3 ± 1.2	—	5.7 ± 1.2	0.789 ± 0.013	(Sánchez-González et al., 2011)
Alginate	Garlic	30.89	—	—	Yellowish	—	—	(Pranoto et al., 2005)
κ-carrageenan	*Satureja hortensis*	0.556 ± 0.032	—	68	Yellowish	15	7.35 ± 0.84	(Shojaee-Aliabadi et al., 2013)
Hake protein	Thyme	5.9	—	20	Clear	—	—	(Pires et al., 2011)
Fish protein	Clove	3.8 ± 0.9	—	12.3 ± 4.7	Darker, yellowish, green	—	7.0 ± 3.3	(Teixeira et al., 2014)
Fish protein	Garlic	4.3 ± 1.0	—	12.8 ± 7.0	Darker, yellowish	—	22.9 ± 2.3	(Teixeira et al., 2014)

incorporated into chitosan and HPMC films (Sánchez-González et al., 2009, 2011), *Satureja hortensis* EO into carrageenan films (Shojaee-Aliabadi et al., 2013), and cinnamon EO into soy films (Table 10.6) (Atarés et al., 2010b).

10.6.4 Chemical Properties

Chemical properties of films that infuse EOs are quantified using Fourier transform infrared spectroscopy (FTIR), which allows for examination and identification of functional groups present in the film matrix. Increase in EO concentration could cause a shift in FTIR bands likely due to altered molecular interactions between EOs and polymer matrix, which can be seen in *Zataria multiflora* EO– and *Rosmarinus officinalis* EO–infused waterborne polyurethane films (Hedayati Rad et al., 2018). Additional chemical parameters that determine the suitability of EOs added to packaging film are migration of active compounds, flavors, and aromas into the food product and antioxidant properties of the EO-enhanced packaging.

10.6.4.1 Migration of Active Compounds

Active packaging, where the active components, such as antimicrobial and antioxidant components, CO_2 emitters, and oxygen and ethylene scavengers, dispense extraneous activities, encourages the interaction between the packaging and food product. Migration of active constituents is measured using chromatographic methods. Eucalyptol, the active ingredient of rosemary EO, migrated the most out of the other active compounds in a whey film infused with a combination of essential oils. During the same study, it was observed that the migration rate of active compounds is faster at higher concentrations of EOs and higher ambient temperatures (Ribeiro-Santos et al., 2017c).

10.6.4.2 Flavors and Aroma Transfer

Flavors and aromas are usually added to the food product to enhance the product; however, unintended flavor and aroma transference from the EO-incorporated active packaging can impart an unwanted taste or odor to the food product. Nanoencapsulating EOs after film drying would be a useful technique in reining in the impact of EOs on the food product as well as improving solubility in water-soluble polymers (Gupta et al., 2016). Nanoencapsulated cinnamon EO included into polyvinyl alcohol nonfibrous films that masked the intense flavor of the EO while still retaining the antibacterial properties has been successful in extending the life of perishable fruits (Ait-Ouazzou et al., 2011). Nanoesncapsulation favors incorporation of lipids from EOs in polymer matrices by creating more interactions between hydrophobic EO and hydrophilic film matrix. This prevents alteration of the microstructure of the packaging material (Buendía-Moreno et al., 2019; Ribeiro-Santos et al., 2017a). *Nanoencapsulation* refers to the capsule sizes between 0.05 and 1 μm. Additionally, larger capsulations, macrocapsules, and microcapsules, between 0 and 0.5 μm, are available (Mohsenabadi et al., 2018).

10.6.4.3 Antioxidant Properties

Oxidation can happen during storage and processing and alters both taste, smell, and nutritive attributes of the food products, ultimately causing deterioration (Viuda-Martos et al., 2011). Lipid oxidation imparts rancid emanations and flavors, reduction in nutritional quality, and potential toxin production in the packaged food products (Yanishlieva et al., 2006). Owing to the plentiful presence of terpenoids and phenolic acids, EOs have antioxidant properties, taking place in two mechanisms: promotion of oxygen barrier capacity and specific antioxidant action after diffusion to the coated product (Bonilla et al., 2013).

Multiple methods can be used to quantify antioxidant activity. Primitive methods to measure the action of phenolic antioxidants utilize 2,2-diphenyl-1-picrylhydrazyl (DDPH) radical. Antioxidants are measured as antiradical efficiency, which is the antioxidant quantity essential for 50% depreciation in initial DDPH concentration measured by a spectrophotometer at 515 nm and the time needed to obtain DDPH equilibrium (Frankel & Meyer, 2000). Green tea EO infused in chitosan films and *Zataria multiflora* EO–infused carboxymethilcellulose films have been tested for antioxidants using this method (Dashipour et al., 2015; Siripatrawan & Harte, 2010). Additionally, the same studies found an increase in antioxidant activity with increasing EO concentration. Another measure of antioxidant activity is the 2-thiobarbituric acid assay (TBARS), which measures lipid oxidation, especially for fish and meat products (Irwin & Hedges, 2004). Reactive substances from thiobarbituric acid are formed when peroxides are oxidized into ketones and aldehydes (Ramezani et al., 2015). Socaciu et al. (2021) demonstrate that fish samples encapsulated with tarragon EO infused with whey protein isolate (WPI)–based films are less susceptible to lipid oxidation (Socaciu et al., 2021).

10.6.5 Thermal Properties

Thermal stability is an important parameter when considering addition of EO to packaging films and coatings, especially considering active packaging. Thermal stability is measured through thermogravimetric assay (TGA) and derivative thermogravimetric analysis (DTG) conducted in thermogravimetric analyzer. Both TG and DTG thermograms elucidate a film's thermal degradation (Xu et al., 2018). TGA indicates the correlation between loss of mass and temperature changes, whereas DTG characterizes the rate of weight loss during heating. Addition of EO overall increased thermal stability of the packaging film, as seen in cinnamon EO–infused chitosan (Xu et al., 2018), bergamot EO in gelatin films (Ahmad et al., 2012), and rosemary EO–infused polyethylene films (Dong et al., 2018). This increase in thermal balance can be ascribed to a stronger matrix network due to the presence of dispersed EO particles. Furthermore, these studies also demonstrate an inverse relationship between concentration of EO and thermal stability; a higher concentration makes the film less thermally stable. This could be due to a disruption of the film matrix, resulting in reduced compatibility of the multicomponent system that caused a reduction in the heat resistance of the film or coating (Noshirvani et al., 2017).

10.6.6 Morphological and Structural Properties

Morphological and structural properties of edible or active films or coatings that have an EO infused are based on qualitative detection of component disposition into the film/coating matrix, usually determined using scanning electron microscopy (SEM). SEM utilizes electron beams to scan the structure of the EO-incorporated film and compare to control film. The film is infused with EOs by applying homogenization techniques, which form an emulsion of EO droplets into the polymer matrix at continuous aqueous phase. EO droplets persist suspended in the film microstructure as detected by microscopic approaches mentioned previously. Destabilization of the matrix due to droplet flocculation, coalescence, and creaming may transpire in the drying stage of the EO-incorporated film. The drying stage can also contribute to loss of EO from the matrix, which can further impact the ultimate microstructure of the film (Atarés & Chiralt, 2016; Sánchez-González et al., 2011). Different types and concentrations of EOs added can have different effects on the polymer matrix. HPMC films incorporating ginger EO created more open matrix structure with thicker films due to EO droplets (Atarés et al., 2011), whereas the addition of multiple EOs such as lemongrass, thyme, and sage can increase roughness in the alginate coatings (Acevedo-Fani et al., 2015). EO droplets cause discontinuities, giving rise to an expansive lattice and more condensed films, which can be observed using SEM. Reducing EO droplet size had corresponding microstructural observable differences due to increased polymer–oil interaction and a less-cohesive polymer matrix (Bonilla et al., 2012). X-ray diffraction (XRD) analysis, utilized for determining the latticework of a matrix by quantifying space between atoms, is another method to investigate the texture of film matrix. Adding EOs to the films made the diffraction peaks more level and less perceptible, indicating an inverse proportionality between crystallinity and concentration. Usually, when films have Eos incorporated in their microstructure, the crystallinity index is lower, probably due to the sturdier interactions between the active components and the polymers (i.e., H-bonds, Van der Waals), resulting in improved crystalline 3D network (Al-Hilifi et al., 2022).

10.6.7 Antimicrobial Properties

The increase in antimicrobial resistance (AMR) to multiple drugs has led to the discovery of new solutions for treatments and disinfections (Yap et al., 2014). Besides, the potential toxic nature of chemical preservatives has also shifted the interest in natural and green alternatives, such as EOs. The antimicrobial ability of different types of EOs and their compounds is widely published and well researched, including its applications in food products (Burt, 2004; Deans & Ritchie, 1987; Mendoza-Yepes et al., 1997; Mourey & Canillac, 2002; Shelef, 1984). This chapter intends to provide a summary on antimicrobial properties and mechanisms of EOs, with a focus on eco-friendly coatings and the challenges associated with it.

Components such as carvacrol, eugenol, cinnamic aldehyde, camphor, and p-cymene are considered the major ones responsible for antimicrobial effects (Chouhan et al., 2017).

Clostridium botulinum and *Clostridium perfringens* were inhibited by EOs of oregano, savory, and thyme (Nevas et al., 2004), while there was 3–5 log CFU/g reduction of spore germination and outgrowth in *C. perfringens* in ground turkey during chilling when carvacrol, thymol, and oregano oil (each at 2% wt/wt) were used (Juneja & Friedman, 2007). EOs derived from lemongrass, oregano, and sage have shown antibacterial activity against *Salmonella typhimurium* with minimum inhibitory concentrations

(MIC) of 1.2 μl/mL, 1.2 μl/mL, and 10–20 μl/mL, respectively (Hammer et al., 1999; Shelef et al., 1984). Specific components of EOs, such as carvacrol, α-terpineol, citral, eugenol, and thymol, have proven antibacterial property against various foodborne pathogens, such as *Escherichia coli*, *Salmonella typhiumurium*, *Staphylococcus aureus*, and *Listeria monocytogenes*, each at different MICs (Cosentino et al., 1999; Kim et al., 1995; Onawunmi, 1989). Besides bacteria, EOs have also been tested against viral and fungal pathogens. Multiple fungal species of *Candida*, *Aspergillus*, *Penicillium*, *Saccharomyces*, and *Trichophyton* are inhibited by different types of EOs and their components (Ebani et al., 2016; Esen et al., 2007; Hammer et al., 2002; Latifah-Munirah et al., 2015; Mekonnen et al., 2016; Omidbaigi et al., 2007). A detailed summary of the antimicrobial efficacy of EOs extracted from a wide range of medicinal and aromatic plants is provided by Swamy et al. (Swamy et al., 2016).

It is important to point out the role of EOs when used in combination with other chemical antimicrobials and food preservation technologies. These strategies are very useful in targeting resistant bacteria due to changes in bacterial mechanisms caused by EOs, thus preventing emergence of AMR strains in the future. One study showed that the combination of oregano EO with commonly used antibiotics such as fluroquinolones, doxycycline, lincomycin, and maquindox flofenicol produced a synergistic effect against extended-spectrum β-lactamase-producing *Escherichia coli* (Si et al., 2008). The dose at which EOs are required for microbial inactivation may affect the sensory properties, and hence, hurdle treatments can be useful to reduce the dose. One study investigated 11 components of EO combined with mild heat or pulsed electric field technology (PEF). The group found out that 0.2 μl/mL of monoterpenes combined with mild heat or PEF led to synergistic inactivation of *E. coli* and *L. monocytogenes* (Ait-Ouazzou et al., 2011). It has been observed that generally, Gram-negative bacteria are less susceptible to EOs than Gram-positive bacteria, and the reasons/mechanisms are discussed later in this chapter.

To minimize the changes in organoleptic properties of foods and to enhance the activity of EOs in an environment (water-rich or liquid–solid interface) where microorganisms prefer to grow, Eos can be encapsulated, thus increasing their concentration (Donsì et al., 2011). Edible coatings made of lipid, proteins, or polysaccharide compounds can be used to encapsulate Eos, each having their unique properties and production methods (Ju et al., 2019). EO coatings have been tested against various microbial species and strains in different food matrices. Mandarin EO emulsified with chitosan combined with high-pressure processing reduced *L. innocua* by 5 log in green beans; coatings containing oregano EO, alginate, fiber, and Tween 80 applied on cheese reduced *S. aureus* and psychrophilic bacteria counts; whey protein–based coating with *Origanum virens* EO reduced the total microbial count on sausages; and galbanum gum coating with *Ziziphora* EO decreased microbial load of aerobic mesophilic, psychrotrophic, *Pseudomonas* spp., Lactic acid bacteria, *L. monocytogenes*, and *Enterobacteriaceae* when chicken fillets were coated with this EO coating (Acevedo Fani et al., 2015; Artiga Artigas et al., 2017; Donsì et al., 2015; Hamedi et al., 2017). These are a few examples of antimicrobial activities of EO coatings, and Ju et al. has summarized the role of various EO coatings and their activity in food models (Ju et al., 2019). In general, many EO coatings have shown stronger antimicrobial activity than free EOs. In one study, alginate-based edible films containing thyme EO on agar plates showed reduction in *E. coli* by 4.71 log, but films containing lemongrass or sage EO did not significantly inactivate *E. coli* (Acevedo Fani et al., 2015). It is important to note that the effectiveness of EO coatings may depend on interactions with coatings, pH of the system, food composition, interaction with food matrix and environment, and release of EOs into the food system.

10.6.7.1 Mechanisms of Action

The antimicrobial mechanism of essential oils depends on the type and quantity of components present in EOs and the target bacterial structure. The hydrophobic compounds of EOs can easily enter the peptidoglycan wall of Gram-positive bacteria because it cannot exclude large molecules, and the lipophilic ends of lipoteichoic acid further facilitate their movement inside the membranes. On the other hand, the hydrophilic lipopolysaccharides present on the outer membrane of Gram-negative bacteria limit the movement of EOs (Hyldgaard et al., 2012; Swamy et al., 2016). Some EOs like thymol can damage the outer membrane of Gram-negative bacteria and enhance its permeability to cytoplasmic membrane (Acevedo Fani et al., 2015). The damage to membrane integrity, leakage of cellular components, disruption of cytoplasmic membrane, change in pH and Ca^{2+} homeostasis, and inhibition of cell wall synthesizing enzymes and cell division are some mechanisms proposed for antimicrobial activity of EOs (Hyldgaard et al., 2012). The encapsulation of essentials oils in coatings and films leads to slow release of the volatile compounds on the food surface, thus ensuring long-lasting antimicrobial activity. The diffusion of EO should take place in a controlled manner, and at any time point, the EO quantity should be at least more than the MICs of target pathogen(s) (Anis et al., 2021). The encapsulation of EO can reduce evaporation in atmosphere or outside of the food matrix, can mask the taste and odor of EO, and hence, higher concentrations can be used, and can provide a uniform distribution in product (Liolios et al., 2009). Oregano EO can modify cell membrane and fatty acid composition of cytoplasmic membrane. However, in the low-fat cut cheese study, antimicrobial activity was highly dependent on the concentration of oregano EO used to prepare nanoemulsions for coating the cheese pieces. A low concentration of 1.5% w/w did not inhibit the growth of *S. aureus* (Artiga Artigas et al., 2017). Chitosan-to-tea-tree EO ratio of 1:2 showed better inactivation against *L. monocytogenes* compared to using the polymer and EO in equal quantities (Sánchez-González et al., 2010). In general, incorporating EO in films or coatings versus films alone or EO alone showed improvement in antimicrobial properties of films. However, exceptions like addition of caraway EO (1% volume) to chitosan did not increase the antimicrobial properties of chitosan (Hromiš et al., 2015), and unencapsulated eugenol was slightly more or as inhibitory as eugenol encapsulated in micelles against *L. monocytogenes* and *E. coli* O157:H7 (Gaysinsky et al., 2007).

10.6.7.2 Shelf Life of Essential Oil Films/Coatings

The benefits of EO coatings, as seen earlier due to its antimicrobial properties, can be utilized in active packaging of food products to improve the shelf life. Chicken breast when covered with alginate coating containing black cumin EO reduced total aerobic mesophilic and psychrotrophic bacteria and lowered the variation in pH and color changes when stored at 4°C for five days (Konuk Takma & Korel, 2019). Active cardboard tray containing EO coating (carvacrol, oregano, and cinnamon encapsulated in β-cyclodextrin) reduced decay incidence, maintained color and firmness, and extended shelf life of cherry tomatoes from 20 to 24 days (Buendía-Moreno et al., 2019). Shelf life can be assessed based on various parameters measured over time, depending on desired qualities and the food product, such as sensory (appearance in color and texture, flavor), microbial load (pathogens, spoilage organisms), texture (firmness), chemical properties (solids, pH, titratable acidity, toxin formation), and color analysis. Table 10.7 in this chapter provides some examples of shelf-life extension of EO coatings/films used in food products and the characteristics analyzed.

Gomes et al. have summarized various parameters that can be analyzed for fruits/vegetables applied with edible coatings containing EO. The total soluble solids may either be maintained or increased because of moisture loss from fruits and vegetables. However, this water loss could be less in products with coatings compared to the products without coating because of the moisture barrier properties of essential oils. This can eventually benefit in reduction of weight loss in fruits and vegetables (Gomes et al., 2017). EO coatings also have an excellent gas barrier property, thus reducing respiration rates and delaying ripening of fruits (Salvia-Trujillo et al., 2015). Bioactive compounds that are sensitive to environmental factors such as heat, light, and oxygen and have potential to degrade over time can be retained by application of EO coatings (Gol et al., 2013). Instead of direct assessment of various parameters over time to calculate shelf life, statistical models such as response surface methodology, Weibull model, Arrhenius model, and accelerated shelf-life testing can be used to obtain predictive values of the desired quality parameters and microbial values. The details of using such approaches for shelf life can be found in the book *Food Packaging and Shelf Life* by Robertson (2010). The statistical methods are commonly used in food industry and can reduce the time and cost involved in shelf-life estimation.

The stability of EO coatings can also be assessed by measuring water vapor transmission rates, oxygen transmission rate, tensile strength, hydrophobicity, water contact angle, optical properties (color, transparency, and gloss), and migration tests (between food and EO coatings), which are useful properties for determining their use as packaging materials (Sharma et al., 2021). EO coatings are also used in fragrances, colorants, and cosmetic products for skin and health care. Their antioxidant, antibiotic, and antiseptic properties assist in antiacne, antiaging, and antidandruff functions.

TABLE 10.7

EO coatings or EO films researched in the food industry to extend shelf life and the characteristics analyzed

Product	EO coating or EO film	Results	References
Dates	Chitosan with 2% citrus EO	Reduction in fungal decay (*A. flavus*) by more than 50% at day 12 stored at 25°C	(Aloui et al., 2014)
Strawberries	Chitosan with limonene EO and Tween 80 chitosan with cinnamon EO	Better visual appearance on day 16 and day 21, lower mold contamination than control on days 8, 10, and 12 (storage at 4°C) No significant increase in fungal decay (*Rhizopus stolonifer*) compared to control stored for 14 days at 10°C	(Vu et al., 2011; Perdones et al., 2014)
Grapes	Pectin with lemon EO	Reduction in weight loss by 1.5%, maintained low pH, and intense red color during storage for 35 days at 4°C	(Breceda-Hernandez et al., 2020)
Sweet peppers	Chitosan–cinnamon oil	Reduced decay incidence, better sensory acceptability, and reduced loss of vitamin C compared to control during 35 days of storage at 8°C	(Xing et al., 2011)
Shrimp	Gelatin–orange leaf EO Chitosan with orange peel EO	Preserved fresh shrimps for 14 days; significant growth inhibition of *Enterobacteriaceae*, psychrotrophs, and total viable counts; reduced peroxide and thiobarbituric acid values; and prevented increase of melanosis (14 days storage at 4°C) Inhibited lipid oxidation and growth of total coliform bacteria (15 days at 4°C)	(Alparslan et al., 2016; Alparslan & Baygar, 2017)
Kiwi fruits	Sodium alginate with lemon EO or orange EO or grapefruit EO	Inhibition of yeasts and mold when stored for 7 days at 0°C	(Valentina Chiabrando & Giovanna Giacalone, 2019)
Oranges	Chitosan–tea tree oil	Antifungal activity against *Penicillium italicum* when stored for 26 days at 25°C	(Cháfer et al., 2012)
Rucola leaves	Chitosan–lemon EO	Prolonged shelf life by 3 to 7 days, reduced total microbial load and yeasts and mold counts, no alterations in color and texture	(Sessa et al., 2015)
Papaya	Mesquite gum–thyme EO or Mexican lime EO	Decrease in disease incidence of *Colletotrichum gloeosporioides* by 100% at the end of 9 days at 20°C	(Bosquez-Molina et al., 2010)

10.8 Economic Viability and Environmental Sustainability of Using Essential Oil Films/Coatings

The materials used for encapsulating essential oils are mostly biodegradable, such as proteins or polysaccharides, thus providing an eco-friendly method for food preservation. Pectin, chitosan, and gelatin are some common materials that you may have noticed from previous examples. Other materials that are researched for EOs are alginate, zein, starch, maltodextrin, soy proteins, wheat gluten, casein, collagen, and inulin (Maurya et al., 2021). The stability of EO coatings can be improved by using nanoemulsion technique (examples: high-pressure homogenization, ultrasonication, phase inversion composition, and spontaneous emulsification) that creates small-size droplets (diameter < 200 nm). This increases their surface area, reduces EO dosage, improves film homogeneity, and enhances antimicrobial activity (Sharma et al., 2022). The life span of EO coatings can be increased, but extracting EOs and creating efficient nanoemulsions require additional instruments that can add cost, thus impacting consumer acceptance.

The solvent casting film is a cost-effective method to prepare EO coatings where the encapsulating material such as chitosan is dissolved in a solvent and plasticizer is added to enhance mechanical properties. After addition of EO and stirring for a long time to obtain uniform solution, the solvent is evaporated (Zhang et al., 2021). However, loss of some volatile components can occur during the drying step. Various other methods for preparing edible coatings using EO are outlined by Ju et al. (2019). As seen in the previous text, the use of EO coatings or EO films can boost the shelf life of food products, thus minimizing losses, and aids in reducing postharvest fruit diseases. The use of chemical additives as food preservatives can be reduced or replaced with EO coatings.

The environmental impact of a new system such as EO coating in our case can be measured by conducting a life cycle assessment (LCA) across the food supply chain. One such example of LCA on EO coating as packaging for fresh beef was published by Zhang et al. They concluded that the use of active packaging solutions proposed in the study could potentially reduce beef losses at the retail of the EU market by up to 147,600 t/year (Zhang et al., 2015). There are more studies published on LCA for other types of active packaging that can be referred for methodology (Settier-Ramirez et al., 2022; Vigil et al., 2020; Zhang et al., 2017).

10.9 Challenges and Limitations of Using Essential Oil Films/Coatings

The strong aroma and its impact on organoleptic properties of food products are a few of the major challenges of using Eos. When Eos are incorporated in polymers to form films or coatings, the challenge with sensory aspect is slightly reduced and the diffusion of antimicrobial compounds in the product is also reduced (Atarés & Chiralt, 2016). Cheese coated with chitosan containing fish EO decreased the mold growth, but the sensory evaluation of coated cheeses received the lowest scores (Yangilar, 2016).

Emulsifying and encapsulating EOs have helped overcome some limitations, such as low solubility, high volatility, and heat and light sensitivity of EOs. In addition, many new techniques, such as halloysite nanotubes, nanofibers, and electrospinning, are being researched to overcome these limitations (Sharma et al., 2021). However, the interaction of EO with macromolecules from emulsions can reduce antimicrobial efficiency of some Eos, and optimizing the coatings is also an important factor (Ju et al., 2019). EOs can cause allergic reactions and irritations to the eye, skin, and mucous membrane. Hence, they should be used in appropriate doses to avoid the risks of toxicity (Ribeiro-Santos et al., 2017a).

10.10 Legal Aspects of Use of Essential Oil Films/Coatings

EOs have been classified as generally recognized as safe (GRAS) in the United States. The European Commission (EC) requires essential oils to be registered as flavoring agents based on regulation (EC) No. 1334/2008 (Sharma et al., 2021). Coatings prepared for food and pharmaceutical products are considered food additives. Regulations will differ depending on whether these coatings are generally consumed or not and differ according to each region (country). The food additives standards for United States are available in 21 CFR Part 172, and definitions are available in 21 CFR Part 170 in the Federal Register (Code of Federal Regulations, 2022).

Edible Coatings and Films to Improve Food Quality by Baldwin et al. provides a detailed regulatory understanding for edible coatings required in various countries and can be used as a reference for EO coatings and EO films (Baldwin et al., 2012).

10.11 Future Perspectives

The regulatory approval allows the use of EO coatings and EO films in place of synthetic preservatives. Research interest to maximize the benefits from these coatings for food preservation and to reduce food loss is increasing. Some patents are granted for use of EOs in food packaging. One example is polyolefin containing eucalyptus, nutmeg, hinoki, cinnamon, and oregano EO for degradable packaging of fruits and vegetables (Yañez Sanchez & Zapata Ramirez, 2019). The extra packaging cost and sensory challenges are good areas of research to explore for EO coatings to be used as preservatives in the food industry. The use of EO coatings will help the environment due to their biodegradable nature and assist with reduction in food loss due to their antimicrobial properties.

10.12 Conclusion

The dawn of active, biodegradable, renewable materials requisites the need for the production of an eco-friendly way for storage, transportation, and extending the shelf life of products in the food industry. Essential oil–based films and coatings are incorporated with various secondary metabolites of higher plants that function as natural antimicrobial agents. Essential oils of plants with natural antibacterial, antiviral, and antifungal components are an appropriate source of bioactive material for producing eco-friendly, active, and edible films and coatings. The safety and quality of foods are of increasing concern because contaminated foods can act as a vehicle for transmission of hazardous diseases. Essential oils have been utilized as alternatives to synthetic preservatives with proven antimicrobial activities. Encapsulation of essential oils into biodegradable polymers, moreover, ensures the maintenance of antioxidant activities, physicochemical properties, and organoleptic attributes, resulting in better shelf life. However, several challenges have to be faced during commercial exploitation of essential oils. For instance, the combinatorial action of essential oil and components in a synergistic system, the judicious selection of essential oils, and more importantly, the safety profile of essential oils should be worked out for large-scale practical recommendation as a novel shelf-life enhancer for stored foods.

REFERENCES

Acevedo-Fani, A., Salvia-Trujillo, L., Rojas-Graü, M. A., & Martín-Belloso, O. (2015). Edible films from essential-oil-loaded nanoemulsions: Physicochemical characterization and antimicrobial properties. *Food Hydrocolloids*, *47*, 168–177. https://doi.org/10.1016/j.foodhyd.2015.01.032

Ahmad, M., Benjakul, S., Prodpran, T., & Agustini, T. W. (2012). Physico-mechanical and antimicrobial properties of gelatin film from the skin of unicorn leatherjacket incorporated with essential oils. *Food Hydrocolloids*, *28*(1), 189–199. https://doi.org/10.1016/j.foodhyd.2011.12.003

Ait-Ouazzou, A., Cherrat, L., Espina, L., Lorán, S., Rota, C., & Pagán, R. (2011). The antimicrobial activity of hydrophobic essential oil constituents acting alone or in combined processes of food preservation. *Innovative Food Science & Emerging Technologies*, *12*(3), 320–329. https://doi.org/10.1016/j.ifset.2011.04.004

Al-Hilifi, S. A., Al-Ali, R. M., & Petkoska, A. T. (2022). Ginger essential oil as an active addition to composite chitosan films: Development and characterization. *Gels (Basel, Switzerland)*, *8*(6), 327. https://doi.org/10.3390/gels8060327

Ali, B., Al-Wabel, N. A., Shams, S., Ahamad, A., Khan, S. A., & Anwar, F. (2015). Essential oils used in aromatherapy: A systemic review. *Asian Pacific Journal of Tropical Biomedicine*, *5*(8), 601–611. https://doi.org/10.1016/j.apjtb.2015.05.007

Aloui, H., Khwaldia, K., Licciardello, F., Mazzaglia, A., Muratore, G., Hamdi, M., & Restuccia, C. (2014). Efficacy of the combined application of chitosan and Locust Bean Gum with different citrus essential oils to control postharvest spoilage caused by *Aspergillus flavus* in dates. *International Journal of Food Microbiology*, *170*, 21–28. https://doi.org/10.1016/j.ijfoodmicro.2013.10.017

Alparslan, Y., & Baygar, T. (2017). Effect of chitosan film coating combined with orange peel essential oil on the shelf life of deepwater pink shrimp. *Food and Bioprocess Technology*, *10*(5), 842–853. https://doi.org/10.1007/s11947-017-1862-y

Alparslan, Y., Yapıcı, H. H., Metin, C., Baygar, T., Günlü, A., & Baygar, T. (2016). Quality assessment of shrimps preserved with orange leaf essential oil incorporated gelatin. *Food Science & Technology*, *72*, 457–466. https://doi.org/10.1016/j.lwt.2016.04.066

Altiok, D., Altiok, E., & Tihminlioglu, F. (2010). Physical, antibacterial and antioxidant properties of chitosan films incorporated with thyme oil for potential wound healing applications. *Journal of Materials Science: Materials in Medicine*, *21*(7), 2227–2236. https://doi.org/10.1007/s10856-010-4065-x

Anis, A., Pal, K., & Al-Zahrani, S. M. (2021). Essential oil-containing polysaccharide-based edible films and coatings for food security applications. *Polymers*, *13*(4), 575. https://doi.org/10.3390/polym13040575

Anwar, R. W., Sugiarto, P., & Warsiki, E. (2018). The comparison of antimicrobial packaging properties with different applications incorporation method of active material. *IOP Conference Series. Earth and Environmental Science, 141*(1). https://doi.org/12002.10.1088/1755–1315/141/1/012002

Arcan, I., & Yemenicioğlu, A. (2011). Incorporating phenolic compounds opens a new perspective to use zein films as flexible bioactive packaging materials. *Food Research International, 44*(2), 550–556. https://doi.org/10.1016/j.foodres.2010.11.034

Artiga Artigas, M., Acevedo Fani, A., & Martín Belloso, O. (2017). Improving the shelf life of low-fat cut cheese using nanoemulsion-based edible coatings containing oregano essential oil and mandarin fiber. *Food Control, 76*, 1–12. https://doi.org/10.1016/j.foodcont.2017.01.001

Atarés, L., Bonilla, J., & Chiralt, A. (2010a). Characterization of sodium caseinate-based edible films incorporated with cinnamon or ginger essential oils. *Journal of Food Engineering, 100*(4), 678–687. https://doi.org/10.1016/j.jfoodeng.2010.05.018

Atarés, L., & Chiralt, A. (2016). Essential oils as additives in biodegradable films and coatings for active food packaging. *Trends in Food Science & Technology, 48*, 51–62. https://doi.org/10.1016/j.tifs.2015.12.001

Atarés, L., De Jesús, C., Talens, P., & Chiralt, A. (2010b). Characterization of SPI-based edible films incorporated with cinnamon or ginger essential oils. *Journal of Food Engineering, 99*(3), 384–391. https://doi.org/10.1016/j.jfoodeng.2010.03.004

Atarés, L., Pérez-Masiá, R., & Chiralt, A. (2011). The role of some antioxidants in the HPMC film properties and lipid protection in coated toasted almonds. *Journal of Food Engineering, 104*(4), 649–656. https://doi.org/10.1016/j.jfoodeng.2011.02.005

Atiwesh, G., Mikhael, A., Parrish, C. C., Banoub, J., & Le, T. T. (2021). Environmental impact of bioplastic use: A review. *Heliyon, 7*(9), e07918. https://doi.org/10.1016/j.heliyon.2021.e07918

Azadbakht, E., Maghsoudlou, Y., Khomiri, M., & Kashiri, M. (2018). Development and structural characterization of chitosan films containing *Eucalyptus globulus* essential oil: Potential as an antimicrobial carrier for packaging of sliced sausage. *Food Packaging and Shelf Life, 17*, 65–72. https://doi.org/10.1016/j.fpsl.2018.03.007

Bakkali, F., Averbeck, S., Averbeck, D., & Idaomar, M. (2008). Biological effects of essential oils—a review. *Food and Chemical Toxicology, 46*(2), 446–475. https://doi.org/10.1016/j.fct.2007.09.106

Baldwin, E. A., Hagenmaier, R., & Bai, J. (2012). *Edible coatings and films to improve food quality* (2nd ed.). CRC Press. https://doi.org/10.1201/b11082

Bhuyan, N., Barua, P. C., Kalita, P., & Saikia, A. (2015). Physico-chemical variation in peel oils of Khasi mandarin (*Citrus reticulata* Blanco) during ripening. *Indian Journal of Plant Physiology, 20*(3), 227–231. https://doi.org/10.1007/s40502-015-0164-5

Biji, K. B., Ravishankar, C. N., Mohan, C. O., & Srinivasa Gopal, T. K. (2015). Smart packaging systems for food applications: A review. *Journal of Food Science and Technology, 52*(10), 6125–6135. https://doi.org/10.1007/s13197-015-1766-7

Bonilla, J., Atarés, L., Vargas, M., & Chiralt, A. (2012). Effect of essential oils and homogenization conditions on properties of chitosan-based films. *Food Hydrocolloids, 26*(1), 9–16. https://doi.org/10.1016/j.foodhyd.2011.03.015

Bonilla, J., Talón, E., Atarés, L., Vargas, M., & Chiralt, A. (2013). Effect of the incorporation of antioxidants on physicochemical and antioxidant properties of wheat starch—chitosan films. *Journal of Food Engineering, 118*(3), 271–278. https://doi.org/10.1016/j.jfoodeng.2013.04.008

Bosquez-Molina, E., Jesús, E. R., Bautista-Baños, S., Verde-Calvo, J. R., & Morales-López, J. (2010). Inhibitory effect of essential oils against *Colletotrichum gloeosporioides* and *Rhizopus stolonifer* in stored papaya fruit and their possible application in coatings. *Postharvest Biology and Technology, 57*(2), 132–137. https://doi.org/10.1016/j.postharvbio.2010.03.008

Breceda-Hernandez, T. G., Martínez-Ruiz, N. R., Serna-Guerra, L., & Hernández-Carrillo, J. G. (2020). Effect of a pectin edible coating obtained from orange peels with lemon essential oil on the shelf life of table grapes (*Vitis vinifera* L. var. Red Globe). *International Food Research Journal, 27*(3), 585–596. https://search.proquest.com/docview/2437454171

Buendía-Moreno, L., Soto-Jover, S., Ros-Chumillas, M., Antolinos, V., Navarro-Segura, L., Sánchez-Martínez, M. J., Martínez-Hernández, G. B., & López-Gómez, A. (2019). Innovative cardboard active packaging with a coating including encapsulated essential oils to extend cherry tomato shelf life. *Food Science & Technology, 116*. https://doi.org/10.1016/j.lwt.2019.108584

Burt, S. (2004). Essential oils: Their antibacterial properties and potential applications in foods—a review. *International Journal of Food Microbiology, 94*(3), 223–253. https://doi.org/10.1016/j.ijfoodmicro.2004.03.022

Castro, J. I., Valencia-Llano, C. H., Valencia Zapata, M. E., Restrepo, Y. J., Mina Hernandez, J. H., Navia-Porras, D. P., Valencia, Y., Valencia, C., & Grande-Tovar, C. D. (2021). Chitosan/polyvinyl alcohol/tea tree essential oil composite films for biomedical applications. *Polymers (Basel), 13*(21). https://doi.org/10.3390/polym13213753

Catarino, M. D., Alves-Silva, J. M., Fernandes, R. P., Gonçalves, M. J., Salgueiro, L. R., Henriques, M. F., & Cardoso, S. M. (2017). Development and performance of whey protein active coatings with *Origanum virens* essential oils in the quality and shelf life improvement of processed meat products. *Food Control, 80*, 273–280. https://doi.org/10.1016/j.foodcont.2017.03.054

Cazón, P., Velazquez, G., Ramírez, J. A., & Vázquez, M. (2017). Polysaccharide-based films and coatings for food packaging: A review. *Food Hydrocolloids, 68*, 136–148. https://doi.org/10.1016/j.foodhyd.2016.09.009

Chaemsanit, S., Sukmas, S., Matan, N., & Matan, N. (2019). Controlled release of peppermint oil from paraffin-coated activated carbon contained in sachets to inhibit mold growth during long term storage of brown rice. *Journal of Food Science, 84*(4), 832–841. https://doi.org/10.1111/1750-3841.14475

Cháfer, M., Sánchez-González, L., González-Martínez, C., & Chiralt, A. (2012). Fungal decay and shelf life of oranges coated with chitosan and bergamot, thyme, and tea tree essential oils. *Journal of Food Science, 77*(8), E182–E187. https://doi.org/10.1111/j.1750-3841.2012.02827.

Chiabrando, V., & Giacalone, G. (2019). Effects of citrus essential oils incorporated in alginate coating on quality of fresh-cut Jintao kiwifruit. *Journal of Food and Nutrition Research, 58*(2), 177. https://search.proquest.com/docview/2222647124

Choi, W. S., Singh, S., & Lee, Y. S. (2016). Characterization of edible film containing essential oils in hydroxypropyl methylcellulose and its effect on quality attributes of 'Formosa' plum (*Prunus salicina* L.). *LWT, 70*, 213–222. https://doi.org/10.1016/j.lwt.2016.02.036

Chouhan, S., Sharma, K., & Guleria, S. (2017). Antimicrobial activity of some essential oils-present status and future perspectives. *Medicines, 4*(3), 58. https://doi.org/10.3390/medicines4030058

Code of Federal Regulations. (2022). *Title 21 Part 170.* www.ecfr.gov/current/title-21/chapter-I/subchapter-B/part-170

Cosentino, S., Tuberoso, C. I. G., Pisano, B., Satta, M., Mascia, V., Arzedi, E., & Palmas, F. (1999). In-vitro antimicrobial activity and chemical composition of *Sardinian Thymus* essential oils. *Letters in Applied Microbiology, 29*(2), 130–135. https://doi.org/10.1046/j.1472-765X.1999.00605.x

Cran, M. J., Rupika, L., Sonneveld, K., Miltz, J., & Bigger, S. W. (2010). Release of naturally derived antimicrobial agents from LDPE films. *Journal of Food Science, 75*(2), E126–E133.

Dajic Stevanovic, Z., Sieniawska, E., Glowniak, K., Obradovic, N., & Pajic-Lijakovic, I. (2020). Natural macromolecules as carriers for essential oils: From extraction to biomedical application. *Frontiers in Bioengineering and Biotechnology, 8*, 563. https://doi.org/10.3389/fbioe.2020.00563

Dashipour, A., Razavilar, V., Hosseini, H., Shojaee-Aliabadi, S., German, J. B., Ghanati, K., Khakpour, M., & Khaksar, R. (2015). Antioxidant and antimicrobial carboxymethyl cellulose films containing *Zataria multiflora* essential oil. *International Journal of Biological Macromolecules, 72*, 606–613. https://doi.org/10.1016/j.ijbiomac.2014.09.006

Deans, S. G., & Ritchie, G. (1987). Antibacterial properties of plant essential oils. *International Journal of Food Microbiology, 5*(2), 165–180. https://doi.org/10.1016/0168-1605(87)90034-1

De Groef, B., Decallonne, B. R., Van der Geyten, S., Darras, V. M., & Bouillon, R. (2006). Perchlorate versus other environmental sodium/iodide symporter inhibitors: Potential thyroid-related health effects. *European Journal of Endocrinology, 155*(1), 17–25. https://doi.org/10.1530/eje.1.02190

De Souza, V. V. M. A., Crippa, B. L., De Almeida, J. M., Iacuzio, R., Setzer, W. N., Sharifi-Rad, J., & Silva, N. C. C. (2020). Synergistic antimicrobial action and effect of active chitosan-gelatin biopolymeric films containing *Thymus vulgaris, Ocimum basilicum* and *Origanum majorana* essential oils against *Escherichia coli* and *Staphylococcus aureus. Cellular and Molecular Biology (Noisy-Le-Grand, France), 66*(4), 223. http://europepmc.org/abstract/MED/32583781

Dong, Z., Xu, F., Ahmed, I., Li, Z., & Lin, H. (2018). Characterization and preservation performance of active polyethylene films containing rosemary and cinnamon essential oils for Pacific white shrimp packaging. *Food Control, 92*, 37–46. https://doi.org/10.1016/j.foodcont.2018.04.052

Dong, Z. J., Touré, A., Jia, C. S., Zhang, X. M., & Xu, S. Y. (2007). Effect of processing parameters on the formation of spherical multinuclear microcapsules encapsulating peppermint oil by coacervation. *Journal of Microencapsulation, 24*(7), 634–646. 10.1080/02652040701500632

Donsì, F., Annunziata, M., Sessa, M., & Ferrari, G. (2011). Nanoencapsulation of essential oils to enhance their antimicrobial activity in foods. *Food Science & Technology, 44*(9), 1908–1914. https://doi.org/10.1016/j.lwt.2011.03.003

Donsì, F., Marchese, E., Maresca, P., Pataro, G., Vu, K. D., Salmieri, S., Lacroix, M., & Ferrari, G. (2015). Green beans preservation by combination of a modified chitosan based-coating containing nanoemulsion of mandarin essential oil with high pressure or pulsed light processing. *Postharvest Biology and Technology, 106*, 21–32. https://doi.org/10.1016/j.postharvbio.2015.02.006

Ebani, V. V., Nardoni, S., Bertelloni, F., Giovanelli, S., Rocchigiani, G., Pistelli, L., & Mancianti, F. (2016). Antibacterial and antifungal activity of essential oils against some pathogenic bacteria and yeasts shed from poultry. *Flavour and Fragrance Journal, 31*(4), 302–309. https://doi.org/10.1002/ffj.3318

Esen, G., Azaz, A. D., Kurkcuoglu, M., Baser, K. H. C., & Tinmaz, A. (2007). Essential oil and antimicrobial activity of wild and cultivated *Origanum vulgare* L. subsp. *hirtum* (Link) letswaart from the Marmara region, Turkey. *Flavour and Fragrance Journal, 22*(5), 371–376. https://doi.org/10.1002/ffj.1808

Essential oils market size, share & trends analysis report by product, by application, by sales channel, by region and segment forecasts, 2021–2028. (2022, January 4). *NASDAQ OMX's News Release Distribution Channel.* https://www.globenewswire.com/news-release/2022/01/04/2360801/0/en/Essential-Oils-Market-Size-Share-Trends-Analysis-Report-By-Product-By-Application-By-Sales-Channel-By-Region-And-Segment-Forecasts-2021-2028.htmlwebsite

Fathi, M., Mozafari, M. R., & Mohebbi, M. (2012). Nanoencapsulation of food ingredients using lipid based delivery systems. *Trends in Food Science & Technology, 23*(1), 13–27. https://doi.org/10.1016/j.tifs.2011.08.003

Fernández-López, J., & Viuda-Martos, M. (2018). Introduction to the special issue: Application of essential oils in food systems. *Foods, 7*(4), 56. https://doi.org/10.3390/foods7040056

Figueroa-Lopez, K. J., Enescu, D., Torres-Giner, S., Cabedo, L., Cerqueira, M. A., Pastrana, L., Fuciños, P., & Lagaron, J. (2020). Development of electrospun active films of poly (3-hydroxybutyrate-co-3-hydroxyvalerate) by the incorporation of cyclodextrin inclusion complexes containing oregano essential oil. *Food Hydrocolloids, 108*, 106013.

Frankel, E. N., & Meyer, A. S. (2000). The problems of using one-dimensional methods to evaluate multifunctional food and biological antioxidants. *Journal of the Science of Food and Agriculture, 80*(13), 1925–1941.

Gaysinsky, S., Taylor, T. M., Davidson, P. M., Bruce, B. D., & Weiss, J. (2007). Antimicrobial efficacy of eugenol microemulsions in milk against *Listeria monocytogenes* and *Escherichia coli* O157:H7. *Journal of Food Protection, 70*(11), 2631–2637. https://doi.org/10.4315/0362–028X-70.11.2631

Gol, N. B., Patel, P. R., & Rao, T. V. R. (2013). Improvement of quality and shelf-life of strawberries with edible coatings enriched with chitosan. *Postharvest Biology and Technology, 85*, 185–195. https://doi.org/10.1016/j.postharvbio.2013.06.008

Gomes, M. D. S., Cardoso, M. D. G., Guimarães, A. C. G., Guerreiro, A. C., Gago, C. M. L., Vilas Boas, Eduardo Valério de Barros, Dias, C. M. B., Manhita, A. C. C., Faleiro, M. L., Miguel, M. G. C., & Antunes, M. D. C. (2017). Effect of edible coatings with essential oils on the quality of red raspberries over shelf-life. *Journal of the Science of Food and Agriculture, 97*(3), 929–938. https://doi.org/10.1002/jsfa.7817

Goulas, A. E., & Kontominas, M. G. (2007). Combined effect of light salting, modified atmosphere packaging and oregano essential oil on the shelf-life of sea bream (*Sparus aurata*): Biochemical and sensory attributes. *Food Chemistry, 100*(1), 287–296.

Guerra, I. C. D., de Oliveira, P. D. L., Santos, M. M. F., Lúcio, A. S. S. C., Tavares, J. F., Barbosa-Filho, J. M., Madruga, M. S., & de Souza, E. L. (2016). The effects of composite coatings containing chitosan and *Mentha* (*piperita* L. or *x villosa* Huds) essential oil on postharvest mold occurrence and quality of table grape cv. *Isabella*. *Innovative Food Science & Emerging Technologies, 34*, 112–121. https://doi.org/10.1016/j.ifset.2016.01.008

Guerreiro, A. C., Gago, C. M. L., Faleiro, M. L., Miguel, M. G. C., & Antunes, M. D. C. (2015). Raspberry fresh fruit quality as affected by pectin- and alginate-based edible coatings enriched with essential oils. *Scientia Horticulturae, 194*, 138–146. https://doi.org/10.1016/j.scienta.2015.08.004

Gupta, S., Khan, S., Muzafar, M., Kushwaha, M., Yadav, A. K., & Gupta, A. P. (2016). Encapsulation: Entrapping essential oil/flavors/aromas in food. In *Encapsulations* (pp. 229–268). Academic Press. https://doi.org/10.1016/b978-0-12-804307-3.00006-5

Hamedi, H., Kargozari, M., Shotorbani, P. M., Mogadam, N. B., & Fahimdanesh, M. (2017). A novel bioactive edible coating based on sodium alginate and galbanum gum incorporated with essential oil of *Ziziphora persica*: The antioxidant and antimicrobial activity, and application in food model. *Food Hydrocolloids*, *72*, 35–46. https://doi.org/10.1016/j.foodhyd.2017.05.014

Hammer, K. A., Carson, C. F., & Riley, T. V. (1999). Antimicrobial activity of essential oils and other plant extracts. *Journal of Applied Microbiology*, *86*(6), 985–990. https://doi.org/10.1046/j.1365-2672.1999.00780.x

Hammer, K. A., Carson, C. F., & Riley, T. V. (2002). In vitro activity of *Melaleuca alternifolia* (tea tree) oil against dermatophytes and other filamentous fungi. https://researchrepository.murdoch.edu.au/id/eprint/35497/

Hassan, B., Chatha, S. A. S., Hussain, A. I., Zia, K. M., & Akhtar, N. (2018). Recent advances on polysaccharides, lipids and protein based edible films and coatings: A review. *International Journal of Biological Macromolecules*, *109*, 1095–1107. https://doi.org/10.1016/j.ijbiomac.2017.11.097

Hedayati Rad, F., Sharifan, A., & Asadi, G. (2018). Physicochemical and antimicrobial properties of kefiran/waterborne polyurethane film incorporated with essential oils on refrigerated ostrich meat. *LWT, 97*, 794–801. https://doi.org/10.1016/j.lwt.2018.08.005

Hosseini, S. F., Rezaei, M., Zandi, M., & Farahmandghavi, F. (2015). Bio-based composite edible films containing *Origanum vulgare* L. essential oil. *Industrial Crops and Products*, *67*, 403–413. https://doi.org/10.1016/j.indcrop.2015.01.062

Hromiš, N. M., Lazić, V. L., Markov, S. L., Vaštag, Ž. G., Popović, S. Z., Šuput, D. Z., Džinić, N. R., Velićanski, A. S., & Popović, L. M. (2015). Optimization of chitosan biofilm properties by addition of caraway essential oil and beeswax. *Journal of Food Engineering*, *158*, 86–93. https://doi.org/10.1016/j.jfoodeng.2015.01.001

Huang, M., Wang, H., Xu, X., Lu, X., Song, X., & Zhou, G. (2020). Effects of nanoemulsion-based edible coatings with composite mixture of rosemary extract and ε-poly-l-lysine on the shelf life of ready-to-eat carbonado chicken. *Food Hydrocolloids*, *102*, 105576. https://doi.org/10.1016/j.foodhyd.2019.105576

Hutchings, J. B. (2011). *Food colour and appearance*. Springer Science & Business Media.

Hyldgaard, M., Mygind, T., & Meyer, R. L. (2012). Essential oils in food preservation: Mode of action, synergies, and interactions with food matrix components. *Frontiers in Microbiology*, *3*, 12. https://doi.org/10.3389/fmicb.2012.00012

Irwin, J., & Hedges, N. (2004). Measuring lipid oxidation. In *Understanding and measuring shelf life of food* (pp. 289–316). Woodhead Publishing in Food Science and Technology.

Jarpa-Parra, M., & Chen, L. (2021). Applications of plant polymer-based solid foams: Current trends in the food industry. *Applied Sciences, 11*(20), 9605. https://doi.org/10.3390/app11209605

Jipa, I. M., Stoica-Guzun, A., & Stroescu, M. (2012). Controlled release of sorbic acid from bacterial cellulose based mono and multilayer antimicrobial films. *Food Science & Technology, 47*(2), 400–406. https://doi.org/10.1016/j.lwt.2012.01.039

Ju, J., Xie, Y., Guo, Y., Cheng, Y., Qian, H., & Yao, W. (2019). Application of edible coating with essential oil in food preservation. *Critical Reviews in Food Science and Nutrition*, *59*(15), 2467–2480. https://doi.org/10.1080/10408398.2018.1456402

Ju, J., Xu, X., Xie, Y., Guo, Y., Cheng, Y., Qian, H., & Yao, W. (2018). Inhibitory effects of cinnamon and clove essential oils on mold growth on baked foods. *Food Chemistry*, *240*, 850–855. https://doi.org/10.1016/j.foodchem.2017.07.120

Juneja, V. K., & Friedman, M. (2007). Carvacrol, cinnamaldehyde, oregano oil, and thymol inhibit *Clostridium perfringens* spore germination and outgrowth in ground turkey during chilling. *Journal of Food Protection, 70*(1), 218–222. https://doi.org/10.4315/0362–028X-70.1.218

Kanatt, S. R., Rao, M., Chawla, S., & Sharma, A. (2012). Active chitosan—polyvinyl alcohol films with natural extracts. *Food Hydrocolloids*, *29*(2), 290–297.

Kaul, P. N., Bhattacharya, A. K., Rajeswara Rao, B. R., Syamasundar, K. V., & Ramesh, S. (2003). Volatile constituents of essential oils isolated from different parts of cinnamon (*Cinnamomum zeylanicum* Blume). *Journal of the Science of Food and Agriculture*, *83*(1), 53–55. https://doi.org/https://doi.org/10.1002/jsfa.1277

Kim, I., Lee, H., Kim, J. E., Song, K. B., Lee, Y. S., Chung, D. S., & Min, S. C. (2013). Plum coatings of lemongrass oil-incorporating carnauba wax-based nanoemulsion. *Journal of Food Science*, *78*(10), E1551–E1559. https://doi.org/10.1111/1750–3841.12244

Kim, J., Marshall, M. R., & Wei, C. (1995). Antibacterial activity of some essential oil components against five foodborne pathogens. *Journal of Agricultural and Food Chemistry, 43*(11), 2839–2845. https://doi.org/10.1021/jf00059a013

Kong, I., Degraeve, P., & Pui, L. P. (2022). Polysaccharide-based edible films incorporated with essential oil nanoemulsions: Physico-chemical, mechanical properties and its application in food preservation-a review. *Foods, 11*(4), 555. https://doi.org/10.3390/foods11040555

Konuk Takma, D., & Korel, F. (2019). Active packaging films as a carrier of black cumin essential oil: Development and effect on quality and shelf-life of chicken breast meat. *Food Packaging and Shelf Life, 19*, 210–217. https://doi.org/10.1016/j.fpsl.2018.11.002

Kumar, N., & Goel, N. (2019). Phenolic acids: Natural versatile molecules with promising therapeutic applications. *Biotechnology Reports, 24*, e00370. https://doi.org/10.1016/j.btre.2019.e00370

Latifah-Munirah, B., Himratul-Aznita, W. H., & Mohd Zain, N. (2015). Eugenol, an essential oil of clove, causes disruption to the cell wall of Candida albicans (ATCC 14053). *Frontiers in Life Science, 8*(3), 231–240. https://doi.org/10.1080/21553769.2015.1045628

Li, J., Ye, F., Lei, L., & Zhao, G. (2018). Combined effects of octenylsuccination and oregano essential oil on sweet potato starch films with an emphasis on water resistance. *International Journal of Biological Macromolecules, 115*, 547–553.

Limpisophon, K., Tanaka, M., & Osako, K. (2010). Characterisation of gelatin—fatty acid emulsion films based on blue shark (*Prionace glauca*) skin gelatin. *Food Chemistry, 122*(4), 1095–1101. https://doi.org/10.1016/j.foodchem.2010.03.090

Liolios, C. C., Gortzi, O., Lalas, S., Tsaknis, J., & Chinou, I. (2009). Liposomal incorporation of carvacrol and thymol isolated from the essential oil of *Origanum dictamnus* L. and *in vitro* antimicrobial activity. *Food Chemistry, 112*(1), 77–83. https://doi.org/10.1016/j.foodchem.2008.05.060

Mancianti, F., & Ebani, V. V. (2020). Biological activity of essential oils. *Molecules, 25*(3), 678. https://doi.org/10.3390/molecules25030678

Maurya, A., Prasad, J., Das, S., & Dwivedy, A. K. (2021). Essential oils and their application in food safety. *Frontiers in Sustainable Food Systems, 5*. https://doi.org/10.3389/fsufs.2021.653420

Mekonnen, A., Yitayew, B., Tesema, A., & Taddese, S. (2016). *In vitro* antimicrobial activity of essential oil of *Thymus schimperi, Matricaria chamomilla, Eucalyptus globulus*, and *Rosmarinus officinalis*. *International Journal of Microbiology, 2016*. https://doi.org/10.1155/2016/9545693

Mendoza-Yepes, M. J., Sanchez-Hidalgo, L. E., Maertens, G., & Marin-Iniesta, F. (1997). Inhibition of *Listeria monocytogenes* and other bacteria by a plant essential oil (DMC) in Spanish soft cheese. *Journal of Food Safety, 17*(1), 47–55. https://doi.org/10.1111/j.1745-4565.1997.tb00175

Miller, K., & Krochta, J. (1997). Oxygen and aroma barrier properties of edible films: A review. *Trends in Food Science & Technology, 8*. https://doi.org/10.1016/S0924-2244(97)01051-0

Miranda, M., Sun, X., Marín, A., dos Santos, L. C., Plotto, A., Bai, J., Benedito Garrido Assis, O., David Ferreira, M., & Baldwin, E. (2022). Nano- and micro-sized carnauba wax emulsions-based coatings incorporated with ginger essential oil and hydroxypropyl methylcellulose on papaya: Preservation of quality and delay of post-harvest fruit decay. *Food Chemistry: X, 13*, 100249. https://doi.org/10.1016/j.fochx.2022.100249

Moghimipour, E., Aghel, N., Zarei Mahmoudabadi, A., Ramezani, Z., & Handali, S. (2012). Preparation and characterization of liposomes containing essential oil of *Eucalyptus camaldulensis* leaf. *Jundishapur Journal of Natural Pharmaceutical Products, 7*(3), 117–122. https://pubmed.ncbi.nlm.nih.gov/24624167; www.ncbi.nlm.nih.gov/pmc/articles/PMC3941848/

Mohsenabadi, N., Rajaei, A., Tabatabaei, M., & Mohsenifar, A. (2018). Physical and antimicrobial properties of starch-carboxy methyl cellulose film containing rosemary essential oils encapsulated in chitosan nanogel. *International Journal of Biological Macromolecules, 112*, 148–155. https://doi.org/10.1016/j.ijbiomac.2018.01.034

Moradi, M., Tajik, H., Razavi Rohani, S. M., & Mahmoudian, A. (2016). Antioxidant and antimicrobial effects of zein edible film impregnated with *Zataria multiflora* Boiss. essential oil and monolaurin. *LWT— Food Science and Technology, 72*, 37–43. https://doi.org/10.1016/j.lwt.2016.04.026

Moradi, M., Tajik, H., Razavi Rohani, S. M., Oromiehie, A. R., Malekinejad, H., Aliakbarlu, J., & Hadian, M. (2012). Characterization of antioxidant chitosan film incorporated with *Zataria multiflora* Boiss essential oil and grape seed extract. *LWT—Food Science and Technology, 46*(2), 477–484. https://doi.org/10.1016/j.lwt.2011.11.020

Mourey, A., & Canillac, N. (2002). Anti-*Listeria monocytogenes* activity of essential oils components of conifers. *Food Control, 13*(4), 289–292. https://doi.org/10.1016/S0956-7135(02)00026-9

Mukurumbira, A. R., Shellie, R. A., Keast, R., Palombo, E. A., & Jadhav, S. R. (2022). Encapsulation of essential oils and their application in antimicrobial active packaging. *Food Control, 136*. https://doi.org/10.1016/j.foodcont.2022.108883

Mustafa, P., Niazi, M. B. K., Jahan, Z., Samin, G., Hussain, A., Ahmed, T., & Naqvi, S. R. (2020). PVA/starch/propolis/anthocyanins rosemary extract composite films as active and intelligent food packaging materials. *Journal of Food Safety, 40*(1), e12725. https://doi.org/10.1111/jfs.12725

Negi, P., & Jayaprakasha, G. (2001). Antibacterial activity of grapefruit (*Citrus paradisi*) peel extracts. *European Food Research and Technology, 213*(6), 484–487.

Nevas, M., Korhonen, A., Lindström, M., Turkki, P., & Korkeala, H. (2004). Antibacterial efficiency of Finnish spice essential oils against pathogenic and spoilage bacteria. *Journal of Food Protection, 67*(1), 199–202.

Ningrum, A., Widyastuti Perdani, A., Supriyadi, Siti Halimatul Munawaroh, H., Aisyah, S., & Susanto, E. (2021). Characterization of tuna skin gelatin edible films with various plasticizers-essential oils and their effect on beef appearance. *Journal of Food Processing and Preservation, 45*(9). https://doi.org/10.1111/jfpp.15701

Noshirvani, N., Ghanbarzadeh, B., Gardrat, C., Rezaei, M. R., Hashemi, M., Le Coz, C., & Coma, V. (2017). Cinnamon and ginger essential oils to improve antifungal, physical and mechanical properties of chitosan-carboxymethyl cellulose films. *Food Hydrocolloids, 70*, 36–45.

Omidbaigi, R., Yahyazadeh, M., Zare, R., & Taheri, H. (2007). The *In vitro* action of essential oils on *Aspergillus flavus*. *Journal of Essential Oil-Bearing Plants (Dehra Dun), 10*(1), 46–52. https://doi.org/10.1080/0972060X.2007.10643518

Onawunmi, G. O. (1989). Evaluation of the antimicrobial activity of citral. *Letters in Applied Microbiology, 9*(3), 105–108. https://agris.fao.org/agris-search/search.do?recordID=US201302680990

Oussalah, M., Caillet, S., & Lacroix, M. (2006). Mechanism of action of Spanish oregano, Chinese cinnamon, and savory essential oils against cell membranes and walls of *Escherichia coli* O157:H7 and *Listeria monocytogenes*. *Journal of Food Protection, 69*, 1046–1055. https://doi.org/10.4315/0362-028x-69.5.1046

Pandey, V. K., Islam, R. U., Shams, R., & Dar, A. H. (2022). A comprehensive review on the application of essential oils as bioactive compounds in nano-emulsion based edible coatings of fruits and vegetables. *Applied Food Research, 2*(1), 100042. https://doi.org/10.1016/j.afres.2022.100042

Parvin, R., Shahrokh, K. O., Mozafar, S., Hassan, E., & Mehrdad, B. (2014). Biosynthesis, regulation and properties of plant monoterpenoids. *Journal of Medicinal Plant Research, 8*(29), 983–991. https://doi.org/10.5897/JMPR2012.387

Pelissari, F. M., Ferreira, D. C., Louzada, L. B., dos Santos, F., Corrêa, A. C., Moreira, F. K. V., & Mattoso, L. H. (2019). Starch-based edible films and coatings. In *Starches for food application* (pp. 359–420). Academic Press. https://doi.org/10.1016/b978-0-12-809440-2.00010-1

Perdones, Á, Vargas, M., Atarés, L., & Chiralt, A. (2014). Physical, antioxidant and antimicrobial properties of chitosan—cinnamon leaf oil films as affected by oleic acid. *Food Hydrocolloids, 36*, 256–264. https://doi.org/10.1016/j.foodhyd.2013.10.003

Perdones, A., Sánchez-González, L., Chiralt, A., & Vargas, M. (2012). Effect of chitosan—lemon essential oil coatings on storage-keeping quality of strawberry. *Postharvest Biology and Technology, 70*, 32–41. https://doi.org/10.1016/j.postharvbio.2012.04.002

Pérez-Recalde, M., Ruiz Arias, I. E., & Hermida, É. B. (2018). Could essential oils enhance biopolymers performance for wound healing? A systematic review. *Phytomedicine (Stuttgart), 38*, 57–65. https://doi.org/10.1016/j.phymed.2017.09.024

Piñeros-Hernandez, D., Medina-Jaramillo, C., López-Córdoba, A., & Goyanes, S. (2017). Edible cassava starch films carrying rosemary antioxidant extracts for potential use as active food packaging. *Food Hydrocolloids, 63*, 488–495. https://doi.org/10.1016/j.foodhyd.2016.09.034

Pires, C., Ramos, C., Teixeira, B., Batista, I., Nunes, M. L., & Marques, A. (2013). Hake proteins edible films incorporated with essential oils: Physical, mechanical, antioxidant and antibacterial properties. *Food Hydrocolloids, 30*(1), 224–231. https://doi.org/10.1016/j.foodhyd.2012.05.019

Pires, C., Ramos, C., Teixeira, G., Batista, I., Mendes, R., Nunes, L., & Marques, A. (2011). Characterization of biodegradable films prepared with hake proteins and thyme oil. *Journal of Food Engineering, 105*(3), 422–428. https://doi.org/10.1016/j.jfoodeng.2011.02.036

Pranoto, Y., Salokhe, V. M., & Rakshit, S. K. (2005). Physical and antibacterial properties of alginate-based edible film incorporated with garlic oil. *Food Research International, 38*(3), 267–272. https://doi.org/10.1016/j.foodres.2004.04.009

Praseptiangga, D., Fatmala, N., Manuhara, G. J., Utami, R., & Khasanah, L. U. (2016). Preparation and preliminary characterization of semi refined kappa carrageenan-based edible film incorporated with cinnamon essential oil. *AIP Conference Proceedings, 1746*(1). https://doi.org/10.1063/1.4953961

Prashar, A., Locke, I. C., & Evans, C. S. (2004). Cytotoxicity of lavender oil and its major components to human skin cells. *Cell Proliferation, 37*(3), 221–229. https://doi.org/10.1111/j.1365–2184.2004.00307.x

Radziejewska-Kubzdela, E., Biegańska-Marecik, R., & Kidoń, M. (2014). Applicability of vacuum impregnation to modify physico-chemical, sensory and nutritive characteristics of plant origin products—a review. *International Journal of Molecular Sciences, 15*(9), 16577–16610. https://doi.org/10.3390/ijms150916577

Ramezani, Z., Zarei, M., & Raminnejad, N. (2015). Comparing the effectiveness of chitosan and nanochitosan coatings on the quality of refrigerated silver carp fillets. *Food Control, 51*, 43–48. https://doi.org/10.1016/j.foodcont.2014.11.015

Ramirez, P. Z., & Sanchez, M. Y. (2019). Degradable packaging film for fruit and vegetables. U.S. Patent Application No. 16/065,428.

Raybaudi-Massilia, R. M., Mosqueda-Melgar, J., & Martín-Belloso, O. (2008a). Edible alginate-based coating as carrier of antimicrobials to improve shelf-life and safety of fresh-cut melon. *International Journal of Food Microbiology, 121*(3), 313–327. https://doi.org/10.1016/j.ijfoodmicro.2007.11.010 [doi]

Raybaudi-Massilia, R. M., Rojas-Graü, M. A., Mosqueda-Melgar, J., & Martín-Belloso, O. (2008b). Comparative study on essential oils incorporated into an alginate-based edible coating to assure the safety and quality of fresh-cut Fuji apples. *Journal of Food Protection, 71*(6), 1150–1161. https://doi.org/10.4315/0362–028x-71.6.1150 [doi]

Ribeiro-Santos, R., Andrade, M., Melo, N. R. D., & Sanches-Silva, A. (2017a). Use of essential oils in active food packaging: Recent advances and future trends. *Trends in Food Science & Technology, 61*, 132–140. https://doi.org/10.1016/j.tifs.2016.11.021

Ribeiro-Santos, R., Andrade, M., & Sanches-Silva, A. (2017b). Application of encapsulated essential oils as antimicrobial agents in food packaging. *Current Opinion in Food Science, 14*, 78–84. https://doi.org/10.1016/j.cofs.2017.01.012

Ribeiro-Santos, R., de Melo, N. R., Andrade, M., & Sanches-Silva, A. (2017c). Potential of migration of active compounds from protein-based films with essential oils to a food and a food simulant. *Packaging Technology and Science, 30*(12), 791–798. https://doi.org/10.1002/pts.2334

Richardson, D. G., & Meheriuk, M. (1982). Controlled atmospheres for storage and transport of perishable agricultural commodities. In *Symposium series/Oregon State University, School of Agriculture (USA)*. Timber Press in cooperation with School of Agriculture, Oregon State University.

Robertson, G. L. (2010). *Food packaging and shelf life*. CRC Press. https://doi.org/10.1201/9781420078459

Sadgrove, N. J., Padilla-González, G. F., Leuner, O., Melnikovova, I., & Fernandez-Cusimamani, E. (2021). Pharmacology of natural volatiles and essential oils in food, therapy, and disease prophylaxis. *Frontiers in Pharmacology, 12*, 740302. https://doi.org/10.3389/fphar.2021.740302

Salvia-Trujillo, L., Rojas-Graü, M. A., Soliva-Fortuny, R., & Martín-Belloso, O. (2015). Use of antimicrobial nanoemulsions as edible coatings: Impact on safety and quality attributes of fresh-cut Fuji apples. *Postharvest Biology and Technology, 105*, 8–16. https://doi.org/10.1016/j.postharvbio.2015.03.009

Sánchez-González, L., Chiralt, A., González-Martínez, C., & Cháfer, M. (2011). Effect of essential oils on properties of film forming emulsions and films based on hydroxypropylmethylcellulose and chitosan. *Journal of Food Engineering, 105*(2), 246–253. https://doi.org/10.1016/j.jfoodeng.2011.02.028

Sánchez-González, L., González-Martínez, C., Chiralt, A., & Cháfer, M. (2010). Physical and antimicrobial properties of chitosan—tea tree essential oil composite films. *Journal of Food Engineering, 98*(4), 443–452. https://doi.org/10.1016/j.jfoodeng.2010.01.026

Sánchez-González, L., Vargas, M., González-Martínez, C., Chiralt, A., & Cháfer, M. (2009). Characterization of edible films based on hydroxypropylmethylcellulose and tea tree essential oil. *Food Hydrocolloids, 23*(8), 2102–2109. https://doi.org/10.1016/j.foodhyd.2009.05.006

Sarkar, S., Gupta, S., Variyar, P. S., Sharma, A., & Singhal, R. S. (2013). Hydrophobic derivatives of guar gum hydrolyzate and gum Arabic as matrices for microencapsulation of mint oil. *Carbohydrate Polymers, 95*(1), 177–182. https://doi.org/10.1016/j.carbpol.2013.02.070

Savoia, D. (2012). Plant-derived antimicrobial compounds: alternatives to antibiotics. *Future Microbiology, 7*(8), 979–990. https://doi.org/10.2217/fmb.12.68

Seifari, F. K., & Ahari, H. (2020). Active edible films and coatings with enhanced properties using nanoemulsion and nanocrystals. *Food & Health Journal, 3*(1), 15–22. https://fh.srbiau.ac.ir/article_16022_d1a57a539d-885d550cf6d732e3d6b975.pdf

Sessa, M., Ferrari, G., & Donsi, F. (2015). Novel edible coating containing essential oil nanoemulsions to prolong the shelf life of vegetable products. *Chemical Engineering Transactions, 43*. https://doi.org/10.3303/CET1543010

Settier-Ramirez, L., López-Carballo, G., Hernandez-Muñoz, P., Tinitana-Bayas, R., Gavara, R., & Sanjuán, N. (2022). Assessing the environmental consequences of shelf life extension: Conventional versus active packaging for pastry cream. *Journal of Cleaner Production, 333*, 130159. https://doi.org/10.1016/j.jclepro.2021.130159

Sharma, K., Babaei, A., Oberoi, K., Aayush, K., Sharma, R., & Sharma, S. (2022). *Essential oil nanoemulsion edible coating in food industry: A review*. Springer Science and Business Media LLC. https://doi.org/10.1007/s11947-022-02811-6

Sharma, S., Barkauskaite, S., Jaiswal, A. K., & Jaiswal, S. (2021). Essential oils as additives in active food packaging. *Food Chemistry, 343*, 128403. https://doi.org/10.1016/j.foodchem.2020.128403

Shelef, L. A. (1984). Antimicrobial effects of spices. *Journal of Food Safety, 6*(1), 29–44. https://doi.org/10.1111/j.1745-4565.1984.tb00477.x

Shelef, L. A., Jyothi, E. K., & Bulgarelli, M. A. (1984). Growth of enteropathogenic and spoilage bacteria in sage-containing broth and foods. *Journal of Food Science, 49*(3), 737–740. https://doi.org/10.1111/j.1365-2621.1984.tb13198.x

Sherry, M., Charcosset, C., Fessi, H., & Greige-Gerges, H. (2013). Essential oils encapsulated in liposomes: A review. *Journal of Liposome Research, 23*(4), 268–275. https://doi.org/10.3109/08982104.2013.819888

Shi, G., Lin, L., Liu, Y., Chen, G., Fu, S., Luo, Y., Yang, A., Zhou, Y., Wu, Y., & Li, H. (2022). Multi-objective optimization and extraction mechanism understanding of ionic liquid assisted in extracting essential oil from Forsythiae fructus. *Alexandria Engineering Journal, 61*(9), 6897–6906. https://doi.org/10.1016/j.aej.2021.12.035

Shiku, Y., Yuca Hamaguchi, P., Benjakul, S., Visessanguan, W., & Tanaka, M. (2004). Effect of surimi quality on properties of edible films based on Alaska pollack. *Food Chemistry, 86*(4), 493–499. https://doi.org/10.1016/j.foodchem.2003.09.022

Shishir, M. R. I., Xie, L., Sun, C., Zheng, X., & Chen, W. (2018). Advances in micro and nano-encapsulation of bioactive compounds using biopolymer and lipid-based transporters. *Trends in Food Science & Technology, 78*, 34–60. https://doi.org/10.1016/j.tifs.2018.05.018

Shojaee-Aliabadi, S., Hosseini, H., Mohammadifar, M. A., Mohammadi, A., Ghasemlou, M., Ojagh, S. M., Hosseini, S. M., & Khaksar, R. (2013). Characterization of antioxidant-antimicrobial kappa-carrageenan films containing *Satureja hortensis* essential oil. *International Journal of Biological Macromolecules, 52*, 116–124. https://doi.org/10.1016/j.ijbiomac.2012.08.026

Si, H., Hu, J., Liu, Z., & Zeng, Z. (2008). Antibacterial effect of oregano essential oil alone and in combination with antibiotics against extended-spectrum β-lactamase-producing *Escherichia coli*. *FEMS Immunology and Medical Microbiology, 53*(2), 190–194. https://doi.org/10.1111/j.1574-695X.2008.00414.x

Siripatrawan, U., & Harte, B. R. (2010). Physical properties and antioxidant activity of an active film from chitosan incorporated with green tea extract. *Food Hydrocolloids, 24*(8), 770–775. https://doi.org/10.1016/j.foodhyd.2010.04.003

Sivarooban, T., Hettiarachchy, N. S., & Johnson, M. G. (2008). Physical and antimicrobial properties of grape seed extract, nisin, and EDTA incorporated soy protein edible films. *Food Research International, 41*(8), 781–785. https://doi.org/10.1016/j.foodres.2008.04.007

Socaciu, M. I., Fogarasi, M., Simon, E. L., Semeniuc, C. A., Socaci, S. A., Podar, A. S., & Vodnar, D. C. (2021). Effects of whey protein isolate-based film incorporated with tarragon essential oil on the quality and shelf-life of refrigerated brook trout. *Foods, 10*(2). https://doi.org/10.3390/foods10020401

Swamy, M. K., Akhtar, M. S., & Sinniah, U. R. (2016). Antimicrobial properties of plant essential oils against human pathogens and their mode of action: An updated review. *Evidence-Based Complementary and Alternative Medicine: eCAM, 2016*. https://doi.org/10.1155/2016/3012462

Talón, E., Lampi, A., Vargas, M., Chiralt, A., Jouppila, K., & González-Martínez, C. (2019). Encapsulation of eugenol by spray-drying using whey protein isolate or lecithin: Release kinetics, antioxidant and antimicrobial properties. *Food Chemistry, 295*, 588–598. https://doi.org/10.1016/j.foodchem.2019.05.115

Tan, L. T. H., Lee, L. H., Yin, W. F., Chan, C. K., Abdul Kadir, H., Chan, K. G., & Goh, B. H. (2015). Traditional uses, phytochemistry, and bioactivities of *Cananga odorata* (Ylang-Ylang). *Evidence-Based Complementary and Alternative Medicine, 2015*, 896314. https://doi.org/10.1155/2015/896314

Teixeira, B., Marques, A., Pires, C., Ramos, C., Batista, I., Saraiva, J. A., & Nunes, M. L. (2014). Characterization of fish protein films incorporated with essential oils of clove, garlic and origanum: Physical, antioxidant and antibacterial properties. *LWT—Food Science and Technology, 59*(1), 533–539. https://doi.org/10.1016/j.lwt.2014.04.024

Tongnuanchan, P., & Benjakul, S. (2014). Essential oils: Extraction, bioactivities, and their uses for food preservation. *Journal of Food Science, 79*(7), R1231–R1249. https://doi.org/10.1111/1750–3841.12492

Tongnuanchan, P., Benjakul, S., & Prodpran, T. (2012). Properties and antioxidant activity of fish skin gelatin film incorporated with citrus essential oils. *Food Chemistry, 134*(3), 1571–1579. https://doi.org/10.1016/j.foodchem.2012.03.094

Tongnuanchan, P., Benjakul, S., Prodpran, T., & Nilsuwan, K. (2015). Emulsion film based on fish skin gelatin and palm oil: Physical, structural and thermal properties. *Food Hydrocolloids, 48*, 248–259. https://doi.org/10.1016/j.foodhyd.2015.02.025

Turasan, H., Sahin, S., & Sumnu, G. (2015). Encapsulation of rosemary essential oil. *LWT—Food Science and Technology, 64*(1), 112–119. https://doi.org/10.1016/j.lwt.2015.05.036

Varona, S., Martín, Á., & Cocero, M. J. (2011). Liposomal incorporation of lavandin essential oil by a thin-film hydration method and by particles from gas-saturated solutions. *Industrial & Engineering Chemistry Research, 50*(4), 2088–2097. https://doi.org/10.1021/ie102016r

Venkatachalam, K., & Lekjing, S. (2020). A chitosan-based edible film with clove essential oil and nisin for improving the quality and shelf life of pork patties in cold storage. *RSC Advances, 1*(3), 17777–17786. https://doi.org/10.1039/d0ra02986f

Vigil, M., Pedrosa-Laza, M., Alvarez Cabal, J., & Ortega-Fernández, F. (2020). Sustainability analysis of active packaging for the fresh cut vegetable industry by means of attributional & consequential life cycle assessment. *Sustainability (Basel, Switzerland), 12*(17), 7207. https://doi.org/10.3390/su12177207

Viuda-Martos, M., Mohamady, M. A., Fernández-López, J., Abd ElRazik, K. A., Omer, E. A., Pérez-Alvarez, J. A., & Sendra, E. (2011). *In vitro* antioxidant and antibacterial activities of essentials oils obtained from Egyptian aromatic plants. *Food Control, 22*(11), 1715–1722. https://doi.org/10.1016/j.foodcont.2011.04.003

Vu, K. D., Hollingsworth, R. G., Leroux, E., Salmieri, S., & Lacroix, M. (2011). Development of edible bio-active coating based on modified chitosan for increasing the shelf life of strawberries. *Food Research International, 44*(1), 198–203. https://doi.org/10.1016/j.foodres.2010.10.037

Wang, B., Yan, S., Qiu, L., Gao, W., Kang, X., Yu, B., Liu, P., Cui, B., & Abd El-Aty, A. M. (2022). Antimicrobial activity, microstructure, mechanical, and barrier properties of cassava starch composite films supplemented with geranium essential oil. *Frontiers in Nutrition, 9*, 882742. https://doi.org/10.3389/fnut.2022.882742

Ward, G., & Nussinovitch, A. (1996). Gloss properties and surface morphology relationships of fruits. *Journal of Food Science, 61*(5), 973–977. https://doi.org/10.1111/j.1365-2621.1996.tb10914.x

Xing, C., Qin, C., Li, X., Zhang, F., Linhardt, R. J., Sun, P., & Zhang, A. (2019). Chemical composition and biological activities of essential oil isolated by HS-SPME and UAHD from fruits of bergamot. *LWT, 104*, 38–44.

Xing, Y., Li, X., Xu, Q., Yun, J., Lu, Y., & Tang, Y. (2011). Effects of chitosan coating enriched with cinnamon oil on qualitative properties of sweet pepper (*Capsicum annuum* L.). *Food Chemistry, 124*(4), 1443–1450. https://doi.org/10.1016/j.foodchem.2010.07.105

Xu, T., Gao, C., Yang, Y., Shen, X., Huang, M., Liu, S., & Tang, X. (2018). Retention and release properties of cinnamon essential oil in antimicrobial films based on chitosan and gum Arabic. *Food Hydrocolloids, 84*, 84–92. https://doi.org/10.1016/j.foodhyd.2018.06.003

Xue, F., Zhao, M., Liu, X., Chu, R., Qiao, Z., Li, C., & Adhikari, B. (2021). Physicochemical properties of chitosan/zein/essential oil emulsion-based active films functionalized by polyphenols. *Future Foods, 3*. https://doi.org/10.1016/j.fufo.2021.100033

Yangilar, F. (2016). Effect of the fish oil fortified chitosan edible film on microbiological, chemical composition and sensory properties of Göbek Kashar cheese during ripening time. *Korean Journal for Food Science of Animal Resources, 36*(3), 377–388. https://doi.org/10.5851/kosfa.2016.36.3.377

Yanishlieva, N. V., Marinova, E., & Pokorný, J. (2006). Natural antioxidants from herbs and spices. *European Journal of Lipid Science and Technology, 108*(9), 776–793. https://doi.org/10.1002/ejlt.200600127

Yao, Z., Chang, M., Ahmad, Z., & Li, J. (2016). Encapsulation of rose hip seed oil into fibrous zein films for ambient and on demand food preservation via coaxial electrospinning. *Journal of Food Engineering, 191*, 115–123. https://doi.org/10.1016/j.jfoodeng.2016.07.012

Yap, P. S. X., Yiap, B. C., Ping, H. C., & Lim, S. H. E. (2014). Essential oils, a new horizon in combating bacterial antibiotic resistance. *The Open Microbiology Journal, 8*(1), 6–14. https://doi.org/10.2174/1874285801408010006

Yeshi, K., & Wangchuk, P. (2022). Chapter 11—essential oils and their bioactive molecules in healthcare. *Herbal Biomolecules in Healthcare Applications* (pp. 215–237). Elsevier Inc. https://doi.org/10.1016/B978-0-323-85852-6.00006-8

Zhang, H., Hortal, M., Dobon, A., Bermudez, J. M., & Lara-Lledo, M. (2015). The effect of active packaging on minimizing food losses: Life cycle assessment (LCA) of essential oil component-enabled packaging for fresh beef. *Packaging Technology & Science, 28*(9), 761–774. https://doi.org/10.1002/pts.2135

Zhang, H., Hortal, M., Dobon, A., Jorda-Beneyto, M., & Bermudez, J. M. (2017). Selection of nanomaterial-based active agents for packaging application: Using life cycle assessment (LCA) as a tool. *Packaging Technology & Science, 30*(9), 575–586. https://doi.org/10.1002/pts.2238

Zhang, X., Ismail, B. B., Cheng, H., Jin, T. Z., Qian, M., Arabi, S. A., Liu, D., & Guo, M. (2021). Emerging chitosan-essential oil films and coatings for food preservation—a review of advances and applications. *Carbohydrate Polymers, 273*, 118616. https://doi.org/10.1016/j.carbpol.2021.118616

Zinoviadou, K. G., Koutsoumanis, K. P., & Biliaderis, C. G. (2009). Physico-chemical properties of whey protein isolate films containing oregano oil and their antimicrobial action against spoilage flora of fresh beef. *Meat Science, 82*(3), 338–345. https://doi.org/10.1016/j.meatsci.2009.02.004

11

Polyhydroxyalkanoates (PHA)

Production, Properties, and Packaging Applications

Purnima Kumari and Anupama Singh

CONTENTS

11.1 Introduction

Polyhydroxyalkanoates (PHAs) belong to the family of biopolymers that are accumulated in various bacterial strains as carbon storage material due to nutrient imbalance and unfavorable growth conditions and are synthesized from the cells (Khanna & Srivastava, 2005; Anderson & Dawes, 1990). The occurrence of polyhydroxyalkanoates in bacterial cells was first observed in 1888 by Dutch microbiologist Martinus Beijerinck (Chowdhury, 1963). PHAs were studied as "lipids" by biochemists until 1925, when a French scientist named Lemoigne isolated it from *Bacillus megaterium*, determined

DOI: 10.1201/9781003303671-11

its composition, and showed that it was a homopolyester containing three hydroxybutyric acids (3-HBs). Since then, 3-hydroxybutyrate (3HB) monomer was considered the only hydroxyalkanoates (HA) accumulated by bacteria as an energy reserve and building block of microbial polymer until 1974, when Wallen and Rohwedder (1974) identified several other HA, including 3-hydroxyvalerate (3HV), 3-hydroxyhexanoate (3HHx) in the chloroform extract of activated sewage sludge. About a decade later, in 1983, DeSmith and fellows revealed that HA was a function of a substrate when *Pseudomonas oleovorans* was cultivated on n-octane. The storage polymer produced by this bacterium was discovered to primarily consist of a 3-hydroxyoctanoate (3HO) unit (De Smet *et al.*, 1983) and a small amount of 3HHx unit (Lageveen *et al.*, 1988). As the research progressed, it has come to the knowledge that these polymers not only accumulate and are synthesized from Gram-positive bacteria but also appear in a wide variety of Gram-negative bacteria, in aerobic and anaerobic bacteria, and in some of archaebacteria as well (Masood *et al.*, 2015; Akiyama *et al.*, 2011; Aarthi & Ramana, 2011; Chee *et al.*, 2010; Li *et al.*, 2007). To date, scientists have identified around 140 types of PHA monomers (Steinbüchel & Valentin, 1995). Consequently, this class of bacterial reserve polymers has been given the more generic polyhydroxyalkanoates (PHA), which encompasses all these components.

PHA synthesis, in general, begins with the formation of hydroxyacyl-CoAs from carbon sources and precursors, followed by polymerization of hydroxyacyl-CoAs into PHAs by PHA synthases. The accumulation of PHA granules inside the bacteria is a natural way of storing carbon and energy, and this will happen during nutritional stress conditions, that is, depletion of nitrogen, phosphorus, trace elements, or dissolved oxygen, whereas surplus carbon supply is still available (Masood *et al.*, 2013; Jendrossek, 2009; Reddy *et al.*, 2003). PHAs are water-insoluble, deposited as intracellular granules generally 0.2–0.5 mm in diameter, and located in the cell cytoplasm (Sudesh *et al.*, 2000). This allows bacteria to store extra nutrients inside the cells. There is a coating of phospholipids and proteins on the PHA surface in which "phasins," a predominant protein, influences the number and size of PHA

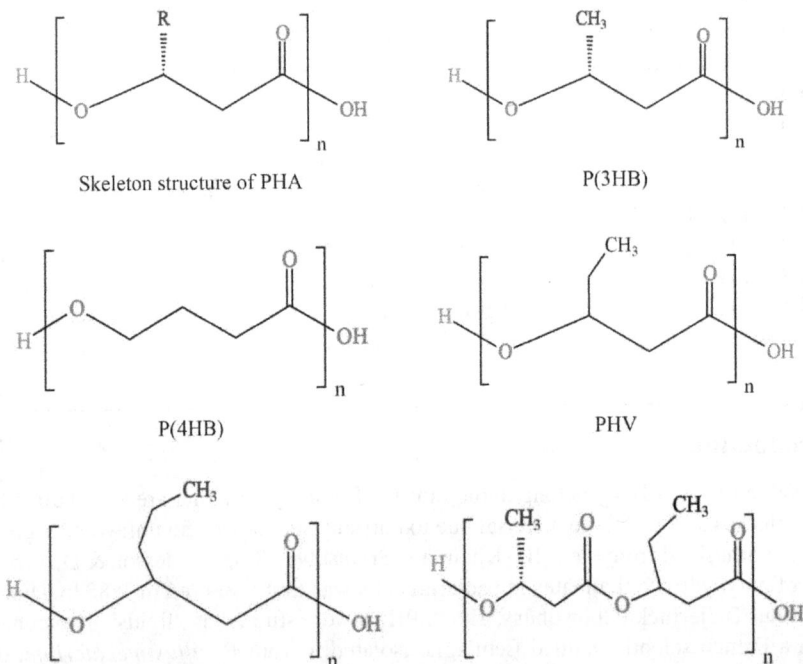

FIGURE 11.1 Molecular structure of different PHA molecules.

molecule (Pötter & Steinbüchel, 2005; Pötter *et al.*, 2002). Carbon substrate, biosynthesis pathways, and microbial strain also influence the molecular structure and copolymer composition of PHAs.

A PHA molecule mainly consists of several hundreds to thousands of (R)-hydroxy fatty acids (HFA) (Khanna & Srivastava, 2005). Each monomer unit is attached to a side chain alkyl group which varies from methyl group to decyl group. The alkyl group need not be saturated, but it can be unsaturated, branched, aromatic, or substituted groups (Lu *et al.*, 2009). Different types of PHA molecules (Figure 11.1) identified have primarily linear structures from one end to another, made up of monomers of 3-hydroxy fatty acids (3-HFA). The polymer is formed by two or more monomer units attaching carboxyl and hydroxyl groups of different units by an ester bond.

PHA can be categorized based on the total number of C-atoms in a monomer unit. Typically, the PHAs are divided into three categories: short-chain length PHA (scl-PHA), which contains 3–5 C-atoms; medium-chain length (mcl-PHA), which contains 6–14 C-atoms; and long-chain length PHA (LCL-PHA), which contain more than 14 C-atoms. Poly-3-hydroxybutyrate (PHB) and poly-3-hydroxybutyrate-co-3 hydroxyvalerate (PHBV) are some of the examples of scl-PHA, whereas poly-3-hydroxyhexanoate (PHHx) and poly-3-hydroxyoctanoate (PHO) are the examples of mcl-PHA. Scl-PHA, such as PHB, has high crystallinity, with a glass transition temperature (T_g) of 4°C and a melting temperature (T_m) of 180°C. Another scl-PHA, that is, PHBV, has somewhat-lower crystallinity and melting temperature. However, the Tg of MCL-PHAs ranges from -25 to 65°C and T_m from 42 to 65°C (Zinn & Hany, 2005). These polymers' average molecular weight (Mw) ranges from 1,104 to 4,106 Da (Agus *et al.*, 2006). Hence, by varying the alkyl group, chain length, composition of side chains, as well as the use of genetically modified organisms (GMOs), a wide range of PHA polymers can be produced, which can have a vast array of functionality and potential application (Escapa *et al.*, 2011; Madison & Huisman, 1999). Moreover, it can be noted that these structural diversities depend on the metabolic pathways of microorganisms for synthesis and the carbon sources (substrate) (Steinbüchel & Lütke-Eversloh, 2003).

Biodegradability, biocompatibility, thermoplasticity, hydrophobicity, gas and water impermeability, non-toxicity, nonlinear optical activity, and piezoelectricity are some of the promising features of produced PHAs (Bugnicourt *et al.*, 2014; Rhim *et al.*, 2013; Thellen *et al.*, 2008; Loo & Sudesh, 2007; Sudesh *et al.*, 2000). Because of their distinctive properties and utilization of renewable resources for their synthesis, PHAs have attracted a lot of commercial and academic interest (Shah *et al.*, 2008). Furthermore, these promising features of PHA make it suitable for various applications, including medical items and biodegradable packaging. However, one of the main reasons for its limited application as a biodegradable commodity plastic is the high cost of production. PHA can be used more widely in daily life if production methods are improved and costs are reduced (Li *et al.*, 2007). This has sparked interest across the globe in the cost-effective and efficient production of PHA using cheaper carbon sources and novel microorganisms.

This chapter deals with the various microorganisms that can produce more PHAs from renewable, less-expensive carbon sources like fatty acids, carbohydrates, or industrial waste. This is followed by a thorough description of the PHAs' characteristics, including their permeability, degradability, thermal properties, and crystallinity. In addition, it has compiled the current data on PHA usage in nanocomposites, multilayer films, paper and cardboard coatings, and active packaging. In the end, some instances of PHAs being used commercially in the food packaging sector are reviewed. A summary and a bright future outlook are provided in conclusion.

11.2 Production of PHAs

As discussed in the previous section, PHAs are synthesized by various bacterial strains using carbon sources. For this purpose, the identification and selection of microorganisms according to the targeted substrate are necessarily known. Further, the production of PHA (Figure 11.2) requires suitable growth conditions for microorganisms by fermentation. After that, the extraction and purification of PHAs from intracellular granules are required to obtain the PHA pallets to characterize. The following section will discuss the PHA production methods in detail.

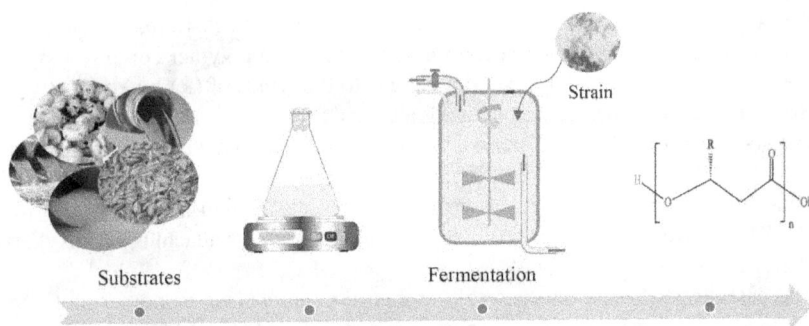

FIGURE 11.2   Schematic diagram of PHA production.

11.2.1 Sources Involved in the Production of PHA

11.2.1.1 Microorganisms for PHA Production

In PHA production, microorganisms play a significant role. Synthesis of PHA can be possible by different living organisms, including bacteria. Scientists have reported about 250 different PHA-accumulating bacteria in which Gram-negative and Gram-positive bacterial species are mainly studied. The following section will discuss the PHA-producing bacteria one by one.

11.2.1.1.1 Role of Gram-Negative Bacteria in PHA Production

Gram-negative bacteria have been discovered to produce the majority of PHA (Lu *et al.*, 2009). *Azohydromonas*, *Burkholderia*, and *Cupriavidus* species are capable of producing large amounts of scl-PHA. According to Tan *et al.* (2014b) and Grothe *et al.* (1999), *A. lata* (ATCC 29714) can synthesize 50–80% of poly(3-hydroxybutyrate) (P(3HB)) from a variety of sugars (glucose, fructose, sucrose), whereas about 69% of P(3HB) can be produced from fatty acids using *Burkholderia* sp. USM (JCM 15050) (Chee *et al.*, 2010). A hydrogen-oxidizing bacterium called *Cupriavidus necator* can make PHA by fixing CO_2 using the Calvin cycle (Khosravi *et al.*, 2013). For growth and PHA synthesis, the well-studied *C. necator* H16 (ATCC 17699) can also switch between autotrophic and heterotrophic modes. Additionally, emerging data revealed that this bacterium could cause concurrent heterotrophic and autotrophic PHA production (Shimizu *et al.*, 2013). Due to its distinct physiology, *C. necator* H16 (ATCC 17699) was able to accumulate P(3HB) in the range of 67 to 88.9% of dry cell mass (DCM) using a variety of chemically diverse carbon substrates, including CO_2, sugars (such as glucose and fructose), n-alkanoic acids (such as 4-hydroxyhexanoic acid), and vegetable oils (such as olive oil, corn oil, avocado oil, and palm oil) (Kamilah *et al.*, 2018; Flores-Sánchez *et al.*, 2017). Because these bacterial species have a high capacity for producing scl-PHA, they are frequently employed in the industrial manufacturing of PHA.

Additionally, Gram-negative methylotroph species can produce scl-PHA. *Methylobacterium* and *Paracoccus* are some of the bacterial species studied previously and reported to produce poly(3-hydroxyhexanoate) (P3HHX) and poly(3-hydroxybutyrate-co-3-hydroxyvalerate) P(HB-co-HV) from alcohols such as ethanol, methanol, and n-pentanol, respectively (Höfer *et al.*, 2011; Chanprateep *et al.*, 2001). Utilizing methylotrophs for commercial scl-PHA synthesis may result in lower PHA costs due to methanol's lower price than pure sugar substrates. Nevertheless, before methylotrophs are regarded as desirable inoculum for industrial production of scl-PHA, more study is needed to increase their PHA content and yield (Khanna & Srivastava, 2005).

Another bacterium species, that is, *Pseudomonas*, is also found to produce mcl-PHA, generally in between 1 and 30% of dry cell mass. *P. putida* mt-2 (NCIMB 10432) and *P. putida* KT2440 (ATCC 47054) were used to produce higher quantities of mcl-PHA, which is approximately 75–77% of DCM (Sun *et al.*, 2007). Besides mcl-PHA, the synthesis of scl-mcl-PHA copolyesters by several *Pseudomonas* species has also been documented. Some of them are *P. marginalis*, *P. mendocina*, *P. oleovorans*, and *P. putida* GPo1, which thrive on carbon sources such as n-alkanoates and 1,3-butanediol. Another well-known characteristic of *Pseudomonas* sp. is that it can biodegrade poisonous and resistant aromatic carbon substrates (Poblete-Castro *et al.*, 2012). These abilities have allowed them to be successfully used in treating polluted soils, exhaust gas, and effluents (Jung & Park, 2005; Greene *et al.*, 2004). Research studies showed that aromatic-degrading microorganisms, including *P. putida* F1, *P. putida* mt-2, and *P. putida* CA-3, are able to biotransform hazardous pollutants such as benzene, ethylbenzene, toluene, styrene, and xylene to mcl-PHA (Nikodinovic-Runic *et al.*, 2011). However, some other *Pseudomonas* sp., including *P. aeruginosa* and *P. frederiksbergensis*, are used for crude pyrolysis by-products of several synthetic polymers (such as polyethylene, polystyrene, polyethylene terephthalate) to produce mcl-PHA, which is advantageous in offsetting the expense of waste treatment as well as PHA production.

Additionally, Gram-negative extremophilic bacteria have been found to accumulate PHA. For this purpose, these bacterial species require specific growth conditions, which can either be high salinity or a high temperature. According to Quillaguaman *et al.* (2005), the thermophilic *Thermus thermophilus* HB8 (ATCC27634) produces up to 35.6% of PHA at a higher growth temperature of 70°C. The bacterial strain uses whey to produce PHA copolymer comprising both short-chain and medium-chain polymers. *Halomonas boliviensis* LC1 (DSM15516), on the other hand, can survive in an environment with high salinity (0.77 M NaCl) and can produce up to 56% of P(3HB), a scl-PHA, utilizing starch hydrolysate as its source of carbon (Pantazaki *et al.*, 2009). Since the extremophiles have the potential to grow under extreme circumstances, they can be used to reduce the cost involved in the pretreatment of waste effluents with a high salt concentration at their initial source and produces PHA.

Although the key concern for Gram-negative bacteria is the existence of lipopolysaccharide endotoxins (LPS), which are present in the outer membrane of bacteria and can be extracted with PHA during processing (Rai *et al.*, 2011) and make the polymer unsuitable for biomedical application, repetitive solvent extractions, solvent extraction, along with purification by activated charcoal or treatment with oxidizing agents (such as benzoyl peroxide ($C_{14}H_{10}O_4$), hydrogen peroxide (H_2O_2), sodium hypochlorite (NaOCl), and ozone (O_3)) are some of the processes used to remove LPS endotoxin from PHA polymer (Rai *et al.*, 2011; Wampfler *et al.*, 2010; Chen & Wu, 2005). However, these techniques result in higher production costs and alter the properties of PHA polymers, like reduction in molecular mass and polydispersity. Using different Gram-negative bacteria for PHA synthesis at the industrial level has been compiled by Chen (2010b) and Chanprateep (2010) in their review articles.

11.2.1.1.2 Role of Gram-Positive Bacteria in PHA Production

In bacteriology, Gram-positive bacteria are defined as bacteria that produce a positive result in the Gram stain test. These bacteria have been shown to synthesize PHA in the genera *Bacillus*, *Clostridium*, *Micrococcus*, *Rhodococcus*, *Staphylococcus*, and *Streptomyces* (Lu *et al.*, 2009). Gram-positive bacteria have been found to produce scl-PHA more frequently than Gram-negative bacteria, but at lower concentrations, between around 2 and 50% DCM, limiting its use for commercial production of PHA (Valappil *et al.*, 2007).

Streptomyces sp. (ATCC 1238) grown on glucose had been reported to have a greater scl-PHA content, that is, 82% of DCM; however, the figure may have been overestimated by the crotonic acid test (Valappil *et al.*, 2007). If appropriate carbon sources and environmental factors are present, some microbial strains can produce mcl-PHA and scl-mcl-PHA copolymers. According to work reported by Shahid *et al.* (2013), *B. megaterium* (DSM 509) generated P(3HB) entirely from glycerol and succinic acid on a nitrogen-supplemented mineral medium. However, the same media began to synthesize scl-mcl-PHA when subcultured without nitrogen. And when the same bacterium was cultivated on nitrogen-deficient octanoic acid, only mcl-PHA (48% DCM) was seen to develop.

Gram-positive bacteria are preferable to Gram-negative bacteria despite producing less PHA on average. This may be because they do not produce LPS and hence are a good source of PHA as a raw material that can be used in the biomedical sector. However, it is known that some Gram-positive bacteria can create lipidated macroamphiphiles, such as lipoglycans and lipoteichoic acids (LTA), which are comparable to LPS in terms of their immunogenicity (Ray *et al.*, 2013). According to reports, the bacterial genera *Corynebacterium*, *Nocardia*, and *Rhodococcus* generate lipoglycans, while *Bacillus*, *Clostridium*, and *Staphylococcus* make LTA (Sutcliffe, 1995; Ruhland & Fiedler, 1990). The lipidated macroamphiphiles in PHA have an unknown effect on the immune system. Thus, the suitability of PHA, generated from Gram-positive bacteria, for biological applications would require in vitro or in vivo examination in the future.

11.2.1.1.3 Role of Archaea Bacteria in PHA Production

Archaea bacteria also contain PHA; however, *haloarchaea* are the only species that have been discovered to produce it (Han *et al.*, 2010). The most halophilic organisms in the *Archaea* domain, referred to as *haloarchaea*, need high salinity for consistent enzyme activity and may survive at maximum concentrations of up to 6M NaCl. According to reports, *haloarchaea* are capable of producing PHA using glucose, volatile fatty acids (VFA), and more complex substrates, like crude glycerol, starch, vinasse, and whey hydrolysate (Hermann-Krauss *et al.*, 2013; Bhattacharyya *et al.*, 2012; Han *et al.*, 2010; Koller *et al.*, 2008a). The only kind of PHA that could be manufactured was either scl-PHA homopolymer or scl-PHA heteropolymer. The homopolymer contains either a 3HB or 3HV monomer unit, whereas the heteropolymer has both 3HB and 3HV monomers in their structural composition (Poli *et al.*, 2011; Han *et al.*, 2010).

There are currently many *haloarchaea* cultures that can produce PHA, but the majority of them can only do so at low cellular contents, typically between 0.8 and 22.9% DCM. The most potent PHA producer is *Haloferax mediterranei* (DSM 1411), which can build up high levels of PHA, ranging from 50 to 76% DCM, and grows best in 2 to 5 M NaCl (Hermann-Krauss *et al.*, 2013; Bhattacharyya *et al.*, 2012). The hypersaline conditions necessary for its development indicate that only a small number of contaminating organisms could survive during PHA cultivation, reducing the necessity of sterilization and its related expense.

In contrast to moderately halophilic bacteria like *H. boliviensis* LC1 (DSM 15516), *haloarchaea* require extremely high salinities, which can be detrimental to PHA production since they increase chemical costs and hasten the corrosive degradation of stainless-steel bioreactors. In terms of ease of PHA recovery, *haloarchaea* surpasses halophilic bacteria. In order to release intracellular PHA granules from halophilic bacteria during the recovery process, it is necessary to use chemical, enzymes, or mechanical procedures to break down the cell wall. These methods account for half or more of the entire cost of PHA production (Chen *et al.*, 2001). Chloroform and acetone, the two most commonly used extraction solvents, could harm the environment if their use and disposal are improperly managed. On the other hand, PHA granules are released when *haloarchaea* cells are lysed in distilled water and can be retrieved by centrifugation at a lower speed. As a result, PHA recovery from *haloarchaea* is an effortless, less energy- and chemical-intensive process, consequently reducing extraction costs and a more negligible ecological impact (Poli *et al.*, 2011).

11.2.1.2 Substrates for PHA Production

Substrates are the material used by microorganisms for their growth and nourishment. For the production of PHA, microorganisms use a carbon source which can be different, including saccharides (such as glucose, fructose, lactose, maltose, arabinose, xylose), alkanes (such as hexane, octane), alkanoic acids (such as acetic acid, butyric acids, lauric acid, propionic acid, oleic acid, valeric acid), alcohols (such as ethanol, methanol, glycerol, octanol), and gases (such as carbon dioxide, methane) (Verlinden *et al.*, 2007; Silva *et al.*, 2004).

Although PHA is profoundly attractive as a replacement for petrochemical-based polymers, the high cost of microbial fermentation makes it five to ten times more expensive than petrochemical-based polymers like polypropylene. It is the major obstacle in producing and applying PHA commercially. The expense of substrates, particularly carbon sources, is the key reason driving up the production cost of PHA. Its high production cost is one of the issues prohibiting PHB from being used commercially. Economically speaking, 50% of the total manufacturing cost is attributable to the cost of the substrates, mostly the carbon source. In order to minimize the cost, research has been done on various inexpensive caron sources and the strains grown by using these cheap carbon substrates.

Agriculture and its associated sectors provide significant amounts of feedstocks and by-products that can be utilized as low-cost substrates for fermentation, in addition to being essential to human life and the well-being of all people. The successful integration of these materials could accelerate the development of a biorefinery business with feedstocks produced from agriculture into commercial processes. Waste streams have also been identified for PHA manufacturing as a free supply of carbons (Koller *et al.*, 2010). These include used cooking oil, used vinegar, used fats, used food, used agricultural waste, used household wastewater, used plant oil mill effluents, used plastic, used landfill gas, etc. The polymer

TABLE 11.1

Inexpensive substrates used by various microorganisms for PHA production

S. No.	Microorganisms	Substrate	PHA concentration	References
1.	*Cupriavidus nector*	Sugarcane vinasse and molasses	11.7 g/L	Dalsasso *et al.*, 2019
2.	*Ralstonia eutropha*	Kenaf biomass	4.81 g/L	Saratale *et al.*, 2019
3.	*Cupriavidus nector*	*Calophyllum inophyllum* oil	10.6 g/L	Arumugam *et al.*, 2018
4.	*Bacillus* sp. ISTVK1	Glycerol	4.44 g/L	Morya *et al.*, 2018
5.	*Serratia* sp.	Municipal secondary wastewater sludge	0.61 g/L	Kumar *et al.*, 2018
6.	*Bacillus safensis* EBT1	Sugarcane bagasse	5.9 g/L	Sakthiselvan & Madhumathi, 2018
7.	*Serratia* sp. ISTVKRI	Wastewater and glucose	0.34 g/L	Gupta *et al.*, 2017
8.	*Ralstonia eutropha* ATCC 17699	Waste biomass hydrolysate	11.4 g/L	Saratale & Oh, 2015
9.	*Bacillus* sp. SV13	Pineapple and sugarcane waste	1.86 g/L	Suwannasing *et al.*, 2015
10.	*Cupriavidus nector* H16	Jatropha oil	15.5 g/L	Mohidin Batcha *et al.*, 2014
11.	*Ralstonia eutropha* MTCC 1472	Paddy straw	19.2 g/L	Sandhya *et al.*, 2013
12.	*Psedomonas oleovorans* NRRL B-14682	Crude glycerol	1.14 g/L	Ashby *et al.*, 2011
13.	*Bacillus megaterium*	Cassava starch by-product	1.48 g/L	Krueger *et al.*, 2012
14.	*Wautersia eutropha*	Wheat hydrolysate	162.8 g/L	Xu *et al.*, 2010
15.	*Cupriavidus nector* DSM 545	Waste glycerol	0.84 g/L	Cavalheiro *et al.*, 2009
16.	*Pseudomonas fluorescens* A2a5	Sugarcane liquor	22 g/L	Jiang *et al.*, 2008
17.	*Pseudomonas hydrogenovora*	Hydrolyzed whey	1.27 g/L	Koller *et al.*, 2008b
18.	*Ralstonia eutropha* NCIMB 11599	Waste potato starch	94 g/L	Haas *et al.*, 2008
19.	*Pseudomonas putida*	Hydrolyzed corn oil	28 g/L	Shang *et al.*, 2008
20.	*Haloferax mediterranei*	Enzymatic extruded starch	20 g/L	Chen *et al.*, 2006

concentration and content were considerably lower than those generated using purified carbon sources, although there are several studies on manufacturing PHAs from inexpensive carbon substrates by wild-type producers. Table 11.1 summarizes various inexpensive substrates used by different microorganisms for PHA production. In order to get a significant amount of PHA from inexpensive carbon sources employing various microbial strains, it is necessary to create more effective fermentation techniques.

Waste biomass has intricate internal structures; they may consist of starch, cellulose, or sugar complex and typically cannot be transformed directly into desirable products by microbial activities. These require some sort of pretreatment to change the complicated molecules into simple sugars that microbes can utilize to produce the desired products. Researchers have used several methods for biomass conversion into fermentable sugars, such as physical methods, like ultrasound or microwave treatment; chemical methods, like acidic or alkaline treatment and enzymatic treatment; and hybrid techniques.

11.2.2 Processing Conditions Involved in PHA Production

11.2.2.1 Microbial Fermentation Techniques

PHA is technologically generated in bioreactors (also known as "fermenters") of various sizes and types under controlled temperature, dissolved oxygen tension (pO_2), and pH conditions. The cylindrical stirred tank reactor (STR), a device well recognized for manufacturing yeast, vinegar, or ethanol, is the most used form. These STRs can be used constantly or intermittently for PHA synthesis in batch or fed-batch mode.

Batch cultivations for the manufacture of PHA are straightforward in operation, but they are inherently underproductive. The producing strain's physiological prerequisites limit the maximum permitted level of carbon and nitrogen supply at the beginning of the batch fermentation. The most significant producing strains frequently begin PHA production processes with fermentable sugar concentrations of 10–30 g/L for simple substrates like sucrose, glucose, or glycerol, and nitrogen concentrations of 2–3 g/L of ammonia sulfate (Koller, 2018).

According to numerous studies, fed-batch technology demonstrated a greater yield than batch production at the industrial level. In the fed-batch fermentation technique, the substrate is pulsed in a bioreactor when its concentration falls below a certain threshold, holding the culture fluid in situ. When it comes to PHA, both carbon and nitrogen sources can be replenished at regular intervals in accordance with the biomass's consumption until the needed level of PHA-deficient biomass is obtained. The supply of nitrogen feeding during the growth phase can be monitored by linking the nitrogen feed with pH changes that occur during the process, since the pH value is inversely proportional to the growth of biomass. The nitrogen concentration is then allowed to fall until it approaches zero to initiate the change from the growth phase to the PHA synthesis phase; this can be achieved by using NaOH solution as an alkaline pH-correcting agent in place of the ammonia solution (da Cruz Pradella *et al.*, 2012). From this point forward, carbon sources are added as substrate pulses up until the process is finished. A slowdown in the pace at which a certain PHA is formed may suggest that the process is approaching the finish.

The main disadvantage of the fed-batch approach is that the addition of substrate during fermentation increases the volume and subsequently dilutes the fermentation broth. Hence, a highly concentrated feed with more than 100 gms per liter of carbon source is utilized to keep this dilution as low as feasible. Glucose feeding solutions with a 500 g carbon source per liter are a classic example. When waste streams (agricultural or industrial wastes) are employed as carbon sources, the substrate concentration is generally lower. For instance, whey obtained from the dairy sector is a very useful carbon source that comprises between 40 and 50 g/L (lactose) only for PHA manufacturing (Koller *et al.*, 2007). Such low substrate concentrations make direct application challenging due to the tremendous increase in volume and excessively low densities of cell production during fermentation. Additionally, toxic molecules with low molecular masses, such as furfural, hydroxymethylfurfural (HMF), carboxylic acids, and aldehydes produced during processing, can be accumulated simultaneously with the desired products and have a detrimental effect on the strain's fermentation ability (Obruca *et al.*, 2015). Hence, a modified semi-continuous cultivation technique, or repeated/cyclic fed-batch fermentation (CFBF), was studied as a substitute approach in which a high-capacity stirred tank fermenter was employed for the processing. In CFBF, the culture broth is partially removed, and the fresh medium is subsequently added to the bioreactor. Thus, it prevents volume growth and increases harmful by-products concentration (Ibrahim & Steinbüchel, 2010).

Continuous processes are often defined as operating under "steady-state" circumstances, in which process parameters such as substrate and product concentrations, as well as components such as pH, pO2, the volume of fermentation broth, and nutrient supply, are maintained at the same level. The substrate is continuously refed in response to the substrate conversion rate by active biomass, and the continuous removal of fermentation broth compensates for volume expansion. The "dilution rate" (D) is an important parameter defined by the flow rate ratio to the fermenter's working volume and is to be considered for continuous operations. Too low D values result in an inadequate supply of substrate to cultures, leading to low growth rates and productivities, whereas D values greater than a threshold value result in a "wash out" of the cultivation reactor. Particular growth rate and production can be smoothened by carefully regulating the D value; hence, the operator causes the system to have a process-engineering parameter to fix a physiological parameter of biomass. To achieve quick cell development and high productivity, D will be chosen slightly below the experimentally found values for max and up. The inverse value of D, known as the "retention period," regulates how much time the cells have to transform the substrate and store PHA in the cells. "Continuous processes" are frequently used interchangeably with "chemostat" processes; however, this is not strictly correct because "continuous" process regimes include, in addition to chemostats ("chemical environment remaining static"), pH-stat, turbido-stat ("turbidity as a parameter for the cell density remaining constant"), redox-stat, or volume-stat processes (Koller & Braunegg, 2015; Koller & Muhr, 2014). To correspond to the existing literature, the phrase "continuous PHA production" is used in the following paragraphs to refer to PHA chemostat studies.

11.2.2.2 Extraction and Purification of PHA

After the accumulation of PHA within cells, centrifugation is used to separate the PHA-containing biomass derived from the fermentation medium. After harvesting, cells are ruptured to recover the PHA. For the recovery of PHAs, numerous approaches are described in the literature in which the use of organic solvents is the oldest method studied. Lemoigne in 1920 used hot alcohol to extract P(3HB) and then purified it with chloroform and diethyl ether. PHAs cannot solubilize in water, but it is soluble in a few organic solvents and form the basis for all solvent extraction techniques. Solvents used in the extraction of PHA are some chlorinated hydrocarbons, including chloroform, 1,2-dichloroethane, and methylene chloride; various cyclic carbonates, including propylene and ethylene carbonates; and also cyclic carbonic esters. Furthermore, it was discovered that these solvents work with most PHAs, especially P(3HB) and other scl-PHA. Acetone is one of many solvents that can be used with mcl-PHA (3-hydroxyoctanoic acid). Mcl-PHA can also be extracted using various non-chlorinated solvents, producing polyesters with shallow endotoxin levels appropriate for medicinal purposes (Madkour *et al.*, 2013). However, the procedure is unfavorable from an economic and environmental standpoint due to the requirement for higher amounts of solvent.

The ease of recovering PHA is a crucial factor in the cost-effective production of PHA. Therefore, researchers have studied different PHA extraction technologies to make the entire process more accessible and less expensive. Enzymatic cell disruption is one of the techniques studied by researchers and has been applied at the commercial level. Imperial Chemical Industries (ICI) has used aqueous enzymatic digestion techniques as an alternative to the traditional solvent extraction method. The steps involved in this process are preheating enzymatic treatment and surfactant treatment. This approach is effective for developing PHB-based goods with lower purity concerns. However, to maximize product purity, this approach typically necessitates further digestion or solvent extraction procedures, which raises the recovery costs (Hahn *et al.*, 1994).

A differential digestion technique using sodium hypochlorite was described by Berger *et al.* (1989) and Ramsay *et al.* (1990). The cells are treated with sodium hypochlorite to extract PHA using this technique. All polymeric non-PHA cell constituents are more or less preferentially degraded by this oxidizing agent at the proper concentrations. In contrast, PHA mostly withstands hypochlorite's chemical attack and remains intact. Filtration or centrifugation can then be used to remove PHA. Although this approach is efficient at breaking down non-PHA cellular materials, it also severely degrades PHB, making it unsuitable for various applications. In order to effectively recover PHB from *C. necator*, Hahn *et al.* (1994) described the technique of using a dispersion of sodium hypochlorite and chloroform. The sodium hypochlorite is used to extract PHB from cells in the aqueous medium. Once released, it quickly transfers to the chloroform phase, where the PHB molecules are partially shielded by chloroform, protected from more degradation by hypochlorite.

Different mechanical disruption methods can also do extraction of PHA. Bead mill, high-pressure homogenization, and ultrasonication are some of the mechanical techniques used to disrupt the cell to isolate PHA. In the bead mill, the cells are disrupted by the force caused due to solid beads. The heat generated during mill operation needs to be efficiently cooled. Several passes are required at a higher speed for complete disruption of cells (Tamer *et al.*, 1998). These techniques are commonly used at the lab scale to extract PHA from bacterial cells, but mechanical disruption techniques are less significant for industrial-scale manufacturing. Other recovery methods studied by various researchers are supercritical fluid disruption, selective dissolution of cell mass, and dissolved air flotation (van Hee *et al.*, 2006; Yu & Chen, 2006; Khosravi-Darani *et al.*, 2004; Hejazi *et al.*, 2003). All these technologies are potential substitutes for solvent extraction.

After extracting PHA from biomass, purification is necessary to raise the degree of purity of the polymer. Commonly used purification techniques include enzymes or chelating agents with a hydrogen peroxide treatment. Horowitz and Brennan (2010) have proposed a new technique, that is, the use of ozone to purify PHAs. Ozone was applied to biomass through an oxygen stream containing 2 to 5% ozone. Impurities are bleached, deodorized, and solubilized by the ozone treatment, making removing them from aqueous polymer suspensions easier. This approach could replace the hydrogen peroxide treatment for purification, which has several disadvantages, including high operating temperatures (80–100°C), the instability of peroxides in the presence of excessive cellular biomass, and a reduction in the molecular weight of the polymer. The use of ammonia at high temperatures was also studied as an effective method of improving PHA purity while preserving the recovery yield (Burniol-Figols *et al.*, 2020).

11.2.2.3 Characterization of PHA

Quantitative analysis of PHA was traditionally done by crotonic acid assay proposed by Law and Slepecky in 1961. In this method, P(3HB) is reacted with concentrated sulfuric acid, and crotonic acid is produced from the reaction mixture. Further, crotonic acid can be detected using a UV spectrophotometer because it strongly absorbs UV light at 235 nm in concentrated sulfuric acid (Law & Slepecky, 1961). Although this method is quick and straightforward, it only determines P(3HB) and tends to overstate P(3HB) content. Modern analytical techniques, including gas chromatography (GC), gel permeation chromatography (GPC), nuclear magnetic resonance (NMR) spectroscopy, and differential scanning calorimetry (DSC), can currently be used to quantify and characterize a variety of intracellular microbial PHAs. It has recently been suggested to determine PHA inside intact cells using flow cytometry and two-dimensional fluorescence spectroscopy.

Following the identification of the fundamental characteristics of PHA polymer, structural and mechanical properties are investigated by methods such as optical microscopy, Fourier transform infrared (FTIR) spectroscopy, dielectric relaxation spectroscopy (DRS), tensile testing, dynamic mechanical thermal analysis (DMTA) (Kulkarni *et al.*, 2011). The three kinds of PHA—scl-PHA, mcl-PHA, and scl-mcl-PHA—found in intact cells or as pure polymers have been identified and distinguished using FTIR. Additionally, FTIR is used as a tool for the quantification of scl-PHA. The FTIR technique's solvent-free and rapid analysis nature removes the risk of exposure to potentially harmful chemicals while offering quick output data generation. However, the detection sensitivity of FTIR-based methods is poor, this method is ineffective in observing the changes in the monomeric structure of PHA, and also, it cannot discriminate between the copolymers and PHA blends. For this reason, FTIR-based approaches are more appropriate for routinely monitoring the PHA production for established bioprocesses.

Another analytical technique involves gas chromatography (GC) and liquid chromatography (LC) based methods, often used for analyzing the quality and quantity of PHA polymer due to their higher separation power and detection sensitivity. Automatic sample analysis makes these techniques provide accurate data. Chromatography-based techniques have higher detection sensitivity than FTIR-based methods, with high-performance liquid chromatography (HPLC) detection sensitivity ranging from 0.014 to 14 µg and GC detection sensitivities ranging from 0.05 pg to 15 mg, respectively, dependent on detector type and chemical derivatization techniques employed. Scl-PHA analysis has been improved with the introduction of automated LC systems. Ion-exchange HPLC combined with UV detection increased the measurement accuracy of crotonic acid produced from P(3HB). Ion chromatography (IC) using a conductivity detector and an anion trap column can also be used to simultaneously analyze and quantify 3HB and 3HV monomers.

A sophisticated liquid chromatography–mass spectrometry (LC-MS) method, which combines tremendous pressure and narrow column packing with particular MS detection, may also be a relatively fast and reliable method for regularly examining PHA monomers. Although, as an alternative approach to analyzing PHA monomers, LC-MS has only been used minimally to date. Through the use of LC-MS to quantify several PHA monomers, there is still much room to increase the potential of the LC system beyond the quantitative study of PHA.

Braunegg *et al.* (1978) have reported the work on the precise and repeatable measurement of P(3HB) concentration in biomass using GC with a flame ionization detector (GC-FID). Later, P(3HV) and mcl-PHA were also analyzed by GC-FID. However, using the proper PHA analytical standards is necessary for the GC-FID analysis to be as robust as possible. On the other hand, linking GC to a mass spectrometry detector (GC-MS) offers more accurate PHA monomer detection, identification, and quantification and allows for the provisional recognition of new PHA monomers while reference standards are unavailable. Without reference standards, this approach enabled the accurate detection and quantification of a wide variety of PHA, ranging from scl-PHA to lcl-PHA. Additionally, increases in sensitivity and specificity for identifying PHA monomers are anticipated from combining GC with tandem mass spectrometry (GC-MS[2]) (Gumel *et al.*, 2012).

Despite their benefits, PHA must first be depolymerized and chemically transformed into acids, diols, or methyl ester derivatives before it can be analyzed using chromatography-based procedures. This meant chromatography-based approaches could not tell if several PHA monomers were contained within a single PHA copolymer or a number of homogenous PHA polymers.

11.3 Properties of PHAs

11.3.1 Physical Properties

The properties of PHA are typically comparable to those of polymers made from petroleum. It has several physical properties, such as being insoluble in water, good UV resistance, stiff, highly polymerized, biodegradable, and thermoplastic (Keskin *et al.*, 2017). PHB's structure is comparable to polypropylene (PP), a synthetic polymer, which has a compact helical structure and a melting point of roughly around 180°C. Although their chemical properties are fundamentally different, PHB exhibits equivalent degrees of crystallinity and glass transition temperature to PP (Tripathi *et al.*, 2019). Hence, most of their applications are intended to replace synthetic plastic in coating and packaging applications.

PHA is a thermoplastic material that can be made from a variety of renewable resources. It exhibits strong UV and oxygen barrier capacities but has poor acid and base resistance. The most significant characteristic of PHA is that it is biocompatible, making it suitable for use in biomedical applications. The fact that PHA is water-insoluble sets it apart from most other biodegradable plastics now available in the market, which are either soluble in water or have high moisture sensitivity. PHA has a molecular weight between 105 and 106 Da. These polymers are partly crystalline, with a crystallinity level between 60 and 80%.

PHB, a popular PHA homopolymer, is a brittle substance with high crystallinity and stiffness that, when spun into fiber, exhibits qualities similar to those of strong elastic material. And when PHB is held at room temperature for several days, it gradually becomes more brittle, which is known as the "aging effect" (De Koning *et al.*, 1994). It has been discovered that creating copolymers by incorporating additional hydroxy acid units is an effective way to change the polymer's melting point, stiffness, and toughness.

PHB-co-PHV copolymers have lower stiffness and brittleness than PHB, while most of PHB's other mechanical properties remain the same. The polymer becomes more robust and flexible as the hydroxyvalerate units (HV) fraction increases. Additionally, the elongation to break also increases. Thermal processing of copolymers without thermal degradation is possible because the increase in 3HV concentration will cause a decrease in the melting temperature of the copolymer. PHB melts at 246.3°C, which is also the temperature at which it thermally degrades. This restricts its ability to be processed through injection molding. PHB-co PHV begins to degrade thermally at 260.4°C. This suggests that the PHBs' thermal stability is increased by copolymerization with valerate (Modi *et al.*, 2011). Young's modulus and tensile strength of PHB are comparable to those of polypropylene; however, PHB's elongation to break is much lower (Ojumu *et al.*, 2004). It has a larger density than polypropylene, which causes PHB to sink in water. The anaerobic biodegradation of PHB in sediments is facilitated by its sinking in water.

11.3.2 Molecular Weight

PHA's molecular weight (Mw) ranges from 50 kDa to 10,000 kDa, and depending on the Mn value, the PDI might be anywhere from 1.1 to 6.0 (Chen, 2010a; Chanprateep, 2010). Despite the presence of various microorganisms, the polymers made from molasses had average molecular weights comparable to those made by frequently used pure cultures $(3.5-4.3) \times 10^5$ g/mol and narrow weight distributions (PDI 1.8–2.1). Because of the strong correlation between mechanical strength and molecular weight, PHA with high molecular weight is chosen. By adjusting the growth conditions for PHA-producing bacteria, such as pH and temperature, the molecular weight of PHA has been altered (Bocanegra *et al.*, 2013; Agus *et al.*, 2006; Choi *et al.*, 2004). These environmental factors impact the four processes that directly regulate molecular weight: PHA synthase concentration, chain transfer reaction occurrence, PHA synthase catalytic activity, and simultaneous PHA degradation during biosynthesis (Tsuge, 2016). Depending on the microorganism and growth circumstances, the monomers are polymerized into high-molecular-weight polymers with molecular weights ranging from 200,000 to 3,000,000 Da (Sudesh *et al.*, 2000). A gel permeation chromatography (GPC) technique with polystyrene as a reference standard could be used to calculate the average molecular mass (Mw), molecular mass distribution (Mn), and polydispersity index (PDI; Mw/Mn) of a PHA polymer.

11.3.3 Crystallinity

PHA polymers can range in crystallinity from 0 to 70%, making them either non-crystalline or highly crystalline (Rai *et al.*, 2011; Chanprateep, 2010). The crystalline morphology and crystal structure of PHAs are directly related to the qualities of PHAs, including mechanical strength, toughness, transparency, storage stability, and permeability. PHAs' crystalline characteristics are significantly correlated with the structural composition of their copolymer units, nucleating agents, and environmental circumstances. Structure analysis tools including FTIR, DSC, and X-ray diffraction can assess crystallinity. PHA exhibits distinctive infrared absorption bands in FTIR analysis at specific wave numbers related to crystallinity (Tan *et al.*, 2014a).

11.3.4 Mechanical Properties

It is known that the type of crystal formation influences the mechanical properties of PHAs. Mechanical properties commonly assessed for PHA polymers include elongation at break (EAB), tensile strength, and Young's modulus. Elongation at break is the amount of a material to be extended before breaking and expressed as a percentage of its initial length. PHA polymers can be hard, stiff, or soft elastomeric, with a broad elongation at break values ranging from 2 to 1,000% (Chen, 2010a). Young's modulus measures the stiffness of PHA, and it varies from highly flexible mcl-PHA (0.008 MPa) to rigid scl-PHA (3.5×10^3 MPa) (Rai *et al.*, 2011). The resistance of a substance to breaking under tension is defined as its tensile strength. PHA polymers typically have tensile strengths ranging from 8.8 to 104 MPa. Standardized test procedures, such as the ASTM standards, can measure the aforementioned mechanical properties using a tensile tester device. It should be noted that PHB is a highly crystalline polymer (55–80%), which makes the material fragile (Domínguez-Díaz & Romo-Uribe, 2012).

11.3.5 Thermal Properties

Thermal properties of PHA material, including glass transition temperature (T_g), melting temperature (T_m), and thermodegradation temperature (T_d), are generally investigated to understand the thermostatic conditions for polymer production and application. PHAs have Tg, Tm, and Td values that typically range from –52 to 4°C, minimal to 177°C, and 227 to 256°C, respectively (Rai *et al.*, 2011; Chen, 2010a). Differential scanning calorimetry (DSC) and differential thermal analysis (DTA) were used to calculate Tg and Tm values. The thermogravimetric analysis (TGA) is used to determine the Td value of PHA. This procedure involves heating a sample at a predetermined rate in a controlled atmosphere while measuring the sample mass loss.

These thermal properties can significantly be affected by the monomer composition of the polymer. It was reported in some studies that as the molar percentage of monomers other than 3HB (non-3HB) increases, the glass transition temperature of the polymer decreases. P(3HB) copolymerization with 90 mol% of 3HV decreased the Tg from 4°C to –16°C.

The polymer with the highest 3HB content had the highest melting temperature. The melting temperature of pure P(3HB) is 174 to 179°C (Song *et al.*, 2001; Savenkova *et al.*, 2000), and the melting enthalpy is around 82 to 88.1 J/g (Wang *et al.*, 2001; Mitomo *et al.*, 1999). Melting temperature, as well as melting enthalpy, decreases when the number of non-3HB monomers increases. The polymer microstructure is also responsible for the elevated Tm and reduced enthalpies.

Polymer thermal stability is a major quality governing their manufacturing and application. The thermal destruction of PHA is known to be caused by a nonradical random chain-scission process. The mechanism involves a progressive drop in the molecular weight of the polymer. This type of procedure is only significant at temperatures equal to or above 200°C. The variation in heat stability could be

TABLE 11.2

Properties of PHA polymers

Properties	PHB	PHV	P(HB-HV)	PHO	PP	PET	Nylon 6,6
Molecular weight ($\times 10^5$)	1–8	2	6	5	2–7		
Density (g/cm)	1.25	1.20	1.20	1.00	0.905	1.385	1.140
Tensile strength (MPa)	40		36.22	6–10	38	70	83
Extension to break (%)	6		8–10	300–450	400	100	60
Crystallinity (%)	80	80	39–69	30	70	30–50	40–60
Glass transition temperature (°C)	5–15	–16	2 to –8	–35	–10	69	50
Water uptake (wt %)	0.2				0.0	0.4	4.5
UV resistance	Good	Good	Good	Good	Good	Poor	
Solvent resistance	Poor	Poor			Good		
Oxygen permeability ($cm^3m^{-2}atm^{-1}day^{-1}$)	45				1700	70	
Biodegradability	+	+	+	+	–	–	–

attributable to distinct structures and molecular weights of the polymer and copolymer due to the various organic matter utilized. Earlier studies discovered that the polymer's molecular weight had a substantial influence on its thermal behavior.

11.3.6 Biodegradability

In addition to their conventional polymeric qualities, PHA also has the benefit of being biodegradable. Microorganisms use PHA hydrolases and PHA depolymerase in nature to break down PHAs. Microorganisms settle on the polymer's surface and secrete enzymes that break down PHB or PHB-co-PHV into HB and HV monomer units. The cell subsequently uses these units as a carbon source for biomass growth. Several factors influence the rate of the polymer degradation process, such as surface area, the microbiological activity of the environment where it is being disposed of, pH, temperature, moisture, pressure, and others. In an aerobic environment, PHA breaks down to CO_2 and water, whereas in an anaerobic environment, methane is created (Ojumu *et al.*, 2004). Numerous researchers have examined the impact of various settings on the rate of degradation of PHB and PHB-co-PHV. PHA-degrading microbes have been identified from various habitats, including soil, river water, compost, sludge, and more. Several bacteria secrete extracellular PHA depolymerase, which hydrolyzes PHA into water-soluble oligomers and monomers. Cells then use the resultant products as nutrition. An extracellular PHB depolymerizes from *Alcaligenes faecalis* and has undergone substantial research into its characteristics. The ASTM and soil burial methods can assess PHA's ability to biodegrade (Kulkarni *et al.*, 2011).

11.4 Applications of PHA

PHA is an environmentally friendly bioplastic used as a packaging material. ICI first developed it in the 1990s under the trade name Biopol and used it to create shampoo bottles. Then they were developed by Procter and Gamble (P&G) under the brand name Nodax into a range of goods, including coated sheets, nonwoven fabrics, and fibers. Another PHA producer, Metabolix, traded their product in the name of Mirel. These initiatives are all geared toward using bioplastic in specific applications. However, its market is still relatively small due to high production costs. The upcoming sections focus on packaging applications covering PHA-based nanocomposites, paper and coatings, multilayer films, active packaging, and food packaging.

11.4.1 PHA Used in Nanocomposites

The term "nanocomposites" refers to hybrids that use at least one component measured in nanometers (nm, 10–9). Maiti and her coworkers reported the first PHB/OMMT (organo-modified montmorillonite) nanocomposites in 2003. In contrast to pure PHB, the resulting nanocomposite had an enhanced storage modulus (>40%) and an intercalated morphology while preserving its biodegradability. Wang *et al.* (2005) investigated the influence of nanoscaled OMMT layers on PHBV crystallization. The addition of OMMT in even a small quantity accelerated the crystallization rate of PHBV; however, the chain mobility was decreased. As a result, the degree of relative crystallinity of the produced PHBV/OMMT nanocomposites decreased, despite an increase in crystallization rate. Increased OMMT concentration expands the range of processing temperature, but it causes a decrease in Tm, enthalpy of fusion, and biodegradability of nanocomposites. The increase in antimicrobial OMMT interaction with PHBV causes a negative impact on the biodegradability of the nanocomposites. Consequently, it was advised to add OMMT in a small amount. By adding bacterial cellulose nanowhiskers (BCNW), Martínez-Sanz *et al.* (2014) created PHBHV nanocomposites. Valerate inclusion in the pure polymer improved water, oxygen, and moisture permeability while decreasing Tm, enthalpy, rigidity, and stiffness.

Pardo-Ibáñez *et al.* (2014) successfully created a new PHBHV-keratin composite material through melt compounding that is both biodegradable and renewable. The optical characteristics were slightly altered, while the dispersion was adequate and suitable for modest additive loadings. Additionally, the water, limonene, and oxygen permeabilities of the composite with keratin were significantly decreased. It was suggested that this particular composition could be used to create entirely biodegradable renewable food packaging materials where improved barrier qualities are required. Contrarily, enhanced additive-loaded composites are recommended for specific packaging applications where transparency of material is not a factor but gas and/or water vapor exchange is preferred.

Kiran *et al.* (2017) produced a PHB–nanomelanin–glycerol (Nm-PHB) polymer film with good flexibility, odor-free, non-toxic, antibacterial, and antioxidant properties. Sonication produced spherical nucleated nanomelanin (Nm) particles that were excellent antimicrobials. The Nm-PHB nanocomposite film was uniform, thermostable up to 281.87°C, and significantly inhibited *Staphylococcus aureus* growth. As a result, it could be used in the food packaging industry to protect against bacterial contamination and oxidation.

The nanocomposite films outperformed pure PHB in terms of network formation and heat stability. Biodegradability of the film is tested via in vitro or in soil burial, which showed that it also degraded more quickly. Significant advancement in PHA nanocomposites and research progress indicates that these materials have the ability to rival the qualities of petro-based polymer, rendering them valuable for the packaging sector. To meet the needs of the current packaging industry from all angles, a deeper analysis of each nanocomposite material is still required, as well as the creation of new ones.

11.4.2 PHA Used in Multilayer Films

A multilayer film comprises three to nine layers of different polymers, with hydrophilic and hydrocolloidal synthetic polymer films (polyolefin) sandwiched between hydrophobic and biodegradable polymer layers, having improved barrier qualities (Fang *et al.*, 2005; Wang *et al.*, 2000). Various research works on producing PHA-based multilayer films have been published; however, they are limited in number. Fabra *et al.* (2013) fabricated a multilayer film using compression molding and casting techniques, which contains electrospun zein nanofibers layered between PHBVs. A multilayer film created using a compression molding technique exhibited superior mechanical characteristics and a better water vapor barrier than the film created using the casting approach. The inclusion of zein in interlayers improved the oxygen and water vapor barrier characteristics of the films produced by both methods. In another work, a multilayer film with inner polyvinyl alcohol (PVOH) layer and an outer PHAs layer was created using the coextrusion method (Thellen *et al.*, 2013). The peel strength of PHA/PVOH-based multilayer films was increased by two times by grafting PHAs with maleic anhydride using dicumyl peroxide (DCP) as an initiator. The multilayered PHA/PVOH film showed an oxygen transmission rate (OTR) of 27, 41, and 52 cc/m²/day at 0, 60, and 90% relative humidity (RH), respectively. In contrast to the non-biodegradable PVOH, the unmodified and malleated PHAs were significantly degradable.

Another multilayer film was developed by Fabra *et al.* (2014) using nanostructured zein in the inner layer with PHB/PHBV as an outer layer. The addition of zein nanostructured interlayers greatly enhanced the oxygen barrier properties and flexibility of PHB/zein and PHBV/zein. However, the film's transparency was decreased. These multilayer films are stable and do not change their properties over storage at even 100% RH, accepting that microorganisms such as *Listeria monocytogenes* can grow.

11.4.3 PHA Used in Paper and Cardboard Coatings

Biodegradable packaging made of paper has also received widespread recognition; its only drawback is hygroscopic. Paper-based packaging is frequently coated with hydrophobic polymers, like orientated PET, PLA, PHB, PVC, poly (ethylene-co-vinyl alcohol), and polyolefin (Rastogi & Samyn, 2015). These composites maintain the structural integrity of the package in a wet environment while reducing disposal costs. Limited information is available about the usage of PHAs in paper coating and paperboard.

In the past, PHB and PHBV were both used successfully to reduce water permeability and absorption over paper and paperboard (Kuusipalo, 2000a, 2000b). Cyras *et al.* (2007) created bilayer films made of PHB and cellulose paper by solvent casting technique. The hydrophobicity, infusion, and overloading of PHB in cellulose fibers significantly improved the barrier characteristics, tensile strength, and modulus of the double-layer films.

In another investigation, compression molding was used to create double-layer films of PHB and cellulose, and the effects of different PHB concentrations on water barrier properties such as absorption, adsorption, and permeability of the films were assessed (Cyras *et al.*, 2009). To improve the cellulose and the PHB adhesion, cellulose cardboards were first acetylated. It was discovered that the mechanical properties of the developed films, such as their rigidity, stiffness, tensile strength, and strain at break, were influenced by the PHB concentration. Using more than 15% of PHB led to a noticeable improvement in uniaxial tensile performance. However, films containing 15% PHB showed a significant reduction in water permeability. This analysis led them to conclude that PHB coatings are an excellent replacement for Tetra Paks.

PHB's and PHBV's potential to produce stable latex was successfully explored in 2010 for paper coatings and sizing (Bourbonnais & Marchessault, 2010). PHB and PHBV granules, both naturally occurring and synthetically created, were employed to size the paper, and the sized paper's resistance to aqueous fluid diffusion was evaluated. Pressing and heating impregnated sheets for a short period resulted in increased paper size due to the formation of a thin coating of melted PHB pellets on the paper's surface. PHB was found to be better suited for paper size than PHBV. The polymer's texture, drying time, temperature, and pressure are additional crucial elements to be considered. In a different study, PHB and PLA layers were coated on the surface of the paper using the dip-coating method by Shawaphun and Manangan (2010). Compared to the PLA that was not coated, the PHB-coated papers showed improved water resistance but low oil resistance.

The creation of scl-PHA-kraft paper composites was attempted by Dagnon *et al.* (2010). It was discovered that uncoated kraft paper lost more weight after eight weeks of soil burial than P(3HB-co-4HB)-coated kraft paper. The thermomechanical characteristics of kraft paper had also been greatly enhanced by P(3HBco-4HB) biopolymer coating. These materials can be used as liners for creating fiberboard boxes for various packing applications where mechanical qualities and environmental disposal are prime considerations.

A superhydrophobic material that is economical, biodegradable, and flexible was created by Obeso *et al.* (2013) using a phase separation technique. PHB was initially precipitated on the cellulose surface of paper fibers for this purpose. A rough surface with a water contact angle of 153.0 degrees was created. Further, argon plasma treatment enhances surface wettability. The researchers claim that this PHB-based material can be used to create open-microfluidic or lab-on-chip systems that are both affordable and useful in a variety of biomedical and environmental applications. In order to create hydrophobic coatings for packing papers, Rastogi and Samyn (2016) looked into the fabrication of structured PHB microparticles (PHB-MP). The carnauba plant wax was additionally utilized as a hydrophobic agent. To create a structured PHB-MP, a phase-separation methodology was used. The wax was used to size the filter papers after dipping them in the particle suspension. The increase in static contact angles represented an improvement in the inherent hydrophobicity of the PHB-MP-coated sheets with wax addition. Contact angle enhancement was encouraged by the highest PHB and nonsolvent concentrations.

11.4.4 PHA Used in Active Packaging

The term "active food packaging" refers to the packaging technology intended to extend the shelf life, enhance the quality attributes, and enhance the sensory quality of packaged food while protecting it against microbiological contamination and mechanical damage. PHA-based active food packaging was first developed by Hany *et al.* (2004) by coating an mcl-PHA, that is, poly (3-hydroxyalkanoate-co-3 hydroxyalkenoate) (PHAE) with zosteric acid, which is a non-toxic and antifouling agent. During contact with sewage sludge, no microbiological growth was observed on the zosteric acid–coated PHAE surface, demonstrating their potential for use as active food packaging. Another active food packaging system was developed by Gonta *et al.* (2012) from PHB films and PHB-coated paper in combination with antimicrobial agents, which are silbiol and benzoic acid. PHB films and PHB/paper systems both showed notable antimicrobial effects against several Gram-positive bacteria (such as *B. cereus* and *S. aureus*) and Gram-negative bacteria (such as *E. coli* and *P. aeruginosa*). Similarly, Kwiecień *et al.* (2014) also created an active packaging system from conjugate of preservative-oligo (3-HB) to stop the migration of antimicrobial agents such as sorbic acid and benzoic acid from the packaging material to the food products. An active packaging material developed from poly(3-hydroxybutyrate-co-4-hydroxybutyrate) or P(3HB-co-4HB) integrating thyme essential oil shows antimicrobial activity, enhancing the shelf life of products (Sharma *et al.*, 2022). The antioxidant capacity of PHA-based packaging material can be enhanced by incorporating different antioxidants, such as apple extract, which makes it an active packaging material (Urbina *et al.*, 2019).

11.4.5 PHA Used in Food Packaging

Only a few research that have provided actual demonstrations of the use of PHAs for food product packaging are discussed throughout this section.

For the first time, Kantola and Helén (2001) investigated quality differences in organic tomatoes kept in Biopol, a paperboard tray coated with poly (HB-co-HV) copolymer, and other biodegradable and LDPE containers, for three weeks at 111°C and 75–85% RH. The biodegradable packaging materials that were evaluated are Mater-Bi type ZF03U (based on cornstarch), Biopol (PHBHV), and cellophane (regenerated cellulose). The various packages tested for suitability included a perforated Mater bag, a Biopol, and a PLA-coated paperboard tray wrapped with a perforated Mater bag and a perforated cellophane bag. After a three-week storage period, LDPE-bagged tomatoes showed a weight loss of 1.7%, but bags made of biodegradable materials showed a weight loss of >2.5%. The sensory attributes of tomatoes were primarily unaffected by the packaging material, but the length of storage had a considerable impact.

According to the study conducted by Haugaard *et al.* (2003), PHB- and PLA-based packing cups may be just as successful as regular HDPE packs for simulating the preservation of orange juice (an acidic food) and dressing (a fatty meal) for ten weeks at 4°C under fluorescent light or complete darkness. Quality changes included color and ascorbic acid degradation for the orange juice simulant. The dressing simulant's quality changes included color, peroxide value, and degradation of α-tocopherols. Due to similar quality differences seen for products packed in PHB, PLA, and HDPE, respectively, it is projected that PHB and PLA will be employed for packaging other commercial juices, acidic beverages, and fatty foods. The study also showed that the change in quality was caused mainly by light and can be controlled by variations in the packages' light permeability.

Bucci *et al.* (2005) used several dimensions and mechanical tests to investigate and compare the utilization of injection-molded P(3HB) with PP (polypropylene). P(3HB) was described as a more rigid material compared to polypropylene, having a 50% lower deformation value. Even while the PHB performed worse than PP under typical freezing and refrigeration settings, it excelled at elevated temperatures. The work strongly emphasized the requirement for creating unique molds and optimizing injection parameters and processing temperature conditions for PHB. The sensory tests with the items (margarine, mayonnaise, and cream cheese) showed no significant difference, demonstrating PHB performance as good as ordinary plastic PP for storing lipophilic items.

In comparison to the traditional dairy packaging materials, such as lean pouches, polyethylene packs, and polystyrene cups, a plasticized PHB material was formulated by Muizniece-Brasava and Dukalska

TABLE 11.3

Globally produced and used PHA polymers

Company name	Origin	Trade name	PHA type	Application
ADM (with Metabolix)	USA	Mirel	Several PHAs	Paper coating, sheet, and foam products
Bio-on	Italy	Minerva-PHA	PHA	—
Boomers	Germany	Boomer	PHB	Packaging and drug delivery
Biocycle	Brazil	Biocycle	PHB	—
Kaneka (with P&G)	Japan	Nodax	Several PHAs	Packaging
Ningbo Tian An, Zhenjiang	China	—	PHBHV	Biomol resin for film and coating
Metabolix	USA	Mirel	Several PHAs	Packaging
Meridian	USA	Nodax	Several PHAs	Raw material
Tianjin Green Bio-science	China	Greenbio	P(3HB)(4HB)	—
Tianan Biological	China	Enmat	PHBHV	—
Yikeman, Shandong	China	—	PHAs	Raw material

in 2006, which is suitable for packaging sour cream with minor quality changes. In order to give thermal treatment to samples of meat and mayonnaise salad, Levkane *et al.* (2008) suggested using plasticized PHB and PLA packaging films. Salad consistency was found to be retained with a lower microbial count at temperatures up to $63 \pm 0.5°C$. Their research focused on the effects of pasteurization methods and packing materials on preserving samples of meat and mayonnaise salad that were vacuum-packed using traditional, plasticized PHB and PLA packaging sheets, respectively. According to Hermida *et al.* (2008), PHB is suitable for gamma radiation and used to sterilize food or packing materials, and its qualities have not significantly changed.

11.5 Conclusion

From the chapter, it can be concluded that PHAs are produced by bacterial fermentation from bio-based feedstocks, such as agricultural and industrial wastes, and so represent a viable substitute for petroleum-based plastics. Nutritional inputs, bacterial strains, fermentation conditions, purification, and recovery processes are all elements that influence PHA production. Because of its unique combination of biocompatibility, biodegradability, and thermal and chemical stability, PHA is an excellent choice for a wide range of applications in the medical industry as well as in the packaging sector, including paper coating, adhesives, film-forming, active packaging. In the near future, higher production of PHA will play a vital role in protecting the environment by replacing non-degradable synthetic polymers.

REFERENCES

Aarthi, N., & Ramana, K. V. (2011). Identification and characterization of polyhydroxybutyrate producing *Bacillus cereus* and *Bacillus mycoides* strains. *International Journal of Environmental Sciences, 1*(5), 744.

Agus, J., Kahar, P., Abe, H., Doi, Y., & Tsuge, T. (2006). Molecular weight characterization of poly [(R)-3-hydroxybutyrate] synthesized by genetically engineered strains of *Escherichia coli. Polymer Degradation and Stability, 91*(5), 1138–1146.

Akiyama, H., Okuhata, H., Onizuka, T., Kanai, S., Hirano, M., Tanaka, S., Sasaki, K., & Miyasaka, H. (2011). Antibiotics-free stable polyhydroxyalkanoate (PHA) production from carbon dioxide by recombinant cyanobacteria. *Bioresource Technology, 102*(23), 11039–11042.

Anderson, A. J., & Dawes, E. (1990). Occurrence, metabolism, metabolic role, and industrial uses of bacterial polyhydroxyalkanoates. *Microbiological Reviews, 54*(4), 450–472.

Arumugam, A., Senthamizhan, S. G., Ponnusami, V., & Sudalai, S. (2018). Production and optimization of polyhydroxyalkanoates from non-edible *Calophyllum inophyllum* oil using *Cupriavidus necator. International Journal of Biological Macromolecules, 112*, 598–607.

Ashby, R. D., Solaiman, D. K., & Strahan, G. D. (2011). Efficient utilization of crude glycerol as fermentation substrate in the synthesis of poly (3-hydroxybutyrate) biopolymers. *Journal of the American Oil Chemists' Society*, *88*(7), 949–959.

Berger, E., Ramsay, B. A., Ramsay, J. A., Chavarie, C., & Braunegg, G. (1989). PHB recovery by hypochlorite digestion of non-PHB biomass. *Biotechnology Techniques*, *3*(4), 227–232.

Bhattacharyya, A., Pramanik, A., Maji, S. K., Haldar, S., Mukhopadhyay, U. K., & Mukherjee, J. (2012). Utilization of vinasse for production of poly-3-(hydroxybutyrate-co-hydroxyvalerate) by Haloferax mediterranei. *AMB Express*, *2*(1), 1–10.

Bocanegra, J. K., da Cruz Pradella, J. G., Da Silva, L. F., Taciro, M. K., & Gomez, J. G. C. (2013). Influence of pH on the molecular weight of poly-3-hydroxybutyric acid (P3HB) produced by recombinant *Escherichia coli*. *Applied Biochemistry and Biotechnology*, *170*(6), 1336–1347.

Bourbonnais, R., & Marchessault, R. H. (2010). Application of polyhydroxyalkanoate granules for sizing of paper. *Biomacromolecules*, *11*(4), 989–993.

Braunegg, G., Sonnleitner, B. Y., & Lafferty, R. M. (1978). A rapid gas chromatographic method for the determination of poly-β-hydroxybutyric acid in microbial biomass. *European Journal of Applied Microbiology and Biotechnology*, *6*(1), 29–37.

Bucci, D. Z., Tavares, L. B. B., & Sell, I. (2005). PHB packaging for the storage of food products. *Polymer Testing*, *24*(5), 564–571.

Bugnicourt, E., Cinelli, P., Lazzeri, A., & Alvarez, V. A. (2014). Polyhydroxyalkanoate (PHA): Review of synthesis, characteristics, processing and potential applications in packaging. *eXPRESS Polymer Letters*, *8*(11), 791–808.

Burniol-Figols, A., Skiadas, I. V., Daugaard, A. E., & Gavala, H. N. (2020). Polyhydroxyalkanoate (PHA) purification through dilute aqueous ammonia digestion at elevated temperatures. *Journal of Chemical Technology & Biotechnology*, *95*(5), 1519–1532.

Cavalheiro, J. M., de Almeida, M. C. M., Grandfils, C., & Da Fonseca, M. M. R. (2009). Poly (3-hydroxybutyrate) production by *Cupriavidus necator* using waste glycerol. *Process Biochemistry*, *44*(5), 509–515.

Chanprateep, S. (2010). Current trends in biodegradable polyhydroxyalkanoates. *Journal of Bioscience and Bioengineering*, *110*(6), 621–632.

Chanprateep, S., Abe, N., Shimizu, H., Yamane, T., & Shioya, S. (2001). Multivariable control of alcohol concentrations in the production of polyhydroxyalkanoates (PHAs) by *Paracoccus denitrificans*. *Biotechnology and Bioengineering*, *74*(2), 116–124.

Chee, J. Y., Tan, Y., Samian, M. R., & Sudesh, K. (2010). Isolation and characterization of a *Burkholderia* sp. USM (JCM15050) capable of producing polyhydroxyalkanoate (PHA) from triglycerides, fatty acids and glycerols. *Journal of Polymers and the Environment*, *18*(4), 584–592.

Chen, C. W., Don, T. M., & Yen, H. F. (2006). Enzymatic extruded starch as a carbon source for the production of poly (3-hydroxybutyrate-co-3-hydroxyvalerate) by *Haloferax mediterranei*. *Process Biochemistry*, *41*(11), 2289–2296.

Chen, G. Q. (2010a). Introduction of bacterial plastics PHA, PLA, PBS, PE, PTT, and PPP. In *Plastics from bacteria* (pp. 1–16). Springer.

Chen, G. Q. (2010b). Plastics completely synthesized by bacteria: Polyhydroxyalkanoates. In *Plastics from bacteria* (pp. 17–37). Springer.

Chen, G. Q., & Wu, Q. (2005). The application of polyhydroxyalkanoates as tissue engineering materials. *Biomaterials*, *26*(33), 6565–6578.

Chen, G. Q., Zhang, G., Park, S., & Lee, S. (2001). Industrial scale production of poly (3-hydroxybutyrate-co-3-hydroxyhexanoate). *Applied Microbiology and Biotechnology*, *57*(1), 50–55.

Choi, J. I., & Lee, S. Y. (2004). High level production of supra molecular weight poly (3-hydroxybutyrate) by metabolically engineered *Escherichia coli*. *Biotechnology and Bioprocess Engineering*, *9*(3), 196–200.

Chowdhury, A. A. (1963). Poly-β-hydroxybuttersäure abbauende Bakterien und Exoenzym. *Archiv für Mikrobiologie*, *47*(2), 167–200.

Cyras, V. P., Commisso, M. S., Mauri, A. N., & Vázquez, A. (2007). Biodegradable double-layer films based on biological resources: Polyhydroxybutyrate and cellulose. *Journal of Applied Polymer Science*, *106*(2), 749–756.

Cyras, V. P., Soledad, C. M., & Analía, V. (2009). Biocomposites based on renewable resource: Acetylated and non acetylated cellulose cardboard coated with polyhydroxybutyrate. *Polymer*, *50*(26), 6274–6280.

da Cruz Pradella, J. G., Ienczak, J. L., Delgado, C. R., & Taciro, M. K. (2012). Carbon source pulsed feeding to attain high yield and high productivity in poly (3-hydroxybutyrate) (PHB) production from soybean oil using *Cupriavidus necator. Biotechnology Letters, 34*(6), 1003–1007.

Dagnon, K. L., Thellen, C., Ratto, J. A., & D'Souza, N. A. (2010). Physical and thermal analysis of the degradation of poly (3-hydroxybutyrate-co-4-hydroxybutyrate) coated paper in a constructed soil medium. *Journal of Polymers and the Environment, 18*(4), 510–522.

Dalsasso, R. R., Pavan, F. A., Bordignon, S. E., de Aragão, G. M. F., & Poletto, P. (2019). Polyhydroxybutyrate (PHB) production by *Cupriavidus necator* from sugarcane vinasse and molasses as mixed substrate. *Process Biochemistry, 85*, 12–18.

De Koning, G. J. M., Scheeren, A. H. C., Lemstra, P. J., Peeters, M., & Reynaers, H. (1994). Crystallization phenomena in bacterial poly [(R)-3-hydroxybutyrate]: 3. Toughening via texture changes. *Polymer, 35*(21), 4598–4605.

De Smet, M. J., Eggink, G., Witholt, B., Kingma, J., & Wynberg, H. (1983). Characterization of intracellular inclusions formed by *Pseudomonas oleovorans* during growth on octane. *Journal of Bacteriology, 154*(2), 870–878.

Domínguez-Díaz, M., & Romo-Uribe, A. (2012). Viscoelastic behavior of biodegradable polyhydroxyalkanoates. *Bioinspired, Biomimetic and Nanobiomaterials, 1*(4), 214–220.

Escapa, I. F., Morales, V., Martino, V. P., Pollet, E., Avérous, L., García, J. L., & Prieto, M. A. (2011). Disruption of β-oxidation pathway in Pseudomonas putida KT2442 to produce new functionalized PHAs with thioester groups. *Applied Microbiology and Biotechnology, 89*(5), 1583–1598.

Fabra, M. J., Lopez-Rubio, A., & Lagaron, J. M. (2013). High barrier polyhydroxyalcanoate food packaging film by means of nanostructured electrospun interlayers of zein. *Food Hydrocolloids, 32*(1), 106–114.

Fabra, M. J., Lopez-Rubio, A., & Lagaron, J. M. (2014). Nanostructured interlayers of zein to improve the barrier properties of high barrier polyhydroxyalkanoates and other polyesters. *Journal of Food Engineering, 127*, 1–9.

Fang, J. M., Fowler, P. A., Escrig, C., Gonzalez, R., Costa, J. A., & Chamudis, L. (2005). Development of biodegradable laminate films derived from naturally occurring carbohydrate polymers. *Carbohydrate Polymers, 60*(1), 39–42.

Flores-Sánchez, A., López-Cuellar, M., Pérez-Guevara, F., Figueroa López, U., Martín-Bufájer, J. M., & Vergara-Porras, B. (2017). Synthesis of poly-(R-hydroxyalkanoates) by *Cupriavidus necator* ATCC 17699 using Mexican avocado (*Persea americana*) oil as a carbon source. *International Journal of Polymer Science, 2017*, 1–10.

Gonta, S., Savenkova, L., Krallish, I., & Kirilova, E. (2012). Antimicrobial activity of PHB based polymeric compositions. *Environmental Engineering and Management Journal, 11*, 99–104.

Greene, E. A., & Voordouw, G. (2004). Biodegradation of c5+ hydrocarbons by a mixed bacterial consortium from a c5+-contaminated site. *Environmental Technology, 25*(3), 355–363.

Grothe, E., Moo-Young, M., & Chisti, Y. (1999). Fermentation optimization for the production of poly (β-hydroxybutyric acid) microbial thermoplastic. *Enzyme and Microbial Technology, 25*(1–2), 132–141.

Gumel, A. M., Annuar, M. S. M., & Heidelberg, T. (2012). Biosynthesis and characterization of polyhydroxyalkanoates copolymers produced by *Pseudomonas putida* Bet001 isolated from palm oil mill effluent. *PLoS ONE, 7*(9), 45214.

Gupta, A., Kumar, M., & Thakur, I. S. (2017). Analysis and optimization of process parameters for production of polyhydroxyalkanoates along with wastewater treatment by *Serratia* sp. ISTVKR1. *Bioresource Technology, 242*, 55–59.

Haas, R., Jin, B., & Zepf, F. T. (2008). Production of poly (3-hydroxybutyrate) from waste potato starch. *Bioscience, Biotechnology, and Biochemistry, 72*(1), 253–256.

Hahn, S. K., Chang, Y. K., Kim, B. S., & Chang, H. N. (1994). Optimization of microbial poly (3-hydroxybutyrate) recover using dispersions of sodium hypochlorite solution and chloroform. *Biotechnology and Bioengineering, 44*(2), 256–261.

Han, J., Hou, J., Liu, H., Cai, S., Feng, B., Zhou, J., & Xiang, H. (2010). Wide distribution among halophilic *archaea* of a novel polyhydroxyalkanoate synthase subtype with homology to bacterial type III synthases. *Applied and Environmental Microbiology, 76*(23), 7811–7819.

Hany, R., Böhlen, C., Geiger, T., Schmid, M., & Zinn, M. (2004). Toward non-toxic antifouling: Synthesis of hydroxy-, cinnamic acid-, sulfate-, and zosteric acid-labeled poly [3-hydroxyalkanoates]. *Biomacromolecules, 5*(4), 1452–1456.

Haugaard, V. K., Danielsen, B., & Bertelsen, G. (2003). Impact of polylactate and poly (hydroxybutyrate) on food quality. *European Food Research and Technology, 216*(3), 233–240.

Hejazi, P., Vasheghani-Farahani, E., & Yamini, Y. (2003). Supercritical fluid disruption of *Ralstonia eutropha* for poly (β-hydroxybutyrate) recovery. *Biotechnology Progress, 19*(5), 1519–1523.

Hermann-Krauss, C., Koller, M., Muhr, A., Fasl, H., Stelzer, F., & Braunegg, G. (2013). Archaeal production of polyhydroxyalkanoate (PHA) co-and terpolyesters from biodiesel industry-derived by-products. *Archaea, 2013*.

Hermida, É. B., Mega, V. I., Yashchuk, O., Fernández, V., Eisenberg, P., & Miyazaki, S. S. (2008). Gamma irradiation effects on mechanical and thermal properties and biodegradation of poly (3-hydroxybutyrate) based films. *Macromolecular Symposia, 263*(1), 102–113.

Höfer, P., Vermette, P., & Groleau, D. (2011). Production and characterization of polyhydroxyalkanoates by recombinant *Methylobacterium extorquens*: Combining desirable thermal properties with functionality. *Biochemical Engineering Journal, 54*(1), 26–33.

Horowitz, D. M., & Brennan, E. M. (2010). Methods for separation and purification of biopolymers. *European Patent EP, 1*, 070–135.

Ibrahim, M. H., & Steinbüchel, A. (2010). High-cell-density cyclic fed-batch fermentation of a poly (3-hydroxybutyrate)-accumulating thermophile, *Chelatococcus* sp. strain MW10. *Applied and Environmental Microbiology, 76*(23), 7890–7895.

Jendrossek, D. (2009). Polyhydroxyalkanoate granules are complex subcellular organelles (carbonosomes). *Journal of Bacteriology, 191*(10), 3195–3202.

Jiang, Y., Song, X., Gong, L., Li, P., Dai, C., & Shao, W. (2008). High poly (β-hydroxybutyrate) production by *Pseudomonas fluorescens* A2a5 from inexpensive substrates. *Enzyme and Microbial Technology, 42*(2), 167–172.

Jung, I. G., & Park, C. H. (2005). Characteristics of styrene degradation by *Rhodococcus pyridinovorans* isolated from a biofilter. *Chemosphere, 61*(4), 451–456.

Kamilah, H., Al-Gheethi, A., Yang, T. A., & Sudesh, K. (2018). The use of palm oil-based waste cooking oil to enhance the production of polyhydroxybutyrate [P (3HB)] by *Cupriavidus necator* H16 strain. *Arabian Journal for Science and Engineering, 43*(7), 3453–3463.

Kantola, M., & Helén, H. (2001). Quality changes in organic tomatoes packaged in biodegradable plastic films. *Journal of Food Quality, 24*(2), 167–176.

Keskin, G., Kızıl, G., Bechelany, M., Pochat-Bohatier, C., & Öner, M. (2017). Potential of polyhydroxyalkanoate (PHA) polymers family as substitutes of petroleum based polymers for packaging applications and solutions brought by their composites to form barrier materials. *Pure and Applied Chemistry, 89*(12), 1841–1848.

Khanna, S., & Srivastava, A. K. (2005). Recent advances in microbial polyhydroxyalkanoates. *Process Biochemistry, 40*(2), 607–619.

Khosravi-Darani, K., Mokhtari, Z. B., Amai, T., & Tanaka, K. (2013). Microbial production of poly (hydroxybutyrate) from C1 carbon sources. *Applied Microbiology and Biotechnology, 97*(4), 1407–1424.

Khosravi-Darani, K., Vasheghani-Farahani, E., Shojaosadati, S. A., & Yamini, Y. (2004). Effect of process variables on supercritical fluid disruption of *Ralstonia eutropha* cells for poly (R-hydroxybutyrate) recovery. *Biotechnology Progress, 20*(6), 1757–1765.

Kiran, G. S., Jackson, S. A., Priyadharsini, S., Dobson, A. D., & Selvin, J. (2017). Synthesis of Nm-PHB (nano-melanin-polyhydroxy butyrate) nanocomposite film and its protective effect against biofilm-forming multi drug resistant *Staphylococcus aureus*. *Scientific Reports, 7*(1), 1–13.

Koller, M. (2018). A review on established and emerging fermentation schemes for microbial production of polyhydroxyalkanoate (PHA) biopolyesters. *Fermentation, 4*(2), 30.

Koller, M., Atlić, A., Dias, M., Reiterer, A., & Braunegg, G. (2010). Microbial PHA production from waste raw materials. In *Plastics from bacteria* (pp. 85–119). Springer.

Koller, M., Atlić, A., Gonzalez-Garcia, Y., Kutschera, C., & Braunegg, G. (2008a). Polyhydroxyalkanoate (PHA) biosynthesis from whey lactose. In *Macromolecular symposia* (Vol. 272, No. 1, pp. 87–92). Wiley-VCH Verlag.

Koller, M., Bona, R., Chiellini, E., Fernandes, E. G., Horvat, P., Kutschera, C., Hesse, P., & Braunegg, G. (2008b). Polyhydroxyalkanoate production from whey by *Pseudomonas hydrogenovora*. *Bioresource Technology, 99*(11), 4854–4863.

Koller, M., & Braunegg, G. (2015). Potential and prospects of continuous polyhydroxyalkanoate (PHA) production. *Bioengineering, 2*(2), 94–121.

Koller, M., Hesse, P., Bona, R., Kutschera, C., Atlić, A., & Braunegg, G. (2007). Potential of various archae- and eubacterial strains as industrial polyhydroxyalkanoate producers from whey. *Macromolecular Bioscience, 7*(2), 218–226.

Koller, M., & Muhr, A. (2014). Continuous production mode as a viable process-engineering tool for efficient poly (hydroxyalkanoate) (PHA) bio-production. *Chemical and Biochemical Engineering Quarterly, 28*(1), 65–77.

Krueger, C. L., Radetski, C. M., Bendia, A. G., Oliveira, I. M., Castro-Silva, M. A., Rambo, C. R., Antonio, R. V., & Lima, A. O. (2012). Bioconversion of cassava starch by-product into *Bacillus* and related bacteria polyhydroxyalkanoates. *Electronic Journal of Biotechnology, 15*(3), 8–8.

Kulkarni, S. O., Kanekar, P. P., Jog, J. P., Patil, P. A., Nilegaonkar, S. S., Sarnaik, S. S., & Kshirsagar, P. R. (2011). Characterisation of copolymer, poly (hydroxybutyrate-co-hydroxyvalerate) (PHB-co-PHV) produced by *Halomonas campisalis* (MCM B-1027), its biodegradability and potential application. *Bioresource Technology, 102*(11), 6625–6628.

Kumar, M., Ghosh, P., Khosla, K., & Thakur, I. S. (2018). Recovery of polyhydroxyalkanoates from municipal secondary wastewater sludge. *Bioresource Technology, 255*, 111–115.

Kuusipalo, J. (2000a). PHB/V in extrusion coating of paper and paperboard: Part I: Study of functional properties. *Journal of Polymers and the Environment, 8*(1), 39–47.

Kuusipalo, J. (2000b). PHB/V in extrusion coating of paper and paperboard—Study of functional properties. Part II. *Journal of Polymers and the Environment, 8*(2), 49–57.

Kwiecień, I., Adamus, G., Bartkowiak, A., & Kowalczuk, M. (2014). Synthesis and structural characterization at the molecular level of oligo (3-hydroxybutyrate) conjugates with antimicrobial agents designed for food packaging materials. *Designed Monomers and Polymers, 17*(4), 311–321.

Lageveen, R. G., Huisman, G. W., Preusting, H., Ketelaar, P., Eggink, G., & Witholt, B. (1988). Formation of polyesters by *Pseudomonas oleovorans*: Effect of substrates on formation and composition of poly-(R)-3-hydroxyalkanoates and poly-(R)-3-hydroxyalkenoates. *Applied and Environmental Microbiology, 54*(12), 2924–2932.

Law, J. H., & Slepecky, R. A. (1961). Assay of poly-β-hydroxybutyric acid. *Journal of Bacteriology, 82*(1), 33–36.

Levkane, V., Muizniece-Brasava, S., & Dukalska, L. (2008). Pasteurization effect to quality of salad with meat in mayonnaise. *Foodbalt, 1*, 69–73.

Li, Z., Lin, H., Ishii, N., Chen, G. Q., & Inoue, Y. (2007). Study of enzymatic degradation of microbial copolyesters consisting of 3-hydroxybutyrate and medium-chain-length 3-hydroxyalkanoates. *Polymer Degradation and Stability, 92*(9), 1708–1714.

Loo, C. Y., & Sudesh, K. (2007). Polyhydroxyalkanoates: Bio-based microbial plastics and their properties. *Malaysian Polymer Journal, 2*(2), 31–57.

Lu, J., Tappel, R. C., & Nomura, C. T. (2009). Mini-review: Biosynthesis of poly (hydroxyalkanoates). *Journal of Macromolecular Science®, Part C: Polymer Reviews, 49*(3), 226–248.

Madison, L. L., & Huisman, G. W. (1999). Metabolic engineering of poly (3-hydroxyalkanoates): From DNA to plastic. *Microbiology and Molecular Biology Reviews, 63*(1), 21–53.

Madkour, M. H., Heinrich, D., Alghamdi, M. A., Shabbaj, I. I., & Steinbüchel, A. (2013). PHA recovery from biomass. *Biomacromolecules, 14*(9), 2963–2972.

Martínez-Sanz, M., Villano, M., Oliveira, C., Albuquerque, M. G., Majone, M., Reis, M., Lopez-Rubio, A., & Lagaron, J. M. (2014). Characterization of polyhydroxyalkanoates synthesized from microbial mixed cultures and of their nanobiocomposites with bacterial cellulose nanowhiskers. *New Biotechnology, 31*(4), 364–376.

Masood, F., Chen, P., Yasin, T., Hasan, F., Ahmad, B., & Hameed, A. (2013). Synthesis of poly-(3-hydroxybutyrate-co-12 mol% 3-hydroxyvalerate) by *Bacillus cereus* FB11: Its characterization and application as a drug carrier. *Journal of Materials Science: Materials in Medicine, 24*(8), 1927–1937.

Masood, F., Yasin, T., & Hameed, A. (2015). Production and characterization of Tailor-made polyhydroxyalkanoates by *Bacillus cereus* FC11. *Pakistan Journal of Zoology, 47*(2), 491–503.

Mitomo, H., Takahashi, T., Ito, H., & Saito, T. (1999). Biosynthesis and characterization of poly (3-hydroxybutyrate-co-3-hydroxyvalerate) produced by *Burkholderia cepacia* D1. *International Journal of Biological Macromolecules, 24*(4), 311–318.

Modi, S., Koelling, K., & Vodovotz, Y. (2011). Assessment of PHB with varying hydroxyvalerate content for potential packaging applications. *European Polymer Journal, 47*(2), 179–186.

Mohidin Batcha, A. F., Prasad, D. M., Khan, M. R., & Abdullah, H. (2014). Biosynthesis of poly (3-hydroxy-butyrate) (PHB) by *Cupriavidus necator* H16 from jatropha oil as carbon source. *Bioprocess and Biosystems Engineering, 37*(5), 943–951.

Morya, R., Kumar, M., & Thakur, I. S. (2018). Utilization of glycerol by *Bacillus* sp. ISTVK1 for production and characterization of polyhydroxyvalerate. *Bioresource Technology Reports, 2*, 1–6.

Muizniece-Brasava, S., & Dukalska, L. (2006). Impact of biodegradable PHB packaging composite materials on dairy product quality. *LLU Raksti, 16*(311), 79–87.

Nikodinovic-Runic, J., Casey, E., Duane, G. F., Mitic, D., Hume, A. R., Kenny, S. T., & O'Connor, K. E. (2011). Process analysis of the conversion of styrene to biomass and medium chain length polyhydroxyalkano-ate in a two-phase bioreactor. *Biotechnology and Bioengineering, 108*(10), 2447–2455.

Obeso, C. G., Sousa, M. P., Song, W., Rodriguez-Pérez, M. A., Bhushan, B., & Mano, J. F. (2013). Modification of paper using polyhydroxybutyrate to obtain biomimetic superhydrophobic substrates. *Colloids and Surfaces A: Physicochemical and Engineering Aspects, 416*, 51–55.

Obruca, S., Benesova, P., Marsalek, L., & Marova, I. (2015). Use of lignocellulosic materials for PHA produc-tion. *Chemical and Biochemical Engineering Quarterly, 29*(2), 135–144.

Ojumu, T. V., Yu, J., & Solomon, A. (2004). Production of polyhydroxyalkanoates, a bacterial biodegradable polymers. *African Journal of Biotechnology, 3*(1), 18–24.

Pantazaki, A. A., Papaneophytou, C. P., Pritsa, A. G., Liakopoulou-Kyriakides, M., & Kyriakidis, D. A. (2009). Production of polyhydroxyalkanoates from whey by *Thermus thermophilus* HB8. *Process Biochemistry, 44*(8), 847–853.

Pardo-Ibáñez, P., Lopez-Rubio, A., Martínez-Sanz, M., Cabedo, L., & Lagaron, J. M. (2014). Keratin–poly-hydroxyalkanoate melt-compounded composites with improved barrier properties of interest in food packaging applications. *Journal of Applied Polymer Science, 131*(4).

Poblete-Castro, I., Becker, J., Dohnt, K., Dos Santos, V. M., & Wittmann, C. (2012). Industrial biotechnology of *Pseudomonas putida* and related species. *Applied Microbiology and Biotechnology, 93*(6), 2279–2290.

Poli, A., Di Donato, P., Abbamondi, G. R., & Nicolaus, B. (2011). Synthesis, production, and biotechnological applications of exopolysaccharides and polyhydroxyalkanoates by *archaea. Archaea, 2011.*

Pötter, M., Madkour, M. H., Mayer, F., & Steinbüchel, A. (2002). Regulation of phasin expression and polyhydroxyalkanoate (PHA) granule formation in Ralstonia eutropha H16. *Microbiology, 148*(8), 2413–2426.

Pötter, M., & Steinbüchel, A. (2005). Poly (3-hydroxybutyrate) granule-associated proteins: Impacts on poly (3-hydroxybutyrate) synthesis and degradation. *Biomacromolecules, 6*(2), 552–560.

Quillaguaman, J., Hashim, S., Bento, F., Mattiasson, B., & Hatti-Kaul, R. (2005). Poly (β-hydroxybutyrate) production by a moderate halophile, *Halomonas boliviensis* LC1 using starch hydrolysate as sub-strate. *Journal of Applied Microbiology, 99*(1), 151–157.

Rai, R., Keshavarz, T., Roether, J. A., Boccaccini, A. R., & Roy, I. (2011). Medium chain length polyhydroxy-alkanoates, promising new biomedical materials for the future. *Materials Science and Engineering: R: Reports, 72*(3), 29–47.

Ramsay, J. A., Berger, E., Ramsay, B. A., & Chavarie, C. (1990). Recovery of poly-3-hydroxyalkanoic acid granules by a surfactant-hypochlorite treatment. *Biotechnology Techniques, 4*(4), 221–226.

Rastogi, V. K., & Samyn, P. (2015). Bio-based coatings for paper applications. *Coatings, 5*(4), 887–930.

Rastogi, V. K., & Samyn, P. (2016). Synthesis of polyhydroxybutyrate particles with micro-to-nanosized struc-tures and application as protective coating for packaging papers. *Nanomaterials, 7*(1), 5.

Ray, A., Cot, M., Puzo, G., Gilleron, M., & Nigou, J. (2013). Bacterial cell wall macroamphiphiles: Pathogen-/microbe-associated molecular patterns detected by mammalian innate immune sys-tem. *Biochimie, 95*(1), 33–42.

Reddy, C. S. K., Ghai, R., & Kalia, V. (2003). Polyhydroxyalkanoates: An overview. *Bioresource Technology, 87*(2), 137–146.

Rhim, J. W., Park, H. M., & Ha, C. S. (2013). Bio-nanocomposites for food packaging applications. *Progress in Polymer Science, 38*(10–11), 1629–1652.

Ruhland, G. J., & Fiedler, F. (1990). Occurrence and structure of lipoteichoic acids in the genus Staphylococcus. *Archives of Microbiology, 154*(4), 375–379.

Sakthiselvan, P., & Madhumathi, R. (2018). Kinetic evaluation on cell growth and biosynthesis of polyhy-droxybutyrate (PHB) by *Bacillus safensis* EBT1 from sugarcane bagasse. *Engineering in Agriculture, Environment and Food, 11*(3), 145–152.

Sandhya, M., Aravind, J., & Kanmani, P. (2013). Production of polyhydroxyalkanoates from *Ralstonia eutropha* using paddy straw as cheap substrate. *International Journal of Environmental Science and Technology, 10*(1), 47–54.

Saratale, G. D., & Oh, M. K. (2015). Characterization of poly-3-hydroxybutyrate (PHB) produced from *Ralstonia eutropha* using an alkali-pretreated biomass feedstock. *International Journal of Biological Macromolecules, 80*, 627–635.

Saratale, R. G., Saratale, G. D., Cho, S. K., Kim, D. S., Ghodake, G. S., Kadam, A., Kumar, G., Bharagava, R. N., Banu, R., & Shin, H. S. (2019). Pretreatment of kenaf (*Hibiscus cannabinus* L.) biomass feedstock for polyhydroxybutyrate (PHB) production and characterization. *Bioresource Technology, 282*, 75–80.

Savenkova, L., Gercberga, Z., Bibers, I., & Kalnin, M. (2000). Effect of 3-hydroxy valerate content on some physical and mechanical properties of polyhydroxyalkanoates produced by *Azotobacter chroococcum*. *Process Biochemistry, 36*(5), 445–450.

Shah, A. A., Hasan, F., Hameed, A., & Ahmed, S. (2008). Biological degradation of plastics: A comprehensive review. *Biotechnology Advances, 26*(3), 246–265.

Shahid, S., Mosrati, R., Ledauphin, J., Amiel, C., Fontaine, P., Gaillard, J. L., & Corroler, D. (2013). Impact of carbon source and variable nitrogen conditions on bacterial biosynthesis of polyhydroxyalkanoates: Evidence of an atypical metabolism in *Bacillus megaterium* DSM 509. *Journal of Bioscience and Bioengineering, 116*(3), 302–308.

Shang, L., Jiang, M., Yun, Z., Yan, H. Q., & Chang, H. N. (2008). Mass production of medium-chain-length poly (3-hydroxyalkanoates) from hydrolyzed corn oil by fed-batch culture of *Pseudomonas putida*. *World Journal of Microbiology and Biotechnology, 24*(12), 2783–2787.

Sharma, P., Ahuja, A., Izrayeel, A. M. D., Samyn, P., & Rastogi, V. K. (2022). Physicochemical and thermal characterization of poly (3-hydroxybutyrate-co-4-hydroxybutyrate) films incorporating thyme essential oil for active packaging of white bread. *Food Control, 133*, 108688.

Shawaphun, S., & Manangan, T. (2010). Paper coating with biodegradable polymer for food packaging. *Science Journal Ubon Ratchathani University, 1*(1), 51–57.

Shimizu, R., Chou, K., Orita, I., Suzuki, Y., Nakamura, S., & Fukui, T. (2013). Detection of phase-dependent transcriptomic changes and Rubisco-mediated CO2 fixation into poly (3-hydroxybutyrate) under heterotrophic condition in *Ralstonia eutropha* H16 based on RNA-seq and gene deletion analyses. *BMC Microbiology, 13*(1), 1–15.

Silva, L. F., Taciro, M. K., Michelin Ramos, M. E., Carter, J. M., Pradella, J. G. C., & Gomez, J. G. C. (2004). Poly-3-hydroxybutyrate (P3HB) production by bacteria from xylose, glucose and sugarcane bagasse hydrolysate. *Journal of Industrial Microbiology and Biotechnology, 31*(6), 245–254.

Song, C., Zhao, L., Ono, S., Shimasaki, C., & Inoue, M. (2001). Production of poly (3-hydroxybutyrate-co-3-hydroxyvalerate) from cottonseed oil and valeric acid in batch culture of *Ralstonia* sp. strain JC-64. *Applied Biochemistry and Biotechnology, 94*(2), 169–178.

Steinbüchel, A., & Lütke-Eversloh, T. (2003). Metabolic engineering and pathway construction for biotechnological production of relevant polyhydroxyalkanoates in microorganisms. *Biochemical Engineering Journal, 16*(2), 81–96.

Steinbüchel, A., & Valentin, H. E. (1995). Diversity of bacterial polyhydroxyalkanoic acids. *FEMS Microbiology Letters, 128*(3), 219–228.

Sudesh, K., Abe, H., & Doi, Y. (2000). Synthesis, structure and properties of polyhydroxyalkanoates: biological polyesters. *Progress in Polymer Science, 25*(10), 1503–1555.

Sun, Z., Ramsay, J. A., Guay, M., & Ramsay, B. A. (2007). Carbon-limited fed-batch production of medium-chain-length polyhydroxyalkanoates from nonanoic acid by *Pseudomonas putida* KT2440. *Applied Microbiology and Biotechnology, 74*(1), 69–77.

Sutcliffe, I. C. (1995). The lipoteichoic acids and lipoglycans of gram-positive bacteria: A chemotaxonomic perspective. *Systematic and Applied Microbiology, 17*(4), 467–480.

Suwannasing, W., Imai, T., & Kaewkannetra, P. (2015). Cost-effective defined medium for the production of polyhydroxyalkanoates using agricultural raw materials. *Bioresource Technology, 194*, 67–74.

Tamer, I. M., Moo-Young, M., & Chisti, Y. (1998). Disruption of alcaligenes latus for recovery of poly (β-hydroxybutyric acid): Comparison of high-pressure homogenization, bead milling, and chemically induced lysis. *Industrial & Engineering Chemistry Research, 37*(5), 1807–1814.

Tan, G. Y. A., Chen, C. L., Ge, L., Li, L., Wang, L., Zhao, L., Mo, Y., Tan, S. N., & Wang, J. Y. (2014a). Enhanced gas chromatography-mass spectrometry method for bacterial polyhydroxyalkanoates analysis. *Journal of Bioscience and Bioengineering, 117*(3), 379–382.

Tan, G. Y. A., Chen, C. L., Li, L., Ge, L., Wang, L., Razaad, I. M. N., Li, Y., Zhao, L., Mo, Y., & Wang, J. Y. (2014b). Start a research on biopolymer polyhydroxyalkanoate (PHA): A review. *Polymers, 6*(3), 706–754.

Thellen, C., Cheney, S., & Ratto, J. A. (2013). Melt processing and characterization of polyvinyl alcohol and polyhydroxyalkanoate multilayer films. *Journal of Applied Polymer Science, 127*(3), 2314–2324.

Thellen, C., Coyne, M., Froio, D., Auerbach, M., Wirsen, C., & Ratto, J. A. (2008). A processing, characterization and marine biodegradation study of melt-extruded polyhydroxyalkanoate (PHA) films. *Journal of Polymers and the Environment, 16*(1), 1–11.

Tripathi, A. D., Raj Joshi, T., Kumar Srivastava, S., Darani, K. K., Khade, S., & Srivastava, J. (2019). Effect of nutritional supplements on bio-plastics (PHB) production utilizing sugar refinery waste with potential application in food packaging. *Preparative Biochemistry and Biotechnology, 49*(6), 567–577.

Tsuge, T. (2016). Fundamental factors determining the molecular weight of polyhydroxyalkanoate during biosynthesis. *Polymer Journal, 48*(11), 1051–1057.

Urbina, L., Eceiza, A., Gabilondo, N., Corcuera, M. Á., & Retegi, A. (2019). Valorization of apple waste for active packaging: Multicomponent polyhydroxyalkanoate coated nanopapers with improved hydrophobicity and antioxidant capacity. *Food Packaging and Shelf Life, 21*, 100356.

Valappil, S. P., Boccaccini, A. R., Bucke, C., & Roy, I. (2007). Polyhydroxyalkanoates in Gram-positive bacteria: insights from the genera Bacillus and Streptomyces. *Antonie Van Leeuwenhoek, 91*(1), 1–17.

van Hee, P., Elumbaring, A. C., van der Lans, R. G., & Van der Wielen, L. A. (2006). Selective recovery of polyhydroxyalkanoate inclusion bodies from fermentation broth by dissolved-air flotation. *Journal of Colloid and Interface Science, 297*(2), 595–606.

Verlinden, R. A., Hill, D. J., Kenward, M. A., Williams, C. D., & Radecka, I. (2007). Bacterial synthesis of biodegradable polyhydroxyalkanoates. *Journal of Applied Microbiology, 102*(6), 1437–1449.

Wallen, L. L., & Rohwedder, W. K. (1974). Poly-. beta.-hydroxyalkanoate from activated sludge. *Environmental Science & Technology, 8*(6), 576–579.

Wampfler, B., Ramsauer, T., Rezzonico, S., Hischier, R., Kohling, R., Thony-Meyer, L., & Zinn, M. (2010). Isolation and purification of medium chain length poly (3-hydroxyalkanoates) (mcl-PHA) for medical applications using nonchlorinated solvents. *Biomacromolecules, 11*(10), 2716–2723.

Wang, L. I. N. F. U., Shogren, R. L., & Carriere, C. (2000). Preparation and properties of thermoplastic starch-polyester laminate sheets by coextrusion. *Polymer Engineering & Science, 40*(2), 499–506.

Wang, S., Song, C., Chen, G., Guo, T., Liu, J., Zhang, B., & Takeuchi, S. (2005). Characteristics and biodegradation properties of poly (3-hydroxybutyrate-co-3-hydroxyvalerate)/organophilic montmorillonite (PHBV/OMMT) nanocomposite. *Polymer Degradation and Stability, 87*(1), 69–76.

Wang, Y., Yamada, S., Asakawa, N., Yamane, T., Yoshie, N., & Inoue, Y. (2001). Comonomer compositional distribution and thermal and morphological characteristics of bacterial poly (3-hydroxybutyrate-co-3-hydroxyvalerate)s with high 3-hydroxyvalerate content. *Biomacromolecules, 2*(4), 1315–1323.

Xu, Y., Wang, R. H., Koutinas, A. A., & Webb, C. (2010). Microbial biodegradable plastic production from a wheat-based biorefining strategy. *Process Biochemistry, 45*(2), 153–163.

Yu, J., & Chen, L. X. (2006). Cost-effective recovery and purification of polyhydroxyalkanoates by selective dissolution of cell mass. *Biotechnology Progress, 22*(2), 547–553.

Zinn, M., & Hany, R. (2005). Tailored material properties of polyhydroxyalkanoates through biosynthesis and chemical modification. *Advanced Engineering Materials, 7*(5), 408–411.

12

Recent Advances in the Development of PHB (Polyhydroxybutyrate)-Based Packaging Materials

Lakshmanan Muthulakshmi, R. Rajam, Shalini Mohan, and P. Karthik

CONTENTS

12.1 Introduction

Packaging materials play an irreplaceable role in ensuring food security and food supply. The plastics used for food packaging have been hugely piled up in the environment, contaminating the surroundings rapidly due to the growing number of consumer products in the market (Ncube et al., 2020). Many forms of the packaging material used for food include rigid and flexible types, like cartons, boxes, sleeves, bottles, and wrappings (Ojha et al., 2015). The type of food packaging that could be adopted for different products is usually based on the characteristics of the food material that has to be packed and the processing operations to be followed (Deshwal et al., 2019). The chronological evolution of packaging material starts with plastics, followed by polymers from renewable sources, then biopolymers, and the latest generation of active packaging material (Shershneva, 2022). The biopolymers are usually classified as synthetically produced polymers, those produced from microbes, and the bio-products from renewable sources. Biodegradable polymers tend to replace conventional polymers in order to reduce the carbon footprint in the environment that occurs due to the accumulation of petroleum-based polymers (Zhong et al., 2020).

Most biopolymers are fragile in nature when they are homopolymers without any other polymers. The desired properties of the polymers could be enhanced by blending them as composite materials with other polymers to improve the required characteristics. The nature of the composite produced is highly dependent on the parent polymers used and the process followed for producing the same (Hamid & Samy, 2021). Polyhydroxybutyrate (PHB) is one of the most commonly studied biopolymers for the utilization of food packaging materials. They are brittle due to their high crystallinity. It was first identified from the *Bacillus* sp. of bacteria as an accumulation to combat environmental stress (Turco et al., 2021). PHB is produced by a variety of both Gram-positive and Gram-negative bacteria. The PHB produced by the bacterial cells as a response to the limitation of nutrient sources in ideal conditions where macronutrients like phosphorous and nitrogen are limited while, carbon and lipid molecules are in excess quantity. They are accumulated within the cells as inclusion bodies and are one of the most common representatives of PHAs produced by the microbes (Arrieta et al., 2017). The degradability of pure PHB is relatively high compared to the composite. It is prone to degradation by a variety of enzymes over a wide range of temperatures and leads to the formation of monomeric units that are completely degradable in nature (Anbukarasu et al., 2015). They also possess good mechanical and gas barrier properties. In virgin form, PHB polymer has a high melting temperature and possesses more tensile strength. The molar mass of the PHB produced depends on the type and conditions opted for polymer production (Turco et al., 2021).

Pure isolates used for PHB production could be screened using the lipophilic stain of Sudan Black B, and the positive colonies on staining tend to appear blue-black in color when observed under the microscope (Thapa et al., 2018). Staining for PHB is done using Nile red solution and observing under the fluorescence microscope at excitation and emission wavelengths of 450–550 nm, respectively (Meixner et al., 2022). In general, PHB is optically active, non-toxic, and biocompatible and usually possesses a yellowish hue shade with improved properties and tends to remain stable even when subjected to exposure under UV irradiation (McAdam et al., 2020). PHB has a wide range of applications in various fields, like tissue engineering, agriculture, biomedical and food packaging (Yeo et al., 2018). Blending is one of the techniques used to improve the applicability of the polymer (Li et al., 2016). Several blends of PHB are analyzed to improve their mechanical and degradable properties. It gets mixed thoroughly with polylactic acid, enhancing the amorphous nature, thereby the flexibility of the produced polymer matrix (Zhang & Thomas, 2011). Likewise, other blends of PHB with polymethacrylate, polyethylene glycol, and ethyl cellulose showed improvements in numerous properties of the packaging material (Abou-Aiad, 2007; Zhang et al., 1998; Chan et al., 2011). PHB possesses strong antimicrobial activity against foodborne pathogens (Slepička et al., 2016). PHB polymer recently exploited for numerous active packaging compared to other types of polymer. Boey et al., (2021) reported that PLA-blended with PHB improved the functional properties of packaging films. PHB has the capacity of releasing active molecules like antibiotics and antioxidants enhancing the quality of packed food (Arrieta et al., 2019). As the thermal and mechanical properties of PHB are similar to polypropylene, it could be used as an effective, eco-friendly biodegradable alternative for the petroleum-based polymeric materials that remain as recalcitrant in the atmosphere for a longer time (Boey et al., 2021).

This chapter focuses on a brief introduction to the petroleum- and bio-based polymers used for packaging, biocomposites, properties of PHBs, blends of PHB, and techniques for the production of PHB biocomposites, antimicrobial activity of PHB blends, the role of PHB in active packaging, and other applications of PHB.

12.2 Packaging Materials

Plastics have occupied a prominent position in the monotonous life schedule of individuals as a better material of choice. But poor waste management of plastics has led a huge amount of it to pollute the environment. The industries manufacturing packaging materials are the largest and a growing consumer of products from fossil fuels to produce plastic (Ncube et al., 2020). A major portion of the packaging work is performed for food products, as it has turned out to be a vital response element for phasing out the environmental footprints that are engraved in the ecosystem due to increased consumption of plastic materials (Guillard et al., 2018). The principal catalysts driving the surge of the utilization of packaging material are the expanding middle class, organized sectors for retail supply, and liberalization. Packaging materials are usually flexible, like foldable cartons, boxes, shrink sleeves, glass bottles, and wrappings (Ojha et al., 2015). Food processing industries select suitable packaging material based on certain factors that are deeply correlated with respect to the food item manufactured in the factory. One of the common environmentally friendly materials used for food packaging is paper. Almost 47% of the paper produced is used for packaging purposes, yet the wide applicability of it is restricted due to its low wear resistance (Deshwal et al., 2019). The paper materials used for packaging include kraft paper, sulfite paper, parchment paper, greaseproof paper, and glassine. Food packaging has numerous emerging innovations, like modified atmosphere packaging, active and intelligent packaging that can abruptly change the internal environment where food is present to protect it from damaging factors like oxidative gases, microbial contamination, etc. (Raheem, 2012). Recently, bio-based products are utilized for eco-friendly packaging materials, and most of the raw materials used for their production are obtained from renewable sources. Though biopolymers have a handful of limitations concerning their cost of production and other physical properties, their recycling stream with respect to the environment has paved the way for numerous research and development of the same (Hong et al., 2021).

12.2.1 Biopolymer-Based Packaging Materials

Biodegradable polymers are extracted from renewable sources in the environment and have an increased rate of degradability compared to non-degradable ones. Biopolymers are extracted from many sources, like microbes, fungi, plants, and waste materials. The units of these polymers could be easily broken down by the biological reactions involving enzymes. It reduces the petroleum products in the environment (Zhong et al., 2020). Packaging material has a humongous task to directly appeal to the consumer in such a way that the person is convinced the information provided relevant to the product is properly conveyed to the consumer. The material for developing the packaging would be desired to possess antibacterial, antimicrobial, and nutrients as an impartation to the product packed (Yuvaraj, 2021). The evolutionary trend of packaging material is divided into four generations that include the synthetic polymers grouped in first generation, composite ones in second generation, biopolymers in third generation, and active packaging material in fourth generation. The products of decomposition formed from the biopolymer are carbon dioxide, water, biomass, and some inorganic compounds (Shershneva, 2022). The sustainable nature of biopolymers makes them prominent candidates for manufacturing packaging materials. Some of the standard test methods employed for testing the quality of the packaging material include the American Society for Testing and Materials (ASTM D5338–15) and International Organization for Standardization (ISO 14855–2:2018). The aforementioned testing methods could evaluate the rate and conditions required for the degradation of the materials (Nanda et al., 2021). The polymers used for food packaging are depicted in Table 12.1.

12.2.2 Types of Bio-Based Packaging Materials

The major classification of biopolymers based on the mode of synthesis is described in this section. The types of bio-based materials that could be used for packaging are shown in Figure 12.1.

12.2.2.1. *Synthetically Produced Biopolymers*

The polymers grouped in this category include those produced by living organisms that could also be produced synthetically using certain chemical modifications (Reichert et al., 2020).

- **Polylactic Acid**

It is derived from starch and is known to be non-toxic and possesses high mechanical strength and plasticity. It is classified generally recognized as safe (GRAS) by the Food and Drug Administration (Süfer

TABLE 12.1

Polymers used for food packaging

S. no.	Polymer	Food material	Reference
Petroleum-based polymers			
1.	Polyethylene terephthalate	Beverage bottles	(Nisticò, 2020)
2.	Low-density polyethylene films	Spice powders	(Niazmand et al., 2020).
3.	High-density polyethylene films	Juice bottles	(Manikantan & Varadharaju, 2011)
4.	Polycarbonate	Ready-to-eat food containers	(Manoli & Voutsa, 2016)
5.	Polyvinyl chloride	Bottle caps, cling films	(Pearson, 1982).
6.	Polypropylene	Teacups	(Wu et al., 2021)
7.	Polyethylene naphthalate	Refillable bottles	(Ewender & Welle, 2019).
8.	Polyamide	Frozen food	(Schweighuber et al., 2021)
9.	Polymethaacrylate	Food wrapping	(Corsaro et al., 2021)
10.	Polyvinyl alcohol	Food films	(Goñi-Ciaurriz et al., 2021)
Bio-based packaging materials			
11.	Cellulose	Films	(Ghaderi et al., 2014)
12.	Starch	Films	(Avella et al., 2005)
13.	Zein protein	Films	(Oymaci & Altinkaya, 2016)
14.	Gluten protein	Films and coatings	(Bibi et al., 2016)
15.	Polyhydroxy alkonoate	Coating, container	(Khosravi-Darani & Bucci, 2015)
16.	Polylactic acid	Films	(Arora & Padua, 2010).
17.	Chitosan	Edible coating	(Flórez et al., 2022)
18.	Polyethylene furanoate	Beverage bottles	(Loos et al., 2020)
19.	Polyhydroxybutyrate co-valerate	Biopaper	(Melendez-Rodriguez et al., 2020)
20.	Pullalan	Coating	(Kraśniewska et al., 2019)

et al., 2017). It is mostly extracted from agricultural feedstock as a result of fermentation of sugars glucose and sucrose refined to the highest purity (McGrath et al., 2013).

- **Polyethylene Furanoate**

It is one of the polymers that resemble the properties of polyethylene terephthalate that is employed as a packaging material for beverages. It is being synthesized using the C6 sugars by selective oxidation of hydroxymethylfurfural residues (Loos et al., 2020). The oxygen diffusion rates have been found to be less for PEF compared to that of PET (Lightfoot et al., 2022).

- **Polybutylene Succinate**

Polybutylene succinate is a biopolymer fabricated from starch and has better properties, like mechanical and barrier. The plasticization process has improved the interfacial bonding that occurs in the polymer matrix (Nazrin et al., 2020). It is a flexible polymer with good clarity, processability, and shiny look (Rafiqah et al., 2021).

12.2.2.2 Non-Synthetically Produced Biopolymers

This class includes polymers produced from living organisms.

- **Polyhydroxyalkanoate**

Polyhydroxyalkanoates have been characterized by manifold variations that are not found in many other polymers. They have a complete green life cycle. They are usually present as intercellular granules

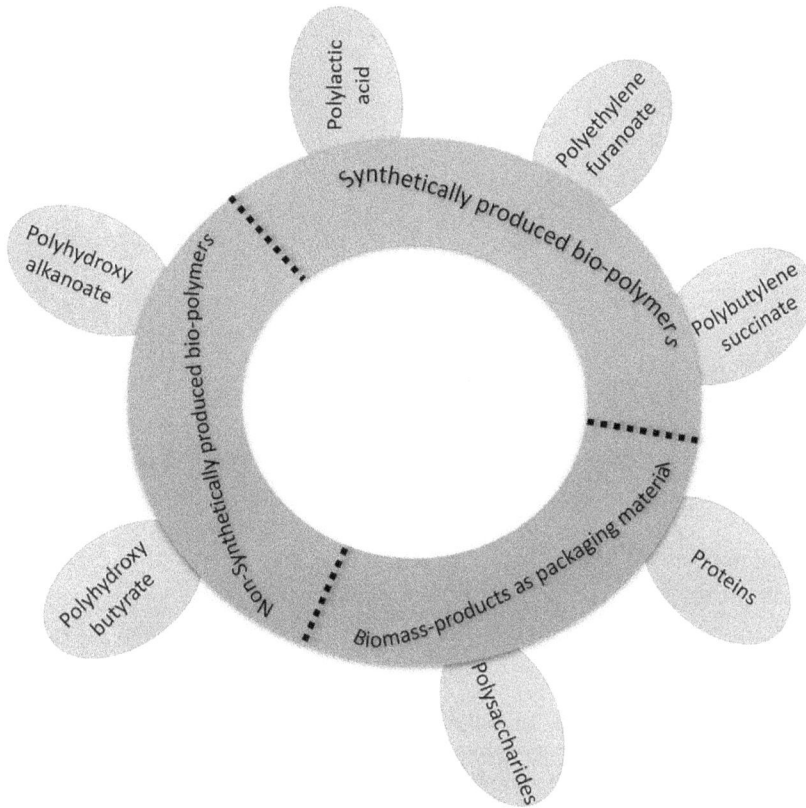

FIGURE 12.1

within the microbial cells (Koller, 2014). PHA is an environment friendly polymer that could be easily purified using water, sodium hydroxide, SDS other than solvents like chloroform that are harmful in nature (Boey et al., 2021).

- **Polyhydroxybutyrate**

Polyhydroxybutyrate is an aliphatic chain of polyester that is of microbial origin and could be an effective substitute for polypropylene (Latos & Masek, 2018). The details about PHB are briefly discussed in later sections of this chapter.

12.2.2.3 Biomass Products as Packaging Material

This class includes products that are obtained directly from biomass for formulating the packaging materials.

- **Polysaccharides**

Polysaccharides have become more appealing as a source of edible packaging. Some of the notable polysaccharides used as packaging materials are cellulose, xylan, glucomannan, amylose, amylopectin, and chitosan (Zhao et al., 2021). Tensile strength and the ability to elongate well make polysaccharides amenable on a large scale to produce packaging (Cazón et al., 2017).

- **Proteins**

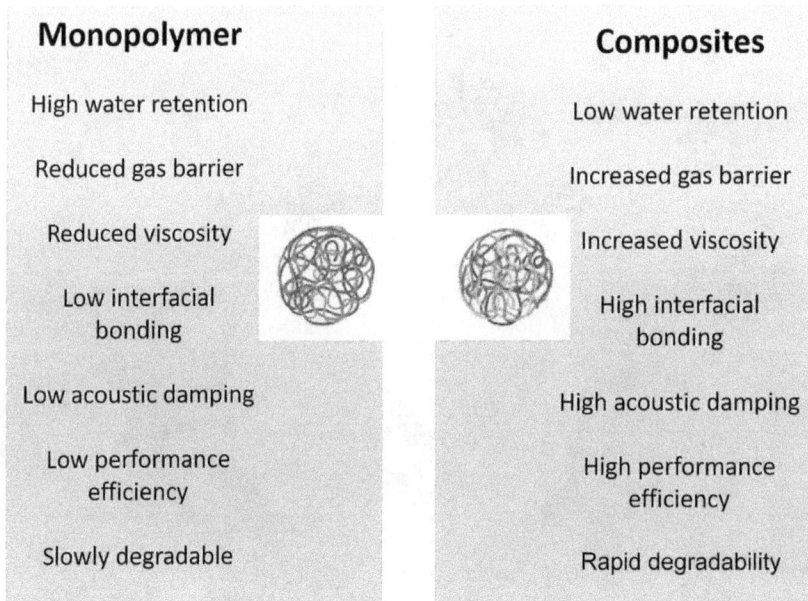

Monopolymer	**Composites**
High water retention	Low water retention
Reduced gas barrier	Increased gas barrier
Reduced viscosity	Increased viscosity
Low interfacial bonding	High interfacial bonding
Low acoustic damping	High acoustic damping
Low performance efficiency	High performance efficiency
Slowly degradable	Rapid degradability

FIGURE 12.2

Proteins can be obtained from plant and animal sources. The gas barrier and wide functional properties of proteins based on the unique structure enable the highest intermolecular binding potential of the protein (Chen et al., 2019). The protein packaging is also edible and is classified as solution-based and mass-based categories, depending on the nature of the material used for preparing the film (Martins et al., 2021).

12.3 Biocomposites

Biocomposites are a material made up of two or more individual constituents of materials, among which at least one of them is derived from natural sources, mainly aimed to increase the performance of the individual polymeric materials. The improved properties increase the applicability of biopolymeric materials (Rudin & Choi, 2012). A polymer composite is a multiphase material integrated with the matrix of a polymer to synergistically enhance the mechanical properties of the component. Biocomposites increase the hardness, gel-forming capacity, and viscosity of the fabricated material (Hamid & Samy, 2021). The uses of biocomposites are shown in Figure 12.2.

Reinforcing the natural fibers within the biopolymer is a markable option for improving the properties of the polymer material. The process of modification or functionalization is of utmost importance to promote interfacial bonding to produce the composite material with high performance efficiency. A class of reinforced biopolymers is termed green composites due to the filler material being a natural fiber (Aaliya et al., 2021). The biodegradable composite–based food packaging material includes cellulose, chitin, PHB, and gelatin (Alim et al., 2022). Natural fiber–reinforced plastics have the ability to replace synthetic fibers in most of the industrial sectors, like automobile, furniture, and packaging industry (Laftah & Wan Abdul Rahman, 2021). Some of the waste materials, like almond shell, rice husk, and seagrass, are sometimes used as fillers for the manufacture of reinforced polymers. The barrier property, thermoforming property, and disintegration property improved than in the individual polymer (Sánchez-Safont et al., 2018). They are also known to possess benefits like low cost, reduced carbon footprint, increased acoustic damping, high availability, low energy consumption for manufacturing, biodegradability, and acceptable ratio of modulus weight (Bajsic et al., 2021). The abrasion resistance property could also be enhanced by the exploitation of the biocomposite material without losing its inherent properties. The limiting factors restricting the use of biocomposites include their incompatibility, hydrophobic nature, sensitivity to temperature, and exacerbation. The methodologies adopted for the production of biocomposites include extrusion, filament welding, extrusion, injection molding, sheet molding, resin transfer molding, and other methods (Edebali, 2021).

12.4 Polyhydroxybutyrate

Polyhydroxybutyrate is a thermoplastic polymer with a relatively low resistance to degradation. The thermal degradation of PHB occurs at temperatures around 190°C. PHB is very brittle and stiff, resulting in poor mechanical properties. PHB was first characterized as early as 1926 within *Bacillus megaterium* cells (Turco et al., 2021). Since then, it is reported to be produced by a variety of Gram-negative bacteria, like *Azobacter* and *Bacillus*, as well as Gram-positive bacteria, like *Streptomyces*, *Nocardia*. They are mostly produced by bacterial cells as a response to limitation of nutrient sources in ideal conditions where macronutrients like phosphorous and nitrogen are limited while carbon and lipid molecules are present in excess quantity. They are accumulated within the cells as inclusion bodies and are one of the most common representatives of PHAs produced by the microbes (Arrieta et al., 2017). The raw materials for the production of PHB are both renewable and sustainable in nature, probably the waste materials from agriculture or biomass. It is a linear-chain structure consisting of both phases that are amorphous and crystalline structures. The general structure of the PHB is composed of a methyl group and an ester linkage, making the material thermoplastic and hydrophobic in nature (McAdam et al., 2020). PHB has a range of applications diversely spread into many fields, like agriculture, food packaging, tissue engineering, water filter, automobile, construction, and biomedical fields (Yeo et al., 2018).

12.5 Properties of PHB

PHB is an unequalled material with two phases that are amorphous and crystalline in nature. The linear structure of it is responsible for the crystal nature of the polymer. It could exist both in virgin form or in a blended form as either composites or blends (McAdam et al., 2020). Most of the PHB produced are crystalline and brittle in structure. In order to overcome this, the crude PHB is blended with other polymers that are capable of reducing the crystallinity of the polymer, making them utilizable (Savenkova et al., 2000). Spherulites are formed by increasing the mechanical strength by the addition of nucleating agents like saccharin. A necking process occurs when the blends form ductile polymers with the progression of plastic deformation (El-Hadi et al., 2002). PHB produced by microorganisms is usually semicrystalline with all the molecules being in R-configuration due to their isotactic steric regular property, which increases their biodegradability (Bugnicourt et al., 2014). The degradability of pure PHB is relatively high compared to the composite. It is prone to degradation by a variety of enzymes over a temperature range of 80°C to 160°C and leads to the formation of monomeric units that are completely degradable in characteristics (Anbukarasu et al., 2015). Besides, they possess fine mechanical and gas barrier properties. In virgin form, PHB polymer has a high melting temperature and consists of high tensile strength. The molar mass of the PHB produced depends on the type and conditions opted for the production of the polymer (Turco et al., 2021). In general, PHB is optically active, non-toxic, biocompatible, and usually possesses a yellowish hue shade with improved properties and tends to remain stable even when subjected to exposure under UV irradiation (McAdam et al., 2020). Hence, the peculiar properties of PHB are listed as follows:

- It possesses the property of thermoplasticity, especially when blended with apt polymers (Olejnik et al., 2021).
- It is biodegradable in nature and does not produce any toxic by-products in the environment (Bucci et al., 2007).

TABLE 12.2

Physical properties of polyhydroxybutyrate

S. no.	Properties	Values of PHB	Reference
1.	Melting point Tm	175°C to 180°C	(Anbukarasu et al., 2015)
2.	Crystallinity	64% to 78%	(Anbukarasu et al., 2015)
3.	Molecular weight	0.5×10^5 Da to 8×10^5	(Liu et al., 2016; Chan et al., 2004, Bourque et al., 1995)
4.	Tensile strength	4 Mpa to 20 Mpa	(Jain & Tiwari, 2015)
5.	UV resistance	High	(Sadi et al., 2010)
6.	Solvent resistance	High	(Babaniyi et al., 2020)
7.	Biodegradability	0.04 mg per day per cm^2 to 0.08 mg per day per cm^2	(Dilkes-Hoffman et al., 2019)

- The synthesized polymer is non-toxic, and it does not produce any toxic by-products as it is highly biocompatible in nature (Adorna et al., 2022).
- PHB is an isotactic stereo polymer with the monomeric units present in the same configuration throughout the polymer chain (dos Santos et al., 2017).
- PHB has higher densities and is completely insoluble in water. It could be depolymerized by heat, pH, and other factors (Gahlawat et al., 2020).
- The crystallinity of PHB varies with respect to the source organism of isolation. Bacterial PHB is highly crystalline, with approximately 65% crystallinity, and those isolated from other sources are amorphous in nature. The brittle nature of the PHB is attributed to its crystalline structure of PHB (Righetti et al., 2019).
- Piezoelectric property is observed in the polymer at glass transition temperature, and relaxations of dielectric nature are also observed in both crystalline and non-crystalline regions of the polymer (Fukada & Ando, 1986).
- PHB is produced mostly from materials of renewable sources, ensuring sustainable goals (de Resende & da Costa, 2020).

The comprehensive properties of PHB are shown in Table 12.2

12.6 Production Strategies of PHB

The most optimal media constituents required for the production of PHB deeply vary with respect to the nature of the organism that produces it. Hence, a different scheme of production is used for varieties of cultures and metabolic pathways involved in producing the same (Trakunjae et al., 2021). It is reported that PHB accumulation influences the c-type cytochrome production, oxidative stress reduction, growth inhibition, perturbation of energy balance, and metabolic changes (Batista et al., 2018). It is also found that the liquid nature of the PHB molecules has the potential of repairing and stabilizing the cell membrane during its readiness to plasmolysis. At hyperosmotic conditions, the levels of pH inside the microbial cell play an important role in the survival of the organism. PHB molecules have the ability to orchestrate the fluid movement into the cell with respect to the internal environments of the microbial cell (Obruca et al., 2017). The production scheme of PHB within microbial cells is represented in Figure 12.3.

Polyhydroxybutyrate is produced as carbon and energy reserve material of the bacterial cells. The yield of PHB produced by fermentation could be enhanced by altering the nutrient and culture conditions of the fermentation process (Trakunjae et al., 2021). A variety of raw materials are used for the synthesis of PHB by biological routes employing suitable microorganisms. Some sources of raw materials used for the production of PHB are mentioned in Table 12.3.

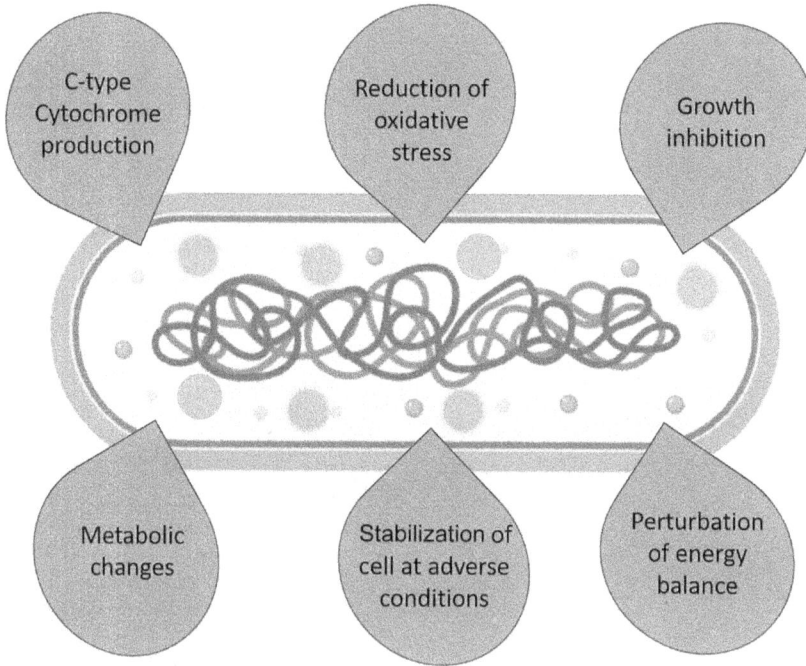

FIGURE 12.3

TABLE 12.3

List of raw materials reported for the production of PHB

S. no	Raw material	Organism	Notable features	Reference
1.	Food and agricultural waste	*Bacillus* sp., *Ralstonia* sp., and *Cupriavidus* sp.	• Cost-effective. • Increased production yield. • Reuse of waste material.	(Sirohi, 2020)
2.	Sugarcane bagasse, corncob, teff straw	*Bacillus* sp.	• Reduced disposal issue of waste. • Modifiable production parameters.	(Getachew & Woldesenbet, 2016)
3.	Dairy waste	*Bacillus megaterium*	• Increased yield of PHB. • Dissolved oxygen is the limiting nutrient.	(Pandian et al., 2010)
4.	Corn bran, corncob, wheat bran, rice bran, dairy waste, and sugarcane molasses	*Bacillus subtilis*	• Rice bran is suitable for PHB production. • Increased thermal stability of PHB.	(Hassan et al., 2019)
5.	Fruit waste	*Bacillus megaterium, Ralstonia eutropha, Bacillus cereus, Zobellella taiwanensis, Pseudomonas chlororaphis*	• Waste of mixed fruit and pineapple showed increased production of PHB. • Alkali-assisted pretreatment with ultrasonication and microwave treatment is better.	(Sirohi et al., 2021)
6.	Agave plant waste	Microbial consortium	• Promising vegetal resource utilization. • Improved properties of PHB.	(Martínez-Herrera et al., 2021)

(*Continued*)

TABLE 12.3 (*Continued*)

S. no	Raw material	Organism	Notable features	Reference
7.	Diary waste	*Brevibacterium casei*	• Green resources for PHB production. • PHB used for nanoparticle synthesis.	(Pandian et al., 2009)
8.	Cane molasses	*Alcaligenes* sp.	• PHB has high molecular weight and crystallinity index. • Increased melting point.	(Tripathi et al., 2019)
9.	Fish solid waste	*Bacillus subtilis*	• Less toxic in nature. • Has favorable structural feature for surface attachment.	(Mohapatra et al., 2017)
10.	Wastewater	*Botryococcus braunii*	• Increased production yield of PHB. • Microalgal production of PHB.	(Kavitha et al., 2016)
11.	Waste starch	*Bacillus arybhattai*	• Cost-effective production of PHB. • Cassava pulp, oil palm trunk starch identified as high-yield substrates.	(Bomrungnok et al., 2020)
12.	Luria Bertani medium	*Bacillus* sp.	• Glucose identified as suitable carbon source. • Ammonium sulfate identified as suitable nitrogen source.	(Alshehrei, 2019)
13.	Waste sludge	Methanotrophs	• Increased yield of PHB. • Type 2 methanotrophs produced increased amount of PHB.	(Fergala et al., 2018)
14.	Wheat waste	*Ralstonia eutropha*	• Ultrasound pretreatment of substrate effective for PHB production. • Sustainable production of biopolymer.	(Saratale et al., 2020)
15.	Pineapple peel	*Bacillus drentensis*	• Low-cost substrate. • Thermally stable PHB with semi-crystalline nature.	(Penkhrue et al., 2020)

12.7 Microbial Production of PHB

The main strategies used for the production of PHB include the ring opening of β-butyrolactone polymerization, transgenic PHB production using plants, and microbial fermentation, in which 90% of the dry weight of the cell is produced as PHB (McAdam et al., 2020). The production strategies for PHB include operation in batch mode, fed-batch, continuous, and two-stage modes of fermentation (Sharma, 2019). Pure isolates used for PHB production could be screened using the lipophilic stain of Sudan Black B, and the positive colonies on staining tend to appear blue-black in color when observed under the microscope (Thapa et al., 2018). Staining for PHB can be performed using Nile red solution and fluorescence microscope at excitation and emission wavelengths of 450–500 nm and 550 nm, respectively (Meixner et al., 2022). The biosynthesis of PHB from acetyl CoA follows three important steps in a sequence of events:

1. Condensation of acetyl-CoA moieties to form acetoacetyl-CoA molecules catalyzed by the enzyme β-ketothiolase.

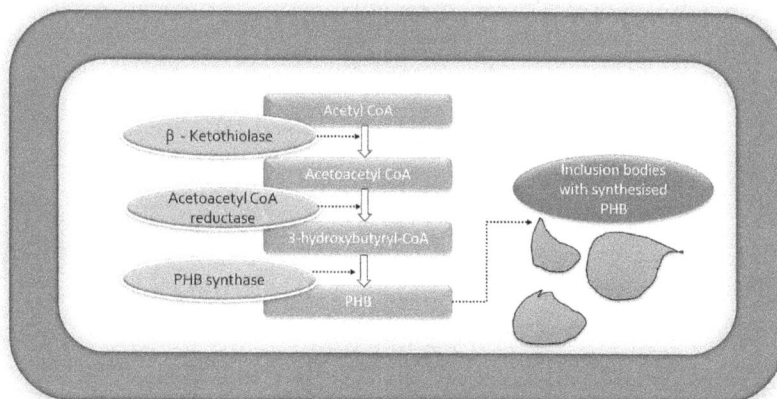

FIGURE 12.4

2. Reduction of acetoacetyl-CoA to 3-hydroxybutyryl-CoA catalyzed by the enzyme acetoacetyl-CoA-reductase.

3. The enzyme PHB synthase polymerizes the monomeric units of 3-hydroxybutyryl-CoA to accumulate PHB (Shrivastav et al., 2013).

PHB is produced as inclusion bodies by the microbes to act as electron sinks and carbon sources for their biosynthetic pathways, as shown in Figure 12.4. Some of the metabolic engineering strategies could improve the accumulation of PHB within microbial cells. By co-expressing certain genes, the enhanced synthesis of PHB in terms of type and quantity could be promoted (Li et al., 2010). PHB production is also increased by modifying the nutrient sources used for the fermentation of the medium (Zhuang et al., 2014).

12.8 Production of PHB by Engineering Microbial Cells

The commercialization of fermentative production of PHB is limited due to numerous factors, most of which could be altered by engineering the microbial cells and increasing their production of the same. The genes coding for the production of PHB include PhaA (β-ketothiolase), PhaB (acetoacetyl coenzyme A reductase), and PhaC (PHB synthase) and are involved in the PHB biosynthetic pathway, as mentioned in the previous section. Therefore, modification of any of the processes in the biosynthetic pathway could enhance the production of PHB (Peoples & Sinskey, 1989). Different sources of raw materials are responsible for generating the various yields of PHB with respect to the culture conditions (García et al., 2014). When PhaC gene was knocked down, complete removal of the PHB synthase enzyme occurred. However, different substrates like octanoate were used as a carbon source for the production of PHB (Fukui & Doi, 1997). On the other hand, recombinants yielded different types of PHBs based on the modified genes and the culture conditions (Höfer et al., 2011; Ma et al., 2018). In addition, certain extremophiles are also known to produce PHB. Genetic modification of *Halomanas* significantly increases the production of PHB (Tan et al., 2014). Similarly, the genetic modification of *Sinorhizobium meliloti* yielded 15% accumulation of PHB material in the cell (Tran & Charles, 2016). Therefore, *Pseudomonas putida* was modeled by the in silico method by mutation of enzymes like (gcd) glucose dehydrogenase and (pgl) 6-phosphogluconolactonase, revealing about 100% accumulation of PHB within the cellular entities (Poblete-Castro et al., 2013). Some of the techniques of processing that could improve the characteristic features of the PHB biocomposites are discussed in this section (Raturi et al., 2021) and are shown in Figure 12.5.

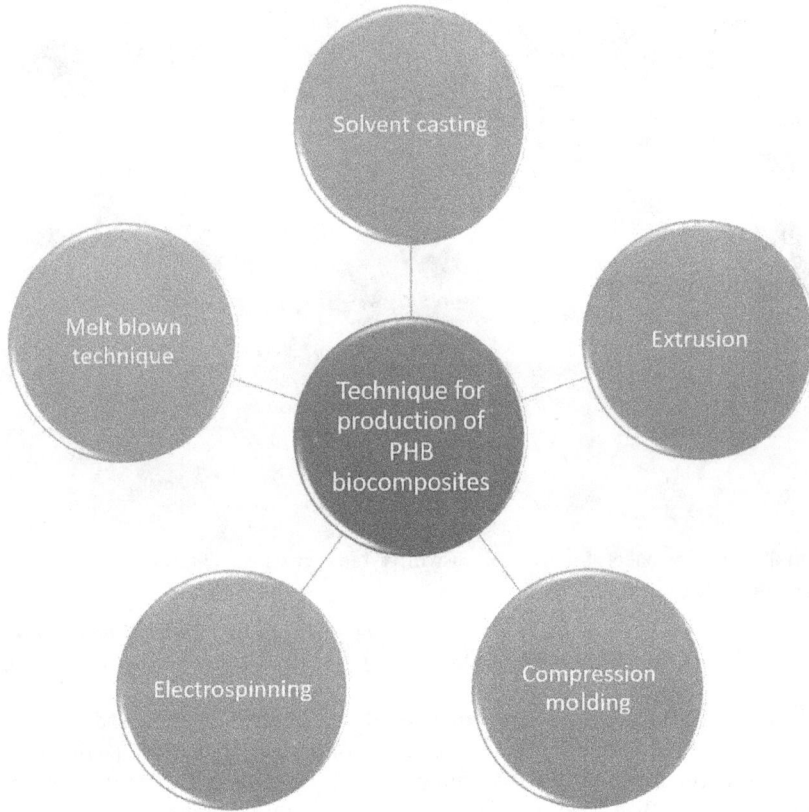

FIGURE 12.5

12.8.1 Solvent Casting

It is one of the most specific ways for the synthesis of biocomposites where the polymer mixture is cast into a common solvent which is allowed to dry under specific conditions. Some advantages of solvent casting include uniformity, optical purity, better transparency, flatness, and feasibility of the procedure (Seimann, 2005). This procedure was adopted for the production of biocomposite of acetylated CNC, poly(3-hydroxybutyrate-co-3-hydroxyhexanoate) (PHBH), and montmorillonite. The produced biocomposite had increased tensile strength, possessed better barrier characteristics, and was readily degraded by thermal properties (Xu et al., 2021).

12.8.2 Extrusion

This process involves conversion of melted polymer into a uniform shape by exerting force under controlled conditions. The notable advantages of the process include uniformity and flexibility of shape alterations that are post-process (Breitenbach, 2002). The extrusion process is adopted for the preparation of PHB composites using filler, like rice husk and almond shell (Sánchez-Safont et al., 2021).

12.8.3 Electrospinning

The electrospinning process can produce biocomposites with modifiable diameters. It is widely used for the synthesis of scaffolds that could be applicable to drug delivery purposes (Zhang et al., 2015). Scaffolds of PHB with gold nanoparticles produced by electrospinning are sought to be used for engineering bone tissue (Iron et al., 2019).

12.8.4 Compression Molding

Compression molding process involves the enforcement of compressive forces onto the heated polymers using molds that could cast the shape of the polymeric material (Park & Lee, 2012). The biocomposite is obtained by curing of material at an adequate time. The process of compression molding has increased reproducibility, improved dimensional stability, and could be automated to a large extent (Mitschang & Hildebrandt, 2012). Biocomposite of PHB and carbon nanotubes produced by this technique have improved mechanical and thermal properties (Seoane et al., 2018).

12.9 Blends of PHB

Blending is one of the versatile approaches used for the production of novel polymers that possess polymers enhanced to the desirable levels. The type and preparation conditions of the initial polymer and the process determine the physical and mechanical properties acquired by the blend (Li et al., 2016). The blends of PHB are represented in Figure 12.6.

PHB is blended with polylactic acid (PLA) to improve the crystallized properties. Though the polymers are immiscible with each other, they interact at the molecular level and increase the distortion temperature at heat. Besides, the PLA/PHB blends had improved tensile strength and biodegradability, and PHB acted as an agent of nucleation and filler in the PLA (Zhang & Thomas, 2011). The miscibility of the two polymers could be enhanced with the use of compatibilizers like propylene and ethylene oxide. Morphological characters were enhanced without much compromise of the rheology, and an enormous increase in the crystallinity and elastic modulus is observed (D'Anna et al., 2019). The plasticizer Lapol 108 tried to enhance only the biodegradability of the blended polymer by selectively decreasing the crystallinity of the polymer (Abdelwahab et al., 2012). Conversely, thermoplastic starch (TPS) is one of the other polymers that drastically improve the properties of the PHB blends. It was

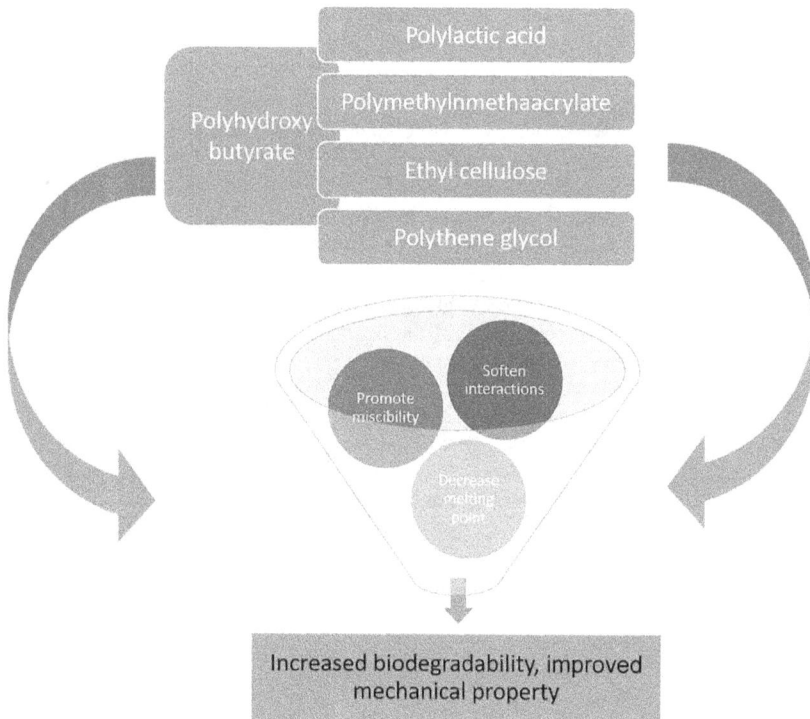

FIGURE 12.6

suggested that morphological property is influenced by the material composition and the processing method (Vanovčanová et al., 2016). Likewise, keratin is added to PLA/PHB blends for increasing the glass transition temperature but restricting the mobility of chains of polymer, thereby softening the interactions. The produced blend had properties similar to a thermoplastic that could extensively be used as packaging material (Mosnáčková et al., 2020).

The brittleness of PHB could be avoided by blending with poly(methylmethacrylate) (PMMA). The blending promoted miscibility of the amorphous phases of both the polymers and increased the compatibility of both. The blend also possessed antimicrobial activity against certain microbes (Abou-Aiad, 2007). Also, polyethylene glycol is one of the other polymers that could decrease the melting point of the blend. The mixture is miscible, and the low melting point signifies the degradable nature of the blend by improving the toughness (Zhang et al., 1998). The polymer ethyl cellulose increases the biocompatibility of blending with PHB. The composite has reduced crystallinity, enhancing the ability to degrade, and improved mechanical strength, with no significant changes in the morphological properties except the elevated surface roughness (Chan et al., 2011). Some more examples of PHB composites are listed in Table 12.4.

TABLE 12.4

Examples of PHB biocomposites

S. no.	PHB biocomposite	Notable properties	Reference
1.	Polylactic acid–PHB, limonene	PHB produced has amber tonality. Limonene increased the composting of PHB and PLA.	(Arrieta et al., 2014)
2.	Polylactic acid–PHB	Loaded with catechin, chitosan. Improved thermal stability.	(Arrieta et al., 2016)
3.	Polylactic acid–PHB	Improved mechanical resistance. Improved barrier performance.	(Arrieta et al., 2017)
4.	PHB–wood flour and kenaf fiber	Increased modulus of strength and elasticity. Higher ability of water absorption.	(Kuciel & Liber-Kneć, 2011)
5.	Cellulose–PHB	Improved uniaxial tensile behavior. Better adhesion property.	(Cyras et al., 2009)
6.	Polylactic acid–PHB	Increased elongation property. Accelerated crystallization process.	(El-Hadi, 2017)
7.	PEG–PHB, starch–PHB	Heavy metal removal. Improved thermal stability.	(Hungund et al., 2018)
8.	PLA–PHB	Films blended with cellulose nanocrystals. Increased oxygen barrier properties.	(Arrieta et al., 2014)
9.	Agave fiber–PHB	Improved mechanical properties. Improved impact and flexural strength.	(Smith et al., 2020)
10.	PLA–PHB	Cellulose nanocrystals and cellulose nanofibers are used as reinforcing agent. Increased crystallinity and thermal stability.	(Frone et al., 2019).
11.	Polypropylene–PHB	Lignophenols used as plasticizing agent. Improved mechanical properties.	(Ren et al., 2015)
12.	Quasi crystal–PHB	Increased hardness of composites. Biodegradable and biocompatible in nature.	(Fernandes et al., 2020)
13.	Cellulose–PHB	Increased tensile strength and modulus. Improved barrier and mechanical properties.	(Cyras et al., 2007)
14.	Polylactic acid–PHB	Increased damping capacity. Degradation ability increased with increase in temperature.	(Ren et al., 2015)
15.	Luffa fiber–PHB	Mercerization process improved blending of the polymer. Degradability increased with fiber content irrespective of treatment methods.	(Avecilla-Ramírez et al., 2020)

12.10 Antimicrobial Activity of PHB

The antimicrobial activity of PHB produced by *Bacillus mycoides* is assessed against a variety of bacterial and fungal agents that are known to contaminate food materials. The concentrations of PHB required to inhibit bacterial cells are > or = to 80 microgram/gram and are > or = to 50 microgram/gram to inhibit fungal cells (Xavier et al., 2015). The antimicrobial activity of PHB fabric could be enhanced with the deposition of silver nanoparticles. The process of direct silver sputtering is followed to deposit silver on the surface of the fabric. The antibacterial potential of the PHB fabric against *Escherichia coli* and *Staphylococcus aureus* proved the antibacterial surface of the fabric (Slepička et al., 2016). PHB is used as a medium that promotes the incorporation of clove essential oil into films. The produced films were assessed against *E. coli*, *E. aerogenes*, and *S. aureus*, which proved to be antibiotic in nature, inhibiting the growth of the aforementioned pathogens. Meanwhile, the flexible nature of films is also enhanced by the addition of clove essential oil (Silva et al., 2020). PHB films on incorporation with phosphoserine phosphatase by techniques like bleached chlorine treatment and electrospinning exhibited antibiotic activity against *E. coli* and *S. aureus*, respectively. The blend of PLA/PHB along with carvacrol inhibited the growth of *L. innocua* and *E. coli* (Hernández-García et al., 2021).

12.11 Techniques for Production of PHB Biocomposites

There are several notable techniques that could be employed for the production of PHB molecules. The constraints involved in the application of PHB include the occurrence of secondary crystallization that might increase the fragile nature of the polymer, low density of nucleation which could cause splitting and cracking, instability at various temperature ranges, and the cost involved in the production of the same (Yeo et al., 2018). To overcome the limitations, the characteristic of PHB is modified by the addition of lubricants, plasticizers, compatibilizers, chemical, and thermal treatment techniques (Al etal., 2018).

12.12 PHB in Active Packaging

Active packing is defined by Farmer as the material that not only contains the food material and protects them but also interacts with them and informs the conditions inside it. Therefore, around 35% of the packaging material available is active packaging material in one or another way (Wyrwa & Barska, 2017). The general classification of the active packaging systems includes active scavenging system and active releasing system with absorber and emitter actions, respectively (Yildirim et al., 2018). Examples of active packaging with PHB are represented in Figure 12.7.

- **Active Scavenging Systems**

The films produced from PHB incorporated with palladium are produced by electrospinning technique, followed by annealing. The dispersion properties of the PHB are enhanced by CTAB (hexadecyltrimethylammonium bromide) and TEOS (tetraethyl orthosilicate). It was reported that the films treated with CTAB exhibited better oxygen-scavenging properties (Cherpinski et al., 2018). Further, the incorporation of palladium into blends of PHB and polycaprolactone (PCL) exhibited better oxygen-scavenging properties compared to the plain PHB. The synthesized nanobiopaper had a significant rate of biodegradability (Cherpinski et al., 2019). Consequently, PHB exhibits better barrier properties against oxygen, carbon dioxide, and moisture. It possesses superior gas barrier property with desirable vapor permeability, which is suitable for food packaging applications (Boey et al., 2021).

- **Active Releasing Systems**

The active releasing systems emit active molecules that are antimicrobial, antioxidant, or any other material capable of increasing the shelf life of food products. Chitosan oligomer on incorporation with

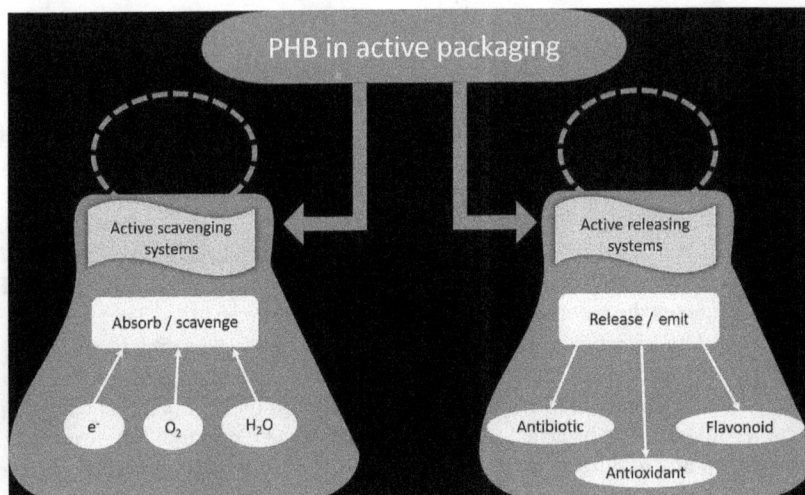

FIGURE 12.7

PHB showed effective antifungal activity, making them suitable packaging material for bread and straw-berries. Besides, catechins are incorporated as an antioxidant for the packaging of fatty food products (Vasile & Baican, 2021). The stretchability, food stimulant, and biodegradable properties could be enhanced by the blend of PHB and PLA incorporation with catechin (Arrieta et al., 2019). The food packaging should also be accounted for the number of molecules that are migrated into the package. The tests of migration performed with both polar and non-polar results suggested that food stimulants had values below the legislated limits (Arrieta et al., 2017).

12.13 Other Applications of PHB

The versatile applications of PHB, like ready availability, biodegradability, compatibility with biological material, and its specific physical properties make them unique materials with suitable sustainability and competitiveness. Blending with other polymers has improved the properties of the fabrications produced by PHB (Yeo et al., 2018). The demand for biopolymers is increasing rapidly with the progress of novel techniques for the production of bioplastics, like PHB using renewable sources of raw material. The miscibility of PHB with other polymers has improved the properties and its potential applications (Holmes, 1985). In addition, the biological properties of PHB are deeply influenced by the presence of water molecules around the implantation site (Bonartsev et al., 2019).

Polypropylene is one of the petroleum-based polymeric materials that is known to remain in the atmosphere as a recalcitrant for a longer time. Since the thermal and mechanical properties of PHB are similar to polypropylene, it could be used as an effective eco-friendly biodegradable alternative for the polymer polypropylene (Boey et al., 2021). The important properties of PHB that make them suitable for biological applications include the controllable properties of retardation. And it could be modulated by varying the parameters of processing and compositional molecular weight of the polymer (Shrivastav et al., 2013). Some of the commercial applications of various fabrications of PHB are listed in Table 12.5.

12.14 Conclusion

This chapter summarizes the wide domains with respect to the applicability of PHB produced from microbial sources. The transition from conventional polymeric materials to bio-based polymers could be mitigated by PHBs. The trade-off implied to the applicability of PHB due to their properties could be

TABLE 12.5

List of some of the commercial applications of PHB

S. no.	Fabrication type of PHB	Advantages	Application	Reference
1.	Composite hydrogel	Thixotropic Biocompatible in nature	Drug delivery	(Liu et al., 2011)
2.	Self-assembled hydrogel	Thixotropic Reversible Favorable for controlled release of molecules	Drug delivery	(Li et al., 2006)
3.	Composite hydrogel with cyclodextrin	Improved stability Biocompatible Have sustained release of molecules	Wound healing	(Pinho et al., 2014)
4.	Composite hydrogel reinforced with chitosan	Increased hydrophilicity Capable of entrapping both hydrophilic and hydrophobic molecules	Injectable carrier molecules	(Kang & Yun, 2022)
5.	PHB fiber mat developed into hybrid scaffold	Fiber acts as backbone and confer strength to scaffold Improved mechanical properties	Bone scaffold	(Sadat-Shojai et al., 2016)
6.	Composite film	Improved optical properties Increased mechanical strength	Packaging material	(Reis et al., 2008)
7.	Composite film	Colored plastic Feasible production method	Cell culture applications	(Jung et al., 2020)
8.	Blend film	Increased crystallization rate Improved structural and mechanical properties	Agricultural applications	(Arrieta et al., 2020)
9.	Film matrix	Suitable for immobilization of fungicide Functionality of the film unaffected	Pesticide matrix	(Savenkova et al., 2002)
10.	Blend film	Suitable oxygen-to-carbon-dioxide ratio Optimal water vapor permeability	Food packaging	(Peterson et al., 2001)
11.	PHB films reinforced with hemp fiber	Increased elastic modulus Increased fading of the film due to photo-oxidation caused by weathering	Packaging material	(Michel & Billington, 2012)
12.	Nanospheres	Possess antibacterial property Improved ability of encapsulation	Drug delivery	(Rodríguez-Contreras et al., 2013)
13.	Nano- and microspheres	Suitable for antibiotic loading On compositing with PEG, developed antifouling property	Biomedical applications	(Rodríguez-Contreras et al., 2016)
14.	Microspheres	Followed sustained-release kinetics Prolonged analgesic effect compared to free form	Drug delivery	(Salman et al., 2003)
15.	Encapsulated within microsphere	Enhance expression of protein of interest Facilitates flow through conversion of substrates within microspheres	Bioengineering applications	(Ogura & Rehm, 2019)

improved by developing biocomposites with materials that aid enhancement of characteristic features of PHB material. The availability of versatile sources, flexibility in synthesis, and fabrication makes PHB an unparalleled source of bio-based polymer. They could also be employed in the fabrication of fourth-generation packaging material that has ability to interact with food material enclosed within the packaging.

Acknowledgments

Ms. Shalini Mohan acknowledges the Department of Biotechnology for the fellowship (DBT/2022-23/KARE/2059). The author (PK) thanks to the Karpagam Academy of Higher Education for the support and help.

REFERENCES

Aaliya, B., Sunooj, K. V., & Lackner, M. (2021). Biopolymer composites: A review. *International Journal of Biobased Plastics, 3*(1), 40–84.

Abdelwahab, M. A., Flynn, A., Chiou, B. S., Imam, S., Orts, W., & Chiellini, E. (2012). Thermal, mechanical and morphological characterization of plasticized PLA—PHB blends. *Polymer Degradation and Stability, 97*(9), 1822–1828.

Abou-Aiad, T. H. M. (2007). Morphology and dielectric properties of polyhydroxybutyrate (PHB)/poly (methylmethacrylate) (PMMA) blends with some antimicrobial applications. *Polymer-Plastics Technology and Engineering, 46*(4), 435–439.

Adorna, J. A., Ventura, R. L. G., Dang, V. D., Doong, R.-A., & Ventura, J.-R. S. (2022). Biodegradable polyhydroxybutyrate/cellulose/calcium carbonate bioplastic composites prepared by heat-assisted solution casting method. *Journal of Applied Polymer Science, 139*(7), 51645.

Al, G., Aydemir, D., Kaygin, B., Ayrilmis, N., & Gunduz, G. (2018). Preparation and characterization of biopolymer nanocomposites from cellulose nanofibrils and nanoclays. *Journal of Composite Materials, 52*(5), 689–700.

Alim, A. A. A., Shirajuddin, S. S. M., & Anuar, F. H. (2022). A review of nonbiodegradable and biodegradable composites for food packaging application. *Journal of Chemistry, 2022*, 1–26.

Alshehrei, F. (2019). Production of polyhydroxybutyrate (PHB) by bacteria isolated from soil of Saudi Arabia. *Journal of Pure and Applied Microbiology, 13*, 897–904.

Anbukarasu, P., Sauvageau, D., & Elias, A. (2015). Tuning the properties of polyhydroxybutyrate films using acetic acid via solvent casting. *Scientific Reports, 5*(1), 1–14.

Arora, A., & Padua, G. W. (2010). Nanocomposites in food packaging. *Journal of Food Science, 75*(1), R43–R49.

Arrieta, M. P., Díez García, A., López, D., Fiori, S., & Peponi, L. (2019). Antioxidant bilayers based on PHBV and plasticized electrospun PLA-PHB fibers encapsulating catechin. *Nanomaterials, 9*(3), 346.

Arrieta, M. P., López, J., Hernández, A., & Rayón, E. (2014). Ternary PLA—PHB—limonene blends intended for biodegradable food packaging applications. *European Polymer Journal, 50*, 255–270.

Arrieta, M. P., López, J., López, D., Kenny, J. M., & Peponi, L. (2016). Effect of chitosan and catechin addition on the structural, thermal, mechanical and disintegration properties of plasticized electrospun PLA-PHB biocomposites. *Polymer Degradation and Stability, 132*, 145–156.

Arrieta, M. P., Perdiguero, M., Fiori, S., Kenny, J. M., & Peponi, L. (2020). Biodegradable electrospun PLA-PHB fibers plasticized with oligomeric lactic acid. *Polymer Degradation and Stability, 179*, 109226.

Arrieta, M. P., Samper, M. D., Aldas, M., & López, J. (2017). On the use of PLA-PHB blends for sustainable food packaging applications. *Materials, 10*(9), 1008.

Avecilla-Ramírez, A. M., del Rocío López-Cuellar, M., Vergara-Porras, B., Rodríguez-Hernández, A. I., & Vázquez-Núñez, E. (2020). Characterization of poly-hydroxybutyrate/luffa fibers composite material. *BioResources, 15*(3), 7159–7177.

Avella, M., De Vlieger, J. J., Errico, M. E., Fischer, S., Vacca, P., & Volpe, M. G. (2005). Biodegradable starch/clay nanocomposite films for food packaging applications. *Food Chemistry, 93*(3), 467–474.

Babaniyi, R. B., Afolabi, F. J., & Obagunwa, M. P. (2020). Recycling of used polyethylene through solvent blending of plasticized polyhydroxybutyrate and its degradation potential. *Composites Part C: Open Access, 2*, 100021.

Bajsic, E. G., Persic, A., Jemric, T., Buhin, J., Kucic Grgic, D., Zdraveva, E., Zizek, K., & Holjevac Grguric, T. (2021). Preparation and characterization of polyethylene biocomposites reinforced by rice husk: Application as potential packaging material. *Chemistry, 3*(4), 1344–1362.

Batista, M. B., Teixeira, C. S., Sfeir, M. Z. T., Alves, L. P. S., Valdameri, G., Pedrosa, F. O., Sassaki, G. L., Steffens, M. B. R., de Souza, E. M., Dixon, R., & Müller-Santos, M. (2018). PHB biosynthesis counteracts redox stress in *Herbaspirillum seropedicae*. *Frontiers in Microbiology, 9*, 1–12. doi: 10.3389/fmicb.2018.00472.

Bibi, F., Guillaume, C., Vena, A., Gontard, N., & Sorli, B. (2016). Wheat gluten, a bio-polymer layer to monitor relative humidity in food packaging: Electric and dielectric characterization. *Sensors and Actuators A: Physical, 247*, 355–367.

Boey, J. Y., Mohamad, L., Khok, Y. S., Tay, G. S., & Baidurah, S. (2021). A review of the applications and biodegradation of polyhydroxyalkanoates and poly (lactic acid) and its composites. *Polymers, 13*(10), 1544.

Bomrungnok, W., Arai, T., Yoshihashi, T., Sudesh, K., Hatta, T., & Kosugi, A. (2020). Direct production of polyhydroxybutyrate from waste starch by newly-isolated *Bacillus aryabhattai* T34-N4. *Environmental Technology, 41*(25), 3318–3328.

Bonartsev, A. P., Bonartseva, G. A., Reshetov, I. V., Kirpichnikov, M. P., & Shaitan, K. V. (2019). Application of polyhydroxyalkanoates in medicine and the biological activity of natural poly (3-hydroxybutyrate). *Acta Naturae (англоязычнаяверсия), 11*(2), 4–16.

Bourque, D., Pomerleau, Y., & Groleau, D. (1995). High-cell-density production of poly-β-hydroxybutyrate (PHB) from methanol by *Methylobacterium extorquens*: Production of high-molecular-mass PHB. *Applied Microbiology and Biotechnology, 44*(3), 367–376.

Breitenbach, J. (2002). Melt extrusion: From process to drug delivery technology. *European Journal of Pharmaceutics and Biopharmaceutics, 54*(2), 107–117.

Bucci, D. Z., Tavares, L. B. B., & Sell, I. (2007). Biodegradation and physical evaluation of PHB packaging. *Polymer Testing, 26*(7), 908–915.

Bugnicourt, E., Cinelli, P., Lazzeri, A., & Alvarez, V. (2014). Polyhydroxyalkanoate (PHA): Review of synthesis, characteristics, processing and potential applications in packaging. *eXPRESS Polymer Letters, 8*(11), 791–808.

Cazón, P., Velazquez, G., Ramírez, J. A., & Vázquez, M. (2017). Polysaccharide-based films and coatings for food packaging: A review. *Food Hydrocolloids, 68*, 136–148.

Chan, C. H., Kummerlöwe, C., & Kammer, H. W. (2004). Crystallization and melting behavior of poly (3-hydroxybutyrate)-based blends. *Macromolecular Chemistry and Physics, 205*(5), 664–675.

Chan, R. T., Garvey, C. J., Marçal, H., Russell, R. A., Holden, P. J., & Foster, L. J. R. (2011). Manipulation of polyhydroxybutyrate properties through blending with ethyl-cellulose for a composite biomaterial. *International Journal of Polymer Science, 2011*.

Chen, H., Wang, J., Cheng, Y., Wang, C., Liu, H., Bian, H., Pan, Y., Sun, J., & Han, W. (2019). Application of protein-based films and coatings for food packaging: A review. *Polymers, 11*(12), 2039.

Cherpinski, A., Gozutok, M., Sasmazel, H. T., Torres-Giner, S., & Lagaron, J. M. (2018). Electrospun oxygen scavenging films of poly (3-hydroxybutyrate) containing palladium nanoparticles for active packaging applications. *Nanomaterials (Basel), 8*(7), 469.

Cherpinski, A., Szewczyk, P. K., Gruszczyński, A., Stachewicz, U., & Lagaron, J. M. (2019). Oxygen-scavenging multilayered biopapers containing palladium nanoparticles obtained by the electrospinning coating technique. *Nanomaterials, 9*(2), 262.

Corsaro, C., Neri, G., Santoro, A., & Fazio, E. (2021). Acrylate and methacrylate polymers' applications: Second life with inexpensive and sustainable recycling approaches. *Materials, 15*(1), 282.

Cyras, V. P., Commisso, M. S., Mauri, A. N., & Vázquez, A. (2007). Biodegradable double-layer films based on biological resources: Polyhydroxybutyrate and cellulose. *Journal of Applied Polymer Science, 106*(2), 749–756.

Cyras, V. P., Soledad, C. M., & Analía, V. (2009). Biocomposites based on renewable resource: Acetylated and non acetylated cellulose cardboard coated with polyhydroxybutyrate. *Polymer, 50*(26), 6274–6280.

D'Anna, A., Arrigo, R., & Frache, A. (2019). PLA/PHB blends: Biocompatibilizer effects. *Polymers, 11*(9), 1416.

de Resende, T. M., & da Costa, M. M. (2020). Biopolymers of sugarcane. In *Sugarcane biorefinery, technology and perspectives* (pp. 229–254). Academic Press.

Deshwal, G. K., Panjagari, N. R., & Alam, T. (2019). An overview of paper and paper based food packaging materials: Health safety and environmental concerns. *Journal of Food Science and Technology, 56*(10), 4391–4403.

Dilkes-Hoffman, L. S., Lant, P. A., Laycock, B., & Pratt, S. (2019). The rate of biodegradation of PHA bioplastics in the marine environment: A meta-study. *Marine Pollution Bulletin, 142*, 15–24.

dos Santos, A. J., Oliveira Dalla Valentina, L. V., Hidalgo Schulz, A. A., & Tomaz Duarte, M. A. (2017). From obtaining to degradation of PHB: Material properties. Part I. *Ingeniería y Ciencia, 13*(26), 269–298.

Edebali, S. (2021). Methods of engineering of biopolymers and biocomposites. In *Advanced green materials* (pp. 351–357). Woodhead Publishing.

El-Hadi, A. M. (2017). Increase the elongation at break of poly (lactic acid) composites for use in food packaging films. *Scientific Reports, 7*(1), 1–14.

El-Hadi, A., Schnabel, R., Straube, E., Müller, G., & Henning, S. (2002). Correlation between degree of crystallinity, morphology, glass temperature, mechanical properties and biodegradation of poly (3-hydroxyalkanoate) PHAs and their blends. *Polymer Testing, 21*(6), 665–674.

Ewender, J., & Welle, F. (2019). Diffusion coefficients of n-alkanes and 1-alcohols in polyethylene naphthalate (PEN). *International Journal of Polymer Science, 2019*.

Fergala, A., AlSayed, A., & Eldyasti, A. (2018). Factors affecting the selection of PHB accumulating methanotrophs from waste activated sludge while utilizing ammonium as their nitrogen source. *Journal of Chemical Technology & Biotechnology, 93*(5), 1359–1369.

Fernandes, M. R. P., França, T. S., Queiroz, I. X., Wanderley, W. F., Cavalcante, D. G. L., Passos, T. A., Melo, D. M. A., & Wellen, R. M. R. (2020). Insights of PHB/QC biocomposites: Thermal, tensile and morphological properties. *Journal of Polymers and the Environment, 28*(9), 2481–2489.

Flórez, M., Guerra-Rodríguez, E., Cazón, P., & Vázquez, M. (2022). Chitosan for food packaging: Recent advances in active and intelligent films. *Food Hydrocolloids, 124*, 107328.

Frone, A. N., Panaitescu, D. M., Chiulan, I., Gabor, A. R., Nicolae, C. A., Oprea, M., Ghiurea, M., Gavrilescu, D., & Puitel, A. C. (2019). Thermal and mechanical behavior of biodegradable polyester films containing cellulose nanofibers. *Journal of Thermal Analysis and Calorimetry, 138*, 2387–2398.

Fukada, E., & Ando, Y. (1986). Piezoelectric properties of poly-β-hydroxybutyrate and copolymers of β-hydroxybutyrate and β-hydroxyvalerate. *International Journal of Biological Macromolecules, 8*(6), 361–366.

Fukui, T., & Doi, Y. (1997). Cloning and analysis of the poly (3-hydroxybutyrate-co-3-hydroxyhexanoate) biosynthesis genes of *Aeromonas caviae*. *Journal of Bacteriology, 179*(15), 4821–4830.

Gahlawat, G., Kumari, P., & Bhagat, N. R. (2020). Technological advances in the production of polyhydroxyalkanoate biopolymers. *Current Sustainable/Renewable Energy Reports, 7*(3), 73–83.

García, A., Segura, D., Espín, G., Galindo, E., Castillo, T., & Peña, C. (2014). High production of poly-β-hydroxybutyrate (PHB) by an *Azotobacter vinelandii* mutant altered in PHB regulation using a fed-batch fermentation process. *Biochemical Engineering Journal, 82*, 117–123.

Getachew, A., & Woldesenbet, F. (2016). Production of biodegradable plastic by polyhydroxybutyrate (PHB) accumulating bacteria using low cost agricultural waste material. *BMC Research Notes, 9*(1), 509.

Ghaderi, M., Mousavi, M., Yousefi, H., & Labbafi, M. (2014). All-cellulose nanocomposite film made from bagasse cellulose nanofibers for food packaging application. *Carbohydrate Polymers, 104*, 59–65.

Goñi-Ciaurriz, L., Senosiain-Nicolay, M., & Vélaz, I. (2021). Aging studies on food packaging films containing β-cyclodextrin-grafted TiO$_2$ nanoparticles. *International Journal of Molecular Sciences, 22*(5), 2257.

Guillard, V., Gaucel, S., Fornaciari, C., Angellier-Coussy, H., Buche, P., & Gontard, N. (2018). The next generation of sustainable food packaging to preserve our environment in a circular economy context. *Frontiers in Nutrition, 5*, 1–13.

Hamid, L., & Samy, I. (2021). Fabricating natural biocomposites for food packaging. In *Fiber-reinforced plastics*. IntechOpen. doi: 10.5772/intechopen.100907

Hassan, M. A., Bakhiet, E. K., Hussein, H. R., & Ali, S. G. (2019). Statistical optimization studies for polyhydroxybutyrate (PHB) production by novel *Bacillus subtilis* using agricultural and industrial wastes. *International Journal of Environmental Science and Technology, 16*(7), 3497–3512.

Hernández-García, E., Vargas, M., González-Martínez, C., & Chiralt, A. (2021). Biodegradable antimicrobial films for food packaging: Effect of antimicrobials on degradation. *Foods, 10*(6), 1256.

Höfer, P., Vermette, P., & Groleau, D. (2011). Production and characterization of polyhydroxyalkanoates by recombinant *Methylobacterium extorquens*: Combining desirable thermal properties with functionality. *Biochemical Engineering Journal, 54*(1), 26–33.

Holmes, P. A. (1985). Applications of PHB-a microbially produced biodegradable thermoplastic. *Physics in Technology, 16*(1), 32.

Hong, L. G., Yuhana, N. Y., & Zawawi, E. Z. E. (2021). Review of bioplastics as food packaging materials. *AIMS Materials Science, 8*(2), 166–184.

Hungund, B. S., Umloti, S. G., Upadhyaya, K. P., Manjanna, J., Yallappa, S., & Ayachit, N. H. (2018). Development and characterization of polyhydroxybutyrate biocomposites and their application in the removal of heavy metals. *Materials Today: Proceedings, 5*(10), 21023–21029.

Iron, R., Mehdikhani, M., Naghashzargar, E., Karbasi, S., & Semnani, D. (2019). Effects of nano-bioactive glass on structural, mechanical and bioactivity properties of poly (3-hydroxybutyrate) electrospun scaffold for bone tissue engineering applications. *Materials Technology, 34*(9), 540–548.

Jain, R., & Tiwari, A. (2015). Biosynthesis of planet friendly bioplastics using renewable carbon source. *Journal of Environmental Health Science and Engineering, 13*(1), 1–5.

Jung, H. R., Choi, T. R., Han, Y. H., Park, Y. L., Park, J. Y., Song, H. S., Yang, S. Y., Bhatia, S. K., Gurav, V., Park. H., Namgung, S., Choi, K. Y., & Yang, Y. H. (2020). Production of blue-colored polyhydroxybutyrate (PHB) by one-pot production and coextraction of indigo and PHB from recombinant *Escherichia coli*. *Dyes and Pigments, 173*, 107889.

Kang, J., & Yun, S. I. (2022). Chitosan-reinforced PHB hydrogel and aerogel monoliths fabricated by phase separation with the solvent-exchange method. *Carbohydrate Polymers, 284*, 119184.

Kavitha, G., Kurinjimalar, C., Sivakumar, K., Kaarthik, M., Aravind, R., Palani, P., & Rengasamy, R. (2016). Optimization of polyhydroxybutyrate production utilizing waste water as nutrient source by *Botryococcus braunii* Kütz using response surface methodology. *International Journal of Biological Macromolecules, 93*, 534–542.

Khosravi-Darani, K., & Bucci, D. Z. (2015). Application of poly (hydroxyalkanoate) in food packaging: Improvements by nanotechnology. *Chemical and Biochemical Engineering Quarterly, 29*(2), 275–285.

Koller, M. (2014). Poly (hydroxyalkanoates) for food packaging: Application and attempts towards implementation. *Applied Food Biotechnology, 1*(1), 3–15.

Kraśniewska, K., Pobiega, K., & Gniewosz, M. (2019). Pullulan—biopolymer with potential for use as food packaging. *International Journal of Food Engineering, 15*(9).

Kuciel, S., & Liber-Kneć, A. (2011). Biocomposites based on PHB filled with wood or kenaf fibers. *Polimery, 56*(3), 218–223.

Laftah, W. A., & Wan Abdul Rahman, W. A. (2021). Rice waste–based polymer composites for packaging applications: A review. *Polymers and Polymer Composites, 29*(9, Suppl.), S1621–S1629.

Latos, M., & Masek, A. (2018). Pro-ecological packaging materials based on polyhydroxybutyrate (PHB). In *E3S web of conferences* (Vol. 44, p. 00092). EDP Sciences.

Li, J., Li, X., Ni, X., Wang, X., Li, H., & Leong, K. W. (2006). Self-assembled supramolecular hydrogels formed by biodegradable PEO—PHB—PEO triblock copolymers and α-cyclodextrin for controlled drug delivery. *Biomaterials, 27*(22), 4132–4140.

Li, Z., Yang, J., & Loh, X. J. (2016). Polyhydroxyalkanoates: Opening doors for a sustainable future. *NPG Asia Materials, 8*(4), e265–e265.

Li, Z. J., Shi, Z. Y., Jian, J., Guo, Y. Y., Wu, Q., & Chen, G. Q. (2010). Production of poly (3-hydroxybutyrate-co-4-hydroxybutyrate) from unrelated carbon sources by metabolically engineered *Escherichia coli*. *Metabolic Engineering, 12*(4), 352–359.

Lightfoot, J. C., Buchard, A., Castro-Dominguez, B., & Parker, S. C. (2022). Comparative study of oxygen diffusion in polyethylene terephthalate and polyethylene furanoate using molecular modeling: Computational insights into the mechanism for gas transport in bulk polymer systems. *Macromolecules, 55*, 498–510.

Liu, K. L., Zhang, Z., & Li, J. (2011). Supramolecular hydrogels based on cyclodextrin—polymer polypseudorotaxanes: Materials design and hydrogel properties. *Soft Matter, 7*(24), 11290–11297.

Liu, X. J., Zhang, J., Hong, P. H., & Li, Z. J. (2016). Microbial production and characterization of poly-3-hydroxybutyrate by *Neptunomonas antarctica*. *PeerJ, 4*, e2291.

Loos, K., Zhang, R., Pereira, I., Agostinho, B., Hu, H., Maniar, D., Sbirrazzuoli, N., Silvestre, A. J. D., Guigo. N., & Sousa, A. F. (2020). A perspective on PEF synthesis, properties, and end-life. *Frontiers in Chemistry, 8*, 585.

Ma, W., Wang, J., Li, Y., Yin, L., & Wang, X. (2018). Poly(3-hydroxybutyrate-co-3-hydroxyvalerate) co-produced with l-isoleucine in *Corynebacterium glutamicum* WM001. *Microbial Cell Factories, 17*. Article no: 93.

Manikantan, M. R., & Varadharaju, N. (2011). Development and evaluation of food packaging related properties of high density polyethylene based nanocomposite films. *Journal of Polymer Materials, 28*(2), 245.

Manoli, E., & Voutsa, D. (2016). Food containers and packaging materials as possible source of hazardous chemicals to food. In H. Takada & H. Karapanagioti (eds) *Hazardous chemicals associated with plastics in the marine environment. The handbook of environmental chemistry* (Vol. 78, pp. 19–50). https://doi.org/10.1007/698_2017_19

Martínez-Herrera, R. E., Rutiaga-Quiñones, O. M., & Alemán-Huerta, M. E. (2021). Integration of Agave plants into the polyhydroxybutyrate (PHB) production: A gift of the ancient Aztecs to the current bioworld. *Industrial Crops and Products, 174*, 114188.

Martins, V. G., Romani, V. P., Martins, P. C., & Nogueira, D. (2021). Protein-based materials for packaging applications. In S. M. Sapuan & R. A. Ilyas (eds) *Bio-based packaging: Material, environmental and economic aspects* (pp. 27–49). https://doi.org/10.1002/9781119381228.ch2

McAdam, B., Brennan Fournet, M., McDonald, P., & Mojicevic, M. (2020). Production of polyhydroxybutyrate (PHB) and factors impacting its chemical and mechanical characteristics. *Polymers, 12*(12), 2908.

McGrath, J. E., Hickner, M. A., & Höfer, R. (2013). Polymers for a sustainable environment and green energy. *Polymer Science, 10*, 849.

Meixner, K., Daffert, C., Bauer, L., Drosg, B., & Fritz, I. (2022). PHB producing cyanobacteria found in the neighborhood—their isolation, purification and performance testing. *Bioengineering, 9*(4), 178.

Melendez-Rodriguez, B., Torres-Giner, S., Lorini, L., Valentino, F., Sammon, C., Cabedo, L., & Lagaron, J. M. (2020). Valorization of municipal biowaste into electrospun poly (3-Hydroxybutyrate-Co-3-Hydroxyvalerate) biopapers for food packaging applications. *ACS Applied Bio Materials, 3*(9), 6110–6123.

Michel, A. T., & Billington, S. L. (2012). Characterization of poly-hydroxybutyrate films and hemp fiber reinforced composites exposed to accelerated weathering. *Polymer Degradation and Stability, 97*(6), 870–878.

Mitschang, P., & Hildebrandt, K. (2012). Polymer and composite moulding technologies for automotive applications. In *Advanced materials in automotive engineering* (pp. 210–229). Woodhead Publishing.

Mohapatra, S., Sarkar, B., Samantaray, D. P., Daware, A., Maity, S., Pattnaik, S., & Bhattacharjee, S. (2017). Bioconversion of fish solid waste into PHB using *Bacillus subtilis* based submerged fermentation process. *Environmental Technology, 38*(24), 3201–3208.

Mosnáčková, K., Opálková Šišková, A., Kleinová, A., Danko, M., & Mosnáček, J. (2020). Properties and degradation of novel fully biodegradable PLA/PHB blends filled with keratin. *International Journal of Molecular Sciences, 21*(24), 9678.

Nanda, S., Patra, B. R., Patel, R., Bakos, J., & Dalai, A. K. (2021). Innovations in applications and prospects of bioplastics and biopolymers: A review. *Environmental Chemistry Letters, 20*, 379–395.

Nazrin, A., Sapuan, S. M., Zuhri, M. Y. M., Ilyas, R. A., Syafiq, R., & Sherwani, S. F. K. (2020). Nanocellulose reinforced thermoplastic starch (TPS), polylactic acid (PLA), and polybutylene succinate (PBS) for food packaging applications. *Frontiers in Chemistry, 8*, https://doi.org/ 10.3389/fchem.2020.00213

Ncube, L. K., Ude, A. U., Ogunmuyiwa, E. N., Zulkifli, R., & Beas, I. N. (2020). Environmental impact of food packaging materials: A review of contemporary development from conventional plastics to polylactic acid based materials. *Materials, 13*(21), 4994.

Niazmand, R., Razavizadeh, B., & Sabbagh, F. (2020). Low-density polyethylene films carrying ferulaasafoetida extract for active food packaging: Thermal, mechanical, optical, barrier, and antifungal properties. *Advances in Polymer Technology, 2020*(65), 1–15.

Nisticò, R. (2020). Polyethylene terephthalate (PET) in the packaging industry. *Polymer Testing, 90*, 1–18. https://doi.org/10.1016/j.polymertesting.2020.106707

Obruca, S., Sedlacek, P., Mravec, F., Krzyzanek, V., Nebesarova, J., Samek, O., Kučera, D., Benešová, P., Hrubanová K., Milerová, M., & Marova, I. (2017). The presence of PHB granules in cytoplasm protects non-halophilic bacterial cells against the harmful impact of hypertonic environments. *New Biotechnology, 39*, 68–80.

Ogura, K., & Rehm, B. H. (2019). Alginate encapsulation of bioengineered protein-coated polyhydroxybutyrate particles: A new platform for multifunctional composite materials. *Advanced Functional Materials, 29*(37), 1901893.

Ojha, A., Sharma, A., Sihag, M., & Ojha, S. (2015). Food packaging—materials and sustainability-A review. *Agricultural Reviews, 36*(3), 241–245.

Olejnik, O., Masek, A., & Zawadziłło, J. (2021). Processability and mechanical properties of thermoplastic polylactide/polyhydroxybutyrate (PLA/PHB) bioblends. *Materials, 14*(4), 898.

Oymaci, P., & Altinkaya, S. A. (2016). Improvement of barrier and mechanical properties of whey protein isolate based food packaging films by incorporation of zein nanoparticles as a novel bionanocomposite. *Food Hydrocolloids, 54*, 1–9.

Pandian, S. R. K., Deepak, V., Kalishwaralal, K., Muniyandi, J., Rameshkumar, N., & Gurunathan, S. (2009). Synthesis of PHB nanoparticles from optimized medium utilizing dairy industrial waste using brevibacterium casei SRKP2: A green chemistry approach. *Colloids and Surfaces B: Biointerfaces, 74*(1), 266–273.

Pandian, S. R. K., Deepak, V., Kalishwaralal, K., Rameshkumar, N., Jeyaraj, M., & Gurunathan, S. (2010). Optimization and fed-batch production of PHB utilizing dairy waste and sea water as nutrient sources by *Bacillus megaterium* SRKP-3. *Bioresource Technology, 101*(2), 705–711.

Park, C. H., & Lee, W. I. (2012). Compression molding in polymer matrix composites. In *Manufacturing techniques for polymer matrix composites (PMCs)* (pp. 47–94). Woodhead Publishing.

Pearson, R. B. (1982). PVC as a food packaging material. *Food Chemistry, 8*(2), 85–96.

Penkhrue, W., Jendrossek, D., Khanongnuch, C., Pathom-Aree, W., Aizawa, T., Behrens, R. L., & Lumyong, S. (2020). Response surface method for polyhydroxybutyrate (PHB) bioplastic accumulation in *Bacillus drentensis* BP17 using pineapple peel. *PLoS One, 15*(3), e0230443.

Peoples, O. P., & Sinskey, A. J. (1989). Poly-beta-hydroxybutyrate (PHB) biosynthesis in *Alcaligenes eutrophus* H16. Identification and characterization of the PHB polymerase gene (phbC). *Journal of Biological Chemistry, 264*(26), 15298–15303.

Petersen, K., Nielsen, P. V., & Olsen, M. B. (2001). Physical and mechanical properties of bio-based materials starch, polylactate and polyhydroxybutyrate. *Starch-Stärke, 53*(8), 356–361.

Pinho, E., Grootveld, M., Soares, G., & Henriques, M. (2014). Cyclodextrin-based hydrogels toward improved wound dressings. *Critical Reviews in Biotechnology, 34*(4), 328–337.

Poblete-Castro, I., Binger, D., Rodrigues, A., Becker, J., Dos Santos, V. A. M., & Wittmann, C. (2013). In-silico-driven metabolic engineering of *Pseudomonas putida* for enhanced production of polyhydroxyalkanoates. *Metabolic Engineering, 15*, 113–123.

Rafiqah, S. A., Khalina, A., Harmaen, A. S., Tawakkal, I. A., Zaman, K., Asim, M., Nurrazi, M. N., & Lee, C. H. (2021). A review on properties and application of bio-based poly (butylene succinate). *Polymers (Basel), 13*(9), 1436.

Raheem, D. (2012). Application of plastics and paper as food packaging materials – An overview. *Emirates Journal of Food and Agriculture, 25*, 177–188.

Raturi, G., Shree, S., Sharma, A., Panesar, P. S., & Goswami, S. (2021). Recent approaches for enhanced production of microbial polyhydroxybutyrate: Preparation of biocomposites and applications. *International Journal of Biological Macromolecules, 182*, 1650–1669.

Reis, K. C., Pereira, J., Smith, A. C., Carvalho, C. W. P., Wellner, N., & Yakimets, I. (2008). Characterization of polyhydroxybutyrate-hydroxyvalerate (PHB-HV)/maize starch blend films. *Journal of Food Engineering, 89*(4), 361–369.

Ren, H., Liu, Z., Zhai, H., Cao, Y., & Omori, S. (2015). Effects of lignophenols on mechanical performance of biocomposites based on polyhydroxybutyrate (PHB) and polypropylene (PP) reinforced with pulp fibers. *BioResources, 10*(1), 432–447.

Ren, H., Zhang, Y., Zhai, H., & Chen, J. (2015). Production and evaluation of biodegradable composites based on polyhydroxybutyrate and polylactic acid reinforced with short and long pulp fibers. *Cellulose Chemistry and Technology, 49*, 641–652.

Righetti, M. C., Aliotta, L., Mallegni, N., Gazzano, M., Passaglia, E., Cinelli, P., & Lazzeri, A. (2019). Constrained amorphous interphase and mechanical properties of poly (3-Hydroxybutyrate-co-3-Hydroxyvalerate). *Frontiers in Chemistry, 7*, 790.

Rodríguez-Contreras, A., Canal, C., Calafell-Monfort, M., Ginebra, M. P., Julio-Moran, G., & Marqués-Calvo, M. S. (2013). Methods for the preparation of doxycycline-loaded phb micro-and nano-spheres. *European Polymer Journal, 49*(11), 3501–3511.

Rodríguez-Contreras, A., Marqués-Calvo, M. S., Gil, F. J., & Manero, J. M. (2016). Modification of titanium surfaces by adding antibiotic-loaded PHB spheres and PEG for biomedical applications. *Journal of Materials Science: Materials in Medicine, 27*(8), 1–15.

Rudin, A., & Choi, P. (2012). *The elements of polymer science and engineering*. Academic Press.

Sadat-Shojai, M., Khorasani, M. T., & Jamshidi, A. (2016). A new strategy for fabrication of bone scaffolds using electrospun nano-HAp/PHB fibers and protein hydrogels. *Chemical Engineering Journal, 289*, 38–47.

Sadi, R. K., Fechine, G. J. M., & Demarquette, N. R. (2010). Photodegradation of poly (3-hydroxybutyrate). *Polymer Degradation and Stability, 95*(12), 2318–2327.

Salman, M. A., Sahin, A., Onur, M. A., Öge, K., Kassab, A., & Aypar, Ü. (2003). Tramadol encapsulated into polyhydroxybutyrate microspheres: In vitro release and epidural analgesic effect in rats. *Acta Anaesthesiologica Scandinavica, 47*(8), 1006–1012.

Sánchez-Safont, E., Aldureid, A., Lagarón, J., Gámez-Pérez, J., & Cabedo, L. (2018). Biocomposites of different lignocellulosic wastes for sustainable food packaging applications. *Composites Part B: Engineering, 145*, 215–225.

Sánchez-Safont, E. L., Aldureid, A., Lagarón, J. M., Gamez-Pérez, J., & Cabedo, L. (2021). Effect of the purification treatment on the valorization of natural cellulosic residues as fillers in PHB-based composites for short shelf life applications. *Waste and Biomass Valorization, 12*(5), 2541–2556.

Saratale, G. D., Saratale, R. G., Varjani, S., Cho, S. K., Ghodake, G. S., Kadam, A., Mulla, S. I., Bharagava, R. N., Kim, D. S., & Shin, H. S. (2020). Development of ultrasound aided chemical pretreatment methods to enrich saccharification of wheat waste biomass for polyhydroxybutyrate production and its characterization. *Industrial Crops and Products*, *150*, 112425.

Savenkova, L., Gercberga, Z., Muter, O., Nikolaeva, V., Dzene, A., & Tupureina, V. (2002). PHB-based films as matrices for pesticides. *Process Biochemistry*, *37*(7), 719–722.

Savenkova, L., Gercberga, Z., Nikolaeva, V. J. P. B., Dzene, A., Bibers, I., & Kalnin, M. (2000). Mechanical properties and biodegradation characteristics of PHB-based films. *Process Biochemistry*, *35*(6), 573–579.

Schweighuber, A., Fischer, J., & Buchberger, W. (2021). Differentiation of polyamide 6, 6.6, and 12 contaminations in polyolefin-recyclates using HPLC coupled to drift-tube ion-mobility quadrupole time-of-flight mass spectrometry. *Polymers*, *13*(12), 2032.

Seoane, I. T., Manfredi, L. B., & Cyras, V. P. (2018). Bilayer biocomposites based on coated cellulose paperboard with films of polyhydroxybutyrate/cellulose nanocrystals. *Cellulose*, *25*(4), 2419–2434.

Sharma, N. (2019). Polyhydroxybutyrate (PHB) production by bacteria and its application as biodegradable plastic in various industries. *Academic Journal of Polymer Science*, *2*(3), 555586.

Shershneva, E. G. (2022). Biodegradable food packaging: Benefits and adverse effects. In *IOP conference series: Earth and environmental science* (Vol. 988, No. 2, p. 022006). IOP Publishing.

Shrivastav, A., Kim, H. Y., & Kim, Y. R. (2013). Advances in the applications of polyhydroxyalkanoate nanoparticles for novel drug delivery system. *BioMed Research International*, *2013*.

Seimann, U. (2005). Solvent cast technology—a versatile tool for thin film production. In *Scattering methods and the properties of polymer materials* (pp. 1–14). Springer.

Silva, I. D. D. L., Andrade, M. F. D., Caetano, V. F., Hallwass, F., Brito, A. M. S. S., & Vinhas, G. M. (2020). Development of active PHB/PEG antimicrobial films incorporating clove essential oil. *Polímeros*, *30*.

Sirohi, R., Gaur, V. K., Pandey, A. K., Sim, S. J., & Kumar, S. (2021). Harnessing fruit waste for poly-3-hydroxybutyrate production: A review. *Bioresource Technology*, *326*, 124734.

Sirohi, R., Pandey, J. P., Gaur, V. K., Gnansounou, E., & Sindhu, R. (2020). Critical overview of biomass feedstocks as sustainable substrates for the production of polyhydroxybutyrate (PHB). *Bioresource Technology*, *311*, 123536.

Slepička, P., Malá, Z., Rimpelová, S., & Švorčík, V. (2016). Antibacterial properties of modified biodegradable PHB non-woven fabric. *Materials Science and Engineering: C*, *65*, 364–368.

Smith, M. K., Paleri, D. M., Abdelwahab, M., Mielewski, D. F., Misra, M., & Mohanty, A. K. (2020). Sustainable composites from poly (3-hydroxybutyrate) (PHB) bioplastic and agave natural fibre. *Green Chemistry*, *22*(12), 3906–3916.

Süfer, Ö., Oz, A. T., & ÇelebiSezer, Y. (2017). Poly (lactic acid) films in food packaging systems. *Food Science and Nutrition Technology*, *2*, 000131.

Tan, D., Wu, Q., Chen, J. C., & Chen, G. Q. (2014). Engineering halomonas TD01 for the low-cost production of polyhydroxyalkanoates. *Metabolic Engineering*, *26*, 34–47.

Thapa, C., Shakya, P., Shrestha, R., Pal, S., & Manandhar, P. (2018). Isolation of polyhydroxybutyrate (PHB) producing bacteria, optimization of culture conditions for PHB production, extraction and characterization of PHB. *Nepal Journal of Biotechnology*, *6*(1), 62–68.

Trakunjae, C., Boondaeng, A., Apiwatanapiwat, W., Kosugi, A., Arai, T., Sudesh, K., & Vaithanomsat, P. (2021). Enhanced polyhydroxybutyrate (PHB) production by newly isolated rare actinomycetes *Rhodococcus* sp. strain BSRT1–1 using response surface methodology. *Scientific Reports*, *11*(1), 1–14.

Tran, T. T., & Charles, T. C. (2016). Genome-engineered *Sinorhizobium meliloti* for the production of poly (lactic-co-3-hydroxybutyric) acid copolymer. *Canadian Journal of Microbiology*, *62*(2), 130–138.

Tripathi, A. D., Raj Joshi, T., Kumar Srivastava, S., Darani, K. K., Khade, S., & Srivastava, J. (2019). Effect of nutritional supplements on bio-plastics (PHB) production utilizing sugar refinery waste with potential application in food packaging. *Preparative Biochemistry and Biotechnology*, *49*(6), 567–577.

Turco, R., Santagata, G., Corrado, I., Pezzella, C., & Di Serio, M. (2021). In vivo and post-synthesis strategies to enhance the properties of PHB-based materials: A review. *Frontiers in Bioengineering and Biotechnology*, *8*, 1454.

Vanovčanová, Z., Alexy, P., Feranc, J., Plavec, R., Bočkaj, J., Kaliňáková, L., Tomanová, K., Perďochová, D., Šariský, D., & Gálisová, I. (2016). Effect of PHB on the properties of biodegradable PLA blends. *Chemical Papers*, *70*(10), 1408–1415.

Vasile, C., & Baican, M. (2021). Progresses in food packaging, food quality, and safety—controlled-release antioxidant and/or antimicrobial packaging. *Molecules, 26*(5), 1263.

Wu, X., Liu, P., Shi, H., Wang, H., Huang, H., Shi, Y., & Gao, S. (2021). Photo aging and fragmentation of polypropylene food packaging materials in artificial seawater. *Water Research, 188*, 116456.

Wyrwa, J., & Barska, A. (2017). Innovations in the food packaging market: Active packaging. *European Food Research and Technology, 243*(10), 1681–1692.

Xavier, J. R., Babusha, S. T., George, J., & Ramana, K. V. (2015). Material properties and antimicrobial activity of polyhydroxybutyrate (PHB) films incorporated with vanillin. *Applied Biochemistry and Biotechnology, 176*(5), 1498–1510.

Xu, X., Ma, X., Li, D., & Dong, J. (2021). Toward sustainable biocomposites based on MMT and PHBH reinforced with acetylated cellulose nanocrystals. *Cellulose, 28*(5), 2981–2993.

Yeo, J. C. C., Muiruri, J. K., Thitsartarn, W., Li, Z., & He, C. (2018). Recent advances in the development of biodegradable PHB-based toughening materials: Approaches, advantages and applications. *Materials Science and Engineering: C, 92*, 1092–1116.

Yildirim, S., Röcker, B., Pettersen, M. K., Nilsen-Nygaard, J., Ayhan, Z., Rutkaite, R., Radusin, T., Suminska, P., Marcos, B., & Coma, V. (2018). Active packaging applications for food. *Comprehensive Reviews in Food Science and Food Safety, 17*(1), 165–199.

Yuvaraj, D., Iyyappan, J., Gnanasekaran, R., Ishwarya, G., Harshini, R. P., Dhithya, V., Chandran, M., Kanishka, V., & Gomathi, K. (2021). Advances in bio food packaging—An overview. *Heliyon, 7*(9), e0799.

Zhang, M., & Thomas, N. L. (2011). Blending polylactic acid with polyhydroxybutyrate: The effect on thermal, mechanical, and biodegradation properties. *Advances in Polymer Technology, 30*(2), 67–79.

Zhang, Q., Zhang, Y., Wang, F., Liu, L., & Wang, C. (1998). Thermal properties of PHB/PEG blends. *Journal of Materials Science and Technology, 14*(1), 95–96.

Zhang, S., Prabhakaran, M. P., Qin, X., & Ramakrishna, S. (2015). Poly-3-hydroxybutyrate-co-3-hydroxyvalerate containing scaffolds and their integration with osteoblasts as a model for bone tissue engineering. *Journal of Biomaterials Applications, 29*(10), 1394–1406.

Zhao, Y., Li, B., Li, C., Xu, Y., Luo, Y., Liang, D., & Huang, C. (2021). Comprehensive Review of polysaccharide-based materials in edible packaging: A sustainable approach. *Foods, 10*(8), 1845.

Zhong, Y., Godwin, P., Jin, Y., & Xiao, H. (2020). Biodegradable polymers and green-based antimicrobial packaging materials: A mini-review. *Advanced Industrial and Engineering Polymer Research, 3*(1), 27–35.

Zhuang, Q., Wang, Q., Liang, Q., & Qi, Q. (2014). Synthesis of polyhydroxyalkanoates from glucose that contain medium-chain-length monomers via the reversed fatty acid β-oxidation cycle in *Escherichia coli*. *Metabolic Engineering, 24*, 78–86.

13

Preparation, Characterization, and Evaluation of Antibacterial Properties of Poly(3-Hydroxybutarate-Co-3-Hydroxyvalerate) (PHBV)-Based Films and Coatings

Pradeep Kumar Panda and Pranjyan Dash

CONTENTS

13.1 Introduction

Poly(3-hydroxybutyrate-co-3-hydroxyvalerate) (PHBV) is a good microbial aliphatic polyester biopolymer used for making different packaging and coating materials. This material is mainly extracted from the 3-hydroxyvalerate (3-HV) components of the biopolymer agent of PHB biopolymer. PHBV has different kinds of eco-friendly properties, including biodegradability and biocompatible. This material also has some beneficial characteristics, including high crystallinity and resistance against ultraviolet (UV) radiation (NCBI, 2022a). In addition, this material has some antimicrobial activity, which makes this a promising material for packaging and biomedical applications.

This material also has a small amount of plasticity, which helps prevent liquids from passing from the material. This biofilm material has some unique characteristics, including high proliferative activity. Due to several cracks on PHBV coating material, it failed to act as a good grease barrier (Sängerlaub et al., 2019). Increasing need and implicational perspectives for replacing plastic have contributed to the increment in the market value of polyhydroxyalkanoates (PHAs). The biopolymer market dominated by PHA and related biopolymers has a market value of $221.08 million. The potential growth of CAGR of the biopolymer market is expected to be 6.29% within 2027 (Redskins, 2022).

PHBV is also used as a coating material for some drugs and controls the release of those encapsulated drugs. In addition, this material also has some thermoplastic properties due to the presence of the linear structure of aliphatic polyester. This promising packaging material can be produced from the recombinant strains of *Escherichia coli* (*E. coli*). At the same time, scaffolds of PHBV make it more suitable for changing the chemical property of it (Rivera-Briso et al., 2018). This chapter is going to shed light on the preparation process of the coating material. Additionally, this chapter also describes the characteristics and antimicrobial properties of this coating film material.

DOI: 10.1201/9781003303671-13

13.2 Preparation of PHBV-Based Films

Electrospinning is one of the prime methods used for preparing PHBV-based films. This method of preparation was especially found in the study of Râpă et al. (2021), where it was revealed that PHBV nanoparticles were subjected to the electrospinning nanosystem (Figure 13.1). This system resulted in the nanoparticles being spun into polymer, which were further reduced to film format (Alhazmi et al., 2022). One of the principal materials which are often implemented in the process is food bio-waste. According to Figueroa-Lopez et al. (2020), the films of PHBV were prepared by collecting fruit waste that was fermented with 1% of doped zinc oxide.

The parameters involved in the electrospinning process of preparation differ concerning the solvent used for the fermentation of the fruit product. According to Kaniuk and Stachewicz (2021), if the solvent used for the preparation process is only chloroform, it results in the film being produced at a flow rate of 3mL/h. On the other hand, when chloroform and dimethylformamide (DMF) are used in 8 wt% concentration during fermentation of the fruit waste, it leads up to 6 m/h of flow rate in a better production process (Bing-Chiuan Shiu et al., 2022).

Another method of preparation that is often implemented for the preparation of PHBV based films is engaging in film blowing of PHBV blends. It was pointed out by Cunha et al. (2016) that often all the PHBV was not implemented for the preparation of films. This material not used in the PHBV-based film manufacturing process was prepared into a blend, thereby being used as a new raw material for the preparation of PHBV-based film (Fajstavr et al., 2022).

The PHBV blends are entered into the extruder. This extruder feeds the blends into the concentration feeder, from which it is sent to the external layer. Here the blends are subjected to pressure orientation, resulting in them being blown from the bi-layered bubble (Khane et al., 2013). As a result of such stress–pressure orientation subjected to the machine, the PHBV-based films are prepared and manufactured in the form of co-extruded bi-layered films. Studies have further shown that using such a preparation method results in good traction resistance (up to 25.9 MPa) and up to 1.25 g/cm³density. However, such a method results in poor interfacial adhesion (Khayrova et al., 2022). According to Scaffaro et al. (2019), implementing such a method in the preparation process results in a two-step approach preparation of fibers for the nanosystem by producing the bi-layered films, followed by the introduction of the film into the electrospinning nanosystem for further processing into a thin film. Righetti et al. (2019) revealed that the properties of the PHBV-based films vary with the method used for preparing the material. It is due to this reason that manufacturers of PHBV-based films are known to determine the approach and method of preparation of the material based on the desired properties to be achieved (Lange et al., 2022). Further, according to Lammi et al. (2019), the electrospinning method of preparation results in an increase in tensile strength and even a reduction in particles, thereby producing a thin layer of the film. Moreover, the flow rate of the PHBV-based films is known to vary with the solvent used, along with other factors, such as the concentration and ratio of the solvent used in the preparation process.

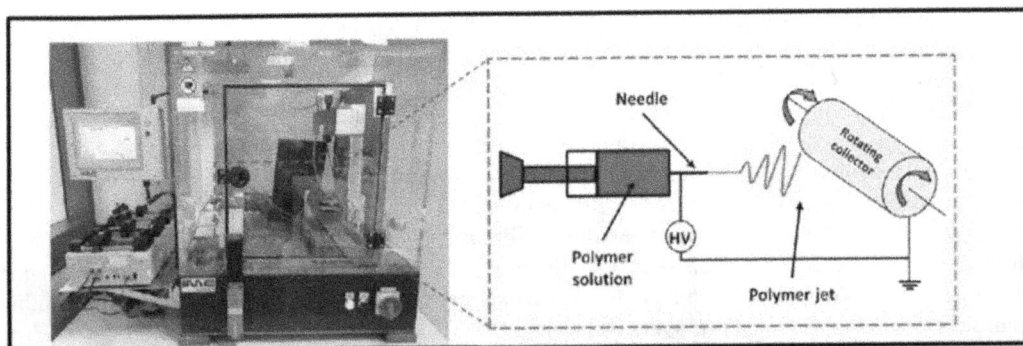

FIGURE 13.1   Electrospinning method of preparation.

Source: Kaniuk and Stachewicz (2021).

13.3 Preparation of PHBV-Based Coatings

PHBV is a microbial biopolymer that is used to prepare for the coating of papers through the extrusion process. As per the perspective of Melendez-Rodriguez et al. (2019), this coating is usually an internal layer in a multilayer system of PHBV films that is applied in the food packaging system. It is further observed that this coating is thoroughly used in packaging food as well as delivering drugs because of its biocompatible and biodegradable nature (Righetti et al., 2019). In this context, the overall preparation associated with the extrusion methods for preparing to coat papers using the PHBV films will be explained briefly. According to the viewpoint of Sängerlaub et al. (2019), PHBV was dried at a flow rate of the air volume of 15 m³/h at 60°C for 24 hours. On the other hand, after drying up the PHBV film, the process of compounding has been done. This is the first step of the preparation method of the PHBV coating. The materials that were used for coating different substances using the chosen biopolymer involve PHBV film, plasticizers, and paper substrate (Sängerlaub et al., 2019). The polymer film that was prepared in the previous process was used for coating purposes. The plasticizers triethyl citrate (TEC) and poly(ethylene glycol) (PEG) of different density and weight were used for the coating process. The combination of these plasticizers along with the film would generate a durable and thick coating (Sängerlaub et al., 2019). The paper substrate was used as the main sample for analyzing the effective coating of the "PHBV film." In the compounding step, the PHBV film is incorporated with TEC and PEG using a twin-screw extruder (Hietala & Oksma, 2018). The mentioned components are integrated into the compounder with the melted PHBV film with the help of a syringe pump. TEC and PEG concentrations were set to 2, 5, 10, and 15 wt.%. However, the delivery rate was different for two different components because of a major difference in their density. The heating and melting temperature of PEG was set to circa 40°C (Mirković et al., 2022). However, distinct temperatures were set for compounding the materials step by step. The temperature set for carrying out the compounding process was "180°C, 170°C, 160°C, 150°C, and 155°C."

The compounder's rotation frequency was set to 100 rpm, with a melting pressure of approximately 40 bars. Furthermore, the strand of the polymer was further cooled and cut into pellets with the help of a granulator. The speed of taking off was set to 20 m per minute, and the length of the strand was 2.5 mm. The diameter of the granule was set to less than 4 mm. The water bath that was used for cooling the "PHBV film" strands was set to 150 cm in length. The extrusion is the next step associated with the preparation of coatings. This step also involves the drying up of the PHBV film (Mohammad & Stiharu, 2022). The sample was further flushed in between each trial. A single-screw extruder was used for carrying out the extrusion process of the PHBV coating (Lewandowski & Wilczyński, 2019). The frequency of rotation for the extrusion process was set to 30 rpm. The extrusion process involves two different temperature profiles, one constant and another constant flow of melting rate. The differences in temperature profiles are further determined (Pahlevanzadeh et al., 2022). The different temperature levels were set for carrying out the extrusion process at a wide range of distinct concentrations. After the process of extrusion, the extruded PHBV film was coated and then laminated using the roll of lamination at 25°C to 40°C. The pressure of the extruder was set to 50–60 bar. During the coating process, the side containing the PHBV film was treated with siliconized paper to avoid sticking the film (Petousis et al., 2022). The speed of taking off from the extruder varied due to the distinct velocity and thickness of the coating. It has been observed that the extrusion process with a high velocity of coating generates a lower coating thickness.

13.4 Characterization Techniques of PHBV-Based Films and Coatings

Nowadays, different coating materials are developed to reduce the use of plastic. In addition to that, the use of different biodegradable materials influenced reducing the negative impact on nature. To develop this material, scientists have started using PHBV-based coating materials (Pradhan et al., 2022). This material also has some character of antimicrobial characteristics. This biofilm material has some unique characteristics, including high proliferative activity (Figure 13.2). In addition to that, the cell-adhesion property of this material results to making this material long-lasting (Insomphun et al., 2017). Additionally, the presence of graphene oxide (GO) influenced the researchers to develop

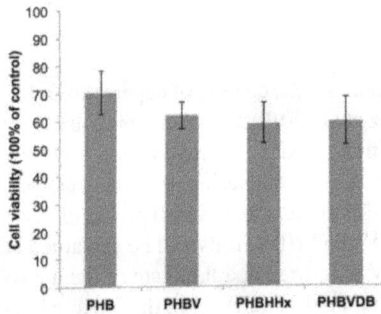

FIGURE 13.2   Viability of PHBV.

Source: Insomphun et al. (2017).

TABLE 13.1

Physical Properties of PHB and PHBV

Properties	PHB	PHBV
Density (g/cm³)	1.25	1.25
Elongation (%)	5.2–8.4	1.4
Fusion temperature (°C)	161	153
Glass transition temperature (°C)	–10	–1
Traction resistance (MPa)	21	25.9
Elasticity module (GPa)	0.93	2.38

Source: Rivera-Briso and Serrano-Aroca (2018).

different nanosheets. The development of nanosheets, along with nanofibers, helped overcome the characteristic brittleness.

Table 13.1 represents changes in the physical and mechanical characteristics of PHBV compared to PHB. The densities of both PHB and PHBV are observed to be identical, 1.25 g/cm³. However, major changes are observed with the other mechanical and physical properties of the biopolymer. The elasticity modulus of PHB is 0.93 GPa, whereas PHBV is 2.38 Gpa (Rivera-Briso & Serrano-Aroca, 2018). This indicates that PHBV is comparatively more elastic than PBH. The comparative surface tension and flexibility of PBHV are also significantly increased compared to PHB. Supporting these characteristics, traction resistance of PHB is identified as 21 Mpa, which is higher in PHBV with a value of 25.9 MPa. Elongation at the point of failure of PHBV is 1.4%, whereas the elongation of PHB is 5.2–8.4% (Vidakis et al., 2022). This justifies the fact that PHBV is stiff and brittle, but with further research and developmental approaches, PHBV is expected to deliver appropriate mechanical characteristics that would replace the use and application of petroleum-based plastics (Wang & Clapper, 2022). The fusion temperature of PBHV is less than PHB, 153°C, whereas the fusion temperature of PBH is 161°C. Glass transition temperature of PHBV is –10°C, and PHBV is –1°C. The higher glass transition temperature (Table 13.1) indicates higher flexibility of the biopolymer compared to PHB (Rivera-Briso & Serrano-Aroca, 2018). 3-HV decreases the biopolymer crystallinity characteristic of the PHBV, and therefore, the degradability of PHBV increases compared to PHB. The present development and production of PHBV depend on its application and use; therefore, the biopolymer is prepared with the inclusion of the required 3-HV (Xia et al., 2022). Although existing development and production of PHBV is expensive, modern application of the biopolymer is significantly limited to a certain application.

The comparison of the values of PHB and PHBV as represented in the previous graph illustrates that PHBV has more flexibility compared to PHB biopolymer. Values such as the traction resistance and the elasticity module of both PHB and PHBV further represent that PHBV is more tensile compared to PHB (Xu et al., 2022). In addition to this, the respective other representing values in the graph show that elongation properties of PHBV are less compared to PHB. This has been evaluated earlier, as PHB is less brittle and stiff compared to PHBVC (Yang et al., 2022). The higher fusion temperatures of PHBV than PHB further shows that efficiency of developing or production of PHB is more convenient compared to PHBV.

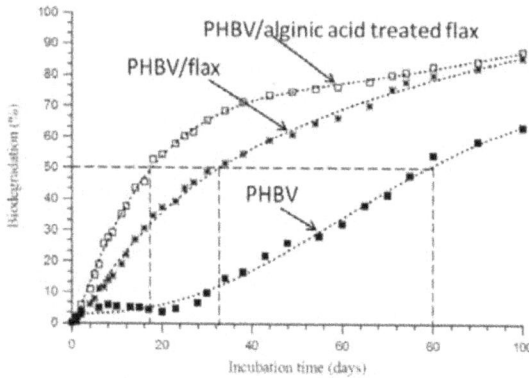

FIGURE 13.3 Biodegradability of PHBV.

Source: Muniyasamy et al. (2019).

The uses of this biodegradable material influenced the researchers to make the coating of some controlled drugs. On the other hand, using this material as a coat reduces the negative impact on nature (Zhou et al., 2022). The presence of a putative blocking of PHBVs allows for using it as an attachment and proliferating material during in vitro cell culture. As per the view of Öner et al. (2018), the use of boron nitride in the coating-making process from PHBV influenced the developers to make it more stable (Zhu et al., 2022). This article has also stated that making changes in the composition of this material and introducing collagen fibers help make it more rigid. On the other hand, John (2022) stated that using alginates helps develop the potentiality of this coating material.

Enhancing the inclusion of biodegradable replacement of plastic, the fusion of natural rubber with PBHV has been conducted to increase application and use efficiency of the biopolymer (Figure 13.3) (Muniyasamy et al., 2019). Blending of PHBV with high-molecular-weight natural rubber (HMWNR) has resulted in promising characteristics that would effectively increase production and use of the polymer for various operations as a replacement for plastic (Abdullah et al., 2022). The PBHV/HMW-NR blends acquired exhibit a 59% increase in flexibility, followed by a 20% increase in toughness of the blended biopolymer product. The potential benefits of the PBHV/HMW-NR blend are bio-plastic, sustainable production, environmental sustainability, and improved packaging services (The Ohio State University, 2022).

The development of PHBV as effectively capable with required stiffness, biodegradability, and related complimentary use for replacing plastic has been conducted following the addition of numerous inorganic materials that modify the tensile strength of PHBV (Almihyawi et al., 2022). Incorporating different agents such as rubber, coagent, peroxide, and a respective combination of the agents with PHBV/NR addresses the changes with the tensile strength of the PHBV/NR. The addition of more rubber is observed to decrease the tensile strength significantly; however, coagent has been observed to increase the tensile strength of PHBV/NR (Alshehri et al., 2022). The other combination included with PHBV/NR is observed to decrease tensile gradually, but the addition of more rubber with PHBV/NR is significantly low.

The fermented product, including PHBV, has shown high thermal plasticity proper due to the presence of "polybutylene terephthalate." This property makes it more heat resistant and makes it more useful for different medical industries. As per the view of Muniyasamy et al. (2019), the presence of flax fibers in the PHBV helps develop the resistance power against heat. In recent days, these materials were used as food packaging materials. Characteristics of this material make it a good packaging material (Arkaban et al., 2022).

In some cases, these fermented microbial products have shown different melting points. Scientists have found that the average melting temperature of PHBV is nearly 100–150°C (NCBI, 2022b). It has also been founded that the ratio between carbon and nitrogen (C:N) also changed the heat resistance capability of this compound. In addition to that, the change in the ratio of C to N also changed the stress resistance power of this compound (Attallah et al., 2022). Hence, evolving some chemical compositions of this compound helps improve the packaging quality of PHBV.

The development and production of PHBV films and coatings are often included with additional biological flax that are obtained from different alginates as biopolymers. The coatings developed from the addition of flax are often further incorporated with different organic and inorganic substances that contribute to the modification of different physical and implicational properties of PHBV films and coatings (Burlou-Nagy et al., 2022). Organic flax acquired from plant-based alginates and other organic sources adds to tensile, strength, and elasticity properties of the PHBV polymer. Adding organic flax for incorporating subsequent supportive characteristics for PHBV is important for the elastic stress of the storage modulus under specified temperature conditions (Burlou-Nagy et al., 2022). The concentration of flax content with the PHBV also adds into the respective implication of storage modulus under the respective temperature conditions. Storage modulus is the elastic strain capability or the ability of the particular material based on the elasticity energy storage capability. The storage modulus changes with influence of temperature on the material. The changes with included organic or inorganic flux with the material under consideration add in to respective changes that affect the storage modulus of the concerned material.

John et al. (2022) evaluated the storage modulus with changing temperature and different flax concentration for PLA and PHBV. The PLA, or polylactic acid, and PHBV are incorporated with acquired flax from organic alginates acquired from plant-based extracts. Flux concentrations included for both PLA and PHBV are 0%, 2%, and 4%, respectively (Dai et al., 2022). Interpretations from the graph show that changes with storage modulus of the respective flux-incorporated PLA and PHBV are almost at a direct proportion. The rate of changes with the respective flax-induced PHBV under 30°C is the highest, indicating significantly strong elasticity energy-storing ability. The lowest storage modulus as observed from the illustration is the PLA storage modulus under 70°C (De Santis et al., 2022). This implies that inclusion of the added organic alginate-based flux in PHBV contributes to acquisition of highly elastic PHBV polymer for use and application.

This microbial fermented product has some different biopolymer characteristics. The density of this compound is 1.25 g/cm³. Additionally, the elasticity modulus of this compound is 2.38 GPa. At the same time, the traction resistance of this compound is 25.9 MPa, and the elongation rate of PHBV is 1.4%. As per the view of Rabbani et al. (2022), changing the character of polydimethylsiloxane (PDMS) by treating it with silica influenced to develop the electrospinning characteristics. In addition, the development of the bonding of the outer coating of this fermented packaging material helps improve its bio-polymeric characteristics (Friščić et al., 2022). PHBV is a non-toxic product that is used as a coating and filming material for different compounds. In addition to that, this material is also used for the coating of different drugs. PHBV also has the characteristics of cytotoxicity against the cancer cell line. According to Álvarez-Álvarez et al. (2019), PHBV encapsulated different chemical compounds, including hydrocortisone (HC). That helps reduce the toxicity level of the packaged materials. Hence, the presence of a low toxicity level makes it a good packaging material.

This packaging material has antimicrobial properties which help develop this material as a good packaging material. As per the view of Figueroa-Lopez et al. (2019), developing the cytotoxicity level of the compound makes it more resistant to different kinds of toxic chemicals. Ojha and Das (2020) stated that developing the ratio of C to N helps develop the antimicrobial property of this packaging material. Thereafter, developing the antimicrobial character helps it become a good packaging material (Frolov et al., 2022). In addition to that, its antifungal property helps keep the food products more secure for a long time. Hence, developing antifungal and antimicrobial properties makes it a good packing material.

13.5 Antibacterial Properties of PHBV-Based Films and Coatings

PHBV has been extensively used in the biomedical field as it has antimicrobial and antibacterial properties. The nanomaterial possesses intrinsic antimicrobial activity which is effective in the formation of the containers in the food packaging industry (Joudeh & Linke, 2022). The experimental method can be used for the detection of the effectiveness of the antimicrobial property of the polymer PHBV. Following the article by Tarrahi et al. (2020), it can be stated that with the help of "the agar well and the minimal inhibitory concentration (MIC) methods," the antimicrobial activity of the component has been evaluated. An experiment to understand the antimicrobial activity of the PHBV has been done with different

TABLE 13.2

Sensitivity of different microorganisms on PHBV

Sample code	Gram-positive bacteria		Gram-negative bacteria	
	Bacillus subtilis	*S. aureus*	*E. coli*	*Pseudomonas aeruginosae*
PHBV	(−)	(−)	(+)	(+)
PHBV-5g-PVP	17	18	29	18
PHBV-22g-PVP	22	27	34	22
PHBV-42g-PVP	25	27	42	33

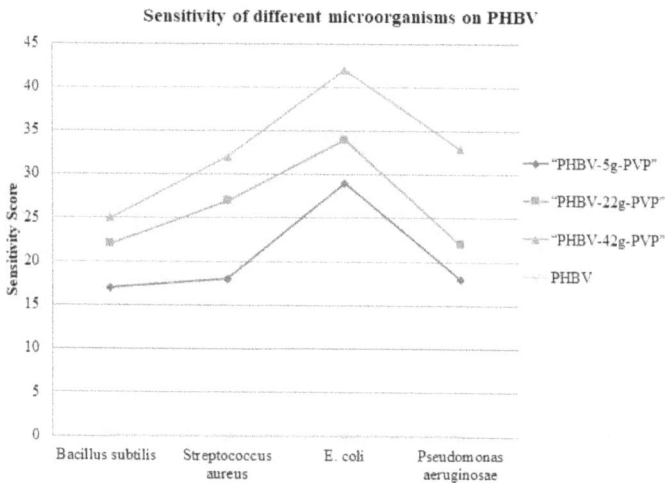

FIGURE 13.4 Sensitivity of different microorganisms on different blends of PHBV.

microorganisms like *Bacillus subtilis, Streptococcus aureus, E. coli,* and *Pseudomonas aeruginosae* (Saad et al., 2012). The effect of the PHBV has been observed on the different microorganisms in the culture media. From the experiment, it has been found that the entire microorganism has been showing sensitivity toward the PHBV (Table 13.2).

Table 13.2 states that PHBV alone is not effective in inhibiting the growth of the microorganism, whereas PHBV-5g-PVP is effective in comparison to PHBV and reduces the growth of the bacteria (Miłek et al., 2022). PHBV-22g-PVP has higher effectivity in comparison to others. PHBV-42g-PVP with the highest concentration is effective in inhibiting the growth of the bacteria. In reference to the result of Zhong et al. (2020), it can be stated that the polymer PHBV, in addition to PVP in higher concentrations, is most effective in the case of Gram-positive bacteria. The Gram-positive bacteria are more sensitive in comparison to the Gram-negative bacteria toward PBHV (Figure 13.4).

The impact and influence of antibacterial and antimicrobial activity of PHBV with different types of microorganism are identified. In addition to this, the impact and influence of the different blends of PHBV are also identified to be differently interacting with the kinds of microorganisms that influence potential uses of the respective blends in replacement of petroleum-based plastic products. The conditional influence of different blends or types of PHBV for antibacterial activities as evaluated earlier complies with the ideas that blends of PHBV are more effective and influential for the successful inhibition bacterial and respective microorganism growth. The development and research based on the respective antibacterial blends of PHBV have shown promising results for potential application of PHBV blends in biomedical applications. The considerable antimicrobial characteristics of PHBV have increased the developmental aspect of the biopolymer to address deliverance of efficiency with biomedical practices and drug application.

The graphical representation of differentiated impact of PHBV blends for antimicrobial activity and influence have been developed based on the tabular data on PHBV actions with different microorganisms. The graph illustrates the action potential and assessed antimicrobial activity of PHBV blends such as PHBV-5g-PVP, PHBV-22g-PVP, and PHBV-42g-PVP. The blends are developed based on crystallization or grafting of the PHBV with PVP polymers. The grafting of the PHBV resulted in the formation of the PHBV-g-PVP polymer. The grafting percentage of the PHBV has been identified to exponentially increase the antimicrobial property of the biopolymer along with increasing the biodegradability rate of the blended PHBV.

The graphical data shows that the influence and contribution of PHBV alone has no significant influence as antimicrobial activity on different microorganisms, but the PVP-copolymer-blended and grafting-crystallization-induced PHBV shows positive values with acting against the growth and development of respective microorganism. Illustrated sensitivity scores of the respective microorganisms *Bacillus subtilis*, *Streptococcus aureus*, *E. coli*, and *Pseudomonas areuginosae* indicate the inhibitory property of the PHBV-g-PVP polymer. It is observed that *E. coli* is the most sensitive bacteria inhibited by the respective PHBV-g-PVP polymers. *E. coli* has 29, 34, and 42 as sensitivity scores for PHBV-5g-PVP, PHBV-22g-PVP, and PHBV-42g-PVP, respectively. This shows that the PHBV blends are significantly efficient in addressing growth of *E. coli*. The graphical representation also highlights that the conditions as inhibition ability of the PHBV-g-PVP blends are more or less the same for other microorganisms assessed against the respective sensitivity with different PHBV-g-PVP polymers.

These elements have effectively broken the components of the cell wall of the bacteria, and the cellular balance has been lost with that, causing the death of the microorganisms. These components are therefore effective against *Staphylococcus aureus*, *Escherichia coli*, *Pseudomonas aeruginosa*, *Listeria monocytogenes*, and *Salmonella enteriditis* (Ferri et al., 2020). However, a drawback to the incorporation of the essential oil for the biodegradable compound is the loss of volatiles during the production of the film. In case the biodegradable activity is lost, then it may cause an environmental impact. Therefore, the most effective way to increase the antimicrobial activity of the PHBV is thermo-compressing the active component of the essential oil on one side of the PHBV films.

On the other hand, there is an effective method of increasing the antimicrobial activity of the component by blending PHBV films with alternative materials like silver nanoparticles. As per the view of Melendez-Rodriguez et al. (2019), the blending of the PHBV with the silver nanoparticles is effective in the case of the *S. aureus* and *Klebsiella pneumonia*. Altogether, the composition is not cytotoxic and has good compatibility with each other. In addition to this, Ciuprina et al. (2020) stated that the composition of PHBV with silver nanoparticles is also effective against some important bacteria, like the *Salmonella enterica* and *L. monocytogenes*. In food particles, most of these microorganisms have originated; therefore, it can be stated that PHBV with silver nanoparticles is effective in reducing toxin-producing microorganisms (Frącz et al., 2021). Therefore, it can also be stated that this has been successfully used in food packaging and the field of medicine for wrapping medicines.

In composition with other biological components like the antimicrobial oxides, the bilayer structure of PHBV is effective against bacteria and viruses. For example, it can be stated that the copper oxide (CuO) nanoparticles formed with the help of the compression process effectively show bactericidal and virucidal performance against a few pathogens like *S. enterica*, *L. monocytogenes*, and *Murine norovirus*grows in the food materials (Castro Mayorga et al., 2018; Scapinello et al., 2018). After compression with the antimicrobial oxides, the PHBV nanoparticle can be used to protect the bacterial growth in the food products. The use of the nanoparticle in the food packaging products inhibits bacterial and fungal growth, and therefore, it has been considered an effective component in the food packaging industry (Figure 13.5).

The previous figure discusses the impact of the PHBV/CEF/SPION nanoparticles on the bacterial culture of *E. coli* in the agarose gel suspension (Rivera-Briso & Serrano-Aroca, 2018). In the test, the nanoparticle has shown a positive antibacterial result of "inhibition halo of 29 mm." On the other hand, PHBV/SPION showed bacterial growth inhibition in the culture media (Melendez-Rodriguez et al., 2019). Therefore, it can be stated that the nanoparticle, in addition to other antimicrobial oxides, is effective in inhibiting the growth of the bacteria and other foreign organisms. This nanoparticle is effective in the reduction of the growth of microorganisms.

On the other hand, recently it has been found that graphene also has a high antimicrobial activity; therefore, the assimilation of the element with the carbon nanomaterials is effective in the production of high

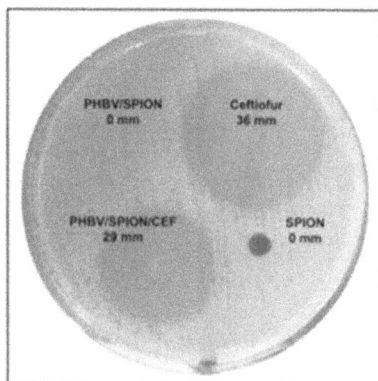

FIGURE 13.5   Response of *E. coli* bacteria to the PHBV nanoparticles.

Source: Rivera-Briso and Serrano-Aroca (2018).

antimicrobial activity (Modi et al., 2022). Thus, by assimilating the antioxidants with the nanoparticles, the 3D scaffold can be prepared, and with the help of this method, porosity can be reduced. The scaffold structure is also effective in the enhancement of the mechanical strength of the nanoparticle (Mazur et al., 2020). In addition to the antimicrobial activity and biodegradability, increasing mechanical strength is effective in the formation of an appropriate nanoparticle that can be used for the covering of the food and the medical components. The well-balanced characteristics of these complex nanoparticles can be effectively used in biomedical applications (Chausali et al., 2022). Due to the presence of high porosity, along with this, the presence of structural resemblance has been effective in the use of PHBV in the biomedical field.

13.6 Applications of PHBV-Based Films and Coatings

13.6.1 Biomedical Applications

Studies have revealed that PHBV-based films and coatings are implemented in the field of tissue engineering (Figure 13.6). It was found that fibers of PHBV-based coatings were often implemented as a strategy of biomedical scaffolding. This application of poly(3-hydroxy-butarate-CO-3-hydroxyvalerate)-based films

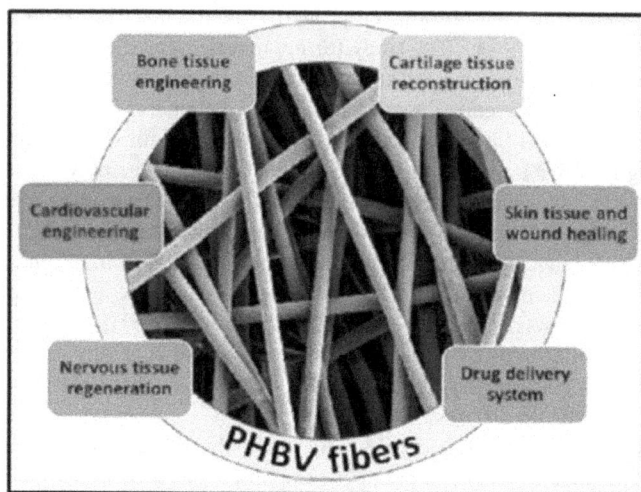

FIGURE 13.6   Implementation of PHBV-based films and coatings in tissue engineering

Source: Kaniuk and Stachewicz (2021).

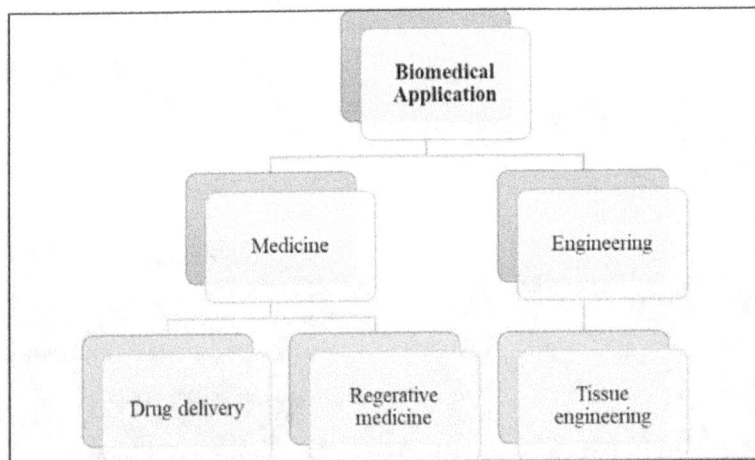

FIGURE 13.7   Biomedical applications of PHBV-based films and coatings.

and coatings was noted to show high antimicrobial activity (Mohammed et al., 2022). Research reveals that the coatings of "PHBV" implemented in various medical and biomedical products added to their antimicrobial properties. This improved proliferation in the activity of the products subjected to "PHBV-based coatings" and even aided in fighting against *Staphylococcus aureus* (Rivera-Briso et al., 2022).

It is further revealed that PHBV-based films and coatings were used for engaging in tissue engineering applications (Figure 13.7). It is often argued that when PHBV-based coating nanofibers were subjected to use in tissue engineering, the material's tensility was increased. This increase in strength was experienced by the use of 6% weight of "PHBV coatings," resulting in an increase in strength by 1%, thereby resulting in biomedical engineers implementing the use of such material in the field of tissue engineering (Pryadko et al., 2021). According to Kaniuk and Stachewicz (2021), the monomers and copolymers involved in PHBV-based films and coats were often subjected to degeneration and electrospinning. These electrospun fibers of PHBV-based coats resulted in a 60% drop in the crystalline property of the item, thereby resulting in increased use in the tissue engineering process (Rakib-Uz-Zaman et al., 2022).

Studies further revealed that the PHYSICOCHEMICAL properties of PHBV-based coatings and films resulted in the increased use of the item in the field of drug delivery. It was revealed by Tebaldi et al. (2019) that the biocompatibility and absorbance properties of the material result in the said material being used in the field of drug delivery. The study further revealed that the use of PHBV-based film and coating for the production of antineoplastic drugs was done, resulting in progress in the field of biomedical applications as well as pharmaceutical medicines (Rufino-Palomares et al., 2022).

Apart from this, PHBV-based films and coatings were even used for the production of regenerative medicines. According to Vahabi et al. (2019), the biodegradability of the polyester-based film and coat was implemented as retardant films and coatings on medicinal products, leading to the counteraction of tumors located in the body. This resulted in PHBV-based films and coats being implemented in the biomedical field of tumor and cancer treatment drug delivery. It was further highlighted by Rahmati et al. (2021) that the use of these chemicals was further done to see to the procedure of scaffolding in the field of biomedical engineering. This resulted in the use of such coatings of PHBV-based film to improve the physicochemical reactions taking place between cells, thereby aiding not only in the field of tissue engineering but even in the field of drug delivery and therapy (Salwa et al., 2022). Table 13.3 summarizes the potential application of PHBV in different fields.

As per the perspective of Dhandapani et al. (2020), this polymer has also been utilized for medical purposes due to its acute inflammatory and prolonged responses. The R-3HAs have been utilized severely for developing blocks of compounds in the pharmaceutical industry. PHBV can be utilized for the development of 3HA that has been widely utilized for the generation of hydroxycarboxylic acids like β-lactones and 2-alkylated. These hydroxycarboxylic acids have been utilized as oral drugs. Sadeghi et al. (2020) have discussed in their study that macrolide and carbapenem antibiotics have been evaluated from PHBV.

TABLE 13.3

Biomedical applications of PHBV

Applicable field	Process of application	Outcomes
Bio-agent	Through metabolizing activity of β-hydroxy SCFAs, which is a component of PHBV in the intestinal tract.	Controlling agents against pathogens, such as controlling the bio-agent giant tiger prawn against the pathogen.
Antibacterial	Development of 3-HA that initiates the generation of hydroxycarboxylic acids like β-lactones and 2-alkylated.	These hydroxycarboxylic acids have been utilized as oral drugs.
Drug carriers	The surface-functional and monodispersity properties of 3HB monomers of PHBV polymers have potential activity as drug carriers.	PHBV is used for producing nanoparticles, scaffolds, and tablets.

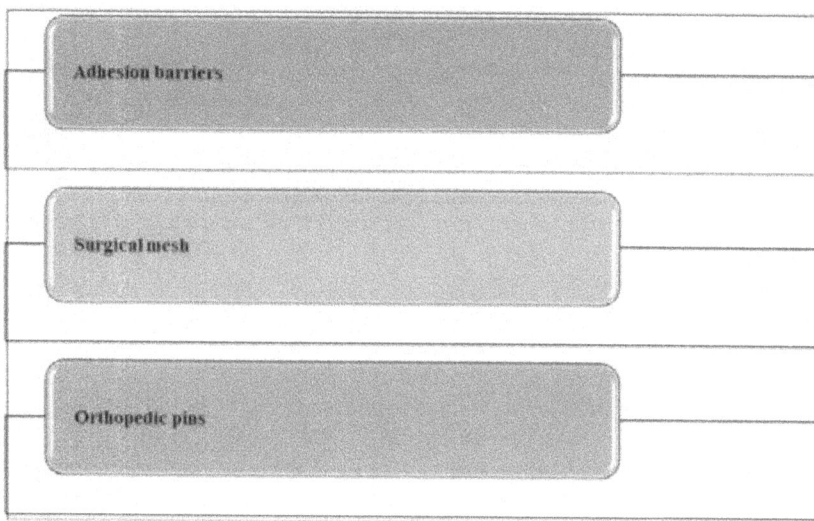

FIGURE 13.8 Application of PHBV films in medical devices.

Pseudomonas fluorescens secrete a depolymerase enzyme that has been utilized for encoding the genre phaZGK13 by the gene GK13. This gene can also depolymerize PHBV polymers into their monomer. These monomeric units have been identified as effective elements for reducing the bacterial infection caused by *Staphylococcus aureus*. The monomeric units also have anti-cancerous properties after conjugating with the *D-peptide*. It has also been assessed that the P4HB and P3HB derived from PHBV also have angiogenic properties that enhance their capabilities of increasing the wound and skin healing processes.

Sadeghi et al. (2020) have stated that PHBV has β-hydroxy SCFAs as a component, and short-chain fatty acids (SCFAs) have been identified as effective elements for controlling agents against pathogens. Therefore, PHBV also has controlling properties over some pathogens due to the presence of β-hydroxy SCFAs. The β-hydroxy SCFAs have been identified to be metabolized in the intestinal tract of the human body and, therefore, associated with the metabolites and have been seen to be acting as biocontrol agents. PHBV has been utilized widely for controlling the growth of bio-agent giant tiger prawn against the pathogen *Penaeus monodon*. The 3-HB monomers have been utilized for the synthesis of polymers like "dendrimers," and the surface-functional and monodispersity properties of these monomers help PHBV to act as a drug carrier (Aguilar-Rabiela et al., 2021).

According to various research and studies, it has been observed that PHBV films and coatings are used in various mechanisms in the biomedical fields. It is used in different medical devices because of its biocompatibility and biodegradability and strong mechanical characteristics. It is commonly used as adhesion barriers, surgical mesh, orthopedic pins, and in many other applications. Due to the biocompatible nature of this biopolymer, it helps resist infection within the host body. On the other hand, these PHBV films are also used in the process of biodegradable implantation. It has been widely used in the implantation technique for fabricating systems and delivering antibiotics.

This biopolymer is also used in enhancing the capability of memory of patients with poor mental health (Figure 13.8). Derivatives of this polymer are used as a potential therapeutic agent for curing diseases such as Alzheimer's and dementia. According to the viewpoint of Ibrahim et al. (2021), it has been observed that the PHBV film is found as a promising component with biocompatible nature that helps in various medical applications.

13.6.2 Food Packaging Applications

The biodegradable characteristics of the polyhydroxyalkanoates (PHA) polymers are increasing the consideration and inducing the importance of food packaging for determining food consumption safety and biodegradability aspects. Modification and research on PHAs have successfully resulted in the development and invention of respective influential polymers that are efficiently environment-friendly and effectively serve the need. Polymers such as polyhydroxybutyrate (PHB) are identified as the most extensively researched PHAs that exhibit exact similar properties shown in modern-day plastics or petroleum-based plastic propylene (PP). The characteristics of PHB are similar to plastic, but they are brittle and stiffer. Adding poly(3-hydroxyvalerate) with PHB develops poly(3-hydroxybutarate-co-3-hydroxyvalerate) (PHBV), which is stated to be expensive in production and brittle but contributes to supportive insights that influence PHBV as a potential biodegradable plastic (Doyle, 2019).

In order to support the inclusion of PHBV for food packaging and related application, rubber was melted and introduced with PHBV along with trimethylolpropane triacrylate (TMPTA) and an organic peroxide. The blending of rubber with the PBHV has been observed to increase flexibility by almost 100% and 75% more tough, indicating a promising future applicability of the blended PHBV as food packaging (Doyle, 2019). The existing concern of non-applicability of blended PHBV in food packaging is the potential increase of toxicity with the biopolymer.

Implications of PHBV's biodegradability and less toxicity of the same been identified to possess the importance of its application in use for food packaging. According to Zhao et al. (2019), the moisture-resistant ability and hot-compression characteristics of PHBV are significantly beneficial for the inclusion of this polymer in food packaging. The brittle and stiff properties of PHBV, however, impose a major issue behind the extensive use of PHBV in place of plastic.

Following this, the separated microorganisms are overfed with fatty acids in order to develop and produce PHBV. PHBV is considered to be fat storage of cells, and these fats are utilized in the production of biodegradable plastic (Ypack, 2018). The implementation of this process is identified to be expensive but supports the effective inclusion of supportive measures that address food safety and environmental safety perspective. Based on the study developed by Ojha and Das (2020), adding in more extensive particle size material (>100nm) could increase the characteristic of PHBV, supporting its subsequent use as a potential replacement of plastic for food packaging and related uses. According to Râpă et al. (2021), PHBV is the most influential and inclusive biodegradable polymer among the PHA polymer family that are potentially indicating the supportive inclusion of the same as replacement of plastic. Addressing this aspect, Requena et al. (2019) have evaluated that the influence of both eugenol and carvacrol on PHBV used for food packaging is significantly effective in addressing the antimicrobial activity of *E. coli*. This signifies that the influence and contributions of PHBV in food packaging would be significantly beneficial to food safety, health, and environmental concerns. Increasing the efficiency of PHBV for food packaging has been observed with developing multilayered active films of the polymer.

The influence of the developmental aspects of PHBV as multilayer films for food packaging would also support the development of efficient biodegradable and food-grade packaging material (Requena et al., 2019). The implications of multilayered PHBV for food packaging would also increase the efficiency of food safety packaging addressing influential impact based on antimicrobial activities. Zhao et al. (2019) have stated that with increasing technological innovation and development of modern environmentally sustainable aspects with plastic use and related food packaging, it is necessary to implement and modify policies that reduced the gap between uses and application of the biodegradable plastic PHBV. The inclusion of PHBV in food packaging and related food safety measures would contribute to supportive implications that would increase sustainability in the environment, in food production, and in human health.

13.7 Conclusion

The chemical properties of the PHBV and, along with this, the extraction of the component have been discussed. The 3-hydroxyvalerate (HV) component is the main factor that has helped in the extraction of PHBV. This material has several properties, like high crystallinity, biodegradability, and resistance against ultraviolet radiation. Due to the presence of these properties, this material has been used for the packaging material of foods and medicines. However, due to the presence of several cracks on PHBV, this coating material has failed to provide a good grease barrier to these products. Due to the presence of aliphatic polyester, the component also shows thermoplastic properties. This biofilm material has some unique characteristics, including high proliferative activity. On the other hand, the presence of graphene has helped the formation of nanosheets from PHBV. One of the prime methods used for the preparation of PHBV-based films is electrospinning. With the help of this process, nanoparticles have been immersed in PHBV. The implementation of this technique is effective in the enhancement of the characteristics of PHBV. The existing characteristics can be altered and enhanced with the help of this technique. Another method of preparation that is often implemented for the preparation of PHBV-based films is engaging in film blowing of PHBV blends. This method is also effective in the preparation of the blended nanoparticle. Further, the method of PHBV-based coatings has been discussed. Through the help of the extrusion process, the coating of the papers has been formed. In the compounding step, the "PHBV film" is incorporated with triethyl citrate (TEC), and polyethylene glycol (PEG) is used as a "twin-screw extruder." Further, the antimicrobial activity of PHBV has been described. The way antimicrobial activity can be increased has been described in this section. The association of the essential components and nanoparticles has been described in the study. Lastly, biomedical applications have been mentioned in the study.

REFERENCES

Abdullah, T., Qurban, R. O., Mohamed Sh, A. W., Salah, N. A., Ammar, A. M., Zamzami, M. A., & Memić, A. (2022). Development of nano-coated filaments for 3D fused deposition modeling of antibacterial and antioxidant materials. *Polymers*, *14*(13), 2645. https://doi.org/10.3390/polym14132645

Aguilar-Rabiela, A. E., Aldo, L. E., Qaisar, N., & Aldo, R. B. (2021). Microspheres as biocompatible composite for drug delivery applications. *Molecules*, *26*(11), 3177. https://doi.org/10.3390/molecules26113177

Alhazmi, A., Aldairi, A. F., Alghamdi, A., Alomery, A., Mujalli, A., Obaid, A. A., . . . Alghamdi, A. (2022). Antibacterial effects of methanolic extract on wound healing. *Molecules*, *27*(10), 3320. https://doi.org/10.3390/molecules27103320

Almihyawi Raed, A. H., Naman, Z. T., Al-Hasani Halah, M. H., Muhseen, Z. T., Sitong, Z., & Chen, G. (2022). Integrated computer-aided drug design and biophysical simulation approaches to determine natural anti-bacterial compounds for *Acinetobacter baumannii*. *Scientific Reports*, *12*(1), 6590. https://doi.org/10.1038/s41598-022-10364-z

Alshehri, M. M., Quispe, C., Herrera-Bravo, J., Sharifi-Rad, J., Tutuncu, S., Aydar, E. F., . . . Cho, W. C. (2022). A review of recent studies on the antioxidant and anti-infectious properties of Seena plants. *Oxidative Medicine and Cellular Longevity*, *2022*, 6025900. https://doi.org/10.1155/2022/6025900

Álvarez-Álvarez, L., Barral, L., Bouza, R., Farrag, Y., Otero-Espinar, F., Feijóo-Bandín, S., & Lago, F. (2019). Hydrocortisone loaded poly-(3-hydroxybutyrate-co-3-hydroxyvalerate) nanoparticles for topical ophthalmic administration: Preparation, characterization and evaluation of ophthalmic toxicity. *International Journal of Pharmaceutics*, *568*, 118519. https://doi.org/10.1016/j.ijpharm.2019.118519

Arkaban, H., Barani, M., Akbarizadeh, M. R., Narendra Pal, S. C., Jadoun, S., Maryam, D. S., & Zarrintaj, P. (2022). Polyacrylic acid nanoplatforms: Antimicrobial, tissue engineering, and cancer theranostic applications. *Polymers*, *14*(6), 1259. https://doi.org/10.3390/polym14061259

Attallah, N. G., Mokhtar, F. A., Elekhnawy, E., Heneidy, S. Z., Ahmed, E., Magdeldin, S., . . . El-Kadem, A. H. (2022). Mechanistic insights on the in vitro antibacterial activity and in vivo hepatoprotective effects of *salvinia auriculata* aubl against methotrexate-induced liver injury. *Pharmaceuticals*, *15*(5), 549. https://doi.org/10.3390/ph15050549

Burlou-Nagy, C., Bănică, F., Jurca, T., Vicaş, L. G., Marian, E., Muresan, M. E., . . . Pallag, A. (2022). *Echinacea purpurea* (L.) Moench: Biological and pharmacological properties. A review. *Plants*, *11*(9), 1244. https://doi.org/10.3390/plants11091244

Castro Mayorga, J. L., Fabra Rovira, M. J., Cabedo Mas, L., Sánchez Moragas, G., & Lagarón Cabello, J. M. (2018). Antimicrobial nanocomposites and electrospun coatings based on poly (3-hydroxybutyrate-co-3-hydroxyvalerate) and copper oxide nanoparticles for active packaging and coating applications. *Journal of Applied Polymer Science*, *135*(2), 45673. https://doi.org/10.1002/app.45673

Chausali, N., Saxena, J., & Prasad, R. (2022). Recent trends in nanotechnology applications of bio-based packaging. *Journal of Agriculture and Food Research*, *7*, 100257. https://doi.org/10.1016/j.jafr.2021.100257

Ciuprina, F., Andrei, L., Stoian, S., Gabor, R., & Panaitescu, D. (2020). Dielectric response and dynamic mechanical analysis of PHBV-TiO$_2$ nanocomposites. In *2020 IEEE 3rd International Conference on Dielectrics (ICD)* (pp. 201–204). IEEE. https://ieeexplore.ieee.org/abstract/document/934

Cunha, M., Fernandes, B., Covas, J. A., Vicente, A. A., & Hilliou, L. (2016). Film blowing of PHBV blends and PHBV-based multilayers for the production of biodegradable packages. *Journal of Applied Polymer Science*, *133*(2). https://doi.org/10.1002/app.42165

Dai, C., Lin, J., Li, H., Shen, Z., Wang, Y., Velkov, T., & Shen, J. (2022). The natural product curcumin as an antibacterial agent: Current achievements and problems. *Antioxidants*, *11*(3), 459. https://doi.org/10.3390/antiox11030459

De Santis, D., Carbone, K., Garzoli, S., Valentina, L. M., & Turchetti, G. (2022). Bioactivity and chemical profile of *Rubus idaeus* L. leaves steam-distillation extract. *Foods*, *11*(10), 1455. https://doi.org/10.3390/foods11101455

Dhandapani, R., Amrutha, M., Swaminathan, S., & Anuradha, S., (2020). Composite nanofiber matrices for biomedical applications. In *Artificial protein and peptide nanofibers* (pp. 241–258). Woodhead Publishing, https://doi.org/10.1016/B978-0-08-102850-6.00011-5

Doyle, A. (2019). *A tougher bioplastic for food packaging*. Retrieved May 26, 2022, from www.thechemicalengineer.com/news/a-tougher-bioplastic-for-food-packaging/

Fajstavr, D., Frýdlová, B., Rimpelová, S., Nikola, S. K., Sajdl, P., Švorčík, V., & Slepička, P. (2022). KrF laser and plasma exposure of PDMS—carbon composite and its antibacterial properties. *Materials*, *15*(3), 839. https://doi.org/10.3390/ma15030839

Ferri, M., Vannini, M., Ehrnell, M., Eliasson, L., Xanthakis, E., Monari, S., . . . Tassoni, A. (2020). From winery waste to bioactive compounds and new polymeric biocomposites: A contribution to the circular economy concept. *Journal of Advanced Research*, *24*, 1–11. https://onlinelibrary.wiley.com/doi/abs/10.1002/appl.202100015

Figueroa-Lopez, K. J., Torres-Giner, S., Enescu, D., Cabedo, L., Cerqueira, M. A., Pastrana, L. M., & Lagaron, J. M. (2020). Electrospun active biopapers of food waste derived poly (3-hydroxybutyrate-co-3-hydroxyvalerate) with short-term and long-term antimicrobial performance. *Nanomaterials*, *10*(3), 506. https://doi.org/10.3390/nano10030506

Figueroa-Lopez, K. J., Vicente, A. A., Reis, M. A., Torres-Giner, S., & Lagaron, J. M. (2019). Antimicrobial and antioxidant performance of various essential oils and natural extracts and their incorporation into bio-waste derived poly (3-hydroxybutyrate-co-3-hydroxyvalerate) layers made from electrospun ultrathin fibers. *Nanomaterials*, *9*(2), 144.

Frącz, W., Janowski, G., Smusz, R., & Szumski, M. (2021). The influence of chosen plant fillers in PHBV composites on the processing conditions, mechanical properties and quality of molded pieces. *Polymers*, *13*(22), 3934. https://doi.org/10.3390/polym13223934

Friščić, M., Petlevski, R., Kosalec, I., Madunić, J., Matulić, M., Bucar, F., . . . Maleš, Ž. (2022). *Globularia alypum* L. and related species: LC-MS profiles and antidiabetic, antioxidant, anti-inflammatory, anti-bacterial and anticancer potential. *Pharmaceuticals*, *15*(5), 506. https://doi.org/10.3390/ph15050506

Frolov, N., Detusheva, E., Fursova, N., Ostashevskaya, I., & Vereshchagin, A. (2022). Microbiological evaluation of novel bis-quaternary ammonium compounds: Clinical strains, biofilms, and resistance study. *Pharmaceuticals, 15*(5), 514. https://doi.org/10.3390/ph15050514

Hietala, M., & Oksman, K. (2018). Pelletized cellulose fibres used in twin-screw extrusion for biocomposite manufacturing: Fibre breakage and dispersion. *Composites Part A: Applied Science and Manufacturing, 109*, 538–545. https://doi.org/10.1016/j.compositesa.2018.04.006

Ibrahim, M. I., Alsafadi, D., Alamry, K. A., & Hussein, M. A. (2021). Properties and applications of poly (3-hydroxybutyrate-co-3-hydroxyvalerate) biocomposites. *Journal of Polymers and the Environment, 29*(4), 1010–1030. https://link.springer.com/article/10.1007/s10924-020-01946-x

Insomphun, C., Chuah, J. A., Kobayashi, S., Fujiki, T., & Numata, K. (2017). Influence of hydroxyl groups on the cell viability of polyhydroxyalkanoate (PHA) scaffolds for tissue engineering. *ACS Biomaterials Science & Engineering, 3*(12), 3064–3075. https://doi.org/10.1021/acsbiomaterials.6b00279

John, M. J. (2022). Biobased alginate treatments on flax fibre reinforced PLA and PHBV composites. *Current Research in Green and Sustainable Chemistry, 5*, 100319. https://doi.org/10.1016/j.crgsc.2022.100319

Joudeh, N., & Linke, D. (2022). Nanoparticle classification, physicochemical properties, characterization, and applications: A comprehensive review for biologists. *Journal of Nanobiotechnology, 20*, 1–29. https://doi.org/10.1186/s12951-022-01477-8

Kaniuk, Ł., & Stachewicz, U. (2021). Development and advantages of biodegradable PHA polymers based on electrospun PHBV fibers for tissue engineering and other biomedical applications. *ACS Biomaterials Science & Engineering, 7*(12), 5339–5362. https://doi.org/10.1021/acsbiomaterials.1c00757

Khane, Y., Benouis, K., Albukhaty, S., Sulaiman, G. M., Abomughaid, M. M., Ali, A. A., . . . Dizge, N. (2022). Green synthesis of silver nanoparticles using aqueous citrus limon zest extract: Characterization and evaluation of their antioxidant and antimicrobial properties. *Nanomaterials, 12*(12), 2013. https://doi.org/10.3390/nano12122013

Khayrova, A., Lopatin, S., Shagdarova, B., Sinitsyna, O., Sinitsyn, A., & Varlamov, V. (2022). Evaluation of antibacterial and antifungal properties of low molecular weight chitosan extracted from *Hermetia illucens* relative to crab chitosan. *Molecules, 27*(2), 577. https://doi.org/10.3390/molecules27020577

Lammi, S., Gastaldi, E., Gaubiac, F., & Angellier-Coussy, H. (2019). How olive pomace can be valorized as fillers to tune the biodegradation of PHBV based composites. *Polymer Degradation and Stability, 166*, 325–333. https://doi.org/10.1016/j.polymdegradstab.2019.06.010

Lange, A., Sawosz, E., Wierzbicki, M., Kutwin, M., Daniluk, K., Strojny, B., . . . Jaworski, S. (2022). Nanocomposites of graphene oxide—silver nanoparticles for enhanced antibacterial activity: Mechanism of action and medical textiles coating. *Materials, 15*(9), 3122. https://doi.org/10.3390/ma15093122

Lewandowski, A., & Wilczyński, K. (2019). Global modeling of single screw extrusion with slip effects. *International Polymer Processing, 34*(1), 81–90. https://doi.org/10.3139/217.3653

Mazur, K., Singh, R., Friedrich, R. P., Genç, H., Unterweger, H., Sałasińska, K., . . . Cicha, I. (2020). The effect of antibacterial particle incorporation on the mechanical properties, biodegradability, and biocompatibility of PLA and PHBV composites. *Macromolecular Materials and Engineering, 305*(9), 2000244. https://doi.org/10.1002/mame.202000244

Melendez-Rodriguez, B., Figueroa-Lopez, K. J., Bernardos, A., Martínez-Máñez, R., Cabedo, L., Torres-Giner, S., & M. Lagaron, J. (2019). Electrospun antimicrobial films of poly (3-hydroxybutyrate-co-3-hydroxyvalerate) containing eugenol essential oil encapsulated in mesoporous silica nanoparticles. *Nanomaterials, 9*(2), 227. https://doi.org/10.3390/nano9020227

Miłek, M., Ciszkowicz, E., Tomczyk, M., Sidor, E., Zaguła, G., Lecka-Szlachta, K., . . . Dżugan, M. (2022). The study of chemical profile and antioxidant properties of poplar-type polish propolis considering local flora diversity in relation to antibacterial and anticancer activities in human breast cancer cells. *Molecules, 27*(3), 725. https://doi.org/10.3390/molecules27030725

Mirković, M., Filipović, S., Kalijadis, A., Mašković, P., Mašković, J., Vlahović, B., & Pavlović, V. (2022). Hydroxyapatite/TiO$_2$ nanomaterial with defined microstructural and good antimicrobial properties. *Antibiotics, 11*(5), 592. https://doi.org/10.3390/antibiotics11050592

Modi, S., Prajapati, R., Inwati, G. K., Deepa, N., Tirth, V., Yadav, V. K., . . . Jeon, B. (2022). Recent trends in fascinating applications of nanotechnology in allied health sciences. *Crystals, 12*(1), 39. doi: https://doi.org/10.3390/cryst12010039

Mohammad, M. D., & Stiharu, I. (2022). Preparing and characterizing novel biodegradable Starch/PVA-based films with nano-sized zinc-oxide particles for wound-dressing applications. *Applied Sciences*, *12*(8), 4001. https://doi.org/10.3390/app12084001

Mohammed, K., Assouguem, A., Fadili, M. E., Benmessaoud, S., Samar, Z. A., Kamaly, O. A., . . . Bahhou, J. (2022). Contribution to the evaluation of physicochemical properties, total phenolic content, antioxidant potential, and antimicrobial activity of vinegar commercialized in Morocco. *Molecules*, *27*(3), 770. https://doi.org/10.3390/molecules27030770

Muniyasamy, S., Ofosu, O., Thulasinathan, B., Rajan, A. S. T., Ramu, S. M., Soorangkattan, S., . . . Alagarsamy, A. (2019). Thermal-chemical and biodegradation behaviour of alginic acid treated flax fibres/poly (hydroxybutyrate-co-valerate) PHBV green composites in compost medium. *Biocatalysis and Agricultural Biotechnology*, *22*, 101394. https://doi.org/10.1016/j.bcab.2019.101394

NCBI.nlm.nih.gov. (2022a). *Properties of PHBV*. Retrieved May 26, 2022, from www.ncbi.nlm.nih.gov/pmc/articles/PMC6403723/

NCBI.nlm.nih.gov. (2022b). *Properties of PHBV*. Retrieved May 26, 2022, from www.ncbi.nlm.nih.gov/pmc/articles/PMC3617235/#:~:text=Because%20the%20melting%20temperatures%20of,polymer%20greatly%20improved%20its%20workability

The Ohio State University. (2022). *Method to improve natural rubber (hevea/guayule)-toughened poly-(3-hydroxybutyrate-co-3-hydroxyvalerate) (PHBV) for food packaging applications*. Retrieved May 26, 2022, from https://oied.osu.edu/technologies/method-improve-natural-rubber-heveaguayule-toughened-poly-3-hydroxybutyrate-co-3

Ojha, N., & Das, N. (2020). Fabrication and characterization of biodegradable PHBV/SiO2 nanocomposite for thermo-mechanical and antibacterial applications in food packaging. *IET Nanobiotechnology*, *14*(9), 785–795. https://doi.org/10.1049/iet-nbt.2020.0066

Öner, M., Kızıl, G., Keskin, G., Pochat-Bohatier, C., & Bechelany, M. (2018). The Effect of Boron Nitride on the Thermal and Mechanical Properties of Poly (3-hydroxybutyrate-co-3-hydroxyvalerate). Nanomaterials, 8(11), 940. https://doi.org/10.3390/nano8110940

Pahlevanzadeh, F., Setayeshmehr, M., Bakhsheshi-Rad, H., Emadi, R., Kharaziha, M., Ali Poursamar, S., . . . Berto, F. (2022). A review on antibacterial biomaterials in biomedical applications: From materials perspective to bioinks design. *Polymers*, *14*(11), 2238. https://doi.org/10.3390/polym14112238

Petousis, M., Vidakis, N., Velidakis, E., Kechagias, J. D., David, C. N., Papadakis, S., & Mountakis, N. (2022). Affordable biocidal ultraviolet cured cuprous oxide filled vat photopolymerization resin nanocomposites with enhanced mechanical properties. *Biomimetics*, *7*(1), 12. https://doi.org/10.3390/biomimetics7010012

Pradhan, B., Nayak, R., Bhuyan, P. P., Patra, S., Behera, C., Sahoo, S., . . . Jena, M. (2022). Algal phlorotannins as novel antibacterial agents with reference to the antioxidant modulation: Current advances and future directions. *Marine Drugs*, *20*(6), 403. https://doi.org/10.3390/md20060403

Pryadko, A., Surmeneva, M. A., & Surmenev, R. A. (2021). Review of hybrid materials based on polyhydroxyalkanoates for tissue engineering applications. *Polymers*, *13*(11), 1738. https://doi.org/10.3390/polym13111738

Rabbani, S., Jafari, R., & Momen, G. (2022). Superhydrophobic micro-nanofibers from PHBV-SiO2 biopolymer composites produced by electrospinning. *Functional Composite Materials*, *3*(1), 1–15. https://doi.org/10.1186/s42252-022-00029-5

Rahmati, M., Mills, D. K., Urbanska, A. M., Saeb, M. R., Venugopal, J. R., Ramakrishna, S., & Mozafari, M. (2021). Electrospinning for tissue engineering applications. *Progress in Materials Science*, *117*, 100721. https://doi.org/10.1016/j.pmatsci.2020.100721

Rakib-Uz-Zaman, S. M., Ehsanul, H. A., Muntasir, M. N., Sadrina, A. M., Mst, G. K., Shah, S. J., . . . Khan, K. (2022). Biosynthesis of silver nanoparticles from cymbopogon citratus leaf extract and evaluation of their antimicrobial properties. *Challenges*, *13*(1), 18. https://doi.org/10.3390/challe13010018

Râpă, M., Stefan, M., Popa, P. A., Toloman, D., Leostean, C., Borodi, G., . . . Matei, E. (2021). Electrospun nanosystems based on PHBV and ZnO for ecological food packaging. *Polymers*, *13*(13), 2123. https://doi.org/10.3390/polym13132123

Redskins. (2022). *Global PHA market research report 2022 professional edition*. Retrieved May 26, 2022, from https://redskins101.com/global-pha-market-research-report-2022-professional-edition-2/

Requena, R., Vargas, M., & Chiralt, A. (2019). Eugenol and carvacrol migration from PHBV films and antibacterial action in different food matrices. *Food Chemistry*, *277*, 38–45. https://doi.org/10.1016/j.foodchem.2018.10.093

Righetti, M. C., Cinelli, P., Mallegni, N., Stäbler, A., & Lazzeri, A. (2019). Thermal and mechanical properties of biocomposites made of poly (3-hydroxybutyrate-co-3-hydroxyvalerate) and potato pulp powder. *Polymers, 11*(2), 308. https://doi.org/10.3390/polym11020308

Rivera-Briso, A., Aparicio-Collado, J., Roser Sabater, i. S., & Serrano-Aroca, Á. (2022). Graphene oxide versus carbon nanofibers in poly(3-hydroxybutyrate-co-3-hydroxyvalerate) films: Degradation in simulated intestinal environments. *Polymers, 14*(2), 348. https://doi.org/10.3390/polym14020348

Rivera-Briso, A. L., & Serrano-Aroca, Á. (2018). Poly (3-Hydroxybutyrate-co-3-Hydroxyvalerate): Enhancement strategies for advanced applications. *Polymers, 10*(7), 732. https://doi.org/10.3390/polym10070732

Rufino-Palomares, E. E., Pérez-Jiménez, A., García-Salguero, L., Mokhtari, K., Reyes-Zurita, F. J., Peragón-Sánchez, J., & Lupiáñez, J. A. (2022). Nutraceutical role of polyphenols and triterpenes present in the extracts of fruits and leaves of olea europaea as antioxidants, anti-infectives and anticancer agents on healthy growth. *Molecules, 27*(7), 2341. https://doi.org/10.3390/molecules27072341

Saad, G. R., Elsawy, M. A., & Elsabee, M. Z. (2012). Preparation, characterization and antimicrobial activity of poly (3-hydroxybutyrate-co-3-hydroxyvalerate)-g-poly (N-vinylpyrrolidone) copolymers. *Polymer-Plastics Technology and Engineering, 51*(11), 1113–1121.

Sadeghi, S., Nourmohammadi, J., Ghaee, A., & Soleimani, N. (2020). Carboxymethyl cellulose-human hair keratin hydrogel with controlled clindamycin release as antibacterial wound dressing. *International Journal of Biological Macromolecules, 147*, 1239–1247. https://doi.org/10.1016/j.ijbiomac.2019.09.251

Salwa, O. B., Baras, B. H., Weir, M. D., & Xu, H. H. K. (2022). Denture acrylic resin material with antibacterial and protein-repelling properties for the prevention of denture stomatitis. *Polymers, 14*(2), 230. https://doi.org/10.3390/polym14020230

Sängerlaub, S., Brüggemann, M., Rodler, N., Jost, V., & Bauer, K. D. (2019). Extrusion coating of paper with poly (3-hydroxybutyrate-co-3-hydroxyvalerate) (PHBV)—packaging related functional properties. *Coatings, 9*(7), 457. https://doi.org/10.3390/coatings9070457

Scaffaro, R., Maio, A., Sutera, F., Gulino, E. F., & Morreale, M. (2019). Degradation and recycling of films based on biodegradable polymers: A short review. *Polymers, 11*(4), 651. https://doi.org/10.3390/polym11040651

Scapinello, J., Aguiar, G. P. S., Dal Magro, C., Capelezzo, A. P., Niero, R., Dal Magro, J., . . . Oliveira, J. V. (2018). Extraction of bioactive compounds from *Philodendron bipinnatifidum* Schott ex Endl and encapsulation in PHBV by SEDS technique. *Industrial Crops and Products, 125*, 65–71. https://doi.org/10.1016/j.indcrop.2018.08.079

Shiu, B. C., Hsu, P. W., Lin, J. H., Chien, L. F., Lin, J. H., & Lou, C. W. (2022). A study on preparation and property evaluations of composites consisting of TPU/triclosan membranes and tencel®/LMPET nonwoven fabrics. *Polymers, 14*(12), 2514. https://doi.org/10.3390/polym14122514

Tarrahi, R., Fathi, Z., Özgür Seydibeyoğlu, M., Doustkhah, E., & Khataee, A. (2020). Polyhydroxyalkanoates (PHA): From production to nanoarchitecture. *International Journal of Biological Macromolecules, 146*, 596–619. www.sciencedirect.com/science/article/pii/S0141813019373143

Tebaldi, M. L., Ana Luiza C. M., Fernanda P., Fabricio V. de A., & Daniel C. F. S. (2019). Poly (-3-hydroxybutyrate-co-3-hydroxyvalerate) (PHBV): Current advances in synthesis methodologies, antitumor applications and biocompatibility. *Journal of Drug Delivery Science and Technology, 51*(2019), 115–126. https://doi.org/10.1016/j.jddst.2019.02.007

Vahabi, H., Rad, E. R., Parpaite, T., Langlois, V., & Saeb, M. R. (2019). Biodegradable polyester thin films and coatings in the line of fire: The time of polyhydroxyalkanoate (PHA)? *Progress in Organic Coatings, 133*, 85–89. https://doi.org/10.1016/j.porgcoat.2019.04.044

Vidakis, N., Petousis, M., Velidakis, E., Korlos, A., Kechagias, J. D., Tsikritzis, D., & Mountakis, N. (2022). Medical-grade polyamide 12 nanocomposite materials for enhanced mechanical and antibacterial performance in 3D printing applications. *Polymers, 14*(3), 440. https://doi.org/10.3390/polym14030440

Wang, W. B., & Clapper, J. C. (2022). Antibacterial activity of electrospun polyacrylonitrile copper nanoparticle nanofibers on antibiotic resistant pathogens and methicillin resistant staphylococcus aureus (MRSA). *Nanomaterials, 12*(13), 2139. https://doi.org/10.3390/nano12132139

Xia, Y., He, L., Feng, J., Xu, S., Yao, L., & Pan, G. (2022). Waterproof and moisture-permeable polyurethane nanofiber membrane with high strength, launderability, and durable antimicrobial properties. *Nanomaterials, 12*(11), 1813. https://doi.org/10.3390/nano12111813

Xu, L., Liu, Y., Zhou, W., & Yu, D. (2022). Electrospun medical sutures for wound healing: A review. *Polymers*, *14*(9), 1637. https://doi.org/10.3390/polym14091637

Yang, F., Song, Y., Hui, A., Kang, Y., Zhou, Y., & Wang, A. (2022). Facile preparation of organo-modified ZnO/Attapulgite nanocomposites loaded with monoammonium glycyrrhizinate via mechanical milling and their synergistic antibacterial effect. *Minerals*, *12*(3), 364. https://doi.org/10.3390/min12030364

Ypack. (2018). *Newsletter article: How do you make food packaging from cheese whey?* Retrieved May 26, 2022, from www.ypack.eu/2019/10/28/article-how-do-you-make-food-packaging-from-cheese-whey/

Zhao, X., Ji, K., Kurt, K., Cornish, K., & Vodovotz, Y. (2019). Optimal mechanical properties of biodegradable natural rubber-toughened PHBV bioplastics intended for food packaging applications. *Food Packaging and Shelf Life*, *21*, 100348. https://doi.org/10.1016/j.fpsl.2019.100348

Zhong, Q., Long, H., Hu, W., Shi, L., Zan, F., Xiao, M., . . . Chen, H. (2020). Dual-function antibacterial micelle via self-assembling block copolymers with various antibacterial nanoparticles. *ACS Omega*, *5*(15), 8523–8533. https://pubs.acs.org/doi/abs/10.1021/acsomega.9b04086

Zhou, F., Wang, D., Zhang, J., Li, J., Lai, D., Lin, S., & Hu, J. (2022). Preparation and characterization of biodegradable κ-carrageenan based anti-bacterial film functionalized with Wells-Dawson polyoxometalate. *Foods*, *11*(4), 586. https://doi.org/10.3390/foods11040586

Zhu, Z., Zhao, S., & Wang, C. (2022). Antibacterial, antifungal, antiviral, and antiparasitic activities of Peganum harmala and its ingredients: A review. *Molecules*, *27*(13), 4161. https://doi.org/10.3390/molecules27134161

14

Pullulan-Based Films and Coatings for Food Packaging

Applications and Challenges

Swati Mitharwal and Rajat Suhag

CONTENTS

14.1 Introduction

Traditional petrochemical polymer–based plastic packaging are non-renewable and non-biodegradable, which pose a serious threat of environmental pollution. This has created a demand for alternative packaging material made from natural edible biopolymers (polysaccharides, proteins, and lipid based), which are environment-friendly and biodegradable (Hassan et al., 2017). The major sources for production of these biopolymers include plants, animal tissues, and various microorganisms. Commonly used biopolymers include cellulose, xanthan gum, chitosan, gellan, sodium alginate, and pullulan (Tang et al., 2012). Polysaccharide-based biopolymers are preferred over others, owing to their higher stability, compatibility, and ease of modifying physicochemical properties (Vuddanda et al., 2017; Pan et al., 2014). Several microorganisms produce exopolysaccharides as amorphous slime attached to the surface of the cell or as an extracellular material (Sutherland, 1998). These exopolysaccharides are categorized as either homopolysaccharides (neutral glucans) or heteropolysaccharides (polyanionic). Microbial-origin polysaccharides are prospective substitute to plant polysaccharides, ascribing their distinct and superior physical characteristics (Singh et al., 2008).

Pullulan is an emerging microbial biopolymer biosynthesized by black yeast like the fungus *Aureobasidium pullulans* through submerged fermentation (Singh et al., 2009; Sugumaran et al., 2014). Pullulan is a linear, unbranched, extracellular exopolysaccharide mainly composed of maltotriose units attached together by linear (1→6) glycosidic linkage (Carolan et al., 1983; Sutherland, 1998). Hayashibara Company Limited began commercial manufacturing of pullulan in the year 1976 in Okayama, Japan, and continues to sell food-grade, deionized pullulan today (Farris et al., 2014). Pullulan is a non-toxic, non-carcinogenic, tasteless, odorless, non-hydroscopic edible polymer. It has wide industrial applications, including cosmetics, pharmaceuticals, agriculture, and food, because of its strong rheological and film-forming properties. The properties, yield, and cost of pullulan depend on the substrate, fermentation method, and fermentation conditions (Sugumaran & Ponnusami, 2017a; Alhaique et al., 2015). The cost of carbon, nitrogen, and other critical resources utilized in the

DOI: 10.1201/9781003303671-14

fermentation of pullulan is high. Thus, in order to bring down the cost, waste from agro-based indus-tries is also being explored as an alternate substrate for pullulan synthesis (Singh et al., 2019). In USA, pullulan is classified as generally recognized as safe (GRAS) status, and the FDA recommends a daily dosage of up to 10 g (Singh & Saini, 2008; Kimoto et al., 1997).

Pullulan is a thickening, stabilizing, binding, texturizing, and gelling agent used to improve sensory characteristics and lengthen food shelf life in the food sector (Singh et al., 2019). Pullulan has a good ability to form films and adhesiveness, because of which it has been used in edible food packaging. The chief methods applied to fabricate pullulan-based edible films or coating include (a) coating by dipping, brushing, or spraying; (b) extrusion; (c) solvent casting; and (d) electrospinning (Islam & Yeum, 2013; Kraśniewska et al., 2019). Pure pullulan films are colorless, are transparent, have no taste or odor, are heat stable, and have high oil and oxygen impermeability (Diab et al., 2001; Silva et al., 2018). The use of pure pullulan films as food packaging, on the other hand, has been limited, owing to the poor strength (breakability), brittleness, and hydrophilic nature of these films (Silva et al., 2018; Vuddanda et al., 2017). High cost and the absence of bioactive functions are other shortcomings of pullulan-based packaging (Tabasum et al., 2018). Thus, in order to obtain pullulan-based edible packaging material with improved mechanical and barrier properties and bioactivity, it co-blended with other functional polymers, bioac-tives, and nanofibers obtained from plant or microbial source (Synowiec et al., 2014; Pattanayaiying et al., 2015; Wu et al., 2013).

14.2 Structure of Pullulan

The chemical formula of pullulan as suggested by elemental analysis is $(C_6H_{10}O_5)_n$ (Singh et al., 2008). The structure of pullulan comprises of maltotriose units (linked to each other by α-1,4 glycosidic bond) linked together by α-1,6 glycosidic bonds, as explained by Bender et al. (1959) (Figure 14.1a). In another study, the enzyme pullulanase was discovered, which hydrolyses α-1,6 links in pullulan to produce maltotriose (Wallenfels et al., 1961).

Pullulan dimeric segments are made up of [→ x)-α-D-glucopyranosyl-(1→4)-α-D-glucopyranosyl-(1→] and [→4)-α-D-glucopyranosyl-(1→6)-α-D-glucopyranosyl-(1→], where *x* value is either 4 or 6 (for (1→ 4) linked segment). Based on previous research results, 1,4 and 1,6 linkages of pullulan are present in a fraction of 2:1 (Bouveng et al., 1962; Wallenfels et al., 1965). However, maltotetraose subunits are also present in the pullulan backbone, distributed randomly throughout the molecule to a maximum extent of approximately 7% (Carolan et al., 1983; Catley et al., 1986). Sometimes, other monomers, such as panose, isopanose, maltose, and isomaltose, are also produced during partial acid hydrolysis of pullulan (Bender et al., 1959; Sowa et al., 1963). As a result, pullulan is frequently referred to as a panose or iso-panose polymer (Singh et al., 2008) (Figures 14.1b and 14.1c).

Arnosti and Repeta (1995) described the structure of pullulan using NMR spectroscopy. They found that every third glucose ring in pullulan has α-(1→6)-linkage. Furthermore, the emergence of a band in the IR spectra at wavelength ($\lambda = 935$ cm^{-1}) indicates the coexistence of α-(1→ 4)-and α-(1→6)-links in the structure of pullulan (Petrov et al., 2002). Pullulan has a structure that is intermediate between that of amylose and dextran because it has both α-1,4 and α-1,6 links, whereas amylose solely has α-1,4 linkages (Farris et al., 2014). Pullulan linear chains have non-uniform segmental mobility, with higher mobility centered on α-1,6 linkages regions (Dais, 1995; Dais et al., 2001). The existence of both α-1,4 and α-1,6 linkages influences conformation of pullulan in aqueous medium, leading to random coiling of pullulan backbone in aqueous solution, resulting in bundled molecular arrangement (Nishinari et al., 1991; Farris et al., 2012).

14.3 Properties of Pullulan

Physicochemical properties of pullulan are presented in Table 14.1. Pullulan is white- to off-white-colored powder and has no taste or odor. When dissolved in water at a concentration of 5–10%, pul-lulan forms a viscous, non-hygroscopic solution (Prajapati et al., 2013). Pullulan at a concentration

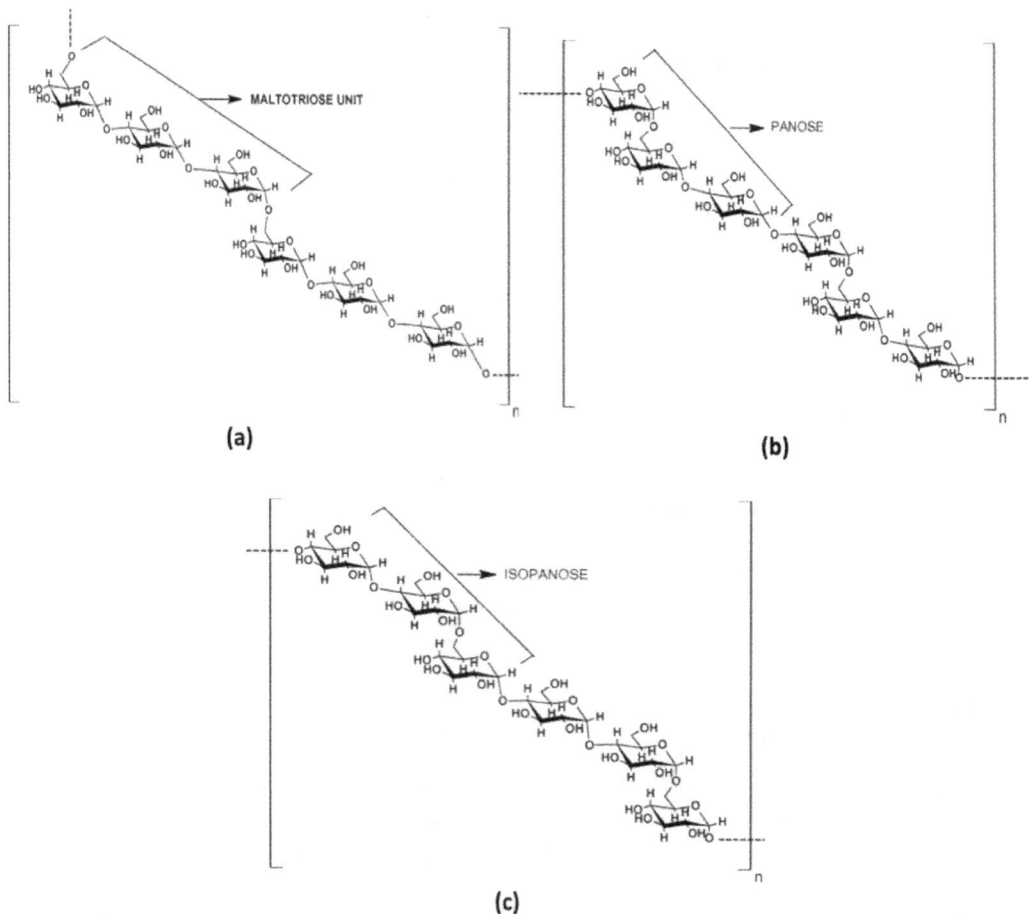

FIGURE 14.1 Structure of pullulan as repeating unit of (a) maltotriose, (b) panose, (c) isopanose.

Source: Reprinted from Singh et al. (2008) with permission from Elsevier.

of <10% forms a viscous solution when dissolved in water. Except for dimethyl sulfoxide and formamide, pullulan is insoluble in organic solvents (Tsujisaka & Mitsuhashi, 1993; Sugumaran & Ponnusami, 2017a). As compared to other polysaccharides, aqueous solutions of pullulan show relatively lower viscosity, because of which it can be used in beverages and sauces (Prajapati et al., 2013). Changes in pH, temperature, or salt concentration have no effect on the viscosity of pullulan solution (Hijiya & Shiosaka, 1975). Decomposition of pullulan starts at 250°C, and it chars at 280°C (Prajapati et al., 2013).

One of the unique characteristics of pullulan is its adhesiveness, and thus, it can be utilized as a binding agent to adhere nuts and edible seeds on the surface of crackers. It can also be used as a thickening agent and to improve the texture of products such as dressings and seasonings (Chaen, 2009; Singh et al., 2019). It is non-mutagenic, non-toxic, edible, and biodegradable in nature (Kimoto et al., 1997). It has film-forming ability—thin films which are transparent, impermeable to oxygen, and oil resistant; as a result, it is suitable for food preservation and packing (Yuen, 1974). Pullulan films' impermeability to oxygen prevents lipids and vitamins in meals from being oxidized. Pullulan films are ingestible and dissolve quickly in water and thus can be cooked along with food, as it disintegrates quickly in the mouth on consumption (Conca & Yang, 1993). Pullulan edible films have antimicrobial properties and help in retaining the moisture of food (Singh et al., 2019; Debeaufort et al., 1998).

TABLE 14.1

Physicochemical properties of pullulan

Parameter	Description
Color	White or off-white
Nature	Powder
Molecular weight	45–600 kDa
Melting point	250°C
Viscosity (10% (w/v) solution, 30°C)	100–180 cP
Solubility	Easily soluble in cold water, hot water, and dilute alkali
Insolubility	Insoluble in organic solvents except formamide and dimethyl sulfoxide
pH (1% (w/v) solution)	5–7
Mineral residue ash (sulfated)	≤3%
Polypeptides	≤0.5%
Moisture content (without tackiness or hygroscopy)	10–15%
Specific optical activity [α] D2O (1% (w/v) in water)	≥+160°

Source: Reprinted from Singh et al. (2019) with permission from Elsevier.

14.4 Production of Pullulan

Pullulan is manufactured as an exopolysaccharide by the fungus *Aureobasidium pullulans* from various sources of carbon (sucrose, mannose, maltose, galactose, maltose, xylose, and others) (Leathers, 2003; Madi et al., 1996). Other microorganisms capable of producing pullulan are presented in Table 14.2.

Pullulan is produced using a multistep biochemical procedure. The transformation of carbon (C) sources to uridine diphosphate glucose (UDPG), a key predecessor for the production of pullulan, is carried out by isomerase and hexokinase. α-phospho glucomutase improves biosynthesis of glucose-1-phosphate (G1P) from glucose-6-phosphate (G6P). UDPG-phyrophosphorylase converts G1P to UDPG (pullulan synthesis precursor). The generation of lipid-associated complexes by means of D-glucose residues is the next step. The last step comprises of a reaction between the generated residues and isomaltosyl to synthesize isopanosyl or pyranosyl subunits, finally polymerizing into pullulan chains with the help of the glucosyl transferase enzyme (Figure 14.2) (Sugumaran & Ponnusami, 2017a; Singh et al., 2017). Antecedents generated in the microbiological system speed up the production of pullulan. Pullulan production is aided by the aggregation of leftover carbohydrate during the early phases of fermentation, within the cell (Sugumaran & Ponnusami, 2017a).

TABLE 14.2

List of microbial sources of pullulan

S. no.	Microorganisms	References
1.	*Aureobasidium pullulans*	Leathers (2003)
2.	*Cytaria darwinii*	Oliva et al. (1986)
3.	*Cytaria harioti*	Oliva et al. (1986)
4.	*Cryphonectria parasitica*	Delben et al. (2006)
5.	*Rhodototula bacarum*	Chi and Zhao (2003)
6.	*Teloschistes flavicans*	Reis et al. (2002)
7.	*Tremella mesenterica*	Fraser and Jennings (1971)

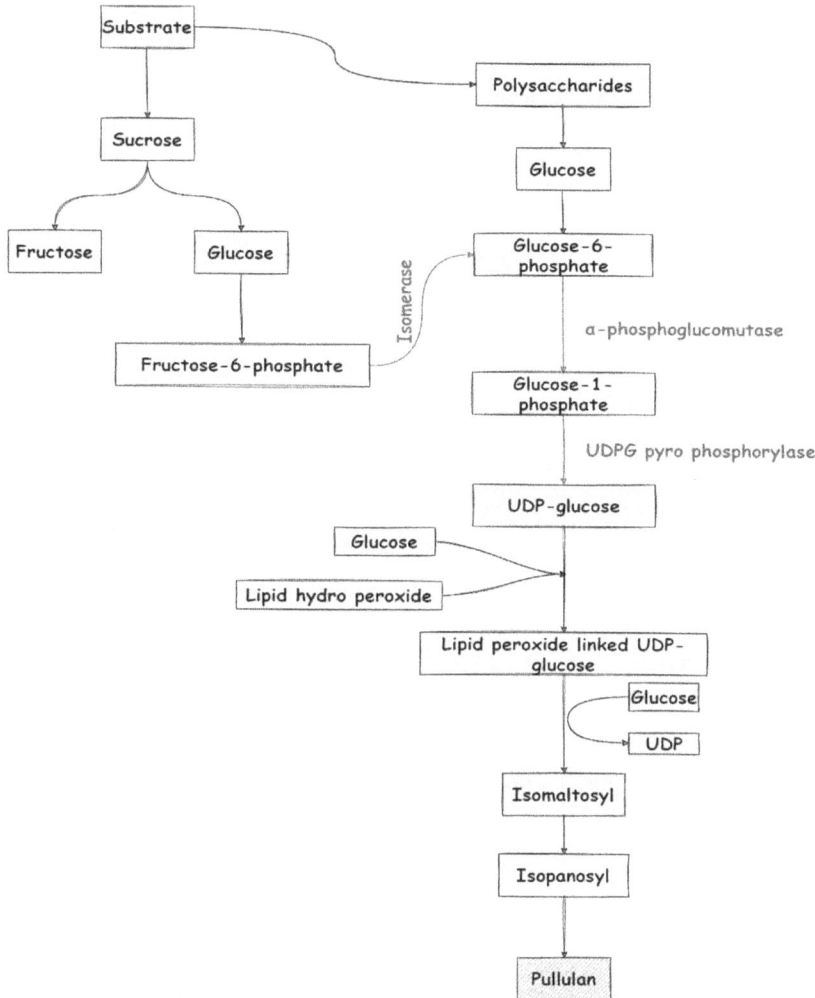

FIGURE 14.2 Pullulan production process.

Pullulan produced inside the cell wall passes across the β-glucan layer in order to get to the cell's surface. The released biopolymer forms a slimy coating on the cell surface that is weakly connected (Simon et al., 1993).

Despite numerous investigations on pullulan synthesis from *Aureobasidium pullulans*, the actual mechanism of pullulan synthesis is yet to be fully known, ascribing to the intricate cytological and physiological properties of this microbe (Cheng et al., 2011a; Singh et al., 2019). Hyashibara, a Japanese manufacturer, started to produce pullulan in 1972. Since then, pullulan has been the company's most important polysaccharide. At the industrial scale, pullulan is manufactured through fermentation by *Aureobasidium pullulans*. Fermentation is a biochemical process that occurs when a microorganism feeds on a specific substrate and yields a product (Wani et al., 2021). According to the class of substrate, pullulan is made using two different fermentation methods: (a) submerged fermentation and (b) solid-state fermentation. For pullulan manufacturing by *Aureobasidium pullulans*, submerged fermentation is commonly used (Singh et al., 2019). Several scientists have recently looked into the possibility of utilizing solid agricultural waste for pullulan synthesis using solid-state fermentation (Sugumaran et al., 2014; Sugumaran & Ponnusami, 2017b; Viveka et al., 2021). Pullulan has been manufactured using a synthetic medium and a variety of industrial waste by-products (Table 14.3). Pullulan made from artificial substrates, on the other hand, is more expensive than those made from

TABLE 14.3

Summary of various agro-industrial waste used for pullulan production

Substrate	Microbial strain	Reference
Beet molasses	*Aureobasidium pullulans* P56	Lazaridou et al. (2002)
Cassava bagasse	*Aureobasidium pullulans* MTCC 2670	Sugumaran et al. (2013b)
Corn steep liquor	*Aureobasidium pullulans* RBF 4A3	Sharma et al. (2013)
Grape skin pulp extract	*Aureobasidium pullulans* NRRLY-6220	Israilides et al. (1994)
Jackfruit seed	*Aureobasidium pullulans* NCIM 1049	Sugumaran et al. (2013b)
Palm kernel	*Aureobasidium pullulans* MTCC 2670	Sugumaran et al. (2013a)
Rice hull hydrolysate	*Aureobasidium pullulans* CCTCC M 2012259	Wang et al. (2014)
Soybean pomace	*Aureobasidium pullulans* HP-2001	Seo et al. (2004)

agricultural wastes, owing to the elevated price of artificial substrate. The kind of bioreactor used, fermentation technique, inoculant age, fermentation process parameters, production medium composition, and working environment all influence pullulan production (Cheng et al., 2011b; Gniewosz & Duszkiewicz-Reinhard, 2008). The characteristics of substrate have a significant impact on product development. Quality control during selection and processing of substrate (upstream process) is still critical to the fermentation process as well as the product recovery phases (downstream process) (Sugumaran & Ponnusami, 2017a).

Recently, Singh et al. (2022) comprehensively explained the downstream processing (Figure 14.3) and techniques to establish the structural confirmation of pullulan. Pullulan processing in the downstream should be efficient and repeatable. Because organic solvents are practically insoluble in pullulan, they are widely used for downstream processing. For pullulan recovery, the proportion of organic solvent to cell-free extract is important. When it comes to manufacturing pullulan, it differs among strains and is also reliant on the available amount of pullulan in the culture supernatant. The time it takes for pullulan to precipitate is influenced by the quantity of pullulan in the cell-free filtrate. Pullulan purity can be increased by repeating the precipitation process. Purification of pullulan can also be done using chromatographic techniques; however, this will increase the expense of the bioprocess. Pullulan has rarely been purified using an aqueous two-phase method. This technology has the potential to be a fast and automated way to recover pure pullulan. As a result, future study should concentrate on developing efficient and automated aqueous two-phase systems. Various advances in instrumental techniques, such as (a) Fourier-transform infrared spectroscopy (FTIR), (b) nuclear magnetic resonance (NMR), (c) high-performance liquid chromatography (HPLC), and (d) high-performance thin-layer chromatography (HPTLC), can all be used to confirm the structural properties of the purified pullulan. FTIR and NMR are two of the most common techniques for determining the structural properties of pullulan. A pullulan purification efficiency can be monitored using an inline liquid chromatography (LC) coupled with high-resolution NMR (LC-NMR) device (Singh et al., 2022).

Pullulan synthesis faces numerous challenges, including viscous fermentation broth, formation of melanin pigment, and product purification, because it entails cellular microbes that produce cellular fragments and residual components (Dailin et al., 2019). Superior pullulan synthesis processes and the discovery of new substrate substitutes emerge from process optimization. Despite the process's intricacy, the reactive spots in the pullulan structure increase the chances of chemical modifications, allowing pullulan to be used across a wide range of industrial settings (Hansen et al., 2015; Singh et al., 2017).

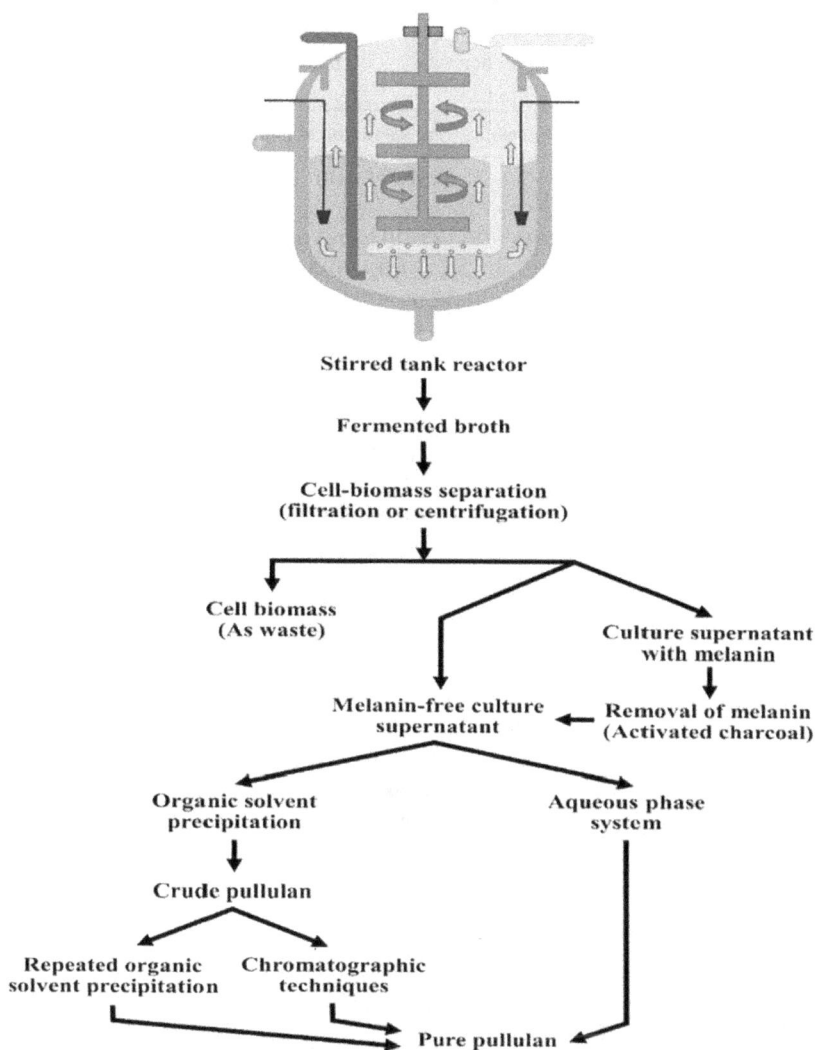

Stirred tank reactor

↓

Fermented broth

↓

Cell-biomass separation
(filtration or centrifugation)

Cell biomass
(As waste)

Culture supernatant
with melanin

↓

Melanin-free culture ← **Removal of melanin**
supernatant **(Activated charcoal)**

Organic solvent **Aqueous phase**
precipitation **system**

↓

Crude pullulan

Repeated organic **Chromatographic**
solvent precipitation **techniques**

Pure pullulan

FIGURE 14.3 Typical layout of pullulan downstream processing.

Source: Reprinted from Singh et al. (2022) with permission from Elsevier.

14.5 Modification of Pullulan

Pullulan can be chemically modified to obtain derivatives with enhanced activity and to broaden its range of applications. The various chemical derivatives of pullulan are presented in Table 14.4. Pullulan is made up of nine hydroxyl groups, each of which can be swapped with a different chemical group. The polarity of the solvent or reagents has an effect on the relativities of hydroxyl groups. Carboxymethylation and sulfation are most commonly used reactions done on pullulan to obtain derivatives with modified properties.

The carboxylation process involves activation of hydroxyl groups of pullulans to enhance chloride substitution (nucleophilic) from monochloroacetic acid as alcoholate in alkaline solution (in aqueous) (Prajapati et al., 2013). Bataille et al. (1997) prepared modified carboxymethyl pullulan by reacting C_{16} alkylamine on carboxylic groups of respective polyacid with degree of substitution (carboxylic groups) varying between 0.76 and 0.84. With increasing polymer concentration, there was a significant rise in

TABLE 14.4

Summary of common pullulan derivatives obtained through various
substitution reactions

Type of reaction	Schematic representation of pullulan derivatives (P-OH)
Etherification	P-O—CH$_3$ (permethylation)
	P-O—(CH$_2$)$_{2-3}$—CH$_3$ (alkylation)
	P-O—CH$_2$—COOH (carboxymethylation)
	P-O—(CH$_2$)$_{2-3}$—CH$_2$—NH$_3^+$ (cationization)
	P-O—CH$_2$—CH$_2$—CN (cyanoethylation)
	P-O—(CH$_2$)1–4-Cl (chloroalkylation)
	P-O—CH$_2$—CH$_2$—(S=O)—CH$_3$ (sulfinyl)
	P-O—CH$_2$—CH$_2$—CH$_2$—SO$_3$Na
	P-O—CH$_2$—CH$_2$—N(CH$_2$CH$_3$)
Esterification	P-O—CO—CH$_2$—CH$_2$—COOH (succinoylation)
	PA-O—CO—CH$_2$—CH$_2$—CO-sulfodimethoxine
	P-O —CO—CH$_2$—CH$_2$—CO-cholesterol
	P-abietate
	P-stearate
	PA-folate
	P-cinnamate
	P-biotin
Urethane derivatives	P-O—CO—NH—CH$_2$—CH(OH)—CH$_3$
	P-O— CO—NH—CH$_2$—CH$_2$—NH$_3$+
	P-O—CO—NH—R (R = phenyl or hexyl)
	P-O—CO—NH-phenyl
Chlorination	P-CH$_2$—Cl (C$_6$ substitution)
Sulfation	P-O—SO3Na
Azido-pullulan	P-CH2—N2
Oxidation	P-COOH (C$_6$ oxidation)
	Glycosidic ring opening (periodate oxidation)
CMP/hydrazone derivative	P-O—CH$_2$—CO—NH-doxorubicin
	P-O—CH2—CO—NH-antibody

Source: Reprinted from Prajapati et al. (2013) with permission from Elsevier.

low-shear viscosity of modified pullulan in 0.1 M NaCl solution. Further, the dilute solution of polymers showed compact globular structure owing to intermolecular aggregation of pullulan. In another study, Souguir et al. (2007) synthesized three alkylated derivatives of pullulans (cationic amphiphilic modified) by first computing the synthesis of diethyl aminoethyl pullulan (degree of substitution 0.80 in water) in anhydroglucose units, followed by application of Hoffmann alkylation reaction to add various alkyl chains, that is, C10, C12, and C16. All the three derivatives contained an amine function (pH dependent) and quaternary ammonium function (hydrophobic alkyl chains). The effect of chain length (alkyl group) on the solubilization and interfacial characteristics of three pullulan derivatives, (a) CMP$_{12}$C$_8$, (b) CMP$_7$C$_{14}$, and (c) CMP$_4$C$_{16}$, was investigated by Henni-Silhadi et al. (2008). Significant differences were observed in interfacial properties between CMP$_{12}$C$_8$ and CMP$_7$C$_{14}$, CMP$_4$C$_{16}$. Further, solubilization of pullulan derivatives increased with an increase in the length of the alkyl chains from 8C to 14C or 16C. Legros et al. (2008) investigated the physicochemical characteristics of amphiphile and cross-linked carboxymethyl pullulan in soluble state. They reported the presence of intermolecular and intramolecular hydrophobic associations in these derivatives along with other chemical links. Further, the nature of these interactions depends on the polymer content and type of solvent used.

Mähner et al. (2001) synthesized sulfated pullulan from pullulan (molar mass between 10^4–10^5 g/ mol) using SO$_3$-pyridine complex. Pullulan derivatives with degree of substitution = 2.0 and 1.4 were obtained. FTIR spectrum revealing peak at 1,260 cm^{-1} relates to the asymmetrical stretching of the S=O functional group, while peak at 820 cm^{-1} corresponds to symmetrical C-O-S group stretching vibration. The reactivity of C-atoms in derivatives as determined by ^{13}C NMR spectroscopy followed the order of C-4 < C-2 < C-3 < C-6. In another study, Alban et al. (2002) synthesized sulfated derivatives of pullulan using the same sulfating agent as used in a previous research by Mähner et al. (2001). Modified pullulan with degree of sulfation between 0.17 and 1.99 and molecular weight ranging between 15 kDa and 250 kDa was obtained. Further, the majority of the sulfate groups (>50%) were connected to carbon atom at 2nd, 3rd, and 4th position in the glucose unit.

14.6 Modification Methods for Pullulan-Based Packaging

The unique linkage ability of pullulan gives it structural flexibility, ability to form fiber, adhesive properties, solubility, and ability to form films (Singh et al., 2008). Native pullulan-based edible films are transparent, colorless, odorless and have low oil and oxygen permeability. However, they have several drawbacks, such as brittleness, hydrophilicity, and high cost (Silva et al., 2018). To develop pullulan-based packaging, pullulan is usually blended with other biopolymers, essential oils, bioactives, plus plant extracts to boost the functionality and properties of the packaging material. The several methods that used to advance the characteristics of pullulan as packaging material are presented in Table 14.5. Physical and chemical cross-linking of pullulan along with the addition of plasticizers such as sorbitol, glycerol, diethylene glycol, etc. have been previously studied (Diab et al., 2001; Gounga et al., 2008;

TABLE 14.5

Modification methods for improving properties of pullulan-based packaging materials

Method	Film matrix	Results	Reference
Blending	Pullulan + pectin	Improved tensile strength, hardness, and thickness in composite film. Decreased elongation at break.	Priyadarshi et al. (2021)
	Pullulan + carboxymethyl-gellan + glycerol	Improved tensile strength, increase in water vapor permeation, and reduction in elongation at break. Reduction in transparency of composite film.	Zhu et al. (2014)
Cross-linking	Pullulan + chitosan	Reduced permeability of water vapor. Elastic modulus and tensile strength of treated film improved, whereas elongation at break reduced.	Qin et al. (2020)
Green synthesis	Pullulan + silver nanoparticles	Improved antimicrobial activity against *Staphylococcus aureus* and *Listeria monocytogenes* in deli meat.	Khalaf et al. (2013)
Metal-based nanoparticles	Pullulan + nano-TiO$_2$	Improved water vapor and UV light barrier properties. Decreased elongation at break of modified film.	Liu et al. (2019)
	Pullulan + chitosan + ZnO	Increased tensile strength, elastic modulus, and decreased elongation at break in modified film. Enhanced antimicrobial activity toward *L. monocytogenes* and *E. coli*.	Roy et al. (2021)
Nanofibers	Pullulan + lysozyme nanofibers	Reduced tensile stress and elongation at break, as well as improved Young's modulus. Improvement in both antioxidant and antibacterial activity toward *Staphylococcus aureus*.	Silva et al. (2018)
	Pullulan + cellulose nanofibers	Increased thermal and mechanical properties (Young's modulus, tensile strength, and elongation at break) of composite film.	Trovatti et al. (2012)

Source: Adapted and modified from Ghosh et al. (2022).

Wang et al., 2019; Tong et al., 2013). Further, the blending of pullulan with other biopolymer such as pectin (Priyadarshi et al., 2021), gelatin (Zhang et al., 2013; Kowalczyk et al., 2020), chitosan (Li et al., 2020), graphene oxide (Unalan et al., 2015), carboxymethyl cellulose (Wu et al., 2013), nanofibrillated cellulose (Trovatti et al., 2012), essential oils and nanoparticles (Morsy et al., 2014), Sakacin A (Trinetta et al., 2010), casein (Tomasula et al., 2016), pea protein (Aguilar-Vázquez et al., 2018), etc. has been investigated.

Priyadarshi et al. (2021) developed an edible packaging film by blending pectin and pullulan in different ratios using glycerol (30% w/w) as plasticizer, using the solvent-casting method. The transparency and lightness of composite films increased while their water vapor permeation declined when supplemented with pullulan (30–70%). Further, the tensile strength (19.5 MPa) and elongation at break (1.8%) of pectin-based films increased with the addition of pullulan (tensile strength, 23.4–23.2 MPa; elongation at break, 2.1–2.9%). In another study, cross-linking of chitosan + pullulan nanofiber films was done by two green processes, that is, (a) heating (120°C for 2 hours) and (b) treatment with cinnamaldehyde (30 minutes) to augment the physicomechanical properties of films. The water vapor barrier characteristics of treated films increased, as shown by the decrease in water vapor permeation values with the rise in incorporation of chitosan. Further, cross-linking treatment and addition of chitosan significantly improved the mechanical properties (elastic modulus, tensile strength, and elongation at break) of developed films, owing to the intermolecular entanglements and cross-linking between pullulan and chitosan chains. FTIR spectra exhibited occurrence of Maillard reaction and Schiff base reaction during thermal and cinnamaldehyde cross-linking process, respectively. Incorporation of metal-based nanoparticles such as TiO_2 on the physicomechanical and optical properties of pullulan-based films was investigated by Liu et al. (2019). The outcomes of the study revealed that supplementing nano-TiO_2 (at 0.04 g addition level) to pullulan augmented the barrier quality (toward water vapors) by reducing the water vapor permissibility to 6.6×10^{-4} g m/m^{2}h KPa, owing to the creation of tortuous pathway, thereby lengthening the time it takes for water vapor molecules to travel through pullulan. Further, tensile strength of pullulan films containing nano-TiO_2 was higher than prepared from pullulan alone, while elongation at break followed an opposite trend, which indicates the enhanced cohesive force and reduced flexibility of films upon addition of nano-TiO_2.

Khalaf et al. (2013) reported the antimicrobial potential of essential oils (oregano or rosemary oil) or metal nanoparticles (Ag or ZnO) in pullulan-based edible films against two pathogens, *Listeria monocytogenes* and *Staphylococcus*. The potential application of nanofibers such as lysozyme to enhance the functionality and characteristics of edible films synthesized from pullulan was explored by Silva et al. (2018). The inclusion of lysozyme nanofibers had a minor effect on the produced films' transmittance (optical transparency), but it enhanced the Young's modulus and elongation at break, which indicate the increased stiffness and rigidity of the film. The addition of lysozyme nanofibers (5–15%) resulted in 0.72 to 3.2 log reduction in the growth of *S. aureus*, conferring antibacterial activity to the developed film.

14.7 Application of Pullulan in Edible Films

The term "edible films and coatings" refers to thin layers of food-grade substances possessing barrier properties (water vapor, lipid, and gases) that are used to protect food or used as carrier medium to deliver active compounds, with the ultimate aim of extending food's shelf life (Yousuf et al., 2018). In spite of the regular functions, edible films and coatings also have certain additional requirements with respect to food applications, such as cost-effectiveness, easy production, safe consumption, good mechanical strength and adhesive property, acceptable sensory attributes, good barrier properties toward oil, water, or gases, and effective carrier for antimicrobial and bioactive compounds (Diab et al., 2001). Solvent casting and compression molding (extrusion) are the two most widely utilized methods for making edible films. Solvent casting technique involves spreading of edible biopolymer solution on a suitable base material, followed by drying the material for an appropriate time period, resulting in the development of film (Dhumal & Sarkar, 2018). However, wrinkling and tearing of film during peeling operations are the main associated drawbacks, which can be reduced by using the right base material and keeping the moisture level of the film at the right level (5–8%) (Tharanathan, 2003). Pullulan is commonly utilized in the production of oral dental strips that disintegrate quickly (<60 seconds) once placed on the tongue (Chatap et al., 2013). It is used for the incorporation of flavor compound, active substances, and colorant or to cover off the taste or odor of the active compound present in a pill or capsule. In recent years, pullulan's potential use in edible films and coatings for food products has been investigated by researchers (Kraśniewska et al., 2019).

The application of pullulan for the development of edible films is presented in Table 14.6. Zhu et al. (2014) investigated the outcome of blending carboxymethyl–gellan and pullulan in different ratios (4:0, 3:1, 2:2, 1:3, 0:4) and their effect on the physicomechanical and barrier characteristics of developed films. FTIR spectra of composite films show a shift in peak positions, signifying the chemical interaction between pullulan and carboxymethyl-gellan. SEM analysis results exhibited rougher matrix with the increase in addition level of carboxymethyl-gellan. The transparency of film did not differ significantly with the incorporation of pullulan to the blend. The addition of pullulan downgraded the tensile strength of palatable films, which can be ascribed to intermolecular hydrogen bonding among pullulan's -OH group to that of -COO group of carboxymethyl-gellan. Further, elongation at break values increased,

TABLE 14.6

Physical, mechanical, and barrier properties of pullulan-based edible films

Film matrix	Properties	Conclusions	Reference
Pullulan + carboxymethyl-gellan (*Plasticizer*: glycerol)	*Thickness:* 0.16 ± 0.01–0.21 ± 0.03 mm *Tensile strength:* 10.60 ± 2.43–27.65 ± 2.06 MPa *Elongation at break:* 9.80 ± 1.20–33.50 ± 1.82 *Water vapor permeability:* 1.34 ± 0.03–$2.64\pm0.05 \times 10^{-6}$ g m/Pa h m^2 *L*:* 90.54 ± 0.32–94.21 ± 0.02 *a*:* -0.94 ± 0.007–1.36 ± 0.014 *b*:* 0.68 ± 0.09–9.65 ± 0.12 *Transparency:* 0.14 ± 0.01–0.51 ± 0.01	• Carboxymethyl-gellan and pullulan have good miscibility, as confirmed by FTIR, TGA, and XRD results. • Adding Pu to CMGe increased the film's barrier characteristics (improved the elongation at break, decreased the tensile strength and values of water vapor permeability).	Zhu et al. (2014)
Pullulan + whey protein isolate (*Plasticizer*: glycerol)	*Transparency:* 3.41–7.42 *Solubility:* 50.58 ± 4.8–$97.98\pm2.2\%$ *Oxygen permeability:* 1.043 ± 0.03 cm^3 mm/m^2 d kPa. *Water vapor permeability:* 30.14 ± 3.5 g mm/m^2 d kPa.	• The whey protein isolate + pullulan composite film had better barrier properties, solubility, and transmittance.	Gounga et al. (2008)
Pullulan + starch (potato/tapioca/corn) + probiotic (*Lactobacillus plantarum* GG ATCC 53103, *Lactobacillus reuteri* ATCC 55730, and *Lactobacillus acidophilus* DSM 20079)	*Thickness:* 0.030 ± 0.008–0.105 ± 0.008 mm *Transparency:* 0.90 ± 0.106–6.48 ± 0.081 *Water solubility:* 52.67 ± 4.3–$79.43\pm1.8\%$ *Water vapor permeability:* 0.142 ± 0.010–0.681 ± 0.012 g mm m^{-2} d^{-1} kPa^{-1}	• Upon storage for a period of two months at 4°C temperature, pullulan and pullulan/potato starch films were found to be the best carriers for probiotics, with relative cell viability between 70 and 80%. • The increase in addition level of starch in pullulan films resulted in decrease in water vapor permeability and water absorption rate.	Kanmani and Lim (2013)
Pullulan + essential oil (oregano/rosemary) or ZnO/Ag nanoparticles *Plasticizer:* glycerin, loctus bean + Tween 20 + lecithin	The edible films were found to exhibit antimicrobial activity.	• The composite film (pullulan-glycerol-xanthan gum: 50:15:1) with added essential oils (2%) and nanoparticles (ZnO 110 nm or Ag 100 nm) was found to be active against *S. aureus*, *L. monocytogenes*, *E. coli* O157:H7, and *S. Typhimurium*.	Morsy et al. (2014)

(Continued)

TABLE 14.6 *(Continued)*

Film matrix	Properties	Conclusions	Reference
Carboxymethyl chitosan + pullulan + galangal essential oil	*Thickness:* 0.105–0.152 mm *Water solubility:* 7.87 × 10⁻⁵–1.03 × 10⁻⁶ g/s *Tensile strength:* 15.32–22.39 MPa	• Carboxymethyl chitosan blended with pullulan in 2.5:2:5 ratio resulted in films with right physicomechanical and barrier characteristics. • The composite blend showed good compatibility, and developed films were thermally stable. • The composite film (carboxymethyl chitosan-pullulan-essential oil) in ratio of 2.5:2.5:8 was found to be effective in preserving mango fruit stored at 25°C for 15 days.	Zhou et al. (2021)
Pullulan + gelatin + potassium sorbate *Plasticizer:* glycerol	*Thickness:* 80.67±2.80–138.80±12.89 µm *Water vapor transmission rate:* 2186.44±103.70–2814.43±135.89 g mm⁻¹ d⁻¹ *Water vapor permeability:* 41.60±0.07–86.79±4.65 g mm m⁻² d⁻¹ kPa⁻¹ *Tensile strength:* 2.54±0.73–61.37±4.11 MPa *Elongation:* 6.14±1.90–430.97±42.65%	• The mechanical properties of the composite film improved with the incorporation level of gelatin. • Pullulan + gelatin blended in the ratio of 50:50 resulted in films with highest opacity.	Kowalczyk et al. (2020)
Pullulan + egg white *Plasticizer:* glycerin	*L*:* 95.21±0.74–95.48±0.18 *a*:* 3.93±0.11–4.66±0.01 *b*:* −1.73±0.04–1.35±0.4 *Tensile strength:* 60.65–329.48 MPa *Elongation at break:* 1.43–10.33%	• Composite films' mechanical characteristics first improved, then deteriorated on increasing the egg white ratio. • XRD and FTIR results demonstrated that pullulan and egg white interacted well.	Han et al. (2020)
Pullulan + lactoferrin	*Tensile strength:* 60.17–20.11 MPa *Elongation at break:* 8.33–2.27% *Water vapor permeability:* 2.43 × 10⁻²–1.83 × 10⁻² g m/Pa h m²	• Addition of lactoferrin to pullulan decreased the mechanical properties of composite films, while barrier and permeation characteristics were improved.	Zhao et al. (2019)
Pullulan + sodium alginate	*Tensile strength:* ~58–97 MPa *Elastic modulus:* ~2200–5100 MPa *Elongation at break:* ~3–14%	• With the incorporation of alginate to pullulan, the mechanical characteristics of developed films increased at low water activity, while it followed a reverse trend at higher water activity level.	Xiao et al. (2012)

indicating that the incorporation of pullulan resulted in the modification of film flexibility. The barrier and permeation characteristics (water vapor) of composite films were higher as compared to films prepared from carboxymethyl-gellan alone. In another study, Gounga et al. (2008) developed various palatable films from whey protein isolates (WPI) and pullulan mixed in different proportions. The films prepared from WPI and pullulan (1:1 ratio) had the highest oxygen and water vapor permeation value of 1.043 cm³ mm/m² d kPa and 42.38 mm/m² d kPa, respectively. Further, the solubility of whey protein isolate-pullulan films (59.11–97.98%) was higher as compared to those prepared from whey protein isolate alone (50.58%), owing to the hygroscopic and water-soluble nature of pullulan. SEM images depicted the presence of several pinholes on composite films.

The effect of potassium sorbate (0.5–2%) and pullulan (25–100%) on the physicochemical characteristics of composite films prepared from pullulan and gelatin was examined by Kowalczyk et al. (2020). The XRD results depicted that films with added pullulan were more amorphous, which further converted to more semi-crystalline-like, owing to recrystallization of potassium sorbate phase. Further, the blending of pullulan and gelatin resulted in uneven films which were characterized by microparticles of different size due to the encapsulation of potassium sorbate in the microspheres. The composite films were thicker and yellow in color as compared to those prepared from gelatin or pullulan alone. Tensile strength and elastic modulus of films decreased with the incorporation level of pullulan from 13.46 MPa and 252.07 MPa to 2.54 MPa and 12.27 MPa, respectively. Further, elongation at break values increased from 65.64% (100% gelatin + 2% potassium sorbate) to 430.97% (100% pullulan + 2% potassium sorbate). In another study, Zhao et al. (2019) prepared edible films by blending pullulan with lactoferrin and characterized the developed films. Increase in lactoferrin concentration significantly decreased the elongation at break as well as the tensile strength of the film, which can be ascribed to aggregation of lactoferrin particles negatively affecting the film consistency. Further, water vapor permeability significantly increased from 2.31×10^{-12} g m/Pa h m^2 (100% pullulan film) to 2.43×10^{-12} g m/Pa h m^2 (0.6% lactoferrin concentration), which can be due to the presence of large amounts of lactoferrin which increased the free volume in film matrix. The SEM analysis results revealed that the composite films morphology consisted of phase-separated spheroidal domains. Xiao et al. (2012) investigated pullulan- and sodium alginate–based composite film for their mechanical properties as a function of water activity. The results revealed that at low water activity level, the addition of alginate to pullulan increased the tensile strength of the composite film; a decrease in elongation at break values was observed, while at higher water activity level, the opposite trend was observed. At water activity level above 0.43, plasticization effect on pullulan-alginate blends was observed. Zhou et al. (2021) worked on developing active composite film (pullulan + carboxymethyl chitosan) incorporated with essential oil (galangal). The films prepared by blending carboxymethyl chitosan and pullulan in 2.5:2.5 ratio resulted in films with appropriate mechanical (tensile strength, elongation at break) and barrier properties (water vapor permeability and oxygen barrier properties). Based on the weight loss, firmness, titratable acidity (TA) values data, the blend containing 8% galangal essential oil was found to be effective in mango preservation stored at 25°C for 15 days.

14.8 Application of Pullulan in Edible Coatings

Edible coating is formed by direct application of coating solution to the food surface that comes into contact with the external environment. Edible coating helps protect the food product against microbes and lipid oxidation, prevent moisture loss, and enhance shelf life. Edible coating can be applied on foods employing several methods, like dipping, spraying, panning, and fluidized-bed coating (Suhag et al., 2020). Pullulan has shown to have significant potential to be used as a coating biopolymer with the purpose of preserving the quality and storage life of the food product. Environmental factors like moisture and temperature influence the characteristics of pullulan-based edible coating; therefore, it is blended with additional biopolymers, extracts, and essential oils to impart protecting characteristics. Chemical alteration of pullulan is one more technique to enhance its operational qualities (Hezarkhani & Yilmaz, 2019). Strawberry fruits were preserved by using pullulan acyl esters substituted with various anhydrides to delay maturation and senescence. Furthermore, derivatives of pullulan have been synthesized through processes like sulfation, esterification, oxidation, etherification, amidification, and other methods, with pullulan butylate exhibiting greater permeation properties (oxygen and water vapor) in comparison to natural pullulan (Niu et al., 2019). Pullulan's n-octenyl succinylation greatly improved the water vapor barrier and coating solution applied on sapota fruits produced using esterified pullulan and aided in delaying maturation and senescence (Shah et al., 2016). To date, findings on the biochemical alteration of pullulan have revealed considerable developments in the quality of biopolymer-based food packaging. Nevertheless, there is no consensus on how to keep pullulan's edible quality following transformation. Researchers suggest using coating solution produced from esterified pullulan for coating fruits; however, until more research is done, various sorts of chemical alterations could be employed in non-edible packing films (Ghosh et al., 2022).

Table 14.7 describes the studies involving pullulan-based edible coating on food products. In a recent work, Zhou et al. (2021) applied carboxymethyl chitosan-pullulan composite edible coating supplemented by different concentrations of galangal essential oil on mango. Findings showed that coating enriched with 8% galangal essential oil maintained the highest firmness value (3.82 ± 0.76 N), TA ($0.185 \pm 0.07\%$),

TABLE 14.7

Pullulan-based edible coatings on food products

Coating	Food product	Findings	Reference
Esterified pullulan + n-octenyl succinic anhydride	Sapota (*Manilkara zapota*)	• Delay in fruit ripening and senescence process.	Shah et al. (2016)
Carboxymethyl chitosan–pullulan coating + galangal essential oil (0%, 4%, 8%, and 12%)	Mango	• Reduction in sample weight loss. • Maintained TSS, TA, and firmness.	Zhou et al. (2021)
Pullulan coating incorporated with water and ethanolic *Satureja hortensis* extracts	Pepper and apple	• Inhibitory properties toward proliferation of Gram-positive and Gram-negative bacteria and *Penicillium expansum*.	Kraśniewska et al. (2014)
Chitosan-pullulan	Pineapple	• Effective in controlling growth of *Listeria monocytogenes* and *Salmonella Typhi*.	Treviño-Garza et al. (2017)
Pectin-pullulan coating enriched with grape seed extract	Peanuts	• 75% reduction in • the peroxide values of oil obtained from coated peanuts.	Priyadarshi et al. (2022)
Pullulan + sweet basil extract	Apples	• Exhibited good antifungal properties toward *Rhizopus arrhizus* and effective against mesophilic bacteria.	Synowiec et al. (2014)
Chitosan incorporated with 1% lemon peel polyphenols	Raw poultry meat	• Improved storage life and sensory quality.	Maru et al. (2021)
Multilayer coating of pullulan-chitosan	Papaya	• Effective in preserving papaya • flavor and overall acceptance. • Best performance was given by four-layer coating.	Zhang et al. (2019)
Pullulan + 1% calcium chloride + 2% lemon juice	Bananas	• Prolonged shelf life of non-climacteric (banana) fruits.	Ganduri (2020)
Whey protein isolate—pullulan coating	Chinese chestnut	• Reduced moisture loss and decay incidences.	Gounga et al. (2008)
Pullulan	Strawberries	• Delayed mold development. • Extended shelf life.	Eroglu et al. (2014)
Pullulan coating	Blueberries	• Retention of L-ascorbic acid. • Reduced in loss of blueberries weight and variations in reducing sugars.	Kraśniewska et al. (2017)
Pullulan (1%) + glutathione (0.8%) + chitooligosaccharides (1%)	Fuji apple slices	• Inhibited enzymatic browning. • Inhibited microbial growth. • Extended shelf life.	Wu and Chen (2013)
Pullulan + sodium carboxymethylcellulose + gallic acid (GA) and/or ε-polylysine hydrochloride (εPH) based edible coating	Sea bass fillets	• Preserved quality of sea bass fillets. • Increased shelf life by 10 days.	Li et al. (2021)
Pullulan and/or nisin	Eggs	• Maintained internal quality of eggs. • Combination of nisin with pullulan was effective against microbes.	Morsy et al. (2015)
Pullulan and cinnamon essential oil nanoemulsion	Strawberries	• Cinnamon essential oil provided antimicrobial effect against bacteria and molds.	Chu et al. (2020)

and total soluble solids (TSS) upon storing for 15 days. This was observed due to the excellent anti-microbial and oxygen barrier characteristics of the galangal essential oil–enriched coating. Similarly, Kraśniewska et al. (2014) applied pullulan-based coating enriched with aqueous and ethanolic extracts of *Satureja hortensis* on sweet peppers (*Capsicum annuum* L.) and apples (*Malus domestica*) cv. Jonagored. Pullulan coating containing aqueous extracts was found to have a strong inhibitory effect against Gram-positive, Gram-negative, and *Penicillium expansum* in comparison to pullulan coating with ethanolic extracts. Furthermore, pullulan coating prevented excessive surface wilting or wrinkling and reduced weight loss, thereby improving the storage life of sweet pepper and apples.

Fresh-cut pineapple can be preserved and extended using a layer-by-layer (LbL) treatment using chitosan-pullulan-based edible coating (Treviño-Garza et al., 2017). Edible coating developed from chitosan and pullulan was found to be effective against *Listeria monocytogenes* and *Salmonella Typhi*. Further, the usage of palatable coating prolonged the storage life up to 14 days when kept at room temperature (Treviño-Garza et al., 2017). Additional work by Zhang et al. (2019) applied pullulan-chitosan-based coating using multilayer LbL (two, four, and six layers) technique on papaya. On the 14th day, papayas coated with 2, 4, and 6 layers (15.53%, 10.41%, and 13.85%, respectively) lost less weight than control papayas (28.86%). Uncoated papayas were entirely softened on day 14 and could not be stored or transported, but coated samples had firmness values of 15.6 N (two-layer coating), 20.2 N (four-layer coating), and 16.9 N (six-layer coating). Multilayered coats preserved the flavor and economic value of papayas for longer durations, according to sensory evaluation. This was attributed to the O_2 and moisture barrier, as well as a drop in the activity of respiratory enzymes. Multilayer coatings of pullulan and chitosan might therefore be employed as a new fruit preservation technique. Application of four layers of coating solution provided the best performance among multilayer coatings.

Recently, pectin-pullulan blend-based coating enriched with extracts of grape seed was put on raw and roasted peanuts (*Arachis hypogaea*). Over the course of 30 days, the antioxidant-rich coating reduced lipid oxidation by 75%, extending the storage life of peanuts kept in an ambient environment. Pullulan's oxygen barrier property and activity of antioxidants-rich grape seed extract played a crucial influence (Priyadarshi et al., 2022). Synowiec et al. (2014) studied the effectiveness of pullulan coating supplemented with sweet basil (extract) on apples. The application of the coating comprising sweet basil extract at a concentration level of 24 mg/cm^2 on apple surface exhibited lower antibacterial activity toward mesophilic bacteria, but significant antifungal activity was observed against *Rhizopus arrhizus*. This coating also resulted in less weight loss and less color and soluble solids changes in the fruits upon storing. Overall, the apple samples coated with pullulan + sweet basil extract had improved preference parameters (Synowiec et al., 2014).

Ganduri (2020) optimized conditions for coating *Rastali* and *Chakkarakeli* bananas with coating solution prepared using pullulan and 1% w/v and 2% v/v calcium chloride and lemon juice, respectively. The results showed that an optimized coating blend with dipping period (10 minutes), temperature (60°C), and pullulan (10% level) resulted in the lowest loss in fruit weight (5.466%). This novel coating preparation had low (64.92) color saturation, a 15% lower peel–pulp ratio, a 19% lower vitamin C content, a 55% higher fruit hardness, a low (212.17) browning index, and a 12–13% higher total and residual sugar content. Gounga et al. (2008) applied WPI-pullulan-based edible coating on Chinese chestnut. SEM images revealed that WPI-pullulan composite edible coating covered the entire surface of the chestnut with excellent adhesion and integrity. Whey protein isolate-pullulan coating had a modest but substantial effect in reducing moisture loss and degradation in chestnuts, delaying exterior color changes.

Maru et al. (2021) applied a pullulan-based active coating incorporated with 1% lemon peel polyphenols on raw poultry meat using the dipping method. Application of coating increased the bacterial log phase and significantly reduced the lipid peroxidation, resulting in the extension of shelf life by a period of six days, whereas no considerable alterations were noted in color, pH, and weight loss and coated raw meat samples had acceptable sensory parameters during the storage period. Kraśniewska et al. (2017) documented that coated blueberries (pullulan-based coatings) were better shielded against undue skin withering and wrinkling. Pullulan-based coating on blueberries was also found to minimize drying and wilting, especially at higher temperatures, allowing it to retain its freshness and appeal for longer. The use of pullulan to coat blueberry fruit resulted in less weight and sugar content loss, which was most likely due to a slower rate of respiration and transpiration. In addition, the coated fruits had a better retention of L-ascorbic acid. Changes in the quantity of bioactives such as polyphenols and anthocyanin (the substances found in the maximum concentrations in the fruit's skin) were then linked to weight loss in the control samples. Uncoated blueberries had a higher phenolic content due to the larger weight loss. Furthermore, the pullulan coating considerably minimized the changes associated with blueberry rotting. Application of coating marked drop in weight loss rate in tomatoes and helped maintaining TSS, pH, acidity, and color. Furthermore, higher content of phenols and flavonoids, as well as higher antioxidant activity, was reported in coated samples compared to control during the storage period. Applying coating improved the storage life of tomatoes by nine days while maintaining the sensory parameters.

Wu and Chen (2013) applied pullulan coating containing anti-browning and antibacterial agents (0.8% glutathione and 1% chitooligosaccharides) on Fuji apple slices. During cold storage, compared to the control, pullulan coating applications substantially delayed enzymatic browning, reduced weight loss, preserved firmness, and suppressed microbiological development and rate of respiration in apple slices. According to the findings, pullulan-based coatings containing glutathione and chitooligosaccharides appears to be a potential strategy for prolonging the storage life of sliced apples. Li et al. (2021) applied pullulan and sodium carboxymethylcellulose-based edible coating containing two additives, that is, (a) gallic acid (GA) and (b) ε-polylysine hydrochloride (εPH), on fish fillets. Findings revealed that applying edible coating enriched with GA and/or εPH inhibited the increase of thiobarbituric acid, total viable counts, pH, as well as retained the immobile water content, capacity to hold water, and sensory attributes of sea bass fish samples. Edible coating containing both GA and εPH was found to be more effective than when used individually. This indicates the synergistic effect of GA and εPH in improving the preserving capacity of pullulan and sodium carboxymethylcellulose–based edible coating. Morsy et al. (2015) used pullulan- and/or nisin-based coating to augment the internal attributes of eggs. Incorporating pullulan and/or nisin resulted in weight loss reduction while the quality of both egg yolk and albumen was retained for three weeks for eggs stored at 25°C temperature and four weeks for those stored at 4°C as compared to control eggs (without coating). Furthermore, a combination of nisin with pullulan was also helpful in reducing the microbial load throughout the storage period.

Strawberries were coated using a pullulan-based emulsion (Eroglu et al., 2014). During storage, the physicochemical characteristics of coated strawberries were more persistent than the untreated control. In comparison to untreated strawberries, the coated strawberries' firmness and mold formation behavior were found to be superior. Coating strawberries with a pullulan suspension (10% concentration) was recommended for a lengthier shelf life from a practical standpoint. Similarly, Chu et al. (2020) coated strawberries using a nanoemulsion prepared using pullulan and cinnamon essential oil. The outcome indicated that edible film blend significantly dropped the loss of firmness, fruit weight, TA, and TSS after six days of storage. Furthermore, when contrasted to the control samples (uncoated) and those coated with pullulan alone, the strawberry samples coated with pullulan + cinnamon essential oil composite blend had the greatest antibacterial efficacy against bacteria (2.544 log CFU/g) and molds (1.958 log CFU/g). The antibacterial effect was attributed to nanosized particles of cinnamon essential oil distributed throughout the pullulan base.

14.9 Other Applications in Food Packaging

Owing to the shortcomings associated with using edible films and coatings created from pullulan alone, in recent years, investigators have been exploring applicability of pullulan as coatings on plastic packaging material. Farris et al. (2011) examined the influence of application of biopolymers (high methoxyl pectin, amidated pectin, pig skin gelatin, shellfish chitosan, and pullulan) as coating material on surface properties (wettability) of polyethylene terephthalate (PET) plastic films (12 ± 0.5 μm). Pullulan surface was found to exhibit highest hydrophilicity among the studied biopolymer samples with lowermost contact angle of 30° and 23° after 0 s and 60 s of droplet deposition, respectively. Further, surface energy analysis results indicated the resemblance between synthetic polymers (polyethyl oxide and polyvinyl alcohol) and pullulan owing to strong electron donor behavior of pullulan. In another work, Introzzi et al. (2012) assessed antifog properties of pullulan coating on low-density polyethylene (LDPE) film substrate as compared to two commercial antifog packaging films. The results revealed that pullulan-coated LDPE films showed no water droplet or stain formation upon refrigeration, but rather, a thin continuous layer of water on the surface which in turn enhanced the transparency of the film, substantiating superior antifog properties of pullulan as compared to commercial antifog films. This wetting-enhancing property of pullulan coating can be ascribed to its lower water contact angle (~24°) and hydrophilic nature of pullulan.

The application of pullulan in combination with tertraethoxysilane was investigated by Farris et al. (2012) with an aim of improving the oxygen barrier properties of PET. Significant decrease in oxygen transmission rate of plastic from 165.20 mL m^{-2} (24 h^{-1}) in control sample to 4.92 mL m^{-2} (24 h^{-1}) and 0.96–82.30 mL m^{-2} (24 h^{-1}) was observed upon coating with pullulan alone or in combination with tertraethoxysilane, respectively, at 0% relative humidity. Significant decrease in oxygen permeability of microfibrillated cellulose-based films from 71.03 mL μm m^{-2} (24 h^{-1}) atm^{-1} to 51.79–34.19 mL μm m^{-2} (24 h^{-1}) atm^{-1} was observed upon addition of pullulan at 0.1–1% level and 0% relative humidity (Cozzolino et al., 2014). Unalan et al. (2015) worked on improving the oxygen barrier properties of edible films made from pullulan by incorporating graphene oxide (2D) nanoplatelets. The oxygen transmission rate of 100% pullulan-based film was 6,337 mL μm m^{-2} (24 h^{-1}) atm^{-1}, which decreased to 1,357 mL μm m^{-2} (24 h^{-1}) atm^{-1} for graphene oxide–added pullulan film (at 0.3% level), which was significantly lower for commodity plastic films. In conclusion, pullulan can be used as a coating material on the commercial plastic-based food packaging to improve the oxygen barrier characteristics of the developed material.

14.10 Future Trends, Opportunities, and Challenges

Pullulan has not been fully utilized despite its numerous intriguing features. Moreover, there are presently only a small number of industrial manufactured goods on the market, with the majority of recommended uses being at the lab or pilot scale. The fundamental cause of such business stagnation is the high cost. Pullulan costs between 25 and 30 USD/kg (Farris et al., 2012), which is much more than other biopolymers derived from polysaccharides and proteins. The higher cost of pullulan is primarily due to its manufacturing. Carbohydrates are used as carbon sources in the commercial bioprocess for the production of pullulan because they enable for high productivity levels and outputs (Singh & Saini, 2008). However, the selling prices for sugar and starch, which are the most often used substrate for the manufacture of many microbial-origin polysaccharides, range around 350–518 USD/ton and 279–310 USD/ton, respectively (Freitas et al., 2014). Because the substrate cost alone (accounting for 40% of production cost) is the highest cost associated with production of any microbial-origin polymer, it seems evident that using low-cost raw materials like wastes and by-products is a good place to start. Despite the fact that considerable efforts are now being taken in this direction, several roadblocks appear to make the road difficult, like the difficulty of ensuring their supply in quantitative and qualitative terms, the diverse nutritional profile of the resources, the occurrence of impurities in the substrates, and the formation of unreacted constituents in the fermentation broth (Farris et al., 2014). Additional factor in pullulan's high pricing is the eradication of the melanin pigment during the fermentation broth's downstream processing. Activated charcoal is commonly used to remove melanin, followed by pullulan purification. Along with innovative alternative fermentation strategies, the selection of strains with higher yield and which are free from melanin could thus constitute another significant strategy for expanding pullulan uses (Choudhury et al., 2012; Gniewosz & Duszkiewicz-Reinhard, 2008).

Various technological concerns must be tackled in the next years to set off new market prospects, particularly in the food packaging industry. Several uses are hampered by drawbacks related to pullulan's strong affinity for water molecules. Although pullulan's hydrophilic characteristic can be advantageous in a variety of applications, it is also the cause of its extreme sensitivity to damp surroundings. If protecting layers from other sources such as polyolefin films are also used along with it, intrinsic or improved qualities such as oxygen barrier properties are not impaired; this would certainly preclude it from being used in conventional packaging. Pullulan's neutrality is another factor that may rule it out for some food packaging applications. Because it has no charge on its backbone, it cannot form ionic connections with oppositely charged molecules, which is necessary for the formation of some complex configurations and release systems (Cozzolino et al., 2013; Trinetta & Cutter, 2016). Chemical modification is a valuable way for imparting new functions to pullulan-based films/coatings among numerous strategies (e.g., mixing, nanotechnology). Functionality of pullulan can be modified by addition of new chemical functional groups onto its -OH group, giving it previously unattainable properties. This type of chemical alteration has the advantage of allowing for the creation of custom structures for each application. The most important chemical modifications in the food packaging field seem to be (a) carboxylation, which introduces negative charges on backbone structure of pullulan, and (b) isocyanate chemistry, which gives novel features to pullulan, such as thermoplasticity and resistance to water (Farris et al., 2014).

REFERENCES

Aguilar-Vázquez, G., Loarca-Piña, G., Figueroa-Cárdenas, J. D., & Mendoza, S. (2018). Electrospun fibers from blends of pea (*Pisum sativum*) protein and pullulan. *Food Hydrocolloids*, *83*, 173–181. https://doi.org/10.1016/j.foodhyd.2018.04.051

Alban, S., Schauerte, A., & Franz, G. (2002). Anticoagulant sulfated polysaccharides: Part I. Synthesis and structure—activity relationships of new pullulan sulfates. *Carbohydrate Polymers*, *47*(3), 267–276. https://doi.org/10.1016/S0144-8617(01)00178-3

Alhaique, F., Matricardi, P., Di Meo, C., Coviello, T., & Montanari, E. (2015). Polysaccharide-based self-assembling nanohydrogels: An overview on 25-years research on pullulan. *Journal of Drug Delivery Science and Technology*, *30*, 300–309. https://doi.org/10.1016/j.jddst.2015.06.005

Arnosti, C., & Repeta, D. J. (1995). Nuclear magnetic resonance spectroscopy of pullulan and isomaltose: Complete assignment of chemical shifts. *Starch-Stärke*, *47*(2), 73–75. https://doi.org/10.1002/star.19950470208

Bataille, I., Huguet, J., Muller, G., Mocanu, G., & Carpov, A. (1997). Associative behaviour of hydrophobically modified carboxymethylpullulan derivatives. *International Journal of Biological Macromolecules*, *20*(3), 179–191. https://doi.org/10.1016/S0141-8130(97)01158-6

Bender, H., Lehmann, J., & Wallenfels, K. (1959). Pullulan, ein extracelluläres Glucan von Pullularia pullulans. *Biochimica et Biophysica Acta*, *36*(2), 309–316.

Bouveng, H. O., Kiessling, H., Lindberg, B., & McKay, J. (1962). Polysaccharides elaborated by *Pullularia pullulans*. I. The neutral glucan synthesized from sucrose solutions. *Acta Chemica Scandinavica*, *16*, 615–622.

Carolan, G., Catley, B. J., & McDougal, F. J. (1983). The location of tetrasaccharide units in pullulan. *Carbohydrate Research*, *114*(2), 237–243. https://doi.org/10.1016/0008-6215(83)88190-7

Catley, B. J., Ramsay, A., & Servis, C. (1986). Observations on the structure of the fungal extracellular polysaccharide, pullulan. *Carbohydrate Research*, *153*, 79–86. https://doi.org/10.1016/S0008-6215(00)90197-6

Chaen, H. (2009). Pullulan. In A. Imeson (Ed.), *Food stabilizers, thickners and gelling agents* (pp. 266–274). John Wiley & Sons, Blackwell Publishing Ltd.

Chatap, V. K., Maurya, A. R., Deshmukh, P. K., & Zawar, L. R. (2013). Formulation and evaluation of nisoldipne sublingual tablets using pullulan & chitosan for rapid oromucosal absorption. *Advances in Pharmacology and Pharmacy*, *1*(1), 18–25. https://doi.org/10.13189/app.2013.010104

Cheng, K. C., Demirci, A., & Catchmark, J. M. (2011a). Pullulan: Biosynthesis, production, and applications. *Applied Microbiology and Biotechnology*, *92*, 29–44. https://doi.org/10.1007/s00253-011-3477-y

Cheng, K. C., Demirci, A., & Catchmark, J. M. (2011b). Evaluation of medium composition and fermentation parameters on pullulan production by *Aureobasidium pullulans*. *Food Science and Technology International*, *17*(2), 99–109. https://doi.org/10.1177/1082013210368719

Chi, Z., & Zhao, S. (2003). Optimization of medium and new cultivation conditions for pullulan production by a new pullulan-producing yeast strain. *Enzyme and Microbial Technology*, *33*, 206–211. https://doi.org/10.1016/S0141-0229(03)00119-4

Choudhury, A. R., Sharma, N., & Prasad, G. S. (2012). Deoiledjatropha seed cake is a useful nutrient for pullulan production. *Microbial Cell Factories*, *11*, 39. https://doi.org/10.1186/1475-2859-11-39

Chu, Y., Gao, C. C., Liu, X., Zhang, N., Xu, T., Feng, X., Yang, Y., Shen, X., & Tang, X. (2020). Improvement of storage quality of strawberries by pullulan coatings incorporated with cinnamon essential oil nanoemulsion. *LWT-Food Science and Technology*, *122*, 109054. https://doi.org/10.1016/j.lwt.2020.109054

Conca, K. R., & Yang, T. C. S. (1993). Edible food barrier coatings. In C. Ching, D. L. Kaplan, & E. L. Thomas (Eds.), *Biodegradable polymers and packaging* (pp. 357–369). Technomic Publishing Company Inc.

Cozzolino, C. A., Cerri, G., Brundu, A., & Farris, S. (2014). Microfibrillated cellulose (MFC): Pullulan bionanocomposite films. *Cellulose, 21*(6), 4323–4335. https://doi.org/10.1007/S10570-014-0433-X/FIGURES/5

Cozzolino, C. A., Nilsson, F., Iotti, M., Sacchi, B., Piga, A., & Farris, S. (2013). Exploiting the nano-sized features of microfibrillated cellulose (MFC) for the development of controlled-release packaging. *Colloids and Surfaces B: Biointerfaces, 110*, 208–216. https://doi.org/10.1016/j.colsurfb.2013.04.046

Dailin, D. J., Mohd Izwan Low, L. Z., Kumar, K., Abd Malek, R., Khairun, H. N., Keat, H. C., Sukmawati, D., & El Enshahy, H. (2019). Agro-industrial waste: A potential feedstock for pullulan production. *Biosciences Biotechnology Research Asia, 16*(2), 229–250. http://dx.doi.org/10.13005/bbra/2740

Dais, P. (1995). 13C nuclear magnetic relaxation and motional behavior of polysaccharides in solution. In T. Theophanides, J. Anastassopoulou, & N. Fotopoulos (Eds.), *Fifth international conference on the spectroscopy of biological molecules* (pp. 217–221). Kluwer.

Dais, P., Vlachou, S., & Taravel, F. (2001). ^{13}C nuclear magnetic relaxation study of segmental dynamics of the heteropolysaccharide pullulan in dilute solutions. *Biomacromolecules, 2*, 1137–1147. https://doi.org/10.1021/bm010073q

Debeaufort, F., Quezada-Gallo, J. A., & Voilley, A. (1998). Edible films and coatings: Tomorrow's packagings: A review. *Critical Reviews in Food Science and Nutrition, 38*, 299–313. https://doi.org/10.1080/10408699891274219

Delben, F., Forabosco, A., Guerrini, M., Liut, G., & Torri, G. (2006). Pullulans produced by strains of *Cryphonectria parasitica*-II. Nuclear magnetic resonance evidence. *Carbohydrate Polymers, 63*, 545–554. https://doi.org/10.1016/j.carbpol.2005.11.012

Dhumal, C. V., & Sarkar, P. (2018). Composite edible films and coatings from food-grade biopolymers. *Journal of Food Science and Technology, 55*(11), 4369–4383. https://doi.org/10.1007/s13197-018-3402-9

Diab, T., Biliaderis, C. G., Gerasopoulos, D., & Sfakiotakis, E. (2001). Physicochemical properties and application of pullulan edible films and coatings in fruit preservation. *Journal of the Science of Food and Agriculture, 81*(10), 988–1000. https://doi.org/10.1002/jsfa.883

Eroglu, E., Torun, M., Dincer, C., & Topuz, A. (2014). Influence of pullulan-based edible coating on some quality properties of strawberry during cold storage. *Packaging Technology and Science, 27*(10), 831–838. https://doi.org/10.1002/pts.2077

Farris, S., Introzzi, L., Biagioni, P., Holz, T., Schiraldi, A., & Piergiovanni, L. (2011). Wetting of biopolymer coatings: Contact angle kinetics and image analysis investigation. *Langmuir, 27*(12), 7563–7574. https://doi.org/10.1021/LA2017006

Farris, S., Introzzi, L., Fuentes-Alventosa, J. M., Santo, N., Rocca, R., & Piergiovanni, L. (2012). Self-assembled pullulan—silica oxygen barrier hybrid coatings for food packaging applications. *Journal of Agricultural and Food Chemistry, 60*(3), 782–790. https://doi.org/10.1021/jf204033d

Farris, S., Unalan, I. U., Introzzi, L., Fuentes-Alventosa, J. M., & Cozzolino, C. A. (2014). Pullulan-based films and coatings for food packaging: Present applications, emerging opportunities, and future challenges. *Journal of Applied Polymer Science, 131*(13). https://doi.org/10.1002/app.40539

Fraser, C. G., & Jennings, H. J. (1971). A glucan from *Tremella mesenterica* NRRL-Y6158. *Canadian Journal of Chemistry, 49*, 1804–1807. https://doi.org/10.1139/v71-297

Freitas, F., Alves, V. D., Reis, M. A., Crespo, J. G., & Coelhoso, I. M. (2014). Microbial polysaccharide-based membranes: Current and future applications. *Journal of Applied Polymer Science, 131*(6). https://doi.org/10.1002/app.40047

Ganduri, V. S. R. (2020). Evaluation of pullulan-based edible active coating methods on Rastali and Chakkarakeli bananas and their shelf-life extension parameters studies. *Journal of Food Processing and Preservation, 44*, e14378. https://doi.org/10.1111/jfpp.14378

Ghosh, T., Priyadarshi, R., Krebs de Souza, C., Angioletti, B. L., & Rhim, J.-W. (2022). Advances in pullulan utilization for sustainable applications in food packaging and preservation: A mini-review. *Trends in Food Science & Technology, 125*, 43–53. https://doi.org/10.1016/j.tifs.2022.05.001

Gniewosz, M., & Duszkiewicz-Reinhard, W. (2008). Comparative studies on pullulan synthesis, melanin synthesis and morphology of white mutant *Aureobasidium pullulans* B-1 and parent strain A.p.-3. *Carbohydrate Polymers, 72*(3), 431–438. https://doi.org/10.1016/j.carbpol.2007.09.009

Gounga, M. E., Xu, S.-Y., Wang, Z., & Yang, W. G. (2008). Effect of whey protein isolate—pullulan edible coatings on the quality and shelf life of freshly roasted and freeze-dried Chinese chestnut. *Journal of Food Science, 73*(4), E155–E161. https://doi.org/10.1111/j.1750-3841.2008.00694.x

Han, K., Liu, Y., Liu, Y., Huang, X., & Sheng, L. (2020). Characterization and film-forming mechanism of egg white/pullulan blend film. *Food Chemistry*, *315*, 126201. https://doi.org/10.1016/j.foodchem.2020.126201

Hansen, G. H., Lübeck, M., Frisvad, J. C., Lübeck, P. S., & Andersen, B. (2015). Production of cellulolytic enzymes from ascomycetes: Comparison of solid state and submerged fermentation. *Process Biochemistry*, *50*(9), 1327–1341. https://doi.org/10.1016/j.procbio.2015.05.017

Hassan, B., Chatha, S. A. S., Hussain, A. I., Zia, K. M., & Akhtar, N. (2018). Recent advances on polysaccharides, lipids and protein based edible films and coatings: A review. *International Journal of Biological Macromolecules*, *109*, 1095–1107. https://doi.org/10.1016/j.ijbiomac.2017.11.097

Henni-Silhadi, W., Deyme, M., de Hoyos, M. R., Le Cerf, D., Picton, L., & Rosilio, V. (2008). Influence of alkyl chains length on the conformation and solubilization properties of amphiphilic carboxymethylpullulans. *Colloid and Polymer Science*, *286*(11), 1299–1305. https://doi.org/10.1007/s00396-008-1896-9

Hezarkhani, M., & Yilmaz, E. (2019). Pullulan modification via poly(N-vinylimidazole) grafting. *International Journal of Biological Macromolecules*, *123*, 149–156. https://doi.org/10.1016/j.ijbiomac.2018.11.022

Hijiya, H., & Shiosaka, M. (1975). Process for the preparation of food containing pullulan and amylase. US Patent Office, Pat. No. 3 872 228.

Introzzi, L., Fuentes-Alventosa, J. M., Cozzolino, C. A., Trabattoni, S., Tavazzi, S., Bianchi, C. L., Schiraldi, A., Piergiovanni, L., & Farris, S. (2012). "Wetting enhancer" pullulan coating for antifog packaging applications. *ACS Applied Materials and Interfaces*, *4*(7), 3692–3700. https://doi.org/10.1021/AM300784

Islam, M. S., & Yeum, J. H. (2013). Electrospun pullulan/poly (vinyl alcohol)/silver hybrid nanofibers: Preparation and property characterization for antibacterial activity. *Colloids and Surfaces A: Physicochemical and Engineering Aspects*, *436*, 279–286. https://doi.org/10.1016/j.colsurfa.2013.07.001

Israilides, C., Bocking, M., Smith, A., & Scanlon, B. (1994). A novel rapid coupled enzyme assay for the estimation of pullulan. *Biotechnology and Applied Biochemistry*, *19*, 285–291.

Kanmani, P., & Lim, S. T. (2013). Development and characterization of novel probiotic-589 residing pullulan/starch edible films. *Food Chemistry*, *141*(2), 1041–1049. https://doi.org/10.1016/j.foodchem.2013.03.103

Khalaf, H. H., Sharoba, A. M., El-Tanahi, H. H., & Morsy, M. K. (2013). Stability of antimicrobial activity of pullulan edible films incorporated with nanoparticles and essential oils and their impact on turkey deli meat quality. *Journal of Food and Dairy Sciences*, *4*(11), 557–573. https://doi.org/10.21608/jfds.2013.72104

Kimoto, T., Shibuya, T., & Shiobara, S. (1997). Safety studies of a novel starch, pullulan: Chronic toxicity in rats and bacterial mutagenecity. *Food and Chemical Toxicology*, *35*, 323–329. https://doi.org/10.1016/S0278-6915(97)00001-X

Kowalczyk, D., Skrzypek, T., Basiura-Cembala, M., Łupina, K., & Mężyńska, M. (2020). The effect of potassium sorbate on the physicochemical properties of edible films based on pullulan, gelatin, and their blends. *Food Hydrocolloids*, *105*, 105837. https://doi.org/10.1016/j.foodhyd.2020.105837

Kraśniewska, K., Gniewosz, M., Synowiec, A., Przybył, J. L., Baczek, K., & Weglarz, Z. (2014). The use of pullulan coating enriched with plant extracts from *Satureja hortensis* L. to maintain pepper and apple quality and safety. *Postharvest Biology and Technology*, *90*, 63–72. https://doi.org/10.1016/j.postharvbio.2013.12.010

Kraśniewska, K., Pobiega, K., & Gniewosz, M. (2019). Pullulan—biopolymer with potential for use as food packaging. *International Journal of Food Engineering*, *15*(9), 20190030. https://doi.org/10.1515/ijfe-2019-0030

Kraśniewska, K., Ścibisz, I., Gniewosz, M., Mitek, M., Pobiega, K., & Cendrowski, A. (2017). Effect of pullulan coating on postharvest quality and shelf-life of highbush blueberry (*Vaccinium corymbosum* L.). *Materials*, *10*, 965. https://doi.org/10.3390/ma10080965

Lazaridou, A., Roukas, T., Biliaderis, C. G., & Vaikousi, H. (2002). Characterization of pullulan produced from beet molasses by Aureobasidium pullulans in a stirred tank reactor under varying agitation. *Enzyme and Microbial Technology*, *31*(1–2), 122–132. https://doi.org/10.1016/S0141-0229(02)00082-0

Leathers, T. D. (2003). Biotechnological production and applications of pullulan. *Applied Microbiology and Biotechnology*, *62*, 468–473. https://doi.org/10.1007/s00253-003-1386-4

Legros, M., Dulong, V., Picton, L., & Le Cerf, D. (2008). Self-organization of water soluble and amphiphile cross-linked carboxymethylpullulan. *Polymer Journal*, *40*(12), 1132–1139. https://doi.org/10.1295/polymj.PJ2008117

Li, Q., Zhang, J., Zhu, J., Lin, H., Sun, T., & Cheng, L. (2021). Effects of gallic acid combined with epsilon-polylysine hydrochloride incorporated in a pullulan-CMC edible coating on the storage quality of sea bass. *RSC Advances*, *47*. https://doi.org/10.1039/d1ra02320a

Li, S., Yi, J., Yu, X., Wang, Z., & Wang, L. (2020). Preparation and characterization of pullulan derivative/chitosan composite film for potential antimicrobial applications. *International Journal of Biological Macromolecules*, *148*, 258–264. https://doi.org/10.1016/j.ijbiomac.2020.01.080

Liu, Y., Liu, Y., Han, K., Cai, Y., Ma, M., Tong, Q., & Sheng, L. (2019). Effect of nano-TiO$_2$ on the physical, mechanical, and optical properties of pullulan film. *Carbohydrate Polymers*, *218*, 95–102. https://doi.org/10.1016/j.carbpol.2019.04.073

Madi, N. S., McNeil, B., & Harvey, L. M. (1996). Influence of culture ph and aeration on ethanol production and pullulan molecular weight by *Aureobasidium pullulans*. *Journal of Chemical Technology & Biotechnology*, *65*(4), 343–350.

Mähner, C., Lechner, M. D., & Nordmeier, E. (2001). Synthesis and characterisation of dextran and pullulan sulphate. *Carbohydrate Research*, *331*(2), 203–208. https://doi.org/10.1016/S0008-6215(00)00315-3

Maru, V. R., Gupta, S., Ranade, V., & Variyar, P. S. (2021). Pullulan or chitosan based active coating by incorporating polyphenols from lemon peel in raw poultry meat. *Journal of Food Science and Technology*, *58*(10), 3807–3816. https://doi.org/10.1007/s13197-020-04841-4

Morsy, M. K., Khalaf, H. H., Sharoba, A. M., Eltanahi, H. H., & Cutter, C. N. (2014). Incorporation of essential oils and nanoparticles in pullulan films to control foodborne pathogens on meat and poultry products. *Journal of Food Science*, *79*(4), M675–M684. https://doi.org/10.1111/1750-3841.12400

Morsy, M. K., Sharoba, A. M., Khalaf, H. H., El-Tanahy, H. H., & Cutter, C. N. (2015). Efficacy of antimicrobial pullulan-based coating to improve internal quality and shelf-life of chicken eggs during storage. *Journal of Food Science*, *80*(5), M1066–M1074. https://doi.org/10.1111/1750-3841.12855

Nishinari, K., Kohyama, K., Williams, P. A., Phillips, G. O., Burchard, W., & Ogino, K. (1991). Solution properties of pullulan. *Macromolecules*, *24*(20), 5590–5593. https://doi.org/10.1021/ma00020a017

Niu, B., Shao, P., Chen, H., & Sun, P. (2019). Structural and physiochemical characterization of novel hydrophobic packaging films based on pullulan derivatives for fruits preservation. *Carbohydrate Polymers*, *208*, 276–284. https://doi.org/10.1016/j.carbpol.2018.12.070

Oliva, E. M., Cirelli, A. F., & De Lederkremer, R. M. (1986). Characterization of a pullulan in *Cyttaria darwinii*. *Carbohydrate Research*, *158*, 262–267. https://doi.org/10.1016/0008-6215(86)84025-3

Pan, H., Jiang, B., Chen, J., & Jin, Z. (2014). Assessment of the physical, mechanical, and moisture-retention properties of pullulan-based ternary co-blended films. *Carbohydrate Polymers*, *112*, 94–101. https://doi.org/10.1016/j.carbpol.2014.05.044

Pattanayaiying, R., Aran, H., & Cutter, C. N. (2015). Incorporation of nisin Z and lauric arginate into pullulan films to inhibit foodborne pathogens associated with fresh and ready-to-eat muscle foods. *International Journal of Food Microbiology*, *207*, 77–82. https://doi.org/10.1016/j.ijfoodmicro.2015.04.045

Petrov, P. T., Shingel, K. I., Scripko, A. D., & Tsarenkov, V. M. (2002). Biosynthesis of pullulan by *Aureobasidium pullulans* strain BMP-97. *Biotechnologia*, *1*, 36–48.

Prajapati, V. D., Jani, G. K., & Khanda, S. M. (2013). Pullulan: An exopolysaccharide and its various applications. *Carbohydrate Polymers*, *95*(1), 540–549. https://doi.org/10.1016/j.carbpol.2013.02.082

Priyadarshi, R., Kim, S.-M., & Rhim, J.-W. (2021). Pectin/pullulan blend films for food packaging: Effect of blending ratio. *Food Chemistry*, *347*, 129022. https://doi.org/10.1016/j.foodchem.2021.129022

Priyadarshi, R., Riahi, Z., & Rhim, J. W. (2022). Antioxidant pectin/pullulan edible coating incorporated with *Vitis vinifera* grape seed extract for extending the shelf life of peanuts. *Postharvest Biology and Technology*, *183*, 111740. https://doi.org/10.1016/j.postharvbio.2021.111740

Qin, Z., Jia, X., Liu, Q., Kong, B., & Wang, H. (2020). Enhancing physical properties of chitosan/pullulan electrospinning nanofibers via green cross-linking strategies. *Carbohydrate Polymers*, *247*, 116734. https://doi.org/10.1016/j.carbpol.2020.116734

Reis, R. A., Tischer, C. A., Gorrin, P. A. J., & Iacomini, M. (2002). A new pullulan and a branched (1→3)-, (1→6)-linked b-glucan from the lichenised ascomycete *Teloschistes flavicans*. *FEMS Microbiology Letters*, *210*, 1–5. https://doi.org/10.1111/j.1574-6968.2002.tb11152.x

Roy, S., Priyadarshi, R., & Rhim, J. W. (2021). Development of multifunctional pullulan/chitosan-based composite films reinforced with ZnO nanoparticles and propolis for meat packaging applications. *Foods*, *10*(11), 2789. https://doi.org/10.3390/foods10112789

Seo, H. P., Son, C. W., Chung, C. H., Jung, D. I., Kim, S. K., Gross, R. A., Kaplan, D. L., & Lee, J. W. (2004). Production of high molecular weight pullulan by *Aureobasidium pullulans* HP-2001 with soybean pomace as a nitrogen source. *Bioresource Technology*, *95*(3), 293–299. https://doi.org/10.1016/j.biortech.2003.02.001

Shah, N. N., Vishwasrao, C., Singhal, R. S., & Ananthanarayan, L. (2016). n-Octenyl succinylation of pullulan: Effect on its physico-mechanical and thermal properties and application as an edible coating on fruits. *Food Hydrocolloids*, *55*, 179–188. https://doi.org/10.1016/j.foodhyd.2015.11.026

Sharma, N., Prasad, G. S., & Choudhury, A. R. (2013). Utilization of corn steep liquor for biosynthesis of pullulan, an important exopolysaccharide. *Carbohydrate Polymers*, *93*(1), 95–101. https://doi.org/10.1016/j.carbpol.2012.06.059

Silva, N. H. C. S., Vilela, C., Almeida, A., Marrucho, I. M., & Freire, C. S. R. (2018). Pullulan-based nanocomposite films for functional food packaging: Exploiting lysozyme nanofibers as antibacterial and antioxidant reinforcing additives. *Food Hydrocolloids*, *692*(77), 921–930. https://doi.org/10.1016/j.foodhyd.2017.11.039

Simon, L., Caye-Vaugien, C., & Bouchonneau, M. (1993). Relation between pullulan production, morphological state and growth conditions in *Aureobasidium pullulans*: New observations. *Journal of General Microbiology*, *139*, 2757–2761. https://doi.org/10.1099/00221287-139-5-979

Singh, R. S., & Saini, G. K. (2008). Production, purification and characterization of pullulan from a novel strain of *Aureobasidium pullulans* FB-1. *Journal of Biotechnology*, *136*, S506–S507. https://doi.org/10.1016/j.jbiotec.2008.07.625

Singh, R. S., Kaur, N., & Kennedy, J. F. (2019). Pullulan production from agro-industrial waste and its applications in food industry: A review. *Carbohydrate Polymers*, *217*, 46–57. https://doi.org/10.1016/j.carbpol.2019.04.050

Singh, R. S., Kaur, N., Rana, V., & Kennedy, J. F. (2017). Pullulan: A novel molecule for biomedical applications. *Carbohydrate Polymers*, *171*, 102–121. https://doi.org/10.1016/j.carbpol.2017.04.089

Singh, R. S., Kaur, N., Singh, D., Bajaj, B. K., & Kennedy, J. F. (2022). Downstream processing and structural confirmation of pullulan—a comprehensive review. *International Journal of Biological Macromolecules*, *208*, 553–564. https://doi.org/10.1016/j.ijbiomac.2022.03.163

Singh, R. S., Saini, G. K., & Kennedy, J. F. (2008). Pullulan: Microbial sources, production and applications. *Carbohydrate Polymers*, *73*(4), 515–531. https://doi.org/10.1016/j.carbpol.2008.01.003

Singh, R. S., Saini, G. K., & Kennedy, J. F. (2009). Downstream processing and characterization of pullulan from a novel colour variant strain of *Aureobasidium pullulans* FB-1. *Carbohydrate Polymers*, *78*(1), 89–94. https://doi.org/10.1016/j.carbpol.2009.03.040

Souguir, Z., Roudesli, S., Picton, E. L., Le Cerf, D., & About-Jaudet, E. (2007). Novel cationic and amphiphilic pullulan derivatives I: Synthesis and characterization. *European Polymer Journal*, *43*(12), 4940–4950. https://doi.org/10.1016/j.eurpolymj.2007.09.017

Sowa, W., Blackwood, A. C., & Adams, G. A. (1963). Neutral extracellular glucan of *Pullularia pullulans* (de Bary) Berkhout. *Canadian Journal of Chemistry*, *41*, 2314–2319.

Sugumaran, K. R., Gowthami, E., Swathi, B., Elakkiya, S., Srivastava, S. N., Ravikumar, R., Gowdhaman, D., & Ponnusami, V. (2013a). Production of pullulan by *Aureobasidium pullulans* from Asian palm kernel: A novel substrate. *Carbohydrate Polymers*, *92*(1), 697–703. https://doi.org/10.1016/j.carbpol.2012.09.062

Sugumaran, K. R., Jothi, P., & Ponnusami, V. (2014b). Bioconversion of industrial solid waste—cassava bagasse for pullulan production in solid state fermentation. *Carbohydrate Polymers*, *99*, 22–30. https://doi.org/10.1016/j.carbpol.2013.08.039

Sugumaran, K. R., Shobana, P., Balaji, P. M., Ponnusami, V., & Gowdhaman, D. (2014a). Statistical optimization of pullulan production from Asian palm kernel and evaluation of its properties. *International Journal of Biological Macromolecules*, *66*, 229–235. https://doi.org/10.1016/j.ijbiomac.2014.02.045

Sugumaran, K. R., Sindhu, R. V., Sukanya, S., Aiswarya, N., & Ponnusami, V. (2013b). Statistical studies on high molecular weight pullulan production in solid state fermentation using jack fruit seed. *Carbohydrate Polymers*, *98*(1), 854–860. https://doi.org/10.1016/j.carbpol.2013.06.071

Sugumaran, K. R., & Ponnusami, V. (2017a). Review on production, downstream processing and characterization of microbial pullulan. *Carbohydrate Polymers*, *173*, 573–591. https://doi.org/10.1016/j.carbpol.2017.06.022

Sugumaran, K. R., & Ponnusami, V. (2017b). Conventional optimization of aqueous extraction of pullulan in solid-state fermentation of cassava bagasse and Asian palm kernel. *Biocatalysis and Agricultural Biotechnology*, *10*, 204–208. https://doi.org/10.1016/j.bcab.2017.03.010

Suhag, R., Kumar, N., Petkoska, A. T., & Upadhyay, A. (2020). Film formation and deposition methods of edible coating on food products: A review. *Food Research International*, *136*, 109582. https://doi.org/10.1016/j.foodres.2020.109582

Sutherland, I. W. (1998). Novel and established applications of microbial polysaccharides. *Trends in Biotechnology, 16*(1), 41–46. https://doi.org/10.1016/S0167-7799(97)01139-6

Synowiec, A., Gniewosz, M., Kraśniewska, K., Przybył, J. L., Bączek, K., & Węglarz, Z. (2014). Antimicrobial and antioxidant properties of pullulan film containing sweet basil extract and an evaluation of coating effectiveness in the prolongation of the shelf life of apples stored in refrigeration conditions. *Innovative Food Science & Emerging Technologies, 23*, 171–181. https://doi.org/10.1016/j.ifset.2014.03.006

Tabasum, S., Noreen, A., Maqsood, M. F., Umar, H., Akram, N., Chatha, S. A. S., & Zia, K. M. (2018). A review on versatile applications of blends and composites of pullulan with natural and synthetic polymers. *International Journal of Biological Macromolecules, 120*, 603–632. https://doi.org/10.1016/j.ijbiomac.2018.07.154

Tang, X. Z., Kumar, P., Alavi, S., & Sandeep, K. P. (2012). Recent advances in biopolymers and biopolymer-based nanocomposites for food packaging materials. *Critical Reviews in Food Science and Nutrition, 52*(5), 426–442. https://doi.org/10.1080/10408398.2010.500508

Tharanathan, R. N. (2003). Biodegradable films and composite coatings: Past, present, and future. *Trends in Food Science & Technology, 14*(3), 71–78. https://doi.org/10.1016/S0924-2244(02)00280-7

Tomasula, P. M., Sousa, A. M. M., Liou, S.-C., Li, R., Bonnaillie, L. M., & Liu, L. S. (2016). Short communication: Electrospinning of casein/pullulan blends for food-grade applications. *Journal of Dairy Science, 99*(3), 1837–1845. https://doi.org/10.3168/jds.2015-10374

Tong, Q., Xiao, Q., & Lim, L. T. (2013). Effects of glycerol, sorbitol, xylitol and fructose plasticisers on mechanical and moisture barrier properties of pullulan—alginate—carboxymethylcellulose blend films. *International Journal of Food Science & Technology, 48*(4), 870–878. https://doi.org/10.1111/ijfs.12039

Treviño-Garza, M. Z., García, S., Heredia, N., Alanís-Guzmán, M. G., & Arévalo-Niño, K. (2017). Layer-by-layer edible coatings based on mucilages, pullulan and chitosan and its effect on quality and preservation of fresh-cut pineapple (*Ananas comosus*). *Postharvest Biology and Technology, 128*, 63–75. https://doi.org/10.1016/j.postharvbio.2017.01.007

Trinetta, V., & Cutter, C. N. (2016). Pullulan: A suitable biopolymer for antimicrobial food packaging applications. In *Antimicrobial food packaging* (pp. 385–397). Academic Press. https://doi.org/10.1016/B978-0-12-800723-5.00030-9

Trinetta, V., Floros, J. D., & Cutter, C. N. (2010). Sakacin a-containing pullulan film: An active packaging system to control epidemic clones of listeria monocytogenes, in ready-to-eat foods. *Journal of Food Safety, 30*(2), 366–381. https://doi.org/10.1111/j.1745-4565.2010.00213.x

Trovatti, E., Fernandes, S. C., Rubatat, L., da Silva Perez, D., Freire, C. S., Silvestre, A. J., & Neto, C. P. (2012). Pullulan—nanofibrillated cellulose composite films with improved thermal and mechanical properties. *Composites Science and Technology, 72*(13), 1556–1561. https://doi.org/10.1016/j.compscitech.2012.06.003

Tsujisaka, Y., & Mitsuhashi, M. (1993). Pullulan. In R. L. Whistler & J. N. BeMiller (Eds.), *Industrial gums. Polysaccharides and their derivatives* (pp. 447–460). Academic Press.

Unalan, I. U., Wan, C., Figiel, Ł. F., Olsson, R. T., Trabattoni, S., & Farris, S. (2015). Exceptional oxygen barrier performance of pullulan nanocomposites with ultra-low loading of graphene oxide. *Nanotechnology, 26*(27), 275703.

Viveka, R., Varjani, S., & Ekambaram, N. (2021). Valorization of cassava waste for pullulan production by *Aureobasidium pullulans* MTCC 1991. *Energy and Environment, 32*(6), 1086–1102. https://doi.org/10.1177/0958305X20908065

Vuddanda, P. R., Montenegro-Nicolini, M., Morales, J. O., & Velaga, S. (2017). Effect of plasticizers on the physico-mechanical properties of pullulan based pharmaceutical oral films. *European Journal of Pharmaceutical Sciences, 96*, 290–298. https://doi.org/10.1016/j.ejps.2016.09.011

Wallenfels, K., Bender, H., Keilich, G., & Bechtler, G. (1961). On pullulan, the glucan of the slime coat of *Pullularia pullulans. Angewandte Chemie, 73*, 245–246.

Wallenfels, K., Keilich, G., Bechtler, G., & Freudenberger, D. (1965). Investigations on pullulan. IV. Resolution of structural problems using physical, chemical, and enzymatic methods. *Biochemische Zeitschrift, 341*, 433–450.

Wang, B., Yang, C., Wang, J., Xia, S., & Wu, Y. (2019). Effects of combined pullulan polysaccharide, glycerol, and trehalose on the mechanical properties and the solubility of casted gelatin-soluble edible membranes. *Journal of Food Processing and Preservation, 43*(1), e13858. https://doi.org/10.1111/jfpp.13858

Wang, D., Ju, X., Zhou, D., & Wei, G. (2014). Efficient production of pullulan using rice hull hydrolysate by adaptive laboratory evolution of *Aureobasidium pullulans*. *Bioresource Technology*, *164*, 12–19. https://doi.org/10.1016/j.biortech.2014.04.036

Wani, S. M., Mir, S. A., Khanday, F. A., & Masoodi, F. A. (2021). Advances in pullulan production from agro-based wastes by *Aureobasidium pullulans* and its applications. *Innovative Food Science & Emerging Technologies*, *74*, 102846. https://doi.org/10.1016/j.ifset.2021.102846

Wu, J., Zhong, F., Li, Y., Shoemaker, C. F., & Xia, W. (2013). Preparation and characterization of pullulan—chitosan and pullulan—carboxymethyl chitosan blended films. *Food Hydrocolloids*, *30*(1), 82–91. https://doi.org/10.1016/j.foodhyd.2012.04.002

Wu, S., & Chen, J. (2013). Using pullulan-based edible coatings to extend shelf-life of fresh-cut "Fuji" apples. *International Journal of Biological Macromolecules*, *55*, 254–257. https://doi.org/10.1016/j.ijbiomac.2013.01.012

Xiao, Q., Tong, Q., & Lim, L.-T. (2012). Pullulan-sodium alginate-based edible films: Rheological properties of film-forming solutions. *Carbohydrate Polymers*, *87*(2), 1689–1695. https://doi.org/10.1016/j.carbpol.2011.09.077

Yousuf, B., Qadri, O. S., & Srivastava, A. K. (2018). Recent developments in shelf-life extension of fresh-cut fruits and vegetables by application of different edible coatings: A review. *LWT-Food Science and Technology*, *89*, 198–209. https://doi.org/10.1016/j.lwt.2017.10.051

Yuen, S. (1974). Pullulan and its applications. *Process Biochemistry*, *9*, 7–22.

Zhang, C., Gao, D., Ma, Y., & Zhao, X. (2013). Effect of gelatin addition on properties of pullulan films. *Journal of Food Science*, *78*(6), C805–C810. https://doi.org/10.1111/j.1750-3841.2012.02925.x

Zhang, L., Huang, C., & Zhao, H. (2019). Application of pullulan and chitosan multilayer coatings in fresh papayas. *Coatings*, *9*, 745. https://doi.org/10.3390/coatings9110745

Zhao, Z., Xiong, X., Zhou, H., & Xiao, Q. (2019). Effect of lactoferrin on physicochemical properties and microstructure of pullulan-based edible films. *Journal of the Science of Food and Agriculture*, *99*(8), 4150–4157. https://doi.org/10.1002/jsfa.9645

Zhou, W., He, Y., Liu, F., Liao, L., Huang, X., Li, R., Zou, Y., Zhou, L., Zou, L., Liu, Y., Ruan, R., & Li, J. (2021). Carboxymethyl chitosan-pullulan edible films enriched with galangal essential oil: Characterization and application in mango preservation. *Carbohydrate Polymers*, *256*, 117579. https://doi.org/10.1016/j.carbpol.2020.117579

Zhu, G., Sheng, L., & Tong, Q. (2014). Preparation and characterization of carboxymethyl-gellan and pullulan blend films. *Food Hydrocolloids*, *35*, 341–347. https://doi.org/10.1016/j.foodhyd.2013.06.009

15

Bionanocomposites in Food and Medicine

Pratiksha Shrestha, Sneh Punia Bangar, and Abebaw Ayele

CONTENTS

DOI: 10.1201/9781003303671-15

15.1 Introduction

Biopolymers are the perfect alternatives of conventional polymers, such as polyethylene terephthalate (PET), polyvinylchloride (PVC), polyethylene (PE), polyamide (PA), polystyrene (PS), and polypropylene (PP), that are not biodegradable (Song et al., 2009). Plant-based compounds, microbial products, chemically synthesized naturally derived monomers acid, and animal by-products are the sources for production of biopolymer (Sharma et al., 2018). The poor mechanical and water vapor barrier nature of biopolymer in comparison to conventional plastics is the constraint that limits its use in commercial scale. Therefore, researchers are working to develop a biopolymer composite material with enhanced mechanical and water vapor barrier properties.

Superior thermal, mechanical, and biodegradable properties of biopolymer-based nanocomposite biocomposite have received considerable interest in recent years (Cesur et al., 2018; Ghelejlu et al., 2016). Bionanocomposite is the combination of biopolymers with inorganic and organic materials that have one or more dimensions on the nanometer scale (<100 nm) into a polymer matrix. These nanomaterials are also called nanofillers, nanoparticles, nanoscale building blocks, or nanoreinforcements. Nanofillers are also unique in the way that they will not affect the clarity of the matrix. They appear transparent since this nanomaterial scale is very minute compared to the wavelength of visual light (Petersson & Oksman, 2006). The use of nanofillers improves functional properties, mechanical strength, barrier properties, and thermal properties of the nanocomposite material (Xiong et al., 2018).

A few decades ago, the use of conventional plastics was the only option for many products. This caused increase in pollution level in soil and water resources, and many people are concerned about this problem. Consequently, the use of biodegradable polymer, and thus its production, has increased significantly in recent years. Increasing consumer awareness toward environmental issues put pressure on industries to switch to green, eco-friendly solution. "Mostly, the mispackages are recognized as green alternatives to synthetic plastics due to their abundance, non-toxicity, biodegradability, biocompatibility, renewability and environmental friendliness" (Ramos et al., 2018). Biopolymers function as a matrix for integrating a diverse list of bioactive compounds that includes antioxidants, antimicrobials, antifungal, nutrients, colors, flavors, as well as reinforcing materials. In addition, the skyrocketing price of crude oil and lack of its sustainability have left manufactures with no option other than to replace conventional polymers with novel biopolymers in the future. Therefore, there is an urgent need to develop packaging materials that offer new functionalities and make less environmental impact but also are economical.

In today's world of globalization, food and medicines are imported and exported throughout the world. The concept of food packaging to keep the food safe and fresh throughout the supply chain is fascinatingly increasing. Similarly, when it comes to packing of medicine, one should be more thoughtful as even a slight failure to meet the quality of the medicine may risk life. Considering the environmental issues of increasing plastic waste and the sustainability of resources, bionanocomposite materials have been developed to replace conventional plastic. Besides, owing to the peculiar characteristics of nanomaterial and biopolymers, they have been extensively studied for use in various food and medical applications.

Bionanocomposites have been equally exploited in food as well as medical applications. Food packaging have advanced to the next level and helped achieve food safety using nanocomposites. Active packaging and smart packaging of food are based on nanocomposites and are already a huge commercial success. biopolymeric nanoparticles (protein, lipid, and carbohydrate) and few metal nanoparticles such as silver have been studied for use in food processing or as edible nanocoating. However, the safety aspect of nanoparticles and the health impact aren the major concerns when used in food itself. The medical field, drug delivery, wound management, implants, and tissue engineering are most benefitted by the use of biopolymer nanocomposite. Antimicrobial property, biocompatibility, massive drug carrying capacity, and sustained drug release are the major characteristics of bionanocomposite film used for medical applications.

15.2 Biopolymers Popularly Used in the Preparation of Bionanocomposites

Biopolymer is a natural substance inferred from the Greek words "bio" and "polymer" and stands for polymer obtained from living things. "They are large macromolecules made up of numerous repeating units" (Baranwal et al., 2022). The biodegradable and biocompatible nature of biopolymer-based bionanocomposite films makes them suitable for various applications in food and pharmaceutical industries. The food industry makes use of incomposite films to prepare edible films, coatings, and packaging materials, whereas pharmaceutical industries use it for developing dressing materials, drug transport materials, wound healing, tissue scaffolds, and medical implants. The monomeric units of the polymeric biomolecules are joined together with the covalent bond to form biopolymers. The term *biopolymer* describes its derivation from biological sources, such as microorganisms, plants, or trees. "Materials such as vegetable oil, sugars, fat, resins, protein, amino acids etc. that are produced synthetically from biological sources can also be referred as biopolymers" (Mohan et al., 2016). Biopolymers are also called renewable polymers. They do not have a destructive impact on the environment as they are produced form microorganisms, polynucleotides, polysaccharides, and biomass and are biodegradable. By incorporating very small amounts of nanofiller into natural and synthetic biopolymers, they can be a better alternative to thermoplastics, such as polyethylene, polypropylene, polystyrene, and many others.

Biopolymers exhibit a good film-forming and coating capability. Films and coating obtained with the use of biopolymers have a capability to preserve the quality and extend the shelf life of packaged food products by acting as a protective barrier to the external environmental and preventing loss of desirable compounds like flavor and other volatiles. Basically, biopolymers function as a matrix base for preparation of bionanocomposite films that are biocompatible, biodegradable, and renewable (Xiong et al., 2018). They are also a good carrier of small molecules, like antioxidants, antimicrobial substances, vitamins, pigments, and flavors, and contribute to improving nutritional and sensory properties (Mihindukulasuriya & Lim, 2014; Otoni et al., 2017) and assure the safety of products.

Generally, bio-based polymers can be classified as polymers (1) obtained from plants and animals, (2) synthesized bio-monomers, (3) produced from microorganism (Ramos et al., 2018). Chitosan, cellulose, pectin, starch, alginate, and gums are carbohydrate-based biopolymers that can be obtained from plants and animals. Similarly, biopolymers can also be obtained from protein biomass, such as soya, gluten, zein, casein, whey, gelatin, etc. and can be obtained from lipid biomass, such as triglyceride. Polylactic acid (PLA) and other polyesters are synthesized from biologically derived monomers. Microorganisms can also be employed for the production of biopolymers, such as bacterial nanocellulose, polyhydroxyalkanoate (PHA), xanthan, and pullulan. Biopolymers popularly used to make bionanocomposites are briefly discussed in the following sections.

15.2.1 Carbohydrate-Based Biopolymers

15.2.1.1 Cellulose

Cellulose is a polymer chain of unbranched 1–4 linked D glucopyranosyl unit, and the length of these 1–4 glucan chain chains varies with the source of cellulose. It has poor water vapor transfer properties

(Azeredo et al., 2009), but its nanofiber can improve permeability and oxygen barrier properties and increase its Young's modulus and tensile strength (Sharma et al., 2018). Therefore, it is widely used in packaging, bio-adsorption, edible coatings, etc.

15.2.1.2 Chitosan

After cellulose, chitin is the most abundant polysaccharide found on Earth. Chitosan is a derivative of N-deacetylated chitin and is also called soluble chitin. Deacetylation is the process of replacing amino (-NH2) to the acetyl (-C2H3O) group in the polymer chain to give N-acetyl-glucosamine and D-glucosamine copolymers. It is inexpensive, commercially available (Cazón & Vázquez, 2019), and biocompatible; possesses antibacterial activity (Yang et al., 2010); and is therefore one of the most studied polysaccharides. The inherent antimicrobial activity of chitosan is because of the chelating property and cationic property (Nouri et al., 2018) of the amino group that can interact with negatively charged bacterial cell membrane (Qu & Luo, 2020). However, the poorer thermal, mechanical, and barrier properties of chitosan make it a material of least choice to use in packaging when compared to petroleum-based polymers.

15.2.1.3 Starch

Starch is a widely available, promising biopolymer in food packaging application (Charles et al., 2003) because of its better film-forming property. When a plasticizer such as glycerol is added, it shows thermoplastic behavior. Gelatinizing granular starch in the presence of the plasticizer, heat, and pressure give rise to thermoplastic starch (TPS). Like other biopolymers, it has poorer barrier properties and is water-sensitive and brittle in nature in its native form. But usually, it is used in combination with other materials (Sadeghizadeh-Yazdi et al., 2019) to overcome its limitations.

15.2.1.4 Alginate

This polysaccharide is the polymer of 1–4 linked β-D mannuronic acids and α-L guluronic acid and has unique gel-forming properties (Norajit et al., 2010). It is derived from the cell wall of the brown algae and is non-toxic. Reinforcing this polysaccharide with nanocellulose (Reddy & Rhim, 2014), nanoclay (Lee et al., 2019) metal particles, and bimetal nanoparticles (Arfat et al., 2017) can improve the properties of the product to get the desired characteristics. Furcellaran is another type of biopolymer obtained from algae. It has gained much popularity in a short period of time because of its excellent film-forming property, biodegradability, and abundance (Júnior et al., 2021).

15.2.1.5 Polylactic Acid (PLA)

It is a polymer of lactic acid with plenty of carboxylic functional groups and exhibits good mechanical and optical properties and is biocompatible and ecofriendly. PLA is considered a renewable material, as the material from which it is made (lactic acid) is produced by lactobacillus fermentation of carbohydrate, which is a renewable resource. Processing PLA nanocomposites is simple and easy, using a range of solvents, like acetone, chloroform, and dimethyl formamide. However, it comes with the drawback that it has low water vapor and gas barrier properties when compared to conventional polymers (Sharma et al., 2018). Environment-friendly, biodegradable food packaging products like cups, trays, tableware, and cutleries are produced from PLA. It is also notably used in tissue engineering and developing biomedical devices and scaffolds.

15.2.1.6 Pectin

Pectin is a polymer chain of methylated galacturonan consisting of rhamnose residues. It is a heterogenous branched polysaccharide and is widely popular among researchers as it is abundant in agricultural waste. Demethylation of pectin is done to modify its nature for preparation of edible film (Tang et al., 2012).

15.2.2 Protein-Based Biopolymers

15.2.2.1 Wheat Gluten

It is an enriched protein and is an intricate conjugation of glutenin and gliadin protein. Glutenin is a water- and ethanol-insoluble glutelin, whereas gliadin is a water-insoluble and ethanol-soluble pro-lamin. Gluten provides channels for interactions, grafting, and cross-linking of chemicals with the polymer (Diao et al., 2014). Basically, for the preparation of gluten-based polymer, by-products of the wheat starch industry are used. Wheat gluten is most popularly used in preparing biodegradable food packaging material.

15.2.2.2 Soy Protein

It is a globulin protein and is obtained from soybean seeds. Soy protein is mainly composed of polar and non-polar amino acids consisting of β-conglycinin (52%) as a major component. This is the material of interest for many researchers for the preparation of edible films and coatings because of its excellent film-forming property, biocompatibility, and biodegradability (Tian et al., 2018).

15.2.2.3 Gelatin

It is a fibrous protein and is water-soluble in nature. It is mainly extracted from the thermal denaturation of collagen found in the connective tissue, skin, and bone of animals. Fish and meat industries by-product are the most potential sources for gelatin production. It dissolves in water at 40°C and has the capability to form into a thermoreversible gel. Gelatin is extremely hygroscopic (Etxabide et al., 2017), such that even storing at high moisture condition can make it swell.

15.2.2.4 Corn Zein

Zein is a prolamin protein obtained from the endosperm of corm. This protein is alcohol-soluble, with good antioxidant activity and oxygen barrier property (Yıldırım & Barutçu Mazı, 2017), making it a desirable food packaging material that can maintain food quality and safety. It also has a strong affinity to nanofillers, as negatively charged nanofillers and positively charged zein protein can make a strong bond together (de Moura et al., 2017). Lately, antimicrobial food packagings developed with zein-based nanocomposites have been reported by many authors (Kasaai, 2018).

15.2.2.5 Whey Protein

Whey protein is mainly comprised of β-lactoglobulin (approximately 57%) and α-lactalbumin (approximately 20%) of total whey protein. It is used in preparing films by heat denaturation of protein in aqueous solution. The water vapor permeability of whey protein is largely dependent on the type and concentration of plasticizer and relative humidity. The major drawback of using whey protein for film preparation is that it shows poorer performance (poorer than that of wheat gluten, soy protein, and zein protein) for water vapor permeability (Tang et al., 2012).

15.2.2.6 Polyhydroxyalkanoate (PHA)

It is a renewable and biocompatible biopolymer having elastic properties. One interesting thing about PHA is, they can be used in biosensor and can form excellent films and coating by using the solution casting method. Like PHA, poly-ε-caprolactone (PCL) is also a synthetic biodegradable biopolymer that is used for biomedical applications because of its easily processable and highly hydrophilic nature (Song et al., 2015). The characteristics of PHA closely resemble those of petroleum-based polymer polypropylene (PP).

15.3 Fillers and Reinforcements Popularly Used in the Preparation of Bionanocomposites

Nanocomposites are polymeric matrixes in which nanoparticles with a scale of 1–100 nm are incorporated. Nanoparticles exhibit extremely high mechanical strength, low density, and high aspect ratio, and most importantly, they are biocompatible. Therefore, researchers have investigated various nanoparticles to reinforce biopolymer for improving its properties for various applications. They are incorporated in bionanocomposites as fillers to obtain the desired optical, electrical, magnetic, thermal, or mechanical properties. The mass fraction of nanoparticles in the bionanocomposite is usually between 0.5% and 5%, which is a very small amount (Padua et al., 2012). The main challenges of biopolymer that researchers are working on are its poor mechanical and barrier properties (Tabatabaei et al., 2018; Youssef & El-Sayed, 2018), to make the material suitable for industrial applications. The nanofillers used in combination with different biopolymers to make bionanocomposites are illustrated in Figure 15.1. Nanofillers have great potential for disabling the shortcomings of biopolymers, and this has drawn the substantial attention of researchers in the recent years (Uysal Ünalan, 2014). Nanoparticles can improve the properties of biopolymer without hindering their biodegradability and non-toxic characters. There are different types of metallic and non-metallic nanofillers used in food and medical applications, such as silver nanoparticles (AgNPs), titanium oxide nanoparticles (TiO$_2$-NPs), montmorillonite (MMT), zinc oxide nanoparticles (ZnONPs), cellulose nanofibers (CNF), cellulose nanocrystals (CNC), and many others.

15.3.1 Non-Metal Nanoparticles

Nanoparticles of biological origin such as starch nanoparticle, protein nanoparticle, and lipid nanoparticle are called biopolymeric nanoparticles. Besides, other non-metal nanoparticles popularly used are discussed next.

15.3.1.1 Cellulose Nanostructures

Commonly used plant-based cellulose nanostructures are cellulose nanocrystals (CNCs) and cellulose nanofibers (CNF). CNCs have an average dimension of less than 25 nm diameter and less than 100 nm length, while CNFs have an average dimension of 10–100 nm diameter. Microcrystalline cellulose

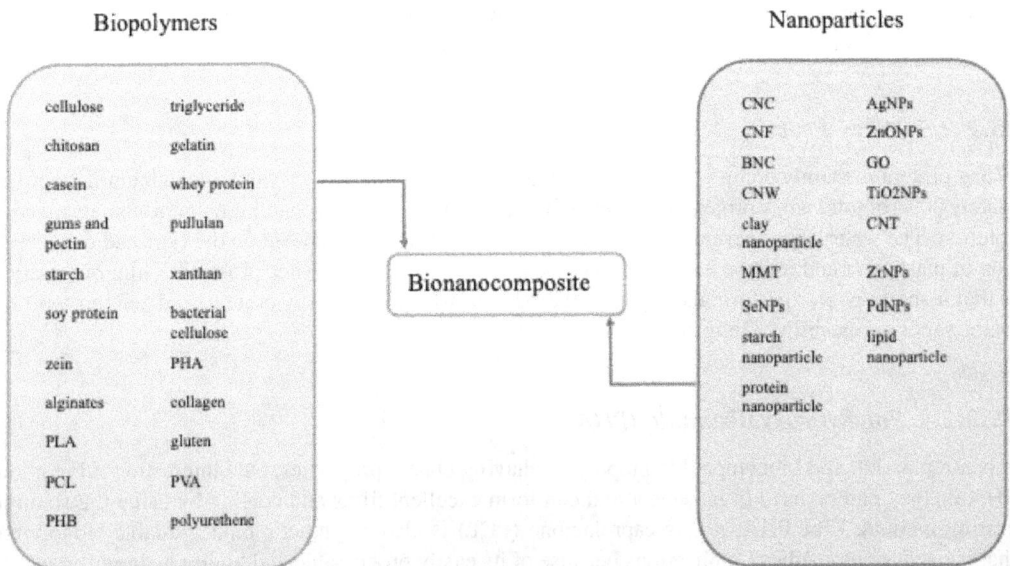

FIGURE 15.1 Options for using various biopolymers and nanoparticles to make suitable bionanocomposites

(MCC) and cellulose nanowhiskers (CNWs) are also closely related products. CNCs can be obtained from the acid hydrolysis of cellulose (Shrestha et al., 2021), while both oxidation or mechanical decomposition of biomass (Shahi et al., 2020) and acid hydrolysis methods can be employed for the extraction of CNF.

Bacterial nanocellulose is popular with the name of BNC, binonanocellulose, microbial cellulose, and bacterial cellulose. It is an unbranched epoxy-polysaccharide of β-1,4 linked glucopyranose units with the molecular formula $(C_6H_{10}O_5)_n$. It is synthesized by Gram-negative bacteria (Lin et al., 2020; Sharma & Bhardwaj, 2019), and its physicochemical properties are different from that of plant-derived nanocellulose. In comparison to other bionanoparticles, BNC possesses better moldability and higher purity, crystallinity, and tensile strength. Cellulose-based nanocomposites most suitable for food packaging are BNC, CNC, and CNF. From an environmental point of view, CNF carries great potential in developing food packaging materials because of its reduced cost and lesser environmental impact (Khalil et al., 2016).

15.3.1.2 Clay Nanoparticles

Clay nanoparticles consist of nanosized compounds, like magnesium and silicates. They are popularly used in developing food packaging with better functionalities, such as thermal resistance during production process, easy transportation, and extended shelf life for storage. "The surface to volume ratio of the nanoparticle may influence the quality of material" (Rhim et al., 2013). Reinforcing polymer composites with clay nanoparticles results in high surface area of bionanocomposites. Like clay nanoparticles, montmorillonite (MMT) is a hydrous aluminosilicate clay packed in between layers of silicon tetrahedron. The silicate layers from which they are modified originally are called organosilicates or nanoclay.

15.3.1.3 Selenium Nanoparticles (SeNPs)

Selenium nanoparticles have expectational free radical scavenging activity and antimicrobial activity. It is mostly used in developing smart packaging material for food industries as it is readily bioavailable and is non-toxic (Nonsuwan et al., 2018).

15.3.2 Metal Nanoparticles

There are various metal nanoparticles, and different nanoparticles have their own unique properties. Mostly, they have good chemical and antibacterial properties and exhibit electrical conductivity and optical polarizability. Metal nanoparticles can also be used for sensing applications as they can underdo reduction in size and enhance functionalization of surface (Faupel et al., 2010). They are synthesized by electrodeposition, spray pyrolysis, chemical vapor deposition (CVD), and other chemical methods.

15.3.2.1 Silver Nanoparticles (AgNPs)

Silver nanoparticles exhibit antimicrobial activity against a broad spectrum of microorganism. It is soluble, biocompatible, and non-toxic (Jamróz et al., 2020) in comparison to other nanoparticles. This makes AgNPs a perfect choice for developing novel food packaging materials that contribute food safety by ensuring extended shelf life of food products. Likewise, reinforcing biopolymers with silver nanoparticles results in improved physical and mechanical properties of bionanocomposites (Kraśniewska et al., 2020).

15.3.2.2 Zinc Oxide Nanoparticles (ZnONPs)

Zinc oxide nanoparticles act as an antimicrobial agent, and hence, its incorporation in biopolymers is being studied to explore its potential for many other applications. Furthermore, ZnONPs maintain greater stability in the polymer matrix, have improved mechanical and barrier properties, and are non-toxic in nature (Indumathi & Rajarajeswari, 2019).

15.3.2.3 Graphene Oxide (GO)

GO is principally a one-atom thick sheet exhibiting high surface area. Besides its excellent mechanical, thermal, and electrical properties, GO is a suitable reinforcing nanofiller that allows plenty of interaction between functional groups (Lyn et al., 2019). During preparation of GO-based nanocomposite films, the ultrasonication technique is used to ensure uniform distribution of GO into polar solvent. Basically, GO possesses an insulating property but, after its reduction, shows increase in conductivity.

15.3.2.4 Titanium Dioxide (TiO$_2$)

TiO$_2$ nanoparticle is an inactive material that exhibits photocatalytic activity. It is a low-cost material that exhibits potential activity against a wide range of microorganisms and is non-toxic. TiO$_2$ nanoparticle can perform better in combination with biopolymers such as PLA and soy protein isolate (Fei et al., 2013). Among various metal oxides, TiO$_2$ shows many distinct characteristics which can potentially be used in the preparation of improved biopolymer-based packaging materials.

15.3.2.5 Carbon Nanotubes (CNTs)

It is a one-dimensional carbon structure with aspect ratio of more than 1,000 (Lee et al., 2008), blending CNTs with Tg of the nanocomposite material when compared to PLA alone. This clearly demonstrates the potential of CNTs for development of nanocomposites with greater fracture toughness and strength and stronger mechanical properties (Zhou et al., 2018).

15.4 Fabrication of Bionanocomposites

The major factor contributing to greenhouse effect is environmental pollution. Petroleum-based plastic waste invites serious environmental issues, and the use of biodegradable biopolymer–based composite materials is the only alternative to replace conventional plastics. However, these biopolymer-based materials are more costly compared to petroleum-based material and have some drawbacks. To overcome these shortcomings, there is a need to find such fabrication techniques that can improve quality and reduce cost. For instance, nanofibers and nanocellulose are used as nanofillers during fabrication to prove the thermal and mechanical properties of bionanocomposites.

In the polymeric matrix, nanofillers are likely to form agglomeration during processing. Chemical modification can help obtain uniform distribution of nanoparticles in the solution matrix, resulting in bionanocomposites with improved performance. In this way, recently developed advanced fabrication methods for different kinds of bionanocomposites can bring solution to the problem by improving mechanical, thermal, and electrical properties of biopolymers. The use of different fabrication methods depends on the size of the nanofiller, the nature of the polymer, the molecular weight of the polymer, and the solvent to be used (Khanam et al., 2015). Preparation techniques can effectively influence the thermal, electrical, and mechanical characteristics of the produced nanocomposite. To prepare bionanocomposites with optimum properties, uniform distribution of nanofillers in the polymer matrix is required (Roy et al., 2012), and the recent fabrication method uses ultrasonication techniques to overcome the problem. Moreover, different structures of nanofillers have been developed, such as nanofibers, nanocrystals, nanoribbons, nanotubes, etc., to achieve the intended result.

Typically, the incorporation of nanofillers into the matrix can be achieved by (1) in situ polymerization, (2) solution intercalation, and (3) melt intercalation (Chivrac et al., 2009). However, recent advances in fabrication technique also recognize electrospinning, extrusion, and compression molding techniques, which are quite successful. Selecting the most applicable method of fabrication is largely dependent on the types of polymers used, the intended quality attribute of the final product, and the volume of production.

15.4.1 In Situ Polymerization

In this method, polymerization is enabled by the incorporation of a photo or thermal catalyst (Okamoto et al., 2000, 2001) that allows homogenous dispersion of nanofillers into the polymer matrix. The catalyst diffusion takes place through inner-layer exchange of cation (Yao et al., 2002). Different types of nanocomposites can be developed using this method (Maser et al., 2003), but it has limited applications because of its struggle in intergallery polymerization.

15.4.2 Solution Intercalation

Another name for the *solution intercalation* technique is solution casting method. This is a traditional process for bionanocomposites film production where all the constituents are mixed to make a solution, and then stirred, heated, and dried. In this fabrication technique, mixing is done by simple shear mixing with a glass rod or magnetic stirrer to obtain a homogenous distribution of nanofillers in the colloidal suspension. The application of ultrasonication during mixing breaks the agglomerations of nanoparticles, thereby enhancing the bionanocomposite film properties (Abral et al., 2018). The homogenously mixed solution is casted on acrylic plates and vacuum-dried (50°C for 48 hours) to remove the solvent. Deionized water, acetone, ethanol, etc. are the preferred solvents (Kalaitzidou et al., 2007) for fabrication of bionanocomposites by solution intercalation method. Eventually, thus prepared films are kept in desiccators (Asad et al., 2018) for further use. Solution intercalation is the common method for small-scale production (Mangaraj et al., 2019) of bionanocomposite materials, such as in research laboratories or pilot plants. This method is most suitable for preparing bionanocomposites with polymers like PCA, PLA, and PVA (Zammarano et al., 2005), but a wide range of nanofillers can be incorporated.

15.4.3 Melt Intercalation

This method does not require a solvent because fillers are directly dispersed into the molten polymer using injection molding and extrusion (Kim et al., 2010). Employing high-shear mixing allows a homogenous mixing of fillers in molten polymers. In this method, an intercalated network is formed between the polymer chain and nanofibers to form a bionanocomposite (Weng et al., 2005). However, this method is only suitable to a limited group of polymers, like PLA, polyolefins, and polyesters (Sharma et al., 2018).

15.4.4 Electrospinning

The electrospinning fabrication method uses high voltage in between the tip of a syringe and the collection plate. The syringe contains a solution of biopolymers that is drawn out from the tip, forming a thin stream which is dried when encountering the cool air to form fiber. In this way, fibers are collected on the surface of the collection plate and are then further dried at higher temperature.

Bionanocomposite films produced by this method are very delicate, possess high specific area, and are extremely porous due to interconnected pores (Zhang et al., 2020b). The electrospinning method is commonly used in laboratories and industries (Khan et al., 2013), and the films produced this way are suited for packaging applications (Zhang et al., 2020a). However, this method is applicable only to those biopolymers that can be charged electrically and have the ability to be dragged out through a nozzle.

15.4.5 Extrusion

This method uses high pressure, temperature, and shear force to plasticize the biopolymer mixture. The plasticized mixture is then pressed through a designated die to give it the required shape. The types of biopolymers and additives used and the processing condition, such as share rate, temperature, and pressure, determine the property of the bionanocomposite films produced (Hyvärinen et al., 2020). However, this method is not applicable for fabricating biopolymers that cannot withstand extreme processing conditions. For better plasticizing property, glycerol is used due to its non-toxic nature and good thermal stability. The extrusion method can work for both small- or large-scale production of biodegradable food packaging materials but is best suited for large-scale production (Krepker et al., 2018) when continuous processes are applied.

15.4.6 Compression Molding

In the compression molding method of fabrication of polymer, a mixture of biopolymer and additives is placed into a suitable mold and compressed to facilitate cross-linking of the biopolymers. Compressing the mold to form film is also called curing, which can be carried out at room temperature (in cold compression) or elevated temperature (in hot compression) (Guerrero et al., 2019). This method is compatible for producing bionanocomposite films at both a small and large scale (Mangaraj et al., 2019; Roy & Rhim, 2021). This method is applicable only to those polymers that can set and form a film by pressing alone or by combined pressing and heating (Mangaraj et al., 2019).

15.5 Bionanocomposite Coating Preparation

Basically, coating is done to maintain the quality of fresh food and to extend the shelf life (Barikloo & Ahmadi, 2018). Coating can be applied on the food surface through two main methods: spraying and dipping. In the spraying process, the solutions are applied to the surface of the foodstuffs using an aerosol sprayer (Qu & Luo, 2021), whereas immersion involves dipping fresh food into a nanocomposite solution, which may include preservatives and plasticizers, to improve the properties of the coating material (Abdollahi et al., 2014). Obtaining coating by immersion is a simple method that provides a good packaging without requiring advanced equipment (Kerch, 2015).

15.6 Popular Bionanocomposites for Food Applications

Nanocomposites consist of two phases that have biopolymer as continuous phase and nanofiller as dispersed phase, with the filler diameter being less than 100 nm (Khalil et al., 2012). "The application of nanoparticles in food industries can be divided into two major groups: food nano-sensing and nanostructured food ingredients" (Singh et al., 2017). Food nanosensing is applicable for improving the safety and quality of food, whereas nanostructured food ingredients are applicable for improving food packaging and processing. Food industries have been using bionanocomposites not only for containment and protection of food but also for nanosensing smart packaging. The nanostructure of biopolymers has also been used for food processing in the form of anticaking agent, chelating agent, and other additives (Primožič et al., 2021).

Despite their being only an alternative for conventional fossil fuel–based packaging, the commercialization of these materials is still in the infantile stage. Their poor barrier and mechanical properties limit their usage for commercial production. Researchers are working to develop bionanocomposite materials that can overcome all the constrains by using different nanoparticles, as it can improve the properties of bionanocomposites (Kraśniewska et al., 2019; Zambrano-Zaragoza et al., 2018). Recently developed bionanocomposite materials for their application in food packaging and processing are listed in Table 15.3.

15.6.1 Food Packaging

The prime objective of food packaging is to protect food from the surrounding environment and extend the shelf life of the food. Other aspects of food packaging are convenience, communication, advertising, etc. The proper packaging of food plays a crucial role in maintaining food safety, as it prevents food spoilage by contamination (He et al., 2019). Many studies have been done regarding the use of nanoparticles in food packaging material, and it is confirmed by many researchers that nanoparticles can improve the performance of the polymer material used for packaging. One such example of using nanoparticles in packaging is nanolaminate film. Such films are prepared by layer-by-layer deposition of nanomaterials on the polymer surface, allowing multiple nanolayers as an interfacial film. Additives such as surfactants, colloidal particles, and electrolytes are used for better performance of nanolaminates (Kuswandi & Moradi, 2019). Similarly, incorporating antibacterial and antioxidant compounds (Galus et al., 2020) helps in increasing the shelf life of the food (Bajpai et al., 2018; Salgado et al., 2015). Biodegradable nanolaminates are prepared by using biopolymers (Ranjan et al., 2014). For food packaging applications, bionanocomposites find their use in developing improved, active, and smart packaging.

15.6.1.1 Improved Packaging

The inherent property of biopolymer can be improved by incorporating CNF (Tibolla et al., 2019) to make bionanocomposite materials that have superior barrier and mechanical properties and are biodegradable. For the preparation of biodegradable food packaging materials, cellulose nanoparticles are used as reinforcing agent. A very small fraction of cellulose nanoparticles is required to achieve the desired result, and overdosing the nanoparticles can act just the opposite due to agglomeration.

Lavoine et al. (2016) developed a biodegradable but strong food packaging material by applying CNF coating on paper (Lavoine et al., 2016). Similarly, improved thermal stability, tensile strength, and oxygen and water vapor barrier properties were achieved by dosing CNF in whey protein–based polymer (Karimi et al., 2020). Song and his coworker (2015) have found in his study that tensile strength was increased to 14.86 MPa with the incorporation of CNC in the polymer matrix (Song et al., 2015). The incorporation of CNF for fabricating cellulose acetate nanocomposite and the application of modified starch (Basumatary et al., 2020) both exhibit an enhanced performance of the nanocomposite material for food packaging. Mechanical and barrier properties of biopolymer-based food packaging material can also be improved by the addition of clay nanoparticles. Besides, such packagings are lightweight, are heat- and shatter-resistant, and promote an extended shelf life of the packaged food (Sharma et al., 2018).

Graphene sheets have excellent gas barrier properties and remarkably high elastic modulus (Uysal Ünalan, 2014), and its coating on biopolymer-based food packaging material can be done to fabricate material with such properties. Sadeghizadeh-Yazdi et al. (2019) used plant polysaccharide (Lallemantia iberica mucilage) polymer reinforced with titanium dioxide (TiO_2) nanoparticle and came up with bionanocomposite films that can withstand high humidity, pressure, and temperature (Sadeghizadeh-Yazdi et al., 2019).

15.6.1.2 Active Packaging

Active packaging makes use of those materials that show antimicrobial, antioxidant, antiscavenging, and absorptive effect (Rodrigues et al., 2021). Such packaging materials are designed to prevent microbial growth and objectionable chemical reactions and enzyme activity (Yildirim et al., 2018), thereby extending the shelf life of the packaged food (Cacciotti et al., 2018). Besides protecting the package from the outside environment, the objective of active packaging is to ensure utilization of the functional ingredients of the food (Alizadeh-Sani et al., 2021; Iordanskii, 2020; Moreirinha et al., 2020).

Some nanomaterials exhibit antimicrobial, UV protection activity, possibility of oxidation prevention, etc. and are therefore incorporated during the fabrication of active food packaging materials. Nanocomposites with active packaging property have been developed for preserving vegetables and dairy and meat products (Jamróz et al., 2019). Similarly, Youssef and coworkers (2016) found that Egyptian soft white cheese can be preserved from being oxidized and from browning by using a ZnONPs-chitosan-CMC film.

Active agents like nanoparticles, antimicrobial and antioxidant compounds, etc. in the active packaging materials are positioned within the film but not in the food, and this prevents the ingestion of objectionable preservatives and additives while consuming the food (Maisanaba et al., 2017; Sun et al., 2017). Based on the mode of action, active packaging systems can be divided into the following three categories (Table 15.1).

TABLE 15.1

Active packaging systems based on their mode of action

	Active packaging system	Mode of action
1	Releasing system	Antimicrobial, antioxidant particles, etc. are released in the packaging system.
2	Absorbing system	Oxygen, carbon dioxide, moisture, or aroma is absorbed by the packaging system.
3	Non-migrating system	Contact between food and packaging material is not required.

TABLE 15.2

Active packaging systems based on the positioning of active agents in the packaging system

	Active packaging system	Positioning of active agent
1	Using active sachets	Sachets containing oxygen or moisture absorbers or ethanol vapor generators are placed inside the packaging system.
2	Coating an active compound	Active compounds are coated in the inner wall of the polymer in the packaging system.
3	Immobilizing active compound	Interacting functional compounds are immobilized on the surface of the polymer in the packaging system. Sustained release of active compound in packaging system.
4	Incorporation into polymer	Active compounds are directly blended throughout the polymer matrix used as a packaging material. High resistance against adverse processing conditions.

Based on the positioning of active agents in the packaging system, active packaging systems are classified into four categories (see Table 15.2).

15.6.1.2.1 Antioxidant and Scavenging Active Packaging

To prevent packaged food from browning and other undesirable physicochemical change, antioxidant- and scavenging agent–loaded bionanocomposites have been developed. According to the nature of food, various food packaging materials that can absorb moisture, carbon dioxide, ethylene, and oxygen from packaged food and can release ethanol and flavor to packaged food (Basumatary et al., 2020) can be fabricated. Usually, nanoparticles are used as carrier of these active agents, as they allow plenty of interactions due to high surface area. In a different study conducted by Chatterjee et al. (2010) and Cherpinski et al. (2018), chitosan-based nanocomposite films reinforced with carbon nanotubes exhibited remarkably good adsorption capability (Chatterjee et al., 2010), and similarly, palladium nanoparticle–reinforced polyhydroxybutyrate (PHB) films showed higher oxygen scavenging property (Cherpinski et al., 2018).

15.6.1.2.2 Antimicrobial Active Packaging

Antimicrobial ability is an important property of packaging materials for food and medicine industries. Food packaging materials can be dosed with antimicrobial compounds for better preservation and shelf life extension of food inside the packet. Dosing antimicrobial compounds is done either by coating or by integrating into the packaging material itself (Mustafa & Andreescu, 2020).

Some biopolymers also exhibit antimicrobial activity. For instance, biopolymer chitosan has antimicrobial property against a wide spectrum of bacterial and fungi microbes (Kravanja et al., 2019; Verlee et al., 2017). To further increase the antibacterial activities, metal nanoparticles are used in combination (Matharu et al., 2018) because of their excellent antimicrobial activity against a wide range of bacteria, protozoa, fungi, and even some virus (Hoseinnejad et al., 2018; Kumar et al., 2017). Nanoparticles such as silver nanoparticles, ZnO nanoparticles, and TiO_2 nanoparticles are most suited (Becerril et al., 2020) for developing such packaging materials. However, essential oils, phytochemicals, or organic biopolymeric (carbohydrate, protein, lipid) nanoparticles can also be used.

Researchers have developed several bionanocomposite films suitable for food packaging that have antibacterial activity, such as silver zeolite–chitosan film (Rhim et al., 2013), ZnONPs-PLA film (Marra et al., 2016), halloysite-clay-starch-nisin film (Meira et al., 2016), PVA-MMT-AgNPs film (Mathew et al., 2019), and mahua oil–based polyurethane-chitosan-ZnONPs film (Indumathi & Rajarajeswari, 2019), ZnONPs-PLA film (Shankar et al., 2018), and CNC-protein-based polymer (Xiao et al., 2020). Nevertheless, not limited to inhibit only bacteria, many bionanocomposites have been developed to inhibit a wide range of microbes, including fungi and protozoa. Some examples of such films are clay-polysaccharide film (R. Sharma et al., 2020) and MMT-starch-clove-essential-oil film (Echeverría et al., 2018).

15.6.1.3 Smart Packaging

Smart packaging is a novel concept that allows monitoring of packaged food in real time. This is one of the prominent innovations in food packaging that help ensure food safety. It can detect physical, chemical, or microbial changes within the packaging system (Alizadeh-Sani et al., 2020; Biji et al., 2015). Many times, smart packaging is referred to as intelligent packaging in literatures, while some researchers have categorized both intelligent packaging and active packaging under smart packaging (Rodrigues et al., 2021).

Such packaging materials are designed to be pH and temperature sensitive, thereby indicating the real-time condition of the food (Chaudhary et al., 2022; Kritchenkov et al., 2021). Change in pH changes the color of the indicator in the packaging material that gives an indication of the level of deterioration of food (Chen et al., 2020). Generally, natural pigments are used as color indicator, as they are environment-friendly and non-toxic (Mohammadalinejhad et al., 2020). Colorimetric-based smart packaging systems are economical and simple (Zhang et al., 2020b).

Smart food packaging materials give information of product history, if it has been exposed to harsh storage condition such as sunlight, oxygen, high temperature, or humidity or has been contaminated by pathogenic microorganisms. Since they provide information about quality and safety of stored food (Merz et al., 2020), they can contribute to safety and quality of the packaged food (Kuswandi et al., 2011). Bionanocomposite-based smart food packaging makes use of nanoparticles owing to its exceptional optical properties and high surface reactivity. The nanoparticles can detect even small changes at a molecular level, and therefore, when used for sensing purposes, they are also called nanosensors. Certain gas nanosensors can detect food spoilage (Madhusudan et al., 2018) and can identify and quantify the pathogenic microorganisms, toxins, and chemicals in the food product (Li & Sheng, 2014; Pramanik et al., 2020; Sharma et al., 2017). There are various types of nanosensors, and among them, optical nanosensors are commercially popular because of their convenience (Bumbudsanpharoke & Ko, 2019).

15.6.2 Food Processing

15.6.2.1 Edible Food Coatings

A layer of edible food-grade material can be coated on the surface of solid food as an intact edible packaging material. Edible coatings can be applicable for fresh fruits and vegetables, meat, fish, cheese, and breads (Azeredo et al., 2019; Nile et al., 2020). Edible nanocoatings can be applied on food surfaces by immersing (dipping), rubbing, or spraying (Zambrano-Zaragoza et al., 2018). Applying nanocoatings by immersion or rubbing does not require specialized equipment, but spray-coating needs an atomizer (Muxika et al., 2017).

Edible coatings preserve the quality of the food and increases the shelf life of food (Barikloo & Ahmadi, 2018; Youssef & Hashim, 2020). It can protect the food from browning and oxidation and can also be fabricated to impart color and/or flavor to the food. It has a huge possibility to flourish as a novel food packaging material in the future (Galus et al., 2020), and therefore, different biopolymers have been exploited for the preparation for edible coatings. Edible coatings are adhered on the surface of the food (Hassan et al., 2018) and can be consumed together with food (Youssef & El-Sayed, 2018). Edible biopolymers are aided with organic or inorganic nanoparticles to enhance its functionality to be used as edible nanocoating. Clay-starch edible bionanocomposites have also been developed with improved mechanical properties (Sharma et al., 2018). BNC is one of the important biopolymers in preparing edible coating. The U.S. Food and Drug Administration (USFDA) has recognized BNC as generally recognized as safe (GRAS) product (Azeredo et al., 2019). It is popularly used in preparing edible food packaging and can also be processed to be consumed as food, such as nata de coco, a traditional dessert in the Philippines (Ludwicka et al., 2020).

15.6.2.2 Nanoparticles Incorporated in Food

Organic and biopolymeric nanoparticles are studied for their incorporation in food for better nutritional or physicochemical properties (Pan & Zhong, 2016). Usually, nanoparticles of carbohydrate, protein and lipid are used as organic biopolymeric nanoparticles. When carbohydrate, protein, and lipid are used together to make a nanocomposite film. Carbohydrates and proteins work as emulsifier and stabilize the lipid nanoparticles in the matrix (Burger & Zhang, 2019). Biopolymeric nanoparticles are fully digested, and hence, researchers are using organic nanoparticles to develop them as a functional ingredient in food (McClements et al., 2017). However, inorganic nanoparticles are also being explored to incorporate in food. In one study, digestion of TiO_2 nanoparticle–incorporated nanoemulsions was studied to find its negative impact on health, and no side effects were observed other than a slight reduction in the digestion of lipid present in nanoemulsions (Li et al., 2017). Recently developed bionanocomposites for application in food packaging and food processing are shown in Table 15.3.

TABLE 15.3

Recently developed bionanocomposites for applications in food packaging and food processing

	Nanoparticle	Biopolymer/Matrix	Objective of bionanocomposite	References
			Improved packaging	
1	CNF	Paper (cellulose)	Strong packaging material	(Lavoine et al., 2016)
2	CNF	Whey protein	Improved thermal stability, tensile strength, and vapor barrier properties of packaging material	(Karimi et al., 2020)
3	CNC	Polyurethane	Increased tensile strength (14.86 MPa) and Young's modulus (344 MPa) of packaging film	(Song et al., 2015)
4	Rice straw derived CNF	Cellulose acetate	Improved thermal, mechanical, and optical properties of packaging material	(Sharma et al., 2021)
			Active packaging	
5	PdNPs	PHB	Oxygen scavenging activity and water and oxygen barrier properties of packaging film	(Cherpinski et al., 2018)
6	CNT	Chitosan	Enhanced adsorption of packaging material	(Chatterjee et al., 2010)
7	CNF	Soy protein	Improved thermal stability, tensile strength, and vapor barrier properties of antibacterial packaging material	(Xiao et al., 2020)
8	Ag zeolite	Chitosan	Antibacterial food packaging film	(Rhim et al., 2013)
9	Clay nano-particle	Polysaccharide	Antimicrobial food packaging film	(Sharma et al., 2020)
10	Halloysite clay	Starch/nisin	Film to protect cheese against *L. monocytogens*	(Meira et al., 2016)
11	MMT	Starch/clove essential oil	Antibacterial and antioxidant packaging film	(Echeverría et al., 2018)
12	MMT, AgNPs	PVA	Chicken sausage packaging to prevent growth of aerobic bacteria	(Mathew et al., 2019)

TABLE 15.3 (*Continued*)

	Nanoparticle	Biopolymer/Matrix	Objective of bionanocomposite	References
13	ZnONPs	Mahua oil–based polyurethane/chitosan	Effective antimicrobial packaging for carrot slices	(Indumathi & Rajarajeswari, 2019)
14	ZnONPs	PLA	Effective packaging against *E. coli* and *L. monocytogens* Smart packaging	(Shankar et al., 2018)
15	Nanosensors	—	Real-time monitoring of freshness of food	(Pramanik et al., 2020)
16	Nanosensors	—	Detection of flavor and color of food	(Li & Sheng, 2014)
17	Nanosensors	—	Detection of food spoilage Food processing	(Madhusudan et al., 2018)
18	Clay	Starch	Edible coating with improved mechanical properties	(Sharma et al., 2018)
19	Lipid nanoparticle	Food matrix	Carrier of small bioactive agents	(Speranza et al., 2013)
20	Protein nanoparticle	Food matrix	Nutrient carrier, texture modifier, fat replacers	(Cho & Jones, 2019)

15.7 Public Acceptance of Nanoparticles in Their Food

Public knowledge about nanoparticle use in food is very limited (Siegrist et al., 2007). People have been consuming nanoparticles in their food for ages, but they are not willing to embrace externally added nanoparticles in their food (Siegrist et al., 2007, 2008). This is one of the reasons that the use of nanoparticles for food packaging has become commercially successful, while its use as a food ingredient is still struggling (Chaudhry et al., 2010). Potential risks of consuming nanoparticles are still uncertain, and this may be the reason for consumers having a negative perception toward the use of nanoparticles in food (Chun, 2009).

15.8 Popular Bionanocomposites in Medical Applications

The application of nanoparticles in the medical field has resulted to many advanced health-care services and products (Pramanik et al., 2020), and along with new, emerging technologies and methods, there are tremendous possibilities for further advancement (Chopdey et al., 2015; Dua et al., 2017; Maheshwari et al., 2015). As shown in Figure 15.2, there are three main categories of health services where nanoparticles in the form of bionanocomposites can possibly be used.

15.8.1 Bionanocomposites in Transdermal Drug Delivery

Drug delivery system is the delivery of medicinal chemical in the body's specific part with minimal risk of toxicity. The most common oral route of drug administration exhibits low bioavailability and low membrane permeability (Gomez-Orellana, 2005). In another hand, drug delivery through the skin provides larger surface area for drug absorption, and the drug is directly absorbed in the body's cells (Kupnik et al., 2020). Chemical penetration enhancers can be used to increase the efficiency of transdermal drug delivery (Maher et al., 2019). Nanomaterials can be used for drug delivery applications. The excellent drug-binding capacity of nanoparticles is due to its large surface area. Biopolymeric nanoparticles have their own distinct ability to deliver to targeted cells in the body (Yang et al., 2006).

Drug release rate from bionanocomposites depends on the nature of the biopolymer used (Kayat et al., 2011; Kurmi et al., 2010), with most used biopolymers being cellulose, chitosan, collagen, gelatin, alginate, and starch (Jacob et al., 2018). Biopolymers are biocompatible and biodegradable and

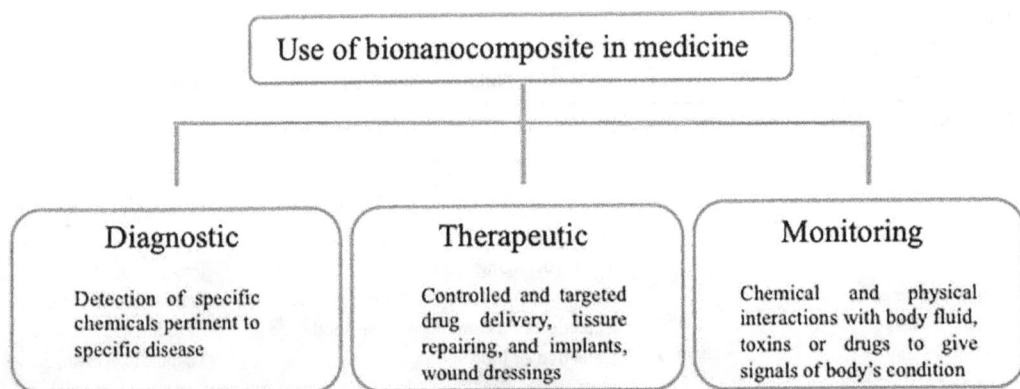

FIGURE 15.2   Applications of bionanocomposites in medical and health-care services.

easily allow surface modification (Tekade et al., 2017) to get the desired properties. Nanocellulose and nanofillers are widely used for drug delivery applications owing to their biocompatibility, biodegradability, larger surface area, and mechanical strength (Huang et al., 2019; Ioelovich, 2016; Salas et al., 2014), which are required characteristics for any drug delivery material. The crystalline nanocellulose obtained from banana pseudo-stem has been studied for the development of an antibacterial patch that shows effective drug release when tested against Gram-negative *E. coli* and Gram-positive *S. aureus* (Shrestha et al., 2021). Nanocellulose can exhibit a sustained release of drug ranging from a few minutes to several days. The use of cellulose in drug products has already been included in the approved list of USFDA (Tran & Wang, 2014).

Likewise, chitosan- (Tekade et al., 2017) and hydrogel-based nanocomposites (Rasoulzadeh & Namazi, 2017) are also being used for drug delivery applications. Chitosan has an excellent bioadhesive property and demonstrates superior ability for sustained release of the drug (Tan et al., 2017). Drug release kinetics may be different for the same nanomaterial from a different source, and this problem can be avoided by using genetically recombinant nanomaterials (Wenk et al., 2011). Bionanocomposites are the recent advancement for efficient drug delivery at target sites (Kukoyi, 2016). Bionanocomposites are fabricated in the form of patches and films for transdermal drug delivery. For different diseases such as diabetes, allergies, inflammation, infection, and cancer, drug-loaded bionanocomposites have already been developed (Nitta & Numata, 2013). Targeted drug delivery of "cisplatin" for treatment of cancerous cells was achieved by using nanocomposites that used protein nanoparticles as a carrier (Zhen et al., 2013).

Besides listed FDA-approved uses, bionanocomposite patches for lowering fever, relieving pain, and controlled nicotine dosing are commercially available (Pastore et al., 2015). Nanocomposites for cancer treatment have also been studied, showing prolonged release of the "doxorubicin" drug with no observed toxicity (Poonguzhali et al., 2017). To prevent the rapid spread of COVID-19 lately, tissue paper having antimicrobial property has also been developed (Patel, 2020). Nanoparticles and nanomaterials are popularly used in skin-care products as well. For instance, ZnO_2 and TiO_2 nanoparticles are incorporated in sunscreen products (Morganti, 2010), and when applied topically on skin, they can make a protective layer against damage caused by the ultraviolet rays of the sun.

15.8.2 Bionanocomposites in Tissue Engineering and Implants

Tissue engineering is an advanced technology with the help of which tissues and organs can be regenerated or reconstructed. Different bionanocomposites have been developed that are suitable scaffolds for tissue engineering.

PVA-based bionanocomposite reinforced with calcium-doped zirconium phosphate (ZrP) nanoparticles and titanium-doped (Ti) zirconium nanoparticles bionanocomposite can be potentially used for scaffold-guided tissue engineering application (Kalita et al., 2017). Hydroxyapatite (HA) nanoparticles are the material of choice for fabricating scaffolds for bone repair, as they can promote differentiation and proliferation of cells (Nazeer et al., 2017). Polyvinylidene fluoride (PVDF)–polymethyl methacrylate (PMMA)-HA-TiO$_2$ bionanocomposite scaffolds have been commended for repairing bone structures, owing to their excellent compatibility with bone (Arumugam et al., 2019). Selenium nanoparticles (SeNPs) have excellent antioxidant, electrical, and mechanical properties, enabling them to be applicable for cardiac tissue engineering, such as the fabrication of cardiac patches (Kalishwaralal et al., 2018).

Nanomaterials are ever more being exploited for developing implantable devices, and its market is predicted to grow with the growing market of nanomaterial-based implantable devices (Pramanik et al., 2020). Several techniques, in particular, physical vapor deposition and chemical vapor deposition (Lukaszkowicz, 2011), can be applied for nanocomposite coating to the surface of implant structures. Various bionanocomposites are designed to offer different structures and properties for use in tissue engineering that act as implantable scaffolds (Stumpf et al., 2018).

However, still more studies and investigations are being carried out for developing bionanocomposites with the most desired characteristics for producing higher-quality prosthetic implants (Schmalz et al., 2017), bone substitute materials, orthopedic implants, tissue implants, and dental implants. PLA is a commonly used biopolymer for biomedical implants and stents, as it is resistant to hydrolysis degradation due to the body's metabolism. It is generally used in combination with hydroxyapatite (HA) nanoparticles (Davachi et al., 2017).

15.8.3 Bionanocomposites in Wound Dressings

Wound dressing materials with nanoscale fabrication have the capability to heal wounds better (Cortivo et al., 2010; Rajendran et al., 2018). Chitosan-gelatin-AgNP bionanocomposites incorporated with *Ganoderma lucidum* can be used for the management of wounds. Graphene oxide–chitosan bionanocomposites have been found to be antibacterial and effective disinfectant (Chowdhuri et al., 2015). Antibacterial nanoparticles can cause lysis of cell by electrostatically binding to the cell wall of microorganisms and changing its membrane potential (Pelgrift & Friedman, 2013).

Improvement in wound healing was observed in albino rats by using nanostarch-reinforced chitosan–polyvinyl propylidene bionanocomposite (Poonguzhali et al., 2017). Similarly, faster wound healing was evinced in male rabbits by using ciprofloxacin-loaded PVA-sodium-alginate-nanofiber bionanocomposite (Kataria et al., 2014). Most interestingly, a coating of gelatin-silver nanoparticle during the fabrication of polycaprolactone-based wound dressing can facilitate easy peeling of the dressing by preventing sticking of dressing to the tissue (Zimmermann et al., 2020). Recently developed bionanocomposite materials for medical applications are listed in Table 15.4.

TABLE 15.4

Recently developed bionanocomposites for medical applications

	Nanoparticle	Biopolymer/Matrix	Objective of bionanocomposite	References
1	Casein protein nanoparticle	Paper (cellulose)	Targeted delivery of "cisplatin" drug for treatment of cancerous cell	(Zhen et al., 2013)
2	CNC	Chitosan/tetracycline	Development of antibacterial patch	(Shrestha et al., 2021)
3	Zirconium nanoparticle (ZrNPs)	PVA	Scaffold for tissue engineering	(Kalita et al., 2017)
4	TiO$_2$NPs	Polyvinylidene/ polymethylmethacrylate	Bone repair	(Arumugam et al., 2019)
5	SeNPs	Chitosan	Cardiac patches	(Kalishwaralal et al., 2018)
6	Nanostarch	Chitosan/polyvinyl propylidene	Enhanced wound healing	(Poonguzhali et al., 2017)
7	Nanofiber	PVA/ciprofloxacin	Wound healing	(Kataria et al., 2014)
8	AgNPs	Gelatin	Easy-to-peel wound dressing patch	(Thanh et al., 2018)

15.9 Regulatory and Safety Issues of Using Bionanocomposites in Food

There is swarming research interest in exploiting nanoparticles (biopolymeric, organic, inorganic) for improving the quality of food. In general, the beneficial effects of nanocomposite materials are well recognized, but fewer studies have been performed for the effects of nanoparticles on human health. The possible routes through which nanoparticles may enter the human body are by skin surface contact, respiration (Ede et al., 2019), and ingestion (He & Hwang, 2016).

Nanoparticle migration in food is a subject of concern, and much research has been done (Bott et al., 2014). But it is found that the migration of nanoparticles used in food packaging, such as TiO_2, ZnO, SiO2, aluminum oxide, to food is insignificantly less (Garcia et al., 2018). For instance, selenium nanoparticles (SeNPs) are safe to be used in food packaging according to regulations of food packaging (Abreu et al., 2015). Edible nanofilms and coatings can be manufactured by following food ingredient–related regulations (Galus et al., 2020). However, the declaration of edible coating/packaging containing allergen (if any) must be labelled on the packaging according to regulations of the country or territory. Nanoparticle exposure for a prolonged period may cause phagocytosis dysfunction, protein denaturation, and reactive oxygen species (ROS) formation (Chaudhary et al., 2020). Nanoparticles of metal and metal oxide are considered non-toxic in nature but can cause oxidative stress. However, TiO_2 nanoparticle is considered all safe and has been approved as a food additive.

Regarding the chemicals they are composed of, bionanocomposites can be equally toxic as compared to petroleum-based plastic. Nevertheless, all-purpose endorsement should not be made regarding safety or toxicity of all types of nanoparticles. Nanomaterial toxicity is greatly determined by nanoparticle size (Mauricio et al., 2018), the nature of the nanoparticle (McClements & Xiao, 2017), biopolymer, food matrix, and the interaction therein. Therefore, all these factors should be studied from safety aspects, and its digestion and absorption should also be closely studied (Neethirajan et al., 2018). There is a concern that nanoparticles from food system can eventually get accumulated in the environment (Youssef & El-Sayed, 2018), thereby increasing the chance of getting back to the food system from the environment. The use of nanoparticle in food is regulated by the USFDA in the United States, Food Standards Australia and New Zealand (FSANZ) in Australia, and the European Union and Novel Food Regulation (EC 258–97) in EU. However, it is necessary to develop universal international protocol for regulating the use of nanoparticles in the food industry throughout the world.

15.10 Conclusion

Medical science has advanced this far with the use of bionanocomposites for diagnostics and the treatment or monitoring of ailments. Medical science has been using bionanocomposites for drug delivery, tissue engineering, implants, and wound dressings. Likewise, they have earned an important place in food packaging in the form of improved, active, and smart packaging and play a vital role in ensuring food safety and quality. They are also being used in food as additives or functional ingredients.

However, nanoparticles have their own distinct characteristics and, when incorporated in food, may not be digested and utilized as other conventional food ingredients. Still, more research and investigations are required for detail and precise understanding of the fate of nanoparticles in human body and environment and its toxicity. While there are many studies to improve the performance of bionanocomposite altogether for food and medical applications, there are even more challenges, and hence more opportunities to work on, in creating better bionanocomposite materials. In the days to come, it is more important to consider the safety aspects of production along with development in the methods and technologies for the fabrication of bionanocomposites with better performance for food packaging and medical applications that are commercially viable.

REFERENCES

Abdollahi, M., Rezaei, M., & Farzi, G. (2014). Influence of chitosan/clay functional bionanocomposite activated with rosemary essential oil on the shelf life of fresh silver carp. *International Journal of Food Science & Technology*, *49*(3), 811–818.

Abral, H., Anugrah, A. S., Hafizulhaq, F., Handayani, D., Sugiarti, E., & Muslimin, A. N. (2018). Effect of nanofibers fraction on properties of the starch based biocomposite prepared in various ultrasonic powers. *International Journal of Biological Macromolecules, 116,* 1214–1221.

Abreu, A. S., Oliveira, M., de Sá, A., Rodrigues, R. M., Cerqueira, M. A., Vicente, A. A., & Machado, A. V. (2015). Antimicrobial nanostructured starch based films for packaging. *Carbohydrate Polymers, 129,* 127–134.

Alizadeh-Sani, M., Moghaddas Kia, E., Ghasempour, Z., & Ehsani, A. (2021). Preparation of active nanocomposite film consisting of sodium caseinate, ZnO nanoparticles and rosemary essential oil for food packaging applications. *Journal of Polymers and the Environment, 29*(2), 588–598.

Alizadeh-Sani, M., Mohammadian, E., Rhim, J.-W., & Jafari, S. M. (2020). pH-sensitive (halochromic) smart packaging films based on natural food colorants for the monitoring of food quality and safety. *Trends in Food Science & Technology, 105,* 93–144.

Arfat, Y. A., Ahmed, J., & Jacob, H. (2017). Preparation and characterization of agar-based nanocomposite films reinforced with bimetallic (Ag-Cu) alloy nanoparticles. *Carbohydrate Polymers, 155,* 382–390.

Arumugam, R., Subramanyam, V., Chinnadurai, R. K., Srinadhu, E. S., Subramanian, B., & Nallani, S. (2019). Development of novel mechanically stable porous nanocomposite (PVDF-PMMA/HAp/TiO$_2$) film scaffold with nanowhiskers surface morphology for bone repair applications. *Materials Letters, 236,* 694–696.

Asad, M., Saba, N., Asiri, A. M., Jawaid, M., Indarti, E., & Wanrosli, W. D. (2018). Preparation and characterization of nanocomposite films from oil palm pulp nanocellulose/poly (Vinyl alcohol) by casting method. *Carbohydrate Polymers, 191,* 103–111.

Azeredo, H. M. C., Barud, H., Farinas, C. S., Vasconcellos, V. M., & Claro, A. M. (2019). Bacterial cellulose as a raw material for food and food packaging applications. *Frontiers in Sustainable Food Systems, 3,* 7.

Azeredo, H. M. C., Mattoso, L. H. C., Wood, D., Williams, T. G., Avena-Bustillos, R. J., & McHugh, T. H. (2009). Nanocomposite edible films from mango puree reinforced with cellulose nanofibers. *Journal of Food Science, 74*(5), N31–N35.

Bajpai, V. K., Kamle, M., Shukla, S., Mahato, D. K., Chandra, P., Hwang, S. K., Kumar, P., Huh, Y. S., & Han, Y.-K. (2018). Prospects of using nanotechnology for food preservation, safety, and security. *Journal of Food and Drug Analysis, 26*(4), 1201–1214.

Baranwal, J., Barse, B., Fais, A., Delogu, G. L., & Kumar, A. (2022). Biopolymer: A sustainable material for food and medical applications. *Polymers, 14*(5), 983.

Barikloo, H., & Ahmadi, E. (2018). Shelf life extension of strawberry by temperatures conditioning, chitosan coating, modified atmosphere, and clay and silica nanocomposite packaging. *Scientia Horticulturae, 240,* 496–508.

Basumatary, I. B., Mukherjee, A., Katiyar, V., & Kumar, S. (2020). Biopolymer-based nanocomposite films and coatings: Recent advances in shelf-life improvement of fruits and vegetables. *Critical Reviews in Food Science and Nutrition, 62,* 1912–1935.

Becerril, R., Nerín, C., & Silva, F. (2020). Encapsulation systems for antimicrobial food packaging components: An update. *Molecules, 25*(5), 1134.

Biji, K. B., Ravishankar, C. N., Mohan, C. O., & Srinivasa Gopal, T. K. (2015). Smart packaging systems for food applications: A review. *Journal of Food Science and Technology, 52*(10), 6125–6135.

Bott, J., Störmer, A., & Franz, R. (2014). Migration of nanoparticles from plastic packaging materials containing carbon black into foodstuffs. *Food Additives & Contaminants: Part A, 31*(10), 1769–1782.

Bumbudsanpharoke, N., & Ko, S. (2019). Nanomaterial-based optical indicators: Promise, opportunities, and challenges in the development of colorimetric systems for intelligent packaging. *Nano Research, 12*(3), 489–500.

Burger, T. G., & Zhang, Y. (2019). Recent progress in the utilization of pea protein as an emulsifier for food applications. *Trends in Food Science & Technology, 86,* 25–33.

Cacciotti, I., Mori, S., Cherubini, V., & Nanni, F. (2018). Eco-sustainable systems based on poly (lactic acid), diatomite and coffee grounds extract for food packaging. *International Journal of Biological Macromolecules, 112,* 567–575.

Cazón, P., & Vázquez, M. (2019). Applications of chitosan as food packaging materials. *Sustainable Agriculture Reviews, 36,* 81–123.

Cesur, S., Köroğlu, C., & Yalçın, H. T. (2018). Antimicrobial and biodegradable food packaging applications of polycaprolactone/organo nanoclay/chitosan polymeric composite films. *Journal of Vinyl and Additive Technology, 24*(4), 376–387.

Charles, A. L., Kao, H.-M., & Huang, T.-C. (2003). Physical investigations of surface membrane–water relationship of intact and gelatinized wheat–starch systems. *Carbohydrate Research*, *338*(22), 2403–2408.

Chatterjee, S., Lee, M. W., & Woo, S. H. (2010). Adsorption of Congo red by chitosan hydrogel beads impregnated with carbon nanotubes. *Bioresource Technology*, *101*(6), 1800–1806.

Chaudhary, P., Fatima, F., & Kumar, A. (2020). Relevance of nanomaterials in food packaging and its advanced future prospects. *Journal of Inorganic and Organometallic Polymers and Materials*, *30*(12), 5180–5192.

Chaudhary, V., Punia Bangar, S., Thakur, N., & Trif, M. (2022). Recent advancements in smart biogenic packaging: Reshaping the future of the food packaging industry. *Polymers*, *14*(4), 829.

Chaudhry, Q., Watkins, R., & Castle, L. (2010). Nanotechnologies in the food arena: New opportunities, new questions, new concerns. *Nanotechnologies in Food*, *14*, 1–17.

Chen, S., Wu, M., Lu, P., Gao, L., Yan, S., & Wang, S. (2020). Development of pH indicator and antimicrobial cellulose nanofibre packaging film based on purple sweet potato anthocyanin and oregano essential oil. *International Journal of Biological Macromolecules*, *149*, 271–280.

Cherpinski, A., Gozutok, M., Sasmazel, H. T., Torres-Giner, S., & Lagaron, J. M. (2018). Electrospun oxygen scavenging films of poly (3-hydroxybutyrate) containing palladium nanoparticles for active packaging applications. *Nanomaterials*, *8*(7), 469.

Chivrac, F., Pollet, E., & Averous, L. (2009). Progress in nano-biocomposites based on polysaccharides and nanoclays. *Materials Science and Engineering: R: Reports*, *67*(1), 1–17.

Cho, Y.-H., & Jones, O. G. (2019). Assembled protein nanoparticles in food or nutrition applications. *Advances in Food and Nutrition Research*, *88*, 47–84.

Chopdey, P. K., Tekade, R. K., Mehra, N. K., Mody, N., & Jain, N. K. (2015). Glycyrrhizin conjugated dendrimer and multi-walled carbon nanotubes for liver specific delivery of doxorubicin. *Journal of Nanoscience and Nanotechnology*, *15*(2), 1088–1100.

Chowdhuri, A. R., Tripathy, S., Chandra, S., Roy, S., & Sahu, S. K. (2015). A ZnO decorated chitosan—graphene oxide nanocomposite shows significantly enhanced antimicrobial activity with ROS generation. *RSC Advances*, *5*(61), 49420–49428.

Chun, A. L. (2009). Will the public swallow nanofood? *Nature Nanotechnology*, *4*(12), 790–791.

Cortivo, R., Vindigni, V., Iacobellis, L., Abatangelo, G., Pinton, P., & Zavan, B. (2010). Nanoscale particle therapies for wounds and ulcers. *Nanomedicine*, *5*(4), 641–656.

Davachi, S. M., Heidari, B. S., Hejazi, I., Seyfi, J., Oliaei, E., Farzaneh, A., & Rashedi, H. (2017). Interface modified polylactic acid/starch/poly ε-caprolactone antibacterial nanocomposite blends for medical applications. *Carbohydrate Polymers*, *155*, 336–344.

de Moura, I. G., de Sá, A. V., Abreu, A. S. L. M., & Machado, A. V. A. (2017). Bioplastics from agro-wastes for food packaging applications. In *Food packaging* (pp. 223–263). Elsevier.

Diao, C., Xia, H., Noshadi, I., Kanjilal, B., & Parnas, R. S. (2014). Wheat gluten blends with a macromolecular cross-linker for improved mechanical properties and reduced water absorption. *ACS Sustainable Chemistry & Engineering*, *2*(11), 2554–2561.

Dua, K., Shukla, S. D., Tekade, R. K., & Hansbro, P. M. (2017). Whether a novel drug delivery system can overcome the problem of biofilms in respiratory diseases? *Drug Delivery and Translational Research*, *7*(1), 179–187.

Echeverría, I., López-Caballero, M. E., Gómez-Guillén, M. C., Mauri, A. N., & Montero, M. P. (2018). Active nanocomposite films based on soy proteins-montmorillonite-clove essential oil for the preservation of refrigerated bluefin tuna (*Thunnus thynnus*) fillets. *International Journal of Food Microbiology*, *266*, 142–149.

Ede, J. D., Ong, K. J., Goergen, M., Rudie, A., Pomeroy-Carter, C. A., & Shatkin, J. A. (2019). Risk analysis of cellulose nanomaterials by inhalation: Current state of science. *Nanomaterials*, *9*(3), 337.

Etxabide, A., Uranga, J., Guerrero, P., & de la Caba, K. (2017). Development of active gelatin films by means of valorisation of food processing waste: A review. *Food Hydrocolloids*, *68*, 192–198.

Faupel, F., Zaporojtchenko, V., Strunskus, T., & Elbahri, M. (2010). Metal-polymer nanocomposites for functional applications. *Advanced Engineering Materials*, *12*(12), 1177–1190.

Fei, P., Shi, Y., Zhou, M., Cai, J., Tang, S., & Xiong, H. (2013). Effects of nano-TiO_2 on the properties and structures of starch/poly (ε-caprolactone) composites. *Journal of Applied Polymer Science*, *130*(6), 4129–4136.

Galus, S., Arik Kibar, E. A., Gniewosz, M., & Kraśniewska, K. (2020). Novel materials in the preparation of edible films and coatings—a review. *Coatings*, *10*(7), 674.

Garcia, C. V., Shin, G. H., & Kim, J. T. (2018). Metal oxide-based nanocomposites in food packaging: Applications, migration, and regulations. *Trends in Food Science & Technology, 82*, 21–31.

Ghelejlu, S. B., Esmaiili, M., & Almasi, H. (2016). Characterization of chitosan—nanoclay bionanocomposite active films containing milk thistle extract. *International Journal of Biological Macromolecules, 86*, 613–621.

Gomez-Orellana, I. (2005). Strategies to improve oral drug bioavailability. *Expert Opinion on Drug Delivery, 2*(3), 419–433.

Guerrero, P., Muxika, A., Zarandona, I., & de La Caba, K. (2019). cross-linking of chitosan films processed by compression molding. *Carbohydrate Polymers, 206*, 820–826.

Hassan, B., Chatha, S. A. S., Hussain, A. I., Zia, K. M., & Akhtar, N. (2018). Recent advances on polysaccharides, lipids and protein based edible films and coatings: A review. *International Journal of Biological Macromolecules, 109*, 1095–1107.

He, X., Deng, H., & Hwang, H. (2019). The current application of nanotechnology in food and agriculture. *Journal of Food and Drug Analysis, 27*(1), 1–21.

He, X., & Hwang, H.-M. (2016). Nanotechnology in food science: Functionality, applicability, and safety assessment. *Journal of Food and Drug Analysis, 24*(4), 671–681.

Hoseinnejad, M., Jafari, S. M., & Katouzian, I. (2018). Inorganic and metal nanoparticles and their antimicrobial activity in food packaging applications. *Critical Reviews in Microbiology, 44*(2), 161–181.

Huang, J., Dufresne, A., & Lin, N. (2019). *Nanocellulose: From fundamentals to advanced materials*. John Wiley & Sons.

Hyvärinen, M., Jabeen, R., & Kärki, T. (2020). The modelling of extrusion processes for polymers—a review. *Polymers, 12*(6), 1306.

Indumathi, M. P., & Rajarajeswari, G. R. (2019). Mahua oil-based polyurethane/chitosan/nano ZnO composite films for biodegradable food packaging applications. *International Journal of Biological Macromolecules, 124*, 163–174.

Ioelovich, M. (2016). Nanocellulose—fabrication, structure, properties, and application in the area of care and cure. In *Fabrication and self-assembly of nanobiomaterials* (pp. 243–288). Elsevier.

Iordanskii, A. (2020). Bio-based and biodegradable plastics: From passive barrier to active packaging behavior. *Polymers, 12*(7), 1537.

Jacob, J., Haponiuk, J. T., Thomas, S., & Gopi, S. (2018). Biopolymer based nanomaterials in drug delivery systems: A review. *Materials Today Chemistry, 9*, 43–55.

Jamróz, E., Khachatryan, G., Kopel, P., Juszczak, L., Kawecka, A., Krzyściak, P., Kucharek, M., Bębenek, Z., & Zimowska, M. (2020). Furcellaran nanocomposite films: The effect of nanofillers on the structural, thermal, mechanical and antimicrobial properties of biopolymer films. *Carbohydrate Polymers, 240*, 116244.

Jamróz, E., Kulawik, P., & Kopel, P. (2019). The effect of nanofillers on the functional properties of biopolymer-based films: A review. *Polymers, 11*(4), 675.

Júnior, L. M., Vieira, R. P., Jamróz, E., & Anjos, C. A. R. (2021). Furcellaran: An innovative biopolymer in the production of films and coatings. *Carbohydrate Polymers, 252*, 117221.

Kalaitzidou, K., Fukushima, H., & Drzal, L. T. (2007). A new compounding method for exfoliated graphite—polypropylene nanocomposites with enhanced flexural properties and lower percolation threshold. *Composites Science and Technology, 67*(10), 2045–2051.

Kalishwaralal, K., Jeyabharathi, S., Sundar, K., Selvamani, S., Prasanna, M., & Muthukumaran, A. (2018). A novel biocompatible chitosan—selenium nanoparticles (SeNPs) film with electrical conductivity for cardiac tissue engineering application. *Materials Science and Engineering: C, 92*, 151–160.

Kalita, H., Pal, P., Dhara, S., & Pathak, A. (2017). Fabrication and characterization of polyvinyl alcohol/metal (Ca, Mg, Ti) doped zirconium phosphate nanocomposite films for scaffold-guided tissue engineering application. *Materials Science and Engineering: C, 71*, 363–371.

Karimi, N., Alizadeh, A., Almasi, H., & Hanifian, S. (2020). Preparation and characterization of whey protein isolate/polydextrose-based nanocomposite film incorporated with cellulose nanofiber and L. plantarum: A new probiotic active packaging system. *LWT, 121*, 108978.

Kasaai, M. R. (2018). Zein and zein-based nano-materials for food and nutrition applications: A review. *Trends in Food Science & Technology, 79*, 184–197.

Kataria, K., Gupta, A., Rath, G., Mathur, R. B., & Dhakate, S. R. (2014). In vivo wound healing performance of drug loaded electrospun composite nanofibers transdermal patch. *International Journal of Pharmaceutics, 469*(1), 102–110.

Kayat, J., Gajbhiye, V., Tekade, R. K., & Jain, N. K. (2011). Pulmonary toxicity of carbon nanotubes: A systematic report. *Nanomedicine: Nanotechnology, Biology and Medicine, 7*(1), 40–49.

Kerch, G. (2015). Chitosan films and coatings prevent losses of fresh fruit nutritional quality: A review. *Trends in Food Science & Technology, 46*(2), 159–166.

Khalil, H. P. S. A., Bhat, A. H., & Yusra, A. F. I. (2012). Green composites from sustainable cellulose nanofibrils: A review. *Carbohydrate Polymers, 87*(2), 963–979.

Khalil, H. P. S. A., Davoudpour, Y., Saurabh, C. K., Hossain, M. S., Adnan, A. S., Dungani, R., Paridah, M. T., Sarker, M. Z. I., Fazita, M. R. N., & Syakir, M. I. (2016). A review on nanocellulosic fibres as new material for sustainable packaging: Process and applications. *Renewable and Sustainable Energy Reviews, 64*, 823–836.

Khan, W. S., Asmatulu, R., Ceylan, M., & Jabbarnia, A. (2013). Recent progress on conventional and non-conventional electrospinning processes. *Fibers and Polymers, 14*(8), 1235–1247.

Khanam, P. N., Ponnamma, D., & Al-Madeed, M. A. (2015). Electrical properties of graphene polymer nanocomposites. In *Graphene-based polymer nanocomposites in electronics* (pp. 25–47). Springer.

Kim, S., Do, I., & Drzal, L. T. (2010). Thermal stability and dynamic mechanical behavior of exfoliated graphite nanoplatelets-LLDPE nanocomposites. *Polymer Composites, 31*(5), 755–761.

Kraśniewska, K., Galus, S., & Gniewosz, M. (2020). Biopolymers-based materials containing silver nanoparticles as active packaging for food applications—a review. *International Journal of Molecular Sciences, 21*(3), 698.

Kraśniewska, K., Pobiega, K., & Gniewosz, M. (2019). Pullulan—biopolymer with potential for use as food packaging. *International Journal of Food Engineering, 15*(9), 20190030.

Kravanja, G., Primožič, M., Knez, Ž., & Leitgeb, M. (2019). Chitosan-based (nano) materials for novel biomedical applications. *Molecules, 24*(10), 1960.

Krepker, M., Zhang, C., Nitzan, N., Prinz-Setter, O., Massad-Ivanir, N., Olah, A., Baer, E., & Segal, E. (2018). Antimicrobial LDPE/EVOH layered films containing carvacrol fabricated by multiplication extrusion. *Polymers, 10*(8), 864.

Kritchenkov, A. S., Egorov, A. R., Volkova, O. V., Artemjev, A. A., Kurliuk, A. V., Le, T. A., Truong, H. H., Le-Nhat-Thuy, G., Thi, T. V. T., & van Tuyen, N. (2021). Novel biopolymer-based nanocomposite food coatings that exhibit active and smart properties due to a single type of nanoparticles. *Food Chemistry, 343*, 128676.

Kukoyi, A. R. (2016). Economic impacts of natural polymers. In *Natural polymers* (pp. 339–362). Springer.

Kumar, S., Bhattacharya, W., Singh, M., Halder, D., & Mitra, A. (2017). Plant latex capped colloidal silver nanoparticles: A potent anti-biofilm and fungicidal formulation. *Journal of Molecular Liquids, 230*, 705–713.

Kupnik, K., Primožič, M., Kokol, V., & Leitgeb, M. (2020). Nanocellulose in drug delivery and antimicrobially active materials. *Polymers, 12*(12), 2825.

Kurmi, B. D., Kayat, J., Gajbhiye, V., Tekade, R. K., & Jain, N. K. (2010). Micro-and nanocarrier-mediated lung targeting. *Expert Opinion on Drug Delivery, 7*(7), 781–794.

Kuswandi, B., & Moradi, M. (2019). Improvement of food packaging based on functional nanomaterial. In *Nanotechnology: Applications in energy, drug and food* (pp. 309–344). Springer.

Kuswandi, B., Wicaksono, Y., Abdullah, A., Heng, L. Y., & Ahmad, M. (2011). Smart packaging: Sensors for monitoring of food quality and safety. *Sensing and Instrumentation for Food Quality and Safety, 5*(3), 137–146.

Lavoine, N., Guillard, V., Desloges, I., Gontard, N., & Bras, J. (2016). Active bio-based food-packaging: Diffusion and release of active substances through and from cellulose nanofiber coating toward food-packaging design. *Carbohydrate Polymers, 149*, 40–50.

Lee, C., Wei, X., Kysar, J. W., & Hone, J. (2008). Measurement of the elastic properties and intrinsic strength of monolayer graphene. *Science, 321*(5887), 385–388.

Lee, H., Rukmanikrishnan, B., & Lee, J. (2019). Rheological, morphological, mechanical, and water-barrier properties of agar/gellan gum/montmorillonite clay composite films. *International Journal of Biological Macromolecules, 141*, 538–544.

Li, Q., Li, T., Liu, C., DeLoid, G., Pyrgiotakis, G., Demokritou, P., Zhang, R., Xiao, H., & McClements, D. J. (2017). Potential impact of inorganic nanoparticles on macronutrient digestion: Titanium dioxide nanoparticles slightly reduce lipid digestion under simulated gastrointestinal conditions. *Nanotoxicology, 11*(9–10), 1087–1101.

Li, Z., & Sheng, C. (2014). Nanosensors for food safety. *Journal of Nanoscience and Nanotechnology, 14*(1), 905–912.

Lin, D., Liu, Z., Shen, R., Chen, S., & Yang, X. (2020). Bacterial cellulose in food industry: Current research and future prospects. *International Journal of Biological Macromolecules, 158,* 1007–1019.

Ludwicka, K., Kaczmarek, M., & Białkowska, A. (2020). Bacterial nanocellulose—a biobased polymer for active and intelligent food packaging applications: Recent advances and developments. *Polymers, 12*(10), 2209.

Lukaszkowicz, K. (2011). Rahman, M. (Eds). Review of nanocomposite thin films and coatings deposited by PVD and CVD technology. In *Intech, Rijeka,* Shanghai, China, (pp. 145–162).

Lyn, F. H., Peng, T. C., Ruzniza, M. Z., & Hanani, Z. A. N. (2019). Effect of oxidation degrees of graphene oxide (GO) on the structure and physical properties of chitosan/GO composite films. *Food Packaging and Shelf Life, 21,* 100373.

Madhusudan, P., Chellukuri, N., & Shivakumar, N. (2018). Smart packaging of food for the 21st century—a review with futuristic trends, their feasibility and economics. *Materials Today: Proceedings, 5*(10), 21018–21022.

Maher, S., Brayden, D. J., Casettari, L., & Illum, L. (2019). Application of permeation enhancers in oral delivery of macromolecules: An update. *Pharmaceutics, 11*(1), 41.

Maheshwari, R., Tekade, M., Sharma, P. A., & Kumar Tekade, R. (2015). Nanocarriers assisted siRNA gene therapy for the management of cardiovascular disorders. *Current Pharmaceutical Design, 21*(30), 4427–4440.

Maisanaba, S., Llana-Ruiz-Cabello, M., Gutiérrez-Praena, D., Pichardo, S., Puerto, M., Prieto, A. I., Jos, A., & Cameán, A. M. (2017). New advances in active packaging incorporated with essential oils or their main components for food preservation. *Food Reviews International, 33*(5), 447–515.

Mangaraj, S., Yadav, A., Bal, L. M., Dash, S. K., & Mahanti, N. K. (2019). Application of biodegradable polymers in food packaging industry: A comprehensive review. *Journal of Packaging Technology and Research, 3*(1), 77–96.

Marra, A., Silvestre, C., Duraccio, D., & Cimmino, S. (2016). Polylactic acid/zinc oxide biocomposite films for food packaging application. *International Journal of Biological Macromolecules, 88,* 254–262.

Maser, W. K., Benito, A. M., Callejas, M. A., Seeger, T., Martınez, M. T., Schreiber, J., Muszynski, J., Chauvet, O., Osváth, Z., & Koós, A. A. (2003). Synthesis and characterization of new polyaniline/nanotube composites. *Materials Science and Engineering: C, 23*(1–2), 87–91.

Matharu, R. K., Ciric, L., & Edirisinghe, M. (2018). Nanocomposites: Suitable alternatives as antimicrobial agents. *Nanotechnology, 29*(28), 282001.

Mathew, S., Snigdha, S., Mathew, J., & Radhakrishnan, E. K. (2019). Biodegradable and active nanocomposite pouches reinforced with silver nanoparticles for improved packaging of chicken sausages. *Food Packaging and Shelf Life, 19,* 155–166.

Mauricio, M. D., Guerra-Ojeda, S., Marchio, P., Valles, S. L., Aldasoro, M., Escribano-Lopez, I., Herance, J. R., Rocha, M., Vila, J. M., & Victor, V. M. (2018). Nanoparticles in medicine: A focus on vascular oxidative stress. *Oxidative Medicine and Cellular Longevity, 2018,* 6231482.

McClements, D. J., & Xiao, H. (2017). Is nano safe in foods? Establishing the factors impacting the gastrointestinal fate and toxicity of organic and inorganic food-grade nanoparticles. *npj Science of Food, 1*(1), 1–13.

McClements, D. J., Xiao, H., & Demokritou, P. (2017). Physicochemical and colloidal aspects of food matrix effects on gastrointestinal fate of ingested inorganic nanoparticles. *Advances in Colloid and Interface Science, 246,* 165–180.

Meira, S. M. M., Zehetmeyer, G., Scheibel, J. M., Werner, J. O., & Brandelli, A. (2016). Starch-halloysite nanocomposites containing nisin: Characterization and inhibition of Listeria monocytogenes in soft cheese. *LWT-Food Science and Technology, 68,* 226–234.

Merz, B., Capello, C., Leandro, G. C., Moritz, D. E., Monteiro, A. R., & Valencia, G. A. (2020). A novel colorimetric indicator film based on chitosan, polyvinyl alcohol and anthocyanins from Jambolan (Syzygium cumini) fruit for monitoring shrimp freshness. *International Journal of Biological Macromolecules, 153,* 625–632.

Mihindukulasuriya, S. D. F., & Lim, L.-T. (2014). Nanotechnology development in food packaging: A review. *Trends in Food Science & Technology, 40*(2), 149–167.

Mohammadalinejhad, S., Almasi, H., & Moradi, M. (2020). Immobilization of *Echium amoenum* anthocyanins into bacterial cellulose film: A novel colorimetric pH indicator for freshness/spoilage monitoring of shrimp. *Food Control, 113,* 107169.

Mohan, S., Oluwafemi, O. S., Kalarikkal, N., Thomas, S., & Songca, S. P. (2016). Biopolymers—application in nanoscience and nanotechnology. *Recent Advances in Biopolymers*, *1*(1), 47–66.

Moreirinha, C., Vilela, C., Silva, N. H. C. S., Pinto, R. J. B., Almeida, A., Rocha, M. A. M., Coelho, E., Coimbra, M. A., Silvestre, A. J. D., & Freire, C. S. R. (2020). Antioxidant and antimicrobial films based on brewers spent grain arabinoxylans, nanocellulose and feruloylated compounds for active packaging. *Food Hydrocolloids*, *108*, 105836.

Morganti, P. (2010). Use and potential of nanotechnology in cosmetic dermatology. *Clinical, Cosmetic and Investigational Dermatology: CCID*, *3*, 5.

Mustafa, F., & Andreescu, S. (2020). Nanotechnology-based approaches for food sensing and packaging applications. *RSC Advances*, *10*(33), 19309–19336.

Muxika, A., Etxabide, A., Uranga, J., Guerrero, P., & de La Caba, K. (2017). Chitosan as a bioactive polymer: Processing, properties and applications. *International Journal of Biological Macromolecules*, *105*, 1358–1368.

Nazeer, M. A., Yilgör, E., & Yilgör, I. (2017). Intercalated chitosan/hydroxyapatite nanocomposites: Promising materials for bone tissue engineering applications. *Carbohydrate Polymers*, *175*, 38–46.

Neethirajan, S., Ragavan, V., Weng, X., & Chand, R. (2018). Biosensors for sustainable food engineering: Challenges and perspectives. *Biosensors*, *8*(1), 23.

Nile, S. H., Baskar, V., Selvaraj, D., Nile, A., Xiao, J., & Kai, G. (2020). Nanotechnologies in food science: Applications, recent trends, and future perspectives. *Nano-Micro Letters*, *12*(1), 1–34.

Nitta, S. K., & Numata, K. (2013). Biopolymer-based nanoparticles for drug/gene delivery and tissue engineering. *International Journal of Molecular Sciences*, *14*(1), 1629–1654.

Nonsuwan, P., Puthong, S., Palaga, T., & Muangsin, N. (2018). Novel organic/inorganic hybrid flower-like structure of selenium nanoparticles stabilized by pullulan derivatives. *Carbohydrate Polymers*, *184*, 9–19.

Norajit, K., Kim, K. M., & Ryu, G. H. (2010). Comparative studies on the characterization and antioxidant properties of biodegradable alginate films containing ginseng extract. *Journal of Food Engineering*, *98*(3), 377–384.

Nouri, A., Yaraki, M. T., Ghorbanpour, M., Agarwal, S., & Gupta, V. K. (2018). Enhanced antibacterial effect of chitosan film using montmorillonite/CuO nanocomposite. *International Journal of Biological Macromolecules*, *109*, 1219–1231.

Okamoto, M., Morita, S., & Kotaka, T. (2001). Dispersed structure and ionic conductivity of smectic clay/polymer nanocomposites. *Polymer*, *42*(6), 2685–2688.

Okamoto, M., Morita, S., Taguchi, H., Kim, Y. H., Kotaka, T., & Tateyama, H. (2000). Synthesis and structure of smectic clay/poly (methyl methacrylate) and clay/polystyrene nanocomposites via in situ intercalative polymerization. *Polymer*, *41*(10), 3887–3890.

Otoni, C. G., Avena-Bustillos, R. J., Azeredo, H. M. C., Lorevice, M. V., Moura, M. R., Mattoso, L. H. C., & McHugh, T. H. (2017). Recent advances on edible films based on fruits and vegetables—a review. *Comprehensive Reviews in Food Science and Food Safety*, *16*(5), 1151–1169.

Padua, G. W., Nonthanum, P., & Arora, A. (2012). Padua, G.W. and Wang, Q. (Eds). Nanocomposites. In *Nanotechnology research methods for foods and bioproducts*, Wiley Blackwell, Oxford, UK (pp. 41–54).

Pan, K., & Zhong, Q. (2016). Organic nanoparticles in foods: Fabrication, characterization, and utilization. *Annual Review of Food Science and Technology*, *7*, 245–266.

Pastore, M. N., Kalia, Y. N., Horstmann, M., & Roberts, M. S. (2015). Transdermal patches: History, development and pharmacology. *British Journal of Pharmacology*, *172*(9), 2179–2209.

Patel, M. (2020). Nanoparticle-based antimicrobial paper as spread-breaker for coronavirus. *Paper Technology International*, *62*, 20–25.

Pelgrift, R. Y., & Friedman, A. J. (2013). Nanotechnology as a therapeutic tool to combat microbial resistance. *Advanced Drug Delivery Reviews*, *65*(13–14), 1803–1815.

Petersson, L., & Oksman, K. (2006). *Preparation and properties of biopolymer-based nanocomposite films using microcrystalline cellulose*. ACS Publications.

Poonguzhali, R., Basha, S. K., & Kumari, V. S. (2017). Synthesis and characterization of chitosan-PVP-nanocellulose composites for in-vitro wound dressing application. *International Journal of Biological Macromolecules*, *105*, 111–120.

Pramanik, P. K. D., Solanki, A., Debnath, A., Nayyar, A., El-Sappagh, S., & Kwak, K.-S. (2020). Advancing modern healthcare with nanotechnology, nanobiosensors, and internet of nano things: Taxonomies, applications, architecture, and challenges. *IEEE Access*, *8*, 65230–65266.

Primožič, M., Knez, Ž., & Leitgeb, M. (2021). (Bio) Nanotechnology in food science—food packaging. *Nanomaterials, 11*(2), 292.

Qu, B., & Luo, Y. (2020). Chitosan-based hydrogel beads: Preparations, modifications and applications in food and agriculture sectors—a review. *International Journal of Biological Macromolecules, 152*, 437–448.

Qu, B., & Luo, Y. (2021). A review on the preparation and characterization of chitosan-clay nanocomposite films and coatings for food packaging applications. *Carbohydrate Polymer Technologies and Applications, 2*, 100102.

Rajendran, N. K., Kumar, S. S. D., Houreld, N. N., & Abrahamse, H. (2018). A review on nanoparticle based treatment for wound healing. *Journal of Drug Delivery Science and Technology, 44*, 421–430.

Ramos, Ó. L., Pereira, R. N., Cerqueira, M. A., Martins, J. R., Teixeira, J. A., Malcata, F. X., & Vicente, A. A. (2018). Bio-based nanocomposites for food packaging and their effect in food quality and safety. In *Food packaging and preservation* (pp. 271–306). Elsevier.

Ranjan, S., Dasgupta, N., Chakraborty, A. R., Melvin Samuel, S., Ramalingam, C., Shanker, R., & Kumar, A. (2014). Nanoscience and nanotechnologies in food industries: Opportunities and research trends. *Journal of Nanoparticle Research, 16*(6), 1–23.

Rasoulzadeh, M., & Namazi, H. (2017). Carboxymethyl cellulose/graphene oxide bio-nanocomposite hydrogel beads as anticancer drug carrier agent. *Carbohydrate Polymers, 168*, 320–326.

Reddy, J. P., & Rhim, J.-W. (2014). Characterization of bionanocomposite films prepared with agar and paper-mulberry pulp nanocellulose. *Carbohydrate Polymers, 110*, 480–488.

Rhim, J.-W., Park, H.-M., & Ha, C.-S. (2013). Bio-nanocomposites for food packaging applications. *Progress in Polymer Science, 38*(10–11), 1629–1652.

Rodrigues, C., Souza, V. G. L., Coelhoso, I., & Fernando, A. L. (2021). Bio-based sensors for smart food packaging—current applications and future trends. *Sensors, 21*(6), 2148.

Roy, N., Sengupta, R., & Bhowmick, A. K. (2012). Modifications of carbon for polymer composites and nanocomposites. *Progress in Polymer Science, 37*(6), 781–819.

Roy, S., & Rhim, J.-W. (2021). Anthocyanin food colorant and its application in pH-responsive color change indicator films. *Critical Reviews in Food Science and Nutrition, 61*(14), 2297–2325.

Sadeghizadeh-Yazdi, J., Habibi, M., Kamali, A. A., & Banaei, M. (2019). Application of edible and biodegradable starch-based films in food packaging: A systematic review and meta-analysis. *Current Research in Nutrition and Food Science Journal, 7*(3), 624–637.

Salas, C., Nypelö, T., Rodriguez-Abreu, C., Carrillo, C., & Rojas, O. J. (2014). Nanocellulose properties and applications in colloids and interfaces. *Current Opinion in Colloid & Interface Science, 19*(5), 383–396.

Salgado, P. R., Ortiz, C. M., Musso, Y. S., di Giorgio, L., & Mauri, A. N. (2015). Edible films and coatings containing bioactives. *Current Opinion in Food Science, 5*, 86–92.

Schmalz, G., Hickel, R., van Landuyt, K. L., & Reichl, F.-X. (2017). Nanoparticles in dentistry. *Dental Materials, 33*(11), 1298–1314.

Shahi, N., Min, B., Sapkota, B., & Rangari, V. K. (2020). Eco-friendly cellulose nanofiber extraction from sugarcane bagasse and film fabrication. *Sustainability, 12*(15), 6015.

Shankar, S., Wang, L.-F., & Rhim, J.-W. (2018). Incorporation of zinc oxide nanoparticles improved the mechanical, water vapor barrier, UV-light barrier, and antibacterial properties of PLA-based nanocomposite films. *Materials Science and Engineering: C, 93*, 289–298.

Sharma, A., Mandal, T., & Goswami, S. (2021). Fabrication of cellulose acetate nanocomposite films with lignocelluosic nanofiber filler for superior effect on thermal, mechanical and optical properties. *Nano-Structures & Nano-Objects, 25*, 100642.

Sharma, B., Malik, P., & Jain, P. (2018). Biopolymer reinforced nanocomposites: A comprehensive review. *Materials Today Communications, 16*, 353–363.

Sharma, C., & Bhardwaj, N. K. (2019). Bacterial nanocellulose: Present status, biomedical applications and future perspectives. *Materials Science and Engineering: C, 104*, 109963.

Sharma, C., Dhiman, R., Rokana, N., & Panwar, H. (2017). Nanotechnology: An untapped resource for food packaging. *Frontiers in Microbiology, 8*, 1735.

Sharma, R., Jafari, S. M., & Sharma, S. (2020). Antimicrobial bio-nanocomposites and their potential applications in food packaging. *Food Control, 112*, 107086.

Shrestha, P., Sadiq, M. B., & Anal, A. K. (2021). Development of antibacterial biocomposites reinforced with cellulose nanocrystals derived from banana pseudostem. *Carbohydrate Polymer Technologies and Applications, 2*, 100112.

Siegrist, M., Cousin, M.-E., Kastenholz, H., & Wiek, A. (2007). Public acceptance of nanotechnology foods and food packaging: The influence of affect and trust. *Appetite, 49*(2), 459–466.

Siegrist, M., Stampfli, N., Kastenholz, H., & Keller, C. (2008). Perceived risks and perceived benefits of different nanotechnology foods and nanotechnology food packaging. *Appetite, 51*(2), 283–290.

Singh, T., Shukla, S., Kumar, P., Wahla, V., Bajpai, V. K., & Rather, I. A. (2017). Application of nanotechnology in food science: Perception and overview. *Frontiers in Microbiology, 8*, 1501.

Song, J. H., Murphy, R. J., Narayan, R., & Davies, G. B. H. (2009). Biodegradable and compostable alternatives to conventional plastics. *Philosophical Transactions of the Royal Society B: Biological Sciences, 364*(1526), 2127–2139.

Song, J., Gao, H., Zhu, G., Cao, X., Shi, X., & Wang, Y. (2015). The preparation and characterization of polycaprolactone/graphene oxide biocomposite nanofiber scaffolds and their application for directing cell behaviors. *Carbon, 95*, 1039–1050.

Speranza, A., Corradini, M. G., Hartman, T. G., Ribnicky, D., Oren, A., & Rogers, M. A. (2013). Influence of emulsifier structure on lipid bioaccessibility in oil—water nanoemulsions. *Journal of Agricultural and Food Chemistry, 61*(26), 6505–6515.

Stumpf, T. R., Yang, X., Zhang, J., & Cao, X. (2018). In situ and ex situ modifications of bacterial cellulose for applications in tissue engineering. *Materials Science and Engineering: C, 82*, 372–383.

Sun, L., Sun, J., Chen, L., Niu, P., Yang, X., & Guo, Y. (2017). Preparation and characterization of chitosan film incorporated with thinned young apple polyphenols as an active packaging material. *Carbohydrate Polymers, 163*, 81–91.

Tabatabaei, R. H., Jafari, S. M., Mirzaei, H., Nafchi, A. M., & Dehnad, D. (2018). Preparation and characterization of nano-SiO2 reinforced gelatin-κ-carrageenan biocomposites. *International Journal of Biological Macromolecules, 111*, 1091–1099.

Tan, G., Yu, S., Pan, H., Li, J., Liu, D., Yuan, K., Yang, X., & Pan, W. (2017). Bioadhesive chitosan-loaded liposomes: A more efficient and higher permeable ocular delivery platform for timolol maleate. *International Journal of Biological Macromolecules, 94*, 355–363.

Tang, X. Z., Kumar, P., Alavi, S., & Sandeep, K. P. (2012). Recent advances in biopolymers and biopolymer-based nanocomposites for food packaging materials. *Critical Reviews in Food Science and Nutrition, 52*(5), 426–442.

Tekade, R. K., Maheshwari, R., & Tekade, M. (2017). Biopolymer-based nanocomposites for transdermal drug delivery. In *Biopolymer-based composites* (pp. 81–106). Elsevier.

Thanh, N. T., Hieu, M. H., Phuong, N. T. M., Thuan, T. D. B., Thu, H. N. T., do Minh, T., Dai, H. N., & Thi, H. N. (2018). Optimization and characterization of electrospun polycaprolactone coated with gelatin-silver nanoparticles for wound healing application. *Materials Science and Engineering: C, 91*, 318–329.

Tian, H., Guo, G., Fu, X., Yao, Y., Yuan, L., & Xiang, A. (2018). Fabrication, properties and applications of soy-protein-based materials: A review. *International Journal of Biological Macromolecules, 120*, 475–490.

Tibolla, H., Pelissari, F. M., Martins, J. T., Lanzoni, E. M., Vicente, A. A., Menegalli, F. C., & Cunha, R. L. (2019). Banana starch nanocomposite with cellulose nanofibers isolated from banana peel by enzymatic treatment: In vitro cytotoxicity assessment. *Carbohydrate Polymers, 207*, 169–179.

Tran, M., & Wang, C. (2014). Semi-solid materials for controlled release drug formulation: Current status and future prospects. *Frontiers of Chemical Science and Engineering, 8*(2), 225–232.

Uysal Ünalan, I. (2014). *Potential use of graphene for the generation of bionanocomposite materials for food packaging applications*. https://air.unimi.it/handle/2434/245012

Verlee, A., Mincke, S., & Stevens, C. V. (2017). Recent developments in antibacterial and antifungal chitosan and its derivatives. *Carbohydrate Polymers, 164*, 268–283.

Weng, W., Chen, G., & Wu, D. (2005). Transport properties of electrically conducting nylon 6/foliated graphite nanocomposites. *Polymer, 46*(16), 6250–6257.

Wenk, E., Merkle, H. P., & Meinel, L. (2011). Silk fibroin as a vehicle for drug delivery applications. *Journal of Controlled Release, 150*(2), 128–141.

Xiao, Y., Liu, Y., Kang, S., Wang, K., & Xu, H. (2020). Development and evaluation of soy protein isolate-based antibacterial nanocomposite films containing cellulose nanocrystals and zinc oxide nanoparticles. *Food Hydrocolloids, 106*, 105898.

Xiong, R., Grant, A. M., Ma, R., Zhang, S., & Tsukruk, V. V. (2018). Naturally derived biopolymer nanocomposites: Interfacial design, properties and emerging applications. *Materials Science and Engineering: R: Reports, 125*, 1–41.

Yang, X., Tu, Y., Li, L., Shang, S., & Tao, X. (2010). Well-dispersed chitosan/graphene oxide nanocomposites. *ACS Applied Materials & Interfaces, 2*(6), 1707–1713.

Yang, Y.-Y., Wang, Y., Powell, R., & Chan, P. (2006). Polymeric core-shell nanoparticles for therapeutics. *Clinical and Experimental Pharmacology and Physiology, 33*(5), 557–562.

Yao, K. J., Song, M., Hourston, D. J., & Luo, D. Z. (2002). Polymer/layered clay nanocomposites: 2 polyurethane nanocomposites. *Polymer, 43*(3), 1017–1020.

Yıldırım, E., & Barutçu Mazı, I. (2017). Effect of zein coating enriched by addition of functional constituents on the lipid oxidation of roasted hazelnuts. *Journal of Food Process Engineering, 40*(4), e12515.

Yildirim, S., Röcker, B., Pettersen, M. K., Nilsen-Nygaard, J., Ayhan, Z., Rutkaite, R., Radusin, T., Suminska, P., Marcos, B., & Coma, V. (2018). Active packaging applications for food. *Comprehensive Reviews in Food Science and Food Safety, 17*(1), 165–199.

Youssef, A. M., & El-Sayed, S. M. (2018). Bionanocomposites materials for food packaging applications: Concepts and future outlook. *Carbohydrate Polymers, 193*, 19–27.

Youssef, A. M., El-Sayed, S. M., El-Sayed, H. S., Salama, H. H., & Dufresne, A. (2016). Enhancement of Egyptian soft white cheese shelf life using a novel chitosan/carboxymethyl cellulose/zinc oxide bionanocomposite film. *Carbohydrate Polymers, 151*, 9–19.

Youssef, K., & Hashim, A. F. (2020). Inhibitory effect of clay/chitosan nanocomposite against penicillium digitatum on citrus and its possible mode of action. *Jordan Journal of Biological Sciences, 13*(3), 349–355.

Zambrano-Zaragoza, M. L., González-Reza, R., Mendoza-Muñoz, N., Miranda-Linares, V., Bernal-Couoh, T. F., Mendoza-Elvira, S., & Quintanar-Guerrero, D. (2018). Nanosystems in edible coatings: A novel strategy for food preservation. *International Journal of Molecular Sciences, 19*(3), 705.

Zammarano, M., Franceschi, M., Bellayer, S., Gilman, J. W., & Meriani, S. (2005). Preparation and flame resistance properties of revolutionary self-extinguishing epoxy nanocomposites based on layered double hydroxides. *Polymer, 46*(22), 9314–9328.

Zhang, C., Li, Y., Wang, P., & Zhang, H. (2020a). Electrospinning of nanofibers: Potentials and perspectives for active food packaging. *Comprehensive Reviews in Food Science and Food Safety, 19*(2), 479–502.

Zhang, C., Sun, G., Cao, L., & Wang, L. (2020b). Accurately intelligent film made from sodium carboxymethyl starch/κ-carrageenan reinforced by mulberry anthocyanins as an indicator. *Food Hydrocolloids, 108*, 106012.

Zhen, X., Wang, X., Xie, C., Wu, W., & Jiang, X. (2013). Cellular uptake, antitumor response and tumor penetration of cisplatin-loaded milk protein nanoparticles. *Biomaterials, 34*(4), 1372–1382.

Zhou, Y., Lei, L., Yang, B., Li, J., & Ren, J. (2018). Preparation and characterization of polylactic acid (PLA) carbon nanotube nanocomposites. *Polymer Testing, 68*, 34–38.

Zimmermann, L., Dombrowski, A., Völker, C., & Wagner, M. (2020). Are bioplastics and plant-based materials safer than conventional plastics? In vitro toxicity and chemical composition. *Environment International, 145*, 106066.

16

Nanostructured Film and Coating Materials

A Novel Approach in Packaging

Advaita, Kinshuk Malik, Verbi P. Bhagabati, and Kritika Sharma

CONTENTS

DOI: 10.1201/9781003303671-16

16.1 Introduction

For almost a period of ten years, nanotechnology has pulled a lot of notice within the research programs of the public as well as in private companies and institutions. Nanotechnology, according to scientists, is a fascinating field to research in since it allows for innovation in the creation of novel substances with purposely induced nanosized designs. Finally, this area will produce unique and novel properties and functions which can be used favorably in day-to-day life. However, a few researchers have shown reservations about nanotechnology, with the major query being if it is a genuine opportunity or only a ruse to raise cash for research (Unalan et al., 2014).

16.1.1 Nanomaterials

Nanomaterials are very fine matter aggregates with diameters of a few to a hundred nanometers. Depending on their composition, they can be crystalline or amorphous, with both conducting and insulating qualities. Metals (like Ag, Au, Pt) or metal oxides (like Ti_2O, FeO), semiconductors, or polymers (like latex) are all typical examples of nanomaterials. The arrangement of atoms is primarily dictated by the reduction of their surface energy, which can lead to a variety of crystal shapes when compared to the bulk arrangement. Because the surface of nanoparticles can contain up to 90% of all atoms, practically the entire nanoparticle can engage in chemical or physical processes, which explains their quick kinetics. Although there are various technologies available to manufacture nanoparticles, "top-down" and "bottom-up" methodologies are the most widely applied all over the world (Ranzoni & Cooper, 2017).

Despite the fact that nanomaterials have been used by humans for many years, users' opinions of nanotechnology remain unfavorable, owing to ongoing uncertainties about the possible harm for not only human beings but also the environment connected with the employment of nanomaterials for sensitive uses, like packaging of food, nutrient delivery, and many other fields in the food service industry (Chun, 2009). One of the most significant advancements in the area of nanotechnology is the invention of nanocomposites (Lagarón et al., 2005). The term "composites" in polymer science usually is used for the combinations of polymers with organic or inorganic fillers or additives with microlength scale and specific geometries (like flakes, fibers, particulates, and spheres), but the utilization of nanolength-scale entities is particularly called "nanocomposites." They are well-developed nanobuilding blocks (NBBs), which are prefabricated items that maintain their integrity in final nanocomposite materials (Sanchez et al., 2001). Nanocore shells, clusters, layered compounds (e.g., clays), and organically post- or pre-functionalized nanostructures are all examples of NBB (e.g., metal oxides) (Sanchez et al., 2005). The beneficial route to form nanocomposites can be found in the "top-down" method, which continuously decreases the measurements of large nanocomposite items by utilizing methods and techniques like etching, milling, lithography, grinding, or even precision engineering (Sanguansri & Augustin, 2006). Nanocomposites have proved to be a great alternate choice to traditional polymer composites as, when correctly managed, they can provide additional advantages, for example, transparency, recyclability, low density, improved surface characteristics even at extremely low filler concentrations (mostly less than 5 wt. %) (Alexandre & Dubois, 2000; Giannelis, 1996; Sinha Ray & Okamoto, 2003; Sorrentino et al., 2007).

16.1.2 Inorganic Fillers: Physicochemical Properties and Recent Innovations in Food Packaging

Inorganic nanofillers have become well-known for their usage in the production of nanocomposite films. Developers and researchers have suggested the use of layered silicates (like clays) mostly as fillers in nanocomposites to date (Duncan, 2011; Liu & Wu, 2001; Pavlidou & Papaspyrides, 2008; Sterky et al., 2010). Further, they have also used carbon-based nanofillers, like expanded graphite (Kalaitzidou et al., 2007), carbon nanotubes (CNTs) (Coleman et al., 2006), carbon black (Maiti et al., 2005), and carbon nanofibers in the development of polymer nanocomposites to attain characteristics that would, in other ways, be unachievable, like thermal and electrical conductivity of the final nanocomposite substance. From the previously discussed nanofillers, carbon nanotubes (CNTs) have shown to be useful as conductive fillers. Except for their high production cost as nanofillers, there is no other as such disadvantage associated with CNTs (Coleman et al., 2006; Biercuk et al., 2002; Price & Tour, 2006; Coleman, 2012).

16.1.3 Other Layered Minerals of Interest

Layered double hydroxides (LDHs), which are known as "non-silicate oxides and hydroxides," exhibit a surprising number of physical and chemical features in common with clay minerals. LDHs are made up of stacked brucite $[Mg(OH)_2]$–type trioctahedral layers which are charged positively because magnesium atoms have been replaced by atoms of aluminum. This is the reason LDHs are dubbed "anionic clays," because anionic exchangeable species are intercalated between the sheets. Several studies and research performed on the manufacture of bionanocomposites based on LDHs suggest that these complexes could be used in food packaging.

Sheet silicates such as muscovite, which are abundant in nature and typically inexpensive, belong to the mica group. Micas, when combined with various biopolymers, could be used as filler in the creation of bionanocomposites for food packaging.

Although there is a large variety of nanofillers available in the market, finding the best fit for a specific biopolymer system and application is not as easy as it appears. This decision necessitates careful consideration of a variety of factors, all of which are based on increasing the bond between the polymer matrix and the filler, with the intention to make them an integrated part of each other.

16.1.4 Metals and Metal Oxide NPs in Films

Silver, due to its antibacterial properties and shelf stability, has been popular for a long time, and that is why it is the most extensively employed metal for the creation of useful and effective bionanocomposite materials (Russell & Hugo, 1994). AgNPs are employed in the food manufacturing industries at large scale, for example, to make biopolymer nanocomposites for packaging of food, in addition to medical and pharmaceutical uses. Nanocomposites based on AgNPs slowly release Ag^+ ions into preserved foods than silver zeolites (which will be explored subsequently) but also lower acute antibacterial reactions. In practice, this implies that although a zeolite-based item may have a stronger immediate effect, the nanocomposite's long-term antibacterial activity would be more beneficial for the packing of foods that must travel large distances or be stored for lengthy periods of time (Egger et al., 2009).

It has been stated that several aspects, like size of particle, distribution of size, extent of particle aggregation, Ag content, and the contact of the Ag surface with the base polymer, must be considered in order for Ag-based bionanocomposite materials to be totally functional. This is due to the antimicrobial effectiveness of AgNPs, like that of any different nanocomposite antimicrobial system, being heavily reliant on the nanosized antimicrobial particles' high surface-to-volume ratio and high surface reactivity, allowing them to deactivate microorganisms in a better way than their micro- or macroscale equivalent.

Copper is another metal with well-known antibacterial characteristics; however, its biocidal potential is lower than that of Ag^+ ions (Kim et al., 2007). Copper ions' activity has been used in very less polymer nanocomposites, like Cu–PE nanocomposite films and Cu–chitosan bionanocomposites films for packaging of food. The main cause behind this restriction is that copper is said to be harmful when it comes into direct contact with food, and its catalytic oxidation action would hasten biochemical deterioration in foods (Llorens et al., 2012; Cárdenas et al., 2009; Fernández et al., 2010).

Nanoscale metal and metal oxides, like zinc oxide (ZnO), copper oxide (CuO), silver nanoparticles (AgNPs), and titanium dioxide (TiO_2), are useful in upgrading the mechanical and restrictive characteristics of chitosan films, besides their inbuilt antimicrobial duties (Bui et al., 2017). Metal oxides have antibacterial characteristics that are effective against a large range of microorganisms. As a result, for packaging of food, TiO_2-, MgO-, and ZnO-based biopolymer nanocomposites have been created. TiO_2 nanocomposites are by far the most commonly studied among them. Cerrada et al. prepared EVOH sheets containing TiO_2 nanoparticles particularly for use in packaging of food (Unalan et al., 2014).

The key practical restriction of TiO$_2$ nanoparticles for the packaging of food is their low photon usage effectiveness and the need for UV rays as a tool for excitation (i.e., antimicrobials based on TiO$_2$ are active in the presence of UV rays only) (Zhang et al., 2008). Aside from antibacterial action, TiO$_2$ nanoparticles may offer additional benefits to the overall quality of the resulting packaging material. Because TiO$_2$ NPs are excellent absorbers of short-wavelength radiations with high photostability, food packaging films, including TiO$_2$ NPs, may provide the added advantage of preserving eatables from the oxidizing results of UV irradiation while keeping good optical clarity. The photoactivity of AgNPs and TiO$_2$ can oxidize ethylene to CO$_2$ and H$_2$O (Llorens et al., 2012). Oxygen scavenger films have also been made with TiO$_2$ nanoparticles. Incorporating TiO$_2$ into a synthetic plastic matrix has also been proven to improve biodegradability (Kubacka et al., 2007; Xiao-e et al., 2004). Another fascinating outcome is the production of smart packaging nanocomposite films, which are developed to check the state of packed food items or the conditions in which it is packaged. Lee et al. created an ultraviolet-activated colorimetric oxygen indicator that employs TiO$_2$ NPs to photosensitize the reduction of methylene blue (MB) by triethanolamine in a polymer-encapsulating medium under UVA light. After ultraviolet exposure, the sensor bleaches and persists bleached until it is provided with oxygen, at which point it reverts to its original blue hue. The amount of exposure to oxygen decides the rate of color restoration (Lee et al., 2002).

The antimicrobial properties of ZnO particles are also well recognized. When compared to AgNPs, ZnO has a few advantages: first, it is less expensive, which is due to the large availability of zinc metal, and secondly, ZnO nanoparticles show UV-blocking capabilities (Dastjerdi & Montazer, 2010). In the last few years, a huge range of nanocomposite films based on ZnO for the packaging of food has been created. For example, PVC films covered with ZnO NPs show antibacterial properties against *S. aureus* and *E. coli*. (Li et al., 2009). In a latest work, Fuji apple slices were tried to be preserved using nanopackaging incorporating ZnO NPs. It was found that quality markers, including ascorbic acid and polyphenol content, were preserved better (Unalan et al., 2014).

16.1.5 Graphene

Graphene/polymer nanocomposites can be used in various applications due to their vast multifunctionality. Due to the projected enhancement of mechanical, thermal, and thermomechanical characteristics of the final materials, graphene-based polymer nanocomposites could also be used in food packaging (Kim et al., 2013; Mortazavi et al., 2013; Kim & Macosko, 2009; Wang et al., 2011). Another advantage of graphene's employment in packaging of food is its gas-impermeable atomic membrane, which allows for the creation of barrier materials that prevent O$_2$, N$_2$, and He from passing through (Kim & Macosko, 2009; Kim et al., 2011). The atomic membrane of graphene is a remarkable controllable, water-permeable atomic membrane, which is critical for its application in packaging of food. In view of recent exciting breakthroughs that may lower graphene production costs, graphene may be the most viable strategy to steer development and approval in the creation of new, innovative packaging materials, although it is still in its early phases (Nair et al., 2012).

16.2 Fabrication of Bionanocomposites

For fully exploiting the enormous capability of bionanocomposites for real-world applications, it is necessary to incorporate specified nano-building blocks in polymer matrices. The final performance of the nanocomposite packaging material may be affected by the level of the interfacial interactions between filler or additives and polymer and the degree of exfoliation of the fillers; their ordering in space, which may be stretching or wrinkling; the structure of the composite; and the dispersion condition in the polymer matrix, which may be stacked or clustered, and so the degree of the interfacial contacts between filler and polymer. Historically, there have been three basic ways for incorporating nanofillers in polymer matrices: (1) polymerization in situ, (2) melt processing, and (3) casting in solution. Methods like high-shear mixing and sonication have recently been suggested as alternatives to traditional methods for preparing bionanocomposite materials (Prolongo et al., 2014)

16.2.1 In Situ Polymerization

This method includes the polymerization of monomers in the existence of multilayer materials. Here, first, in a liquid monomer or solution of monomer, nanoclays are swelled and then polymerized in between the intercalated layers. Polymerization can be started off by various ways, like heat or light, diffusion of an appropriate initiator, or an organic catalyst inserted inside the interlayer by cationic exchange before the engorgement process, if necessary. The main drawback of this method is the capability of inorganic species to cause phase separation and silt fast from the organic polymer. Specific groups must be attached onto their surface to stabilize the connection at the solvent/filler contact (Sinha Ray & Okamoto, 2003).

16.2.2 Melt Processing

In this method, the NPs are integrated with the polymer in liquid condition. The particles are combined with the polymers, and the combination is heated over the polymer's softening point, either statically or by shear interlayers (Vaia et al., 1996; Vaia & Giannelis, 1997). There is no need for solvent while performing this process, which is one of the great advantages of this method. Also, it is consistent with present industrial processes, such as injection molding and extrusion, which is another significant benefit of the melt processing approach. However, melt processing approach method is unsuccessful in case of maximum bionanocomposites, as these polymers deteriorate, owing to mechanical shearing force or temperature conditions used while processing.

16.2.3 Solution Casting

The process of solution casting utilizes a system of solvent where the polymer and any other mixture constituent (such as surfactants) are soluble. Before the two are combined together to create a homogeneous dispersion, the polymer is normally added and mixed in a suitable solvent; on the other hand, the nanofillers are distributed in the same or a separate solvent (Figure 5c). The major benefit of this procedure is that it permits relatively quick exfoliation of the stacked layers, credits for which go to the use of a suitable solvent (Vaia et al., 1996; Sinha Ray & Okamoto, 2003). This process requires an enormous amount of solvent for its working due to which when organic solvents are used, the process becomes dangerous and environmentally unfriendly (Reddy et al., 2013). On the other side, this process is gaining popularity for water-soluble polymers like PVOH, especially as thin coatings, which minimizes the quantity of water utilized in the method. Because of the inherent features of most biopolymers, the processes of melt intercalation and in situ polymerization both are typically unsuited for the development of bionanocomposites, so the method of solution casting has recently been suggested for the formation of bionanocomposites. From a practical standpoint, the solution casting process for fabricating bionanocomposite films and coatings necessitates extra caution during the solvent removal (evaporation) step. Indeed, a decreased interfacial engagement between the polymer and the filler can occur if a little quantity of solvent stays trapped in the finished product (Jin et al., 2002). Due to this, the ideal technique for avoiding this potential downside is to combine infrared lamps with high-efficiency air ovens (Farris, 2009).

16.2.4 Sonication

In the top-down production of nanoparticles, sonication is becoming more popular. This is accomplished by using sound waves (more commonly, ultrasound waves) to deagglomerate and reduce micro-sized particles (like tactoids) as an impact of the mechanical outcomes of the phenomenon known as cavitation. This phenomenon relates to the creation, growth, and implosive collapse of bubbles in a liquid (Hielscher, 2007). More recently, the technique of ultrasonication has been found useful for the formation of nanocomposites, which are graphene-based from graphite flakes or particles distributed in liquid media, both aqueous and non-aqueous systems, starting from graphite flakes scattered in liquid solvent system.

16.2.5 High-Shear Mixing Methods

This method has been developed more recently than previous methods in an attempt to give an alternate strategy that can speed up the transition from the lab to commercial applications. This method involves exfoliation of multilayer materials. Although still in its infancy, high-shear mixing methods have shown significant promise for the exfoliation of graphene and other 2D substances, paving the way for large-scale industrial uses based on these layered crystals (Costantino et al., 2009; Mangiacapra et al., 2006). Graphite can be exfoliated into graphene by creating shear in an extremely thin liquid layer in a speedily spinning tube, according to a recent study. This process, however, produces very small amounts of graphene and is intrinsically unscalable. In terms of yield, this approach is said to be many times faster than sonication, allowing for the production of great amounts of defect-free, unoxidized graphene. Furthermore, the procedure can be developed up to an industrial scale.

Another frequent high-shear mixing process is ball milling. It involves solid-state mixing at surrounding temperature (Sorrentino et al., 2005). This process is an effective and simple approach that breaks the Van der Waals connections between layers by using pure shear between balls of varied diameters (Guo & Chen, 2014). Among the current processes, ball milling has several significant advantages, like it doesn't require extreme temperature conditions or solvent treatments, making composites formation more environmentally friendly, suitable, and successful (Tammaro et al., 2014). Bionanocomposites such as MMT/pectin (Mangiacapra et al., 2006) and LDH/PCL (Costantino et al., 2009) were created using the ball milling process (Unalan et al., 2014).

16.3 Aspects of Manufacturing Bionanocomposite Coatings from a Technological Perspective

The effectiveness of bionanocomposite materials is heavily reliant on a number of key elements during the design and development phases. Aside from high expenses, technological considerations may serve as a "go-no-go" gate in the development of market applications. These considerations apply to the coating system before and after deposition on the chosen substrate.

16.3.1 Compatibility of the Polymer with the Filler

The bond between inorganic and organic components at the hybrid interface determines how well bionanocomposite polymer systems operate in the end. The source material to be chosen must be the best one to maximize the benefits of the interface effect between polymer and filler. There is a substantial difference among petrol-based polymers and biopolymers in this regard.

In practice, increasing the affinity of organic and inorganic constituents allows for better intercalation, swelling, and exfoliation of nanoclays, as well as a reduction in phase during processing. As a result of the depression of the "interface effect," this will result in balanced dispersal and distribution of the filler inside the polymer matrix, preventing the creation of typical filled polymers with lesser utilizations. Eventually, efficient filler dispersion in the hosting matrix should allow for low-cost nanocomposite materials that are both sustainable and lightweight (Kim et al., 2010; Cao et al., 2011; Cai & Song, 2010; Potts et al., 2011).

16.3.2 Rheological Properties

Rheological measures are extensively used by manufacturers of nanocomposite polymer systems because they provide an indirect way to examine the status of distribution of nano-objects (i.e., fillers), like the extent of intercalation, exfoliation, and dispersion in the polymer matrix (Rezadoust et al., 2013). Since maximum preparation methods to achieve the best system of nanocomposite polymer (i.e., melt processing, mesophase mediated processing, solution processing, and in situ polymerization) require flow, understanding the rheological behavior of polymer nanocomposites can be beneficial to get important data on the structure–characteristics relationship in the finished products (Sinha Ray & Okamoto, 2003; Chen et al., 2005).

16.3.3 Optical Characteristics

Optical qualities of substances are very essential in specific industries, as they might influence the finished material's performance or the consumer's decision. Both issues are important to consider when building a new bionanocomposite coating in the food packaging industry. On the one hand, ultraviolet light (of wavelengths less than 340 nm) must be avoided since it can promote photo-oxidation in photosensitive foods like meat, beer, and milk, leading in color, flavor, and taste alterations. On the other hand, high visible light transmittance (of wavelengths b/w 340 nm and 800 nm) must be ensured as this enables people to see packed food items through the package (Unalan et al., 2014).

16.3.4 Surface Characteristics and the Scalping Effect

The introduction of additives or fillers to coating formulations may have an effect on the finished material's surface characteristics. Surface qualities are important in numerous converting procedures in food packaging, including printing, laminating, and coextrusion. In general, the incorporation of an inorganic filler reflects the modifications in the surface topography, that is, rise in surface coarseness, especially for a high filler volume percentage, which is connected to the arrangement of the additives within the coating thickness (Introzzi et al., 2012; Vartiainen et al., 2010; Rhim, 2011).

After solvent removal, platy clays with huge surface areas, for example, frequently show a "house-of-cards" or "cell-like" arrangement (Introzzi et al., 2012). According to both Wenzel's and Cassie-Baxter's theories, an increase in roughness can keep a notable effect on the wettability quality of the surface, in addition to increasing the haze of the final material (Cassie & Baxter, 1944; Wenzel, 1936). Furthermore, very porous fillers (like zeolites) and fillers with compatibility for water molecules (such as natural MMT) may enhance absorption at the solid–liquid interface. A rise in roughness can affect not only surface qualities but also mechanical properties (Vartiainen et al., 2010).

Finally, but certainly not least, there is the potential for bionanocomposite coatings to have a notable effect on the standard and condition of packed edibles. The sorption of food ingredients, especially fragrance compounds, by materials used in packaging is known as "scalping." In the construction of a bionanocomposite coating, both chemical affinity and porosity of the filler for smells and volatile molecules should be considered, especially if it is going to face the inside of the package and will be in close contact to the edibles. The process of scalping has gotten a lot of attention in the food packaging business because they might affect consumer approval of a food product by reducing scent strength or creating an uneven flavor profile (Nielsen & Jagerstad, 1994).

16.3.5 Gas Barrier Properties

The penetration and distribution of gas molecules through the polymer membrane determine the gas barrier qualities. The mechanism is fairly complicated, and it involves the polymer chains' free volume holes, which are caused by Brownian movements or thermal perturbations. As a result, the most common technique for improving barrier qualities is to limit mass flow by incorporating non-permeable barriers that produce a tortuous track. Three distinct approaches have been taken to this winding road: because crystallites are impermeable to tiny molecules, increasing the crystallinity of the polymer, adding nanofillers, and building multilayers are all options.

Even today, the link between PLA crystallinity and gas barrier characteristics is a hot topic of discussion. Although early investigations found that increasing crystallinity decreased the permeability of PLA films, now latest studies did not find any influence on the permeability of helium and oxygen or water. The results were associated with either crystal polymorphism and PLA stereochemistry or crystalline and amorphous phase structures, according to the authors (Armentano et al., 2013).

16.3.6 Antioxidant Characteristics

The antioxidant potential of food packaging films is determined in general by two factors. The first is their ability to protect against environmental stresses, like UV rays and oxygen. The second goal is to preserve antioxidant capabilities in products or to slow lipid oxidation by protecting antioxidant components in packaging from oxidants (Qu & Luo, 2021). Barrier qualities could be improved with the inclusion of nanoclays, similar to how chitosan films have improved antioxidant function. Pires et al. (2018) found that adding MMT to chitosan film improved its UV light barrier capabilities and provided further protection against food deterioration. Nanoclays have the ability to improve barrier performance in addition to increasing surface area and width-to-height ratio. Nanoclays possess the ability to decrease UV rays by absorption or reflection, in addition to barrier augmentation because of their large surface area and width-to-height ratio (Rehan et al., 2018).

16.4 Applications of Nanocoatings and Nanocomposite Films

Nanocoatings and films have a large number of uses in areas like aerospace, packaging, and manufacturing. A *nanocoating* is defined as a coating that is nanostructured or has a thickness in the nanometer range. Nanocoatings, in particular, act as a product's interface with the environment, affecting not only the design of materials but also crucially specific qualities, including self-cleaning, chemical and scratch resistance, and anticorrosion. Nanostructuring is a technique for improving the hydrophobicity, radiation hardness, and resistance to corrosion of materials while also making them more flexible. Coatings and ultrathin films are used extensively in micro- and nanoelectronics, machine building, automotive, and aviation manufacturing.

Free electrons and conducting systems can move only in the x–y plane in nanocoatings and films since they are two-dimensional systems. In the context of electronic materials, restricting in the z-direction can provide a number of unique features. To decrease stiction and light reflection, change surfaces in tough settings, and increase dirt release properties, ultrathin films and nanocoatings with appropriate design can be used. These days, nanophase thermal barrier coatings with exceptionally low heat conductivity are becoming more widespread. Decorative nanocoatings made from special paints and inks are also gaining popularity (Krishnamoorthy & Chidambaram, 2018).

16.4.1 Packaging Industry

The notion of packaging is constantly being revised and improved, as it is critical to attaining a competitive advantage in today's industry. Packaging's original purpose as a measure of protection and preservation against external contamination is now simply one of multiple market demands. A novel pack design can open up new distribution channels while also improving presentation, lowering costs, assuring safety and reliability, and raising brand awareness. As a result, the packaging industry is always challenged to provide cost-effective pack performance while keeping health and safety in mind. Simultaneously, regulations and political pressure are being applied to minimize both the amount of packaging used and the amount of packaging waste. Therefore, Quality, production, engineering, marketing, purchasing, legal difficulties, finance, logistics, and environmental management are all factors that influence the development of new types of packaging (Sorrentino, 2011).

Single films, multilayer flexible structures, sheets, coatings, adhesives, foams, laminations, and rigid or semi-rigid containers are all examples of polymers used in packaging. Low density polyethylene (LDPE), linear low density polyethylene (LLDPE), and polypropylene are the most often used plastics in flexible packaging (PP). For rigid containers, high density polyethylene (HDPE) is commonly used. Polyethylene terephthalate (PET) and its copolymers, on the other hand, are the fastest-growing rigid container plastics. Other resins with particular applications as high barrier materials include ethylene vinyl alcohol (EVOH), polyvinyl chloride (PVC) copolymers, and nylons. However, the increased use of plastic in the packaging industry has raised many environmental issues because of their non-biodegradable nature due to which scientists are looking for better alternatives and found that nanomaterials might be one of the solutions to this problem (Sorrentino, 2011).

16.4.1.1 Packaging Using Nanocoatings and Ultrathin Films

Nanomaterials are a new class of materials that outperform their traditional or bulk equivalents in terms of physical and mechanical qualities (Thostenson et al., 2005; Markarian, 2005). Regulating the order in which matter is arranged at the atomic or molecular scale, their qualities can be achieved (Sorrentino, 2011; Ray et al., 2006) and, hence, gaining advantage over conventional packaging. Nanotechnology has the capability to change the nature of traditional packaging material to the preferred nature. Due to the application of nanotechnology, traditional packaging will get swapped with multifunctional intelligent packaging. Materials with variable amounts of gas/water vapor permeability can be created using various nanostructures, depending on the requirements. Packaging can be made lighter and stronger with greater thermal performance and reduced gas absorption by using nanocoatings or nanoparticles (Sanguansri & Augustin, 2006). Furthermore, nanostructured packages can effectively prevent bacteria and pathogens from invading and ensure product safety. Consumers will be allowed to see the exact status of the contents thanks to nanosensors embedded in the packaging (Krishnamoorthy & Chidambaram, 2018). Packages with dirt-repellent coatings are also being developed (Sorrentino, 2011).

A nanostructured composite is a substance that is made up of two or more separate material components, at least one of which has a nanometer-scale dimension (Sorrentino, 2011; Alexandre & Dubois, 2000). Polymer/inorganic particles, polymer/polymer coatings, metal/ceramic, and inorganic-based composites are among the various forms of nanostructured composites now being researched and developed (Sorrentino, 2011; Johnston et al., 2008; Jordan et al., 2005; Wiley et al., 2005). One of the methods to prepare nanostructured composites is to dissolve both complex and polymer in solvents (Zeng et al., 2006). The melt mixing approach is another method. The use of shear during compounding aids exfoliation and dispersion in this scenario. There is also an alternate method, called "in situ polymerization," in which the nanolayer is placed in the form of liquid monomers that must be polymerized directly on or inside the sheets (Sinha Ray & Okamoto, 2003; Lepoittevin et al., 2002). Solid-state mixing at ambient temperature is a method for making nanocomposites that does not necessitate the use of high temperatures or solvents. The energy transmission between the milling tools and the composite particle mixture promotes the dispersion of nanometric particles in this scenario. One of the most significant advantages of these systems is that, because of the nanometric scale dimension, only a little quantity of material is required, typically 0.1–5 wt.% (Lan et al., 1994). As a result, these systems bypass many of the time-consuming and expensive fabrication procedures used in traditional composites. They can also be modified for application in films, fibers, and monoliths (Sorrentino, 2011).

16.4.1.1.1 Food Packaging

Some criteria for packaging of food items usually involve function as a wall between the environment and the food to avoid the passage of light, bacteria, insects, heat, moisture, and other contaminants (Krishnamoorthy & Chidambaram, 2018). A suitable packing material should manage internal gas compositions, like oxygen, ethylene, and carbon dioxide, and water loss in the packaged food. It should give physical protection and provide the correct physicochemical conditions for items, both of which are necessary for a long shelf life. It is also expected that the material should biodegrade in a short time without causing harm to the environment. The development of "nanocomposites" is a major success of nanotechnology in packaging. The growing effort to substitute polymers with polymers based on oils derived partially or entirely from renewable sources has spawned a novel subclass of nanocomposites. The word "bionanocomposites" is used for substances where the polymer matrix having the nano-sized fillers is either a biopolymer (like polysaccharides and proteins) or a natural polymer received through synthetic or biotechnological pathways. The utilization of NBB (nano-building blocks) for the formation of bionanocomposites offers immense promise for eliminating natural polymer limitations, like poor thermal and mechanical capabilities, sensitivity to damp conditions, and insufficient gas and vapor barrier properties. As a result, bionanocomposites have gotten considerable attention in recent times. The majority of cases, on the other hand, involve incorporating the inorganic phase directly into the bulky biopolymer, which is discussed later (Unalan et al., 2014). Biopolymer-based packaging materials, in this respect, have specific advantages as packaging substances in terms of bettering quality of food items and increasing shelf life by minimizing microbial development in the item (Hun & Gennadios, 2005; Petersen et al., 1999; Rhim & Ng, 2007; Benbettaïeb et al., 2019). Indeed, biopolymers have grown in popularity as a consequence of their environmental friendliness and potential for usage in the area of packaging (of food items) (Briassoulis, 2004; Petersen et al., 1999; Debeaufort, 2010; Rhim & Ng, 2007).

16.4.1.2 Nanocomposite Films Based on Natural Biopolymers

Biopolymers made from agricultural or animal goods. Biopolymers derived from natural materials like starch, protein, and cellulose have been proposed as potential substitutes to non-biodegradable petroleum-based plastics due to their abundance, renewable nature, low cost, ecologically friendly nature, and biodegradability. Only high-molecular-weight polymers may produce such a film structure due to their strong cohesive strength and coalescence tendency. The extent of cohesiveness of the polymer matrix influences film characteristics, like density and compactness, porosity and permeability, flexibility and brittleness. Almost any natural biopolymer can be utilized to make films. Film-forming substances include polysaccharides (starch, cellulose derivatives, chitosan, pullulan, and natural gums), proteins (casein, whey protein, collagen, gelatin, keratin, soy protein, wheat gluten, and corn zein), and lipids (neutral lipids, fatty acids, and wax). Plasticizers (glycerol, propylene glycol, polyethylene glycol, etc.) and additives (glycerol, propylene glycol, polyethylene glycol, etc.) are another component (Rhim & Ng, 2007). Bionanocomposite coatings and films also show strong antimicrobial properties. This has proved their ability as potential use in packaging of food for increasing shelf life and enhancing food safety (Dubas et al., 2006; Zhu et al., 2009).

- **Polysaccharide Films**

Carbohydrate-based films are another name for them. Because they are hydrophilic, they have poor moisture barrier qualities. The majority of the efforts in this area were first concentrated on starch and cellulose. These polysaccharides have attracted considerable attention as biopolymers due to economic and availability reasons, but their low elasticity is a key limitation to their use. A number of polysaccharides and their derivatives have been utilized as biodegradable film-developing matrixes. Among them are starch and starch derivatives, cellulose derivatives, alginate, and other gums. The main method of film making in polysaccharide films is hydrophilic and hydrogen bonding, which is created by breaking away polymer segments and reconstructing the polymer chain into a film matrix or gel by solvent evaporation (Rhim & Ng, 2007).

- **Protein Films**

Proteins are huge macromolecules that are generated by condensation polymerization of different amino acid combinations. Proteins from diverse plant sources contain several amino acid compilations and so have varied characteristics. Proteins have characteristics such as network formation, flexibility, and elasticity that make them suitable in the production of packaging biomaterials. The tendency of various protein compounds to form films has been employed in industrial use for a long time (Bhawani et al., 2017). Because of its water resilience, earlier Egyptians, Chinese, Greeks, and Romans utilized casein in glue.

Soy protein, wheat gluten, cottonseed protein, and protein extracted from sorghum kafrin, rice bran, peanuts, corn zein, and pea protein have all been studied for their film-forming potential (Rhim & Ng, 2007; Brandenburg et al., 1993; Hernández-Muñoz et al., 2003; Rayas et al., 1997). Protein film-forming components used in the formation of biodegradable/edible films involve collagen, gelatin, fish myofibrillar protein, keratin, egg white, and whey protein (Rhim & Ng, 2007).

Protein films are generated when polypeptide chains are partially denatured by adding a solvent, changing the pH, adding an electrolyte to make cross-linking, and/or applying heat. The partially denatured peptide chains form a protein matrix, which leads in the formation of films, owing to hydrophobic and hydrogen interactions.

- **Lipid Films**

Direct coating, dipping, and pan coating are all methods for applying lipid ingredients and waxes to food goods (Guilbert et al., 1995). Edible film-forming ingredients include lipids like beeswax, cadellila wax, triglycerides, fatty acids, and alcohols, as well as resins like terpene resin (Rhim & Ng, 2007; Skoryakov, 1958). The good thing about lipid edible films is that they have great barrier qualities because of their low polarity. However, they possess poor oxygen-restrictive properties, and employing fats as a guard coating for foods may have downsides, like lipids becoming rancid and an oily surface layer being imparted by the fats (Rhim & Ng, 2007). These lipid and resin components do not form cohesive stand-alone films because they are not polymers. In addition to generating a desirable shine, they can be utilized to cover a drug or food layer to give restricted moisture or to give a moisture-restrictive constituent of a composite film.

- **Composite Films**

Protein- and polysaccharide-based films are effective carbon dioxide and oxygen barriers, although their hydrophilic property limits their resistance to water vapor transmission. The majority of these films possess good mechanical characteristics, enabling them to be valuable for strengthening the structural integrity of delicate items. To take advantage of the properties of both lipid and hydrocolloid, composite films comprise both of them as its components (proteins or polysaccharides). When a water vapor barrier is required, the lipid part can do so while the hydrocolloid part gives the essential mechanical strength.

Polylactic acid (PLA) has emerged as one of the major crucial biopolymers for replacing petroleum-based polymers in industries, among many others. PLA has intriguing physical properties, as well as biocompatibility and biodegradability, all of which are controlled by its stereochemistry and molecular mass. PLA has a lot of potential in consumer items like packaging because of its clarity, lesser toxicity, and environmental friendliness.

Here are the advantages that make PLA so special for packaging purposes:

1. *Transparency* is commonly described as visible light transmission in the region of 540–560 nm, which is somewhat greater than that of poly(ethylene terephthalate) (PET) and poly(styrene) (PS).
2. PLA has been shown to be bio-friendly and decompose into non-toxic parts, and it has been cleared for utilization in the human system by the Food and Drug Administration (FDA).
3. Melt processing is the most common PLA conversion strategy. High-molecular-mass PLA melt viscosities range from 500 to 1,000 Pa•s at shear rates of 10–50 s^{-1}, and commercial grades of PLA may normally be handled by a conventional twin-screw extruder. In fact, NatureWorks offers a variety of PLA grades customized to specific processing and applications, like extrusion, thermoforming, injection stretch blow molding, or the fabrication of film, fiber, and foam. However, the greatest disadvantage of treating it in a molten form is its heat deterioration. PLA is a sustainable substitute to petrochemical-based synthetic polymers as a packaging substance because of all these characteristics.

Commercial PLA, on the other hand, has some limitations that limit its today's application in packaging of food items, like its high brittleness, which restricts its applications in flexible films, or sheets with large impact strength and limits its processability range. Its thermomechanical resistance is limited by its weak crystallization behavior. PLA is inappropriate for hot filling liquid food packing due to its larger rate of hydrolysis and limited thermal resilience. It has a low gas restriction to oxygen, carbon dioxide, or water, all of which can interact with or harm food (Armentano et al., 2013).

16.4.1.3 Additives and Inorganic Fillers

This is another component used in making biopolymer-based nanocomposite films. To adjust physical qualities or add functionality to films, plasticizers, inorganic fillers, and other substances are mixed with the film-forming biopolymers. Plasticizers are low-molecular-mass chemicals that are mixed with polymer film-forming materials to make the films more flexible and processable. They raise the ratio of crystalline to amorphous areas, the glass transition temperature, and the free volume of a polymer structure or the molecular mobility of a polymer matrix (Rhim & Ng, 2007; Guilbert et al., 1997; Sothornvit & Krochta, 2000). Extra compounds like antioxidants, antimicrobials, nutraceuticals, fragrances, and colorants are occasionally incorporated to film-forming solutions to provide active packaging functions (Rhim & Ng, 2007; Kloeckner, 2013). Natural biopolymer films, a new group of eco-friendly plastics, can be made from these plentiful, renewable, and biodegradable natural basic materials. They have limited application, however, due to their mechanical and water vapor–restricting characteristics, which are of lower quality to those of synthetic biopolymer–based plastics (Rhim & Ng, 2007).

Inorganic nanosized building blocks have also piqued interest in recent years as fillers in the creation of polymer/inorganic nanocomposites for food packaging, because of their unique features that have multiple usage in a variety of industries. The preceding part went over this in great depth.

Biopolymer-based nanocomposite films have many advantages, but their poor resistance against water doesn't allow them to be used at a large scale. Researchers are continuously working to improve these properties and make them better for easy use.

16.4.1.4 Potential Applications of Nanofilms and Coatings in Active Packaging of Food

Active food packaging systems play an active role in the standard of food and/or stability, either by scavenging or inactivating harmful substances or by releasing suitable constituents like antimicrobial compounds. This method is an alternative to introducing antimicrobials directly into bulk food, with the benefit of being able to manage antimicrobial dispersion toward food (Azeredo et al., 2019) and being more cost-effective, as food surfaces are more sensitive to microbial growth than the bulk itself. The usage of nanofilms with antimicrobial activities in active packaging helps in managing the growth of pathogens and spoilage of microorganisms.

It has been suggested and/or tested to use organic acids like sorbate, propionate, and benzoate or their acid anhydrides, bacteriocins like nisin and pediocin, enzymes like lysozyme, metals like copper, and fungicides like benomyl and imazalil. Active components in these antibacterial systems may be organic or inorganic. Metal ions like silver, copper, and platinum are particularly common in inorganic systems. Applications for silver ions as an antibacterial include appliances, building materials, medical equipment, water filtration, delivery systems, and of course, food processing and packaging. Ag-modified zeolite is a very common antibacterial agent used in polymers.

The way antimicrobial agents are used in food systems is crucial to their success. And hence, keeping in mind the problems related to direct spraying of antimicrobial agents, the use of antimicrobial polymer–based packaging film could be more effective in terms of retaining high concentrations on the surface of food while limiting active substance migration.

Due to its possible antibacterial action and good film-forming capabilities, chitosan has emerged as an attractive ingredient for preparation of antimicrobial films. Because of this, chitosan-based nanocomposite films and coatings are being developed as alternatives to traditional antimicrobial food packaging (Rhim & Ng, 2007).

16.4.1.5 Antimicrobial Activity of Chitosan

Chitosan is a natural biopolymer with antibacterial activity against bacteria, filamentous fungus, and yeast (Hosseinnejad & Jafari, 2016). Many researches have proved chitosan's antibacterial properties, but the mechanism behind them is still unknown (Kumar et al., 2020). In order to be employed in films and coatings and to enhance its performance as a food packaging substance, chitosan must be blended with other biopolymers like polysaccharides, proteins, and lipids (Zhu et al., 2019). In comparison to blends with proteins and/or lipids, polysaccharide blends have a number of benefits, such as inexpensive material costs, easy access to resources, relative stability, and greater heat sealability and water solubility (Li et al., 2019).

- **Chitosan and Different Carbohydrate Blend**s

Chitosan is combined with a variety of polysaccharides, including alginates, starch, carrageenan, pectin, and bacterial cellulose. Alginate and chitosan are compatible polysaccharides of aquatic origin, and their mixes are better suited for use in food packaging than those that combine chitosan with protein or lipid polymers (Li et al., 2019).

Starch from various sources, including rice, corn, and cassava, has been researched the most when combined with chitosan. Higher-amylose content starch blends well because of the stable hydrogen bond formation with chitosan. Both research utilized starch gelatinization and solution casting in chitosan solution to create the combined film. Plasticizers of 40% sorbitol or 25% (w/w) glycerin were incorporated to the blended polymer to enhance the mechanical properties of the blended films.

Carrageenan is a class of anionic linear polysaccharides produced from red seaweed. Fresh fruits and vegetables have already been coated with this polysaccharide to stop oxidative deterioration and moisture loss. Along with chitosan, carrageenan creates a polyelectrolyte complex.

Chitosan–pectin blend is another mixture that has been researched. In addition to this, the possible use(s) of chitosan-blended films with some of the third-generation carbohydrate polymers, like polylactic acid and microbial cellulose, in food packaging are being researched (Kumar et al., 2020).

- **Chitosan-Based Polymeric Antimicrobial Composites and Their Blend**

Chitosan has antibacterial properties against a wide range of pathogens (Verlee et al., 2017). But its antibacterial characteristics should be improved further, notably for preservation of food and packaging applications. Natural antimicrobials and antimicrobial nanostructures were used to improve its antimicrobial capabilities.

16.4.1.5.1 Natural Antimicrobials

Animal, plant, and microbial sources are the most common sources of natural antimicrobials. Biocompatible and biodegradable natural antimicrobials have minimal toxicity and antimicrobial property comparable to synthetic antimicrobials, with antibacterial action against a broad spectrum of pathogens, including rotting germs. In order to manage natural spoiling processes (i.e., food preservation), preserve or enhance nutritional value (i.e., quality of food), and decrease or cease the growth of microbes, natural antimicrobials are utilized in food (i.e., food safety) (Kumar et al., 2020).

Plant extracts, such as essential oil, offer high antioxidant properties as well as antibacterial capabilities and are utilized to make antimicrobial composites for packaging of food and storage (Pisoschi et al., 2018). It has been demonstrated that adding essential oils (EOs) like cinnamon, clove, oregano, or eucalyptus to chitosan (CS) can enhance the antibacterial effect in food packaging applications (Yuan et al., 2016). This antimicrobial effect led to the forming of chitosan-based active packaging ideas that decreased, prevented, or postponed the growth of bacteria on food surfaces or in food, hence prolonging postharvest/postmanufacturing shelf life (Ponce et al., 2016).

16.4.1.5.2 Metal- and Metal Oxide Nanostructure–Based Antimicrobials

Metal NPs, including silver AgNPs and CuNPs, are well-known broad-spectrum antibacterial agents because of their strong efficacy against Gram-negative and Gram-positive bacteria, fungus, protozoa, and some viruses. AgNPs have been produced from the bottom up for use in culinary applications using various chemical and biological techniques, including green synthesis. AgNPs may also be synthesized in a chitosan matrix in situ. Chitosan has a strong affinity to Ag^+ ions because it contains amine and hydroxyl groups, and in alkaline conditions, it can convert Ag^+ ions into AgNPs (Kumar et al., 2020).

16.4.1.6 Nanoedible Films in the Food Packaging Industry

Packaging of food is typically done to preserve foods and drinks at a low cost, to save them from the impact of nature or harm while in transit, and to maintain food quality and standard from the time of packing through the consumption time (Marsh & Bugusu, 2007; Petersen et al., 1999).

According to FICCI, the most commonly used packaging materials are plastics (42%), paperboard (31%), metals (15%), glass (7%), and other substances (5%).

Due to their excellent cost-to-performance ratio, lightweight, and ease, plastics have become the most widely used material for packaging (FICCI, 2020). The largest and fastest-growing area of the plastic packaging market is food packaging. For the years 2003–2009, it was predicted that the food and beverage industries will utilize 54%, 40%, and 14% of all plastic packaging, respectively (WPO, 2013). Plastics utilized in food packaging are typically thrown post-use, with just a small percentage being recycled. If plastic garbage is recycled, landfill space could be reduced. However, only a small percentage of plastic is recovered by recycling. Either plastics are burned for energy recovery (incineration), which releases toxic gases into the atmosphere, or they are discarded in landfills, where they will remain indefinitely. In many countries, especially in heavily populated areas, the availability of landfills is a major problem.

Plastics relying on petroleum reserves is also a worry (Hopewell et al., 2009). On the other hand, biodegradable plastics break down into methane, water, carbon dioxide, and other by-products as they break down. Biodegradability is harmed by landfills because they lessen the presence of moisture and oxygen, both of which are necessary for biodegradation. Furthermore, land space is still needed to store wastes until they disintegrate completely. As a result, biodegradable materials have a minor impact on landfill reduction (Kale et al., 2007; Marsh & Bugusu, 2007).

A thin layer (film or coating) applied to a food product's surface that may be consumed in its whole is known as edible packaging. The term "coating" describes the quick creation of a thin layer on the food surface. When it is manufactured independently and later wound around the surface of food, it is referred to as film. Edible films and coatings can only be produced when the material (used for packaging) is capable of forming a continuous and cohesive structure. Composites and polymers that can produce such continuous cohesive structures and are edible are considered edible packaging materials (Guilbert et al., 1995). Nothing will be left out for trash because these edible packagings can be consumed along with the food items. Even if they are not consumed, they do not affect the environment. Edible packaging decomposes more quickly than synthetic and biodegradable packaging materials since it is made from food ingredients. Consequently, edible packaging materials have piqued the interest of researchers as a promising option that has the capability to take the place of both synthetic and biodegradable plastics (Shit & Shah, 2014).

Edible films are made from polysaccharides (like, starch, cellulose, chitosan), proteins (e.g., zein, soy protein, gelatin), and lipids (e.g., resins, essential oils and extracts, animal oils, and fats), which are all edible polymers.

The food and beverage business is very interested in incorporating nanotechnology's benefits. Nanotechnology allows for the creation of lightweight packaging materials with improved barrier qualities that protect food quality throughout packaging, transit, and consumption, as well as the preservation of meat and poultry products from rotting pathogens. Nanomaterials with diameters in between 10 and 100 nm can be added into a range of substances used for packaging to create new materials for packaging with better film qualities (for example, barrier capabilities and mechanical strength) (Lagarón et al., 2016; Bumbudsanpharoke & Ko, 2015; Tsagkaris et al., 2018). Nanomaterials have several unique qualities that make them useful for the packaging sector, such as a huge surface-area-to-volume ratio, peculiar optical behavior, and greater mechanical strength. When the correct nanoparticles are incorporated into compatible polymers, the packaging materials can have better obstacle and mechanical qualities, thermal stability, and optical characteristics than conventional packaging substances (Eleftheriadou et al., 2017; Jeevahan & Chandrasekaran, 2019; Pan & Zhong, 2016; Cerqueira et al., 2018).

- **Edible Films with Nanostarch Reinforcement**

Growth rings that make up the starch structure are contained in white powder known as starch granules, which range in size from 2 to 100 lm. The blocklets (20–50 nm) that make up these growth rings are made up of crystalline and amorphous lamellae, which contain amylose chains and amylopectin (Jeevahan & Chandrasekaran, 2019). Dufresne et al. (Havens, 1984), in 1996, conducted a study on the isolation of starch nanocrystals which was latter published. His studies found that when potato starch was acid hydrolyzed, it yielded so-called "microcrystals" with a few tens of nanometer in diameter.

The effects of waxy maize starch edible films upheld by waxy maize starch nanocrystals on their aging were studied (Angellier et al., 2006). After two weeks, the waxy maize starch–enhanced edible films' mechanical properties underwent a significant change. With age, percent E fell, while YM and TS increased. The mechanical characteristics of aged starch films can be compared to non-aged starch films containing 10% starch nanocrystals. Significant YM and TS (after aging) developed from the presence of nanocrystals greater than 10% by weight, but percent E remained rather stable at around 20%. Starch nanoparticles were used to make maize starch edible films, and Shi et al. (2013) studied how different drying techniques affected the films' physical and mechanical characteristics. The results revealed that when starch nanoparticles, prepared using drying techniques, were added to cornstarch films, the surface roughness increased by 23.5%, the degree of crystallinity decreased by 44%, the WVP decreased by 44%, and the glass transition temperature decreased by 4.3°C when compared to control films.

Ice starch edible films were created by Piyada et al. (2013) and distributed with rice starch nanocrystals (5 to 30%). It was discovered that the rice starch nanocrystal distribution increased the WVP, thermal stability, and mechanical characteristics of the rice starch edible films. The best edible films in terms of mechanical properties had TS (16.43 MPa) and percent E (5.76%) when up to 20% of rice starch nanoparticles were present. Liu et al. (2016) created waxy maize starch–based nanocomposite films upheld by cornstarch NPs (0–25%). The incorporation of cornstarch nanoparticles significantly increased mechanical strength, water vapor barrier, and thermal stability in contrast to standard cornstarch–based films. Gonza'lez and Igarzabal (Jeevahan & Chandrasekaran, 2019) demonstrated that adding nanostarch improved mechanical and physical characteristics, like swelling, WVP, water solubility, and TS, by using a casting procedure to create edible films made of soy protein isolate mixed with natural cornstarch nanocrystals (5–40%). The films became more homogenous and clearer when the quantity of nanostarch particles put to them was increased. However, the degree of crystallinity and the opacity rose slightly with increasing levels of starch nanocrystals.

Jiang et al. (2016) developed potato starch nanoparticle–coated pea starch–based nanoedible films. In SEM examination, the potato nanoparticles showed spherical-shaped NPs with a diameter of about 15–30 nm. Potato starch NP–enhanced nanoedible films were shown to have a greater relative crystallinity than native starch films. The mechanical characteristics were improved by adding nanoparticles. In addition, the nanocomposite films outperformed native pea starch–based films in terms of thermal stability and water resistance.

- **Edible Films with Nanocellulose Reinforcement**

Due to its distinctive physicochemical characteristics, like light weight, less density (approximately 1.6 g/cc), transparency, and large strength-to-weight ratio, non-toxicity nanocellulose is one of the most discussed novel biopolymers in recent years. Nanocellulose can be produced from various cellulosic sources, including agricultural plants and leftovers, lignocellulosic biomass and wastes, microcrystalline cellulose, animals, bacteria, and algae. Nanocellulose has been separated from different cellulose sources, including rice husk, rice straw, banana pseudostem, jute stem, pineapple leaf, wheat straw, kenaf bast, potato pulp, corncob, mulberry, wood, cotton, ramie, sisal, soybean straw, sugarcane bagasse, tunicates, bacteria, and fruits and vegetable wastes. Acid hydrolysis, steam explosion, ultrasonication, high-pressure homogenization, and enzymatic hydrolysis are some of the methods for separating such nanocellulose that have been documented.

Cellulose nanocrystals, bacterial cellulose, and nanofibrillated cellulose are three varieties of nanocellulose structures that can be made. Nanocrystals of cellulose, with a width of roughly 5 nm, are the smallest fundamental unit. With an aspect ratio spanning from 11 to 67, cellulose nanocrystals are excellent for thin-film applications (Jeevahan & Chandrasekaran, 2019).

Nanocomposite films made of chitosan reinforced with nanocellulose were created by Azeredo et al. (2010), and the resulting films, which contained 18% glycerol and 15% nanocellulose, displayed larger YM and TS than synthetic polymers. On the other hand, their WVP and percent E were lower, enabling it only workable for uses that didn't need flexibility or WVP. *Gluconacetobacter xylinus* was used to make bacterial cellulose nanocrystals and a bionanocomposite with gelatin by George and Siddaramaiah (2012).

- **Edible Films with Nanochitosan/Nanochitin Reinforcement**

The second most frequent semi-crystalline biopolymer after cellulose is the polysaccharide chitin, and chitosan. It was discovered in the cell walls of fungi, yeast, and mushrooms and also in crabs, shellfish, shrimp, tortoises, and other animals (Jeevahan & Chandrasekaran, 2019).

Ifuku et al. (2013, 2014; de Moura et al., 2009) used chitosan and surface-deacetylated chitin nanofibers to make transparent nanoedible films. The mechanical strength of the nanoedible films enhanced by 65% and 94%, respectively, with a 10% incorporation of chitin nanofibers, whereas the coefficient of thermal expansion was reduced. Antifungal action against *Alternaria alternata* was also demonstrated by these films.

Cinnamon essential oil and chitosan nanoparticles (4%) were added to zein-based nanofilms by Vahedikia et al. [D125] (2%). The produced films, especially those with both chitosan NPs and cinnamon essential oil, improved TS by 112% and decreased percent E by about 45% while lowering WVP by 41%. Additionally, it was discovered that the composite nanofilms were efficient against both Gram-negative bacteria (*E. coli*) and Gram-positive bacteria (*S. aureus*).

- **Edible Films with Nanoprotein Reinforcement**

A *protein* is defined as a linear heterogeneous biopolymer which consists of various types of amino acids. It forms a three-dimensional network structure. Protein nanoparticles with easily changeable surfaces, biodegradability, and non-antigenicity are typical. The size, shape, and weight of protein nanoparticles may be easily controlled and manufactured. The most popular techniques for producing protein nanoparticles include nanospray drying, desolvation, emulsification, and thermal gelation. The separation of protein nanoparticles from protein sources like maize zein, gelatin, whey protein, egg albumin, soy protein, and casein has been reported in the literature (Jeevahan & Chandrasekaran, 2019).

Oymaci and Altinkaya (2016) used solution casting to create edible self-standing whey protein isolate and maize zein nanoparticle–based films. Zein is a food-safe substance with high hydrophobicity, making it an excellent option for enhancing WVP.

To attain homogeneous dispersion, sodium caseinate was applied on the films. The consequences clearly revealed that the mechanical (303% greater TS) and barrier characteristics were enhanced by the inclusion of zein nanoparticles (84% lower). Using maize starch that included zein–rutin nanocomposite at varying degrees, Zhang and Zhao (2017) created edible active films. NPs improved the physical and mechanical characteristics of the cornstarch films and their antioxidant activities.

- **Edible Films with Nanolipid Reinforcement**

Edible films and coatings are developed with lipids obtained from vegetable and animal fats, like acylglycerols, fatty acids, and waxes. Water barrier characteristics of lipid films are excellent (Jeevahan & Chandrasekaran, 2019). Nanolipids have a lot of potential as self-assembling nanofilms and other nanostructures.

To increase antibacterial activity against *Escherichia coli*, Cui et al. (2017) created an edible film which was agar-based with artemisia annua oil nanoliposomes and chitosan. It was found that 191.8 nm was the average diameter of liposomes. When agar-based films were tested for their antibacterial effectiveness on cherry tomatoes, it was revealed that the films were more effective at combating germs when chitosan and nanoliposomes were added.

To learn about the antibacterial properties of the nanoemulsion droplet size, Hashemi-Gahruie et al. developed edible films based on basil seed gum that were reinforced with nanoemulsions of *Zataria multiflora* essential oil (Hashemi Gahruie et al., 2017). It has been demonstrated that as the size of the nanoemulsion particle was decreased, the antibacterial property against Gram-negative and Gram-positive bacteria increased.

Due to availability of large varieties of edible films (EFs) and coatings (ECs), it has become relatively easy in present times to choose them for different food products for maintaining quality. So far, they have been implemented on meat, poultry, grains, nuts, confectionaries, fruits, and vegetables.

- When used on fresh or frozen poultry and seafood products, EFs and Ecs have a number of benefits in terms of quality of product and safety as well as the capacity to increase shelf life. Efs or Ecs limit texture degradation by cutting down on moisture loss, preventing unattractive product juice drips, and cutting down on financial losses (Umaraw & Verma, 2017). Efs and Ecs may help lessen biochemical product deterioration by inhibiting irritating proteolytic enzymes, protecting lipids and proteins from oxidation, delaying rancidity, and preventing unwanted change in color (Gennadios et al., 1997).

- Grains and nuts have also been protected using Ecs and Efs from external environmental factors. They help in protecting in this way: Ecs aid in the regulation of an unwanted "mass" exchange across products and the atmosphere. Water absorption from moisture transfer from the atmosphere lowers the sensory quality of objects and speeds up deterioration (Arnon-Rips & Poverenov, 2016). Also, grains and nuts are fragile products, so ECs also provide protection and prevent their breaking.

- The utilization of ECs to the candies' and confectioneries' surfaces can aid in lessening the unfavorable traits that are associated with these food items, like stickiness, agglomeration, moisture absorption, and in the case of confectioneries that include fat, oil migration (Debeaufort et al., 2010). Hydrocolloid-based EFs can prevent the bleaching or blooming of chocolate because of reduced lipid permeability (Arnon-Rips & Poverenov, 2016).

- Fruits and vegetables are also protected using ECs and EFs. Although there are many factors that are considered before using them on fruits and vegetables, they provide protection by avoiding moisture loss, preventing moisture loss, and maintaining perfect aroma.

Nanoparticles have improved edible films' overall performance in comparison to plain edible films by increasing their thermal, restrictive, and mechanical (without nanomaterials) capabilities. As a result, using nanomaterials to increase the effectiveness of food packaging sheets is a viable option. In addition to passively transporting food, some nanoedible films also include active or intelligent features, like antibacterial activities, versatility, oxygen-scavenging capacity, enzyme immobilization, and biosensing to detect food breakdown. However, lack of research and evidence on biodegradability and digestibility is a problem here. Nanoedible films are an excellent choice for packaging materials in which food comes into direct touch with the packaging. Various nano-based edible films have been proposed by the majority of researchers. These studies, however, did not evaluate the edibility of such films using the edibility test. Instead, based on the edibility of the components used in their manufacture, the edibility of such films was assumed. The finished films were deemed edible if all the ingredients used to create them were edible. Compared to synthetic packaging materials, edible packaging materials' barrier and mechanical properties are less durable, and their biodegradability is more susceptible. As a result, more research is needed into the features of the anticipated functions' safety and stability (Jeevahan & Chandrasekaran, 2019).

16.5 Conclusion

Nanotechnology has gained a lot of attention from scientists in past decades. Although nanotechnology has applications in various fields like aerospace industries, the packaging industry has shown a lot of interest in its usage, and hence, numerous studies and research work have been going on to figure out and develop new and better alternatives, especially in food packaging.

The creation of nanocomposites has been a major accomplishment since, when properly handled, they can provide various advantages, including low density, enhanced flow, clarity, improved surface features, and recyclability, even at extremely low filler concentrations, which is especially useful for packaging. Nanocomposites are a revolutionary alternative to typical polymer composites. The desire of consumers for high-quality food products and worries about environmental waste brought on by non-biodegradable packaging materials which are petrochemical-based have attracted people's attention toward the creation of packaging materials which are environmentally safe and made from naturally renewable natural biopolymers like proteins and polysaccharides. Natural packaging materials based on polymers have certain inherent shortcomings, though, like poor mechanical properties and a lack of water resistance, which restrict their industrial use. Additionally, interest in the application of natural biopolymers (polysaccharides, lipids, and proteins) in food packaging has been rekindled by the recent advancements in the development of nanocomposite. Polymer nanocomposites, especially natural biopolymer nanocomposites, show noticeably better packing characteristics due to their nanoscale size distribution. Gains involve improved reduced gas permeability, improved water resistance, and mechanical strength. It has also been discussed to utilize zeolites, metals and metal oxides, graphene, and inorganic fillers. The creation of nanoedible films is another step in resolving problems with food packaging. A nanoedible film is created when an edible nanofiller is supported with an edible matrix. When compared to ordinary edible films, the addition of nanofillers enhanced the film color, mechanical characteristics, and WVP of the final edible films. However, these qualities are not enough to rival those of synthetic plastic. More research is therefore necessary.

Potential uses of nanofilms and coatings are also explored because they offer superior protection and are more affordable than techniques for adding antimicrobials directly into bulk.

The scope and need of further study is also explored for delivering better and enhanced solutions to environmental challenges.

REFERENCES

Alexandre, M., & Dubois, P. (2000). Polymer-layered silicate nanocomposites: Preparation, properties and uses of a new class of materials. *Materials Science and Engineering R: Reports, 28*(1), 1–63. https://doi.org/10.1016/S0927-796X(00)00012-7

Angellier, H., Molina-Boisseau, S., Dole, P., & Dufresne, A. (2006). Thermoplastic starch—waxy maize starch nanocrystals nanocomposites. *Biomacromolecules, 7*(2), 531–539. https://doi.org/10.1021/bm050797s

Armentano, I., Bitinis, N., Fortunati, E., Mattioli, S., Rescignano, N., Verdejo, R., Lopez-Manchado, M. A., & Kenny, J. M. (2013). Multifunctional nanostructured PLA materials for packaging and tissue engineering. *Progress in Polymer Science, 38*(10–11), 1720–1747. https://doi.org/10.1016/j.progpolymsci.2013.05.010

Arnon-Rips, H., & Poverenov, E. (2016). Biopolymers-embedded nanoemulsions and other nanotechnological approaches for safety, quality, and storability enhancement of food products: Active edible coatings and films. In *Emulsions*. Elsevier Inc. https://doi.org/10.1016/b978-0-12-804306-6.00010-6

Azeredo, H. M. C., Mattoso, L. H. C., Avena-Bustillos, R. J., Filho, G. C., Munford, M. L., Wood, D., & McHugh, T. H. (2010). Nanocellulose reinforced chitosan composite films as affected by nanofiller loading and plasticizer content. *Journal of Food Science, 75*(1), 1–7. https://doi.org/10.1111/j.1750-3841.2009.01386.x

Azeredo, H. M. C., Otoni, C. G., Corrêa, D. S., Assis, O. B. G., de Moura, M. R., & Mattoso, L. H. C. (2019). Nanostructured antimicrobials in food packaging—recent advances. *Biotechnology Journal, 14*(12), 1–9. https://doi.org/10.1002/biot.201900068

Benbettaïeb, N., Debeaufort, F., & Karbowiak, T. (2019). Bioactive edible films for food applications: Mechanisms of antimicrobial and antioxidant activity. *Critical Reviews in Food Science and Nutrition, 59*(21), 3431–3455. https://doi.org/10.1080/10408398.2018.1494132

Bhawani, S. A., Hussain, H., Bojo, O., & Fong, S. S. (2017). Proteins as agricultural polymers for packaging production. In *Bionanocomposites for packaging applications* (pp. 243–267). https://doi.org/10.1007/978-3-319-67319-6_13

Biercuk, M. J., Llaguno, M. C., Radosavljevic, M., Hyun, J. K., Johnson, A. T., & Fischer, J. E. (2002). Carbon nanotube composites for thermal management. *Applied Physics Letters, 80*(15), 2767–2769. https://doi.org/10.1063/1.1469696

Brandenburg, A. H., Weller, C. L., & Testin, R. F. (1993). Edible films and coatings from soy protein. *Journal of Food Science, 58*(5), 1086–1089. https://doi.org/10.1111/j.1365-2621.1993.tb06120.x

Briassoulis, D. (2004). An overview on the mechanical behaviour of biodegradable agricultural films. *Journal of Polymers and the Environment, 12*(2), 65–81. https://doi.org/10.1023/B:JOOE.0000010052.86786.ef

Bui, V. K. H., Park, D., & Lee, Y. C. (2017). Chitosan combined with ZnO, TiO$_2$ and Ag nanoparticles for antimicrobial wound healing applications: A mini review of the research trends. *Polymers, 9*(1). https://doi.org/10.3390/polym9010021

Bumbudsanpharoke, N., & Ko, S. (2015). Nano-food packaging: An overview of market, migration research, and safety regulations. *Journal of Food Science, 80*(5), R910–R923. https://doi.org/10.1111/1750-3841.12861

Cai, D., & Song, M. (2010). Recent advance in functionalized graphene/polymer nanocomposites. *Journal of Materials Chemistry, 20*(37), 7906–7915. https://doi.org/10.1039/c0jm00530d

Cao, Y., Lai, Z., Feng, J., & Wu, P. (2011). Graphene oxide sheets covalently functionalized with block copolymers via click chemistry as reinforcing fillers. *Journal of Materials Chemistry, 21*(25), 9271–9278. https://doi.org/10.1039/c1jm10420a

Cárdenas, G., Díaz, V. J., Meléndrez, M. F., Cruzat, C. C., & García Cancino, A. (2009). Colloidal Cu nanoparticles/chitosan composite film obtained by microwave heating for food package applications. *Polymer Bulletin, 62*(4), 511–524. https://doi.org/10.1007/s00289-008-0031-x

Cassie, B. A., & Baxter, S. (1994). Wettability of porous surfaces. *Transactions of the Faraday Society, 40*, 546–551. https://doi.org/10.1039/TF9444000546

Cerqueira, M. A., Vicente, A. A., & Pastrana, L. M. (2018). Nanotechnology in food packaging: Opportunities and challenges. In *Nanomaterials for food packaging: Materials, processing technologies, and safety issues*. Elsevier Inc. https://doi.org/10.1016/B978-0-323-51271-8.00001-2

Chen, D., Yang, H., He, P., & Zhang, W. (2005). Rheological and extrusion behavior of intercalated high-impact polystyrene/organomontmorillonite nanocomposites. *Composites Science and Technology, 65*(10), 1593–1600. https://doi.org/10.1016/j.compscitech.2005.01.011

Chun, A. L. (2009). Will the public swallow nanofood? *Nature Nanotechnology, 4*(12), 790–791. https://doi.org/10.1038/nnano.2009.359

Coleman, J. N., Khan, U., & Gun'ko, Y. K. (2006). Mechanical reinforcement of polymers using carbon nanotubes. *Advanced Materials, 18*(6), 689–706. https://doi.org/10.1002/adma.200501851

Coleman, K. S. (2012). Nanotubes. *Annual Reports on the Progress of Chemistry – Section A, 108*, 478–492. https://doi.org/10.1039/c2ic90014a

Costantino, U., Bugatti, V., Gorrasi, G., Montanari, F., Nocchetti, M., Tammaro, L., & Vittoria, V. (2009). New polymeric composites based on poly(ε-caprolactone) and layered double hydroxides containing antimicrobial species. *ACS Applied Materials and Interfaces, 1*(3), 668–677. https://doi.org/10.1021/am8001988

Cui, H., Yuan, L., Li, W., & Lin, L. (2017). Edible film incorporated with chitosan and *Artemisia annua* oil nanoliposomes for inactivation of *Escherichia coli* O157:H7 on cherry tomato. *International Journal of Food Science and Technology, 52*(3), 687–698. https://doi.org/10.1111/ijfs.13322

Dastjerdi, R., & Montazer, M. (2010). A review on the application of inorganic nano-structured materials in the modification of textiles: Focus on anti-microbial properties. *Colloids and Surfaces B: Biointerfaces, 79*(1), 5–18. https://doi.org/10.1016/j.colsurfb.2010.03.029

Debeaufort, F., Quezada-Gallo, J.-A., & Voilley, A. (2010). Edible films and coatings: Tomorrow's packagings: A review edible films and coatings. *Food Science and Nutrition, 38*(4), 37–41. https://doi.org/10.1080/10408699891274219

de Moura, M. R., Aouada, F. A., Avena-Bustillos, R. J., McHugh, T. H., Krochta, J. M., & Mattoso, L. H. C. (2009). Improved barrier and mechanical properties of novel hydroxypropyl methylcellulose edible films with chitosan/tripolyphosphate nanoparticles. *Journal of Food Engineering, 92*(4), 448–453. https://doi.org/10.1016/j.jfoodeng.2008.12.015

Dubas, S. T., Kumlangdudsana, P., & Potiyaraj, P. (2006). Layer-by-layer deposition of antimicrobial silver nanoparticles on textile fibers. *Colloids and Surfaces A: Physicochemical and Engineering Aspects, 289*(1–3), 105–109. https://doi.org/10.1016/j.colsurfa.2006.04.012

Duncan, T. V. (2011). Applications of nanotechnology in food packaging and food safety: Barrier materials, antimicrobials and sensors. *Journal of Colloid and Interface Science, 363*(1), 1–24. https://doi.org/10.1016/j.jcis.2011.07.017

Egger, S., Lehmann, R. P., Height, M. J., Loessner, M. J., & Schuppler, M. (2009). Antimicrobial properties of a novel silver-silica nanocomposite material. *Applied and Environmental Microbiology, 75*(9), 2973–2976. https://doi.org/10.1128/AEM.01658-08

Eleftheriadou, M., Pyrgiotakis, G., & Demokritou, P. (2017). Nanotechnology to the rescue: Using nano-enabled approaches in microbiological food safety and quality. *Current Opinion in Biotechnology, 44*, 87–93. https://doi.org/10.1016/j.copbio.2016.11.012

Farris, S. (2009). *The Wiley encyclopedia of packaging technology* (K. L. Yam, Ed., 3rd ed., pp. 285–294). Wiley & Sons.

Fernández, A., Soriano, E., Hernández-Muñoz, P., & Gavara, R. (2010). Migration of antimicrobial silver from composites of polylactide with silver zeolites. *Journal of Food Science, 75*(3). https://doi.org/10.1111/j.1750-3841.2010.01549.x

FICCI. (2020). *A report on India's plastic industry. January*, 44. http://ficci.in/spdocument/20690/plastic-packaging-report.pdf

Gennadios, A., Hanna, M. A., & Kurth, L. B. (1997). Lebensm.-Wiss. u.-Technol. 30, 337–350.pdf. *LWT – Food Science and Technology, 350*, 337–350. https://doi.org/10.1006/fstl.1996.0202

George, J., & Siddaramaiah. (2012). High performance edible nanocomposite films containing bacterial cellulose nanocrystals. *Carbohydrate Polymers, 87*(3), 2031–2037. https://doi.org/10.1016/j.carbpol.2011.10.019

Giannelis, E. P. (1996). Polymer layered silicate nanocomposites. *Advanced Materials, 8*(1), 29–35. https://doi.org/10.1002/adma.19960080104

Guilbert, S., Cuq, B., & Gontard, N. (1997). Recent innovations in edible and/or biodegradable packaging materials. *Food Additives and Contaminants, 14*(6–7), 741–751. https://doi.org/10.1080/02652039709374585

Guilbert, S., Gontard, N., & Cuq, B. (1995). Technology and applications of edible protective films. *Packaging Technology and Science, 8*(6), 339–346. https://doi.org/10.1002/pts.2770080607

Guo, W., & Chen, G. (2014). Fabrication of graphene/epoxy resin composites with much enhanced thermal conductivity via ball milling technique. *Journal of Applied Polymer Science, 131*(15). https://doi.org/10.1002/app.40565

Hashemi Gahruie, H., Ziaee, E., Eskandari, M. H., & Hosseini, S. M. H. (2017). Characterization of basil seed gum-based edible films incorporated with *Zataria multiflora* essential oil nanoemulsion. *Carbohydrate Polymers, 166*, 93–103. https://doi.org/10.1016/j.carbpol.2017.02.103

Havens, T. (1984). Communications to the editor. *Journal of Asian Studies, 43*(3), 499. https://doi.org/10.1017/S0021911800071588

Hernández-Muñoz, P., Kanavouras, A., Ng, P. K. W., & Gavara, R. (2003). Development and characterization of biodegradable films made from wheat gluten protein fractions. *Journal of Agricultural and Food Chemistry, 51*(26), 7647–7654. https://doi.org/10.1021/jf034646x

Hielscher, T. (2007). *Ultrasonic production of nano-size dispersions and emulsions* (pp. 138–143). ENS 200. hal-00166996.

Hopewell, J., Dvorak, R., & Kosior, E. (2009). Plastics recycling: Challenges and opportunities. *Philosophical Transactions of the Royal Society B: Biological Sciences, 364*(1526), 2115–2126. https://doi.org/10.1098/rstb.2008.0311

Hosseinnejad, M., & Jafari, S. M. (2016). Evaluation of different factors affecting antimicrobial properties of chitosan. *International Journal of Biological Macromolecules, 85*, 467–475. https://doi.org/10.1016/j.ijbiomac.2016.01.022

Hun, J. H., & Gennadios, A. (2005). Edible films and coatings: A review. *Innovations in Food Packaging*, 239–262. https://doi.org/10.1016/B978-012311632-1/50047-4

Ifuku, S., Ikuta, A., Egusa, M., Kaminaka, H., Izawa, H., Morimoto, M., & Saimoto, H. (2013). Preparation of high-strength transparent chitosan film reinforced with surface-deacetylated chitin nanofibers. *Carbohydrate Polymers, 98*(1), 1198–1202. https://doi.org/10.1016/j.carbpol.2013.07.033

Ifuku, S., Ikuta, A., Izawa, H., Morimoto, M., & Saimoto, H. (2014). Control of mechanical properties of chitin nanofiber film using glycerol without losing its characteristics. *Carbohydrate Polymers, 101*(1), 714–717. https://doi.org/10.1016/j.carbpol.2013.09.076

Introzzi, L., Blomfeldt, T. O. J., Trabattoni, S., Tavazzi, S., Santo, N., Schiraldi, A., Piergiovanni, L., & Farris, S. (2012). Ultrasound-assisted pullulan/montmorillonite bionanocomposite coating with high oxygen barrier properties. *Langmuir, 28*(30), 11206–11214. https://doi.org/10.1021/la301781n

Jeevahan, J., & Chandrasekaran, M. (2019). Nanoedible films for food packaging: A review. *Journal of Materials Science, 54*(19), 12290–12318. https://doi.org/10.1007/s10853-019-03742-y

Jiang, S., Liu, C., Wang, X., Xiong, L., & Sun, Q. (2016). Physicochemical properties of starch nanocomposite films enhanced by self-assembled potato starch nanoparticles. *LWT – Food Science and Technology, 69*, 251–257. https://doi.org/10.1016/j.lwt.2016.01.053

Jin, Y.-H., Park, H.-J., Im, S.S., Kwak, S.-Y., & Kwak, S. (2002). Polyethylene/clay nanocomposite by in-situ exfoliation of montmorillonite during Ziegler-Natta polymerization of ethylene. *Macromolecular Rapid Communications, 23*, 135–140. https://doi.org/10.1002/1521-3927(20020101)23:2%3C135::AID-MARC135%3E3.0.CO;2-T

Johnston, J. H., Borrmann, T., Rankin, D., Cairns, M., Grindrod, J. E., & Mcfarlane, A. (2008). Nanostructured composite calcium silicate and some novel applications. *Current Applied Physics, 8*(3–4), 504–507. https://doi.org/10.1016/j.cap.2007.10.060

Jordan, J., Jacob, K. I., Tannenbaum, R., Sharaf, M. A., & Jasiuk, I. (2005). Experimental trends in polymer nanocomposites – A review. *Materials Science and Engineering A, 393*(1–2), 1–11. https://doi.org/10.1016/j.msea.2004.09.044

Kalaitzidou, K., Fukushima, H., & Drzal, L. T. (2007). Mechanical properties and morphological characterization of exfoliated graphite-polypropylene nanocomposites. *Composites Part A: Applied Science and Manufacturing, 38*(7), 1675–1682. https://doi.org/10.1016/j.compositesa.2007.02.003

Kale, G., Kijchavengkul, T., Auras, R., Rubino, M., Selke, S. E., & Singh, S. P. (2007). Compostability of bioplastic packaging materials: An overview. *Macromolecular Bioscience, 7*(3), 255–277. https://doi.org/10.1002/mabi.200600168

Kim, H., & Macosko, C. W. (2009). Processing-property relationships of polycarbonate/graphene composites. *Polymer, 50*(15), 3797–3809. https://doi.org/10.1016/j.polymer.2009.05.038

Kim, H., Abdala, A. A., & MacOsko, C. W. (2010). Graphene/polymer nanocomposites. *Macromolecules, 43*(16), 6515–6530. https://doi.org/10.1021/ma100572e

Kim, H., Kobayashi, S., Abdurrahim, M. A., Zhang, M. J., Khusainova, A., Hillmyer, M. A., Abdala, A. A., & MacOsko, C. W. (2011). Graphene/polyethylene nanocomposites: Effect of polyethylene functionalization and blending methods. *Polymer, 52*(8), 1837–1846. https://doi.org/10.1016/j.polymer.2011.02.017

Kim, J. S., Kuk, E., Yu, K. N., Kim, J. H., Park, S. J., Lee, H. J., Kim, S. H., Park, Y. K., Park, Y. H., Hwang, C. Y., Kim, Y. K., Lee, Y. S., Jeong, D. H., & Cho, M. H. (2007). Antimicrobial effects of silver nanoparticles. *Nanomedicine: Nanotechnology, Biology, and Medicine, 3*(1), 95–101. https://doi.org/10.1016/j.nano.2006.12.001

Kim, N. H., Kuila, T., & Lee, J. H. (2013). Simultaneous reduction, functionalization and stitching of graphene oxide with ethylenediamine for composites application. *Journal of Materials Chemistry A, 1*(4), 1349–1358. https://doi.org/10.1039/c2ta00853j

Kloeckner, B. (2013). Optimal transport and dynamics of expanding circle maps acting on measures. *Ergodic Theory and Dynamical Systems, 33*(2), 529–548. https://doi.org/10.1017/S014338571100109X

Krishnamoorthy, C., & Chidambaram, R. (2018). Nanostructured thin films and nanocoatings. In *Emerging applications of nanoparticles and architectural nanostructures: Current prospects and future trends*. Elsevier Inc. https://doi.org/10.1016/B978-0-323-51254-1.00017-8

Kubacka, A., Serrano, C., Ferrer, M., Lunsdorf, H., Bielecki, P., Cerrada, M. L., Fernández-García, M., & Fernández-García, M. (2007). High-performance dual-action polymer-TiO$_2$ nanocomposite films via melting processing. *Nano Letters, 7*(8), 2529–2534. https://doi.org/10.1021/nl0709569

Kumar, S., Mukherjee, A., & Dutta, J. (2020). Chitosan based nanocomposite films and coatings: Emerging antimicrobial food packaging alternatives. *Trends in Food Science and Technology, 97*, 196–209. https://doi.org/10.1016/j.tifs.2020.01.002

Lagarón, J. M., Cabedo, L., Cava, D., Feijoo, J. L., Gavara, R., & Gimenez, E. (2005). Improving packaged food quality and safety. Part 2: Nanocomposites. *Food Additives and Contaminants, 22*(10), 994–998. https://doi.org/10.1080/02652030500239656

Lagarón, J. M., López-Rubio, A., & José Fabra, M. (2016). Bio-based packaging. *Journal of Applied Polymer Science, 133*(2). https://doi.org/10.1002/app.42971

Lan, T., Kaviratna, P., & Pinnavaia, T. (1994). On the nature of polyimide-clay hybrid composites. *Chemistry of Materials, 5*(6), 573–575. https://doi.org/10.1021/cm00041a002

Lee, S-W., Mao, C., Flynn, C. E., Belcher, A. M., (2002). Ordering of quantum dots using genetically engineered viruses. *Science, 296*, 892. http://dx.doi.org/10.1126/science.1068054

Lepoittevin, B., Pantoustier, N., Alexandre, M., Calberg, C., Jérôme, R., & Dubois, P. (2002). Polyester layered silicate nanohybrids by controlled grafting polymerization. *Journal of Materials Chemistry, 12*(12), 3528–3532. https://doi.org/10.1039/b205787e

Li, K., Zhu, J., Guan, G., & Wu, H. (2019). Preparation of chitosan-sodium alginate films through layer-by-layer assembly and ferulic acid cross-linking: Film properties, characterization, and formation mechanism. *International Journal of Biological Macromolecules, 122*, 485–492. https://doi.org/10.1016/j.ijbiomac.2018.10.188

Li, X., Xing, Y., Jiang, Y., Ding, Y., & Li, W. (2009). Antimicrobial activities of ZnO powder-coated PVC film to inactivate food pathogens. *International Journal of Food Science and Technology, 44*(11), 2161–2168. https://doi.org/10.1111/j.1365-2621.2009.02055.x

Liu, C., Jiang, S., Zhang, S., Xi, T., Sun, Q., & Xiong, L. (2016). Characterization of edible cornstarch nanocomposite films: The effect of self-assembled starch nanoparticles. *Starch/Staerke, 68*(3–4), 239–248. https://doi.org/10.1002/star.201500252

Liu, X., & Wu, Q. (2001). PP/clay nanocomposites prepared by grafting-melt intercalation. *Polymer, 42*, 10013–10019.

Llorens, A., Lloret, E., Picouet, P. A., Trbojevich, R., & Fernandez, A. (2012). Metallic-based micro and nanocomposites in food contact materials and active food packaging. *Trends in Food Science and Technology, 24*(1), 19–29. https://doi.org/10.1016/j.tifs.2011.10.001

Maiti, M., Sadhu, S., & Bhowmick, A. K. (2005). Effect of carbon black on properties of rubber nanocomposites. *Journal of Applied Polymer Science, 96*(2), 443–451. https://doi.org/10.1002/app.21463

Mangiacapra, P., Gorrasi, G., Sorrentino, A., & Vittoria, V. (2006). Biodegradable nanocomposites obtained by ball milling of pectin and montmorillonites. *Carbohydrate Polymers, 64*(4), 516–523. https://doi.org/10.1016/j.carbpol.2005.11.003

Markarian, J. (2005). Automotive and packaging offer growth opportunities for nanocomposites. *Plastics, Additives and Compounding, 6*(7), 18–21. https://doi.org/10.1016/S1464-391X(05)70485-2

Marsh, K., & Bugusu, B. (2007). Food packaging – Roles, materials, and environmental issues: Scientific status summary. *Journal of Food Science, 72*(3). https://doi.org/10.1111/j.1750-3841.2007.00301.x

Mortazavi, B., Hassouna, F., Laachachi, A., Rajabpour, A., Ahzi, S., Chapron, D., Toniazzo, V., & Ruch, D. (2013). Experimental and multiscale modeling of thermal conductivity and elastic properties of PLA/expanded graphite polymer nanocomposites. *Thermochimica Acta, 552*, 106–113. https://doi.org/10.1016/j.tca.2012.11.017

Nair, R. R., Wu, H. A., Jayaram, P. N., Grigorieva, I. V., & Geim, A. K. (2012). Unimpeded permeation of water through helium-leak-tight graphene-based membranes. *Science, 335*(6067), 442–444. https://doi.org/10.1126/science.1211694

Nielsen, T., & Jagerstad, M. (1994). Flavor scalping by food technology. *Trends in Food Science and Technology, 5*(11), 353–356.

Oymaci, P., & Altinkaya, S. A. (2016). Improvement of barrier and mechanical properties of whey protein isolate based food packaging films by incorporation of zein nanoparticles as a novel bionanocomposite. *Food Hydrocolloids, 54*, 1–9. https://doi.org/10.1016/j.foodhyd.2015.08.030

Pan, K., & Zhong, Q. (2016). Organic nanoparticles in foods: Fabrication, characterization, and utilization. *Annual Review of Food Science and Technology, 7*(December 2015), 245–266. https://doi.org/10.1146/annurev-food-041715-033215

Pavlidou, S., & Papaspyrides, C. D. (2008). A review on polymer-layered silicate nanocomposites. *Progress in Polymer Science (Oxford), 33*(12), 1119–1198. https://doi.org/10.1016/j.progpolymsci.2008.07.008

Petersen, K., Væggemose Nielsen, P., Bertelsen, G., Lawther, M., Olsen, M. B., Nilsson, N. H., & Mortensen, G. (1999). Potential of biobased materials for food packaging. *Trends in Food Science and Technology, 10*(2), 52–68. https://doi.org/10.1016/S0924-2244(99)00019-9

Pires, J. R. A., de Souza, V. G. L., & Fernando, A. L. (2018). Chitosan/montmorillonite bionanocomposites incorporated with rosemary and ginger essential oil as packaging for fresh poultry meat. *Food Packaging and Shelf Life, 17*(December 2017), 142–149. https://doi.org/10.1016/j.fpsl.2018.06.011

Pisoschi, A. M., Pop, A., Georgescu, C., Turcuş, V., Olah, N. K., & Mathe, E. (2018). An overview of natural antimicrobials role in food. *European Journal of Medicinal Chemistry, 143*, 922–935. https://doi.org/10.1016/j.ejmech.2017.11.095

Piyada, K., Waranyou, S., & Thawien, W. (2013). Mechanical, thermal and structural properties of rice starch films reinforced with rice starch nanocrystals. *International Food Research Journal, 20*(1), 439–449.

Ponce, A., Roura, S. I., & Moreira, M. R. (2016). Casein and chitosan polymers: Use in antimicrobial packaging. In *Antimicrobial food packaging* (pp. 455–466). https://doi.org/10.1016/B978-0-12-800723-5.00037-1

Potts, J. R., Dreyer, D. R., Bielawski, C. W., & Ruoff, R. S. (2011). Graphene-based polymer nanocomposites. *Polymer, 52*(1), 5–25. https://doi.org/10.1016/j.polymer.2010.11.042

Price, B. K., & Tour, J. M. (2006). Functionalization of single-walled carbon nanotubes "on water." *Journal of the American Chemical Society, 128*(39), 12899–12904. https://doi.org/10.1021/ja063609u

Prolongo, S. G., Moriche, R., Jiménez-Suárez, A., Sánchez, M., & Ureña, A. (2014). Advantages and disadvantages of the addition of graphene nanoplatelets to epoxy resins. *European Polymer Journal, 61*, 206–214. https://doi.org/10.1016/j.eurpolymj.2014.09.022

Qu, B., & Luo, Y. (2021). A review on the preparation and characterization of chitosan-clay nanocomposite films and coatings for food packaging applications. *Carbohydrate Polymer Technologies and Applications, 2*, 100102. https://doi.org/10.1016/j.carpta.2021.100102

Ranzoni, A., & Cooper, M. A. (2017). The growing influence of nanotechnology in our lives. In *Micro- and nanotechnology in vaccine development*. Elsevier Inc. https://doi.org/10.1016/B978-0-323-39981-4.00001-4

Ray, S., Quek, S., Easteal, A., & Chen, X. D., (2006). The potential use of polymer-clay nanocomposites in food packaging. *International Journal of Food Engineering, 4*(2). https://doi.org/10.2202/1556-3758.1149

Rayas, L. M., Hernandez, R. J., & Ng, P. K. W. (1997). Development and characterization of bio-degradable/edible wheat protein films. *Journal of Food Science, 62*(1), 160–162. https://doi.org/10.1111/j.1365-2621.1997.tb04390.x

Reddy, M. M., Vivekanandhan, S., Misra, M., Bhatia, S. K., & Mohanty, A. (2013). Biobased plastics and bionanocomposites: Current status and future opportunities. *Progress in Polymer Science, 38*(10–11), 1653–1689. http://dx.doi.org/10.1016/j.progpolymsci.2013.05.006

Regev, O., ElKati, P. N. B., Loos, J., & Koning, C. E. (2004). Preparation of conductive nanotube-polymer composites using latex technology. *Advanced Materials, 16*(3), 248–251. https://doi.org/10.1002/adma.200305728

Rehan, M., El-Naggar M. E., Mashlay H. M., & Wilken R. (2018). Nanocomposites based on chitosan/silver/clay for durable multi-functional properties of cotton fabrics. *Carbohydrate Polymers Journal, 182,* 29–41. https://doi.org/10.1016/j.carbpol.2017.11.007

Rezadoust, A. M., Esfandeh, M., Beheshty, M. H., & Heinrich, G. (2013). Effect of the nanoclay types on the rheological response of unsaturated polyester—Clay nanocomposites. *Polymer Engineering and Science, 53,* 809–817.

Rhim, J. W. (2011). Effect of clay contents on mechanical and water vapor barrier properties of agar-based nanocomposite films. *Carbohydrate Polymers, 86*(2), 691–699. https://doi.org/10.1016/j.carbpol.2011.05.010

Rhim, J. W., & Ng, P. K. W. (2007). Natural biopolymer-based nanocomposite films for packaging applications. *Critical Reviews in Food Science and Nutrition, 4*(47), 411–433. https://doi.org/10.1080/10408390600846366

Russell, A. D., & Hugo, W. B. (1994). Antimicrobial activity and action of silver. *Progress in Medicinal Chemistry, 31*(C), 351–370. https://doi.org/10.1016/S0079-6468(08)70024-9

Sanchez, C., Julián, B., Belleville, P., & Popall, M. (2005). Applications of hybrid organic-inorganic nanocomposites. *Journal of Materials Chemistry, 15*(35–36), 3559–3592. https://doi.org/10.1039/b509097k

Sanchez, C., Soler-Illia, G. J. D. A. A., Ribot, F., Lalot, T., Mayer, C. R., & Cabuil, V. (2001). Designed hybrid organic-inorganic nanocomposites from functional nanobuilding blocks. *Chemistry of Materials, 13*(10), 3061–3083. https://doi.org/10.1021/cm011061e

Sanguansri, P., & Augustin, M. A. (2006). Nanoscale materials development—a food industry perspective. *Trends in Food Science and Technology, 17*(10), 547–556. https://doi.org/10.1016/j.tifs.2006.04.010

Shi, A. M., Wang, L. J., Li, D., & Adhikari, B. (2013). Characterization of starch films containing starch nanoparticles Part 1: Physical and mechanical properties. *Carbohydrate Polymers, 96*(2), 593–601. https://doi.org/10.1016/j.carbpol.2012.12.042

Shit, S. C., & Shah, P. M. (2014). Edible polymers: Challenges and opportunities. *Journal of Polymers, 2014,* 1–13. https://doi.org/10.1155/2014/427259

Sinha Ray, S., & Okamoto, M. (2003). Polymer/layered silicate nanocomposites: A review from preparation to processing. *Progress in Polymer Science (Oxford), 28*(11), 1539–1641. https://doi.org/10.1016/j.progpolymsci.2003.08.002

Skoryakov, M. M. (1958). Chemical polishing of crystal ware. *Glass and Ceramics, 15*(2), 61–64. https://doi.org/10.1007/BF00668465

Sorrentino, A. (2011). *Nanocoatings and ultra-thin films* (pp. 203–234). Woodhead Publishing Limited. http://dx.doi.org/10.1533/9780857094902.2.203

Sorrentino, A., Gorrasi, G., Tortora, M., Vittoria, V., Costantino, U., Marmottini, F., & Padella, F. (2005). Incorporation of Mg-Al hydrotalcite into a biodegradable Poly(ε-caprolactone) by high energy ball milling. *Polymer, 46*(5), 1601–1608. https://doi.org/10.1016/j.polymer.2004.12.018

Sorrentino, A., Gorrasi, G., & Vittoria, V. (2007). Potential perspectives of bio-nanocomposites for food packaging applications. *Trends in Food Science and Technology, 18*(2), 84–95. https://doi.org/10.1016/j.tifs.2006.09.004

Sothornvit, R., & Krochta, J. M. (2000). Plasticizer effect on oxygen permeability of β-lactoglobulin films. *Journal of Agricultural and Food Chemistry, 48*(12), 6298–6302. https://doi.org/10.1021/jf0008361

Sterky, K., Jacobsen, H., Jakubowicz, I., Yarahmadi, N., & Hjertberg, T. (2010). Influence of processing technique on morphology and mechanical properties of PVC nanocomposites. *European Polymer Journal, 46*(6), 1203–1209. https://doi.org/10.1016/j.eurpolymj.2010.03.021

Tammaro, L., Vittoria, V., & Bugatti V. (2014). Dispersion of modified layered double hydroxides in poly(ethylene terephthalate) by high energy ball milling for food packaging applications. *European Polymer Journal, 52,* 172–180. https://doi.org/10.1016/j.eurpolymj.2014.01.001

Thostenson, E. T., Li, C., Chou, T.-W. (2005). Nanocomposites in context. *Composites Science and Technology, 3–4*(65). https://doi.org/10.1016/j.compscitech.2004.11.003

Tsagkaris, A. S., Tzegkas, S. G., & Danezis, G. P. (2018). Nanomaterials in food packaging: state of the art and analysis. *Journal of Food Science and Technology, 55*(8), 2862–2870. https://doi.org/10.1007/s13197-018-3266-z

Umaraw, P., & Verma, A. K. (2017). Comprehensive review on application of edible film on meat and meat products: An eco-friendly approach. In *Critical Reviews in Food Science and Nutrition* (Vol. 57, Issue 6). https://doi.org/10.1080/10408398.2014.986563

Unalan, I. U., Cerri, G., Marcuzzo, E., Cozzolino, C. A., & Farris, S. (2014). Nanocomposite films and coatings using inorganic nanobuilding blocks (NBB): Current applications and future opportunities in the food packaging sector. *RSC Advances, 4*(56), 29393–29428. https://doi.org/10.1039/c4ra01778a

Vaia, R. A., Jandt, K. D., Kramer, E. J., & Giannelis, E. P. (1996). Microstructural evolution of melt intercalated polymer-organically modified layered silicates nanocomposites. *Chemistry of Materials, 8*, 2628–2635. https://doi.org/10.1021/cm960102h

Vaia, R. A., & Giannelis, E. P. (1997). Polymer melt intercalation in organically-modified layered silicates: model predictions and experiment. *Macromolecules, 30*, 8000–8009. https://doi.org/10.1021/ma9603488

Vartiainen, J., Tuominen, M., & Nättinen, K. (2010). Bio-hybrid nanocomposite coatings from sonicated chitosan and nanoclay. *Journal of Applied Polymer Science, 116*(6), 3638–3647. https://doi.org/10.1002/app.31922

Verlee, A., Mincke, S., & Stevens, C. V. (2017). Recent developments in antibacterial and antifungal chitosan and its derivatives. *Carbohydrate Polymers, 164*, 268–283. https://doi.org/10.1016/j.carbpol.2017.02.001

Wang, J., Wang, X., Xu, C., Zhang, M., & Shang, X. (2011). Preparation of graphene/poly(vinyl alcohol) nanocomposites with enhanced mechanical properties and water resistance. *Polymer International, 60*(5), 816–822. https://doi.org/10.1002/pi.3025

Wenzel, R. N. (1936). Resistance of Solid Surfaces to wetting by water. *Industrial and Engineering Chemistry, 28*(8), 988–994. https://doi.org/10.1021/ie50320a024

Wiley, B., Sun, Y., Mayers, B., & Xia, Y. (2005). Shape-controlled synthesis of metal nanostructures: The case of silver. *Chemistry – A European Journal, 11*(2), 454–463. https://doi.org/10.1002/chem.200400927

WPO. (2013). Market statistics and future trends in global packaging. *Journal of Chemical Information and Modeling, 53*(9), 1689–1699.

Xiao-e, L., Green, A. N. M., Haque, S. A., Mills, A., & Durrant, J. R. (2004). Light-driven oxygen scavenging by titania/polymer nanocomposite films. *Journal of Photochemistry and Photobiology A: Chemistry, 162*(2–3), 253–259. https://doi.org/10.1016/j.nainr.2003.08.010

Yuan, G., Chen, X., & Li, D. (2016). Chitosan films and coatings containing essential oils: The antioxidant and antimicrobial activity, and application in food systems. *Food Research International, 89*, 117–128. https://doi.org/10.1016/j.foodres.2016.10.004

Zhang, S., & Zhao, H. (2017). Preparation and properties of zein–rutin composite nanoparticle/cornstarch films. *Carbohydrate Polymers, 169*, 385–392. https://doi.org/10.1016/j.carbpol.2017.04.044

Zhang, W., Chen, Y., Yu, S., Chen, S., & Yin, Y. (2008). Preparation and antibacterial behavior of Fe^{3+}-doped nanostructured TiO_2 thin films. *Thin Solid Films, 516*(15), 4690–4694. https://doi.org/10.1016/j.tsf.2007.08.053

Zeng, H., Gao, C., Wang, Y., Watts, C. P. P., Kong, H., Cui, X., & Yan, D. (2006). In situ polymerization approach to multiwalled carbon nanotubes-reinforced nylon 1010 composites: Mechanical properties and crystallization behavior. *Polymer, 1*(47), 113-122. DOI: 10.1016/j.polymer.2005.11.009

Zhu, C., Xue, J., & He, J. (2009). Controlled in-situ synthesis of silver nanoparticles in natural cellulose fibers toward highly efficient antimicrobial materials. *Journal of Nanoscience and Nanotechnology, 9*(5), 3067–3074. https://doi.org/10.1166/jnn.2009.212

Zhu, J., Wu, H., & Sun, Q. (2019). Preparation of cross-linked active bilayer film based on chitosan and alginate for regulating ascorbate-glutathione cycle of postharvest cherry tomato (Lycopersicon esculentum). *International Journal of Biological Macromolecules, 130*, 584–594. https://doi.org/10.1016/j.ijbiomac.2019.03.006

17

Biopolymer Production Methods and Regulatory Aspects

Anand Kishore, Anupama Singh, Pradeep Kumar,
Khushbu Kumari, and Rohan Jitendra Patil

CONTENTS

17.1 Introduction

Food packaging is the enclosure of food products in a different packaging material (plastics, paper, metal, glass, and paperboard) for the effective delivery of high-quality, safe food throughout the supply chain. Synthetic petrochemical-based plastic, such as low-density polyethylene (LDPE), high-density polyethylene (HDPE), polypropylene (PP), polystyrene (PS), and polyvinyl chloride (PVC), has been widely used in food packaging in different formats due to their excellent properties (Cheng et al., 2021). However, migration and the non-biodegradable nature of plastics affect the environmental concerns and safety of the food products. Considering the environmental threats, biopolymers made from renewable materials and agro-industry waste have gained popularity as an alternative to conventional plastics. Biopolymers made from polysaccharides, starch, cellulose, protein, and lipids with reinforcement agents require additional modification in the production method to improve the properties of eco-friendly packaging material. The biopolymer production methods help convert different bio-sourced material into film, bag, bottle, trays, cup, and other packaging forms. The processing methods significantly affect the chemical bond, intermolecular force, spatial arrangements, and molecular orientation, and hence the properties, of the final product (Lee et al., 2017). The biopolymer processing is more complex as compared to synthetic plastic production method due to its moisture sensitivity and narrow gap between melting and decomposition point of bio-sourced materials. The type of material, additives, and heat resistance of raw material must be considered to develop bio-based films. The preparation method of the biopolymer is also vital for the feasibility of film for commercialization. There are various processes available in making the bio-based film. For bio-based packaging, traditional techniques like solvent casting are simple, but not suitable for industrial application due to lengthy processing times and non-uniformity of film quality (Buanz et al., 2015). The production of biopolymer from different sources has been improved using novel techniques, such as reactive extrusion, electrospinning, nanotechnology, multilayer films, coatings, etc. Moreover, a single layer of bio-based material sometimes does not meet all the food packaging requirements, like preservation (from moisture, gas, light, and flavor), tensile strength, and other functional properties, such as antimicrobial and antioxidant properties (Tyagi et al., 2021). Considering this, in recent years, the concept of multilayer packaging, active and smart packaging has been increased to improve the functional properties of the film. Recently, many studies have been conducted to develop a multifunctional packaging that improves the mechanical strength and barrier properties and also indicates the quality of packed food. Traditionally, functional packaging were prepared from coating, lamination, and co-injection, and recently, novel approaches for multilayer packaging, such as nanotechnology, electrospinning, atomic layer deposition, electrostatic powder coating, and spray coating have been widely used in food packaging systems (Khwaldia et al., 2010; Wang et al., 2022). There are also concerns about the migration of additive used in the production methods; they can have a toxic effect on consumers and the environment. So, several countries have introduced regulations to assure the biodegradability and safety of the product. This chapter provides information on the fundamentals of conventional and novel biopolymer production methods, coating strategies, and regulatory considerations in food packaging.

17.2 Biopolymer Preparation Methods

There are different traditional and advanced techniques used for the preparation of bio-based packaging materials, shown in Figure 17.1. The common biopolymer materials, production methods, and their applications are summarized in Table 17.1.

17.3 Traditional Techniques for Biopolymer Preparation

17.3.1 Film Solution Casting Methods

Solution casting is a traditional film-forming method to prepare biopolymers. Typically, the processes involved are mixing, heating and stirring, degassing, pouring, and drying. However, it may change with the types of additives used in the film to improve the physical and functional properties. A step of the general process and their control points for film preparation is shown in Figure 17.2. During preparation of films, the base matrix, along with other additives, like plasticizers, active materials, nanoparticles, and fillers, must be dispersed and dissolved in a suitable solvent, like water. The degree

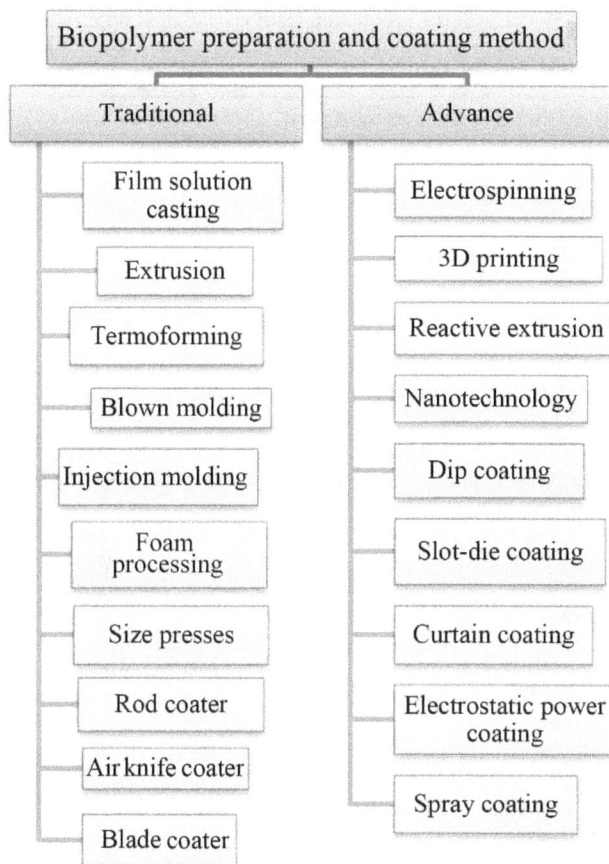

FIGURE 17.1 Biopolymer preparation and coating methods.

TABLE 17.1

Common biopolymer material production methods and their application

S. no.	Biopolymer material	Method	Application	References
1	Zein/alginate, basil oil	Electrospinning	Edible antibacterial packaging	Dede et al. (2022)
2	Plantain flour/poly(ε-caprolactone) (PCL) blends	Reactive extrusion, thermocompression molding	Antimicrobial packaging	Gutiérrez et al. (2017)
3	Cassava starch, anthocyanin	Extrusion	Smart packaging for meat	Vedove et al. (2021)
4	PHBV/PBAT, nanoclay	Melt extrusion, compression molding, and cast film extrusion	Flexible packaging	Pal et al. (2020)
5	Starch-based biopolymer BIOPar-S, alpha cellulose fibers	Injection molding	Improved mechanical properties in packaging	Belhassen et al. (2008)
6	Cornstarch, nanoclay fillers	Solution casting, thermoforming	Better mechanical and thermal properties in packaging	Mohan et al. (2015)
7	Chitosan-tetraethyl orthosilicate (TEOS) and chitosan-glutaraldehyde	Slot-die film casting	Biodegradable replacements for plastics and packaging	Pemble et al. (2021)

(Continued)

TABLE 17.1 Common biopolymer material production methods and their application (*Continued*)

S. no.	Biopolymer material	Method	Application	References
8	Nanofibrillated cellulose (NFC), beeswax latex, acrylated epoxidized soybean oil (AESO), and 3-aminopropyltriethoxysilane (APTS)	Solvent casting, rod coating	Moisture-resistant packaging	Lu et al. (2014)
9	Cellulose nanofibril (CNF), bleached bagasse soda pulp (Paper)	Auto bar coating	Better mechanical and improved barrier properties in paper packaging	Afra et al. (2016)
10	Acetylated hemicellulose (AH)-nanocellulose (ACNC), polycaprolactone (PCL)	Solvent casting, dip coating	Active packaging for fatty foods	Mugwagwa et al. (2020)
11	Silanized castor oil (SCO) and Silanized methyl ricinoleate (SMR) bio-resins, cellulosic paper	Dip coating	Alternate to LDPE liner for paper-based products	Parvathy et al. (2021)
12	Carboxymethyl cellulose/cellulose nanocrystals immobilized silver nanoparticles (CMC/CNC@ AgNPs), cellulosic paper	Dip coating	Antimicrobial packaging for strawberries	He et al. (2021)
13	Poly(3-hydroxybutyrate) (PHB) and PLA in 2,2,2-trifuoroethanol and a trichloromethane/N,N-dimethyl formamide mixture, commercial bleached kraft eucalyptus pulp (paper)	Electrospinning, annealing (hydraulic press)	Best performing multilayer paper packaging	Cherpinski et al. (2017)
14	Polylactic acid (PLA), magnesium oxide (MgO) nanoparticles	Blown film extrusion	Antimicrobial food packaging	Swaroop et al. (2019)
15	Polypropylene non-woven fabric (NWF), PVA (polyvinyl alcohol)	Dip coating	Better antifouling property	Zhang et al. (2008)
16	Polystyrene (PS) dielectric layer and TIPS-pentacene	Slot-die coating	Organic thin film transistors and logic circuits on the plastic substrates	Lin et al. (2017)
17	Agar (*Gracilariavermiculophyl*), Glycerol	Knife coating	Edible packaging for cherry	Sousa et al. (2010)

of mixing depends on the solvent, temperature, time, and additives, which affects the intra- and intermolecular associations of biopolymers (Cheng et al., 2021). After mixing, the film solution is heated and stirred to a suitable temperature. After that, it is important to degas the film solution before pouring onto the plate. Ultrasonic bath, vacuum pump, and strainer are generally used to remove trapped air bubbles from the solution and ensure homogeneity of the gel. It is also important to select the right casting plate for the film-forming process so that the film can be easily removed after drying. Teflon and glass plates have been widely used as casting surface because dried films could be easily peeled off (stripped) without any tears and wrinkles (Vartiainen et al., 2014). The next step involved in the film-forming process is drying of the solution at a suitable temperature and time in an oven to evaporate the solvent and, at last, peeling of the film from the plate for further application. The film solution casting method has been used for the preparation of different biopolymer films from starch-protein-laver-fiber composites (Chen et al., 2019), cornstarch film (Nordin et al., 2020), and potato starch/apple peel pectin–based composite films (Sani et al., 2021). Basically, solvent casting with slight modifications is a popular method for preparing films. However, the use of this technique for biopolymer production on a commercial scale is restricted due to non-uniform thickness, high energy consumption, and long drying time.

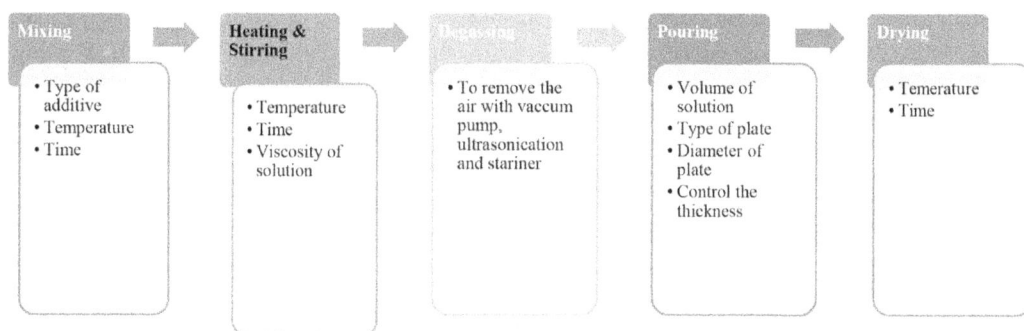

FIGURE 17.2 Casting process and control points.

17.3.2 Extrusion

Extrusion processing is commonly used for the preparation of biopolymer films. It is a dry method which forms films by high temperature and mechanical forces without any solvent. The extrusion process is carried out by using an extruder, which has four major components: a hopper, a conveying section, a die, and an auxiliary equipment. The first step of extrusion processing is to feed the film-forming material in a hopper and continuously supply into the conveying section. The conveying section comprises the barrel and the screw(s). The heat from the barrel is responsible for softening the material, while the rotating screw is responsible for mixing, melting, and transporting the material. A screw pushes the molten material into a die, forming the material into the desired shape. The extrudate is then processed by cooling, cutting, and collecting with the help of different auxiliary equipment to form a required film. The properties of the final film are influenced by various factors, like plasticizers, lubricants, feeding rate, the temperature in the heating zones, time, shear force, screw speed, and moisture content (Cheng et al., 2021; Anukiruthika et al., 2020). Extruded films have different physical properties and higher production efficiency than those produced by solution casting method. There are three main types of extrusion processing: extrusion film blowing, compression, and injection molding.

Blown-film extrusion is a continuous and efficient method for industrial-scale film production. It is also known as tubular extrusion and a potential alternative to solvent casting method to produce biopolymers. Blowing, pressing, and cooling are the general steps involved in blown film. In the blown film process, a thermoplastic is extruded through an annular slit die, followed by the blowing of air through the center of the die to blow up the polymer melted into a large tube of film. Following the inflation of the tube of film, a thin bubble is formed, which is then cooled and flattened (Lee et al., 2017). Thermal degradation, moisture sensitivity, and stickiness are the major processing limitations for biopolymer production in blown film extrusion. This process requires a specific viscosity and strain hardening, which are hardly found in bio-based materials. Therefore, blending with other polymers/fillers and plasticizers is required to produce the material with thermoplastic properties and can be converted into a different shape without thermal degradation. Huntrakul et al. (2020) found that combining plant-derived pea protein and cassava starch significantly enhanced the blown-film processability and barrier properties of film for packaging food products containing oil. Similarly, a blend of PLA and polybutylene adipate terephthalate (PBAT) reinforced with chitin nanofibers effectively enhanced the processability and barrier qualities of blown films (Herrera et al., 2016). Also, films made by extrusion blowing of polylactic acid–based (PLA) thermoplastic resin mixed with natural fillers from date palm leaves have also been reported (Kharrat et al., 2020). Moreover, the amount of moisture content and plasticizers during the blown-film process significantly affects the viscosity, extrusion efficiency, and physical properties of bio-based films. In this context, Gao et al. (2021) investigated the effects of water (W) and glycerol (G) on the physicochemical and structural characteristics of starch-based films produced by extrusion blowing. It was found that the addition of 15% G–15% W co-plasticizers was optimal for film blowing and could lower the cost of film production. In recent years, the trend of multilayer biopolymer packaging has emerged to improve the film's physical and functional properties. The solvent casting and multistep techniques are often used for the production of multilayer packaging. It is, however, time-consuming and less cost-effective and produces film with non-uniform thickness. Although in coextrusion film blowing, several operations like melting, mixing, stretching, conveying, and shaping take place at the same time. A number of studies have been performed to develop multilayer packaging via coextrusion film blowing for industrial scale. This technique has been used for the preparation of bilayer biodegradable, polylactic acid, and bio-flex-based eco-friendly films (Scaffaro et al., 2020).

Compression molding is the oldest industrial technique for material processing. Extrusion compression molding forms the film in different forms through mixing, heating, and cooling. The basic processes involve feeding material into a heated mold cavity. The mold is then closed with a top plug and compressed with large hydraulic presses. This process gives a desired shape to biopolymers/plastics with the use of mold shape, high temperature, and pressure (Teil et al., 2021). Extrusion and compression molding have been used for the preparation of PHBV/PBAT-based nanocomposite films (Pal et al., 2020), fish gelatin–based edible films (Krishna et al., 2012), starch-gelatin-based antimicrobial packaging (Moreno et al., 2018), and active and smart film–based on yerba mate extract (Ceballos et al., 2020). In recent years, active and intelligent packaging has been a market trend, and several studies reported to increase the shelf life and safety of food products. However, the production methods affect the stability and interaction of active agents in film with food materials. In this context, Ciannamea et al. (2016) compared the antioxidant activity of soy protein concentrate–based films incorporated with red grape extract and processed by casting and compression molding. The study indicated that the film processed by compression molding has higher antioxidant activity than solvent casting film and concluded that in compression molding, the red grape extract produced a redistribution of the interactions, interfering with the formation of disulfide bonds and favoring hydrophobic and hydrogen interactions, which help improve the functional properties of the film.

Cast film extrusion involves the feeding of a biopolymer material from a gravimetric feeding system to one or more extruders. Then the materials melt and mix by slit extruder die, and the melt is drawn around two or more chill rolls to form a film. The rotating speed of chill rolls and stretching affect the properties of the film. Cast film extrusion has the advantage of transparent appearance, high production rate, uniform film thickness, and improved barrier properties. Recently, Pal et al. (2020) developed the PHBV/PBAT-based nanocomposite films with nano clay by compression molding and cast film extrusion. The film from cast film extrusion shows improved oxygen and water vapor permeability compared to a film produced from compression molding. This is affected by the difference in operation involved; as in compression molding, materials are pressed between two heated plates to form a film without stretching. However, cast film extrusion involves stretching of biopolymer chains, and it helps improve the dispersion of nanoparticles and, hence, enhances the barrier properties of the film. Cast film extrusion has been used for the preparation of PHBV-based films (Jost et al., 2018), polylactic acid–based films (Wang et al., 2011), and poly(butylene succinate)-based biodegradable films (Wang et al., 2013).

17.3.3 Injection Molding

Injection molding is the most common processing method for synthetic polymer film for different shapes. It consists of heating thermoplastic materials until they melt and injecting the polymer melt into the steel mold, where it cools and solidifies to take its final shape. The processes involved in injection molding are extrusion (clamping), injection, cooling, and ejection. The final film quality depends on the type of bio-sourced material, melting point of mix material, temperature profile, screw speed, injection pressure, mold temperature, and residence time. The way of processing, like the injection molding process and extrusion-injection molding process, also affects the film quality. In injection molding, the bio-sourced material and additive are directly added to the hopper of the injection molding machine, and in extruded injection molding, an extruder is used to extrude the bio-sourced material and then put it into the injection molding for further processing. Siva et al. (2021) compare the method of preparation of PLA-based hemp fiber composite by both extrusion-injection molding process and injection molding process and recommended that the extrusion-injection molding technique for PLA matrix composite preparation has higher mechanical strength and better bonding with the matrix than the injection-molded composite. Biopolymer production in injection molding requires additional care compared to synthetic polymer because of the variant properties. The selection of suitable material for injection molding is challenging. The decision model, like multi-objective optimization by ratio analysis (MOORA) technique, may be helpful in the selection of a suitable material based on the required properties (Ramkumar et al., 2021). The ratio of plasticizer and water is also essential for developing biopolymers, as water is required for the mixing stage, especially for starch gelatinization. Moreover, the moisture content should be low enough for processing adequately in injection molding, as excess water may cause voids and cracks in the film if the mold temperature is above 100°C (Alonso-González et al., 2021). In addition, a standardized ratio of plasticizer and suitable temperature is required to obtain an adequate viscosity for further processing; a higher amount led to the higher viscosity of the blend, which made problematic the injection of the polymer melt. Rheology and DSC measurement of bio-sourced material is also required to select the processing condition for injection molding. It has a good potential for the commercial production of biopolymer from albumen/soy bio-based (Félix et al., 2014), rice-protein-based film (Félix et al., 2016), eco-friendly bioplastic from rice bran (Alonso-González et al., 2022), pea protein–based bioplastics (Perez et al., 2016), and in bio-based polymer nanocomposites (Mistretta et al., 2018).

17.3.4 Thermoforming

Thermoforming is a process for the conversion of polymeric materials into finished parts with high production speed and low cost. The steps involved in this process are clamping the thermoplastic sheet, softening it with heat, applying vacuum or pressure to stretch the sheet into or over a temperature-controlled mold, cooling, and demolding the formed part. The vacuum pump is used to supply the vacuum to suck it on the surface of the mold, and in pressure forming, compressed air pushes the heated material against the tool's surface. Thermoformed material has been used in food packaging applications for monolayer, multilayer, or laminated structures in containers, cups, and trays. The thermal stability of biopolymer, moisture, thickness of the sheet, pressure, and holding time affect the properties of the final bioplastics. The thermoforming process has been used for the development of PLA/PBS-based bioplastic (Barletta & Puopolo, 2020), polybutylene succinate (PBS)/starch-based tray (Ayu & Khalina, 2021) for low-cost industrial application in food packaging.

17.3.5 Foaming Process

Foaming produces biodegradable foam, which is used for food packaging and helps in food delivery. The process involved in foaming is gas dissolution, cell nucleation, growth and stabilization of gas bubbles in the polymer matrix. Biopolymer-based foam is developed by extrusion, baking, molding, and supercritical fluid foaming. The properties of the foam are influenced by the biopolymer source, process conditions, type, and amount of plasticizer and water. The foaming properties, like mechanical properties, thickness, density, water absorption capacity, biodegradability, and morphology, must be studied to optimize the process condition of foaming methods. Extrusion foaming is a continuous process which uses temperature, pressure, and shear rate to produce different foams. This process involves heating and mixing the raw material and other additives under pressure. Then, a melted polymer matrix exists from the die. Due to pressure drop, dissolved gas forms from the bubble, and these bubbles expand, grow, and get trapped in the polymer matrix. The extrusion operating condition, like pressure, temperature zone from feeding to die zone, type of screw, screw speed, feeding rate, and raw material type, plasticizer, nucleating and blowing agent affect the foam quality (Cheng et al., 2021). Density is an important quality parameter affecting the transportation cost of packaging foam. Cha et al. (2001) observed that the bulk density of starch-based foams decreased as the extrusion temperature increased. Recently, novel starch-based foams with a calcium peroxide–embedded coating of PVA have been formed using extrusion and coating techniques. The prepared foam can potentially preserve guava during transportation and storage. The foam showed improved moisture resistance, superior buffering ability, and self-regulation of the storage atmosphere (Zhang et al., 2022). Baking is another method to develop bio-based foam to replace the synthetic plastic–based foam. The cellulose-based lightweight foam was developed to replace the petrochemical-based plastic foam with good mechanical strength and water stability (Liao et al., 2022). It involves mixing bio-sourced material, water, foaming agent, and additives until a homogeneous mass is obtained. After mixing, the paste/batter is added to the preheated mold and then baked in an oven. During heating, water is evaporated, which helps the formation of foam. In this process, the viscosity of the paste, mold dimension, drying temperature, and time affect foam quality (Engel et al., 2019). Cassava starch–based foam mixed with water hyacinth fiber and polyvinyl alcohol has been developed from the baking technique, and the prepared foam has improved physical and mechanical properties (Nugroho et al., 2022). Recently, baking has been used to develop the foam with the incorporation of sugarcane fiber in plantain flour–based foam (Moreno et al., 2018) and starch-based foam with eggshell and shrimp shell (Kaewtatip et al., 2018) to improve the physical and functional properties of foam for food packaging application. Compression molding is similar to a baking technique which produces biodegradable foam. In this method, the bio-based material is placed in a hydraulic press and foam is formed using heating and pressure. Cassava starch is mixed with guar gum, water hyacinth (WH) powder, glycerol, and water in a mixer. Then the batter was poured into a mold and pressed at 200°C for 5 minutes in a compression machine to develop the biodegradable composite foam for food packaging (Chaireh et al., 2020). In order to improve physical and functional properties and meet consumer requirements, some functionalized foam materials have been developed. For example, Cruz-Tirado et al. (2020) reported that oregano (OEO) or thyme (TEO) essential oil was used as an antimicrobial agent in developing starch-based foam as a food container by thermo-compressing. Citric acid, cotton fibers, cotton microfibers, coconut residue, and cellulose fibers are used as fillers to improve the thermal and mechanical properties of starch-based foams in compression molding (Nansu et al., 2021).

17.4 Novel Techniques for Biopolymer Preparation

17.4.1 Electrospinning

Electrospinning is a novel method for producing nanofiber-based mats from biopolymer solution or melt using a high electric field. It is widely used in tissue engineering, textiles, biosensors, and treatment of the environment. Recently, the application of electrospinning in packaging material has been increasing to improve the physical and functional properties of the film (Gutierrez-Gonzalez et al., 2020). There are three major critical requirements for electrospinning: high voltage, spinneret (needle), and collector. In this process, the polymer solution is stored in the pump, which controls the flow rate of the solution. Then the polymer solution is pumped into a needle which is connected to a high-voltage power source. At the high voltage, the power solution melts and comes out through the tip of the needle in the form of a Taylor cone, the solvent evaporates while the solution comes out through the needle, and finally, nanofibers are collected on plates of the collector (Angel et al., 2022). The production and quality of nanofiber mat in electrospinning are affected by the flow rate, voltage, collector type (flat or round), distance between needle tip and collector, rheological properties, conductivity, surface tension of solution, and environmental conditions like temperature, humidity, and air velocity (Li et al., 2018). It can make fibers from a broad range of materials, including biopolymers, such as carbon polysaccharides and proteins, composite metals, metal oxides, and ceramics. Bio-based materials have great potential for electrospun as a reinforcement of fillers to enhance the barrier and mechanical properties of biopolymers (Thenmozhi et al., 2017). In addition, it can be used to immobilize or modify the surface of active agents, such as antimicrobials, antioxidants, oxygen scavengers, carbon dioxide emitters, and ethylene scavengers in an active food packaging system. In intelligent packaging, electrospun nanofibers can be combined with different product deterioration indicators and used for real-time monitoring during packed food storage (Zhang et al., 2020). The fibers produced from electrospinning are more communicative to changes in the atmosphere and controlled release of bioactive compounds than the conventional method. Li et al. (2018) developed butylated hydroxyanisole–based (BHA) gelatin fiber mats by electrospinning technique for better antibacterial activity. It was interesting to find the controlled release of volatile from fiber mat, and the result indicated that the shelf life of strawberry increased during storage. Electrospinning has excellent potential to develop biopolymers with improved properties. However, it requires toxic solvents, harsh chemicals, and other additives to complete the process (Leidy & Maria Ximena, 2019), so developing an efficient and eco-friendly electrospinning method requires more research in the future.

17.4.2 3D Printing

Three-dimensional printing (3D printing) is a digital technique that involves multiple layer deposition of a viscous solution, paste, slurry, and melted material through a controlled nozzle to create the food material or design the suitable material (Nachal et al., 2019; Armstrong et al., 2022). This technique involves heating the raw material, extruding it through a print nozzle that can move in three directions (x, y, and z), and then cooling it after deposition. The components in 3D printing technology include an extruder assembly with a motor to break and mix the biopolymer compound, nozzle tip to inject out mixed polymer, compressor unit, and printing platform, over which printing takes place. Recently, 3D printing has been emerging in food packaging to create different geometric shapes of material, sensors, indicators, and electronics tags to meet the specific need of consumers. It is a feasible technique to develop the biopolymer and to create smart components in an intelligent packaging system to monitor food quality. The composition and properties of raw materials are essential factors for developing 3D printing technology in the food packaging industry. It is important to understand the rheological and thermal properties of raw material and machine characteristics such as pressure, nozzle diameter, speed of the motor, printing speed, printing rate, and nozzle height to develop the final product. In this direction, Nida et al. (2021), optimize the process parameter for 3D-printed food package casings from sugarcane bagasse. The result indicated that the material was extrudable with pressure of 3.2 bar, nozzle diameter of 1.28 mm, speed of the motor of 240 rpm, printing speed of 500 mm/min, printing rate of 0.304g/min, and nozzle height of 0.450 mm, and the developed package casings were used to pack cake, suggesting that it may be an alternative to single-use plastic. The spreading characteristics, printability, gelling ability, mechanical properties, and heat resistance of bio-based material affect the final product's properties in 3D printing technology. The 3D printing technology may be essential in developing a sustainable packaging system, although low-cost print processes with excellent printing quality to improve the properties of the film remain a serious issue that must be addressed in the future.

17.4.3 Reactive Extrusion

Conventional extrusion involves simple mixing of additives without any chemical reaction, which affects the properties of the final biopolymeric material and limits their further application. In this context, reactive extrusion is a novel polymer extrusion manufacturing process that uses chemical or enzymatic processes to change the physical and functional properties of the material by polymerization, grafting, cross-linking, polycondensation, functionalizing, compatibilizing, and controlling the degradation during processing (Meng et al., 2019). In this method, an extruder works as a reactor, where reaction can happen to give better biopolymer with good properties. It has multiple internal screws for the continuous conveyance of reactant molecules. The control of chemical reactions in the extruder is essential, and it is crucial to understand the parameters of extruder and reaction kinetics. The type of reactants, catalysts, the temperature profile of the extruder from the feeding zone to the die zone, and the residence time of the extruder affect the reaction progress and speed of reaction (Beyer & Hopmann, 2018). A wide range of bio-sourced materials has been used to develop biopolymers at a low cost in a shorter time with higher efficiency. Various coampatelizers, like glycidyl methacrylate, dicumyl peroxide, glycidyl methacrylate grafted poly(caprolactone), diethyl maleate grafted poly(caprolactone), citric acid, dicumyl peroxide, etc., are used in reactive extrusion for manufacturing bio-based materials. Reactive extrusion has been successfully used for the preparation of biodegradable films from poly(3-hydroxybutyrate-co-3-hydroxyvalerate), gluten/PCL-based food packaging films (Gutiérrez & Alvarez, 2017), and phosphate starch-based food packaging films (Gutiérrez & Valencia, 2021). It is also effective in the inclusion of bioactive compounds in active packaging development. Mücke et al. (2021) reported that curcumin is effectively incorporated in starch/poly (butylene adipate-co-terephthalate) blend films in the reactive extrusion process, and citric acid is used as a compatibilizer. The data indicated that the curcumin-based film could protect chia oil from oxidation and suggested that reactive extrusion is a promising approach to developing a functional food packaging system.

17.4.4 Nanotechnology

Nanotechnology is a novel method for manufacturing nanoscaled materials (1 to 100 nm) using mechanical and chemical forces. Milling, electrospinning, lithography, sputtering, etching, and laser ablation are common techniques to convert larger compounds to nanoscaled (Ashfaq et al., 2022). Nanomaterials have better physical and functional properties as compared to conventional materials. Therefore, nanotechnology is widely used in agriculture, food, medical, auto, aerospace, packaging, and other industries. In recent years, research on nano-based biopolymers in food packaging has increased due to the growing interest in enhancing the qualities of bio-based materials for sustainable development. The application of nanotechnology used in biopolymer packaging is in two ways: (1) nanoparticles as nanofillers and (2) nanostructured materials in which nonmaterial are distributed into the biopolymer matrix as nanocomposites for improving the physical and functional properties of the packaging film. The three most common methods to use the nanocomposite in bipolymer packaging are solution blending, melt blending, and in situ polymerization (Alfei et al., 2020). Various organic nanoparticles (nanocellulose, nano starch, protein nanoparticles, chitosan nanoparticles), inorganic nanoparticles (silver nanoparticles, zinc nanoparticles, titanium dioxide nanoparticles), and nano clay are commonly used in biopolymer matrix for adequate packaging (Ashfaq et al., 2022). The addition of nanomaterials changes the polymeric structure, grain size, and shape, thus improving water vapor, gas barrier, dimension stability, heat resistance, and strength of biopolymers. In addition, it is also used as an active material, like antimicrobial and antioxidant, to control releases in food packaging to improve the product's shelf life. Ren et al. (2022) have developed a nanocellulose fibril–based (isolated from bamboo parenchyma cell) composite film, which was embedded with silver nanoparticles (AgNPs). The results show that AgNPs (0.1 wt%) improves the tensile strength and Young's modulus of the film. In addition to improved mechanical properties, it was also found that the high amount of AgNPs enhances the antimicrobial activity against *Salmonella Typhi* (*S. Typhi*) and *Escherichia coli* (*E. coli*), which indicates their potential as antimicrobial food packaging. Beikzadeh et al. (2021) also reported that optimal gelatin nanofibers with electrospun ethylcellulose/polycaprolactone-based film have the highest biocompatibility, antioxidant activity, and antifungal properties for food packaging. Recently, nanosensors and nanoindicators have been used as freshness indicators in intelligent packaging to communicate food quality. As relative humidity affects the quality of perishable food, considering this more recently, Rahman et al. (2022) fabricated guar-gum-sodium-alginate-based film blended with carbon dots nanocomposite to check the freshness of food. It was interesting to find that the film acts as a smart sensor based on the fluorescence "on–off" mechanisms against humidity which monitor the food quality. Despite so many benefits, more study is needed on the migration of nanomaterials from packaged food items and their impact on human health, as well as transparent labelling and regulation, to ensure public acceptance.

17.5 Traditional Techniques for Coatings in Biopolymers

17.5.1 Size Presses

It is an old method used for coating biopolymer or cellulosic-based packaging material and is also known as puddle presses. This method uses two rolls to create a nip that can be vertical, horizontal, or inclined to transfer the coating solution to the bio-based material for coating. The nip load profile is critical, as an uneven load profile can result in various concerns, including tenability problems, profile issues, unequal surface degradation, and web breakage (Dal et al., 2021). The viscosity of the solution, temperature, and gap between roll affect the quality of the final film. Bio-sourced functional additives are used for coating in this method to improve film properties (Tyagi et al., 2021).

17.5.2 Rod Coating

It is a simple and low-cost method for coating on paper and bio-based material. The wet applicator rod and wire bar coater are widely used for coating the cellulose-based packaging system. In the wet film applicator, the coating solution is poured on the paper sheet and spread over the sheet using a wet film applicator which is tightly wound with stainless steel wire and then dried of coated paper at a suitable temperature. In this technique, coating thickness can be changed as per application without modifying the properties of the coating material. The thickness of the coating can be standardized as per the requirement with a selection of wire-bound coating bars diameters. Shankar et al. (2018) developed an antimicrobial paper with three different thickness coating bars (no. 20, 40, 60) and observed that the water vapor permeability coefficient of paper was significantly decreased when the paper was coated with no. 40. Similarly, the wire bar coater was used to coat the hydroxypropyl methylcellulose on paper to improve the mechanical and water vapor permeability of paper (Sothornvit, 2009). The flex rod coater is an innovation in rod coaters where air pressure controls the coating layer on a packaging sheet.

17.5.3 Air Knife Coating

It is a method for coating paper, plastic, and bio-based sheet material to improve film properties. In this technique, the coating is applied by a roll applicator, and then coating thickness is smoothed by passing the web through a doctor knife (an angled metal blade) or an air knife. The knife is the critical element of this method to remove the extra substrate. In air knife coating, high pressure is directed onto the web, and an air separation pan collector is used to collect the excess coating material. The air knife angle and distance between the knife blade and rolls affect the properties of the coating. The material's viscosity, the knife's pressure, the distance between the knife and the substrate, and the web's speed all influence the wet thickness (Roth et al., 2015). The knife coating technique was used to develop an agar-based edible coating on biodegradable film to improve the quality of fresh fruits and vegetables (Sousa et al., 2010).

17.5.4 Blade Coating

It is the most popular and conventional method to apply a thin coating on cellulosic-based packaging material. In this technique, the blade moves over the substrate or the substrate flows underneath the blade. Various blade coaters, such as roll blade coater, fountain blade coater, and Vari-dwell fountain blade coater, have been used for coating in the paper as per requirement. This coating technique is efficient and inexpensive for creating industrial, scalable thin film. Blade coater was used to coat the cellulose nanofibrils on paperboard. The result shows that the barrier properties of paperboard improved as high weights coat and full surface coverage were achieved using this coating technique (Mousavi et al., 2018). Similarly, the blade coating method was used to prepare zinc oxide (ZnO) and polycarbonate (PC) based nanocomposite film for different food packaging as an effective antibacterial with optical properties (Dhapte et al., 2015). The rheological properties of the coating solution, speed of coating, and gap between substrate and blade affect the properties and thickness of the film.

17.6 Advance Techniques for Coatings

17.6.1 Dip Coating

It is a novel and effective method for thin layer coating onto a substrate surface. In this technique, the substrate is dipped in a solution and withdrawn from a solution at a controlled rate using a dip coater. The process involved in dip coating is immersion, dwelling, withdrawal, and drying. The final film's quality is influenced by inertia, drag, gravitational force, and surface tension in the coating solution during the process. In the dip-coating technique, the speed of dipping the substrate into the solution and pulling it out of the solution is regulated by the continuous motor. This method creates thin films through self-assembly and the sol-gel process (Tyagi et al., 2021). The viscosity of the solution, the evaporation rate of the solvent and the angle at which the substrate is taken out from the solution, the withdrawal velocity, and drying are the critical factors that affect the thickness of the coating. Due to its low cost and simple processing methods, the dip coating technique was used to prepare titania-based thin films, and it was observed that the dip-coating process gives a good degree of crystallinity and homogeneity under controlled conditions (Dastan et al., 2016).

17.6.2 Slot-Die Coating

Slot-die coating is an efficient technique for preparing a thin coating on paper, glass, metal, and biopolymer packaging material. Recently, this method has been adopted in industries due to its uniform thickness, minimum waste, and low production cost. The slot-die coating technique is recently used in the food packaging and medical areas for preparing film coatings with better barrier characteristics (Kyaw et al., 2016). The slot-die coater is mainly used for producing films with uniform thickness in the slot-die coating method. In this coating process, a liquid is conveyed through a stationary aperture onto a movable substrate, filling the gap between the substrate and the die. The liquid present produces the coating bead in the gap due to the downstream and upstream meniscus. The downstream meniscus transports a layer of liquid over the moving substrate. Then, after evaporation, the uniform dried film was obtained. Coating speed and flow rate of liquid passing through the die directly influence the thickness of the film. In precise coating applications, the slot-die coating method is efficiently helpful in different biopolymers (Ding et al., 2016). The operating parameters, such as flow rate, speed of coating, die diameter, pressure, coating solution viscosity, and surface tension, are important for optimizing the coating process in bio-based packaging.

17.6.3 Curtain Coating

Curtain coating is a type of pre-metered coating operation in which the coating layer is formed before it contacts the substrate, and it has a strong potential for biopolymer coating. The spacing between the coating head and the substrate distinguishes curtain coating from slot coating. The slot coating method has no noticeable curtain forms (Wang et al., 2022). A curtain coater coats the substrate with a fluid curtain located between two conveyors. The coating fluid falls at a constant rate between the two conveyors from a reservoir tank. The coating thickness is determined by the conveyer's speed and the material flow rate from the tank (Tyagi et al., 2021). This coating technique is most efficient for barrier applications due to its tendency to deliver consistent coating layers with no pinholes and high coverage at low coat weights (Makhlouf, 2011). The curtain coating method was used for coating chitosan on paperboard to improve the mechanical and barrier properties of the material (Gällstedt et al., 2005).

17.6.4 Electrostatic Powder Coating

It is a novel dry coating method involving particle charging, spraying or atomization, and deposition of the charged particle onto a grounded substrate and then heating deposited particles to form a film. In these techniques, particles can be charged using either turboelectric or corona charging mechanisms. Turboelectric charging refers to the accumulation of charge by friction, and in corona charging, particles charge by passing an electric field in the availability of free ions. The significant advantage of the electrostatic powder coating is to coat without using organic or aqueous solvents (Prasad et al., 2016), as in the conventional method solvent used for coating, which may cause toxicity problems in the environment. Solvent coating is not suitable for moisture-sensitive materials, and electrostatic powder coating has good potential to prepare barrier film coating to protect the product from moisture. Yang et al. (2020) prepare the barrier film for herbal medicine by electrostatic coating with hydroxypropyl cellulose to protect from moisture. Andrade-Del Olmo et al. (2020) also reported that biopolymer chitosan was effectively linked to a poly(l-lactic acid) (PLLA-ZnO) surface by electrostatic to improve the physical properties with antimicrobial characteristics for use as active packaging of food.

17.6.5 Spray Coating

Spray coating is one of the recent advancements for coating paper and bio-based materials. Spraying has numerous advantages, including contour and non-contact coating with the base substrate (Shanmugam et al., 2017). It is a low-cost, high-volume manufacturing method and has the advantages of a more significant deposition rate and repeatability, large-quality targets, and also, a vacuum is not required for the process (Prabeesh et al., 2021). In this method, the number of spray nozzles controls the coat weight and thickness of the coating layer on the material. The quality of coating material can be enhanced by arranging the nozzles in such a way that the spray fans overlap. Low-moisture foods, spices, dried fruits, and nuts can be packed in packaging materials made by spray-coating techniques (Vartiainen et al., 2014). Spray coating was used to coat the biopolymer solution of gelatin, enzyme, glycerol, and citric acid on a paper sheet to improve the barrier and functional properties of paper for beef packaging (Battisti et al., 2017).

17.7 Regulatory Aspects of Bio-Based Packaging

17.7.1 Definition of Biodegradable and Compostable

The term *biodegradable* is defined as substances that can undergo aerobic or anaerobic degradation, resulting in biomass, carbon dioxide, methane, water, and mineral salts, depending on the process's climatic conditions. Microorganisms, which are found in the environment and are primarily fed by organic waste, play a key role in biodegradation. The plastics which have susceptibility to biodegradation are called biodegradable plastics. Biomass-based plastics, excluding fossilized biomass, are bio-based plastics. In contrast, composting creates compost by microbial digestion from the disintegration of organic waste. Compostable plastics are the plastics that biodegrade in the composting cycle's conditions and time frame.

17.7.2 Standards of Biodegradability

The interest in developing biodegradable plastics that can be safely used for applications in various areas, such as agriculture, packaging, etc., has resulted in the creation of intense work at the standardization level. In the case of biodegradable plastics, standardization is critical. *Standardization* can be defined, according to the ISO (International Organization for Standardization), as the "activity of establishing, with regard to actual or potential problems, provisions for common and repeated use, aimed at the achievement of the optimum degree of order in a given context" (Briassoulis & Degli Innocenti, 2017).

17.7.3 The Norms and Standards for Soil Biodegradability Testing

a. **French Normalization Organization (AFNOR) NF U52–001**

   The standards for biodegradability testing are based on the biodegradable agricultural and horticultural materials—mulching products—requirements and testing methods.

b. **International Organization for Standardization (ISO)-ISO/PRF 17556**

   The standards for testing biodegradability, which are based on the plastics—determining the utmost aerobic biodegradability in soil by measuring the oxygen requirement in a respirometer or the amount of CO2 evolved.

c. **American Society for Testing and Materials International (ASTM)-ASTM D 5988–96**

   The standards for testing biodegradability which are based on the standard test method to determine the aerobic biodegradation of plastic materials or residual plastic materials in the soil after composting.

17.7.4 Biodegradable Plastic Testing

The implementation of recycling processes and the development of biodegradable plastics are the alternative solution for plastic waste reduction. The ISO gives a series of tests for testing the biodegradability of biodegradable plastics in different environments, like seawater, soil, etc., and these tests are regularly updated. The supporting body for the ISO is the United Nations Standards Coordinating Committee (UNSCC). A material is considered biodegradable if it degrades to at least 70% in 14 days using dissolved organic carbon or biological oxygen demand measurements. A readily biodegradable material, on the other hand, requires at least 60% in a 10-day window within 28 days. The test duration is dependent on the ecosystem and could last up to six months in marine water (Filiciotto & Rothenberg, 2020). There is a general problem while testing the deterioration phenomena of plastics in the environment in terms of the type of test to be used and the conclusions that can be drawn. The tests for biodegradable plastics are classified into laboratory, simulation, and field tests.

17.8 Conclusion

Recently, the industry, regulatory bodies, and consumers have preferences toward bio-based packaging systems. Therefore, academic and industry research has focused on the different techniques to develop films from natural sources to improve the properties of food packaging. The selection of a suitable method for biopolymer production is essential for film quality. Hence, understanding the effect of process parameters on gelatinization, retrogradation, denaturation, conformational change, hydrogen bonding, and texturization of different materials is required to manufacture bio-based films. Similarly, the production time and cost of biopolymer production determine the feasibility of the methods. The traditional and advanced techniques of production methods have their merit and demerits. In conventional techniques, solution casting is simple, but it is good for only lab-scale and not suitable for commercial-scale production. Extrusion techniques are a fast and continuous process, hence are beneficial for commercial production, but its processing condition should be optimized for using bio-sourced material for the production of biopolymer films. Advanced preparation techniques like electrospinning, coating techniques, reactive extrusion, and nanotechnology improve the film's physical and functional properties. However, advanced techniques have a complex operation and higher production costs than conventional preparation techniques. More study is needed to understand the balance between biopolymer production's cost, processing, and efficiency. Food packaging producers can also choose suitable techniques of biopolymer production as per their requirements. In addition, it will be essential to understand the regulatory aspects for the selection of bio-based materials, additive requirement, migration effect, and biodegradability for the sustainable packaging system.

Acknowledgments

The authors are thankful to the National Institute of Food Technology Entrepreneurship and Management, Kundli, Sonepat (Haryana), India, for providing the infrastructure and other support for this work.

REFERENCES

Afra, E., Mohammadnejad, S., & Saraeyan, A. (2016). Cellulose nanofibils as coating material and its effects on paper properties. *Progress in Organic Coatings*, *101*, 455–460. https://doi.org/10.1016/j.porgcoat.2016.09.018

Alfei, S., Marengo, B., & Zuccari, G. (2020). Nanotechnology application in food packaging: A plethora of opportunities versus pending risks assessment and public concerns. *Food Research International*, *137*, 109664. https://doi.org/10.1016/j.foodres.2020.109664

Alonso-González, M., Felix, M., Guerrero, A., & Romero, A. (2021). Effects of mould temperature on rice bran-based bioplastics obtained by injection moulding. *Polymers*, *13*(3), 398. https://doi.org/10.3390/polym13030398

Alonso-González, M., Felix, M., & Romero, A. (2022). Influence of the plasticizer on rice bran-based eco-friendly bioplastics obtained by injection moulding. *Industrial Crops and Products*, *180*, 114767. https://doi.org/10.1016/j.indcrop.2022.114767

Andrade-Del Olmo, J., Pérez-Álvarez, L., Ruiz-Rubio, L., & Vilas-Vilela, J. L. (2020). Antibacterial chitosan electrostatic/covalent coating onto biodegradable poly(l-lactic acid). *Food Hydrocolloids*, *105*, 105835.

Angel, N., Li, S., Yan, F., & Kong, L. (2022). Recent advances in electrospinning of nanofibers from bio-based carbohydrate polymers and their applications. *Trends in Food Science & Technology*, *120*, 308–324. https://doi.org/10.1016/j.tifs.2022.01.003

Anukiruthika, T., Sethupathy, P., Wilson, A., Kashampur, K., Moses, J. A., &Anandharamakrishnan, C. (2020). Multilayer packaging: Advances in preparation techniques and emerging food applications. *Comprehensive Reviews in Food Science and Food Safety*, *19*(3), 1156–1186. https://doi.org/10.1111/1541-4337.12556

Armstrong, C. D., Yue, L., Deng, Y., & Qi, H. J. (2022). Enabling direct ink write edible 3D printing of food purees with cellulose nanocrystals. *Journal of Food Engineering*, *330*, 111086. https://doi.org/10.1016/j.jfoodeng.2022.111086

Ashfaq, A., Khursheed, N., Fatima, S., Anjum, Z., & Younis, K. (2022). Application of nanotechnology in food packaging: Pros and Cons. *Journal of Agriculture and Food Research*, *7*, 100270. https://doi.org/10.1016/j.jafr.2022.100270

Ayu, R. S., & Khalina, A. (2021). Effect of different polybutylene succinate (PBS)/starch formulation on food tray by thermoforming process. In *Biopolymers and biocomposites from agro-waste for packaging applications* (pp. 85–100). Woodhead Publishing.

Barletta, M., & Puopolo, M. (2020). Thermoforming of compostable PLA/PBS blends reinforced with highly hygroscopic calcium carbonate. *Journal of Manufacturing Processes*, *56*, 1185–1192. https://doi.org/10.1016/j.jmapro.2020.06.008

Battisti, R., Fronza, N., Júnior, Á. V., da Silveira, S. M., Damas, M. S. P., & Quadri, M. G. N. (2017). Gelatin-coated paper with antimicrobial and antioxidant effect for beef packaging. *Food Packaging and Shelf Life*, *11*, 115–124.

Beikzadeh, S., Hosseini, S. M., Mofid, V., Ramezani, S., Ghorbani, M., Ehsani, A., & Mortazavian, A. M. (2021). Electrospun ethyl cellulose/poly caprolactone/gelatin nanofibers: The investigation of mechanical, antioxidant, and antifungal properties for food packaging. *International Journal of Biological Macromolecules*, *191*, 457–464. https://doi.org/10.1016/j.ijbiomac.2021.09.065

Belhassen, R., Boufi, S., Vilaseca, F., López, J. P., Méndez, J. A., Franco, E., Pèlach, M. A., & Mutjé, P. (2008). Biocomposites based on *Alfa* fibers and starch-based biopolymer. *Polymers for Advanced Technologies*, *20*(12), 1068–1075. https://doi.org/10.1002/pat.1364

Beyer, G., & Hopmann, C. (2018). *Reactive extrusion: Principles and applications*. John Wiley & Sons.

Briassoulis, D., & Degli Innocenti, F. (2017). Standards for soil biodegradable plastics. *Green Chemistry and Sustainable Technology*, *139*–*168*. https://doi.org/10.1007/978-3-662-54130-2_6

Buanz, A. B., Belaunde, C. C., Soutari, N., Tuleu, C., Gul, M. O., & Gaisford, S. (2015). Ink-jet printing versus solvent casting to prepare oral films: Effect on mechanical properties and physical stability. *International Journal of Pharmaceutics*, *494*(2), 611–618. https://doi.org/10.1016/j.ijpharm.2014.12.032

Ceballos, R. L., Ochoa-Yepes, O., Goyanes, S., Bernal, C., & Famá, L. (2020). Effect of yerbamate extract on the performance of starch films obtained by extrusion and compression molding as active and smart packaging. *Carbohydrate Polymers*, *244*, 116495. https://doi.org/10.1016/j.carbpol.2020.116495

Cha, J., Chung, D., Seib, P., Flores, R., & Hanna, M. (2001). Physical properties of starch-based foams as affected by extrusion temperature and moisture content. *Industrial Crops and Products*, *14*(1), 23–30. https://doi.org/10.1016/s0926-6690(00)00085-6

Chaireh, S., Ngasatool, P., & Kaewtatip, K. (2020). Novel composite foam made from starch and water hyacinth with beeswax coating for food packaging applications. *International Journal of Biological Macromolecules*, *165*, 1382–1391. https://doi.org/10.1016/j.ijbiomac.2020.10.007

Chen, Y., Yu, L., Ge, X., Liu, H., Ali, A., Wang, Y., & Chen, L. (2019). Preparation and characterization of edible starch film reinforced by laver. *International Journal of Biological Macromolecules*, *129*, 944–951. https://doi.org/10.1016/j.ijbiomac.2019.02.045

Cheng, H., Chen, L., McClements, D. J., Yang, T., Zhang, Z., Ren, F., Miao, M., Tian, Y., & Jin, Z. (2021). Starch-based biodegradable packaging materials: A review of their preparation, characterization and diverse applications in the food industry. *Trends in Food Science & Technology*, *114*, 70–82. https://doi.org/10.1016/j.tifs.2021.05.017

Cherpinski, A., Torres-Giner, S., Cabedo, L., Méndez, J. A., & Lagaron, J. M. (2017). Multilayer structures based on annealed electrospun biopolymer coatings of interest in water and aroma barrier fiber-based food packaging applications. *Journal of Applied Polymer Science*, *135*(24), 45501. https://doi.org/10.1002/app.45501

Ciannamea, E. M., Stefani, P. M., & Ruseckaite, R. A. (2016). Properties and antioxidant activity of soy protein concentrate films incorporated with red grape extract processed by casting and compression molding. *LWT*, *74*, 353–362. https://doi.org/10.1016/j.lwt.2016.07.073

Cruz-Tirado, J., Barros Ferreira, R. S., Lizárraga, E., Tapia-Blácido, D. R., Silva, N., Angelats-Silva, L., & Siche, R. (2020). Bioactive Andean sweet potato starch-based foam incorporated with oregano or thyme essential oil. *Food Packaging and Shelf Life, 23*, 100457. https://doi.org/10.1016/j.fpsl.2019.100457

Dal, A. B., & Hubbe, M. A. (2021). Hydrophobic copolymers added with starch at the size press of a paper machine: A review of findings and likely mechanisms. *BioResources, 16*(1), 2138

Dastan, D., Panahi, S. L., &Chaure, N. B. (2016). Characterization of titania thin films grown by dip-coating technique. *Journal of Materials Science: Materials in Electronics, 27*(12), 12291–12296. https://doi.org/10.1007/s10854-016-4985-4

Dede, S., Sadak, O., Didin, M., & Gunasekaran, S. (2022). Basil oil-loaded electrospunbiofibers: Edible food packaging material. *Journal of Food Engineering, 319*, 110914. https://doi.org/10.1016/j.jfoodeng.2021.110914

Dhapte, V., Gaikwad, N., More, P. V., Banerjee, S., Dhapte, V. V., Kadam, S., & Khanna, P. K. (2015). Transparent ZnO/polycarbonate nanocomposite for food packaging application. *Nanocomposites, 1*(2), 106–112. https://doi.org/10.1179/2055033215y.0000000004

Ding, X., Liu, J., & Harris, T. A. L. (2016). A review of the operating limits in slot die coating processes. *AIChE Journal, 62*(7), 2508–2524. https://doi.org/10.1002/aic.15268

Engel, J. B., Ambrosi, A., & Tessaro, I. C. (2019). Development of biodegradable starch-based foams incorporated with grape stalks for food packaging. *Carbohydrate Polymers, 225*, 115234. https://doi.org/10.1016/j.carbpol.2019.115234

Félix, M., Lucio-Villegas, A., Romero, A., & Guerrero, A. (2016). Development of rice protein bio-based plastic materials processed by injection molding. *Industrial Crops and Products, 79*, 152–159. https://doi.org/10.1016/j.indcrop.2015.11.028

Félix, M., Martín-Alfonso, J., Romero, A., & Guerrero, A. (2014). Development of albumen/soy biobased plastic materials processed by injection molding. *Journal of Food Engineering, 125*, 7–16. https://doi.org/10.1016/j.jfoodeng.2013.10.018

Filiciotto, L., & Rothenberg, G. (2020). Biodegradable plastics: Standards, policies, and impacts. *ChemSusChem, 14*(1), 56–72. https://doi.org/10.1002/cssc.202002044

Gällstedt, M., Brottman, A., & Hedenqvist, M. S. (2005). Packaging-related properties of protein-and chitosan-coated paper. *Packaging Technology and Science: An International Journal, 18*(4), 161–170.

Gao, W., Zhu, J., Kang, X., Wang, B., Liu, P., Cui, B., & Abd El-Aty, A. (2021). Development and characterization of starch films prepared by extrusion blowing: The synergistic plasticizing effect of water and glycerol. *LWT, 148*, 111820. https://doi.org/10.1016/j.lwt.2021.111820

Gutierrez-Gonzalez, J., Garcia-Cela, E., Magan, N., & Rahatekar, S. S. (2020). Electrospinning alginate/polyethylene oxide and curcumin composite nanofibers. *Materials Letters, 270*, 127662. https://doi.org/10.1016/j.matlet.2020.127662

Gutiérrez, T. J., & Alvarez, V. A. (2017). Eco-friendly films prepared from plantain flour/PCL blends under reactive extrusion conditions using zirconium octanoate as a catalyst. *Carbohydrate Polymers, 178*, 260–269. https://doi.org/10.1016/j.carbpol.2017.09.026

Gutiérrez, T. J., & Valencia, G. A. (2021). Reactive extrusion-processed native and phosphated starch-based food packaging films governed by the hierarchical structure. *International Journal of Biological Macromolecules, 172*, 439–451. https://doi.org/10.1016/j.ijbiomac.2021.01.048

He, Y., Li, H., Fei, X., & Peng, L. (2021). Carboxymethyl cellulose/cellulose nanocrystals immobilized silver nanoparticles as an effective coating to improve barrier and antibacterial properties of paper for food packaging applications. *Carbohydrate Polymers, 252*, 117156. https://doi.org/10.1016/j.carbpol.2020.117156

Herrera, N., Roch, H., Salaberria, A. M., Pino-Orellana, M. A., Labidi, J., Fernandes, S. C., Radic, D., Leiva, A., & Oksman, K. (2016). Functionalized blown films of plasticized polylactic acid/chitin nanocomposite: Preparation and characterization. *Materials & Design, 92*, 846–852. https://doi.org/10.1016/j.matdes.2015.12.083

Huntrakul, K., Yoksan, R., Sane, A., & Harnkarnsujarit, N. (2020). Effects of pea protein on properties of cassava starch edible films produced by blown-film extrusion for oil packaging. *Food Packaging and Shelf Life, 24*, 100480. https://doi.org/10.1016/j.fpsl.2020.100480

Jost, V., & Miesbauer, O. (2017). Effect of different biopolymers and polymers on the mechanical and permeation properties of extruded PHBV cast films. *Journal of Applied Polymer Science, 135*(15), 46153. https://doi.org/10.1002/app.46153

Kaewtatip, K., Chiarathanakrit, C., & Riyajan, S. A. (2018). The effects of egg shell and shrimp shell on the properties of baked starch foam. *Powder Technology, 335*, 354–359. https://doi.org/10.1016/j. powtec.2018.05.030

Kharrat, F., Khlif, M., Hilliou, L., Haboussi, M., Covas, J. A., Nouri, H., & Bradai, C. (2020). Minimally processed date palm (Phoenix dactylifera L.) leaves as natural fillers and processing aids in poly (lactic acid) composites designed for the extrusion film blowing of thin packages. *Industrial Crops and Products, 154*, 112637.

Khwaldia, K., Arab-Tehrany, E., & Desobry, S. (2010). Biopolymer coatings on paper packaging materials. *Comprehensive Reviews in Food Science and Food Safety, 9*(1), 82–91. https://doi. org/10.1111/j.1541-4337.2009.00095.x

Krishna, M., Nindo, C. I., & Min, S. C. (2012). Development of fish gelatin edible films using extrusion and compression molding. *Journal of Food Engineering, 108*(2), 337–344. https://doi.org/10.1016/j. jfoodeng.2011.08.002

Kyaw, A. K. K., Lay, L. S., Peng, G. W., Changyun, J., & Jie, Z. (2016). A nanogroove-guided slot-die coating technique for highly ordered polymer films and high-mobility transistors. *Chemical Communications, 52*(2), 358–361. https://doi.org/10.1039/c5cc05247e

Lee, D. S., Yam, K. L., & Piergiovanni, L. (2017). *Food packaging science and technology.* CRC Press.

Leidy, R., & Maria Ximena, Q. C. (2019). Use of electrospinning technique to produce nanofibres for food industries: A perspective from regulations to characterisations. *Trends in Food Science & Technology, 85*, 92–106. https://doi.org/10.1016/j.tifs.2019.01.006

Li, L., Wang, H., Chen, M., Jiang, S., Jiang, S., Li, X., & Wang, Q. (2018). Butylated hydroxyanisole encapsulated in gelatin fiber mats: Volatile release kinetics, functional effectiveness and application to strawberry preservation. *Food Chemistry, 269*, 142–149. https://doi.org/10.1016/j.foodchem.2018.06.150

Liao, J., Luan, P., Zhang, Y., Chen, L., Huang, L., Mo, L., . . . & Xiong, Q. (2022). A lightweight, biodegradable, and recyclable cellulose-based bio-foam with good mechanical strength and water stability. *Journal of Environmental Chemical Engineering, 10*(3), 107788.

Lin, Z., Guo, X., Zhou, L., Zhang, C., Chang, J., Wu, J., & Zhang, J. (2018). Solution-processed high performance organic thin film transistors enabled by roll-to-roll slot die coating technique. *Organic Electronics, 54*, 80–88. https://doi.org/10.1016/j.orgel.2017.12.030

Lu, P., Xiao, H., Zhang, W., & Gong, G. (2014). Reactive coating of soybean oil-based polymer on nanofibrillated cellulose film for water vapor barrier packaging. *Carbohydrate Polymers, 111*, 524–529. https:// doi.org/10.1016/j.carbpol.2014.04.071

Makhlouf, A. (2011). Current and advanced coating technologies for industrial applications. *Nanocoatings and Ultra-Thin Films*, 3–23. https://doi.org/10.1533/9780857094902.1.3

Meng, L., Liu, H., Yu, L., Duan, Q., Chen, L., Liu, F., . . . & Lin, X. (2019). How water acting as both blowing agent and plasticizer affect on starch-based foam. *Industrial Crops and Products, 134*, 43–49.

Mistretta, M., Botta, L., Morreale, M., Rifici, S., Ceraulo, M., & La Mantia, F. (2018). Injection molding and mechanical properties of bio-based polymer nanocomposites. *Materials, 11*(4), 613. https://doi. org/10.3390/ma11040613

Mohan, T., & Kanny, K. (2015). Thermoforming studies of cornstarch-derived biopolymer film filled with nanoclays. *Journal of Plastic Film & Sheeting, 32*(2), 163–188. https://doi.org/10.1177/8756087915590846

Moreno, O., Atarés, L., Chiralt, A., Cruz-Romero, M. C., & Kerry, J. (2018). Starch-gelatin antimicrobial packaging materials to extend the shelf life of chicken breast fillets. *LWT, 97*, 483–490. https://doi. org/10.1016/j.lwt.2018.07.005

Mousavi, S. M. M., Afra, E., Tajvidi, M., Bousfield, D. W., & Dehghani-Firouzabadi, M. (2018). Application of cellulose nanofibril (CNF) as coating on paperboard at moderate solids content and high coating speed using blade coater. *Progress in Organic Coatings, 122*, 207–218.

Mücke, N., da Silva, T. B. V., de Oliveira, A., Moreira, T. F. M., Venancio, C. D. S., Marques, L. L. M., Valderrama, P., Gonçalves, O. H., da Silva-Buzanello, R. A., Yamashita, F., Shirai, M. A., Genena, A. K., & Leimann, F. V. (2021). Use of water-soluble curcumin in TPS/PBAT packaging material: Interference on reactive extrusion and oxidative stability of chia oil. *Food and Bioprocess Technology, 14*(3), 471–482. https://doi.org/10.1007/s11947-021-02584-4

Mugwagwa, L. R., & Chimphango, A. F. (2020). Enhancing the functional properties of acetylated hemicellulose films for active food packaging using acetylated nanocellulose reinforcement and polycaprolactone coating. *Food Packaging and Shelf Life, 24*, 100481. https://doi.org/10.1016/j.fpsl.2020.100481

Nachal, N., Moses, J. A., Karthik, P., & Anandharamakrishnan, C. (2019). Applications of 3D printing in food processing. *Food Engineering Reviews, 11*(3), 123–141. https://doi.org/10.1007/s12393-019-09199-8

Nansu, W., Ross, S., Ross, G., & Mahasaranon, S. (2021). Coconut residue fiber and modified coconut residue fiber on biodegradable composite foam properties. *Materials Today: Proceedings, 47,* 3594–3599. https://doi.org/10.1016/j.matpr.2021.03.623

Nida, S., Moses, J. A., & Anandharamakrishnan, C. (2021). 3D printed food package casings from sugarcane bagasse: a waste valorization study. *Biomass Conversion and Biorefinery,* 1–11.

Nordin, N., Othman, S. H., Rashid, S. A., & Basha, R. K. (2020). Effects of glycerol and thymol on physical, mechanical, and thermal properties of cornstarch films. *Food Hydrocolloids, 106,* 105884. https://doi.org/10.1016/j.foodhyd.2020.105884

Nugroho, A., Maharani, D. M., Legowo, A. C., Hadi, S., & Purba, F. (2022). Enhanced mechanical and physical properties of starch foam from the combination of water hyacinth fiber (*Eichhornia crassipes*) and polyvinyl alcohol. *Industrial Crops and Products, 183,* 114936. https://doi.org/10.1016/j.indcrop.2022.114936

Pal, A. K., Wu, F., Misra, M., & Mohanty, A. K. (2020). Reactive extrusion of sustainable PHBV/PBAT-based nanocomposite films with organically modified nanoclay for packaging applications: Compression moulding vs. cast film extrusion. *Composites Part B: Engineering, 198,* 108141. https://doi.org/10.1016/j.compositesb.2020.108141

Parvathy, P., & Sahoo, S. K. (2021). Hydrophobic, moisture resistant and biorenewable paper coating derived from castor oil based epoxy methyl ricinoleate with repulpable potential. *Progress in Organic Coatings, 158,* 106347. https://doi.org/10.1016/j.porgcoat.2021.106347

Pemble, O. J., Bardosova, M., Povey, I. M., & Pemble, M. E. (2021). A slot-die technique for the preparation of continuous, high-area, chitosan-based thin films. *Polymers, 13*(10), 1566. https://doi.org/10.3390/polym13101566

Perez, V., Felix, M., Romero, A., & Guerrero, A. (2016). Characterization of pea protein-based bioplastics processed by injection moulding. *Food and Bioproducts Processing, 97,* 100–108. https://doi.org/10.1016/j.fbp.2015.12.004

Prabeesh, P., Sajeesh, V. G., Selvam, I. P., & Potty, S. N. (2021). Influence of thiourea in the precursor solution on the structural, optical and electrical properties of CZTS thin films deposited via spray coating technique. *Journal of Materials Science: Materials in Electronics, 32*(4), 4146–4156. https://doi.org/10.1007/s10854-020-05156-y

Prasad, L. K., McGinity, J. W., & Williams, R. O. (2016). Electrostatic powder coating: Principles and pharmaceutical applications. *International Journal of Pharmaceutics, 505*(1–2), 289–302. https://doi.org/10.1016/j.ijpharm.2016.04.016

Rahman, S., & Chowdhury, D. (2022). Guar gum-sodium alginate nanocomposite film as a smart fluorescence-based humidity sensor: A smart packaging material. *International Journal of Biological Macromolecules, 216,* 571–582. https://doi.org/10.1016/j.ijbiomac.2022.07.008

Ramkumar, P. L., Gupta, N., & Shukla, A. (2021). Bio-polymer selection for injection molding process using multi objective optimization by ratio analysis method. *Materials Today: Proceedings, 45,* 4447–4450.

Ren, D., Wang, Y., Wang, H., Xu, D., & Wu, X. (2022). Fabrication of nanocellulose fibril-based composite film from bamboo parenchyma cell for antimicrobial food packaging. *International Journal of Biological Macromolecules, 210,* 152–160. https://doi.org/10.1016/j.ijbiomac.2022.04.171

Román-Moreno, J. L., Radilla-Serrano, G. P., Flores-Castro, A., Berrios, J. D. J., Glenn, G., Salgado-Delgado, A., Palma-Rodríguez, H. M., & Vargas-Torres, A. (2020). Effect of size and amount of sugarcane fibers on the properties of baked foams based on plantain flour. *Heliyon, 6*(9), e04927. https://doi.org/10.1016/j.heliyon.2020.e04927

Roth, B., Søndergaard, R. R., & Krebs, F. C. (2015). Roll-to-roll printing and coating techniques for manufacturing large-area flexible organic electronics. In *Handbook of flexible organic electronics: Materials, manufacturing and applications* (pp. 171–197). Elsevier. https://doi.org/10.1016/B978-1-78242-035-4.00007-5

Sani, I. K., Geshlaghi, S. P., Pirsa, S., & Asdagh, A. (2021). Composite film based on potato starch/apple peel pectin/ZrO2 nanoparticles/microencapsulated Zataria multiflora essential oil; Investigation of physicochemical properties and use in quail meat packaging. *Food Hydrocolloids, 117,* 106719. https://doi.org/10.1016/j.foodhyd.2021.106719

Scaffaro, R., Maio, A., Gulino, F. E., Di Salvo, C., & Arcarisi, A. (2020). Bilayer biodegradable films prepared by co-extrusion film blowing: Mechanical performance, release kinetics of an antimicrobial agent and hydrolytic degradation. *Composites Part A: Applied Science and Manufacturing, 132,* 105836. https://doi.org/10.1016/j.compositesa.2020.105836

Shankar, S., & Rhim, J. W. (2018). Antimicrobial wrapping paper coated with a ternary blend of carbohydrates (alginate, carboxymethyl cellulose, carrageenan) and grapefruit seed extract. *Carbohydrate Polymers, 196*, 92–101. https://doi.org/10.1016/j.carbpol.2018.04.128

Shanmugam, K., Varanasi, S., Garnier, G., & Batchelor, W. (2017). Rapid preparation of smooth nanocellulose films using spray coating. *Cellulose, 24*(7), 2669–2676. https://doi.org/10.1007/s10570-017-1328-4

Siva, R., Sundar Reddy Nemali, S., Kishore kunchapu, S., Gokul, K., & Arun kumar, T. (2021). Comparison of mechanical properties and water absorption test on injection molding and extrusion – Injection molding thermoplastic hemp fiber composite. *Materials Today: Proceedings, 47*, 4382–4386. https://doi.org/10.1016/j.matpr.2021.05.189

Sothornvit, R. (2009). Effect of hydroxypropyl methylcellulose and lipid on mechanical properties and water vapor permeability of coated paper. *Food Research International, 42*(2), 307–311. https://doi.org/10.1016/j.foodres.2008.12.003

Sousa, A. M., Sereno, A. M., Hilliou, L., & Gonçalves, M. P. (2010). Biodegradable agar extracted from *GracilariaVermiculophylla*: Film properties and application to edible coating. *Materials Science Forum, 636–637*, 739–744. https://doi.org/10.4028/www.scientific.net/msf.636-637.739

Swaroop, C., & Shukla, M. (2019). Development of blown polylactic acid-MgO nanocomposite films for food packaging. *Composites Part A: Applied Science and Manufacturing, 124*, 105482. https://doi.org/10.1016/j.compositesa.2019.105482

Teil, M., Regazzi, A., Harthong, B., Dumont, P., Imbault, D., Putaux, J. L., & Peyroux, R. (2021). Manufacturing of starch-based materials using ultrasonic compression moulding (UCM): Toward a structural application. *Heliyon, 7*(3), e06482. https://doi.org/10.1016/j.heliyon.2021.e06482

Thenmozhi, S., Dharmaraj, N., Kadirvelu, K., & Kim, H. Y. (2017). Electrospun nanofibers: New generation materials for advanced applications. *Materials Science and Engineering: B, 217*, 36–48. https://doi.org/10.1016/j.mseb.2017.01.001

Tyagi, P., Salem, K. S., Hubbe, M. A., & Pal, L. (2021). Advances in barrier coatings and film technologies for achieving sustainable packaging of food products—a review. *Trends in Food Science & Technology, 115*, 461–485. https://doi.org/10.1016/j.tifs.2021.06.036

Vartiainen, J., Vähä-Nissi, M., & Harlin, A. (2014). Biopolymer films and coatings in packaging applications—a review of recent developments. *Materials Sciences and Applications, 5*(10), 708–718. https://doi.org/10.4236/msa.2014.510072

Vedove, T. M., Maniglia, B. C., & Tadini, C. C. (2021). Production of sustainable smart packaging based on cassava starch and anthocyanin by an extrusion process. *Journal of Food Engineering, 289*, 110274. https://doi.org/10.1016/j.jfoodeng.2020.110274

Wang, K., Jiao, T., Wang, Y., Li, M., Li, Q., & Shen, C. (2013). The microstructures of extrusion cast biodegradable poly(butylene succinate) films investigated by X-ray diffraction. *Materials Letters, 92*, 334–337. https://doi.org/10.1016/j.matlet.2012.10.121

Wang, Q., Chen, W., Zhu, W., McClements, D. J., Liu, X., & Liu, F. (2022). A review of multilayer and composite films and coatings for active biodegradable packaging. *Npj Science of Food, 6*(1). https://doi.org/10.1038/s41538-022-00132-8

Wang, Y., Li, M., & Shen, C. (2011). Effect of constrained annealing on the microstructures of extrusion cast polylactic acid films. *Materials Letters, 65*(23–24), 3525–3528. https://doi.org/10.1016/j.matlet.2011.07.090

Yang, Q., Yuan, F., Xu, L., Zhong, W., Yang, Y., Shi, K., Yang, G., & Zhu, J. (2020). Moisture barrier films for herbal medicines fabricated by electrostatic dry coating with ultrafine powders. *Powder Technology, 366*, 701–708. https://doi.org/10.1016/j.powtec.2020.03.023

Zhang, C., Li, Y., Wang, P., & Zhang, H. (2020). Electrospinning of nanofibers: Potentials and perspectives for active food packaging. *Comprehensive Reviews in Food Science and Food Safety, 19*(2), 479–502. https://doi.org/10.1111/1541-4337.12536

Zhang, C. H., Yang, F. L., Wang, W. J., & Chen, B. (2008). Preparation and characterization of hydrophilic modification of polypropylene non-woven fabric by dip-coating PVA (polyvinyl alcohol). *Separation and Purification Technology, 61*(3), 276–286. https://doi.org/10.1016/j.seppur.2007.10.019

Zhang, Y., Li, C., Fu, X., Ma, N., Bao, X., & Liu, H. (2022). Characterization of a novel starch-based foam with a tunable release of oxygen. *Food Chemistry, 389*, 133062.

18

Food Contamination from Packaging Material

Kartik Soni, Rizwana, Aparna Agarwal, and Abhishek Dutt Tripathi

CONTENTS

DOI: 10.1201/9781003303671-18

18.1 Introduction

The term *food contamination* refers to a state of presence of some undesirable material in the food in more quantities than is considered safe. Contamination can occur right from the harvest and can end up in being consumed by a consumer, causing health hazards. These foreign materials in food can be incorporated from the pesticides sprayed on crops, from the equipment used while processing, and from packaging materials used in packaging of the final food products. Food contamination brings harm to human health, be it accidental or intentional. *Intentional contamination* refers to the act of adding substances that profit the seller or the manufacturer. Contamination can also happen due to microplastics and hazardous chemical coatings which can leach out of the packaging material into the food. This movement of substances from packaging into food is called migration. Packaging materials give a way to safeguard, secure, product, and showcase appropriate food varieties. They perform a very important role in ensuring the quality of the products while they reach the customer in a healthy and protected system. As a consequence, packaging has become a vital feature in the food manufacturing process. To fulfill the enormous demand of the food industry, there has been a phenomenal surge in the development of improved packaging in recent decades. Various additives, including antioxidants, stabilizers, lubricants, antistatic and antiblocking agents, have also been designed to improve the functionality of these polymeric packaging materials during processing and fabrication as well as during application. Nonetheless, awareness about the wholesomeness and safety of foods has recently escalated. The changes that take place during the time lag between this transport are mostly caused by the contact of food items with the packaging material. It is hence significant that few variables are viewed while picking the correct material for a specific food item.

Plastic is one of the most used packaging materials. It was discovered during the 19th century. Most plastics were saved for difficult times, like war, to be used by soldiers. In 1831, styrene was first refined from an amber tree. Protection and padding materials, just like froth boxes, cups, and meat plates, for the food business became famous. In 1835, vinyl chloride was discovered and gave the chance to bring advancement of elastic science. Plastic packaging materials consist of huge natural (carbon-containing) chemical components and can be shaped into a variety of valuable items; they are flexible, heat sealable, and easy to print on and can be used in certain equipment where the formation, filling, and fixing of the packaging take place in a similar creation line (Marsh & Bugusu, 2007). The quality of variable porousness to gases, odors, particles with low atomic weights, and light is an important attribute. Mechanical properties are easily given by polymers, such as polyethylene and polypropylene, while hindrance polymers, for example, polyvinylidene chloride and ethylene vinyl liquor, provide a barrier against the move of gaseous compounds, like flavors and scents, past the packaging.

18.1.1 Plastic Packaging

More than 30 plastics are in use as packaging materials now. These macromolecules are polymers because they are made up of numerous repeating subunits. Plastic packaging materials are primarily composed of polymers (70–99%) with varying amounts of additives, such as plasticizers, antioxidants, colors, antistatic, fillers, and several other substances. Since these compounds are necessary to deliver the required functionalities, the final products are not typically polymers. Likewise, various sorts of added substances, like cell reinforcements, stabilizers, greases, antiblocking and antistatic agents, have been created to positively enhance the efficiency either during handling and manufacture or to improve the functional attributes of these polymeric packaging materials. Still, the quality and well-being of packaged food items have become a concern of great importance recently. Most concern ordinarily centers on food-added substances, both those added deliberately to the food items and those entering the food from the processing equipment or the packaging material. In this field, since the mid-eighties, the movement of plasticizers from food contact materials into food has brought numerous worries. Uncovered by toxicological investigations of a few generally used plasticizers, they were shown to have a cancer-causing impact in rodents and possible estrogenic impact in humans. Such studies showed that the packaging could itself act as a potential contaminant because of the migration or movement of chemical compounds from the packaging material into food. Therefore, governing bodies all throughout the planet have standardized that it is important to keep a check on such contamination, and many have put limits on the levels of allowed contamination from such polymers. For the same, a lot of examinations for migration of monomers, oligomers, volatiles, and other added substances from plastic packaging materials into food were performed (Lau & Wong, 2000).

18.1.2 Paper and Board Packaging

Using an interwoven network of cellulose strands obtained from natural wood, sheet materials like paper and paperboard are produced by making use of sulfate and sulfite. These strands are then pulped and additionally dyed and treated with synthetics, for example, slimicides and fortifying specialists, to deliver the paper item. Layered boxes, milk containers, collapsible containers, packs and sacks, and wrapping paper are some packages that are made from paper and paperboard. Paperboards and paper are biodegradable, environment-friendly, and have great printability and also possess mechanical strength. Waxes or polymeric materials can be utilized in the form of coatings to improve their weak obstruction properties. Usually, paper is not in direct contact with the food items, as they are coated with waxes or polymers, which provide good barrier properties and result in more efficient packaging material. When the food is in direct contact with a paper or paperboard packaging material, migration of printed inks and other microscopic impurities of paper can take place into food and hamper the quality and cause health hazards. Two major chemical substances that leach out from the paper and board packaging into the food are formaldehyde and bisphenol A (BPA) (Ungureanu et al., 2020).

18.2 Migration Effect

The movement of synthetic chemical components by diffusion from a contact surface, subject to both kinetic and thermodynamic control, is termed migration effect.

18.2.1 Stages of the Migration Effect

The migration or movement of added substances or toxins into food from polymeric food packaging might be differentiated into three relative stages: diffusion inside the polymer, solvation at the polymer–food interface, and scattering into bulk food.

18.2.1.1 Diffusion in the Polymer

The migration of added substances or components is caused by diffusion, which is a naturally visible appearance of an arbitrary movement or Brownian movement of individual moving particles inside the polymer grid. The thickness of the polymer packaging has an effect on the pace of migration of particles. Anyhow, the pace stops changing after a specific thickness is obtained. This thickness was characterized as the limiting thickness (Figge, 1988).

18.2.1.2 Solvation at the Polymer–Food Interface

The movement of polymers or monomers taking place due to their solubility in foods causes food contamination due to solvent migration. It takes place at the contact surface, where the packaging is in direct contact with the food item. If the migrant particle is more soluble in food as compared to the base polymer, the rate of migration is facilitated and contamination is increased. Whereas if the migrant particle is less soluble in food, there will be retardation in the migration of polymer into the food. Hence, fatty foods are more susceptible to polymer migration, as most additives and other contaminants are fat-soluble (Lau & Wong, 2000).

18.2.1.3 Dispersion in Bulk Food

Just past the interface between the food and the polymer packaging, the solvated particles diffuse past the interface and migrate into the mass of food. The movement at this stage just as that for the past two phases is driven essentially by entropy, a proportion of irregularity. Blending could escalate the migration deeper into food since blending upgrades kinetic migrant solvation by eliminating solvated migrants from the interface, consequently lessening reprecipitation (Limm & Hollifield, 1996).

18.2.2 Food and Migration

Migration in food substances from food contact materials and other packaging materials depends on various factors that can either support or cause hindrance in the movement of migrants.

18.2.2.1 Compatibility

Some packaging materials can be made of substances that can leach out into the food material faster because of increased incompatibility. For instance, some oils and fats can cause swelling of certain plastics, which leads to a higher diffusion of more plastics into the food products. Also, when metals leach into acidic foods from uncoated metal packaging due to reactions in between the food and packaging.

18.2.2.2 Solubility

The amount of migration of a certain packaging can be determined by the solubility of packaging materials into a food mass. The higher the affinity of packaging material for the food, the higher will be the degree of migration into the food. The nature of food influences the chemical migration of packaging material.

18.2.3 Factors Affecting the Rate of Migration

The rate of movement of migrant components is affected by some environmental and physical parameters.

18.2.3.1 Duration of Food Contact

Some packaging materials are only safe for a limited contact time. These should not be subjected to longer service durations. The value of migration increases with an increase in the square root of the duration of contact: $M \propto t^{1/2}$ (Barnes et al., 2006). Some durations for the contact of common packaging materials are as follows:

1. Minutes (takeaway foods)
2. Hours (ready-to-eat bakery)
3. Days (milk, fruits, and vegetables)
4. Weeks (butter, cheese)
5. Months or years (frozen, dried, canned food items)

18.2.3.2 Temperature of Food

Just like most physicochemical reactions, the migration rate increases with an increase in heat. In food packaging or processing applications, temperatures ranging from freezing to boiling all are specific. Hence, a packaging material used for storing frozen meat mustn't be used for baking bread or storing chilled beverages.

18.2.3.3 Mobility of Chemical Substances

The mobility of chemical substances in the form of molecules, atoms, or compounds depends on their size, shape, interaction with food, and resistance offered by the food to the material. If the food is highly incompatible with the packaging material, it can lead to a higher intensity of migration of plastic and other migrants on the food surface, like a bloom (Barnes et al., 2007).

18.3 Majorly Used Plastics and Their Hazards

Plastic additives, monomers, oligomers, and contaminants are among the many chemicals that migrate from packaging materials to foodstuffs. During the processing and manufacture of the polymeric packaging materials, a wide range of additives are produced and utilized. Plasticizers, antioxidants, light stabilizers, thermal stabilizers, lubricants, antistatic agents, and slip additives are common additives in a variety of plastics. Migrating solvents such as adipic acid, toluene, butanone-2, ethyl acetate, and hexane, as well as pigments like molybdate orange, are also a threat. Monomers are oxidizable substances, regarding living creatures, and thus pretty much harmful. Consequently, regulatory guidelines typically confine the limit of remaining monomers in the raw materials, plastics, and articles made thereby. Their harmfulness is discussed in the following sections.

18.3.1 Polyvinyl Chloride

Being that this oligomer is an organochlorine compound, its migration is of great concern as it has high toxicity. Residual monomer levels in food packaging made with PVC and the migration level to foods are closely kept in check. The gas chromatographic headspace technique is broadly used to quantify the residual monomer in food and packaging. It is known to possess carcinogenic properties and targets the lungs (if inhaled) and liver (when consumed), which is also known to have adverse effects on the human brain or the lymphohematopoietic system (Wagoner, 1983).

18.3.2 Polystyrene

Even though the severe toxicology of styrene isn't high, its digestion includes a mutagenic compound, phenyloxirane (Bond & Bolt, 1989; European Center for Ecotoxicology and Toxicology of Chemicals

TABLE 18.1

Abbreviations of plastics used

Abbreviations	Full form
PS	Polystyrene
PA	Polyamide
PP	Polypropylene
PET or PETE	Polyethylene terephthalate
PEN	Polyethylene naphthalene dicarboxylate
PC	Polycarbonate
EVA	Ethylene vinyl acetate
PE	Polyethylene
PVC	Polyvinyl chloride
HMT	Hexamethylenetetramine
HNP	High-nitrile polymers
TPX	Polymethyl pentene
EVO	Ethylene vinyl alcohol
SB	Styrene butadiene
PVA	Polyvinyl alcohol
ABS	Acrylonitrile butadiene styrene
PVdC	Polyvinylidene chloride

Technical, 1993). Moreover, even at low doses (200–500 ppb in yogurt, and 40–730 ppb in water), styrene may impact sensory attributes (Jenkins, 1978). The most used technique for quantification of styrene is headspace gas chromatography (Varner & Breder, 1981; Nerín et al., 1998). Styrene oxide can cause stomach cancers in humans at very high dosages. A low dosage (<200 ppb), the migration of this compound from packaging material into food is of less concern, as the human body can detoxify it (Roe, 1994).

18.3.3 Bisphenol A Diglycidyl Ether

It is a chemical plasticizer added in polycarbonate to enhance its rigidity. The level of harmfulness of epoxy compounds relies significantly upon the presence of free epoxy groups. These free groups are alkylating agents and can cause certain cytotoxic activities in tissues relating to rapid division of cells. Coatings on food cans and food stockpiling vessels are some applications of such epoxy resins. On account of their harmfulness, it is important to reduce the amount of free radicals to inhibit their migration into food (Lau & Wong, 2000). BADGE is seen to react with unidentified food components. It has a higher potential reaction with proteins like methionine and cysteine-forming derivatives or diminishing the amount of protein in the food (Petersen et al., 2008). The effects of long-term consumption of these compounds are relative to the detrimental effects on sperm count and fertility in males and can be carcinogenic, a precursor for breast cancer in females (Lawley et al., 2008).

18.3.4 Isocyanate

As isocyanates are considered harmful mixtures and their health hazards are all around reported (Woolrich, 1982), their amount of use in the assembling of plastic packaging and the articles expected to come into direct contact with food sources is controlled (Commission of the European Communities, Directive 90, 1990a,b). As of now, for use in food-contact materials, 12 isocyanates are allowed. Leftover amount in the completed packaging materials should not surpass 1.0 mg/kg. Ingestion can cause irritation in the mouth, pharynx, and GI tract, while long-term toxicity can cause nephrotoxicity (Sheftel, 2000).

18.3.5 Caprolactam

Polyamides, usually known as "nylons," are one sort of food packaging material which is used to carry food during flame processing. There was proof showing that remaining caprolactam and moderately a lot of nylon 6 oligomers, monomers of nylon, can possibly diffuse into hot water (Barkby & Lawson, 1993). In spite of the fact that caprolactam isn't particularly harmful to the oral organization, it might cause a minor extended impact on thermoregulation and unpleasant sensory attributes in cooked food items (Stepek et al., 1987). Disintegration-and-precipitation technique was used to control the remaining oligomers and caprolactam in food packagings made of nylon and were measured by HPLC with UV recognition at 210 nm (Begley et al., 1995).

18.3.6 Polyethylene Terephthalate Oligomer

The copolymer of ethylene glycol with either terephthalic acid or dimethyl terephthalate is called polyethylene terephthalate (PET). For fluids and oils, it is usually used as a packaging, as it won't thermally degrade at a temperature less than 220°C. PET is additionally utilized as plate and dishes for conventional and microwave processing. In any case, PET usually contains limited quantities of oligomers ranging from dimer to pentamer of low molecular weight. Based upon the kind of PET, these cyclic mixtures were discovered with levels going from 0.06 to 1.0%. HPLC strategy has been utilized to decide the movement levels of oligomers from PET (Begley et al., 1995). DEHP, a derivative of PET, may have reproductive issues that can be acquired by upcoming generations.

18.3.7 Polyolefins

Linear low-density polyethylene (LLDPE), which is the main compound used, is synthesized by copolymerizing ethylene with other olefins such as propylene, 1-butene, 4-methyl-1-pentene, 1-hexene, or 1-octene, or a combination of these alkenes. A group of antioxidants that are present in polyolefins also works in unison to inhibit polymer breakdown through diverse and complementary processes. The restricted phenols and the phosphite-based compounds are the antioxidants that are most typically found in polyolefin films. Typically, these substances are mixed into the polymer at a concentration of 0.1% (weight for weight). Antistatic compounds, slip agents, and light stabilizers are some additional ingredients in the polymer (Lawson et al., 1996). Polyolefins pose no major health hazard when used within limits, but the additives, when migrated, can pose certain ill effects (Boone et al., 1993).

18.4 Migration from Paper and Paperboards

18.4.1 Dioxins

A large variety of polychlorinated dibenzo-p-dioxins and polychlorinated dibenzofurans, which are used to make paper food packaging, are together referred to as "dioxin." The majority of research show that dioxins have significant toxicity, with 2,3,7,8-tetrachlorodibenzo-p-dioxin being the most dangerous isomer (Ackermann et al., 2006).

18.4.2 Benzophenone

The primary application of benzophenone is as a photoinitiator for UV-treated inks, varnishes, and lacquers. In addition to the uses listed previously, benzophenone works perfectly as a wetting agent for pigments or to increase the flow rate of inks. These inks typically include 5 to 10% photoinitiator (Anderson & Castle, 2003). Since continuous cutting and folding are made possible by UV treatment of printed cardboard inks, it is frequently utilized to produce finished packaging quickly. Benzophenone is not completely eliminated from the printing medium during treatment because only a little amount of the initiator is used up, and migration via the open structure of cartonboard is probable.

Additionally, it may be present if the cartonboard is made from recycled fibers that were once components of printed materials. In addition to benzophenone, 4-methoxybenzophenone may be applied. Studies on this product's hazardous effects show that they can be very damaging to genes and can behave as carcinogens in addition to having estrogenic effects (Muncke, 2009).

18.4.3 Nitrosamines

Nitrosamines are genotoxic carcinogens that are often present in food and other media and can develop endogenously in the human body (Tricker & Preussmann, 1991). N-nitrosamines have been discovered in a variety of meals and drinks (Robertson, 2006). Various food-contact items, such as papers and waxed containers, are potential sources of origin. N-nitrosomorpholine and morpholine are contaminants that can get into foods if they come in touch with certain packing materials directly for a specific amount of time. By combining with salivary or ingested nitrite, ingested amines can cause the body to produce nitrosamines.

18.5 Migration from Metal Packaging

18.5.1 Tin

Tin is presently present in small amounts in a wide range of canned goods, especially those that are packaged in tinplate cans that are either unlacquered or only partly lacquered. Although there have been a few case reports of acute gastrointestinal disturbances following consumption of foods with a level of 100–500 mg/kg tin (Benoy et al., 1971; Omori et al., 1973), these studies have a number of limitations. There is a threshold concentration of >730 mg/kg for adverse effects, according to controlled clinical tests on the acute effects of tin consumed after migration from packages (Boogaard et al., 2003). Since tin can effectively scavenge oxygen, it provides substantial protection against corrosion and is frequently used instead of coating.

18.5.2 Chromium

Passivation is a treatment that is frequently used to increase the adhesion of the enamel and reduce the susceptibility of the tin layer in tinplate cans to oxidation degradation (Kim et al., 2008). The presence of chromium as Cr(VI) can have serious impacts on living beings because it is known to be both carcinogenic and mutagenic (Skrzydlewska et al., 2003). Chromium is characterized by relative toxicity and undesirable organoleptic qualities.

18.5.3 Lead

Lead is frequently utilized in metal packaging, despite its toxicity. It has been established that one of the most significant pollutants originating from packing materials is lead. Lead is very poisonous, and certain organolead species (especially neonates) can have a negative impact on the central nervous system. It is sometimes argued that children are far more vulnerable to the harmful effects of lead than adults are, with a larger concentration of lead remaining in their bones and brains. Children who consume lead subacutely may experience encephalopathy, convulsions, and mental impairment (Skrzydlewska et al., 2003; Robertson, 2006).

18.5.4 Aluminum

High intake and higher tissue levels of this metal have been associated with a number of illnesses, including dialysis encephalopathy, osteodystrophy, and microcytic anemia. In addition to the everyday consumption of Al from food, aluminum can migrate from cooking equipment, storage containers, and packaging (Arvanitoyannis & Kotsanopoulos, 2013).

18.6 Other Contaminants

Aside from added substances and monomer deposits present in the packaging materials, there are various other sources of food contamination. Disintegration items from added substances or monomers are likely to move into the food when suitable conditions are formed. Deposits of synthetic compounds that were utilized in the manufacture of packaging materials can cause contamination through direct contact with food items too. It was determined that some additives include substances that seem to be toxic to human health and the environment based on assessments of several additives. It is not always the case that the additives are pure substances. They are usually blended with resins and waxes (rosin), oils (paraffin), and other substances which are not in the safe category for food contact. These compounds are commonly observed in fat or liquid additives.

18.6.1 Plasticizers

Plasticizers are a class of additives used to enhance the characteristics of polymers in plastic materials. The most frequent forms of plasticizers include butyl stearate, acetyl tributyl citrate, alkyl sebacates, and adipates, which have minimal toxicity but may be carcinogenic and estrogenic (Shiota & Nishimura, 1982). Monomeric plasticizers migrate more readily when they come into direct contact with fatty foods and when the temperature rises. Plasticizer migration from plastics into food has been documented (Hammarling et al., 1998).

18.6.2 Thermal Stabilizers

The use of thermal stabilizers in polymers is also common. Epoxidized seed and vegetable oils are often used in food contact polymers, heat stabilizers, lubricants, and plasticizers, among other things. Because residual ethylene oxide is extremely hazardous, their purity has a significant impact on their toxicity (Lau & Wong, 2000). The initial phase in stabilizer migration, namely, the diffusion of the stabilizer from the polymer matrix to the surface in contact with a food-mimicking solvent or other environments, has been the subject of various studies over the last few decades.

18.6.3 Slip Additives

Fatty acid amides are employed as slip additives in a range of packaging plastics, including polyolefins, polystyrene, and polyvinyl chloride (PVC). Plastic compositions include slip additives, but they gradually develop and tend to bloom to the surface. They provide beneficial features, such as lubrication to prevent films from adhering together or creating conglomerates, as well as reducing static charge (Cooper & Tice, 1995).

18.6.4 Light Stabilizers

Plastics, particularly polyolefins, are treated with light stabilizers to improve their long-term weathering qualities. Polymeric hindered amines (HALs) are extensively employed as light stabilizers in polyolefins.

TABLE 18.2

Types of contaminants

S. no.	Packaging additive	Examples/Metabolite	Toxicity reported
1.	Antioxidants	Aryl substituted phospites, triphenyl phosphate	Can be highly toxic
2.	Plasticizer	Butyl stearate, acetyl tributyl citrate, alkyl sebacates, adipates	Estrogenic and carcinogenic effects
3.	Thermal Stabilizers	Polyvinyl chloride, polyvinylidene chloride, and polystyrene	Causes toxicity in the cells
4.	Light stabilizers	Tinuvin 770 and Chimasorb 944	Effect on cardiac muscles
5.	Slip additives	Polyolefins, polystyrene, and polyvinyl chloride	Depends on the residual migration

18.6.5 Antioxidants

Antioxidants are used to delay the oxidation process that polymers go through when exposed to light. The most often utilized antioxidants are BHT and Irganox 1010. The majority of antioxidants were found to be non-toxic and to have a good stabilizing effect. For many polymers, migration rates of Irganox 1010 and Irganox 1076 were determined (Arvanitoyannis & Bosnea, 2004).

18.6.6 Decomposition Products

For the manufacture of PVC film, a thermal stabilizer called diphenylthiourea is used. The stabilizer and its decay items, including isothiocyanatobenzene, aniline, and diphenylurea, were recognized in the food and the film used to pack the food. With dichloromethane–ethanol as the mobile phase, the mixtures are controlled by HPLC on a Nucleosil amine section. These products are considered potential carcinogens.

18.6.7 Benzene and Other Volatiles

It was exhibited that benzene and alkyl-benzene could be produced and migrated from some food contact plastics in high-temperature applications (Jickells et al., 1990, 1993). Also, benzene may move into food from PET packaging which was preliminarily contaminated with benzene (Komolprasert et al., 1994).

18.6.8 Vapor Contaminants

It was concluded that milk can be contaminated by migration of naphthalene which is consumed from the surroundings by the LDPE packaging (Lau et al., 1994). A numerical model was approved which was created to portray the movement of naphthalene from the surroundings into milk. Likewise, the degree of naphthalene migration into food was found to be directly proportional to the fat content of milk (Lau et al., 1995).

18.6.9 Inks

Major health-risking chronic conditions have been associated with the migration of inks into food products over prolonged storage. Inks contain various aromatic hydrocarbons which are linked to cancerous conditions in human. Several laws have been made and put into action across the globe to safely handle inks and avoid their migration into foodstuffs (Jadhav et al., 2020).

18.7 Permissible Quantities

The human body has a very strong defense mechanism where it can fight against the toxicity caused by overdosage of any specific additive or packaging material. The detox function of the liver reduces the toxicity and prevents the harm that could be caused if not kept in check. Keeping the body's capacity to overcome toxicity and fight the harmful components that have been ingested from the migration of packaging material into the food, certain limits of permissible quantities of these compounds have been set by regulatory bodies for specific packaging materials.

18.7.1 Indian Standards

As per Indian standards of maximum migration limits of packaging material into the food, the limits on migration in foodstuffs, liquid foodstuffs, and overall migration for packaging materials like polyethylene (LDPE, HDPE, LLDPE), polystyrene, polyvinyl chloride (PVC), polypropylene, ionomer resins, ethylene acrylic acid (EAA), polyalkylene terephthalates (PET and PBT), nylon 6 polymer, ethylene vinyl acetate (EVA), ethylene meta acrylic acid (EMAA), polycarbonate resins, polyalkylene terephthalates (PET and PBT), and melamine-formaldehyde resins are up to 60 mg/kg or 10 mg/dm² (maximum).

Most international standards are similar but are more concerned with controlling the daily intake of these materials to a daily acceptable/tolerable limit of 1 mg/kg of body weight, which reduces the consumption-related toxicity of these migrations on human health and resulting risk factors (Food Safety Standards Authority of India, 2011; Bureau of Indian Standards, 1982).

18.7.2 European Union Standards

Migration testing are only required for plastic packaging in the EU. The migrants are measured in food simulants in this system, and the total amount of migrants that could be absorbed each day from 1 kg of food in contact with the plastic is calculated. It should be noted, however, that this amount of food is assumed to be composed of either 1 kg of fatty food or 1 kg of one of the other types of food, never by the sum of different food types. To calculate exposure assessment, the highest migration among several simulants is selected. A cube containing 1 kg (or 1 liter) of food has a surface area of 6 dm² (600 cm²).

For all packaging materials, there is a 60 mg/kg total limit on migration; any migration of chemicals from plastic into food beyond this limit is prohibited (equivalent to 10 mg of substances per 1 dm² of surface area of the plastic material).

18.7.3 USFDA Standards

As long as the following conditions are met, a material used in a food-contact object (such as food packaging or food processing equipment) that migrates into food or could reasonably be expected to migrate into food is exempt from regulation as a food additive:

1. The compound has not been proven to cause cancer in humans or animals, and there is no reason to believe it is a carcinogen based on its chemical makeup. The substance must also not contain a carcinogenic impurity, or if it does, it must not contain a carcinogenic impurity with a TD50 value of less than 6.25 mg/kg body weight per day based on chronic feeding studies reported in the scientific literature or otherwise available to the Food and Drug Administration. (The TD50, for the purposes of this section, is the feeding dose that causes cancer in 50% of the test animals when tumors found are corrected for. If more than one TD50 value for a substance has been reported in the scientific literature, the FDA will utilize the lowest suitable TD50 value in its review.)

2. There are no significant health or safety risks with this chemical because:
   - The use in question has been proven to result in dietary concentrations of less than 0.5 parts per billion, equivalent to dietary exposure levels of less than 1.5 micrograms/person/ day (based on a diet of 1,500 g of solid food and 1,500 g of liquid food per person per day).
   - The drug is currently regulated for direct addition to food, and the proposed use's dietary exposure to the substance is at or below 1% of the permissible daily intake, as determined by safety data in the Food and Drug Administration's files or other relevant sources.

3. The chemical has no technical impact on the food it migrates to.

4. The use of the material has no substantial negative effects on the environment.

18.8 Effects of Processing on Migration

18.8.1 High-Pressure Processing

The sorption of food ingredients or the entry of packaging materials into food is limited by the compression of the polymer matrix during HP processing. The polymer progressively returns to its initial condition after treatment, and as a result, migration continues as anticipated before treatment. However, it should be noted that there aren't enough documented trials to make conclusive results. In light of these investigations, attention should be paid to the extremely minimal impact of HP processing on the food/ packaging interactions of typical plastic materials. This is not surprising, because HP treatments are used to lessen the effects of traditional stabilization procedures on food, which in turn limits their influence on conventional packing materials (Guillard et al., 2010).

18.8.2 Microwave Processing

It is important to distinguish between two dimensional effects when analyzing how microwave treatment affects migration phenomena from packing materials: the direct influence of microwave radiation and the indirect effect of temperature rise brought on by product heating.

In an attempt to distinguish between the two effects of microwave processing, due to the microwaves themselves and due to the increase in temperature by microwave heating, it was concluded that the migration increase during and after the microwave treatment is thus exclusively brought on by the influence of the temperature rise. Measurement of the total migration of food stimulants was the researcher's first method for figuring out how microwave treatments affected interactions between food and packaging (Jickells et al., 1991). It is true that temperature acts as an activating factor for a wide range of reactions, particularly those involving the degradation of elements that were originally present in the packaging, such as the polymer itself or its additives.

- *Plasticizers* A well-known and well-documented instance of food/packaging interaction taking place during microwave treatment is the migration of plasticizers out of packaging into food. The direct contact between the packaging and the food may significantly increase the migration of plasticizers. It was found that migration was substantially influenced by heating duration, product's fat content, and certainly, the plasticizer's original concentration in the film (Badeka & Kontominas, 1999; Badeka et al., 1999). Additional research emphasized on the movement of plasticizers that aren't in the layer that come into direct contact with food. For instance, to achieve the appropriate adhesive performance characteristics, certain plasticizers are added to acrylic- or vinyl-acetate-based adhesives that are used to laminate PET film to paperboard. Several investigations (Begley & Hollifield, 1990a; Begley et al., 1991; Sharman et al., 1995) have demonstrated that these plasticizers may migrate into the food at the temperatures reached by microwave cooking.

- *Monomers and oligomers* A non-negligible amount of migrated monomers and oligomers of certain types of microwaveable materials was observed (nylon 6,6, nylon 6, PET) in fatty foodstuffs like oils, popcorn, pizza, and french fries, all of which are sometimes cooked in a microwave (Rijk & Dekruijf, 1993; Soto-Valdez et al., 1997). These studies support the notion that migration is greater in fried foods than in non-fried foods. This raises severe concerns regarding the household usage of fatty foods intended for microwave cooking, especially given the fact that these foods are occasionally reheated or overheated. Another case of BPA (bisphenol A) was studied, and it showed high migration amounts in polycarbonate containers. Usually, polycarbonate contains BPA, and it is also created by constant degradation of polycarbonate by microwave heating (Biedermann-Brem et al., 2008).

- *Inks* Since inks are not a natural component of the food contact layer that comes into touch with food, they shouldn't, in general, be present there. The well-known instance of ink migration from paper, cardboard, or plastic materials is highly intriguing because of this. Foods can include ink molecules or residues because the inner layer that is in direct contact with food is not always a functional barrier. The migration of benzophenone, a substance found in paperboard's inks, into food was investigated (Johns et al., 1995, 2000). Benzophenone was found in less than a week in the frozen food of storage at 20°C, indicating that this compound might move even during frozen storage from packaging into food (potato chips and patties).

- *Benzene* During microwave treatment, the chemical benzene often leaks from packing. T-butyl perbenzoate, a catalyst for the polymerization of polymers, particularly polyester, is degraded to produce benzoene. During microwave treatment, benzene migration from multiple thermoset polyester samples into various food items was assessed. The same experiment was completed using PVC, PS, and expanded PVC. Food that came into contact with PS had the greatest levels of benzene (up to 0.96 mg/dm^2) in it. As a confirmation that benzene is present in significant amounts in polyester, which uses t-butyl perbenzoate as a polymerization initiator, PVC proved surprisingly resistant to benzene migration (rate less than 0.01 mg/dm^2) (Guillard et al., 2010).

18.8.3 Irradiation

Plasticizers and other additives could potentially be unstable. Goulas et al. (1998) investigated the impact of high-dose irradiation on the migration of the plasticizers dioctyl adipate (DOA) and acetyl tributyl citrate (ATBC) from food-grade polyvinyl chloride (PVC) and poly(vinylidene chloride/vinyl chloride) copolymer (PVdC/PVC) copolymer. These authors found no statistically significant differences in the amount of ATBC that migrated between non-irradiated samples and samples that were irradiated, but they did find a significantly higher amount of DOA that migrated into olive oil from irradiated (20 kGy) versus non-irradiated samples (Guillard et al., 2010).

18.9 Migration Testing and Analysis

To guarantee that the food isn't contaminated with chemicals or synthetic compounds from the food contact materials, bringing about undesirable additives in the food, a wide range of guidelines for food contact materials is set up. For the manufacturers and merchants of the food contact materials, it is difficult to recall with which specific guidelines the material should consent and how to guarantee that it conforms to the specific guidelines and suggested use.

For practicing these guidelines, testing and analysis for all the food contact materials have been standardized by the responsible authorities. Testing of migration from food contact materials mimics food contact by carrying a food simulant into similar or direct contact with the food contact materials (at a certain temperature and time conditions). Afterward, the amount of migrated chemical substances from the food contact materials into the food is identified by testing how much altogether migrates (overall migration). Residual content or specific migration should be analyzed for some particular chemical compounds. For more accuracy, extra trials like purity of ingredients, sensory analysis, change or release of colors, and quantitative analysis of volatiles should likewise be carried out (Barnes et al., 2007).

18.9.1 Mimicking Food Contact

Five food simulants have been standardized by the authorities to avoid any discrepancies or irregularities in obtaining results while performing migration testing.

TABLE 18.3

Food simulants and their descriptions

Abbr.	Simulant	Description	Examples
A	Water	Non-acid, non-fat, aqueous	Mineral water, syrups, skimmed milk, molasses, yeast paster
B	3% AA	Non-fat, acidic, aqueous (pH ≤5)	Fruit (juices, squashes, purees, pastes, or chunks), vinegars, jams, jellies, carbonated beverages, rennet, sauces, broths
C^1	10% Ethanol	Alcohol <10%	All alcoholic beverages, arrack, ppharmaceutical
C^2	50% Ethanol	Alcohol ≥10%	syrups
D	N-heptane	Fatty foods—oils and fats	Oils, ghee, cocoa butter, lard, fatty/fried savory snacks, nuts and seeds
A & D	Water & N-heptane	Emulsions	Bakery items (breads, cakes), ice creams, mayonnaise, butters, milk-based sweets
B & D	3% AA & N-heptane	Purees and pickles	Fresh processed meat and fish, frozen foods, pickles, fatty sauces, cheese, ketchup
No end test	None	Dry items	Oats, dried yeast, dehydrated vegetables and fruits, vermicelli, spaghetti, salt, sugar, milling products (cereals and pulses)

Source: Bureau of Indian Standards (1982).

18.9.2 Overall Migration Testing

To check on the undesirable additives addition in food due to migration from food contact materials, overall migration testing is carried out.

1. Measure the weight of the simulant before contact.
2. Keep the simulant in contact with the food contact material for a particular time and temperature combination.
3. Remove the food contact material sample from the simulant and take out the maximum possible quantity of the simulant (sometimes it is done by evaporating the simulant).
4. Measure the weight of the simulant after separation.
5. Calculate overall migration by subtracting the original weight.

(Barnes et al., 2007)

18.9.3 Specific Migration Testing

Specific migration should be calculated experimentally by suggested procedures. After the simulant is isolated from the sample, specific migration can be calculated by a huge set of analytical experiments, like liquid chromatography–mass spectrometry (LCMS), ultraperformance liquid chromatography (UPLC), high-performance liquid chromatography (HPLC), and gas chromatography (GC), with a wide scope of other detection techniques. The detection technique depends on the type of sample used (Barnes et al., 2007).

18.9.4 Residual Migration Testing

Residual migration limits are set for some specific components which are too low to be calculated. This happens when the component is very reactive, volatile, or impossible to analyze by experimental techniques. If the amount of the migrant is so low that it does not touch the overall migration limit, residual migration can be calculated by the amount added in the ingredients or the chemical specifications.

When the calculated amount is in excess and exceeds the limits or restrictions put on the specified substance, calculation of actual residual content is carried out. For this, firstly, the food contact material is made free of any residual components. Then, precipitation of polymer in a poor solvent can be done by firstly using techniques like Soxhlet, reflux, or dissolution in a good solvent (Barnes et al., 2007).

18.9.5 Other Tests

1. *Organoleptic or sensory evaluation.* Food contact material testing cannot provide results for organoleptic changes in the food due to migration. For this, changes in the actual food product have to be noticed by mimicking the actual conditions of packaging, storage, and transportation. Trained panel members are chosen for these tests.
2. *Physical parameters.* Even for this, food contact material testing isn't sufficient to obtain results; actual conditions are used on food items to check for the changes in physical parameters of food, like viscosity, density, etc.
3. *Color changes.* Any color release from the packaging into/onto the food is completely unacceptable. To check for color release, a white absorbent paper is wetted with the food simulant and kept in contact with the food contact material under certain temperature and pressure. Then it is compared to the other paper which was kept under similar conditions but without any contact surface.
4. *Volatiles.* Some volatile components cannot be determined by testing techniques; for analyzing them, silicone material is used as it can trap volatiles, and then it can be dried and the volatiles can be evaporated. The difference in the weights measured gives the amount of volatiles migrated from the food contact materials (Barnes et al., 2007).

18.10 Prevention of Migration

For non-reactive materials (tempered steel, earthenware, glass), synthetic substances from within the surface, straightforwardly in contact with the food, can migrate. They move from the internal surface to the food by surface movement. Chemical migration from inside the packaging material or from an external perspective (printing inks, cements) is unfeasible. This reactiveness is because of the molecular structure, with pores that are little and keep particles or single atoms from going through, whereas glass-packed oil food varieties can be contaminated by movement of plasticizers (like epoxidized soybean oil [ESBO] or phthalates) from the conclusion. Movement can be decreased via cautious assembling or the utilization of uncommonly grown low relocation terminations (Food Packaging Forum).

Successful guideline of food-contact materials is conceivable with the assistance of the latest logical strategies and accomplishments of present-day experimental and administrative toxicology. All applicable data ought to be coordinated into the danger assessment measure dependent upon the situation. A right procedure in toxicology of plastics in most of the situations includes exact chemical investigation of possible contamination of food, water, or simulant media under particular conditions. The results so obtained must be compared with the existing toxicology data and specified safety standards (Barnes et al., 2007).

18.10.1 Multilayer Packaging

At the point when migration limits for substances are set, a conventional framework is applied to determine the consumption. It is estimated that a 60 kg individual will feed on at least 1 kg of packed food each day, although an alternate framework is important for specific conditions. One such example is on account of lipophilic substances. Lipophilic substances migrate promptly into fat-rich food items. The consumption pattern of fat-rich food varieties is typically just 200 g or less each day. For these compounds, a reduction factor is hence set for use in compliance testing, considering the lower intake of fat.

In multilayer materials, a layer can work as a hindrance to migration of compounds into food. At the point when such a useful boundary layer is applied that guarantees no movement into food, it may not be important to approve the substances behind that layer if the substance isn't cancer-causing, genotoxic, or harmful for reproduction. Reusing of plastic materials has come into highlight as sustainability of manufacture and ecological issues becomes more significant (Barnes et al., 2007).

18.10.2 Use of Functional Barriers

A packaging enhancer that restricts the degree of migration of a compound from the packaging to food in levels under a safe limit can be characterized as a functional barrier. Functional barrier substances can be productively applied to accomplish a similar prevention level of "super-clean" recycling innovations which eliminate, or considerably lessen, the level of post-consumer chemicals and toxicants in the polymer down to comparative levels to those in virgin polymers. Complete or approximately all migration residues of any undesirable unfamiliar chemical compounds can be restricted (Piringer et al., 1998). As per this idea, the migration of any non-cancer-causing compound resulting in dietary focuses equivalent or lower than the limit isn't viewed as a huge well-being hazard.

It is, for the most part, known that only a restricted amount of packaging materials gives complete assurance properties from layers behind or from the environment when concerning the migration of chemical compounds, for example, glass or metal. When talking of multilayers with functional barriers primarily made with plastic materials, there happens, somewhat, an unavoidable migration from the plastic layers into the food item. This should be perceived as a particular amount which should conform to the food safety guidelines. In this manner, it is important initially to comprehend functional barrier qualities and systems and, furthermore, to characterize the functional barrier effectiveness corresponding to safety and to form suitable testing strategies. This is particularly significant where reused plastics are covered by plastic functional barriers in food packaging applications (Piringer & Baner, 2000).

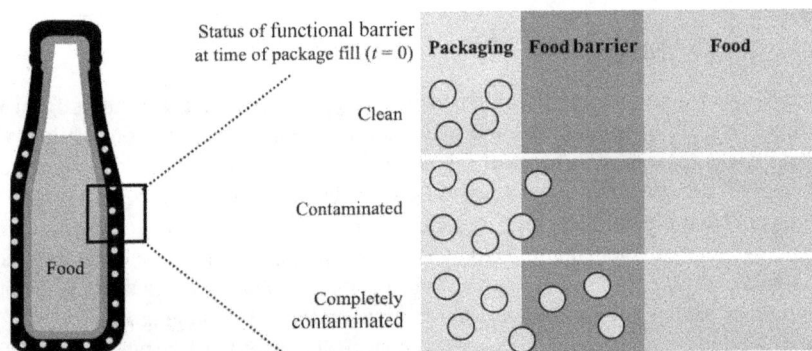

FIGURE 18.1 Possible levels of contamination of functional barrier packaging structures at time of package fill ($t = 0$).

In the past, various publications have managed the idea of functional barriers according to various aspects and with various logical and scientific intentions. One significant point was not, or not adequately, considered: it was accepted that the functional barrier layer was clean without any pollutants right after manufacturing of the package which was produced using virgin materials. Since manufacture of multilayer plastic structures is generally under co-extrusion conditions, the temperatures are usually beyond the level at which a plastic starts to melt; a significant inter-diffusion between the in situ framed layers of polymer takes place in actuality.

Considering temperatures up to 280°C relative to co-extrusion, it may very well be assessed, corresponding to the thickness and the type of polymer, that within a time ranging from a fraction of a second to a few seconds, the contaminants from the middle layer are infiltrating the functional barrier layer halfway or totally. As an outcome, the "virgin" functional barrier layer is probably going to be contaminated during its production. This deteriorates the initially planned functional barrier properties to a decreased or less-pronounced barrier effect and could even bring about the chance of complete contamination, resulting in direct contact of food with the impurities beginning from the middle layer toward the advent of migration, that is, after when the packaging is filled with the food product. Different papers have researched this inquiry tentatively and considered the in situ pollution for their analytical approach. Any analytical approach that does not give importance to the in situ contamination from the packaging material into the functional barrier would give incorrect estimates of migration amounts into the food. It ought to be noticed that a similar impact, that is, the effect of functional barrier, decreases; this could happen during the extensive time durations before the multilayer packaging is utilized to pack food when packaging sheets are put away.

As a result, three main circumstances for a functional barrier packaging material are possible, as portrayed in Figure 18.1. The relative active migration traits are illustrated in Figure 18.2, which sorts the circumstantial active migration for a compound at time $t = 0$ (for example, the time of package fill) into three particular cases:

1. Clean functional barrier: Complete lag by the functional barrier is observed.
2. Contaminated functional barrier: Relying on the level of defilement of the barrier material, the lag is reduced.
3. Completely contaminated barrier: No lag, direct contact with packaging material.

Rather than complete barriers like an aluminum sheet ranging from 6 to 7 um, the efficiency of functional barrier material is identified with a "functional" amount referring to mass transfer, which is subject to the technological (Piringer & Baner, 2000) and application-related attributes of the individual food packaging framework. These attributes are:

1. Production process of the package (e.g., high temperatures applied)
2. Kind of functional barrier material

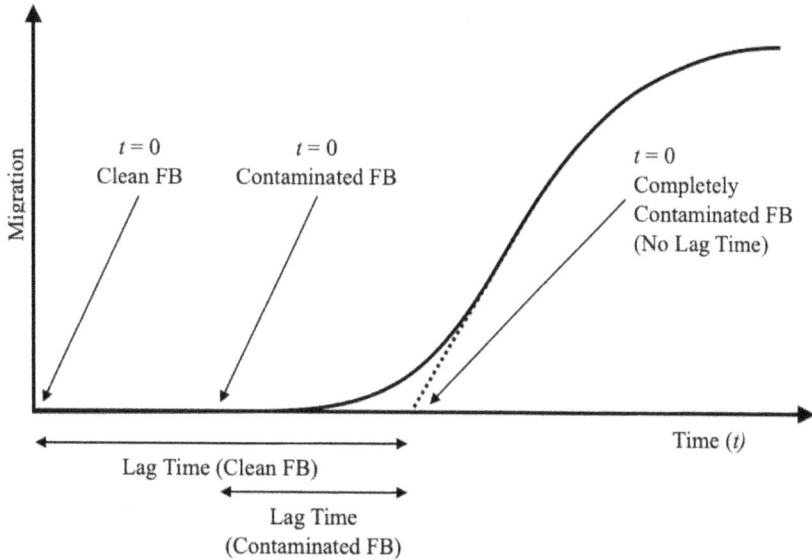

FIGURE 18.2 Possible migration behavior characteristics of migrants from functional barrier packaging structures dependent on status of FB at time of package fill ($t = 0$).

3. Functional barrier layer thickness
4. Chemical structure of contaminants and molecular weight
5. Mobility and concentration of contaminants in the packaging behind the functional barrier
6. Time difference between package preparation and filling
7. Type of food item, that is, fat content, polarity, etc.
8. Filling conditions and storage (time, temperature) of the packed food items

18.10.3 Future Patterns

Future patterns in the field of food contact plastics are destined to be aimed at growing all environment-friendly or durable substances, for example, biodegradable and plant-inferred substances, to diminish the ill effects of plastics made via landfill and incineration on the nature.

The trend in the application and development of "active and intelligent" packaging with the related advantages of enhanced food safety concerning the consumers has been escalating. Specifically, the presence of oxygen scavengers like squalene (Lestido-Cardama et al., 2020) that can be fused into internal layers has been a critical component in the plan of new PET brew bottles (Barnes et al., 2007).

To target the microbes present on the food surface, technology developers are creating frameworks of antimicrobial food packaging. For stopping any microbial development on food particles caught in tiny crevices in the surface contact layer, the antimicrobial synthetic compound is expected to be immobilized in the packaging film, thus not moving into food (bread or cheddar, for instance). Just those antimicrobials that are successful in this specific sort of activity and are on a positive biocides list for food contact will be allowed for use. Suitable naming of food packaging containing the biocides will relatively be needed to conduct their wise use (Cooksey, 2005).

18.10.4 What's Next?

1. Research efforts have been increasing to develop detection technologies of contaminants present in food, but information on toxicity caused by the degradation products and the degradation processes of these contaminants is still not available.

2. The risks that are created because of the new logistic methods of transportation and their standard norms are lacking.

3. New technological substances that enhance the stability or durability of the packaging have been added to packaging materials, and they can possess a potential threat to human health.

The statements mentioned showcase the urgent need to develop new detection techniques and to standardize the latest changes in the industry to ensure consumer health and safety (Li et al., 2020).

18.11 Plant-Based or Edible Packaging

In the food packaging industry, edible packaging is viewed as a sustainable and biodegradable alternative that improves food quality when compared to conventional packaging. The value of edible packaging can be demonstrated in its ability to preserve food quality, prolong shelf life, reduce waste, and contribute to packing material efficiency. Because of their adaptability, ability to be manufactured from a range of materials, and ability to carry diverse active chemicals such as antioxidants and/or antibacterial agents, edible films are one of the most promising disciplines in food science. Over the last decade, there has been a major increase in research activities in this field, with various challenges identified for addressing before adequate and safe industrial scale-up of edible food packaging (Aguirre-Joya et al., 2018; Restrepo et al., 2018).

Food packaging materials are made from edible elements, such natural polymers, which may be ingested by people without posing any health risks. These materials may be formed into different types of films and coatings by adjusting their thicknesses rather than changing their material makeup. Wraps, pouches, bags, capsules, and casings are commonly made with films, whereas coatings are placed directly onto the food surface. The coatings, in contrast to the films, are regarded an intrinsic element of the food product and are normally not meant to be removed (Aguirre-Joya et al., 2018) As a result, the right selection of edible packaging components is primarily determined by the food product to be packaged and the nature of the material used to create the edible packaging, as well as the method of processing. Furthermore, the packaging must be sensory compatible with the food (Restrepo et al., 2018).

Novel packaging options for a range of food goods are generating a lot of buzz across the world. New packaging methods have made it possible for freshly designed items to function better than before by offering confinement and physical protection. The future of edible packaging materials is bright, as increased food industry innovation is both forthcoming and happening now. Global consumer demand is pushing revolutionary material research and development in order to identify alternatives to fossil-based packaging materials. Consumers and the food business both want recyclable, biodegradable, or edible materials made from renewable and sustainable resources to replace them (Trajkovska Petkoska et al., 2021).

REFERENCES

Ackermann, P., Herrmann, T., Stehr, C., & Ball, M. (2006). Status of the PCDD and PCDF contamination of commercial milk caused by milk cartons. *Chemosphere, 63*(4), 670–675. https://doi.org/10.1016/j.chemosphere.2005.08.001

Aguirre-Joya, J. A., de Leon-Zapata, M. A., Alvarez-Perez, O. B., Torres-León, C., Nieto-Oropeza, D. E., Ventura-Sobrevilla, J. M., Aguilar, M. A., Ruelas-Chacón, X., Rojas, R., Ramos-Aguiñaga, M. E., & Aguilar, C. N. (2018). Basic and applied concepts of edible packaging for foods. *Food Packaging and Preservation*, 1–61. https://doi.org/10.1016/b978-0-12-811516-9.00001-4

Anderson, W. A. C., & Castle, L. (2003). Benzophenone in cartonboard packaging materials and the factors that influence its migration into food. *Food Additives and Contaminants, 20*(6), 607–618. https://doi.org/10.1080/0265203031000109486

Arvanitoyannis, I. S., & Bosnea, L. (2004). Migration of substances from food packaging materials to foods. *Critical Reviews in Food Science and Nutrition, 44*(2), 63–76. https://doi.org/10.1080/10408690490424621

Arvanitoyannis, I. S., & Kotsanopoulos, K. V. (2013). Migration phenomenon in food packaging. Food–package interactions, mechanisms, types of migrants, testing and relative legislation—a review. *Food and Bioprocess Technology, 7*(1), 21–36. https://doi.org/10.1007/s11947-013-1106-8

Badeka, A. B., & Kontominas, M. G. (1999). Effect of microwave heating on the migration of dioctyladipate and acetyltributylcitrate plasticizers from food-grade PVC and PVDC/PVC films into olive oil and water. *Zeitschrift Fur Lebensmittel-Untersuchung Und-Forschung, 202*(4), 313–317.

Barkby, C. T., & Lawson, G. (1993). Analysis of migrants from nylon 6 packaging films into boiling water. *Food Additives and Contaminants, 10*(5), 541–553. https://doi.org/10.1080/02652039309374177

Barnes, K. A., Sinclair, C. R., & Watson, D. H. (Eds.). (2006). *Chemical migration and food contact materials.* Woodhead Publishing.

Barnes, K. A., Sinclair, C. R., & Watson, D. H. (2007). *Chemical migration and food contact materials.* Woodhead Publishing.

Begley, T. H., Biles, J. E., & Hollifield, H. C. (1991). Migration of an epoxy adhesive compound into a food-simulating liquid and food from microwave susceptor packaging. *Journal of Agricultural and Food Chemistry, 39*(11), 1944–1945.

Begley, T. H., Gay, M. L., & Hollifield, H. C. (1995). Determination of migrants in and migration from nylon food packaging. *Food Additives and Contaminants, 12*(5), 671–676. https://doi.org/10.1080/02652039509374355

Begley, T. H., & Hollifield, H. C. (1990a). Migration of dibenzoate plasticizers and polyethylene terephthalate cyclic oligomers from microwave susceptor packaging into food-simulating liquids and food. *Journal of Food Protection, 53*(12), 1062–1066.

Benoy, C. J., Hooper, P. A., & Schneider, R. (1971). The toxicity of tin in canned fruit juices and solid foods. *Food and Cosmetics Toxicology, 9*(5), 645–656. https://doi.org/10.1016/0015-6264(71)90152-0

Biedermann-Brem, S., Grob, K., & Fjeldal, P. (2008). Release of bisphenol A from polycarbonate baby bottles: Mechanisms of formation and investigation of worst case scenarios. *European Food Research and Technology, 227*(4), 1053–1060. https://doi.org/10.1007/s00217-008-0819-9

Bond, J. A., & Bolt, H. M. (1989). Review of the toxicology of styrene. *CRC Critical Reviews in Toxicology, 19*(3), 227–249. https://doi.org/10.3109/10408448909037472

Boogaard, P. J., Boisset, M., Blunden, S., Davies, S., Ong, T. J., & Taverne, J. P. (2003). Comparative assessment of gastrointestinal irritant potency in man of tin(II) chloride and tin migrated from packaging. *Food and Chemical Toxicology, 41*(12), 1663–1670. https://doi.org/10.1016/s0278-6915(03)00216-3

Boone, J., Lox, F., & Pottie, S. (1993). Deficiencies of polypropylene in its use as a food-packaging material—a review. *Packaging Technology and Science, 6*(5), 277–281. https://doi.org/10.1002/pts.2770060508

Commission of the European Communities. (1990a). Commission Directive 90/128/EEC of February 1990. Relating to plastics materials and articles intended to come into contact with foodstuffs. *Official Journal of the European Communities, L349*, 26–47.

Commission of the European Communities. (1990b). Directive 90/128/EEC of 23 February 1990. Relating to plastics materials and articles intended to come into contact with foodstuffs. *Official Journal of the European Communities, L75*, 1–19.

Cooksey, K. (2005). Effectiveness of antimicrobial food packaging materials. *Food Additives and Contaminants, 22*(10), 980–987. https://doi.org/10.1080/02652030500246164

Cooper, I., & Tice, P. A. (1995). Migration studies on fatty acid amide slip additives from plastics into food simulants. *Food Additives and Contaminants, 12*(2), 235–244. https://doi.org/10.1080/02652039509374298

European Centre for Ecotoxicology and Toxicology of Chemicals. (1993). *Technical report no. 56: aquatic toxicity data evaluation.* Author.

Figge, K. (1988). Dependence of the migration out of mass plastics on the thickness and sampling of the material. *Food Additives & Contaminants, 5*(S1), 397–420.

Goulas, A. E., Riganakos, K. A., Ehlermann, D. A. E., Demertzis, P. G., & Kontominas, M. G. (1998). Effect of high-dose electron beam irradiation on the migration of DOA and ATBC plasticizers from food-grade PVC and PVDC/PVC films, respectively, into olive oil. *Journal of Food Protection, 61*(6), 720–724. https://doi.org/10.4315/0362-028x-61.6.720

Guillard, V., Mauricio-Iglesias, M., & Gontard, N. (2010). Effect of novel food processing methods on packaging: Structure, composition, and migration properties. *Critical Reviews in Food Science and Nutrition, 50*(10), 969–988. https://doi.org/10.1080/10408390903001768

Hammarling, L., Gustavsson, H., Svensson, K., Karlsson, S., & Oskarsson, A. (1998). Migration of epoxidized soya bean oil from plasticized PVC gaskets into baby food*. *Food Additives and Contaminants, 15*(2), 203–208. https://doi.org/10.1080/02652039809374631

Jadhav, S., Sonone, S. S., Sankhla, M. S., & Kumar, R. (2020). Health risks of newspaper ink when used as food packaging material. *Letters in Applied NanoBioScience, 10*, 2614–2623.

Jenkins, S. (1978). Compilation of odour threshold values in air and water by G. J. van Gemert and A. H. Nettenbreijer. Joint publication of the Central Institute for Nutrition and Food Research TNO (CIVO) and the National Institute for Water Supply (RID). National Institute for Water Supply, P.O. Box 150, Leidschendam, The Netherlands. Price: Dfl. 22. *Water Research, 12*(7), 503. https://doi.org/10.1016/0043-1354(78)90158-6

Jickells, S. M., Crews, C., Castle, L., & Gilbert, J. (1990). Headspace analysis of benzene in food contact materials and its migration into foods from plastics cookware. *Food Additives and Contaminants, 7*(2), 197–205. https://doi.org/10.1080/02652039009373884

Jickells, S. M., Gancedo, P., Nerin, C., Castle, L., & Gilbert, J. (1993). Migration of styrene monomer from thermoset polyester cookware into foods during high temperature applications. *Food Additives and Contaminants, 10*(5), 567–573. https://doi.org/10.1080/02652039309374179

Jickells, S. M., Gramshaw, J. W., Gilbert, J., & Castle, L. (1991). Migration into food during microwave and conventional oven heating. *ACS Symposium Series, 473*, 11–21.

Johns, S. M., Gramshaw, J. W., Castle, L., & Jickells, S. M. (1995). Studies on functional barriers to migration. 1. Transfer of benzophenone from printed paperboard to microwaved food. *Deutsche Lebensmittel-Rundschau, 91*(3), 69–73.

Johns, S. M., Jickells, S. M., Read, W. A., & Castle, L. (2000). Studies on functional barriers to migration. 3. Migration of benzophenone and model ink components from cartonboard to food during frozen storage and microwave heating. *Packaging Technology and Science, 13*(3), 99–104.

Kim, K. C., Park, Y. B., Lee, M. J., Kim, J. B., Huh, J. W., Kim, D. H., Lee, J. B., & Kim, J. C. (2008). Levels of heavy metals in candy packages and candies likely to be consumed by small children. *Food Research International, 41*(4), 411–418. https://doi.org/10.1016/j.foodres.2008.01.004

Komolprasert, V., Hargraves, W. A., & Armstrong, D. J. (1994). Determination of benzene residues in recycled polyethylene terephthalate (PETE) by dynamic headspace-gas chromatography. *Food Additives & Contaminants, 11*(5), 605–614. https://doi.org/10.1080/02652039409374260

Lau, O. W., & Wong, S. K. (2000). Contamination in food from packaging material. *Journal of Chromatography A, 82*(1–2), 255–270. https://doi.org/10.1016/s0021-9673(00)00356-3

Lau, O. W., Wong, S. K., & Leung, K. S. (1994). Naphthalene contamination of sterilized milk drinks contained in low-density polyethylene bottles. Part 1. *The Analyst, 119*(5), 1037. https://doi.org/10.1039/an9941901037

Lau, O. W., Wong, S. K., & Leung, K. S. (1995). Naphthalene contamination of sterilized milk drinks contained in low-density polyethylene bottles. Part 2. Effect of naphthalene vapour in air. *The Analyst, 120*(4), 1125. https://doi.org/10.1039/an9952001125

Lawley, R., Curtis, L., & Davis, J. (2008). *Food safety hazard guidebook* (1st ed.). Royal Society of Chemistry.

Lawson, G., Barkby, C. T., & Lawson, C. (1996). Contaminant migration from food packaging laminates used for heat and eat meals. *Analytical and Bioanalytical Chemistry, 354*(4), 483–489. https://doi.org/10.1007/s0021663540483

Lestido-Cardama, A., Rodríguez Bernaldo De Quirós, A., Bustos, J., Lomo, M. L., Paseiro Losada, P., & Sendón, R. (2020). Estimation of dietary exposure to contaminants transferred from the packaging in fatty dry foods based on cereals. *Foods, 9*(8), 1038. https://doi.org/10.3390/foods9081038

Li, C., Li, C., Yu, H., Cheng, Y., Xie, Y., Yao, W., Guo, Y., & Qian, H. (2020). Chemical food contaminants during food processing: Sources and control. *Critical Reviews in Food Science and Nutrition, 61*(9), 1545–1555. https://doi.org/10.1080/10408398.2020.1762069

Limm, W., & Hollifield, H. C. (1996). Modelling of additive diffusion in polyolefins. *Food Additives & Contaminants, 13*(8), 949–967.

Marsh, K., & Bugusu, B. (2007). Food packaging roles, materials, and environmental issues. *Journal of Food Science, 72*(3), R39–R55. https://doi.org/10.1111/j.1750-3841.2007.00301.x

Ministry of Agriculture, Fisheries and Food. (1980). *The third report of the steering group on food surveillance* (The Working Party on Vinylidene Chloride, Food Surveillance Paper No. 3). HMSO.

Muncke, J. (2009). Food contamination and human exposure to endocrine disrupting compounds from food packaging. *Epidemiology, 20*, S103. https://doi.org/10.1097/01.ede.0000362360.51281.7f

Nerín, C., Rubio, C., Cacho, J., & Salafranca, J. (1998). Parts-per-trillion determination of styrene in yoghurt by purge-and-trap gas chromatography with mass spectrometry detection. *Food Additives and Contaminants, 15*(3), 346–354. https://doi.org/10.1080/02652039809374650

Omori, Y., Takanaka, A., Tanaka, S., Ikeda, Y., & Furuya, T. (1973). Experimental studies on toxicity of tin in canned orange juice. *Food Hygiene and Safety Science (Shokuhin Eiseigaku Zasshi), 14*(1), 69–74. https://doi.org/10.3358/shokueishi.14.69

Petersen, H., Biereichel, A., Burseg, K., Simat, T., & Steinhart, H. (2008). Bisphenol A diglycidyl ether (BADGE) migrating from packaging material "disappears" in food: Reaction with food components. *Food Additives & Contaminants: Part A, 25*(7), 911–920. https://doi.org/10.1080/02652030701837399

Piringer, O., & Baner, A. L. (Eds.). (2000). *Plastic packaging materials for food: Barrier function, mass transport, quality assurance, and legislation* (1st ed.). https://doi.org/10.1002/9783527613281

Piringer, O., Franz, R., Huber, M., Begley, T. H., & McNeal, T. P. (1998). Migration from food packaging containing a functional barrier: Mathematical and experimental evaluation. *Journal of Agricultural and Food Chemistry, 46*(4), 1532–1538. https://doi.org/10.1021/jf970771v

Restrepo, A. E., Rojas, J. D., García, O. R., Sánchez, L. T., Pinzón, M. I., & Villa, C. C. (2018). Mechanical, barrier, and color properties of banana starch edible films incorporated with nanoemulsions of lemongrass (*Cymbopogon citratus*) and rosemary (*Rosmarinus officinalis*) essential oils. *Food Science and Technology International, 24*(8), 705–712. https://doi.org/10.1177/1082013218792133

Rijk, R., & de Kruijf, N. (1993). Migration testing with olive oil in a microwave oven. *Food Additives & Contaminants, 10*(6), 631–645. https://doi.org/10.1080/02652039309374190

Robertson, G. L. (2006). Safety and legislative aspects of packaging. In *Food packaging: Principles and practice*, G. L. Robertson (ed.) (pp. 473–502). Taylor & Francis Group.

Roe, F. J. C. (1994). Styrene: Toxicity studies—what do they show? *Critical Reviews in Toxicology, 24*(Supp. 1), s117–s125. https://doi.org/10.3109/10408449409020144

Sharman, M., Honeybone, C. A., Jickells, S. M., & Castle, L. (1995). Detection of residues of the epoxy adhesive component bisphenol A diglycidyl ether (BADGE) in microwave susceptors and its migration into food. *Food Additives and Contaminants, 12*(6), 779–787. https://doi.org/10.1080/02652039509374370

Sheftel, V. O. (2000). *Indirect food additives and polymers: Migration and toxicology* (1st ed.). CRC Press.

Shiota, K., & Nishimura, H. (1982). Teratogenicity of di(2-ethylhexyl) phthalate (DEHP) and di-n-butyl phthalate (DBP) in mice. *Environmental Health Perspectives, 45*, 65–70. https://doi.org/10.1289/ehp.824565

Skrzydlewska, E., Balcerzak, M., & Vanhaecke, F. (2003). Determination of chromium, cadmium and lead in food-packaging materials by axial inductively coupled plasma time-of-flight mass spectrometry. *Analytica Chimica Acta, 479*(2), 191–202. https://doi.org/10.1016/s0003-2670(02)01527-1

Soto-Valdez, H., Gramshaw, J. W., & Vandenburg, H. J. (1997). Determination of potential migrants present in Nylon "microwave and roasting bags" and migration into olive oil. *Food Additives and Contaminants, 14*(3), 309–318. https://doi.org/10.1080/02652039709374529

Stepek, J., Duchacek, V., Curda, D., Horacek, J., & Sipek, M. (1987) Polymers as materials for packaging. Ellis Horwood Ltd., Halsted Press; A division of John Wiley & Sons, 1987, 489 pages. £65.00. (1988). *Packaging Technology and Science, 1*(3), 171–172. https://doi.org/10.1002/pts.2770010309

Trajkovska Petkoska, A., Daniloski, D., D'Cunha, N. M., Naumovski, N., & Broach, A. T. (2021). Edible packaging: Sustainable solutions and novel trends in food packaging. *Food Research International, 140*, 109981. https://doi.org/10.1016/j.foodres.2020.109981

Tricker, A., & Preussmann, R. (1991). Carcinogenic N-nitrosamines in the diet: Occurrence, formation, mechanisms and carcinogenic potential. *Mutation Research/Genetic Toxicology, 259*(3–4), 277–289. https://doi.org/10.1016/0165-1218(91)90123-4

Ungureanu, E. L., Mustatea, G., & Popa, M. E. (2020). Chemical contaminants migration from food contact materials into aqueous extracts. *E3S Web of Conferences, 215*, 01007. https://doi.org/10.1051/e3sconf/202021501007

Ungureanu, E. L., Mustatea, G., & Popa, M. E. (2020). BPA Incidence in Babies Drinking Water Available on Romanian Market. *J. Agroaliment. Proc. Technol, 26*, 353-356.

Varner, S. L., & Breder, C. V. (1981). Headspace sampling and gas chromatographic determination of styrene migration from food-contact polystyrene cups into beverages and food simulants. *Journal of AOAC International, 64*(5), 1122–1130. https://doi.org/10.1093/jaoac/64.5.1122

Wagoner, J. K. (1983). Toxicity of vinyl chloride and poly (vinyl chloride): A critical review. *Environmental Health Perspectives, 52,* 61–66. https://doi.org/10.1289/ehp.835261

Woolrich, P. F. (1982). Toxicology, industrial hygiene and medical control of TDI, MDI and PMPPI. *American Industrial Hygiene Association Journal, 43*(2), 89–97. https://doi.org/10.1080/15298668291409415

For Product Safety Concerns and Information please contact our EU
representative GPSR@taylorandfrancis.com
Taylor & Francis Verlag GmbH, Kaufingerstraße 24, 80331 München, Germany